Nuclear and Radiochemistry

Nuclear and Radiochemistry

Third Edition

Gerhart Friedlander

Senior Chemist, Brookhaven National Laboratory

Joseph W. Kennedy

Late Professor of Chemistry, Washington University, St. Louis

Edward S. Macias

Associate Professor of Chemistry, Washington University, St. Louis

Julian Malcolm Miller

Late Professor of Chemistry, Columbia University

A Wiley-Interscience Publication

JOHN WILEY & SONS

New York · Chichester · Brisbane · Toronto

A NOTE TO THE READER:
This book has been electronically reproduced from digital information stored at John Wiley & Sons, Inc. We are pleased that the use of this new technology will enable us to keep works of enduring scholarly value in print as long as there is a reasonable demand for them. The content of this book is identical to previous printings.

Library of Congress Cataloging in Publication Data:

Main entry under title:

Nuclear and radiochemistry.

Second ed. by Gerhart Friedlander, Joseph W. Kennedy, and Julian Malcolm Miller.
"A Wiley-Interscience publication."
Bibliography: p.
Includes index.
1. Radiochemistry. 2. Nuclear chemistry.
I. Friedlander, Gerhart. II. Friedlander, Gerhart. Nuclear and radiochemistry.
QD601.2.N83 1981 541.3'8 81-1000
ISBN 0-471-28021-6
ISBN 0-471-86255-X pbk.

Printed in the United States of America

20 19 18 17 16 15

Preface

Over 30 years have passed since the forerunner of this text appeared under the title *Introduction to Radiochemistry*. Casual comparison will reveal little resemblance between the slender 1949 volume and the present work. Yet through the several editions cur purpose has remained the same: to provide a textbook for advanced undergraduate and beginning graduate students who have some chemical but little or no nuclear physics background and to make available a ready reference source for practitioners of nuclear chemistry, radiochemistry, and related fields.

In adopting the present title of the book in 1955 we gave explicit recognition to a dichotomy in the field and in the audience addressed, a dichotomy that has probably become even more pronounced since then. The book is written as an introductory text for two broad groups: nuclear chemists, that is, scientists with chemical background and chemical orientation whose prime interest is the study of nuclear properties and nuclear reactions; and radiochemists, that is, chemists concerned with the chemical manipulation of radioactive sources and with the application of radioactivity and other nuclear phenomena to chemical problems (whether in basic chemistry or in biology, medicine, earth and space sciences, etc.). Despite the apparently growing division between these two audiences, individuals have always moved fairly freely from one field to the other, and we continue to feel that nuclear chemistry and radiochemistry interact strongly with each other and indeed are so interdependent that their discussion together is almost necessary in an introductory text.

In choosing and arranging the subject matter we have been guided by the firm conviction that a good grounding in the fundamentals of radioactivity and other nuclear phenomena is equally essential for both of the broad groups we are trying to address. At the same time we are fully aware that for those who wish to do active research in nuclear chemistry this book can serve only as an introduction; they will certainly want to go on to more advanced nuclear physics books. We have assumed that students using this book have had basic courses in chemistry and physics. While we do not expect that the reader has a rigorous quantum-mechanics background, some aquaintance with the language of quantum mechanics is needed. This is often covered in a modern physics course dealing with the basics of atomic structure or even in many first-year "general" chemistry and physics courses.

In keeping with the general purposes outlined, the material in this third edition has been somewhat rearranged. After the essentially unchanged, largely historical chapter 1 we have put all the basic material on nuclear properties, radioactivity, and nuclear reactions in chapters 2–5, and we hope that these chapters will form the backbone of practically any course in which this book is used at all. The relative emphasis on different parts of this material may, of course, differ widely, and we have put some sections in small print to indicate that they might be considered too advanced and detailed for some courses. The choice among the remaining 10 chapters will depend very much on the predilections of the instructor, the length and purpose of the course, and the preparation of the students. One can pick and choose among the chapters, and in most instances the order is not crucial, although it seems advisable, for example, to precede any discussion of Radiation Detection and Instruments (chapter 7) with coverage of Interactions of Radiation with Matter (chapter 6). In a course oriented toward nuclear studies, chapter 10 (Nuclear Models) would probably be taken up right after chapters 3 and 4, whereas a chemically oriented course might at that point jump to chapters 11 (Radiochemical Applications) and 12 (Nuclear Processes as Chemical Probes).

In the 17 years since the preparation of the preceding edition the field covered has greatly changed and expanded. Well over half of the text in this edition is newly written. The temptation to expand the book considerably was great, but in order to keep the size and price within reasonable bounds we have also excised much old material that no longer seemed as essential or appropriate as it did 15 or 20 years ago. As a result the text has grown only modestly. No attempt has been made to convert to consistent use of SI units in this edition; however, they are introduced occasionally in the body of the text and definitions of the relevant SI units are given in appendix A.

The preparation of this new edition has been in progress for a long time. It was interrupted by the untimely death of one of the authors. Julian Malcolm Miller died suddenly in December 1976 after participating actively in the planning and writing of the revision. We have greatly missed his knowledgeable, stimulating, perceptive participation in the latter stages of our work on the manuscript, but we have attempted to complete the task in the spirit with which he approached it. Although our other coauthor, Joseph W. Kennedy, has been dead for over 20 years, his contributions to concept and content are also still in evidence.

We have, as in previous editions, given a set of exercises and a list of references at the end of each chapter. To quote from the original 1949 preface, the exercises "are intended as an integral part of the course, and only with them does the text contain the variety of specific examples which we consider necessary for an effective presentation." Some old exercises have been retained, some new ones added. The references are listed by initial of first author and a serial number, and most of them are referred to in the text. Some selected, specific research papers, including some of historical interest, are cited, but a particular effort has been made to give references to comprehensive review articles and books that provide thorough coverage of a subject area and can serve as a guide to the literature. General references to works that cover broad areas, such as the subject matter of a whole chapter, are marked with an asterisk.

We are very much indebted to C. M. Lederer and V. S. Shirley for preparing the Table of Nuclides in appendix D from their much more extensive *Table of Isotopes*, 7th edition (Wiley, New York, 1978). Throughout the text we have relied on their book as the primary source of half-life and radiation energy data.

We have had the benefit of helpful advice from many colleagues. The suggestions concerning general subject matter to be covered were not always mutually compatible, since they ranged from pleas for much more rigorous nuclear physics to appeals for much greater emphasis on applications. This divergence of opinion reflects, we believe, the breadth of the field and the wide spectrum of courses for which our book has been found useful in the past. We can only hope that this new edition will again find such widespread use and, for the reasons outlined earlier, we have again chosen a middle ground between the two extreme positions mentioned.

We are grateful to those colleagues who were kind enough to read parts of the manuscript; they include J. H. Barker, J. P. Blewett, J. B. Cumming, W. Faubel, P. P. Gaspar, G. E. Gordon, V. P. Guinn, P. Gütlich, L. A. Haskin, M. Kaplan, P. J. Karol, P. Peuser, L. P. Remsberg, D. G. Sarantites, A. M. Schmitt, A. C. Wahl, M. J. Welch, J. Weneser, and M. A. Yates; each of them offered useful comments and called our attention to errors and inaccuracies. Our very special thanks go to H. N. Erten and G. Herrmann, both of whom read almost the entire manuscript with great care and made innumerable helpful suggestions. Whatever errors and misstatements remain in the book are, of course, our responsibility. We earnestly request any reader who finds a mistake, no matter how trivial, to communicate it to us, so that it can be eradicated in later printings.

Both of us had the benefit of sabbatical leaves from our home institutions during the major writing effort. One of us (G. F.) was the recipient of a Senior Scientist Award from the Alexander von Humboldt Foundation that enabled him to spend the year 1978–1979 at the Institut für Kernchemie at the University of Mainz; he is deeply grateful both to the Humboldt Foundation and to the host institute and its Director, Professor Günter Herrmann, for providing an ideal atmosphere for this writing effort. Special thanks go to Mr. W. Kelp for the meticulous care with which he prepared many of the illustrations.

The younger author (E. S. M.), who was five years old when the first version of the text appeared, spent his sabbatical year at the California Institute of Technology in the Division of Chemistry and Chemical

Engineering as a guest of Professor Sheldon K. Friedlander. He is very grateful for the hospitality and stimulating environment that he enjoyed during that year with "the other Friedlander." He would also like to thank Professors John H. Seinfeld and Harry B. Gray for making his stay in the Division so pleasant. William Wilson and the USEPA deserve special thanks for providing financial support that made this sabbatical leave possible. We would like to acknowledge the help of several people who typed portions of the manuscript: Elaine E. Granger at Cal Tech; and Betty Henley, Mary Baetz, and Karen Klein at Washington University.

We are grateful to Brookhaven National Laboratory and Washington University for making it possible for us to undertake the time-consuming task of preparing this new edition.

Blue Point, New York
St. Louis, Missouri
May 1981

GERHART FRIEDLANDER
EDWARD S. MACIAS

Contents

Nuclear and Radiochemistry

Chapter 1

Introduction

A. EARLY HISTORY OF RADIOACTIVITY

Becquerel's Discovery. The more or less accidental series of events that led to the discovery of radioactivity depended on two especially significant factors: (1) the mysterious X rays discovered about one year earlier by W. C. Roentgen produced fluorescence (the term phosphorescence was preferred at that time) in the glass walls of X-ray tubes and in some other materials; and (2) Henri Becquerel had inherited an interest in phosphorescence from both his father and grandfather. The father, Edmund Becquerel (1820–1891), had actually studied phosphorescence of uranium salts, and about 1880 Henri Becquerel prepared potassium uranyl sulfate, $K_2UO_2(SO_4)_2 \cdot 2H_2O$, and noted its pronounced phosphorescence excited by ultraviolet light. Thus in 1895 and 1896, when several scientists were seeking the connection between X rays and phosphorescence and were looking for penetrating radiation from phosphorescent substances, it was natural for Becquerel to experiment along this line with the potassium uranyl sulfate.

It was on February 24, 1896, that Henri Becquerel reported his first results: after exposure to bright sunlight, crystals of the uranyl double sulfate emitted a radiation that blackened a photographic plate after penetrating black paper, glass, and other substances. During the next few months he continued the experiments, obtaining more and more puzzling results. The effect was as strong with weak light as with bright sunlight; it was found in complete darkness and even for crystals prepared and always kept in the dark. The penetrating radiation was emitted by other uranyl and also uranous salts, by solutions of uranium salts, and even by what was believed to be metallic uranium, and in each case with an intensity proportional to the uranium content. Proceeding by analogy with a known property of X rays, Becquerel observed that the penetrating rays from uranium would discharge an electroscope. All these results were obtained in the early part of 1896 (B1). Although Becquerel and others continued investigations for several years, the knowledge gained in this phase of the new science was summarized in 1898, when Pierre and Marie Sklodowska Curie concluded that the uranium rays were an atomic phenomenon characteristic of the element, and not related to its chemical or physical state, and introduced the name "radioactivity" for the phenomenon (C1).

The Curies. Much new information appeared during the year 1898, mostly through the work of the Curies. Examination of other elements led to the discovery, independently by Mme. Curie and G. C. Schmidt, that compounds of thorium emitted rays similar to those from uranium. A very important observation was that some natural uranium ores were even more radioactive than pure uranium, and more active than a chemically similar "ore" prepared synthetically. The chemical decomposition and fractionation of such ores constituted the first exercise in radiochemistry and led immediately to the discovery of polonium—as a new substance observed only through its intense radioactivity—and of radium, a highly radioactive substance recognized as a new element and soon identified spectroscopically. The Curies and their co-workers had found radium in the barium fraction separated chemically from pitchblende (a dark almost black ore containing about 75 percent U_3O_8), and they learned that it could be concentrated from the barium by repeated fractional crystallization of the chlorides, the radium salt remaining preferentially in the mother liquor. By 1902 Mme. Curie reported the isolation of 100 mg of radium chloride spectroscopically free from barium and gave 225 as the approximate atomic weight of the element. (The work had started with about two tons of pitchblende, and the radium isolated represented about a 25 percent yield.) Still later Mme. Curie redetermined the atomic weight to be 226.5 (within 0.2 percent of the present best value) and also prepared radium metal by electrolysis of the fused salt.

Becquerel in his experiments had shown that uranium, in the dark and not supplied with energy in any known way, continued for years to emit rays in undiminished intensity. E. Rutherford had made some rough estimates of the energy associated with the radioactive rays; the source of this energy was quite unknown. With concentrated radium samples the Curies made measurements of the resulting heating effect, which they found to be about 100 cal h^{-1} per gram of radium. The evidence for so large a store of energy not only caused a controversy among the scientists of that time but also helped to create a great popular interest in radium and radioactivity. (An interesting article in the *St. Louis Post-Dispatch* of October 4, 1903, speculated on this inconceivable new power, its use in war, and as an instrument for destruction of the world.)

Early Characterization of the Rays. The effect of radioactive radiations in discharging an electroscope was soon understood in terms of the ionization of the air molecules, as J. J. Thomson and others were developing a knowledge of this subject in their studies of X rays. The use of the amount of ionization in air as a measure of the intensity of radiations was developed into a more precise technique than the photographic blackening, and this technique was employed in the Curie laboratory, where ionization currents were measured with an electrometer. In 1899 Rutherford began a study of the properties of the rays themselves, using a similar instrument. Measurements of the absorption of the rays in metal foils showed that there were two components. One component was absorbed in the first few thousandths of a centimeter of aluminum and was named α radiation; the other was absorbed considerably in roughly 100 times this thickness of aluminum and was named β radiation. For the β rays Rutherford found that the ionization effect was reduced to the fraction $e^{-\mu d}$ of its original value when d cm of absorber were interposed; the absorption coefficient μ was about $15\ \mathrm{cm}^{-1}$ for aluminum and increased with atomic weight for other metal foils.

Rutherford at that time believed that the absorption of the α radiation also followed an exponential law and listed for its absorption coefficient in aluminum the value $\mu = 1600\ \mathrm{cm}^{-1}$. About a year later Mme. Curie found that μ was not constant for α rays but increased as the rays proceeded through the absorber. This was a surprising fact, since one would have expected that any inhomogeneity of the radiation would result in early absorption of the less penetrating components with a corresponding decrease in absorption coefficient with distance. In 1904 the concept of a definite range for the α particles (they were recognized as particles by that time) was proposed and demonstrated by W. H. Bragg. He found that several radioactive substances emitted α rays with different characteristic ranges.

The recognition of the character of the α and β rays as streams of high-speed particles came largely as a result of magnetic and electrostatic deflection experiments. In this way the β rays were seen to be electrons moving with almost the velocity of light. At first the α rays were thought to be undeviated by these fields. More refined experiments did show deflections, from which the ratio of charge to mass was calculated to be about half that of the hydrogen ion, with the charge positive, and the velocity was calculated to be about one tenth that of light. The suggestion that α particles were helium ions was immediately made, and it was confirmed after much more study. The presence of helium in uranium and thorium ores had already been noticed and was seen to be significant in this connection. A striking demonstration was later made, in which α rays were allowed to pass through a very thin glass wall into an evacuated glass vessel; within a few days sufficient helium gas appeared in the vessel to be detected spectroscopically.

Before the completion of these studies of the α and β rays, an even more penetrating radiation, not deviated by a magnetic field, was found in the rays from radioactive preparations. The recognition of this γ radiation as electromagnetic waves, like X rays in character if not in energy, came rather soon. For a long time no distinction was made between the nuclear γ rays and some extranuclear X rays that often accompany radioactive transformations.

Rutherford and Soddy Transformation Hypothesis. In the course of radioactivity measurements on thorium salts, Rutherford observed that the electrometer readings were sometimes quite erratic. During 1899 it was determined that the cause of this effect was the diffusion through the ionization chamber of a radioactive substance emanating from the thorium compound. Similar effects were obtained with radium compounds. Subsequent studies, principally by Rutherford and F. Soddy, showed that these emanations were inert gases of high molecular weight, subject to condensation at about −150°C. Another radioactive substance, actinium, had been separated from pitchblende in 1899, and it too was found to give off an active emanation.

The presence of the radioactive emanations from thorium, radium, and actinium preparations was a very fortunate circumstance for advancement of knowledge of the real nature of radioactivity. Essentially the inert-gas character of these substances made radiochemical separations not only an easy process but also one that forced itself on the attentions of these early investigators. Two very significant consequences of the early study of the emanations were (1) the realization that the activity of radioactive substances did not continue forever but diminished in intensity with a time scale characteristic of the substance; and (2) the knowledge that the radioactive processes were accompanied by a change in chemical properties of the active atoms. In 1900 and the succeeding years the application of chemical separation procedures, especially by W. Crookes and by Rutherford and Soddy, revealed the existence of other activities with characteristic decay rates and radiations. These included uranium X (now known as ^{234}Th), separable from uranium by precipitation with excess ammonium carbonate (the uranyl carbonate redissolves in excess carbonate through formation of a complex ion), and thorium X (now known as ^{224}Ra), which remains in solution when thorium is precipitated as the hydroxide with ammonium hydroxide. In each case it was found that the activity of the X body decayed appreciably in a matter of days and that a new supply of the X body appeared in the parent substance in a similar time. It was also shown that both uranium and thorium, when effectively purified of the X bodies and other products, emitted only α rays and that uranium X and thorium X emitted β rays.

By the spring of 1903 Rutherford and Soddy had reached an excellent understanding of the nature of radioactivity and published their conclusions that the radioactive elements were undergoing spontaneous transformation from one chemical atom into another, that the radioactive radiations were an accompaniment of these changes, and that the radioactive process was a subatomic change within the atom. However, it should be remembered that the idea of the atomic nucleus did not emerge until eight years later and that in 1904 Bragg was attempting to understand the α particle as a flying cluster of thousands of more or less independent electrons.

Statistical Aspect of Radioactivity. In 1905 E. von Schweidler used the foregoing conclusions as to the nature of radioactivity and formulated a new description of the process in terms of disintegration probabilities. His fundamental assumption was that the probability p of a particular atom of a radioactive element disintegrating in a time interval Δt is independent of the past history and the present circumstances of the atom; it depends only on the length of the time interval Δt and for sufficiently short intervals is just proportional to Δt; thus $p = \lambda \, \Delta t$, where λ is the proportionality constant characteristic of the particular species of radioactive atoms. The probability of the given atom not disintegrating during the short interval Δt is $1 - p = 1 - \lambda \, \Delta t$. If the atom has survived this interval, then its probability of not disintegrating in the next like interval is again $1 - \lambda \, \Delta t$, and so forth for each successive interval Δt. Thus the probability that the given atom will survive n such intervals is $(1 - \lambda \, \Delta t)^n$. Setting $n \, \Delta t = t$, the total time, we obtain for the survival probability $(1 - \lambda t/n)^n$. Now the probability that the atom will remain unchanged after time t is just the value of this quantity when Δt is made indefinitely small; that is, it is the limit of $(1 - \lambda t/n)^n$ as n approaches infinity. Recalling that $e^x = \lim_{n \to \infty} (1 + x/n)^n$, we obtain $e^{-\lambda t}$ for the limiting value. If we consider not one atom but a large initial number N_0 of the radioactive atoms, then we may take the fraction remaining unchanged after time t to be $N/N_0 = e^{-\lambda t}$, where N is the number of unchanged atoms at time t. This exponential law of decay is just that which Rutherford had already found experimentally for the simple isolated radioactivities.

A more detailed discussion of the statistical nature of radioactivity is presented in chapter 9.

B. RADIOACTIVE DECAY AND GROWTH

The exponential law just derived for the decay of a radioactive species, $N = N_0 e^{-\lambda t}$, has the form of the rate law for any unimolecular reaction, as should be expected in view of the nature of the radioactive process. It may alternatively be derived if the decay rate $-dN/dt$ is set proportional to the number of atoms present: $-dN/dt = \lambda N$, which merely expresses the expectation that twice as many disintegrations occur per unit time in a sample containing twice as many atoms, and so forth. Upon integration and letting $N = N_0$ at $t = 0$, we obtain $N = N_0 e^{-\lambda t}$.

The constant λ is known as the **decay constant** for the radioactive species and has the dimensions of reciprocal time. We should note that for most radioactive substances no attempt to alter λ by variation of ordinary experimental conditions, such as temperature, chemical change, pressure, and gravitational, magnetic, or electric fields, has given a detectable effect.[1]

[1] The exceptional cases in which slight changes of λ have been achieved are considered in chapter 12, p. 458.

The characteristic rate of a radioactive decay may very conveniently be given in terms of the **half life** $t_{1/2}$, which is the time required for an initial (large) number of atoms to be reduced on the average to half that number by transformations. Thus at the time $t = t_{1/2}$, we have $N/N_0 = \frac{1}{2}$, and therefore we can write $\ln \frac{1}{2} = -\lambda t_{1/2}$ or

$$t_{1/2} = \frac{\ln 2}{\lambda} \approx \frac{0.693}{\lambda}.$$

In practical work with radioactive materials the number of atoms N is not directly evaluated, and even the rate of change dN/dt is usually not measured absolutely. The usual procedure is to determine a quantity proportional to λN; we call this quantity the **activity A**, with $\mathbf{A} = c\lambda N = c(-dN/dt)$. The coefficient c, which we call the detection coefficient, will depend on the nature of the detection instrument, the efficiency for the recording of the particular radiation in that particular instrument, and the geometrical arrangement of sample and detector. Care must be taken to keep these factors constant throughout a series of measurements. We may now write the decay law as it is commonly observed: $\mathbf{A} = \mathbf{A}_0 e^{-\lambda t}$.

The usual procedure for treating measured values of **A** at successive times is to plot $\log \mathbf{A}$ versus t; for this purpose semilog paper (with a suitable number of decades) is most convenient. Now λ could be found from the slope of the resulting straight line corresponding to the simple decay law. However, it is more convenient to read from the plot on semilog paper the time required for the activity to fall from any value to half that value; this is the half life $t_{1/2}$.

In this discussion we have considered only the radioactivity corresponding to the transformation of a single atomic species; however, the daughter substance resulting from the transformation may itself be radioactive, with its own characteristic radiation and half life as well as its own chemical identity. Indeed, among the naturally occurring radioactive substances this is the more common situation, and in chapter 5 we treat quite complicated combinations of radioactive growth and decay. For the moment consider the decay of the substance ^{238}U (uranium I). This species of uranium is an α-particle emitter, with $t_{1/2} = 4.47 \times 10^9$ y. The immediate product of its transformation is the radioactive substance ^{234}Th (uranium X_1), a β emitter with half life 24.1 d (cf. figure 1-1). As already mentioned, the parent uranium may be separated from the daughter atoms by precipitation of the daughter with excess ammonium carbonate. The daughter precipitate will show a characteristic activity, which will decay with the rate indicated; that is, it will be half gone in 24.1 d, three fourths gone in 48.2 d, and so on. The parent fraction will, of course, continue its α activity as before but will for the moment be free of the β radiations associated with the daughter. However, in time new daughter atoms will be formed, and the daughter activity in the parent fraction will return to its initial value, with a time scale corresponding to the rate of decay of the isolated daughter fraction.

In an undisturbed sample containing N_1 atoms of ^{238}U, a steady state is established in which the rate of formation of the daughter ^{234}Th atoms (number N_2) is just equal to their rate of decay. This means that $-dN_1/dt = \lambda_2 N_2$ in this situation, because the rate of formation of the daughter atoms is just the rate of decay of the parent atoms. Using the earlier relation, we have then $\lambda_1 N_1 = \lambda_2 N_2$, with λ_1 and λ_2 the respective disintegration constants. This result only applies when $\lambda_2 \gg \lambda_1$ (this is discussed in more detail in chapter 5).

Units of Radioactivity. The **curie** (Ci) is a unit of radioactivity originally based on the disintegration rate of 1 g of radium but now defined as the quantity of any radioactive nuclide in which the number of disintegrations per second (dis s^{-1}) is 3.700×10^{10}. The SI unit (cf. Appendix A) of radioactivity is the **becquerel**, defined as 1 dis s^{-1}. The millicurie (mCi) and the microcurie (μCi) are practical units also in common use. The megacurie (MCi = 10^6 Ci) finds use in reactor technology.

As an illustration, we calculate the weight W in grams of 1.00 mCi of ^{14}C from its half life of 5730 y:

$$\lambda = \frac{0.693}{5730 \times 3.156 \times 10^7} = 3.83 \times 10^{-12}\ s^{-1},$$

$$-\frac{dN}{dt} = \lambda N = \lambda \frac{W}{14} \times 6.02252 \times 10^{23} \approx 1.65 W \times 10^{11}\ s^{-1};$$

with

$$-\frac{dN}{dt} = 3.700 \times 10^7\ \text{dis}\ s^{-1}\ (1\ \text{mCi}),$$

$$W = \frac{3.700 \times 10^7}{1.65 \times 10^{11}} = 0.224 \times 10^{-3}\ g.$$

The **rad** is a quantitative measure of radiation energy *absorption* (usually called the dose). A dose of 1 rad deposits 100 ergs g^{-1} of material. In the SI system the unit of dose is the **gray** (Gy), which is defined as 1 J kg^{-1}. Note that 1 Gy = 100 rad. A unit of radiation *exposure* is the **roentgen** (R). The roentgen is defined as "that quantity of X or γ radiation such that the associated corpuscular emission per 0.001293 g of air[2] produces, in air, ions carrying 1 esu of electricity of either sign." This means that 1 R produces 1.61×10^{12} ion pairs per gram of air, which corresponds to the absorption of 84 ergs of energy per gram of air. In water the energy absorption corresponding to 1 R is about 93 ergs g^{-1} or 0.93 rad for all X- or γ-ray energies above about 50 keV.

[2] This is the weight of 1 cm^3 of dry air at 0°C and 760 mm pressure.

C. NATURALLY OCCURRING RADIOACTIVE SUBSTANCES

Uranium, Thorium, and Actinium Series. All elements found in natural sources with atomic number greater than 83 (bismuth) are radioactive. They belong to chains of successive decays, and all the species in one such chain constitute a radioactive family or series. Three of these families include all the natural activities in this region of the periodic chart. One family has ^{238}U as the parent substance, and after 14 transformations (8 of them by α-particle emission and 6 by β-particle emission) reaches a stable end product, ^{206}Pb; this is known as the uranium series (which includes radium and its decay products). Since the mass is changed by four units in α decay and by only a small fraction of one unit in β decay, the various mass numbers found in members of the family differ by multiples of 4, and a general formula for the mass numbers is $4n+2$, where n is an integer. Therefore the uranium series is also known as the $4n+2$ series. Figure 1-1 shows the members and transformations of the uranium series.

Thorium (^{232}Th) is the parent substance of the $4n$ or thorium series with ^{208}Pb as the stable end product. This series is shown in figure 1-2. The $4n+3$ or actinium series has ^{235}U (formerly known as actino-uranium) as the parent and ^{207}Pb as the stable end product. This series is shown in figure 1-3.

The fairly close similarity between the three series and their relations to the periodic chart are interesting and helpful in remembering the decay modes of and nomenclature for the active bodies. Actually these historical names

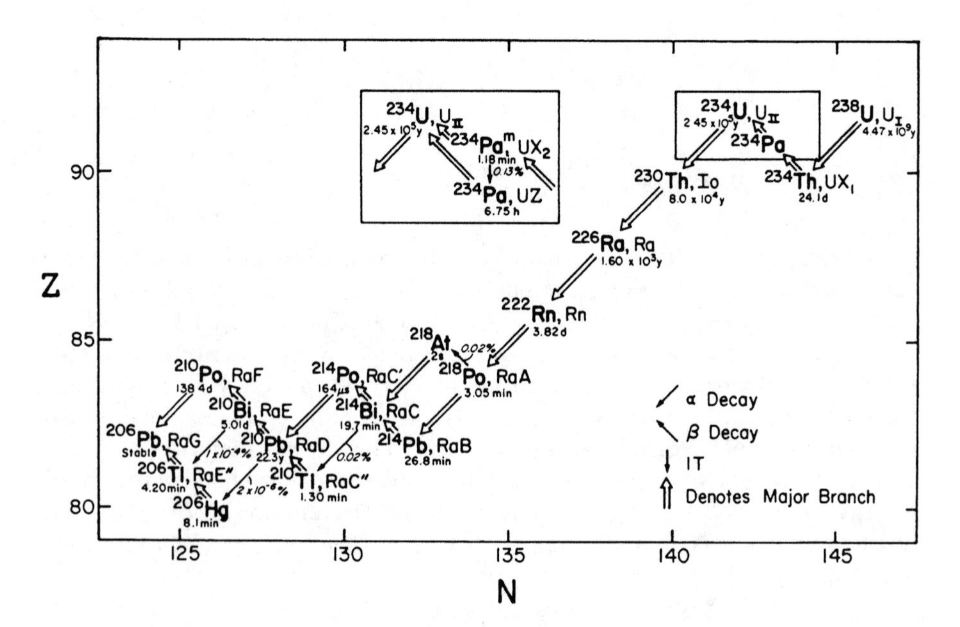

Fig. 1-1 The uranium series. IT stands for isomeric transition.

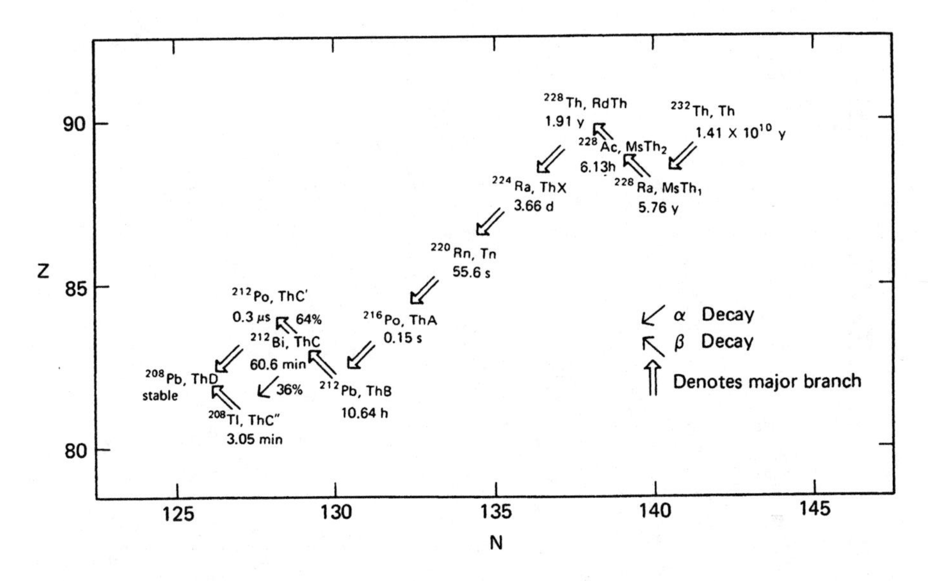

Fig. 1-2 The thorium series.

have become almost obsolete, and the designations of chemical element and mass number are now standard; we are more familiar with ^{238}U, ^{235}U, and ^{234}U than with U_I, AcU, and U_{II}. However, some of the historical names like RaA and RaB indicate immediately positions in the decay chains.

The existence of branching decays in each of the three series should be noticed. As more sensitive means for the detection of low-intensity branches have become available, more branching decays have been discovered. For example, the occurrence of astatine of mass number 219 in a 5×10^{-3} percent branch of the actinium series was recognized as late as 1953. With further refinements in technique additional branchings may be found.

One important result of the unraveling of the radioactive decay series was the conclusion reached as early as 1910, notably by Soddy, that different radioactive species of different mass numbers could nevertheless have identical chemical properties. This is the origin of the concept of isotopes, which we have already used implicitly in writing such symbols as ^{235}U and ^{238}U for uranium of mass numbers 235 and 238. Further discussion of isotopes is deferred until chapter 2, section B.

In each of the three families there is an isotope of element number 86, known as radon (sometimes called emanation). These radioactive rare-gas isotopes, referred to as radon, thoron, and actinon, are the emanations that we mentioned earlier; they were very important for the early understanding of radioactivity. It is because of the gaseous character of these substances that their descendants, A, B, C products and so on, of the three families,

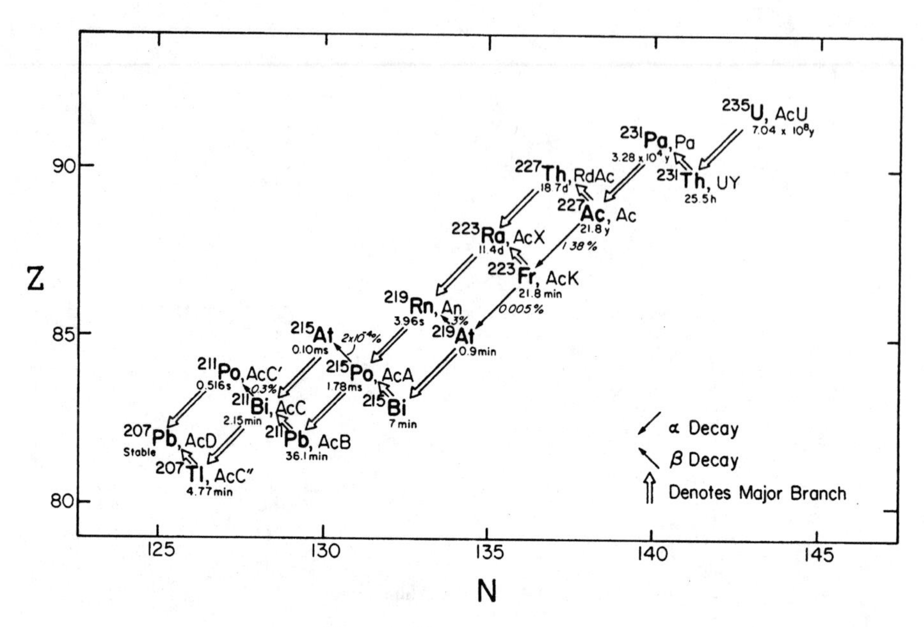

Fig. 1-3 The actinium series.

can be so readily isolated from their longer-lived precursors. These descendants of the emanations are referred to as active deposits. The active deposit from any of the three radioactive series may be collected by exposure of any object, or more efficiently of a negatively charged electrode, to the emanation.

Other Naturally Occurring Radioactivities. Since the discovery of radioactivity nearly every known element has at one time or another been examined for evidences of naturally occurring radioactivity. In 1906 N. R. Campbell and A. Wood discovered weak β radioactivity in both potassium and rubidium, and for about 25 years they remained the only known radioactive elements outside the three decay series. In 1932 G. Hevesy and M. Pahl reported a radioactivity in samarium, and more recently several other naturally occurring radioactivities have been found. The presently known natural radioactivities, other than those of the uranium, thorium, and actinium series, are listed in table 1-1, along with some of their properties. In some of these elements the particular isotope responsible for the radioactivity occurs in very low abundance, and in other cases the half lives are extremely long. Either of these factors makes such activities difficult to detect. With improvement in detection techniques additional natural radioactivities will almost certainly be discovered among the apparently stable species.

In attempts to extend the search for new radioactivities to very low intensity levels, difficulty arises from the general background of radiations

Table 1-1 Additional Naturally Occurring Radioactive Substances

Active Substance[a]	Type of Disintegration[b]	Half Life (y)	Isotopic Abundance (%)	Stable Disintegration Products
^{40}K	β^-, EC, β^+	1.28×10^{9}	0.0117	^{40}Ca, ^{40}Ar
^{87}Rb	β^-	4.8×10^{10}	27.83	^{87}Sr
^{113}Cd	β^-	9×10^{15}	12.2	^{113}In
^{115}In	β^-	5.1×10^{14}	95.7	^{115}Sn
^{138}La	EC, β^-	1.1×10^{11}	0.089	^{138}Ba, ^{138}Ce
^{144}Nd	α	2.1×10^{15}	23.8	^{140}Ce
^{147}Sm	α	1.06×10^{11}	15.1	^{143}Nd
^{148}Sm	α	8×10^{15}	11.3	^{144}Nd
^{152}Gd	α	1.1×10^{14}	0.20	^{148}Sm
^{176}Lu	β^-	3.6×10^{10}	2.61	^{176}Hf
^{174}Hf	α	2.0×10^{15}	0.16	^{170}Yb
^{187}Re	β^-	4×10^{10}	62.60	^{187}Os
^{190}Pt	α	6×10^{11}	0.013	^{186}Os

[a] There is some evidence that ^{50}V ($t_{1/2} > 7\times10^{16}$ y), ^{123}Te ($t_{1/2} > 1\times10^{15}$ y), and ^{156}Dy ($t_{1/2} > 1\times10^{18}$ y) are naturally occurring radioactive species.
[b] The symbols EC, β^-, and β^+ stand for electron capture, negatron decay and positron decay, respectively; these decay modes are described in section D, and more fully treated in chapter 3. Positron decay has been found in nature only in a very small branch of ^{40}K (0.001%).

present in every laboratory. This general background is due in part to the presence of traces of uranium, thorium, potassium, and so on, and in large part to the cosmic radiation, which is discussed in chapter 13. The cosmic rays reach every portion of the earth's surface; their intensity is greater at high altitudes but persists measurably even in deep caves and mines. The magnitude of the background effect is indicated in the discussion of radiation-detection instruments in chapter 7. In recent years there have been occasional temporary increases in background radiation due to scattered residues from large-scale atomic and thermonuclear explosions.

D. ARTIFICIALLY PRODUCED RADIOACTIVE SUBSTANCES

Historical Development. The naturally occurring radioactive substances were the only ones available for study until 1934. In January of that year I. Curie (daughter of Pierre and Marie Curie) and F. Joliot announced that boron and aluminum could be made radioactive by bombardment with the α rays from polonium (J1). This very important discovery of artificially produced radioactivity came in the course of their experiments on the production of positrons by bombardment of these elements with α particles. The positron had been discovered only two years earlier by C. D. Anderson as a component of the cosmic radiation; it is a particle much like the electron but positively charged. A number of laboratories quickly found that positrons could be produced in light elements by α-ray bombardment. The Curie-Joliot discovery was that the boron and aluminum targets continued to emit positrons after removal of the α source and that the induced radioactivity in each case decayed with a characteristic half life (reported as 14 min for B, 3.25 min for Al).

Much earlier, in 1919, Rutherford had produced nuclear transmutations by α-particle bombardment (R1), and the new phenomenon of induced radioactivity was therefore quickly understood in terms of the production of new unstable nuclei. The unstable nucleus ^{13}N is produced from boron (the stable nitrogen nuclei are ^{14}N and ^{15}N); from aluminum the product is ^{30}P (the only stable phosphorus is ^{31}P). These are examples of but one of the many types of nuclear reactions now known to produce radioactive products (cf. chapter 4).

At the time that artificial radioactivity was discovered several laboratories had developed and put into operation devices for the acceleration of hydrogen ions and helium ions to energies at which nuclear transmutations were produced. Furthermore, the discovery of the neutron in 1932 and the isolation of deuterium in 1933 made available two additional bombarding particles that turned out to be especially useful for the production of induced activities. In the 30 years following Curie's and Joliot's discovery there occurred an almost unbelievably rapid growth of the new field. Within 3 years the number of known artificially produced radioactive species reached 200, within 20 years it was about 1000, and by 1978 over 2500, with new species still being reported almost every month. The measured half lives range from microseconds[3] to many million of years. The discovery of nuclear fission by O. Hahn and F. Strassmann in 1938 (H1) gave further strong impetus to the study of new radioactive products. The subsequent development of nuclear chain reactors opened the way for the production of many radioactive substances in large quantities and for their widespread applications in such diverse fields as chemistry, physics, biology, medicine, agriculture, and engineering. Charts (G1) and tabulations (L1, N1) of the properties of radioactive species are available.

Types of Radioactive Decay. Although the first artificial radioactive substances decayed by positron emission, this is not the only or even the most common type of decay. Alpha-particle emitters are found also, but only among the heavier elements. In **α decay** two protons and two neutrons

[3] Modern electronic techniques have made possible the measurement of even much shorter half lives, down to about 10^{-11} s. Many γ emitters with such short half lives are known but are not considered as separate radioactive species for the present purpose (cf. chapter 2, section B, and chapter 3, section E).

are emitted from the nucleus together as an α particle. A typical example is

$$^{226}_{88}\text{Ra} \rightarrow {}^{222}_{86}\text{Rn} + {}^{4}_{2}\alpha + Q.$$

The energy released in the disintegration is represented by Q.

Ordinary **β decay**, as in the natural radioactive series, is commonly found throughout the range of the periodic table. In this type of decay a negative electron is emitted by the nucleus and the atomic number is *increased* by one unit, as illustrated below:

$$^{14}_{6}\text{C} \rightarrow {}^{14}_{7}\text{N} + \beta^- + \bar{\nu} + Q,$$

where $\bar{\nu}$ stands for the antineutrino (described in chapter 3, section D). **Positron emission**, also a β-decay process, results in a *decrease* by one unit in atomic number as in the following example:

$$^{22}_{11}\text{Na} \rightarrow {}^{22}_{10}\text{Ne} + \beta^+ + \nu + Q,$$

where ν represents the neutrino. **Electron capture** (EC) is a third type of β decay in which the atomic number is *decreased* by one unit, as in positron emission, but in this case by spontaneous incorporation into the nucleus of one of the atomic electrons (most often one from the K shell of the atom). In these three β-decay processes the atomic mass decreases only very slightly, the mass *number* remaining unchanged.

Spontaneous fission is a decay mode of some heavy nuclei in which the nucleus breaks up into two intermediate-mass fragments and several neutrons. An example is

$$^{252}_{98}\text{Cf} \rightarrow {}^{140}_{54}\text{Xe} + {}^{108}_{44}\text{Ru} + 4n + Q.$$

Alpha decay, β decay, and spontaneous fission may leave the nucleus in an excited state, which may de-excite by the emission of electromagnetic radiation called γ rays. All of these decay modes are described in more detail in chapter 3.

Synthetic Elements and the $4n+1$ Series. Not only has it been possible by transmutation techniques to produce radioactive isotopes of every known element, but also a number of elements not found in nature have been synthesized. In each case the new element has first been recognized in unweighably small amounts detectable only by its radioactivity; however, macroquantities of many of these new elements have now been prepared. The best known of the synthetic elements is probably plutonium, an element that within less than five years of its discovery was available in sufficient quantities—kilogram amounts—to serve as an ingredient in atomic bombs. Up to 1980 14 new elements beyond uranium in the periodic table had been produced by artificial transmutations, as had the elements technetium (atomic number 43) and promethium (number 61), which are not known to occur naturally on the earth. Among the artificially produced radioactivities in the heavy-element region are many members of

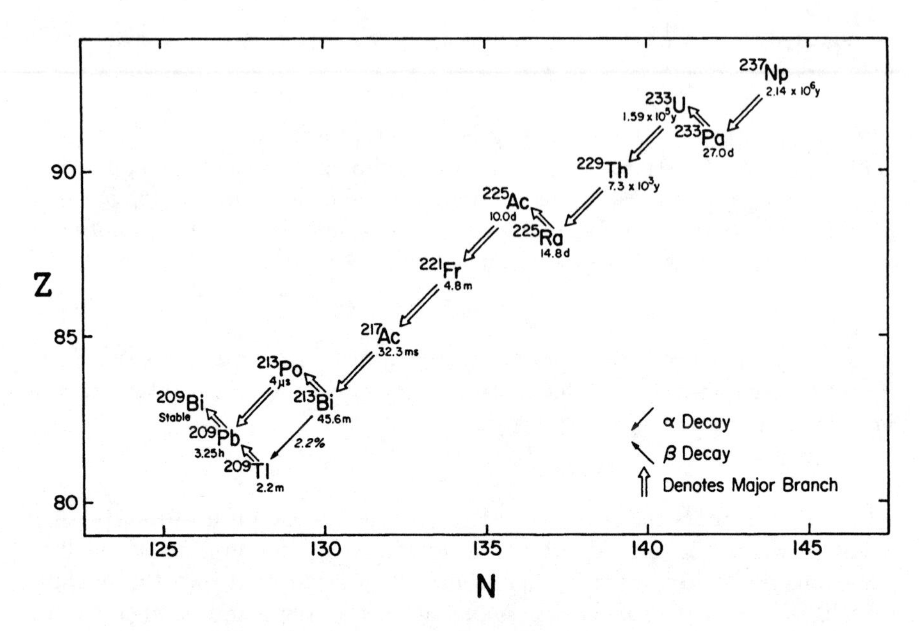

Fig. 1-4 The $4n + 1$ series.

the $4n + 1$ series—the one family missing in nature (see figure 1-4). This well developed series generally resembles the three naturally occurring families and has ^{209}Bi as its stable end product. The reason it does not occur in nature is that its longest-lived member, ^{237}Np, has a half life of only 2.1×10^6 y, far too short to have survived since the time of the formation of the elements.

Superheavy Elements. None of the known species beyond uranium have sufficiently long half lives to have survived in nature over the span of several billion years since the elements in the solar system were formed (cf. chapter 13, section A). A possible exception is the longest-lived known transuranium species, ^{244}Pu (half life 8.1×10^7 y), which has been reported in minute concentrations.

In general, as we progress beyond uranium to heavier elements, the half lives become shorter. However, in recent years there has been much speculation, based on theoretical considerations of nuclear structure, about the possible existence of a new "island of stability" far beyond uranium, near atomic numbers 110–114 and mass number 300. Much effort has gone into attempts to find such superheavy elements in nature as well as to produce them by nuclear reactions. Up to 1980 none of these attempts had been successful (H2, S1). The subject of superheavy elements is further discussed in chapters 3, 4, and 10.

REFERENCES

B1 H. Becquerel, *Compt. Rend.* **122**, 420, 501, 559, 689, 762, 1086 (1896).

C1 P. Curie and M. Sklodowska Curie, "Sur une Substance Nouvelle Radio-active, Contenue dans la Pechblende," *Compt. Rend.* **127**, 175 (1898).

G1 General Electric Co., *Chart of the Nuclides*, 12th ed., Schnectady, New York, 1977.

H1 O. Hahn and F. Strassmann, "Ueber den Nachweis und das Verhalten der bei der Bestrahlung des Urans mittels Neutronen entstehenden Erdalkalimetalle," *Naturwissenschaften* **27**, 11 (1939).

H2 G. Herrmann, "Superheavy Element Research," *Nature* **280**, 543 (1979).

J1 F. Joliot and I. Curie, "Artificial Production of a New Kind of Radio-Element," *Nature* **133**, 201 (1934).

*J2 G. E. M. Jauncey, "The Early Years of Radioactivity," *Am. J. Phys.* **14**, 226 (1946).

L1 M. Lederer and V. Shirley, Eds., *Table of Isotopes*, 7th ed., Wiley, New York, 1978.

M1 S. Meyer and E. Schweidler, *Radioaktivität*, B. G. Teubner, Berlin, 1927.

N1 *Nuclear Data Sheets*, Academic Press, New York.

R1 E. Rutherford, "Collision of α Particles with Light Atoms, IV. An Anomalous Effect in Nitrogen," *Phil. Mag.* **37**, 581 (1919).

*R2 E. Rutherford, J. Chadwick, and C. D. Ellis, *Radiations from Radioactive Substances*, Cambridge University Press, New York, 1930.

S1 G. T. Seaborg, W. Loveland, and D. J. Morrissey, "Superheavy Elements: A Crossroads," *Science* **213**, 711 (1979).

EXERCISES

1. One hundred milligrams of ^{226}Ra would represent what percentage yield from 2000 kg of a pitchblende ore containing 75 percent U_3O_8? *Answer:* 23 percent.
2. Calculate the rate of energy liberation (in calories per hour) for 1.00 g of pure radium (^{226}Ra) free of its decay products. What can you say about the actual heating effect of an old radium preparation? *Answer to first part:* 25 cal h^{-1}.
3. A certain active substance (which has no radioactive parent) has a half life of 8.0 d. What fraction of the initial amount will be left after (a) 16 d, (b) 32 d, (c) 4 d, (d) 83 d? *Answer:* (a) 0.25, (c) 0.707.
4. How long would a sample of ^{226}Ra have to be observed before the decay amounted to 1 percent? (Neglect effects of radium A, B, C, and so on, on the detector.)
5. Find the number of disintegrations of ^{238}U atoms occurring per minute in 1 mg of ordinary uranium, from the half life of ^{238}U, $t_{1/2} = 4.47 \times 10^9$ y.
6. How many β disintegrations occur per second in 1.00 g of pitchblende containing 70 percent uranium? You may assume that there has been no loss of radon from the ore. *Answer:* $5.3 \times 10^4\ s^{-1}$.
7. If 1 g of radium (^{226}Ra) is separated from its decay products and then placed in a sealed vessel, how much helium will accumulate in the vessel in 60 days? Express the answer in cubic centimeters at STP. *Answer:* 0.028 cm^3.

8. What is the natural radioactivity in disintegrations per minute per milligram of ordinary potassium chloride (KCl)? *Answer:* 1.0 dis min^{-1} mg^{-1}.

9. Compute (a) the weight of 1 Ci of ^{222}Rn, (b) the weight of 1 Ci of ^{32}P, (c) the disintegration rate of 1 cm^3 of tritium ($^3H_2$) at STP. *Answer:* (a) 6.50 μg.

Chapter 2

Atomic Nuclei

A. ATOMIC STRUCTURE

Early Views. At the time the phenomenon of radioactivity was discovered the chemical elements were regarded as unalterable; they were thought to retain their identities throughout all chemical and physical processes. This view became untenable when it was recognized that radioactive disintegration involved the transformation of one element into another. As a result of J. J. Thomson's discovery of the electron in 1897, it had already become clear that atoms, until then regarded as the indivisible building blocks of matter, must have some structure. From experiments on the scattering of X rays and electrons by matter, Thomson and others concluded that the number of electrons per atom was approximately equal to the atomic weight.[1] This conclusion, together with Thomson's determination of the electron mass as approximately one two-thousandth of the mass of a hydrogen atom, led to the assumption that most of the mass of an atom must reside in its positively charged parts.

The problem that remained to be solved was: how are the positive and negative charges distributed inside the atom?

Alpha-Particle Scattering. The question just posed was eventually answered by E. Rutherford as a result of experiments on the scattering of α particles by thin metal foils carried out in his laboratory. If a collimated beam of α particles is allowed to strike a thin film of matter, some of the particles are deflected from their original direction in passing through the film. This scattering is clearly caused by the electrostatic forces between the positively charged α particle and the positive and negative charges in the atoms of the scattering material. In the then current model (due to Thomson, 1910) atoms were considered to consist of electrons imbedded in a positively charged mass distributed uniformly over the volume of the atom. The scattering experiments of H. Geiger and E. Marsden proved that deflections through large angles, ranging well above 90°, occurred much

[1] It is, as we now know, more nearly equal to half the atomic weight, and this was recognized as early as 1911 by C. G. Barkla.

more frequently than could be accounted for by either single or multiple scattering from such atoms. These results led Rutherford to propose a different atomic model, which rapidly became universally accepted.

The Nuclear Model of the Atom. In his classic paper of 1911 (R1) Rutherford postulated that the observed large-angle scattering was due to single scattering processes, and that these large-angle scatterings could be produced only by an intense electric field; consequently the positive charge and most of the mass of the atom had to be concentrated in a very small region, later known as the nucleus. A number of electrons sufficient to balance the positive charge was thought to be distributed over a sphere of atomic dimensions. Rutherford then proceeded to show that for large-angle scatterings of α particles by this sort of atom the effect of the electrons was negligible compared to that of the central charge. Considering this central charge (Ze) of the atom and the charge ($Z_\alpha e = 2e$) of the α particle as point charges, Rutherford then merely assumed the force between them at any distance d to be given by Coulomb's law: $F = Ze \cdot Z_\alpha e/d^2$. On this basis, and with the additional simplifying assumption that the nucleus is sufficiently heavy to be considered at rest during the encounter, Rutherford showed that the path of an α particle in the field of a nucleus is a hyperbola with the nucleus at the external focus. From the conditions of conservation of momentum and energy and from the geometric properties of the hyperbola he then derived[2] his celebrated scattering formula, which relates the number $n(\theta)$ of α particles falling on a unit area at a distance r from the scattering point to the scattering angle θ (the angle between the directions of incident and scattered particle):

$$n(\theta) = n_0 \frac{Nt}{16r^2}\left(\frac{Ze \cdot Z_\alpha e}{\frac{1}{2}M_\alpha v_\alpha^2}\right)^2 \frac{1}{\sin^4(\theta/2)}, \tag{2-1}$$

where n_0 is the number of incident α particles, t is the thickness of scatterer, N is the number of nuclei per unit volume of scatterer, and M_α and v_α are the mass and initial velocity of the α particle.

The specific predictions of the Rutherford formula were quickly subjected to experimental test, principally by Geiger and Marsden. They verified that, for heavy-element scatterers, the number of scattered particles detected per unit area was indeed inversely proportional both to the fourth power of the sine of half the scattering angle and to the square of the α-particle energy. For light-element scatterers agreement was found to be excellent also, provided the theory was suitably modified to take into account the fact that the nuclei cannot be assumed to be at rest during impact.

Nuclear Charge and Atomic Number. The experimental verification of the scattering formula led to a general acceptance of Rutherford's

[2] For detailed derivations see, for example, references R2 (p. 191) and S1 (p. 22).

picture of the atom as consisting of a small positively charged nucleus containing nearly the entire mass of the atom, and surrounding it, a distribution of negatively charged electrons. In addition, it was now possible to study the magnitude of the nuclear charge in the atoms of a given element through scattering experiments since, according to the scattering law, the scattered intensity depends on the square of the nuclear charge. It was by the method of α-particle scattering that nuclear charges were first determined, and this work led to the suggestion that the atomic number Z of an element, until then merely a number indicating its position in the periodic table, was identical with the nuclear charge (expressed in units of the electronic charge e). This suggestion was subsequently confirmed by H. G. J. Moseley's work on the X-ray spectra of the elements (M1). Moseley showed that the frequencies of the K X-ray emission lines increase regularly from element to element when the elements are arranged in the order of their appearance in the periodic system. The relation between frequency and atomic weight showed irregular variations, but when each element was assigned an "atomic number" Z, according to its position in the periodic table, Moseley noticed that the square root of the K X-ray frequency was proportional to $Z-1$. He identified the atomic number with the number of unit charges on the nucleus. This number (which is also the number of extranuclear electrons in the neutral atoms) was thus shown to be closely related to the chemical properties of an element.

Following the acceptance of Rutherford's nuclear model of the atom, the further understanding of atomic structure developed rapidly through the study of X-ray and optical spectra and culminated in N. Bohr's theory of 1913 and E. Schroedinger's and W. Heisenberg's quantum-mechanical description of the atom in 1926. Further discussion of the extranuclear features of atomic structure is, however, outside the scope of this book.

B. COMPOSITION OF NUCLEI

Nuclear Size and Density. The α-particle scattering experiments of Rutherford and his school not only confirmed the nuclear model and led to the determination of nuclear charges but, as we see in Section C, also gave the first information of the sizes of nuclei, establishing that their dimensions are of the order of 10^{-12} cm, roughly 10^{-4} times the sizes of atoms. Since the nucleus contains nearly the entire mass of an atom, it follows that nuclei are very much denser than ordinary matter; the density of nuclear matter is in the neighborhood of 10^{14} g cm^{-3} or 10^{8} tons cm^{-3}.

Once it was established that almost the entire mass of an atom resides in its nucleus, and that the atoms of each element have nuclei of characteristic charge, it became evident that radioactive transformations are, in fact, nuclear processes. The newly discovered nuclei thus could not be regarded as indivisible entities, but they had to have some structure of their own. The forces holding nuclei together had to be strong and of short range; their exact nature was not well understood and is still a subject of lively research (see chapter 10, section A).

Isotopes and Integral Atomic Weights. The existence of isotopes, as we saw briefly in chapter 1, became evident when different radioactive bodies in the naturally occurring decay series, for example, RaB, AcB, and ThB, were found to exhibit identical chemical properties (in the case mentioned, the properties of lead). This discovery led to a search for the existence of isotopes in nonradioactive elements. In early experiments with ion deflections in magnetic and electric fields Thomson showed in 1913 that neon consisted of two isotopes with atomic weights about 20 and 22 (later a third neon isotope of atomic weight 21 was found). Subsequently, it was established, principally as a result of F. W. Aston's pioneering work with his mass spectrograph, that most elements consist of mixtures of isotopes and that the atomic weights of the individual isotopes are almost exactly integers (the integer nearest the atomic weight of an isotope now being called its mass number A). This "whole number rule" of Aston's naturally led to a revival, in modern terms, of the hypothesis proposed a hundred years earlier by W. Prout, namely that all elements are built up of hydrogen.

The nucleus of the common hydrogen atom, called a proton, is the simplest known nucleus. Its positive charge is equal in magnitude to the negative charge on an electron, 4.80325×10^{-10} electrostatic units (esu). The mass of a proton is approximately equal to that of a hydrogen atom and therefore nearly equal to 1 on the atomic weight scale.

Proton-Electron Hypothesis. With the charges of nuclei known to be exact multiples and the masses very nearly exact multiples of the charge and mass, respectively, of the proton, it was natural to suppose that all nuclei were built up of protons. Prior to the discovery of the neutron it was generally thought that a nucleus of mass number A and atomic number Z contained A protons (accounting for its mass) and $A-Z$ electrons (to make the net positive charge Z).

This proton–electron hypothesis posed a number of difficulties. In order to be contained in a nucleus an electron would presumably have to have a **de Broglie wavelength** $\lambda = h/mv$ no larger than nuclear dimensions ($\sim 10^{-12}$ cm). The kinetic energy corresponding to such a de Broglie wavelength is more than an order of magnitude larger than the energies of β particles emitted by nuclei, which cast grave doubts on the idea of free electrons as constituents in nuclei. Other difficulties stemmed from considerations of angular-momentum conservation and statistics in nuclei with odd Z and even A, such as ^{14}N (see section C).

To circumvent some of the problems encountered by the hypothesis of free electrons in nuclei, Rutherford suggested as early as 1920 the existence in nuclei of the "neutron," a close combination of a proton and an electron.

Discovery of the Neutron. Many fruitless attempts were made to find evidence for the neutron postulated by Rutherford. Success finally came in 1932 to J. Chadwick (C1) in the course of his investigations of a very penetrating radiation previously observed by other experimenters when they bombarded beryllium and boron with α particles. When this radiation was found to be capable of ejecting energetic protons from hydrogen-containing substances such as paraffin, the previously held view that one was dealing with high-energy γ rays became untenable. Chadwick showed that all the evidence was compatible with the assumption that the radiation consisted of neutrons, that is, neutral particles of zero charge and of approximately the mass of protons. Later, more precise measurements showed the neutron mass to be about 0.08 percent larger than the mass of a hydrogen atom.

Being electrically neutral, neutrons do not cause any primary ionization in passing through matter and are therefore not so readily detected as charged particles. Furthermore, they are not stable in the free state but undergo β-decay, disintegrating into protons and electrons with a half life of about 11 min. It was probably for these reasons that neutrons escaped discovery for so long.

Proton-Neutron Hypothesis. Because of the difficulties we mentioned earlier, the proton-electron hypothesis was quickly discarded after the discovery of the neutron and replaced by the now accepted proton-neutron hypothesis of nuclear composition. According to this picture, the number of protons in a nucleus equals its atomic number Z, and the total number of neutrons and protons (collectively called **nucleons**) equals its mass number A. Therefore the neutron number N equals $A - Z$. Thus the nucleus of ^{14}N is thought to contain seven protons and seven neutrons.

The atomic numbers of the known elements range from 1 for hydrogen to 106 for the heaviest known transuranium element. Nuclei with neutron numbers 0 to 159 are known, and known mass numbers range from 1 to 263. The difference $N - Z$ (or $A - 2Z$) between the numbers of neutrons and protons in a nucleus is referred to as its **neutron excess** or **isotopic number**.

The symbol used to denote a nuclear species is the chemical symbol of the element with the atomic number as a left subscript and the mass number as a left superscript (in the older U.S. literature a right superscript was more common), for example, $^{4}_{2}He$, $^{59}_{27}Co$, $^{235}_{92}U$. The atomic number is often omitted because it is uniquely determined by the chemical symbol.

Isotopes and Nuclides. As we have already mentioned, atomic species of the same atomic number, that is, belonging to the same element but having different mass numbers, are called **isotopes**. In the nuclei of the different isotopes of a given element the number of protons characteristic of that element is combined with different numbers of neutrons. For example, a $^{35}_{17}Cl$ nucleus contains 17 protons and 18 neutrons, whereas a $^{37}_{17}Cl$ nucleus contains 17 protons and 20 neutrons. Deuterium, a rare isotope of hydrogen, has a nucleus containing one proton and one neutron.

As a result of mass-spectrographic investigations we now know that the elements with atomic numbers between 1 and 83 have on the average more than three stable isotopes each. Some elements, such as beryllium, phosphorus, arsenic, and bismuth, each have a single stable nuclear species, whereas tin, for example, has as many as 10 stable isotopes.

The stable isotopes of a given element generally occur together in constant proportions. This accounts for the fact that atomic weight determinations on samples of a given element from widely different sources generally agree within experimental errors. However, there are some notable exceptions to this rule of constant isotopic composition. One is the variation in the abundances of lead isotopes, especially in ores containing uranium and thorium. Depending on the age and composition of such ores, the end products of the three radioactive families, ^{206}Pb, ^{207}Pb, and ^{208}Pb, and the nonradiogenic ^{204}Pb may occur in different proportions. Similarly, the isotope ^{87}Sr has been found to have an abnormally high abundance in rocks that contain rubidium; this is explained by the β decay of the naturally occurring ^{87}Rb.

Helium from gas wells probably has its origin in radioactive processes (α disintegrations) and contains a much smaller proportion of the rare isotope ^{3}He than does atmospheric helium. Water from various sources shows slight variations in the $^{1}H/^{2}H$ ratio. This is in some cases due to the slightly lower vapor pressure of heavy water compared to ordinary water. The enrichment of ^{2}H in the water of the Dead Sea and in certain vegetables is ascribed to this cause. The waters that show abnormally high ^{2}H concentrations usually also have slightly higher than normal $^{18}O/^{16}O$ ratios. Another cause for small variations in isotopic composition is that chemical equilibria are slightly dependent on the molecular weights of the reactants, and this may lead to isotopic enrichments in the course of reactions occurring in nature. For example, the slight enrichment of ^{13}C in limestones relative to some other sources of carbon comes about because the equilibrium in the reaction between CO_2 and water to form bicarbonate ion lies somewhat further toward the side of bicarbonate for $^{13}CO_2$ than for $^{12}CO_2$. The effects of isotopic substitution on equilibria and rates of chemical reactions are discussed in chapter 11, section A.

The word isotope has also been used in a broader sense to signify any particular nuclear species characterized by its A and Z values. In this meaning it should be, and now generally is, replaced by the word **nuclide**, defined as a species of atom characterized by the constitution of its nucleus, in particular by the numbers of protons and neutrons in its nucleus.

Isobars, Isotones, and Isomers. Atomic species having the same mass number but different atomic numbers are called **isobars**. Examples among stable nuclei are: $^{76}_{32}Ge$ and $^{76}_{34}Se$; $^{130}_{52}Te$; $^{130}_{54}Xe$ and $^{130}_{56}Ba$.

Atomic species having the same number of neutrons but different mass numbers are sometimes referred to as **isotones**. For example, $^{30}_{14}Si$, $^{31}_{15}P$, and $^{32}_{16}S$ are isotones because they all contain 16 neutrons per nucleus.

As early as 1922 Hahn was able to prove through careful radiochemical work that two of the naturally occurring radioactive bodies, UX_2 and UZ, had to be assigned the same mass number (234) as well as the same atomic number (91) although they differ in their radioactive properties (cf. figure 1-1). This was the first example of nuclear **isomerism**, a phenomenon that, despite this early discovery, received little attention until 15 years later. Then another pair of isomers was found among artificially radioactive species, in ^{80}Br. Since then it has become clear that nuclear isomerism is by no means a rare phenomenon—about 500 pairs of isomers have been characterized. Nuclear isomers are different energy states of the same nucleus, each having a different measurable lifetime (except that the ground state may be stable).[3] In a number of cases more than two isomeric states have been found for a given A and Z. For example, three radioactive species of half lives 60 d, 1.6 min, and 20 min have been assigned to ^{124}Sb. The notation that has become standard for representing isomeric states other than the ground state is a right superscript m (for metastable); if there are two or more excited isomeric states, they are labeled m_1, m_2 and so on, in order of increasing excitation energy. Thus the isomers of ^{124}Sb are denoted as $^{124}Sb^g$ (ground state, 60-day half life), $^{124}Sb^{m_1}$ (1.6 min), and $^{124}Sb^{m_2}$ (20 min). We do not consider each isomeric state as an individual nuclide.

C. NUCLEAR PROPERTIES

In this section we are primarily concerned with the static properties of nuclei in their *ground*, or lowest-energy, states. At the end of the section we briefly mention excited-state properties, but these are chiefly dealt with in connection with radioactive decay processes (chapter 3) and nuclear models (chapter 10).

1. *Mass and Energy*

Mass Scales and Units. Masses of atomic nuclei are so small when stated in ordinary units (less than 10^{-21} g) that they are generally expressed

[3] To specify what constitutes a "measurable lifetime" has become difficult through continued extension of lifetime-measurement techniques to shorter and shorter time scales (cf. discussion of isomerism in chapter 3, section E). The number of known isomers quoted above includes half lives down to 10^{-6} s.

on a different scale. The scale that is now universally used is based on the mass of an atom of ^{12}C taken as exactly 12.000000 units.[4]

It should be noted that *mass tables always give atomic rather than nuclear masses*; in other words, the masses quoted include the masses and the binding energies of the extranuclear electrons in the neutral atoms. This convention, as we shall see, turns out to have some advantages in the treatment of nuclear reactions and energy relations. More importantly, however, it arises from the fact that it is always *atomic* masses or differences between *atomic* masses that are measured experimentally.

The experimental determination of exact atomic masses involves the use of a mass spectrograph or mass spectrometer. In most of these instruments the charge-to-mass ratio of positive ions is determined from the amount of deflection in a combination of magnetic and electric fields; different arrangements are used for bringing about velocity focusing or directional focusing, or both, for ions of a given e/M. Instruments that use photographic plates for recording the mass spectra are called mass spectrographs; those that make use of collection and measurement of ion currents are referred to as mass spectrometers. The fact that ions of the same kinetic energy and different masses require different times to traverse a given path length has been utilized in the design of several types of so-called time-of-flight (ToF) mass spectrometers. These devices have proved particularly useful for the determination of accurate mass values.

Mass determinations throughout the mass range from hydrogen to bismuth have been made with precisions varying between about 0.01 and 1 part per million (ppm). For precision mass determinations the method generally used is the so-called doublet method. This substitutes the measurement of the difference between two almost identical masses for the direct measurement of absolute masses. All measurements must, of course, eventually be related to the standard ^{12}C. But for convenience the masses of ^1H, ^2H, and ^{16}O have been adopted as secondary standards and for this purpose have been carefully measured by determinations of the fundamental doublets:

$(^{12}C^1H_4)^+$ and $(^{16}O)^+$ at mass-to-charge ratio 16,
$(^2H_3)^+$ and $(^{12}C)^{2+}$ at mass-to-charge ratio 6,
$(^2H)^+$ and $(^1H_2)^+$ at mass-to-charge ratio 2.

On the ^{12}C scale the mass of a hydrogen atom (sometimes loosely called

[4] Prior to the adoption of the ^{12}C scale two different scales were used: the physical atomic-weight scale, based on the mass of ^{16}O taken as exactly 16.00000 units; and the chemical atomic-weight scale, in which the natural isotopic mixture of oxygen (containing small amounts of ^{17}O and ^{18}O) was assigned the value 16.00000. Care is thus indicated in the use of the older literature. The ^{12}C scale and the old chemical scale differ by only 0.005 percent so that chemical atomic weights have remained virtually unaffected. However, on the old physical ^{16}O scale all atomic masses were about 0.0318 percent larger than on the ^{12}C scale.

the proton mass) is 1.007825037(10),[5] the mass of a neutron 1.008665012(37), and that of an electron 0.00054858026(21) mass units. One mass unit equals $1.6605655(86) \times 10^{-24}$ g.

Mass Versus Energy. One of the important consequences of A. Einstein's special theory of relativity[6] is the equivalence of mass and energy. The total energy content E of a system of mass M is given by the relation

$$E = Mc^2,$$

where c is the velocity of light (2.9979246×10^{10} cm s^{-1}). Therefore the mass of a nucleus is a direct measure of its energy content. The measured mass of a nucleus is always smaller than the combined masses of its constituent nucleons, and the difference between the two is called the **binding energy** of the nucleus.

To find the energy equivalent to 1 mass unit we put $M = 1.660566 \times 10^{-24}$ g and find $E = Mc^2 = 1.492442 \times 10^{-3}$ erg. However, energy units much more useful in nuclear work than the erg are the electron volt (eV), the kiloelectron volt (keV; 1 keV = 1000 eV), and the million electron volt (MeV; 1 MeV = 10^6 eV). The **electron volt** is defined as the energy necessary to raise one electron through a potential difference of 1 V.

$$1 \text{ eV} = 1.602189(5) \times 10^{-12} \text{ erg}.$$

Using these new units we find

$$1 \text{ atomic mass unit (amu)} = 931.502(3) \text{ MeV},$$

$$1 \text{ electron mass } (m_e) = 0.5110034(2) \text{ MeV}.$$

As an example, we calculate the binding energy of ^{4}He. The mass of ^{4}He is 4.0026033 amu; the combined mass of two hydrogen atoms[7] and two neutrons is 4.0329801 amu. Thus the difference between these two numbers, the binding energy of ^{4}He, is 0.0303768 amu, or $0.0303768 \times 931.502 = 28.2960$ MeV. The binding energy per nucleon in ^{4}He is therefore approximately 7.1 MeV. The binding energy of the deuteron calculated by the same method is found to be 2.2246 MeV.[8]

Binding Energies. The average binding energy per nucleon is remarkably constant in all nuclei except for a few of the lightest ones. For $A > 11$ it ranges between 7.4 and 8.8 MeV throughout the table of elements, with

[5] The number in parentheses gives the uncertainty in the last digits.

[6] A summary of the most frequently used relativistic equations may be found in appendix B.

[7] Since the mass of ^{4}He includes the mass of two electrons, it is clear that it is also the *atomic* mass of ^{1}H that must be used.

[8] The deuteron binding energy is, in fact, an experimentally determined quantity (from the minimum photon energy required to disintegrate a deuteron into a proton and a neutron). Together with the measured masses of proton and deuteron, this binding energy is used to derive the neutron mass.

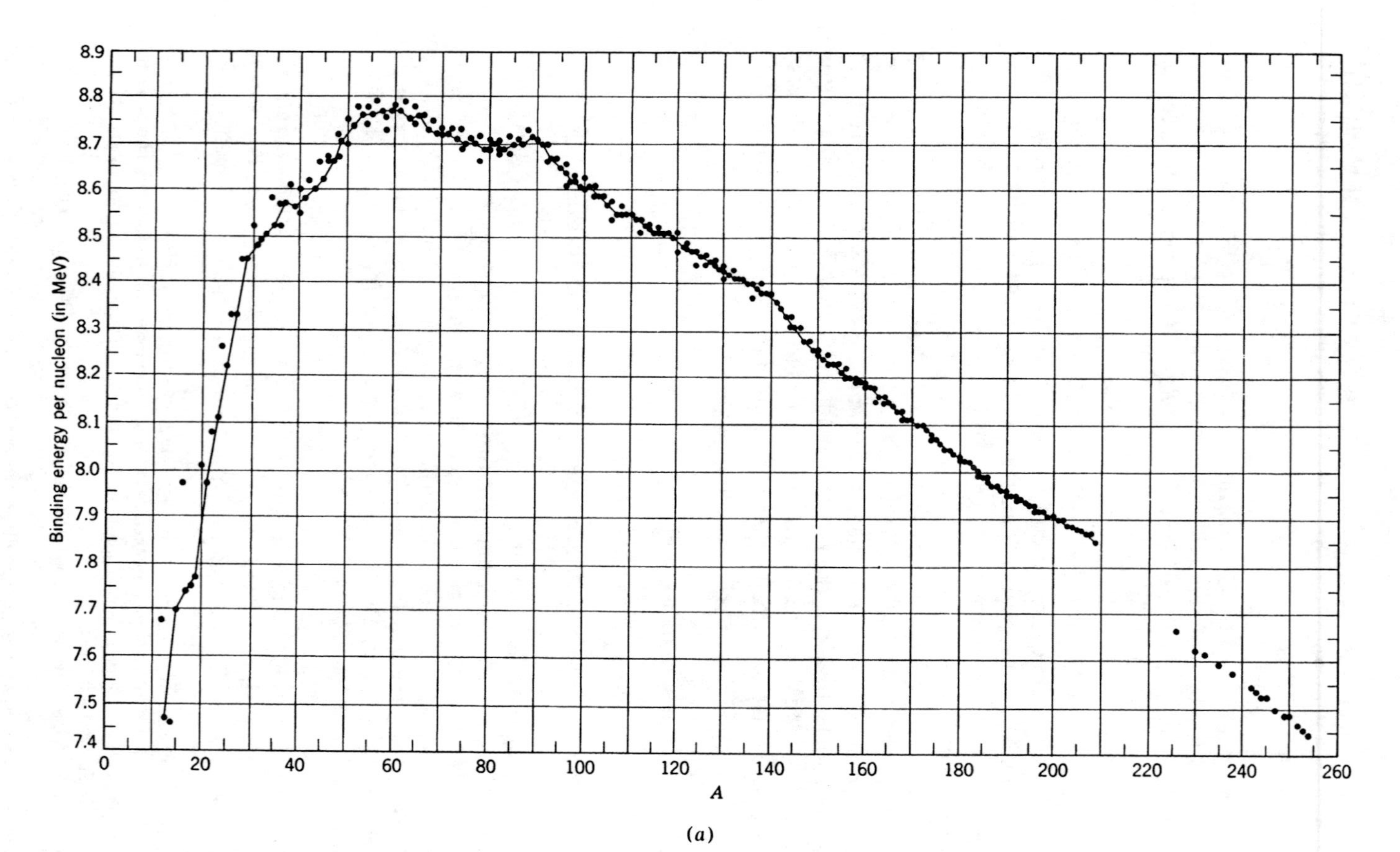

Fig. 2-1 Average binding energy per nucleon as a function of A per stable nuclei. (a) $12 \leq A \leq 250$, with line connecting the odd-A points.

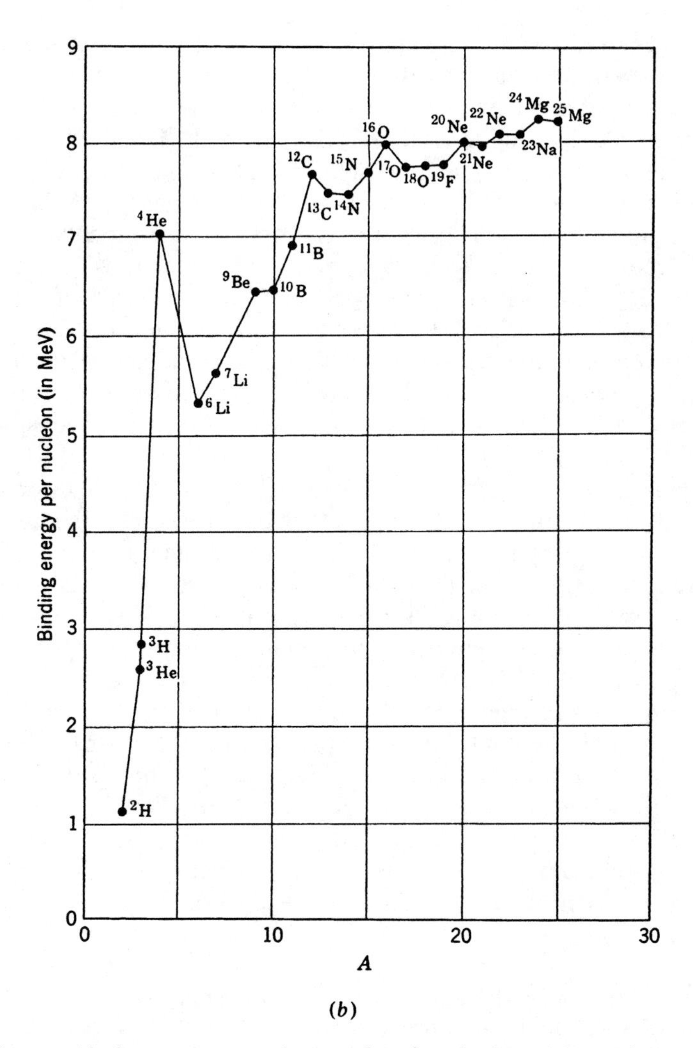

(*b*)

Fig. 2-1 Average binding energy per nucleon as a function of A for stable nuclei. (*b*) $2 \leq A \leq 25$.

the maximum values (near 8.8 MeV) occurring in the vicinity of $A = 60$ (for iron and nickel nuclei). In figure 2-1 the average binding energy per nucleon is plotted as a function of A for all stable and some of the heavy radioactive nuclei. From the maximum in the iron region the values are seen to decrease more slowly towards the high-A than towards the low-A side. Despite the near-constancy of the average binding energy, some interesting details can be discerned. Among the lighter nuclei the value for a nucleus of even A is generally higher than the average of the values for the adjacent odd-A nuclei. The same is true at higher mass numbers when the most stable nucleus of a given even A (there are often two and occasionally three) is compared with the neighboring odd-A nuclei. The slight deviations from a completely smooth curve (e.g., at $A \cong 88$) are real and well established and are discussed later in connection with nuclear shell structure. A number of irregularities occur among the lightest nuclei; in particular, the binding energies of $^{4}_{2}He$, $^{12}_{6}C$, and $^{16}_{8}O$ are very high.

The behavior of the binding-energy curve as a function of A has several important consequences. Thus the very exoergic nature of the fusion of hydrogen atoms to form helium—the process that presumably gives rise to the sun's radiant energy—follows immediately from the very large binding energy of ^{4}He. Similarly, the energy released in the fission of the heaviest nuclei is large because nuclei near the middle of the periodic table have higher binding energies per nucleon. Finally the maximum in the nuclear stability curve in the iron-nickel region is thought to be responsible for the abnormally high natural abundances of these elements (cf. chapter 13, p. 510).

The quantity most frequently tabulated is neither the total atomic mass M nor the total binding energy, but rather the **mass excess** (sometimes also called the mass defect) $\Delta = M - A$, where A is the mass number. The older literature frequently refers also to the packing fraction $f = \Delta/A$. Tables of mass excesses are available in the literature (e.g., *W*1, which lists both mass excess and binding energy). The table of nuclides in appendix D gives mass excesses.

Although the *average* binding energy per nucleon is a rather slowly varying function, the contribution to the binding energy from the addition of *one* more proton or neutron shows large fluctuations from one nucleus to the next. (Chemists may enjoy thinking of the binding energy of one additional nucleon as a sort of partial molal binding energy.) The quantity may be defined here as the mass of the nucleus plus the mass of the additional nucleon minus the mass of the resulting nucleus, expressed in energy units.

As an illustration of the fluctuations in this quantity consider the binding energies for an additional neutron to ^{45}Ti, ^{46}Ti, ^{47}Ti, ^{48}Ti, ^{49}Ti, and ^{50}Ti, which have values of 13.19, 8.88, 11.63, 8.15, 10.94, and 6.38 MeV, respectively; the even-odd effect is much more pronounced here than with the *average* binding energy per nucleon. Similar trends are found in the binding energies for additional protons; for example, proton addition to the nuclei $^{122}_{50}Sn$, $^{123}_{51}Sb$, $^{124}_{52}Te$, $^{125}_{53}I$, and $^{126}_{54}Xe$ involves the liberation of 6.57, 8.57, 5.67, 7.56, and 4.32 MeV, respectively.

For some purposes it is convenient to consider the binding energies of nuclear aggregates, such as α particles, in particular nuclei.

The binding energy of an α particle (^{4}He, mass 4.002603) in ^{235}U (mass 235.043915) may be obtained from these masses and the mass of ^{231}Th (231.036291): $231.036291 + 4.002623 - 235.043915 = -0.005001$ amu $=$ -4.658 MeV. The negative binding energy means that the ^{235}U atom is, as we already know, thermodynamically unstable with respect to decomposition into ^{231}Th and ^{4}He.

Alpha-particle binding energies are, in fact, negative in all "stable" nuclei with $A \gtrsim 140$; the reason for the apparent stability are discussed in chapter 3.

The masses of some radioactive nuclei can be determined from an accurate knowledge of the energy balance in nuclear reactions involving these nuclei and from their disintegration energies. This subject is discussed in chapter 4, section A.

2. *Radius* (B1)

We have already mentioned that nuclei have dimensions of the order of 10^{-12} cm. The unit of length that is usually used in discussing nuclear radii is the **fermi** (fm)[9]: 1 fm = 10^{-13} cm.

All experiments designed for the investigation of nuclear radii lead to the conclusion that, in crude approximation at least, nuclear radii can be represented by the simple formula

$$R = r_0 A^{1/3}, \tag{2-2}$$

where r_0 is a constant independent of A. In other words, nuclear volumes are very nearly proportional to nuclear masses, and thus all nuclei have approximately the same density. We note that, although nuclear densities are high compared to ordinary matter (cf. p. 19), nuclei are by no means densely packed with nucleons; this is an important factor in the success of the nuclear shell model (see section E and chapter 10, section D).]

Different experimental methods lead to somewhat different values of r_0, ranging between ≈ 1.1 and ≈ 1.6 fm, and also differ in the degree to which their results are fitted by (2-2). This should not be surprising, since the term "nuclear radius" can, in fact, have different meanings, and different experiments measure quite different quantities. Thus we can think of the radius of the nuclear force field, the radius of the distribution of charges (protons), or the radius of the nuclear mass distribution. Experimental approaches are available for the measurement of the first two of these quantities, whereas the third can generally be inferred only from less direct evidence.

Nuclear-Force Radii. The earliest information on nuclear sizes came from α-particle scattering experiments. The agreement between experimental results and the predictions of the Rutherford formula (2-1) for α

[9] By coincidence 1 fermi equals 1 femtometer (10^{-15} meter), so the abbreviation fm has a double meaning.

particles from radioactive sources scattered from medium and heavy elements immediately showed that the Coulomb force law holds around the nuclei of these elements out to the distance of closest approach of the α particles; in other words, the radius of the nuclear force field must be less than the distance of closest approach. To estimate the distance of closest approach d_0 consider an α particle of charge $2e$ and initial kinetic energy T coming within the Coulomb field of a nucleus of charge Ze. At distance d from the center of the nucleus the α particle's kinetic energy T' is given by conservation of energy as $T' = T - 2Ze^2/d$. The distance of closest approach d_0 is reached in a head-on collision at the point at which the α particle reverses its direction and where therefore $T' = 0$. Hence

$$d_0 = \frac{2Ze^2}{T},$$

and if T is in millions of electron volts,

$$d_0 = \frac{2.6Z}{T} \text{ fm}.$$

For typical α-particle energies from radioactive sources (4–8 MeV) d_0 thus turns out to be 10–20 fm for copper and 30–60 fm for uranium.

With lighter-element scatterers, such as aluminum, deviations from Rutherford scattering *were* observed, and the distances at which the deviations from Coulomb's law appeared (~7 fm for aluminum) were taken to represent nuclear radii.

Square-Well and Woods-Saxon Potentials. Any positively charged particle subject to nuclear forces can similarly be used to probe the distance from the center of a nucleus within which the nuclear (attractive) forces become significant relative to the Coulombic (repulsive) force.[10] Protons clearly constitute the simplest of such probes and have been used extensively. In figure 2-2*a* the potential energy between a nucleus and a proton is shown schematically as a function of the distance r between their centers. The solid curve depicts the crudest approximation to the data, with a so-called square well of radius R representing the region within which the nuclear forces act, and with the Coulomb potential with its r^{-1} dependence acting at $r > R$. Analyzed in terms of this picture the proton (and other hadron) scattering experiments lead to only approximate agreement with (2-2) and to r_0 values ranging from about 1.35 to 1.6 fm. Many other data indicate that nuclear potential wells have somewhat sloping rather than vertical walls. The most widely accepted analytical form used to describe

[10] The particle need, in fact, not be charged as long as it is subject to nuclear forces. Particles subject to nuclear forces are collectively known as strongly interacting particles, or **hadrons**.

their shape is one due to R. D. Woods and D. S. Saxon:

$$V = \frac{V_0}{1 + \exp[(r - R)/a]}, \tag{2-3}$$

where V_0 is essentially the potential at the center of the nucleus, a is a constant (≈ 0.5 fm), and R is now that distance from the center at which $V = V_0/2$. The dashed curve in figure 2-2a represents the sum of the Woods-Saxon and Coulomb potentials.[11] When analyzed in terms of the so-called optical model (cf. chapter 4, section D) with Woods-Saxon potentials, proton scattering experiments lead to half-potential radii that are well represented by (2-2), with $r_0 \approx 1.25$ fm and with a drop-off from 90 to 10 percent of the full potential within a distance of ≈ 2.2 fm. This "skin thickness" of the potential well is in fact 2 ln 9 times the constant a, as can be readily verified by solution of (2-3) for $V/V_0 = 0.9$ and $V/V_0 = 0.1$.

Since neutrons are not subject to Coulomb forces, we might expect neutron scattering and absorption experiments to be easier to interpret in terms of nuclear force radii than charged-hadron experiments. However, they too are fraught with problems: the neutrons must be of sufficiently high energy to have de Broglie wavelengths small compared to nuclear dimensions (say $\gtrsim 10$ MeV); but at still higher energies nuclei, especially those of small A, become quite transparent to neutrons. To the extent that a nucleus can be considered a completely opaque sphere of radius R, it will present a cross-sectional area of πR^2 for absorption of a beam of neutrons.[12] Fast-neutron measurements have on this basis (which corresponds to a square-well potential as depicted in figure 2-2b, solid curve) led to r_0 values in the neighborhood of 1.4 fm. If a Woods-Saxon potential (dashed curve in figure 2-2b) is used, some nuclear absorption of neutrons must be expected to take place at distances beyond the half-potential radius R and therefore if interpreted with such a potential shape, smaller r_0 values (now referring to the half-potential radius) are deduced, in general agreement with the charged-hadron results.

The radius of the inner wall of the "Coulomb barrier" depicted in figure 2-2a may be deduced by analysis, not only of the interaction between the nucleus and charged particles *approaching* it, but also of charged particles

[11] Note that the Coulomb potential *inside* the nucleus (considered as a uniformly charged sphere) is given by $V_c(r) = (Ze^2/2R)\,[3 - (r/R)^2]$. This is shown by the dot-dashed curve in figure 2-2a. At the surface, where $r = R$, the Coulomb potential has the well-known value Ze^2/R; at the center ($r = 0$) it takes on 3/2 that value.

[12] In most experiments of this type, in which the transmitted neutron beam is observed some distance behind a target, the total cross section actually measured is $2\pi R^2$ because it includes, in addition to the cross section for absorption, that for "shadow scattering," which is also πR^2. This phenomenon, the scattering of radiation of wavelength λ by a black object of radius R, which causes a shadow to extend only a finite distance behind the object, is a general result of wave optics, but it is not observed with light waves incident on macroscopic objects because the scattering is confined to angles $\lesssim \lambda/2\pi R$. For a further discussion of total cross sections, cf. chapter 4 section B.

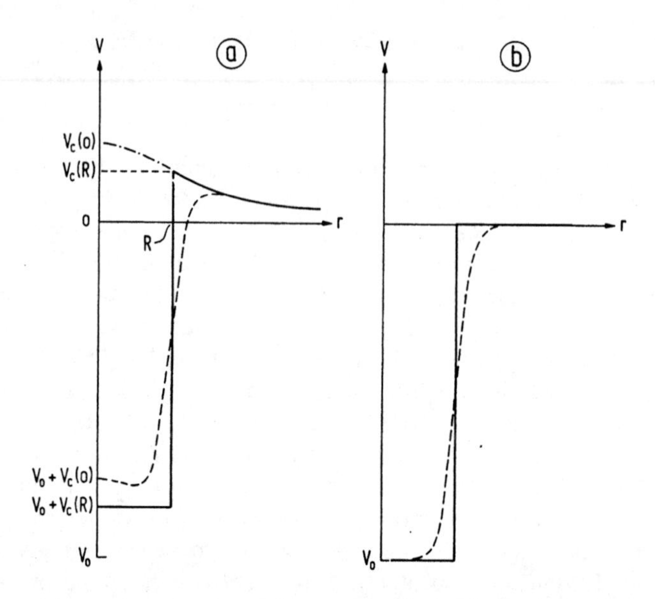

Fig. 2-2 Potential energy as a function of distance from the center of a nucleus for (*a*) proton, (*b*) neutron. The solid curves represent square-well potentials, the dashed curves Woods-Saxon potentials. In (*a*) the dot-dash curve is the Coulomb potential V_c inside the nucleus.

leaving the nucleus. The emission of α particles from heavy nuclei involves a transition from a state in which the α particle is inside the nuclear potential to one in which it is outside the range of nuclear forces. The probability for this process and therefore the lifetime for α decay depends very strongly on the height of the potential barrier that the α particle must penetrate, and therefore on the nuclear radius R beyond which the repulsive Coulomb potential is not compensated by the attractive nuclear potential. The quantum-mechanical theory of α decay accounts very well for the relation between lifetimes for α decay and α-particle energies (see chapter 3, section B) and allows nuclear radii to be deduced from the experimental half lives and decay energies. Values between 8.4 and 9.8 fm are obtained in this way for the square-well radii of α-emitting nuclei with $A > 208$, corresponding to r_0 values in the range of 1.4–1.5 fm.

Nuclear-Charge Radii. We now turn to an entirely different class of experimental methods, which use as probes of nuclear dimensions particles that are not affected by specific nuclear forces but are sensitive to the electric charges of nuclei. Scattering of electrons (H1) is the most widely used of these techniques and the only one that we briefly discuss here. Others involve fine-structure splitting in atomic spectra due to the finite nuclear size, and measurement of the transition energies between energy levels of so-called mesonic atoms, that is, atoms in which an orbital electron is replaced by a π- or μ-meson (cf. chapter 12, section D).

Because of their much greater masses, these mesons penetrate far inside the nucleus in their orbits, and their energy levels are therefore very sensitive to the distribution of nuclear charge.

Scattering data obtained with electrons of moderate energies (<100 MeV) are compatible with nuclei being spheres of uniformly distributed charges, but with radii distinctly smaller than indicated by the methods that determine nuclear force radii. Equation 2-2 is, in fact, not quite adequate to represent these electron-scattering results, since they indicate r_0 values varying from about 1.4 fm for light nuclei to about 1.2 fm for heavy ones.

When electrons of higher energy are used the angular distribution of the scattered electrons leads to more detailed information about the charge *distribution* in the scattering nuclei. Specifically one finds that the results are no longer compatible with nuclei as uniformly charged spheres, but that the charge density drops off gradually at the edge of the nucleus. Two parameters can generally be deduced from the data: the half-density radius R_e, defined as the distance from the center at which the charge density has fallen to half its value at the center; and the skin thickness d_e, usually given as the distance over which the charge density drops from 90 to 10 percent of its central value. The R_e values are very well represented by (2-2), with $r_{0e} = 1.07$ fm. The skin thickness d_e is approximately 2.4 fm for all but the very lightest nuclei. The specific functional form of the drop-off cannot be deduced directly from the experiments, but the most commonly used representation is the so-called Fermi shape, which has the same functional form as the Woods-Saxon potential:

$$\rho(r) = \frac{\rho_0}{1 + \exp\left[(r_e - R_e)/a_e\right]}. \tag{2-4}$$

Again, the skin thickness d_e as defined above is $d_e = 2a_e \ln 9 \approx 4.4a_e$. The Fermi shape is shown schematically in figure 2-3 and also used in figure 2-4 to represent charge distributions for the nuclei of several elements as deduced from electron scattering. Whether the distributions are, in fact, as flat in the central regions as shown cannot be ascertained from the scattering data. Detailed theoretical calculations of charge distributions based on models of nuclear structure predict more complex shapes of $\rho(r)$ as a function of r in various nuclei, and these calculated distributions are often just as compatible with the experimental results as the centrally flat distributions shown in figure 2-4.

While the charge density results give information on how protons are distributed in nuclei, there are, as mentioned earlier, no experimental techniques for determining the total nucleon distribution. It is generally assumed that neutrons are distributed in roughly the same way as protons. However, some differences are predicted in at least some nuclei by theoretical calculations. In particular, it appears that neutron distributions may extend to slightly larger distances from the centers than do proton

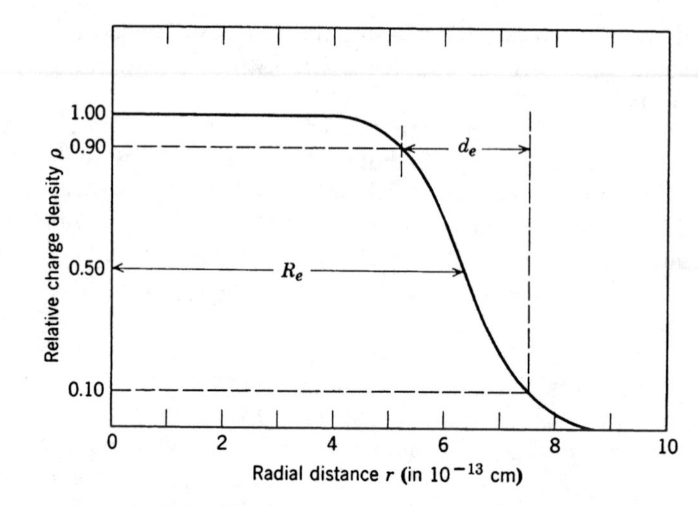

Fig. 2-3 Typical charge distribution in a nucleus, as determined by electron-scattering experiments. The half-density radius R_e and the skin thickness d_e are indicated. The particular distribution shown is that of the gold nucleus. (Data from R. Hofstadter, reference H1.)

distributions, and some experimental results, based, for example, on meson interaction with nuclear surfaces, corroborate this conclusion.

The general picture that emerges is that nuclear-potential radii are about 0.2 fm larger than the radii of the charge (and matter) distributions, and that in both instances we deal not with sharp cutoffs, but with tapering distributions. That the potential wells extend farther out than the nucleons is entirely understandable in terms of the finite, though short, range of nuclear forces (cf. chapter 10, section A).

Our entire discussion tacitly assumes that nuclei are spherical. As we see later (see following section and chapter 10), many nuclei are in fact not strictly spheres, but rather spheroids or ellipsoids. In those cases our discussion of radii may be taken to apply to the mean semi-axes of these more complex shapes.

3. *Spins and Moments.* (S2, Y1)

Spin. That nuclei possess angular momenta was first suggested by W. Pauli in 1924 in order to explain the hyperfine structure (hfs) in atomic spectra of monoisotopic elements.[13] The angular momentum of a nucleus is always expressible as $Ih/2\pi$ or $I\hbar$, where I is an integral or half-integral number known as the nuclear spin. Both neutron and proton have intrinsic

[13] In spectra of elements having more than one isotope, an additional source of hfs is the so-called isotope shift, that is, the splitting of spectral lines due to the different masses of the isotopic nuclei.

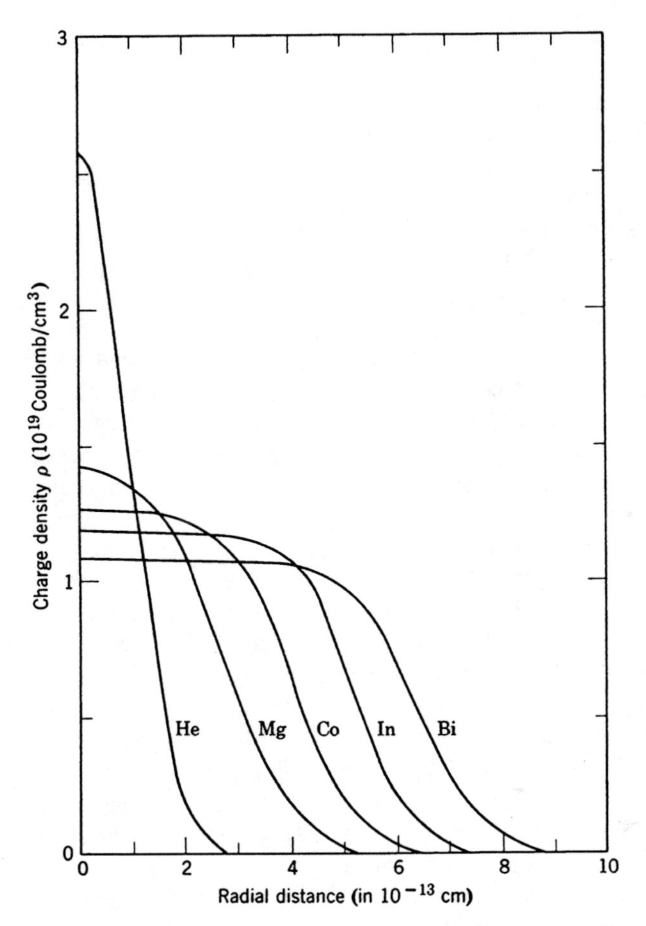

Fig. 2-4 Nuclear charge distributions for a number of elements as deduced from electron scattering. (From reference HI.)

spin $I = \frac{1}{2}$, and the nucleons in the nucleus, just like the electrons in an atom, contribute some orbital angular momentum (which is an integral multiple of $\hbar$) as well as their intrinsic spins. Thus since each nucleon can only add or subtract its intrinsic spin $\frac{1}{2}$ and its integral orbital angular momentum, the spin of any nucleus of even A must be zero or integral, and that of any odd-A nucleus must be half-integral. All spin measurements have confirmed this rule; furthermore, it appears that all nuclei of even A and even Z have $I = 0$ in their normal, or ground, states.

Magnetic Moments. Nuclei with nonzero angular momenta have magnetic moments. The prediction of Dirac's theory for the magnetic moment of an electron (charge e, mass m_e), namely $eh/4\pi m_e c = 0.9274 \times 10^{-20}\ \text{erg G}^{-1} \equiv 1$ Bohr magneton (μ_B), agrees so well with the experimentally determined value that similar success might be expected in the case of the proton (charge e, mass $M_p = 1836\,m_e$). However, the magnetic moment of the proton is not equal to 1/1836 Bohr magneton, but about 2.79 times this value. Nevertheless, the quantity $\mu_B m_e/M_p$ is used as the unit of nuclear magnetic moments and called a nuclear magneton ($\approx 5.05 \times 10^{-24}$ erg G^{-1}).

The observation that the proton has a magnetic moment very different from that expected from the theory for a simple structureless charged particle indicates that the proton is, in fact, not such a simple entity. Perhaps even more startling is the observation that the neutron has a magnetic moment of -1.91 nuclear magnetons (the negative sign indicates that spin and magnetic moment are in opposite directions). This magnetic moment presumably results from a distribution of charges in the neutron, with negative charge (perhaps due to negative mesons) concentrated near the periphery and overbalancing the effect of an equal positive charge nearer the center.

In general, the magnetic moments of nuclei differ from values calculated by any simple theory. Magnetic moments are often expressed in terms of gyromagnetic ratios (nuclear g factors); the magnetic moment is then $g \cdot I$ nuclear magnetons, with g positive or negative, depending on whether spin and magnetic moment are in the same or opposite directions.

Methods of Measurement. Nuclear spins and magnetic moments can sometimes be determined from hyperfine structure in atomic spectra. Hyperfine structure derives from the fact that the energy of an atom is slightly different for different (quantized) orientations between nuclear spin and angular momentum of the electrons because of the interaction between the nuclear magnetic moment and the magnetic field of the electrons. From the number of lines in a spectroscopic "hypermultiplet," the nuclear spin I can be determined under suitable conditions, and once the nuclear spin is known, the magnetic moment can be calculated from the magnitude of the splitting (which is typically of the order of 1 Å). Hyperfine structure occurs also in molecular spectra, and the hfs of transitions between rotational states can be observed in microwave spectra and used to deduce nuclear spins, quite in analogy to the method outlined for atomic spectra.

A second method for the determination of nuclear magnetic moments and spins is the atomic-beam method of I. I. Rabi and co-workers, an extension of the Stern-Gerlach experiment for the determination of magnetic moments of atoms. A beam of atoms (or molecules) is sent through an inhomogeneous magnetic field. The nuclear spin I, uncoupled from the electron angular momentum J by the external field, orients itself with respect to the field. This orientation is governed by the usual quantum conditions, and the beam is therefore split into $2I + 1$ components whose separations are dependent on the nuclear magnetic moment. The energies of these splittings may be found in terms of characteristic alternating magnetic-field frequencies, which induce transitions between components. The magnetic moment of the neutron was directly determined by a suitable (and rather drastic) modification of this principle.

Several resonance techniques are useful for spin and magnetic-moment determinations. In nuclear resonance absorption the magnetic dipoles of spin I are aligned with a strong external magnetic field in $2I+1$ different orientations. The energy differences between the resulting $2I+1$ energy states (which lie in the radio-frequency region) depend on the gyromagnetic ratio. Resonance absorption of radio-frequency radiation will, therefore, take place at a frequency corresponding to these transitions; the resonance frequency is a measure of the gyromagnetic ratio and, if I is known, of the magnetic moment. What is observed in the paramagnetic resonance method is the resonance absorption frequency for a paramagnetic substance in a radio-frequency field and the splitting of this frequency caused by the interaction between the nuclear spin and the electronic angular momentum of the molecule or ion.

Information on spins of radioactive nuclei can be inferred from detailed studies of β- and γ-decay processes. This subject is discussed in chapter 3, as are some methods for the determination of spins and moments of excited states of nuclei. Spin information is also obtainable from nuclear reaction data (cf chapter 4).

Quadrupole Moments. In addition to its magnetic dipole moment a nucleus may have an electric quadrupole moment. This property may be thought of as arising from an elliptic charge distribution in the nucleus. The quadrupole moment q is given by the equation $q=\frac{2}{5}Z(a^2-b^2)$, where a is the semiaxis of rotation of the ellipsoid and b is the semiaxis perpendicular to a; q has the dimensions of area. For the deuteron $q = +2.74\times10^{-27}\ \text{cm}^2$, the plus sign denoting a prolate (cigar-shaped) charge distribution. A negative quadrupole moment corresponds to an oblate (flattened) charge distribution. Quadrupole moments, including both positive and negative values, have been determined for quite a number of nuclei with $I>\frac{1}{2}$. (Nuclei with $I=0$ or $I=\frac{1}{2}$ cannot have quadrupole moments.) The interactions of nuclear quadrupole moments with the electric fields produced by electrons in atoms and molecules give rise to abnormal hyperfine splittings in spectra, and the methods for quadrupole-moment measurements are therefore the ones already discussed: optical spectroscopy, microwave spectroscopy, nuclear resonance absorption, and some modified molecular-beam techniques.

4. *Other Quantum-Mechanical Properties*

Statistics. This is a quantum-mechanical property of particles that becomes important when large numbers of them occur together in a system. For detailed discussions of the concept the reader is referred to other works (B2, B3). Here we merely indicate the nature of this property and give some useful results.

All nuclei and elementary particles are known to obey one of two kinds of statistics: **Bose-Einstein** or **Fermi-Dirac.** If all the coordinates describing a particle in a system (including three space coordinates and the spin) are interchanged with those describing another identical particle in the system, the absolute magnitude of the wave function representing the system must remain the same, but the wave function may or may not change sign. If it does not change sign (the wave function is then called symmetrical), Bose statistics applies. If the particle wave function does change sign with the interchange of coordinates (antisymmetrical wave function), the particles obey Fermi statistics. In Fermi statistics each completely specified quantum state can be occupied by only one particle; that is, the Pauli exclusion principle applies to all particles obeying Fermi statistics. For particles obeying Bose statistics no such restriction exists. Protons, neutrons, electrons (and some other elementary particles such as positrons, neutrinos, and some types of mesons) all obey Fermi statistics. A nucleus will obey Bose or Fermi statistics, depending on whether it contains an even or odd number of nucleons.

The statistics of nuclei can be deduced from the alternating intensities in rotational bands of the spectra of diatomic homonuclear molecules. With Bose statistics the even-rotational states and with Fermi statistics the odd-rotational states are more populated. This can be illustrated by the rotational spectra of hydrogen and deuterium. In normal hydrogen, H_2, the ratio of the populations in the states of odd- and even-rotational quantum numbers is 3:1 corresponding to spin $\frac{1}{2}$ and Fermi statistics; in deuterium, D_2, the ratio is 1:2 corresponding to spin 1 and Bose statistics.

Parity (Y1). Another nuclear property connected with symmetry properties of wave functions is parity. A system is said to have odd or even parity according to whether or not the wave function for the system changes sign when the signs of all the space coordinates are changed. We make some use of the concept of parity in our discussions of nuclear reactions and radioactive decay processes because the parity of an isolated system, like its total energy, momentum, angular momentum, and statistics, is conserved.[14] We require merely the very simple rules of combination for parity. Two particles in states of even parity or two particles in states of odd parity can combine to form a state of even parity only. A particle of

[14] As postulated in 1956 by T. D. Lee and C. N. Yang and subsequently verified by many experiments, parity is *not* conserved in the so-called weak interactions (e.g., β-decay). This discovery has had profound impact on the development of some areas of modern physics and it is discussed briefly in chapter 3, section D. For most of the considerations of nuclear phenomena in this book we need, however, not be concerned with nonconservation of parity.

Table 2-1 Properties of Some "Elementary" Particles[a]

Symbol	Name	Charge[b]	Rest Mass[c]	Spin[d]	Magnetic Moment[e]	Statistics[f]
e^-, β^-	Electron	−1	0.0005486	$\frac{1}{2}$	−1836	F
e^+, β^+	Positron	+1	0.0005486	$\frac{1}{2}$	+1836	F
γ	Photon	0	0	1	0	B
ν	Neutrino	0	$<2\times10^{-7}$	$\frac{1}{2}$	<0.3	F
n	Neutron	0	1.0086650	$\frac{1}{2}$	−1.913	F
$\mu^\pm$	Mu-meson (muon)	±1	0.1134	$\frac{1}{2}$	±8.891	F
$\pi^\pm$	Pi-meson (pion)	±1	0.1498	0		B
π^0	Pi-meson (pion)	0	0.1449	0		B
p	Proton	+1	1.0072765[g]	$\frac{1}{2}$	+2.793	F

[a] According to current views only the first four of the particles listed (the so-called leptons) are truly elementary, all the others (hadrons) are composed of quarks.
[b] In units of the elementary charge $e = 4.80324 \times 10^{-10}$ esu.
[c] In atomic mass units ($^{12}C = 12.000000$).
[d] In units of $\hbar$.
[e] In units of the nuclear magneton ($e\hbar/2M_pc$), where M_p is the proton mass. Positive values indicate moment orientations with respect to spin orientations that would result from spinning positive charges.
[f] F means Fermi and B means Bose statistics.
[g] In contrast to the usual convention the mass given here is that of the bare proton, not the hydrogen atom.

even parity and one of odd parity result in a system of odd parity. We may illustrate this by an example from atomic spectroscopy: allowed transitions in atoms occur only between an atomic state of even and one of odd parity, not between two even or two odd states, because the quanta of ordinary dipole radiation are characterized by odd parity.

In discussing nuclear energy states we make use of the fact that parity is connected with the angular-momentum quantum number l. States with even l ($s, d, g, \ldots$ states) have even parity, those with odd l ($p, f, h, \ldots$ states) have odd parity.

Now that we have briefly discussed the principal properties by which nuclei are characterized, we list in table 2-1 the values of some of these properties for those elementary particles that are of prime importance in nuclear science.

5. *Excited States*

Most of the discussion of nuclear properties in this section has dealt implicitly or explicitly with the properties of *ground* states of nuclei.

However, much of nuclear chemistry and physics is in fact concerned with the detailed investigation of the various *excited* states of nuclei, the systematization of their properties, and an understanding of these systematics in terms of nuclear models. The particular static properties of most interest in this context are the energies (usually given in terms of energy differences from the ground state, rather than in absolute terms), spins, and parities. Magnetic moments are becoming more accessible and of increasing importance. Radii of excited states have been determined in very few instances.

In addition to these static properties the transition probabilities for transitions between excited states or from an excited state to the ground state are of prime importance for the understanding of nuclear structure. They are usually expressed in terms of half lives. However, even though the techniques for measuring half lives have been extended to shorter and shorter times (see chapter 8, section D), there are still vast numbers of nuclear excited states whose lifetimes have not been measured. The relative transition probabilities for transitions from a given state to two or more other states are much easier to determine and are always of interest.

In chapter 3 we deal in considerable detail with excited-state properties, their determination, and their importance in radioactive decay processes; in

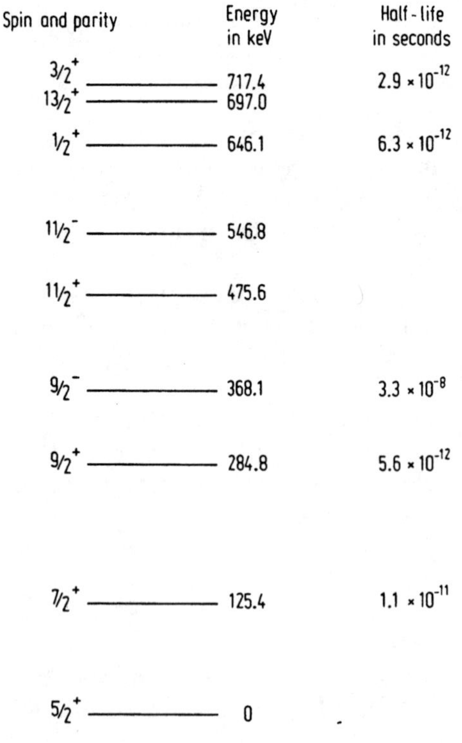

Spin and parity	Energy in keV	Half-life in seconds
$3/2^+$	717.4	2.9×10^{-12}
$13/2^+$	697.0	
$1/2^+$	646.1	6.3×10^{-12}
$11/2^-$	546.8	
$11/2^+$	475.6	
$9/2^-$	368.1	3.3×10^{-8}
$9/2^+$	284.8	5.6×10^{-12}
$7/2^+$	125.4	1.1×10^{-11}
$5/2^+$	0	

Fig. 2-5 The first few energy levels of ^{185}Re.

chapter 4 and again in chapter 8, section F, we touch on the role of nuclear reaction studies in determining these properties; in chapter 10 we discuss the nuclear models that have been devised to account for the enormous body of data that has been accumulated. For many nuclei dozens or even hundreds of excited states have been characterized.

Here we merely call attention to the existence of this vast subject and, as an illustration, show in figure 2–5 the first few excited states of ^{185}Re with their energies, spins, parities, and, where known, half lives.

D. MASS AND BINDING-ENERGY SYSTEMATICS

Binding-Energy Equation. We have seen in preceding sections that both the volumes and the total binding energies of nuclei are very nearly proportional to the numbers of nucleons present. From the first of these observations we can conclude that nuclear matter is quite incompressible, from the second that the nuclear forces must have a saturation character; that is, a nucleon in a nucleus can apparently interact with only a small number of other nucleons, just as an atom in a liquid or solid is strongly bound to only a small number of neighboring atoms. These characteristics of nuclei suggest a similarity with drops of liquid and have prompted attempts to account for the binding energies of nuclei in terms of a model in which nuclei are considered as charged liquid drops with surface tension. It is then possible to express the binding energy of the total mass of such a drop as the sum of terms that individually correspond to the volume, surface, and Coulomb energies (and possibly other contributions), and that each has a simple functional dependence on the mass and charge (A and Z) of the nucleus. An equation of this form, with coefficients for the various terms fitted empirically, was first given by C. F. von Weizsäcker (W2) in 1935. Many authors have since revised and refined the Weizsäcker or semiempirical binding-energy equation through the addition of further terms and changes in the coefficients, but the functional form of the principal terms has remained essentially the same.

We base our discussion on the form of the liquid-drop binding-energy equation given by W. D. Myers and W. J. Swiatecki (M2):

$$E_B = c_1 A\left[1 - k\left(\frac{N-Z}{A}\right)^2\right] - c_2 A^{2/3}\left[1 - k\left(\frac{N-Z}{A}\right)^2\right] - c_3 Z^2 A^{-1/3} + c_4 Z^2 A^{-1} + \delta, \tag{2-5}$$

where E_B is the binding energy, that is, the energy required to dissociate the nucleus into its constituent nucleons, and A, Z, and N have the usual meanings. With E_B expressed in MeV the coefficients take on the following values: $c_1 = 15.677$ MeV, $c_2 = 18.56$ MeV, $c_3 = 0.717$ MeV, $c_4 = 1.211$ MeV, and $k = 1.79$. The δ term is discussed below.

Equation 2–5 contains only six empirically adjusted parameters, yet this simple equation yields binding energies that agree with experimental values for each of the approximately 1200 nuclides of known mass to within less than 10 MeV, and very much better than that for most.[15] This represents a remarkable success indeed for the liquid-drop model of the nucleus.

Volume Energy. We now proceed to discuss the individual terms in (2-5). The first and dominant term, proportional to A and thus to the nuclear volume, expresses the fact already discussed that the binding energy is in first approximation proportional to the number of nucleons. This is a direct consequence of the short range and saturation character of the nuclear forces. The saturation is almost, though not entirely, complete when four nucleons, two protons and two neutrons, interact, as is indicated by the large observed binding energies of ^{4}He, ^{12}C, ^{16}O, and so on (see figure 2-1). The correction term proportional to $(N - Z)^2/A$, which in our representation is included with the volume energy, is referred to as the **symmetry energy.** It reflects the observation that for a given A the binding energy due to *nuclear* forces (i.e., disregarding the Coulomb effect discussed below) is greatest for the nucleus with equal numbers of neutrons and protons and decreases symmetrically on both sides of $N = Z$. The simplest functional form expressing these empirical facts is a term in $(N - Z)^2$. The A^{-1} dependence of the symmetry energy comes about because the binding-energy contribution per neutron-proton pair is proportional to the probability of having such a pair within a certain volume (determined by the range of nuclear forces), and this probability in turn is inversely proportional to the nuclear volume. The extra stability of $N = Z$ nuclei comes about at least in part through the Pauli exclusion principle: since two identical nucleons cannot be in the same energy state, the lowest state for a given number of nucleons is attained (in the absence of Coulomb forces) for equal numbers of neutrons and protons.

Surface Energy. The nucleons at the surface of a nucleus can be expected to have unsaturated forces, and consequently a reduction in the binding energy proportional to the nuclear surface should be taken into account. This effect gives rise to the second (negative) term; it contains $A^{2/3}$, which is a measure of the surface (since A is proportional to the volume). With increasing nuclear size, the surface-to-volume ratio decreases, and therefore this term becomes relatively less important. The correction term to the surface energy, $k[(N - Z)/A]^2$, does not appear in most conventional binding-energy equations. While it is not needed to account for measured binding energies,[16] it is included by Myers and

[15] The major deviations occur as a result of shell-structure effects discussed in section E (see figure 2-9).

[16] However, it should be noted that, if this term is omitted, the coefficients of other terms must be readjusted to obtain agreement with experimental values.

Swiatecki (M2), and expressed in the same functional form as the symmetry correction term to the volume energy, in order to ensure that nuclei with values of $|N - Z|$ large enough to make the volume energy go to zero have their surface tension vanish also.

Coulomb Energy. The third term, $c_3 Z^2 A^{-1/3}$, represents the electrostatic energy that arises from the Coulomb repulsion between the protons. This electrostatic repulsion, of course, lowers the binding energy—hence the negative sign. The electrostatic energy of a uniformly charged sphere of charge q and radius R is $\frac{3}{5}q^2/R$ and, since $q = Ze$ and $R = r_0 A^{1/3}$ for a nucleus of radius R and atomic number Z, we can write its electrostatic energy as $(3e^2/5r_0) Z^2 A^{-1/3}$. The coefficient $c_3 = 0.717$ MeV corresponds to an r_0 value of 1.205 fm. Because of its Z^2 dependence, the Coulomb energy becomes increasingly important as Z increases and accounts for the fact that all stable nuclei with $Z > 20$ contain more neutrons than protons (see figure 2-6) despite the symmetry energy that maximizes *nuclear* binding for $N = Z$.

We already know from our discussion of nuclear radii that nuclei are not uniformly charged, but have charge distributions with diffuse boundaries given by (2-4). The diffuse boundary gives rise to a correction to the Coulomb energy (lowering it), and this is expressed by the fourth term in (2-5). Without deriving its functional form (a derivation may be found in

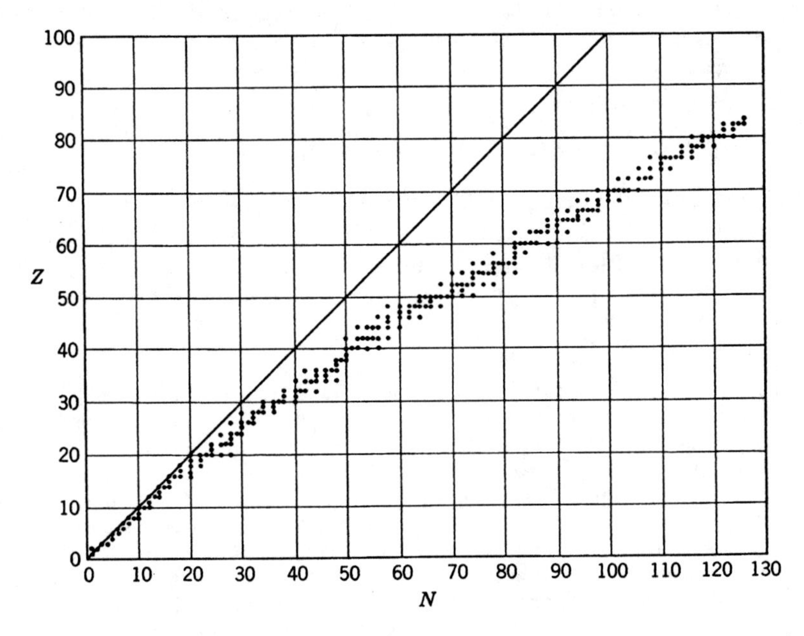

Fig. 2-6 The known stable nuclei on a plot of Z versus N. Note the gradual increase in the neutron–proton ratio; the 45° line indicates a neutron–proton ratio of 1.

M2) we merely state that the value of the coefficient $c_4 = 1.211$ MeV corresponds to a skin thickness $d_e = 2.4$ fm, in conformity with electron-scattering results.

Pairing Energy. The final term in (2-5) is a quantitive expression of the fact (already noted on p. 27) that binding energies for a given A depend somewhat on whether N and Z are even or odd. So-called even-even nuclei (Z and N even) are the stablest, and for them δ in (2-5) may be taken as $+11/A^{1/2}$; for even-odd (Z even, N odd) and odd-even (Z odd, N even) nuclei $\delta = 0$; for odd-odd nuclei $\delta = -11/A^{1/2}$. The difference in the stabilities of the four types of nuclei is manifested in the distribution of the known stable nuclides among them: 157 even-even, 55 even-odd, 50 odd-even, and 4 odd-odd. The striking preponderance of even-even nuclei and the complete absence of stable odd-odd nuclei outside the region of the lightest elements[17] can be explained in terms of a tendency of two like particles to complete an energy level by pairing opposite spins. The δ term in the binding-energy equation is therefore often called the pairing term.[18]

The greater stability of nuclei with filled energy states is apparent not only in the larger number of even-even nuclei but also in their greater abundance relative to the other types of nuclei. On the average, elements of even Z are much more abundant than those of odd Z (by a factor of about 10). For elements of even Z the isotopes of even mass (even N) account in general for about 70 to 100 percent of the element (beryllium, xenon, and dysprosium being exceptions). The general shape of the binding-energy curve (figure 2-1) with the maximum at $A \cong 60$ comes about through the opposing trends with mass number of the relative contributions of surface energy (decreasing with A) and Coulomb and symmetry energies (increasing with A).

Nuclear Energy Surface and Mass Parabolas. The binding energies of all nuclei can be represented as a function of A and Z by means of a three-dimensional plot of an equation such as (2-5). Without attempting to construct this nuclear energy surface in three dimensions, we can obtain

[17] The four odd-odd nuclei are ^2_1H, ^6_3Li, $^{10}_5B$, and $^{14}_7N$.

[18] The pairing energy for a neutron-proton pair in the same energy state is actually larger than that for a pair of like nucleons because of the spin-dependent character of nuclear forces (see chapter 10, section A): these forces are stronger between two nucleons of parallel spin than between two nucleons of opposite spin, and the Pauli principle prevents two like nucleons with parallel spins from being in the same energy state. It is this large neutron proton pairing energy that stabilizes the odd-odd nuclei ^2H, ^6Li, ^{10}B, and ^{14}N relative to their even-even isobars, the di-neutron, ^6He, ^{10}Be, and ^{14}C. With increasing Z the Coulomb effect keeps increasing and soon prevents the most loosely bound protons from occupying the same energy state as the most loosely bound neutrons; thus the neutron-proton pairing energy is no longer observable in heavier nuclei, whereas the pairing of like nucleons (expressed by the δ term in the binding-energy equation) is manifest throughout the table of nuclides. (Cf. reference B2, pp. 211–225.)

useful information about some of its features. For this purpose it is more convenient to consider the total atomic mass M rather than the binding energy E_B. According to the definition of binding energy we can write

$$M = ZM_H + (A - Z)M_N - E_B, \tag{2-6}$$

where M_H and M_N are the masses of the hydrogen atom (938.791 MeV) and the neutron (939.573 MeV), respectively. By combining (2-5) and (2-6) we obtain the semiempirical mass equation:

$$M = 939.573A - 0.782Z - (c_1A - c_2A^{2/3})\,[1 - k(1 - 2Z/A)^2] + Z^2(c_3A^{-1/3} - c_4A^{-1}) - \delta. \tag{2-7}$$

Equation 2-7 is quadratic in Z and can be written in the form

$$M = f_1(A)Z^2 + f_2(A)Z + f_3(A) - \delta, \tag{2-8}$$

with the three coefficients being functions of A:

$$f_1(A) = 0.717A^{-1/3} + 111.036A^{-1} - 132.89A^{-4/3},$$
$$f_2(A) = 132.89A^{-1/3} - 113.029,$$
$$f_3(A) = 951.958A - 14.66A^{2/3}.$$

Thus for a given A the coefficients are constants and (2-8) then represents a parabola when A is odd ($\delta = 0$) and a set of two parabolas when A is even ($\delta = \pm 11A^{-1/2}$). These mass (or energy) parabolas, which are sections through the nuclear energy surface along planes of constant A, are very useful in β-decay systematics because values of the energy available for β decay between neighboring isobars can be read directly from them. For illustrative purposes the parabolas for $A = 157$ and $A = 75$ are shown in figure 2-7 and the pair of parabolas for $A = 156$ in figure 2-8.

The vertex of each mass parabola gives, for the given value of A, the minimum mass or maximum binding energy. To find the nuclear charge Z_A corresponding to this minimum mass, we differentiate (2–8) with respect to Z, considering A constant, and set the derivative $\partial M/\partial Z$ equal to zero. This gives

$$Z_A = \frac{-f_2(A)}{2f_1(A)}. \tag{2-9}$$

Since we have treated Z as a continuous function, we should expect to find nonintegral values for Z_A. For $A = 157$, for example, we get $Z_A = 64.69$; for $A = 156$, $Z_A = 64.32$; and for $A = 75$, $Z_A = 33.13$.

For the purpose of plotting energy parabolas we can now use (2-9) to rewrite (2-8) in the following form:

$$M = f_1(A)(Z - Z_A)^2 - \delta + f(A),$$

where $f(A) = f_3(A) - f_2(A)^2/4f_1(A)$ is a function of A only and does not need to be evaluated, since we are usually concerned only with *differences* among a group of isobars. Thus in figures 2-7 and 2-8 the ordinate is plotted as

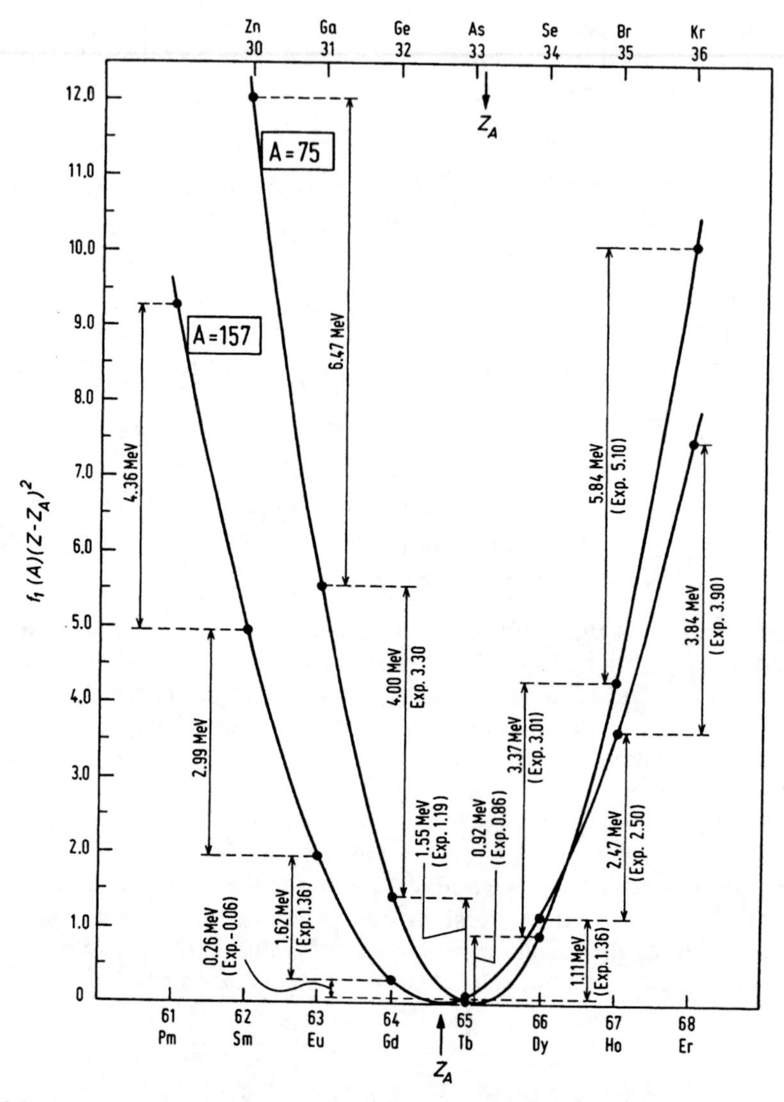

Fig. 2-7 Mass parabolas for $A = 75$ and $A = 157$, as calculated from (2-8). Calculated mass differences between neighboring isobars are indicated, with experimentally determined values shown in parentheses for comparison. The top Z scale refers to $A = 75$, the bottom one to $A = 157$.

$f_1(A)(Z - Z_A)^2$; for odd A the mass corresponding to $Z = Z_A$ is then the zero on the ordinate scale, and for even A the zero is the mass halfway between the vertices of the even-even and odd-odd parabolas. The widths of the energy parabolas are determined by the values of $f_1(A)$ that decrease with increasing A. The **stability valley** in the nuclear energy surface thus broadens with increasing A as is illustrated in figure 2-7.

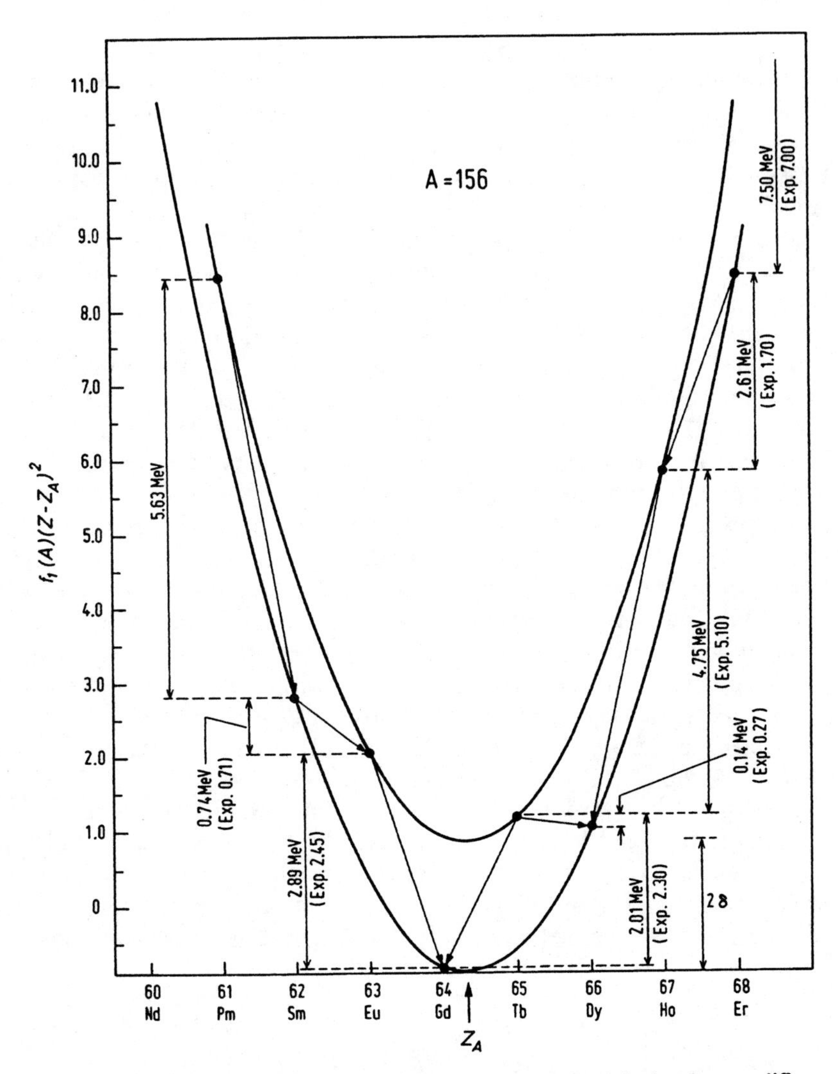

Fig. 2-8 Mass parabolas for $A = 156$ as calculated from (2-8). Calculated mass differences between neighboring isobars are indicated, with experimentally determined values shown in parentheses for comparison.

By considering the parabolic curves for a set of isobars, we can draw several important conclusions about nuclear stability. For example, it is immediately clear that for any given odd A there can be only one β-stable nuclide, that nearest the minimum of the parabola. For even A there are usually two and sometimes three possible β-stable isobars, all of the even-even type. In figure 2-8 both ^{156}Gd and ^{156}Dy are indicated as stable, since both have smaller masses than their odd-odd neighbor ^{156}Tb. Strictly speaking ^{156}Dy, with its mass larger than that of ^{156}Gd, is thermodynamically not really stable. However, its decay to ^{156}Gd requires a so-called double β-decay process, involving simultaneous emission of two β particles (in this case β^+) or simultaneous capture of two electrons. Such processes are expected to have exceedingly long half lives, and only two double β decays [^{130}Te to ^{130}Xe, with $t_{1/2} \approx 2 \times 10^{21}$ y and ^{82}Se to ^{82}Kr, with $t_{1/2} \approx 1 \times 10^{20}$ y (K1)] have been established experimentally.

It also becomes immediately evident from figures 2-7 and 2-8 why, in isobaric decay chains of even A, the β-decay energies alternate between small and large values, whereas in odd-A chains they increase monotonically toward either side of Z_A. We also note that an odd-odd nucleus ($^{156}_{65}Tb$ is an example) may decay to both its even-even isobaric neighbors, by β^- emission and by electron capture (EC) (and possibly β^+ emission), respectively. For ^{156}Tb the β^- branch has, in fact, not been detected, probably because the available decay energy is so small, but there are a number of examples of branching decays (e.g., ^{64}Cu).

In figures 2-7 and 2-8 the experimentally determined energy differences between neighboring isobars have been included for comparison with those obtained from the binding-energy equation. The agreement is seen to be within a few hundred keV in the particular mass regions shown. Closer agreement may be obtained by local adjustment of $f_1(A)$ and Z_A to fit known points in a particular region of A and Z. For example, according to our universal equation ^{157}Tb is the stable isobar at $A = 157$ and ^{157}Gd would be expected to decay by β^- emission to ^{157}Tb, with a decay energy of 0.26 MeV. Actually ^{157}Gd is β stable, and ^{157}Tb decays to it by EC, with a decay energy of 0.06 MeV. To obtain agreement with this experimental fact the value of Z_A would have to be decreased by ~0.2 unit, and that change would also improve the fit to other experimentally determined decay energies among the $A = 157$ isobars and presumably give more reliable predictions for the as yet unknown decay energies (e.g., of ^{157}Sm and ^{157}Pm).

E. NUCLEAR SHELL STRUCTURE

Magic Numbers. The liquid-drop model, in which nuclei are treated essentially as statistical assemblies of neutrons and protons, is successful in accounting for many of the gross properties of nuclei. For example, as we have seen in the preceding section, the liquid-drop approach does very well in correlating the overall behavior of nuclear masses and binding energies. However, if the differences between experimentally determined masses and those obtained from a mass formula such as (2-7) are plotted against the neutron or proton numbers, as is done in figure 2-9, we find that these differences are greatest at certain values of N and Z: 28, 50, 82, and 126. In other words, nuclei with these neutron and proton numbers exhibit unusual stability. Such extra stability has also long been known for nuclei with

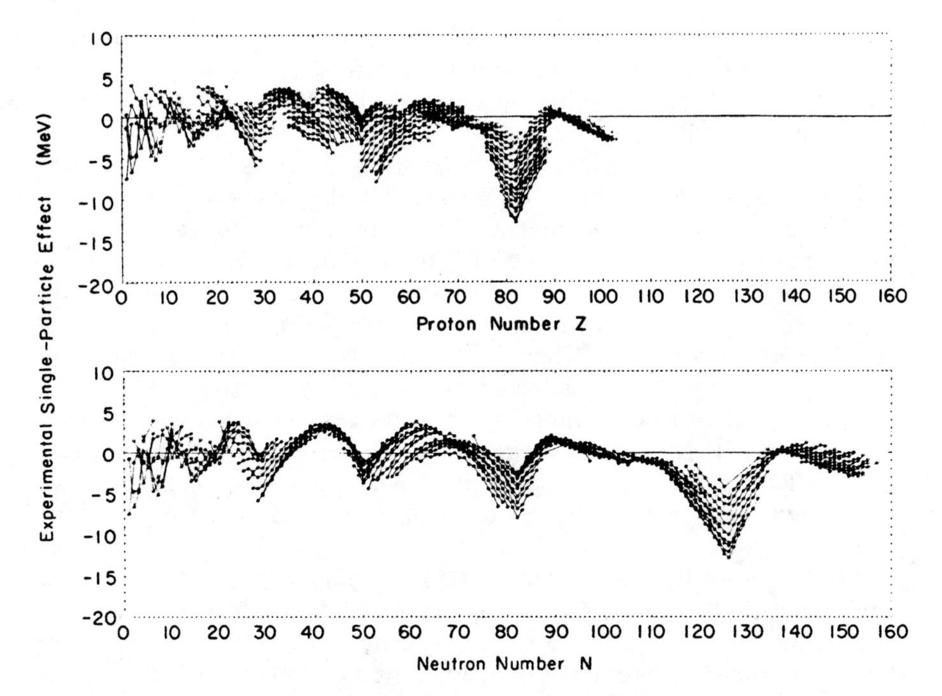

Fig. 2-9 Differences between experimental and liquid-drop-formula masses. In top graph isotones, in bottom graph isotopes are connected by lines. (From reference M2; drawing made available by J. R. Nix.)

N or Z values of 2, 8, and 20, although this is not so readily discernible from figure 2-9.

This special stability associated with certain values of Z and N led, through analogy with the special stability of the atoms of the noble gases, to the concept of closed shells in nuclei. However, early attempts (by W. M. Elsasser, 1934) to account for the stable configurations in terms of nucleons in a potential well failed for N and Z values above 20 and received little attention until much more evidence for the special stability of certain configurations was amassed. Because the unusual properties of the numbers 2, 8, 20, 28, 50, 82, and 126 remained unexplained for so long, they became known as "magic numbers." Much of the empirical evidence for these magic numbers came, as we mentioned, from masses and binding energies. Other indications stemmed from elemental and isotopic abundances, numbers of species with given N or Z, and α-particle energies. We briefly summarize some of the pertinent facts.

Above $Z = 28$ the only nuclides of even Z that have isotopic abundances exceeding 60 percent are ^{88}Sr (with $N = 50$), ^{138}Ba ($N = 82$), and ^{140}Ce ($N = 82$). No more than five isotones occur in nature for any N except $N = 50$, where there are six, and $N = 82$, where there are seven. Similarly, the largest number of stable isotopes (10) occurs at tin, $Z = 50$, and in both calcium ($Z = 20$) and tin the stable isotopes span an unusually large mass range. The fact that all the heavy natural radioactive chains end in lead ($Z = 82$) is significant, as is the neutron number 126 of the two heaviest stable nuclides, ^{208}Pb and ^{209}Bi.

The particularly weak binding of the first nucleon outside a closed shell (analogous to the low ionization potential for the valence electron in an alkali atom) is shown by the unusually low probabilities for the capture of neutrons by nuclides having $N = 50$, 82, and 126. Also, in nuclei such as ^{87}Kr ($N = 51$) and ^{137}Xe ($N = 83$) one neutron is bound so loosely that it can be emitted spontaneously when these nuclei are formed in states of high excitation by β decay from ^{87}Br and ^{137}I, respectively. Much evidence for the $N = 126$ shell has been accumulated from α-decay systematics. Alpha-decay energies are rather smooth functions of A for a given Z, but show striking discontinuities at $N = 126$ (see figure 3-4). Finally the occurrence of long-lived nuclear isomers is correlated with magic numbers: islands of such isomerism occur for N and Z values just below 50, 82, and 126.

The Single-Particle Shell Model (M3). By 1948 the evidence for the magic numbers had become so strong that an explanation in terms of some sort of nuclear shell structure was sought by a number of scientists. As discussed in much more detail in chapter 10, two important insights were essential in enabling M. G. Mayer in the United States and J. H. D. Jensen and co-workers in Germany to arrive independently in 1949 at an explanation of the magic numbers in terms of single-particle orbits. One was the realization that collisions between nucleons in a nucleus are greatly suppressed by the Pauli exclusion principle, so that an individual nucleon can travel rather freely through nuclear matter. This then makes it plausible to consider an individual nucleon as moving independently in an effective potential due to the presence of all the other nucleons.

Choosing for the nuclear potential a spherically symmetric harmonic oscillator, one can solve the Schrödinger equation for a nucleon moving in such a potential and thus arrive at the energy levels of nucleons in that potential. The numbers of nucleons of one kind required to fill all the levels up to and including the first, second, and third levels, respectively, indeed turn out to be 2, 8, and 20. But beyond the third harmonic oscillator level the numbers for completed shells deviate from the magic numbers. This dilemma was resolved by the second important insight due to Mayer and Jensen, namely the strong effect of spin-orbit interactions. They found that, if the orbital angular momentum l and the spin of a nucleon interact in such a way that the state with total angular momentum $l + \frac{1}{2}$ lies at a significantly lower energy than that with $l - \frac{1}{2}$, large energy gaps occur above nucleon numbers 28, 50, 82, and 126. For further details see chapter 10.

As we see in subsequent chapters, the single-particle or independent-particle shell model accounts for much more than just the existence of the magic numbers. It immediately says that the ground states of closed-shell nuclei must have 0 spin and even parity and that the ground-state spins and parities of nuclei with one nucleon above (or below) a closed shell are those of the single extra (or missing) nucleon energy level. These considerations can in fact be extended to subshell closures. Furthermore, from the sequence of energy levels in the nuclear potential the spins and parities of excited states corresponding to the excitation of an individual nucleon can be predicted.

Nonspherical Nuclei and Extensions of the Shell Model. The single-particle shell model as described above is clearly an oversimplification. Various extensions that make the shell model more widely applicable are discussed in chapter 10. Here we merely indicate some of the directions in which these extensions lead.

For one thing, not all nuclei are strictly spherical. It then becomes necessary to consider the modifications in single-particle level sequence that result from a nonspherical potential. Furthermore, in nonspherical nuclei collective motions—rotations and vibrations—of the nucleus as a whole become possible, and these lead to new classes of excited states (analogous to rotational and vibrational excitations of molecules) in addition to the single-particle excitations. Couplings between the collective and single-particle modes are considered in a more sophisticated, so-called unified, model of nuclei.

Even in spherical nuclei, the extreme single-particle model is too naive, except in the immediate vicinity of closed shells. When several nucleons are present outside a closed shell, the residual interaction among them, although relatively small, must be taken into account.

REFERENCES

*B1 R. C. Barrett and D. F. Jackson, *Nuclear Sizes and Structure*, Clarendon, New York, 1977.

*B2 J. M. Blatt and V. F. Weisskopf, *Theoretical Nuclear Physics*, Wiley, New York, 1952.

*B3 H. A. Bethe and P. Morrison, *Elementary Nuclear Theory*, 2nd ed., Wiley, New York, 1956.

C1 J. Chadwick, "The Existence of a Neutron," *Proc. Roy. Soc. (London)* **A136**, 692 (1932).

*F1 H. Frauenfelder and E. Henley, *Subatomic Physics*, Prentice-Hall, Englewood Cliffs, N.J., 1974.

H1 R. Hofstadter, "Nuclear and Nucleon Scattering of High-Energy Electrons," *Ann. Rev. Nucl. Sci.* **7**, 231 (1957).

K1 T. Kirsten, "Nachweis des Doppelten Betazerfalls," *Fortschr. Phys.* **18**, 449 (1970).

*K2 I. Kaplan, *Nuclear Physics*, 2nd ed. Addison Wesley, Cambridge, Mass., 1963.

M1 H. G. J. Moseley, "The High-Frequency Spectra of the Elements,"*Phil. Mag.* **26**, 1024 (1913); **27**, 703 (1914).

M2 W. D. Myers and W. J. Swiatecki, "Nuclear Masses and Deformations," *Nucl. Phys.* **81**, 1 (1966).

*M3 M. G. Mayer and J. H. D. Jensen, *Elementary Theory of Nuclear Shell Structure*, Wiley, New York, 1955.

R1 E. Rutherford, "The scattering of α and β Particles by Matter and the Structure of the Atom," *Phil. Mag.* **21**, 669 (1911).

R2 E. Rutherford, J. Chadwick, and C. D. Ellis, *Radiations from Radioactive Substances*, Cambridge University Press, New York, 1930.

*S1 E. Segrè, *Nuclei and Particles*, 2nd ed., Benjamin, Reading, MA, 1978.

S2 K. F. Smith, "Nuclear Moments and Spins," *Prog. Nucl. Phys.* **6**, 52 (1957).

W1 A. H. Wapstra and K. Bos, "The 1977 Atomic Mass Evaluation, Part I. Atomic Mass Table," *At. Data Nucl. Data Tables* **19**, 177 (1977).

W2 C. F. von Weizsäcker, "Zur Theorie der Kernmassen," *Z. Phys.* **96**, 431 (1935).

Y1 L. C. L. Yuan and C. S. Wu (Eds.), *Methods of Experimental Physics*, Vol. 5B, *Nuclear Physics*, "Determination of Spin, Parity, and Nuclear Moments," Academic, New York, 1963, pp. 44–213.

EXERCISES

1. Show that $h(=6.626\times10^{-27}$ erg s) has the dimensions of angular momentum.
2. Calculate the binding energy per nucleon for ^{6}Li, ^{31}P, ^{60}Ni, ^{108}Pd, ^{195}Pt, and ^{238}U both from masses given in appendix D and from the semiempirical binding-energy equation.
3. An α-particle beam is directed at a thin gold foil. A 1-cm^2 detector placed at an angle of 20° to the beam direction and 50 cm from the foil receives 10^5 scattered α particles per second. At what angle should the same detector be placed (keeping the distance the same) if a rate of 100 particles per second is desired? What would be the counting rate at this new location if the gold foil were replaced by a silver foil of the same linear thickness?
4. Why are the measured spin of zero and the spectroscopically determined Bose statistics of ^{14}N inconsistent with protons and electrons being the constituents of nuclei?
5. Using (2-2), estimate the central density of nuclei (a) in g cm^{-3}, (b) in nucleons fm^{-3}.
6. From the masses given in appendix D, find (a) the binding energy for an additional neutron to ^{16}O, ^{50}V, ^{239}Pu; (b) the binding energy for an additional proton to ^{10}B, ^{52}Mn, ^{234}Th. *Answers:* ^{239}Pu 6.53 MeV; (b) ^{52}Mn 7.53 MeV.
7. What is the kinetic energy of (a) an electron, (b) a proton, (c) a π meson with a de Broglie wavelength of 1.5×10^{-13} cm? (You may want to refer to the relativistic relations in appendix B.) *Answer:* (a): 826 MeV.
8. Estimate what might be the heaviest element for which deviations from the predictions of the Rutherford scattering formula would be observed with 12 MeV ^{4}He ions. *Answer:* In the vicinity of Co.
9. The three fundamental mass doublets have been found to have the following separations:

$$(^{12}CH_4)^+ - (^{16}O)^+ = 36.385 \text{ millimass units at } A/q = 16,$$

$$H_2^+ - D^+ = 1.548 \text{ millimass units at } A/q = 2,$$

$$D_3^+ - (^{12}C)^{2+} = 42.307 \text{ millimass units at } A/q = 6,$$

where q is the charge in units of the electronic charge. Calculate the atomic masses of H, D, and ^{16}O.

10. With the aid of the semiempirical mass or binding-energy equations estimate (a) the energy liberated when one additional neutron is added to ^{235}U, (b) the energy liberated when one additional neutron is added to ^{238}U, (c) the amount of energy by which ^{129}I is unstable with respect to β decay to ^{129}Xe. Compare your answers with values obtained from mass excesses given in appendix D.

Answer: (a) 6.7 MeV.

11. Determine from the semiempirical mass equations the atomic number Z_A corresponding to maximum stability for $A = 27$, $A = 131$, and $A = 204$. Compare your results with the experimental data as listed, for example, in appendix D.

Answer: $Z_{27} = 12.68$.

12. Estimate as best you can the energy available for (a) α decay, (b) spontaneous fission into equal fragments, of ^{192}Pt, ^{238}U, and ^{252}Cf.

Answer: ^{252}Cf: (a) 6.22 MeV, (b) 232 MeV.

13. Estimate the radii at which the charge density has fallen to 0.5 and 0.1 of its central value for nuclei of ^{59}Co and ^{209}Bi.

Chapter 3

Radioactive Decay Processes

A. INSTABILITY OF NUCLEI

Sources of Instability. In discussing mass parabolas in chapter 2 we concluded that all but one of the isobars of a given A must be unstable toward β decay (or, in the case of even-even nuclides, at least toward double β decay) since only the isobar of lowest atomic mass is truly stable in the thermodynamic sense. At the same time we pointed out that the rates of such decay processes can be exceedingly slow—for double β decay the half lives are $\gtrsim 10^{20}$ y. For most purposes a nuclide with such a long half life may be considered stable.

We may generalize and state the following condition for the stability of any nuclide toward spontaneous (radioactive) decay: a nuclide will be energetically stable toward decay by some specified mode, such as α emission, β emission, spontaneous fission into two fragments, and so on, if its atomic mass is smaller than the sum of the masses of the products that would be formed in that decay mode. Thus for example, all so-called stable nuclides with $A \gtrsim 140$ are in fact unstable toward α emission, but have half lives so long that their decay has remained unobservable.

The condition for nuclear stability just stated makes it clear that the various modes of radioactive decay can be discussed in terms of the properties of the nuclear energy surface, which in turn can be understood as resulting from the interplay of the terms in the binding-energy equation (2-5): volume, surface, Coulomb, symmetry, and pairing energy. Thus for example, the instability of heavy nuclei toward α emission comes about because the emission of an α particle lowers the Coulomb energy, the principal negative contribution to the binding energy of heavy nuclei, but changes the nuclear binding very little, since the α particle is almost as tightly bound as a heavy nucleus. On the other hand, proton decay is not normally expected to occur because, although it too would result in a reduction in Coulomb energy, it would generally also lead to substantial reduction in nuclear binding.

We have seen that the binding energy per nucleon is largest for nuclei in the region of $A = 60$ (figure 2-1). This maximum in the binding-energy curve comes about through the different A and Z dependence of the negative terms in the binding-energy equation: the surface energy per nucleon [i.e., $1/A$ times the surface energy term in (2-5)] decreases, whereas the Coulomb energy per nucleon increases in magnitude with increasing A. The magnitude of the symmetry correction term also increases as the β-stable nuclei deviate more and more from $N = Z$. A consequence of the maximum in the binding energy per nucleon near $A = 60$ is that all nuclides with $A \gtrsim 100$ are, in fact, unstable with respect to spontaneous fission. However, because of the high Coulomb barriers for the emission of fission fragments, measurable rates of spontaneous fission are found among the heaviest elements ($A > 230$) only.

Nuclear Spectroscopy. The several examples given in the preceding paragraphs show that, in nuclear as in chemical systems, a statement about thermodynamic stability tells only part of the story. For any system that is energetically *not* stable, we are usually interested in the rates of the possible processes. As we have already seen, a thermodynamically unstable system may, for all practical purposes, behave as if it were stable.

In considering the various forms of radioactive decay we thus always inquire about the decay *rates* or half lives, and in this chapter we are chiefly concerned with the factors that affect these decay rates. In other words, we outline the theoretical framework within which each decay mode is described, explore the predictions the theory makes about the dependence of the decay rate on such factors as the energy change ΔE, the spin change ΔI, and the parity change $\Delta\Pi$ involved in the transition, and when possible, compare these predictions with experimental data. In addition to striving for a basic understanding of the decay processes themselves, we are interested in the information that can be obtained about the properties of nuclear energy levels (energy spacings, spins, and parities[1]) via the study of decay processes. Such knowledge of nuclear spectroscopy is vital for any systematic understanding of nuclei, and forms the basis of the various nuclear models discussed in chapter 10. Needless to say, the development of each of these models has in turn stimulated much work in nuclear spectroscopy designed to test model predictions.

B. ALPHA DECAY

Alpha-Particle Spectra. Much empirical information on α decay was accumulated in the early decades of radioactivity research. The three naturally occurring radioactive series provided a sizable number of α emitters spanning a large range of half lives. As we saw in chapter 1, the identity of α particles as $^4He^{2+}$ ions was established as early as 1903 and the monoenergetic nature of α rays was also soon recognized. Until 1929 it was thought that each α-emitting species had only one α-particle energy associated with it; only then was the so-called fine structure of α spectra

[1] The notation I^+ (or I^-) is used to denote a state of spin I and even (or odd) parity.

discovered. This phenomenon had escaped attention for so long because of the strong dependence of transition probability on decay energy, which is discussed later and which usually leads to a marked preference for transitions to the product ground state, with smaller transition probabilities to the lowest lying excited states; the latter had not been resolved from the dominant ground-state branches until magnetic spectrometry was used.

Since the 1930's the increasing sophistication of measuring and analyzing instruments and the discovery of over 350 artificially produced α-emitting nuclides have led to an enormous increase in experimental information on α decay. From data on the energies of the different α groups emitted by a given nuclide an energy-level diagram of the daughter product can be constructed. It must be noted that the **decay** or **disintegration energy** E_α for a given α transition (defined as the energy difference between the two nuclear states involved in the transition)[2] exceeds the *kinetic energy* T_α of the corresponding α-particle group by the recoil energy T_f of the product nucleus. Conservation of momentum requires the recoil momentum p_f and the α-particle momentum p_α to be equal in magnitude and opposite in direction; thus since nonrelativistic mechanics applies so that $p^2 = 2MT$, it follows that $M_\alpha T_\alpha = M_f T_f$. For heavy-element α emitters T_f is of the order of 0.1 MeV.

In addition to the energies of α particles and γ rays, their intensities also bear a definite relation with one another. To illustrate these energy and intensity considerations figure 3-1*a* shows an α-particle spectrum of ^{228}Th and figure 3-1*b* shows the energy-level diagram of ^{224}Ra derived from these α-decay data and the associated γ spectrum. The selection rules governing the depopulation of a given level by γ emission are discussed in section E.

Half Life versus Alpha Energy. The energies of α particles emitted by radioactive nuclides range from 1.8 MeV (^{144}Nd) to 11.7 MeV (^{212}Po^m), and most of them lie between 4 and 8 MeV. This relatively small range in energies is associated with an enormous range in half lives, from about 10^{-7} s (e.g., ^{213}At) to nearly 10^{16} y (^{148}Sm), a factor of over 10^{30}. A qualitative inverse correlation between energy release and half life was recognized by Rutherford in 1906, and in 1911 Geiger and J. M. Nuttall formulated a quantitative relation between decay constant λ and range in air r:

$$\log \lambda = a + b \log r,$$

where b is a constant and a takes on a different value for each of the three decay series (G1).

The systematic variation of α-decay half lives with decay energy can be expressed in a variety of ways. Figure 3-2 shows this smooth behavior for the even-even α emitters from polonium to nobelium. Ground-state decay

[2] When the decay takes place between ground states the decay energy is called the ground-state decay energy, and is denoted by Q_α. It can be obtained from the atomic masses M_i, M_f, and M_α of initial nuclide, final nuclide, and ^{4}He.

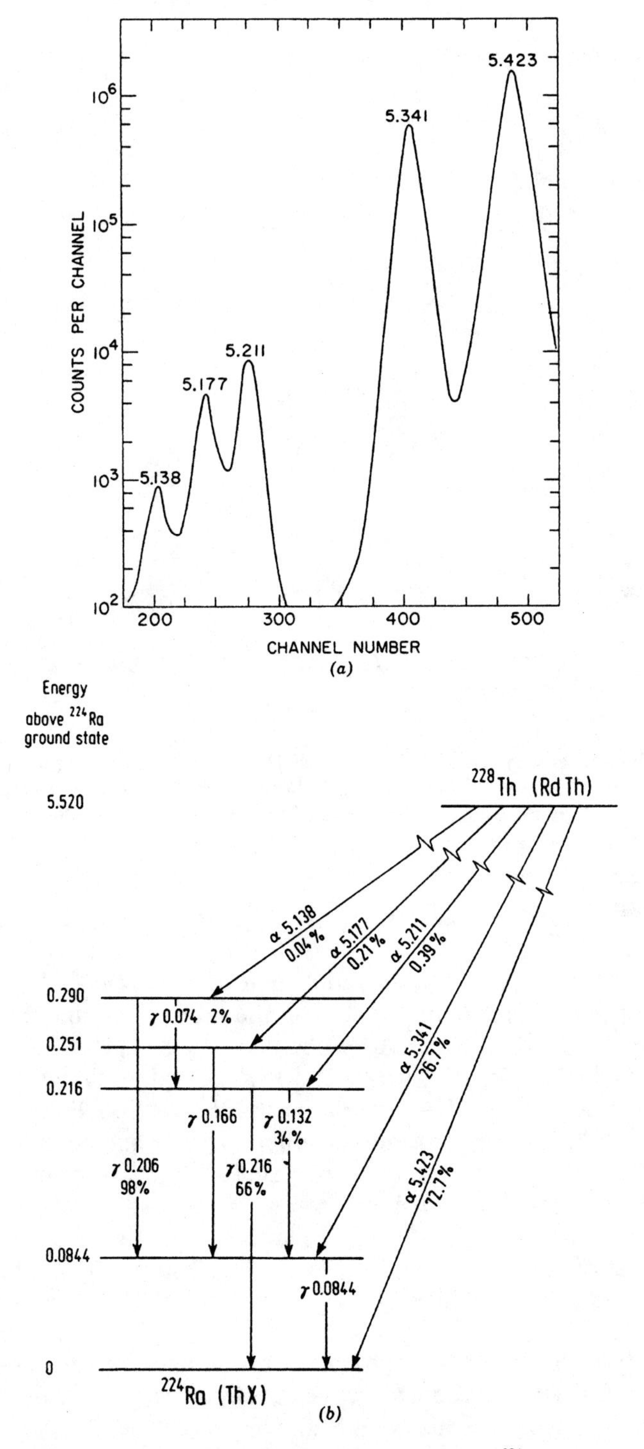

Fig. 3-1 (*a*) A ^{228}Th α spectrum. (*b*) Energy level diagram for ^{224}Ra as obtained from the observed α and γ radiations of ^{228}Th. Energies are in MeV. For each α-particle group the kinetic energy (not the disintegration energy) is given.

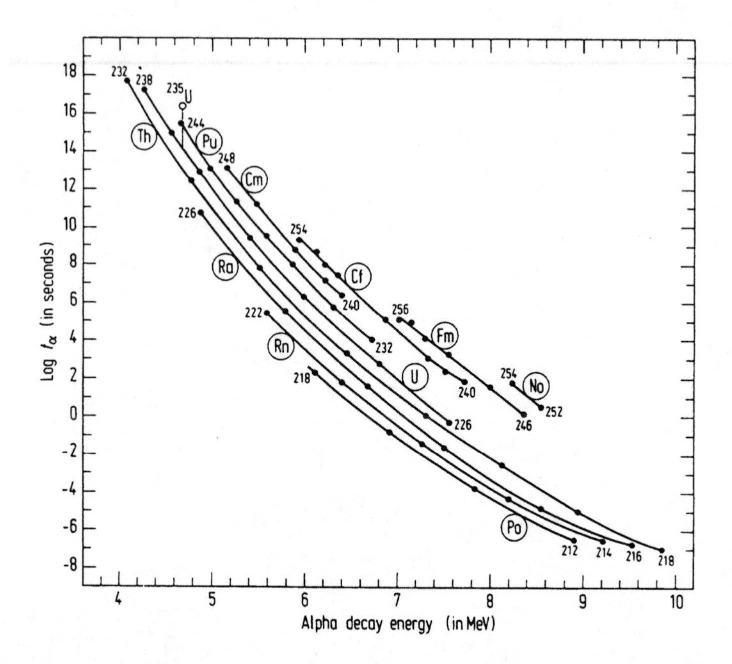

Fig. 3-2 Relation between partial α half life and Q_α for even-even nuclides. Points for each element are connected, and the mass numbers of the heaviest and lightest isotopes are given. The even-odd ^{235}U is also shown. (The data are from reference L1; the representation is adapted from reference P1.)

energies (Q_α values) are plotted against the logarithms of the partial α half lives.

A theoretical basis for understanding α decay was lacking until the advent of quantum mechanics. It was all the more gratifying that the basic quantum-mechanical theory, developed in 1928 independently by G. Gamow (G2) and by R. W. Gurney and E. U. Condon (G3), was brilliantly successful in accounting for the relationship between half lives and energies. With relatively minor modifications and refinements this theory remains the cornerstone of our understanding of α-decay rates, even though the body of experimental data has been vastly expanded, both by great improvements in measuring techniques and by the discovery and characterization of a large number of new α emitters, particularly in the rare-earth and transuranium regions.

Penetration of Potential Barriers. In outline the theory takes the following form. The Schrödinger wave equation for an α particle of energy E inside the nuclear potential well is set up and solved. The wave function representing the α particle does not go abruptly to zero at the wall of the potential barrier (R_1 in figure 3-3) and has finite, although small, values outside the radial distance R_1. By applying the boundary condition that the

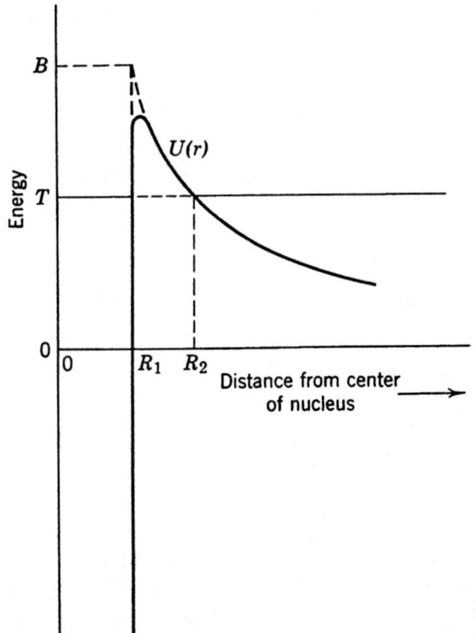

Fig. 3-3 Potential energy for a nucleus-α particle system.

wave function and its first derivative must be continuous at R_1 and R_2, we can solve the wave equation for the region between R_1 and R_2, that is, inside the barrier where the potential energy $U(r)$ is greater than the total kinetic energy T (sum of kinetic energies of α particle and recoil nucleus). The probability P for the α particle of mass M_α to penetrate that region, the so-called barrier penetrability factor, is given by the square of the wave function and turns out to be

$$P = \exp\left(-\frac{4\pi}{h}\sqrt{2\mu}\int_{R_1}^{R_2}\sqrt{U(r)-T}\,dr\right) \tag{3-1}$$

where

$$\mu = \frac{M_\alpha M_R}{M_\alpha + M_R}$$

is the reduced mass of α particle and recoil nucleus. [For a derivation of (3-1), see, for example, E1, pp. 45–74.] It is clear from (3-1) that the probability for barrier penetration decreases with increasing value of the integral in the exponent, that is, with increasing barrier height and width. (The higher the barrier, the larger the difference $U(r) - T$, and the wider the barrier, the greater the range of the integration.)

The decay constant λ may be considered as the product of P and the frequency f with which an α particle strikes the potential barrier; the order of magnitude of f may be estimated as follows. The de Broglie wavelength

$h/\mu v$ of the α particle of velocity v and momentum μv inside the nucleus is taken comparable to R_1, thus

$$\frac{h}{\mu v} \approx R_1, \qquad \text{or} \qquad v \approx \frac{h}{\mu R_1}.$$

If the α particle is considered as bouncing back and forth between the potential walls,

$$f = \frac{v}{2R_1}, \qquad \text{or} \qquad f \approx \frac{h}{2\mu R_1^2}.$$

Therefore the decay constant is

$$\lambda \approx \frac{h}{2\mu R_1^2} \exp\left[-\frac{4\pi}{h}\sqrt{2\mu}\int_{R_1}^{R_2}\sqrt{U(r)-T}\,dr\right]. \tag{3-2}$$

By a more elaborate treatment more accurate expressions for f and λ are obtained.

For certain simple forms of the potential energy $U(r)$, the integral in the exponential of (3-1) and (3-2) can be solved in closed form, and explicit expressions for λ in terms of R_1, Z, and E can thus be obtained. For more realistic shapes of the nuclear potential, (3-1) has to be solved by numerical integration.

For a square-well nuclear potential of radius R_1 and a Coulomb potential[3] $U(r) = Zze^2/r$ for $r > R_1$ (the heavy dashed line in figure 3-3), the integral in (3-1) and (3-2) becomes

$$\text{Int.} = \int_{R_1}^{R_2} (Zze^2 - Tr)^{1/2}\frac{dr}{r^{1/2}}, \tag{3-3}$$

which, by the substitutions $x = r^{1/2}$ and $a^2 = Zze^2/T$, turns into the readily integrable form $2\sqrt{T}\int_{\sqrt{R_1}}^{\sqrt{R_2}}\sqrt{a^2 - x^2}\,dx$, with the solution

$$\text{Int.} = \sqrt{T}\left[x(a^2 - x^2)^{1/2} + a^2 \arcsin\frac{x}{a}\right]_{\sqrt{R_1}}^{\sqrt{R_2}}. \tag{3-4}$$

Values of the radii R_1 and R_2 are obtained from the expressions for the total kinetic energy T and the barrier height B (see figure 3-3):

$$T = \frac{Zze^2}{R_2} \qquad \text{and} \qquad B = \frac{Zze^2}{R_1}.$$

After substitution of the integration limits and some algebraic manipulations we obtain

$$\text{Int.} = \frac{Zze^2}{\sqrt{T}}\left[\arccos\left(\frac{T}{B}\right)^{1/2} - \left(\frac{T}{B}\right)^{1/2}\left(1 - \frac{T}{B}\right)^{1/2}\right]. \tag{3-5}$$

[3] Here Z is the atomic number of the daughter nucleus.

Table 3-1 Comparison of Decay Constants Calculated from (3-6) with Experimental Data

Alpha Emitter	T (*MeV*)	$R_1 \times 10^{13}$ cm[a]	λ_{calc} (s^{-1})	λ_{exp}[b] (s^{-1})
^{144}Nd	1.9	7.950	2.7×10^{-24}	1.0×10^{-23}
^{148}Gd	3.27	8.014	2.6×10^{-10}	2.2×10^{-10}
^{210}Po	5.408	8.878	1.0×10^{-6}	5.80×10^{-8}
^{214}Po	7.835	8.927	4.9×10^{3}	4.23×10^{3}
^{226}Th	6.448	9.072	2.6×10^{-4}	2.95×10^{-4}
^{228}Th	5.521	9.095	8.0×10^{-9}	8.35×10^{-9}
^{230}Th	4.767	9.118	1.7×10^{-13}	2.09×10^{-13}
^{232}Th	4.080	9.142	7.8×10^{-19}	1.20×10^{-18}
^{254}Fm	7.310	9.390	1.3×10^{-4}	5.1×10^{-5}

[a] Radii were calculated from the formula $R_1 = (1.30A^{1/3} + 1.20) \times 10^{-13}$ cm (see text).
[b] The λ values listed are *partial decay constants* for the ground-state α transitions only.

Finally, remembering that $T = \frac{1}{2}\mu v^2$, substitution of (3-5) in (3-2) gives

$$\lambda \approx \frac{h}{2\mu R_1^2} \exp\left\{-\frac{8\pi Zze^2}{hv}\left[\arccos\left(\frac{T}{B}\right)^{1/2} - \left(\frac{T}{B}\right)^{1/2}\left(1-\frac{T}{B}\right)^{1/2}\right]\right\}. \quad (3\text{-}6)$$

As an illustration of the remarkable success of the Gamow-Gurney-Condon approach, we show in table 3-1 a few values calculated with (3-6) and the corresponding experimental values. The calculations were done with $R_1 = (1.30 \times A^{1/3} + 1.20) \times 10^{-13}$ cm and with no other adjustable parameters. The agreement between calculated and measured values is seen to be within a factor of 4 in all cases except ^{210}Po (a nucleus with 126 neutrons that decays to a nucleus with 82 protons—both tightly bound closed-shell nuclei that may have abnormally small radii), even though the λ values extend over a range of about 10^{27}. The absolute values λ_{calc} depend sensitively on the nuclear radii assumed, each increase by 0.03 fm in the nuclear radius parameter r_0 giving rise to an approximate doubling of all λ values. The fact that (3-6) gives good agreement with experimental data when r_0 is taken as 1.30 fm should not be considered significant, since (1) we used a square-well potential rather than a more realistic potential shape with tapering sides, (2) the effective α-particle radius $\rho_\alpha = 1.20$ fm was chosen somewhat arbitrarily, and (3) the pre-exponential factor was derived on the rather naïve assumption that α particles pre-exist and oscillate in nuclei ("one-body model"). The important point is that none of these assumptions strongly affects the spectacular dependence of λ on T, which stems entirely from the exponential in (3-6),[4] but they have a significant effect on the relation between λ and R_1, the effective nuclear radius. Even within the framework of the one-body model, much can be and has been done to refine the derivation and form of the pre-exponential factor in the expression for λ (see, e.g., H1). As mentioned in chapter 2, section C2, nuclear radii deduced from α-decay data can be expressed in the form $R_1 = r_0' A^{1/3}$, with r_0' values between 1.45 and 1.57 fm, depending on the particular form of the theory used. Since R_1 is an effective interaction radius that includes the radius of the α particle, it is more appropriate to express it in the form

$$R_1 = r_0 A^{1/3} + \rho_\alpha, \tag{3-7}$$

as we did above. Here ρ_α is the α-particle radius, and is usually taken as 1.20 fm (B1, p. 357).[5] The various theories then lead to r_0 values in the range 1.25–1.35 fm. We have already mentioned that apparent anomalies in the half life-versus-energy relationships at shell crossings ($N = 126$, $Z = 82$, and $N = 82$) have been interpreted in terms of abnormally small radii for closed-shell nuclei. However, this is probably too simplistic an explanation, and other consequences of closed shells probably play a significant part, as indicated below.

Hindered Alpha Decay. The α emitters listed in table 3-1 all have even Z and even N, and the dependence of half life on decay energy predicted by the barrier-penetration theory in its simple form applies to even-even α emitters only. In fact, good agreement with experimentally determined partial half lives is generally found only for transitions to ground states and first excited states of even-even nuclei. All other α transitions (those to higher excited states in even-even nuclei and most transitions in other nuclear types) tend to be slower (by factors up to about 10^4) than would be predicted by the simple theory. This is illustrated in figure 3-2 by the point for the odd-A nucleus ^{235}U.

Since the ground states of even-even nuclei have 0 spin and even parity (first excited states have spin 2 and even parity) and since (3-6) was derived without regard to angular-momentum effects (that is, for emission of s-wave α particles only), we might at first sight be tempted to ascribe the relative slowness of other α transitions to angular-momentum changes. However, the so-called **hindrance factors**—the ratios of calculated (for $\Delta l = 0$) to observed transition probabilities—are larger than can be ac-

[4] The pre-exponential factor representing the "striking frequency" of the α particle hitting the potential barrier is of the order of $10^{20}\ s^{-1}$ and varies only by about 30 percent for different nuclei.

[5] The best estimates of ρ_α come from excitation functions for α-induced reactions; these involve barrier penetration by α particles also and can be investigated over a wider range of Z (and thus R) than can α radioactivity, and therefore give more definite information on ρ_α (see chapter 4, section A).

counted for by inclusion of angular-momentum effects in the theory.[6] In even-even nuclei the transitions to states other than the ground and first excited states (mostly 4^+, 6^+, 8^+, and 1^- states, and a few other odd-parity states) have hindrance factors ranging from unity to about 12,000. There is some general trend toward larger hindrance factors with increasing ΔI, and for some transition types ($0^+ \rightarrow 4^+$ and $0^+ \rightarrow 1^-$) there is a fairly regular progression with Z or N.

Among the even-odd, odd-even, and odd-odd nuclides, the situation appears to be even more complex. Hindrance factors range from unity to $\sim 3 \times 10^4$, with systematic trends difficult to discern. One striking feature is that the ground-state transitions, especially for strongly deformed odd-A nuclei, are highly hindered even when there is no spin change involved, whereas some transition to an excited state is usually almost unhindered. For example, the α transitions of ^{241}Am ($5/2^-$) to the ground and first excited states of ^{237}Np ($5/2^+$ and $7/2^+$) have hindrance factors of ~ 500, and the main transition, almost unhindered, is to the second excited state at 60 keV above ground ($5/2^-$). Similarly, the transitions of the even-odd ^{235}U to the ground and first excited states of ^{231}Th are hindered by factors of about 10^3 (the ground-state transition is indicated on figure 3-2); if it were not for this circumstance, this important nuclide would be much too short lived to be found on earth. The possibility of large hindrance factors for ground-state transitions makes it sometimes difficult to decide whether the ground-state transition has in fact been found.

Nuclear-Structure Effects. We should certainly not be surprised that the simple one-body theory that assumes the existence of preformed α particles does not fully account for α-decay rates in different types of nuclei and to various kinds of excited states. While the Gamow-Gurney-Condon theory does a remarkable job of explaining the barrier penetrability once an α particle is formed, it stands to reason that the probability of formation of an α particle in a nucleus should depend on details of nuclear structure. Much progress has been made in calculating the relative probabilities for assembling α particles in different nuclei on the basis of the shell and collective models (see chapter 10 for a description of these models), thus accounting for many of the observed hindrance factors. A review of this approach, for both spherical and deformed nuclei, may be found in M1. Here we can give only some qualitative ideas, following the

[6] To take account of an α particle carrying away l units of angular momentum, one has to use for $U(r)$ in the barrier penetration factor (3-1) the expression $Zze^2/r + \hbar^2 l(l+1)/2\mu r^2$ instead of just the Coulomb potential Zze^2/r as before. The added term is referred to as a **centrifugal barrier** and is relatively small. The integral can still be solved after suitable expansion of the square root (H1). Barrier penetrability decreases with increasing l, but only by moderate factors, severalfold for $l = 4$, a few hundredfold for $l = 8$, for example. The pre-exponential factor is also affected by l, in the opposite direction from the penetrability, so that the overall effect is rather small.

treatment in P1. It seems evident that a ground-state transition from a nucleus containing an odd nucleon in the highest filled state can take place only if that nucleon becomes part of the α particle and therefore if another nucleon pair is broken; this is certainly a less favorable situation than the formation of an α particle from already existing pairs in an even-even nucleus and may give rise to the observed hindrance. If, on the other hand, the α particle *is* assembled from existing pairs in such a nucleus, the product nucleus will be in an excited state, and this may explain the "favored" transitions to excited states. Detailed data obtained from other evidence about the spins, parities, and other quantum numbers of the particular states between which these favored transitions take place appear to confirm this interpretation.

Actually the detailed study of α-particle spectra, along with the associated γ-ray spectroscopy and $\alpha\gamma$-coincidence and angular-correlation measurements, produced much of the data on energy levels of deformed nuclei that gave impetus to the development of the collective model discussed in chapter 10.

Alpha-Decay Energies. In addition to the regularities in lifetimes, α-particle emitters exhibit some interesting systematic trends in their Q_α values. Most of these features can be derived from the general properties of the nuclear-energy surface discussed in chapter 2. By obtaining, with the aid of (2-5), a general expression for the energy difference between ground states of two nuclei with $\Delta A = 4$ and $\Delta Z = 2$ and examining the properties of the first and second partial derivatives of this expression with respect to A and Z, we arrive at the following conclusions. For the isotopes of any element in the region of the α emitters the α-decay energies are expected to decrease monotonically with increasing A, and for a series of isobars they will increase with increasing Z.

These predictions of the liquid-drop model of nuclei are largely borne out by experimental data as shown in figure 3-4, in which α-decay energies for ground-state transitions are plotted against the mass number of the α emitter. Points belonging to one element are joined. A few of the points were obtained not by direct measurements but by the method of closed decay cycles, which is illustrated for ^{242}Am in figure 3-5.

In a diagram such as figure 3-4 the abrupt interruption of the predicted regularities in the neighborhood of $A = 210$ is as striking as the regular trends themselves. Had we plotted neutron number rather than mass number as the abscissa,[7] it would have been more immediately evident that the sharp drop in decay energy occurs for each element between the α emitter with 128 and that with 126 neutrons. This indicates exceptionally large binding energies (small masses) just below neutron number 126 and

[7] This was not done because the data for different elements would be less well separated on such a plot.

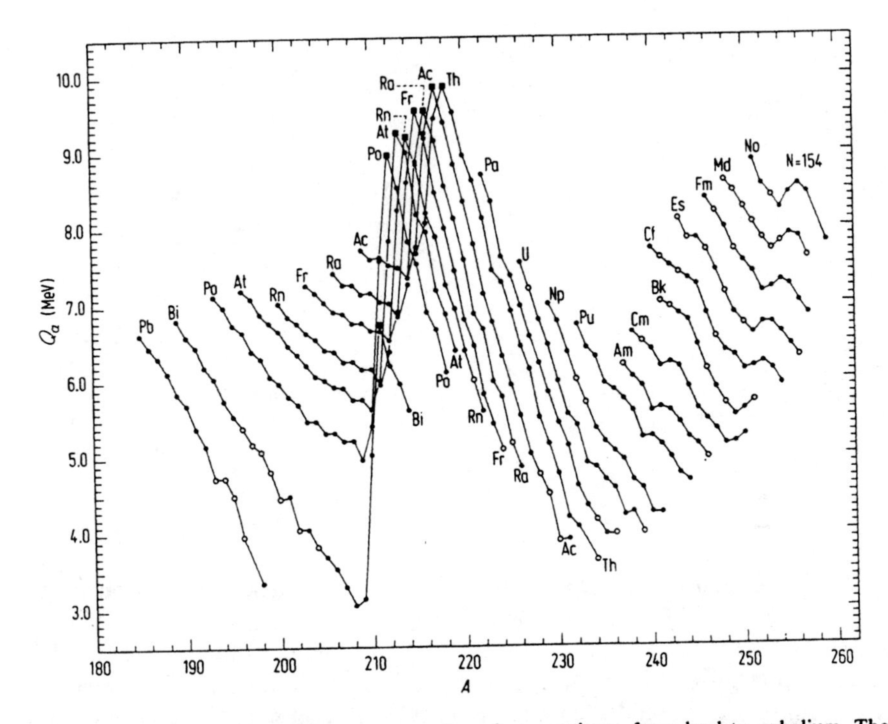

Fig. 3-4 Plot of Q_α values versus mass numbers for α emitters from lead to nobelium. The maxima (filled squares, with sharp drop-off to the left) occur at $N = 128$. The less pronounced maxima at $N = 154$ indicate a subshell closure at $N = 152$. Decay energies estimated from systematics are shown as open circles. (Data from reference L1; the representation is adapted from reference P1.)

is one of the strongest pieces of evidence for a closed shell at $N = 126$. Similarly, the sharp decrease in Q_α in going from the heavy polonium to the heavy bismuth isotopes is a consequence of the closed proton shell at $Z = 82$. Evidently this shell effect is responsible for the absence of observable α decay in lead and thallium isotopes. The slight breaks in the curves of figure 3-4 below ^{252}Cf, ^{253}Es, ^{254}Fm, etc. have been interpreted as evidence for a closed neutron subshell at $N = 152$.

In the rare-earth region, just above neutron number 82, an "island" of α emitters has been found that includes several naturally occurring nuclides (see table 1-1) and a few dozen artificially produced ones, most of them neutron-deficient relative to the stable isotopes. The highest decay energies in this region occur for emitters with $N = 84$ because the daughters are closed-shell nuclei with $N = 82$. More recently another island of (very short-lived) α emitters has been found among the highly neutron-deficient nuclei just above the doubly magic ^{100}Sn; this group includes ^{107}Te, ^{108}Te, and ^{111}Xe. Many other α emitters are found among the very neutron-deficient isotopes of the elements from osmium to mercury.

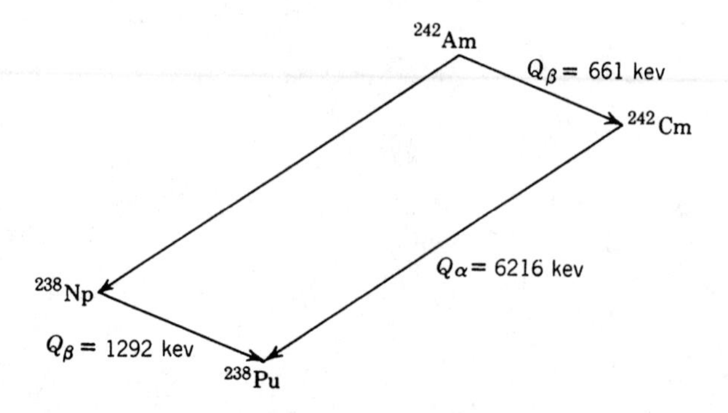

Fig. 3-5 Closed decay cycle for determination of the ground-state decay energy Q_α of ^{242}Am. From the measured Q values shown, Q_α of ^{242}Am is computed to be $661 + 6216 - 1292 = 5585$ keV.

Other Heavy-Particle Decay Modes. We may ask why proton decay with measurable lifetimes does not appear to occur from nuclear ground states, whereas α radioactivity is such a common phenomenon. The reasons can be readily given in the light of our discussions of α decay. In the vicinity of the β-stable region proton emission is, in contrast to α emission, energetically impossible over the entire mass range. The difference lies simply in the large binding energy of the four nucleons in the free α particle; they are more tightly bound than the most loosely bound nucleons in any heavy nucleus. On the other hand, a proton to be emitted must be supplied with an energy of several million electron volts—its own binding energy to the residual nucleus.

Far on the proton-excess side of the β-stability valley nuclei do indeed become unstable with respect to proton emission, largely as a result of the increasing importance of the Coulomb and symmetry terms in the binding-energy equation (2-5). Just inside the limit of stability toward instantaneous proton emission we thus might expect proton emission with measurable lifetimes. However, for nuclei so far removed from β stability (e.g., ^{45}Mn or ^{60}Se) the β-decay energies will be extremely high, hence, as we see in section D, the half lives for positron decay or electron capture very short; proton emission can therefore be observable only if it, too, has comparably short half lives (say <1 s). The half life for proton emission can be estimated very roughly by the use of (3-6), and on that basis a half-life range of $10^{-10}\ \mathrm{s} \leq t_{1/2} \leq 1$ s turns out to correspond to a very narrow range of decay energies—about 30–80 keV at $Z = 10$ and about 0.2–0.5 MeV at $Z = 30$. Although these estimates cannot be considered quantitatively significant, they indicate that it will be a fortuitous circumstance to find a nuclide with a proton-decay energy in the required narrow interval and with a sufficiently favorable ratio of proton- to positron-emission probability for observation. Nevertheless, a case of proton radioactivity has been reported in the decay of $^{53}Co^m$ ($E_p = 1.56$ MeV, $t_{1/2} = 0.24$ s).[8]

Delayed proton emission of a different sort has been observed in proton-rich nuclei, principally among low-Z reaction products. The process is quite analogous to the well-known delayed neutron emission first observed among fission products (cf. p. 166): a β^+ decay leads to a proton-unstable excited state that instantaneously (in $<10^{-12}$ s) emits a proton. The observed half life for proton emission is just the half life for the preceding β^+ decay. A series of β^+ emitters from ^9C to ^{41}Ti, all with $N = Z - 3$, are such delayed-proton precursors, with half lives between a few milliseconds and about 0.5 s.

Binding-energy considerations alone would certainly permit spontaneous emission of tightly bound nuclear entities other than α particles. Deuterons are not sufficiently tightly bound for their emission from nuclei anywhere near the β-stability valley. But many α-unstable nuclei could, on purely energetic grounds, emit nuclei such as ^{12}C, ^{16}O, or ^{20}Ne also. However, the barrier heights for the emission of such nuclei are so high that the expected half lives are far beyond presently detectable limits.

Two-Proton Radioactivity. The possibility of an interesting additional class of radioactive decay has been pointed out by V. I. Goldanskii (G4, G5): the simultaneous emission of two protons. This process can presumably occur among light ($Z \lesssim 50$) nuclides of even Z that lie near the proton-stability limit. Such nuclei, though slightly stable toward proton emission, may be unstable with respect to the emission of two protons, for as a result of the pairing energy [last term in (2-5)] the last (even) proton in the nucleus ^AZ may be positively bound, whereas the last (odd) proton in the nucleus $^{A-1}(Z-1)$ may not be. Furthermore, if the absolute value of the positive binding energy of the Zth proton is less than that of the negative binding energy of the $(Z-1)$th proton, the energetic condition for two-proton radioactivity is fulfilled. Proton pairing energies are of the order of 1–3 MeV, and the decay energy for two-proton emission must always be less than that pairing energy.

On the basis of detailed binding-energy considerations[9] Goldanskii (G5) has predicted the nuclides that can be expected to undergo two-proton decay and their decay energies. With the usual barrier-penetration formulas we can then estimate half lives and find that they are expected to span an appreciably wider range than in single-proton emission. Furthermore, the

[8] The half life given is not the partial half life for p decay, but the total half life, largely determined by the β^+-decay branch, which is estimated to account for ~98.5 percent of the $^{53}Co^m$ decays. Even after correction for this large branch the energy—half life relationship observed for the p decay does not agree well with the prediction of (3-6). This is not too surprising since a large hindrance factor may be expected from single-particle effects.

[9] Our equation (2-5) is probably not sufficiently reliable at large distances from β stability to be useful for this purpose.

total barrier penetration probability for a pair of protons turns out to be sharply peaked when they are emitted with equal energies. The emission of two protons of nearly equal energies should be a fairly characteristic event observable in emulsions or cloud chambers, even when β^+ emission is the predominant decay mode. Among promising candidates as two-proton-radioactive species are ^{16}Ne, ^{38}Ti, and ^{67}Kr. No case of two-proton radioactivity has yet been observed.

C. SPONTANEOUS FISSION

Within about a year of the discovery of nuclear fission under neutron bombardment by Hahn and Strassmann, K. A. Petrzhak and G. N. Flerov reported that ^{238}U undergoes fission spontaneously, although with a half life of about 10^{16} y, very long compared to the α-decay half life. Since that discovery in 1940 the spontaneous fission rates of several dozen nuclides, all with $Z \geq 90$, have been measured, and spontaneous fission is thus firmly established as another mode of radioactive decay. The observed partial half lives for fission extend from fractions of a nanosecond to about 2×10^{17} y.

Energetics and Half Lives. We have already seen (p. 55) that the breakup of any nucleus of $A \gtrsim 100$ into two nuclei of approximately equal size is exoergic. We may then ask why spontaneous fission has been observed only for nuclei with $A \geq 230$ and what factors govern the half lives for the process. The answers to these questions are to be found in considerations somewhat analogous to those used in the discussion of α decay. Clearly the separation of a heavy nucleus into two positively charged fragments is hindered by a Coulomb barrier, and fission can therefore be treated as a barrier penetration problem. The height of the so-called **fission barrier** is, in first approximation, the difference between the Coulomb energy between the two fragments when they are just touching and the energy released in the fission process. For elements in the region of uranium both these quantities have values near 200 MeV, and fission barriers are therefore rather low.

The Coulomb energy between two spherical nuclei $^{A_1}Z_1$ and $^{A_2}Z_2$ in contact is given by

$$V_c = \frac{Z_1 Z_2 e^2}{R_1 + R_2}. \tag{3-8}$$

where R_1 and R_2 are the nuclear radii. Using (2-2) for nuclear radii and setting $r_0 = 1.5$ fm, we get

$$V_c = 0.96 \frac{Z_1 Z_2}{A_1^{1/3} + A_2^{1/3}} \text{ MeV}. \tag{3-9}$$

Using this simple formula, we obtain 206 MeV for the Coulomb energy between two nuclei in contact, each with one half the A and Z of ^{238}U. This can be compared to 193 MeV, the energy release in the symmetric fission of ^{238}U as calculated from the semiempirical mass equation (2-7). Analogous calculations for the symmetric splitting of ^{200}Hg give 165 MeV for the Coulomb energy between the fragments and 139 MeV for the energy release. Neither the Coulomb energies[10] nor the energy-release estimates should be considered quantitatively significant. What is important to note is that the barrier height increases more slowly with increasing nuclear size than does the decay energy for fission. In view of the steep dependence of barrier penetrability on the ratio of available energy to barrier height, it is thus not surprising that spontaneous fission is observed only among the very heaviest elements and that spontaneous-fission half lives, in general, decrease rapidly with increasing Z.

It is instructive to pursue the energetics of fission a little further with the aid of the semiempirical mass equation (2-7). The energy release Q_f in the fission of nucleus ^{A}Z into two equal fragments $^{A/2}(Z/2)$ is given by

$$Q_f = M_{Z,A} - 2M_{Z/2,A/2},$$

and from (2-7) we obtain

$$Q_f = 0.265Z^2A^{-1/3} - 34.54ZA^{-1/3}\left(1 - \frac{Z}{A}\right) + 3.81A^{2/3}. \qquad (3\text{-}10)$$

If Q_f exceeds the height of the Coulomb barrier between the two fragments, we may expect breakup within a few nuclear vibrations since no barrier penetration is required. Thus the condition for stability against such instantaneous breakup may be stated as $Q_f < V_c$. Considering still the simple case of breakup of ^{A}Z into two equal fragments $^{A/2}Z/2$, we get from (3-9) that $V_c = 0.151Z^2/A^{1/3}$ MeV. Combining this with (3-10) and rearranging, we can express the condition $Q_f < V_c$ as follows:

$$\frac{Z^2}{A} < 303\frac{Z}{A}\left(1 - \frac{Z}{A}\right) - 33.4.$$

Since Z/A varies only between about 0.38 and 0.42 throughout the heavy-element region, the condition for stability against instantaneous fission can be expressed as $Z^2/A \lesssim (39.2 \pm 1.2)$. Again, the numerical value is not to be taken seriously—recall that it is based on $r_0 = 1.5$ fm, and with different r_0 values we would obtain different results for the critical value of Z^2/A; for example, with $r_0 = 1.4$ fm, $(Z^2/A)_{crit} = 43.3 \pm 1.3$. The important point is that, according to the liquid-drop model, there *is* a critical value of Z^2/A above which nuclei cannot be expected to hold together for more than about

[10] The Coulomb energy estimates are presumably high because the fragments are surely not spherical at the moment of separation. Also, breakup into *equal* fragments does not necessarily give the largest energy release (and is, in fact, much less likely than asymmetric mass splits, as discussed in chapter 4, section F), but for the rather crude estimates given here this is not important.

10^{-22} s. Further, we might expect that, the closer the Z^2/A value of a nuclide is to $(Z^2/A)_{crit}$, the shorter will be its half life for spontaneous fission.

In figure 3-6 the logarithm of spontaneous-fission half life is plotted against Z^2/A. We see immediately that, while there is a general trend with Z^2/A, the simple expectation of the liquid-drop model is not borne out. For each even-Z element the spontaneous-fission half lives of the even-even isotopes go through a maximum, and the data suggest that the widths of the distributions decrease with increasing Z. Further, as with α decay, the half lives of the odd-A and odd-odd nuclides, a few of which are shown in figure 3-6, are considerably longer than interpolation between neighboring, even species would suggest. To put these half-life data and the deviations from a simple dependence on Z^2/A in perspective, we should remember that the entire range of half lives covered in figure 3-6, 32 orders of magnitude, corresponds to a range of only a few million electron volts in the height of the fission barrier. Thus we are dealing with very subtle effects indeed.

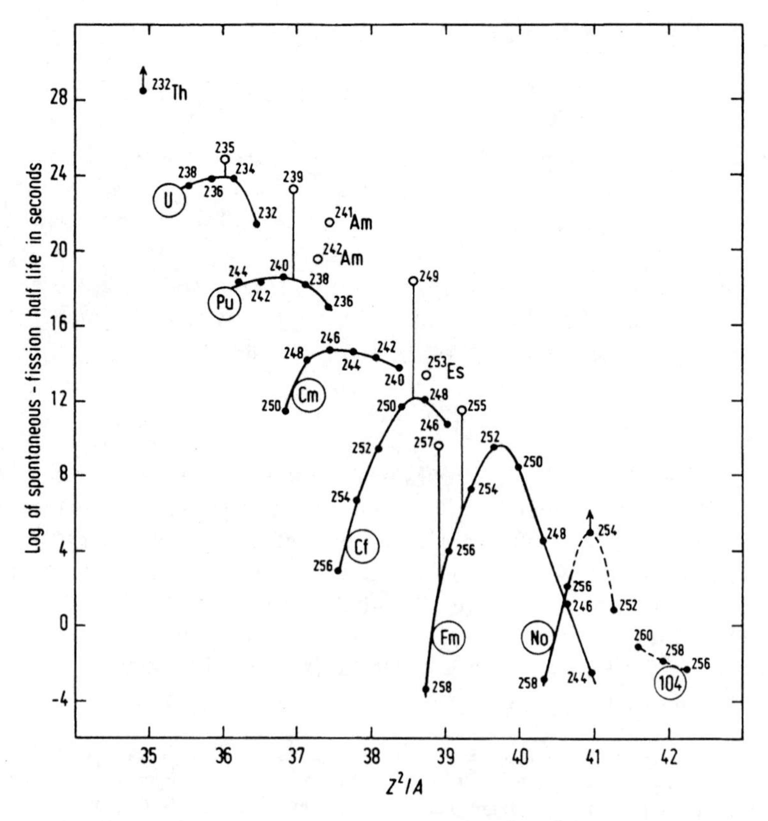

Fig. 3-6 Partial half lives for spontaneous fission versus Z^2/A. Even-even isotopes of each element are connected. A few odd-A nuclides and the odd-odd ^{242}Am are also shown.

Dynamical Considerations. In discussing fission barriers we have so far used only static considerations. But how are we to picture the dynamics of the fission process? Theoretical treatments of fission, whether spontaneous or induced, have largely followed the basic approach developed in 1939 in a pioneering paper by N. Bohr and J. A. Wheeler (B2). Using the liquid-drop model, they treated the fission process in a manner quite analogous to transition-state theory of chemical reactions. Essentially the same ideas were independently and almost simultaneously put forth by J. I. Frenkel in the USSR (F1).

Consider the initial nucleus as a spherical, uniformly charged, incompressible drop. Any small deformation of such a drop leads to an increase in surface area and therefore in surface energy (a negative term in the binding-energy equation), thus reducing the overall binding. However, at the same time such a deformation also increases the average spacing between the protons and thus decreases the Coulomb repulsion term, thus tending to increase total binding. As long as the change ΔE_S in the surface energy exceeds the change ΔE_c in the Coulomb energy,[11] there is a net restoring force that tends to return the nucleus to its spherical shape. However, for some deformations the magnitude of ΔE_c can be greater than that of ΔE_S, and the nucleus then becomes unstable toward some breakup process if it reaches such a deformation.[12] The task of a general theory then is to map potential-energy surfaces for various kinds of deformation from initially spherical or other stable shapes, to locate the saddle points (transition states) on these surfaces, to determine the barrier heights (activation energies), and to trace the trajectories from initial to final states.

Following Bohr and Wheeler, only axially symmetric drop shapes have until recently been considered, and they have most often been parameterized in terms of the radius vector $R(\theta)$, and expanded in Legendre polynomials of $\cos(\theta)$, where θ is the angle between the radius vector and the symmetry axis:

$$R(\theta) = R_0\left[1 + \sum_{n=1}^{\infty} \alpha_n P_n(\cos\theta)\right]. \tag{3-11}$$

Using only the first few terms and considering only very small distortions, Bohr and Wheeler showed that $\Delta E_c = -\frac{1}{5}\alpha_2^2 E_c^0$ and $\Delta E_S = \frac{2}{5}\alpha_2^2 E_S^0$, where E_c^0 and E_S^0 are the Coulomb and surface energies of the undistorted sphere.

[11] Note that the Coulomb energy E_c referred to here (the Coulomb repulsion term in the binding-energy equation) is quite different from the Coulomb repulsion V_c between two touching spheres in (3-9).

[12] This statement can be translated into an alternative way of deriving an expression for $(Z^2/A)_{crit}$, one that was in fact used by Bohr and Wheeler. From the expressions for ΔE_c and ΔE_S given below we see that there will be no net restoring force to return the drop to its spherical shape if $E_c^0/2E_S^0 > 1$. Using for E_c^0 and E_S^0 the Coulomb and surface terms of the semiempirical mass equation (2-7) and setting $(E_c^0/2E_S^0)_{crit} = 1$, we obtain a value for $(Z^2/A)_{crit}$. Again, the particular numerical value derived depends on the parameters used.

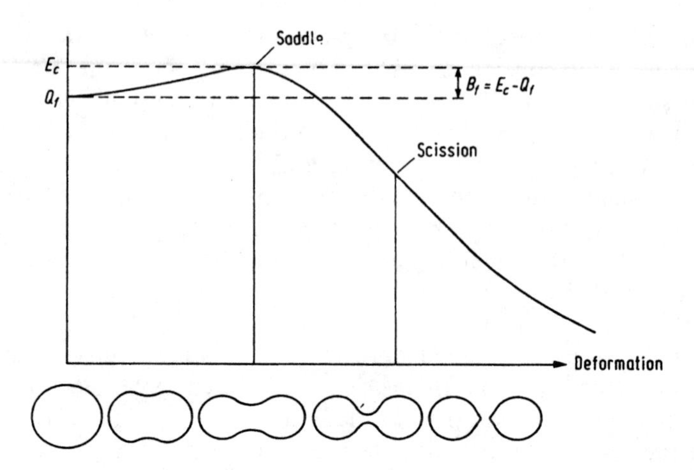

Fig. 3-7 Potential energy as a function of deformation in a simple liquid-drop picture. The fission barrier B_f, the saddle point (critical deformation), and the scission point (separation into two fragments) are indicated. The distortion of an initially spherical nucleus is schematically shown beneath the potential-energy diagram.

Thus the total distortion energy $\Delta E = \Delta E_S + \Delta E_c = \frac{1}{5}\alpha_2^2(2E_S^0 - E_c^0)$. The coefficient α_2 is a measure of the axial stretching of the drop. Qualitatively the progression of drop shapes and the corresponding potential-energy changes can be pictured for this simple case as sketched in figure 3-7.

Although the original Bohr-Wheeler formulation proved qualitatively very successful, it failed in many quantitative respects. The observed mass splits, especially the preference for asymmetric splits in the uranium region, could not be accounted for, nor could absolute barrier heights be predicted. The latter difficulty is not surprising in view of the functional form of the distortion energy shown above. We know the distortion energy at the saddle point to be a few million electron volts, yet it is proportional to the difference between two very large quantities, $2E_S^0$ and E_c^0, each of which has a value of several hundred million electron volts and would thus have to be known extremely accurately.

Considerable progress has been made in recent years in refining the theory by using additional terms in the Legendre polynomial expansion (3-11), by using other expansions more suited to the transition from one distorted drop to two drops, by considering other than spherical ground-state shapes, by taking account of a possible slight compressibility of nuclear matter, and so on. The most significant advance, however, came from the so-called shell-correction approach of V. M. Strutinsky (S1, B3) in which single-particle effects are combined with the average liquid-drop properties. We postpone discussion of this theory until further details of the fission process have been covered in chapter 4. Here we merely note one of its important successes (N1): the prediction of a double potential barrier toward fission in some regions of A and Z as shown in figure 3-8.

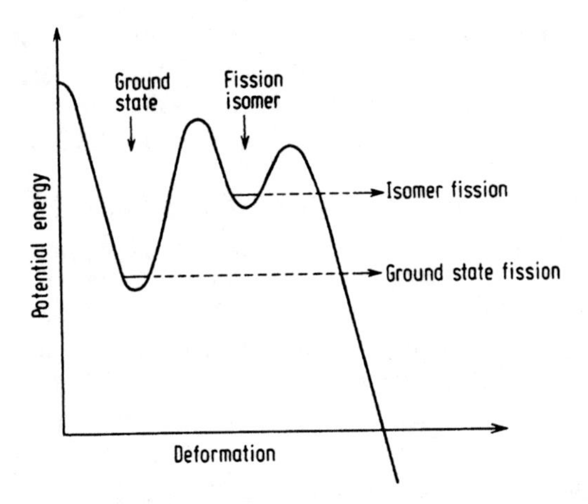

Fig. 3-8 Potential-energy diagram showing double-humped fission barrier.

The second minimum readily explains the existence and properties of the spontaneously fissioning isomers discussed below.

Spontaneously Fissioning Isomers (V1). In 1962 S. M. Polikanov and co-workers (P2) discovered some isomeric states in heavy nuclei that decay by spontaneous fission with very short (nanosecond to microsecond) half lives. Initially the properties of these isomers were quite puzzling, but when the Strutinsky model mentioned above was developed several years later and predicted the existence of a second minimum in the potential-energy surfaces for certain nuclei, it became clear that the fissioning isomers are states in these second potential wells (figure 3-8). The fission isomers are thus so-called **shape isomers**, that is, they exist not by virtue of their spins being very different from those of the ground states as in ordinary isomers (see section E), but because, at a nuclear shape different from that of the ground state, the potential-energy surface has another minimum.[13] The short half lives for spontaneous fission are readily accounted for, since only the outer barrier has to be tunneled through.

Some 30 fission isomers are now known, spanning the region from uranium to berkelium. The half lives range from 0.03 ns ($^{236}Pu^m$) to 14 ms ($^{242}Am^m$). They have been produced by a wide variety of nuclear reactions induced by neutrons, protons, deuterons, and α particles. From measurements of the minimum energies needed for the production of ground and isomeric states, the energy differences between these states can be deter-

[13] One of the characteristic differences between the two types of isomerism appears in the relative production cross sections of ground and excited states: for spin isomers this cross-section ratio changes steeply with the angular momentum brought into the reaction by the bombarding particle, whereas for shape isomers it is nearly independent of angular momentum.

mined. These values are in good agreement with the energy differences between first and second potential-energy minima calculated from the Strutinsky shell-correction theory. There is also reasonable agreement between the calculated and experimentally deduced barrier heights.

Superheavy Elements. As we have seen, spontaneous-fission half lives get shorter and shorter as Z (and hence Z^2/A) increases. However, an interruption in this trend is expected at shell closures, because the greater binding energy of closed-shell nuclei will lead to less energy release in fission (smaller Q_f) and hence larger fission barriers than would otherwise be expected. The next closed proton shell after 82 is most likely 114, the next neutron shell after 126 is expected to be 184. In the vicinity of the doubly magic nucleus $^{298}_{184}114$ an island of relatively stable nuclei has therefore been predicted. This prediction will again be mentioned in connection with shell structure in chapter 10 and in connection with attempts to produce these exotic nuclei by nuclear reactions in chapter 4. Here we merely point out that potential-energy calculations with the shell-correction approach predict in this region spherical ground states, inner barriers of the order of 10 MeV, rather shallow second wells, and low outer barriers (N1). The predicted half lives for spontaneous fission are of the order of 10^{15} y for $^{298}114$, and decrease rapidly as nucleon numbers move away from the closed shells. The actually expected half lives are considerably shorter because of instabilities toward α and β decay and, although half lives as long as 10^9 y have been predicted for some superheavy nuclei such as $^{294}110$, the predictions may be uncertain by several orders of magnitude. In any case, despite vigorous searches in many types of materials, no superheavy elements have been found in nature to date.

D. BETA DECAY

Energetic Conditions. Any radioactive decay process in which the mass number A remains unchanged but the atomic number Z changes is classed as a β decay. We concluded from the parabolic cross section of the nuclear-mass surface at constant A [see (2-8)] that for each odd A there can be only one, and for each even A at most three, β-stable nuclides. On the neutron-rich side of the β-stability valley decay occurs by β^- (electron) emission, on the proton-rich side by β^+ (positron) emission or by electron capture (EC). Odd-odd nuclei near the stability valley (e.g., ^{64}Cu) can decay in both directions, to the neighboring, stable, even-even nuclei.

The energetic conditions for the three types of β decay of a nuclide of atomic number Z and atomic mass M_Z are:

β^- decay:	$M_Z > M_{Z+1}$,
Electron capture:	$M_Z > M_{Z-1}$,
β^+ decay:	$M_Z > M_{Z-1} + 2m_e$,

where m_e is the electron mass.[14] Thus we see that β^+ decay is energetically possible only if the **decay energy** (mass difference between decaying and product atoms) exceeds $2m_ec^2$ (= 1.02 MeV). For lower decay energies EC is the only possible process for $A_Z \rightarrow A_{Z-1}$ and, with increasing energy above 1.02 MeV, β^+ emission competes more and more effectively with EC.

The decay energies of β-unstable nuclei vary rather systematically with distance from the β-stability line, as predicted by the mass parabolas, except for shell-edge perturbations. On the other hand, although there is an obvious qualitative connection between half life and decay energy (large decay energies being generally associated with short lifetimes), the quantitative relations are not nearly so simple as in the case of α decay. As we discuss in detail below, β-decay half lives depend strongly on spin and parity changes as well as on available energy.

Beta Spectra and Conservation Laws. In contrast to α particles β particles[15] from a given radioactive nuclide are not emitted in discrete energy groups, but with a continuous energy distribution extending from zero to a maximum value. It is this maximum energy E_{max} that corresponds to the energy difference between initial and final states. Values of E_{max} for known β emitters range from a few thousand electron volts to about 15 MeV. Beta-ray spectra have been studied in detail by magnetic deflection methods. Typical shapes of a β spectrum in terms of momentum and energy are shown in figure 3-9. The average energy is about one third the maximum.

From 1914, when Chadwick established the continuous nature of β spectra, until nearly two decades later, β decay presented a great puzzle, because the transition from one discrete energy state to another with the emission of β particles of variable kinetic energy appeared to violate the law of conservation of energy.

Furthermore, the observations show discrepancies with other conservation laws. As we have seen in chapter 2, section C, all nuclei of even mass number have integral spins and obey Bose statistics; all nuclei of odd mass number have half-integral spins and obey Fermi statistics. Since the mass number remains unchanged in β decay, the spins of initial and final nuclei should belong to the same class, either integral or half-integral, and the statistics should remain the same. Yet electrons (and positrons) have

[14] To understand why the condition for β^+ decay involves two electron masses while the other two decay modes do not, we must remember that the masses are *atomic* masses and include the masses of the extranuclear electrons in the neutral atoms of parent and daughter nuclides. (See exercise 22.)

[15] By β particles we mean only electrons, positive or negative, emitted from *nuclei.* Electrons originating in the *extranuclear shells* (see later) should not be referred to as β particles; they are often represented by the symbol e^-. In the early literature any electrons emitted in radioactive processes were usually called β particles.

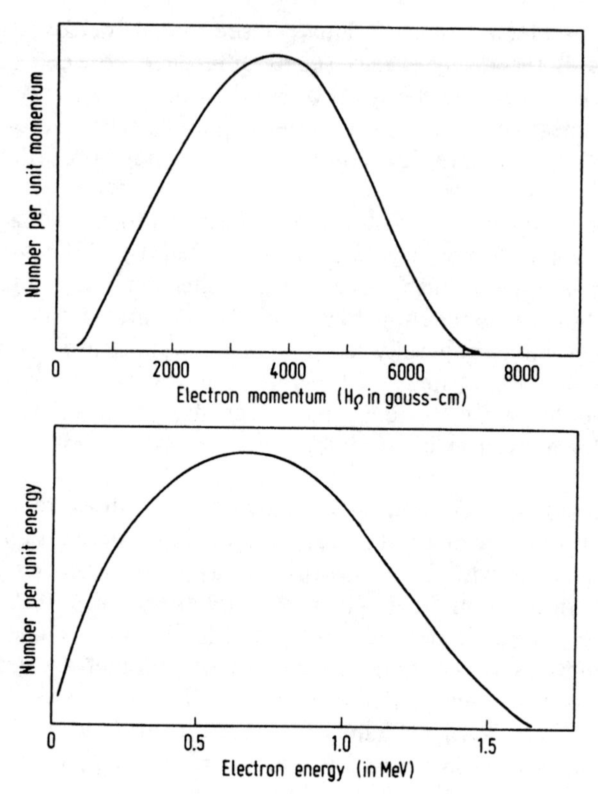

Fig. 3-9 Beta spectrum of ^{32}P, shown as function both of electron momentum and of electron energy. [Data from E. N. Jensen et al., *Phys. Rev.* **85**, 112 (1952).]

one half unit of spin and obey Fermi statistics. Thus angular momentum and statistics appear not to be conserved in β decay.

Neutrinos. To avoid the necessity of abandoning all these conservation laws for the case of β-decay processes, Pauli postulated in 1930 that in each β disintegration an additional unobserved particle is emitted. The properties attributed to this hypothetical particle, which has come to be known as the neutrino, are such that the conservation difficulties are eliminated. The neutrino is assigned zero charge, spin $\frac{1}{2}$, and Fermi statistics, and carries away the appropriate amount of energy and momentum in each β process to conserve these quantities. To account for the fact that neutrinos are almost undetectable, it is in addition necessary to assume that they have a very small or zero rest mass and a very small or zero magnetic moment. By careful measurements of the maximum energy of a β spectrum and determination of the masses of the corresponding β emitter and product atom, an upper limit can be obtained for the rest mass of the neutrino. Best suited for such measurements is the β decay of ^{3}H, and here the most accurate data give an upper limit of about 200 eV (0.0004 times the electron rest mass) for the neutrino rest mass.

In recent years the existence of neutrinos has, in fact, been proved by the observation of their capture by protons to give neutrons and positrons. This is an example of so-called inverse β processes that take place with extremely small probability and are therefore exceedingly hard to observe. Yet it has now been possible, by investigation of inverse processes, to establish with certainty that the neutrinos emitted in β^+ decay are *not* identical with those emitted in β^- decay; the latter are called antineutrinos. This nonidentity of neutrino and antineutrino also follows from the presently accepted theories of β decay that include nonconservation of parity. However, the properties of neutrinos and antineutrinos are indistinguishable (except in the capture reactions just mentioned), and we sometimes use "neutrinos" as a generic term for both.

In EC, as in other β-decay processes, conservation of momentum, angular momentum, and statistics require that a neutrino be emitted. However, since the electron is captured from a definite energy state, the neutrinos emitted in this process are monoenergetic.

In 1934 E. Fermi (F2) formulated a quantitative theory of β decay incorporating Pauli's hypothesis of neutrinos, and this theory is still the cornerstone in our understanding of β decay. It is in many ways analogous to the theory of light emission from atoms: as light quanta are "created" at the moment of emission, so electrons (+ or −) and neutrinos are thought to be created when a nucleon in the nucleus makes the transition from the neutron to the proton state or vice versa. The elementary processes can be written

$$n \rightarrow p + \beta^- + \bar{\nu} \qquad \text{and} \qquad p \rightarrow n + \beta^+ + \nu, \tag{3-12}$$

where ν and $\bar{\nu}$ are the symbols for neutrino and antineutrino. We note that in (3-12) n and p are to be considered not as free particles but as bound in the nucleus. In the free state the neutron has a mass that exceeds the mass of a hydrogen atom by 0.782 MeV. Free neutrons therefore decay by the process shown in (3-12), with a half life of about 11 min, whereas free protons are stable. Inside a nucleus, however, protons, too, can decay as shown without violation of energy conservation because the nucleus as a whole can supply the energy necessary to drive the reaction.

Before going further into the theory of β decay we digress to discuss some properties of positrons.

Positrons. The existence of positrons was postulated in 1931 by P. A. M. Dirac on purely theoretical grounds. He had found that his relativistic wave equations for electrons had solutions corresponding to electrons in negative as well as positive energy states, but with the magnitude of the energy always greater than mc^2 (where m is the electron mass). As to the physical meaning of the unobserved negative-energy states of electrons,

Dirac suggested that normally all the negative-energy states are filled. The raising of an electron from a negative- to a positive-energy state (by the addition of an amount of energy necessarily greater than $2mc^2$) should then be observable not only in the appearance of an ordinary electron but also in the simultaneous appearance of a "hole" in the infinite "sea" of electrons of negative energy. This hole would have the properties of a positively charged particle, otherwise identical with an ordinary electron. The subsequent discovery of positrons, first in cosmic rays and then in radioactive disintegrations, was soon followed by discoveries of the processes of pair production and positron-electron annihilation, which may be regarded as experimental verifications of Dirac's theory.

Pair production is the name for a process that involves the creation of a positron-electron pair by a photon of at least 1.02 MeV ($2mc^2$). It can be shown that in this process both momentum and energy cannot be conserved in empty space; however, the pair production may take place in the field of a nucleus that can then carry off some momentum and energy. The cross section for pair production goes up with increasing Z and with increasing photon energy. Pair production may be thought of as the lifting of an electron from a negative- to a positive-energy state. The reverse process, the falling of an ordinary electron into a hole in the sea of electrons of negative energy, with the simultaneous emission of the corresponding amount of energy in the form of radiation, is observed in the so-called **positron-electron annihilation** process. This process accounts for the very short lifetime of positrons; whenever a hole in the sea of electrons is created, it is quickly filled again by an electron. The energy corresponding to the annihilation of a positron and electron is usually released in the form of two γ quanta, although a very much rarer mode involving the emission of three quanta is also known (see chapter 12, section B). The two-quantum annihilation occurs almost always after the positrons have been slowed by ionization processes to essentially thermal energies. Momentum conservation thus demands that the two γ quanta have equal and opposite momenta; each carries off an energy of $mc^2 = 511.0$ keV. This radiation is referred to as annihilation radiation.

The Weak Interaction. As we have already mentioned Fermi's theory of β decay was patterned after the electromagnetic theory of light emission. The well-known electromagnetic interaction, characterized by the electronic charge e, has to be replaced by a new type of interaction characterized by a new universal constant, the Fermi constant g, whose magnitude must be determined by experiment. The probability $P(p_e)\,dp_e$ that an electron of momentum between p_e and $p_e + dp_e$ is emitted per unit time may be written as

$$P(p_e)\,dp_e = \frac{4\pi^2}{h}\,|\psi_e(0)|^2\,|\psi_\nu(0)|^2\,|M_{if}|^2\,g^2\,\frac{dn}{dE_0}. \tag{3-13}$$

Here ψ_e and ψ_ν are the electron and neutrino wave functions (plane waves in Fermi's theory), and $|\psi_e(0)|^2$ and $|\psi_\nu(0)|^2$ are the probabilities of finding electron and neutrino, respectively, at the nucleus. M_{if} represents the matrix element characterizing the transition from the initial to the final nuclear state; the square of its magnitude $|M_{if}|^2$ is a measure of the amount of overlap between the wave functions of initial and final nuclear states. The so-called statistical factor dn/dE_0 is the density of final states (number of states of the final system per unit decay energy) with the electron in the specified momentum interval $p_e \to p_e + dp_e$.[16]

The interaction constant g governs not only β-decay processes but many other interactions, such as μ-meson decay ($\mu^\pm \to e^\pm + \nu + \bar{\nu}$), π-meson decay ($\pi^+ \to \mu^+ + \nu$, $\pi^- \to \mu^- + \bar{\nu}$), and neutrino-electron scattering ($\nu + e \to \nu + e$). (The distinction between electron-neutrinos, ν_e, and muon-neutrinos, ν_μ [see footnote 20, p. 93] is ignored for the present.) All interactions of this type are classed as weak interactions to distinguish them from the very much stronger interactions governed by nuclear forces and from the electromagnetic interactions, which are of intermediate strength. The fourth type of fundamental interaction known, gravitation, is governed by a still weaker force. The magnitude of the Fermi constant g is about 10^{-49} erg cm^3 as determined from the transition probabilities of simple β decays such as that of the free neutron.

Energy Spectrum. Returning to a consideration of (3-13) we now sketch the derivation of the shape of β-energy (or β-momentum) spectra. Integration over all electron momenta from zero to the maximum possible momentum should then give transition probabilities or lifetimes.

Let us consider so-called **allowed transitions**, that is, transitions in which both electron and neutrino are emitted with zero orbital angular momentum or, what is classically equivalent, with zero impact parameter. The magnitudes of $|\psi_e|^2$ and $|\psi_\nu|^2$ at the position of the nucleus will certainly be much larger for these s-wave neutrinos and electrons than for electrons and neutrinos emitted with larger orbital angular momenta. Therefore the largest transition probabilities will be associated with s-wave electron and neutrino emission. The treatment of these allowed transitions is relatively simple. The magnitudes of $|\psi_\nu(0)|$ and $|M_{if}|$ are independent of the division of energy between electron and neutrino, and the spectrum shape is thus determined entirely by $|\psi_e(0)|$ and dn/dE_0. The first of these factors enters only through the Coulomb interaction between nucleus and emitted electron, and we begin by neglecting this effect (a good approximation at low Z) and evaluating the statistical factor alone.

[16] Despite its appearance, the statistical factor dn/dE_0 is an infinitesimal quantity, as is required for (3-13) to be correct as written (with dp_e on the left side). This can be understood when we remember that dn is the number of states with *both* electron and neutrino in specified momentum intervals [see (3-16)]. Perhaps it would be preferable to write the statistical factor as $(dn)^2/dE_0$ and the number of states in (3-16) as $(dn)^2$.

The density of final states of the system, dn/dE_0, with the electron in the momentum interval $p_e \rightarrow p_e + dp_e$, can be found as follows. Consider an infinitesimal interval dE_0 of the total (electron plus neutrino) kinetic energy E_0. An electron with kinetic energy E_e has associated with it a neutrino with kinetic energy $E_\nu = E_0 - E_e$ (if we neglect the minute amount of recoil energy given to the nucleus), and the range of E_ν is dE_0. For the neutrino of zero rest mass the relation between momentum and kinetic energy is

$$p_\nu = \frac{E_\nu}{c} = \frac{(E_0 - E_e)}{c}. \tag{3-14}$$

Therefore for a given electron energy E_e we have

$$dp_\nu = \frac{dE_0}{c}. \tag{3-15}$$

The number of neutrino states with neutrino momentum between p_ν and $p_\nu + dp_\nu$ is[17]

$$\frac{4\pi p_\nu^2\, dp_\nu}{h^3}.$$

This, we should emphasize, is the number of neutrino momentum states associated with a *given* electron momentum. However, the number of electron states in the momentum interval $p_e \rightarrow p_e + dp_e$ is $4\pi p_e^2\, dp_e/h^3$, and with each of these electron states the number of neutrino states given above can be associated. Therefore the total number of states of the system in the interval dE_0 and with electron momentum in the range $p_e \rightarrow p_e + dp_e$ is

$$dn = \frac{4\pi p_\nu^2\, dp_\nu}{h^3} \cdot \frac{4\pi p_e^2\, dp_e}{h^3}. \tag{3-16}$$

We can now substitute (3-14) and (3-15) into (3-16) and obtain

$$dn = \frac{16\pi^2}{h^6 c^3} p_e^2 (E_0 - E_e)^2\, dp_e\, dE_0. \tag{3-17}$$

[17] The number of translational states of a particle in a certain momentum interval is derived from the quantum-mechanical treatment of the particle in a box. The form of the expression can be made plausible by recourse to the uncertainty principle: for a particle whose position is specified by the Cartesian coordinates x, y, z, and whose momentum is given by the momentum components p_x, p_y, p_z, the product of the uncertainty in a space coordinate and the uncertainty in the corresponding momentum component is of the order of Planck's constant. Thus

$$\Delta x\, \Delta p_x \cong h; \qquad \Delta y\, \Delta p_y \cong h; \qquad \Delta z\, \Delta p_z \cong h.$$

In the six-dimensional **phase space** characterized by three space coordinates and three momentum coordinates, a particle can therefore be specified only as being in a volume $\Delta x\, \Delta y\, \Delta z\, \Delta p_x\, \Delta p_y\, \Delta p_z \cong h^3$; called the unit cell in phase space. The number of translational states available to a particle in a certain volume and in a certain momentum interval is taken as the number of unit cells in the corresponding volume of phase space. For further discussions of these concepts the reader is referred to standard works on statistical mechanics.

In (3-16) we have taken the volume in coordinate space as unity. Any arbitrary volume could have been used; but the wave functions ψ_e and ψ_ν in (3-13) would then have to be normalized over the same volume, and the volume used would subsequently cancel out of the equations.

The volume in (three-dimensional) momentum space corresponding to momentum between p and $p + dp$ is given by the spherical shell with inner radius p and outer radius $p + dp$; this is equal to $4\pi p^2\, dp$.

It is customary to express momentum in units of m_0c and *total* energy W (kinetic plus rest energy) in units of m_0c^2, that is, to set $p_e/m_0c = \eta$, $(E_e/m_0c^2)+1 = W$, and $(E_0/m_0c^2)+1 = W_0$. We can then rewrite (3-17) as

$$\frac{dn}{dE_0} = \frac{16\pi^2 m_0^5 c^4}{h^6}\eta^2(W_0 - W)^2\,d\eta, \tag{3-18}$$

or, making use of the relativistic relation (appendix B) $\eta^2 = W^2 - 1$ and therefore $\eta\,d\eta = W\,dW$,

$$\frac{dn}{dE_0} = \frac{16\pi^2 m_0^5 c^4}{h^6} W(W^2-1)^{1/2}(W_0 - W)^2\,dW. \tag{3-19}$$

The expression for the statistical factor (3-19) is worth examining briefly even if the derivation was not followed. Recapitulating that W is the total (kinetic plus rest) energy of an electron (in units of m_0c^2) and that W_0 is the maximum value of W, we readily see from (3-19) that dn/dE_0 goes to zero both at $W = 1$ and $W = W_0$. The characteristic bell shape of β spectra (figure 3-9) is thus reproduced, at least qualitatively, by the statistical factor. For β emitters of low Z the agreement with experimental spectrum shapes is almost quantitative.

Coulomb Correction. So far we have neglected the Coulomb interaction between the nucleus and the emitted electron. The effect of this interaction is to decelerate negatrons and to accelerate positrons, so that negatron spectra may be expected to contain more, positron spectra fewer, low-energy particles than predicted by the purely statistical considerations of the preceding paragraphs. This indeed corresponds to experimental observations, as shown, for example, by the measured shapes of the negatron and positron spectra of ^{64}Cu, which happen to have similar endpoint energies (0.57 and 0.65 MeV). They are displayed in figure 3-10. Formally, the Coulomb interaction may be treated as a perturbation on the electron wave function $\psi_e(0)$; the entire spectrum (3-19) then has to be multiplied by a Coulomb correction factor $F(Z, W)$, also known as the **Fermi function** and defined as the ratio of $|\psi_e(0)|^2_{\text{Coul}}$ to $|\psi_e(0)|^2_{\text{free}}$. The nonrelativistic result for $F(Z, W)$ is

$$F(Z, W) = \frac{2\pi x}{1 - \exp(-2\pi x)}, \tag{3-20}$$

where $x = \pm Ze^2/\hbar v$, with the + sign applicable to negatrons, the − sign to positrons, v the velocity of the β particle far from the nucleus, and Z the atomic number of the *product* nucleus.

Since the Coulomb effect is most important for the lowest-energy electrons emitted, the nonrelativistic Coulomb correction (3-20) is a fairly useful approximation in many cases. For precise computations, however, the much more complex relativistic form of $F(Z, W)$ given by Fermi (F2) must be used. Values of this relativistic Coulomb correction factor for a wide range of Z and W are available in tabular form (B4). Additional

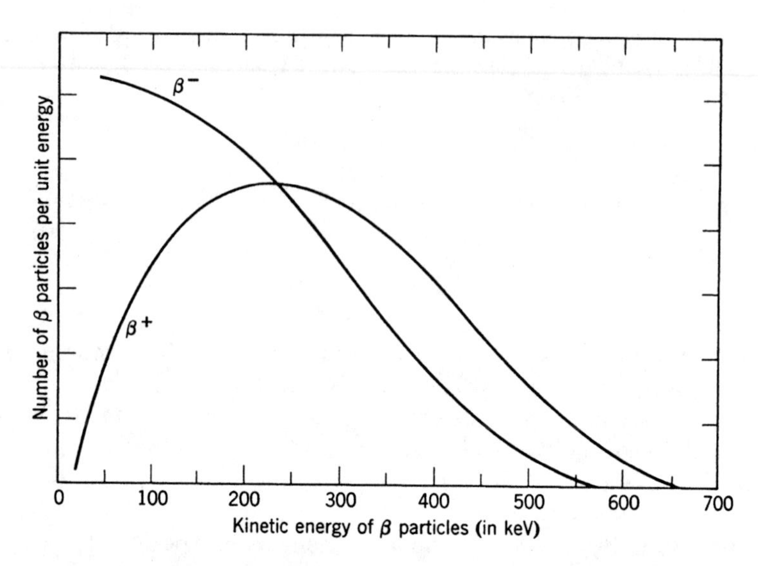

Fig. 3-10 Energy spectra of the positrons and negatrons emitted by ^{64}Cu. The pronounced difference between the two spectral shapes results largely from the Coulomb effect. [Data from J. R. Reitz, *Phys. Rev.* **77**, 10 (1950).]

correction terms for the screening effects of extranuclear electrons have also been calculated (B4) and may be important, especially in enhancing the emission of very-low-energy positrons.

Kurie Plots. Much effort has been expended by experimenters in checking the theoretical predictions about spectrum shapes. In magnetic β-ray spectrometers β particles are analyzed according to their momenta, and the quantity measured is the (relative) number of β particles per unit momentum. Therefore, for comparisons between theory and experiment, the spectrum as given by (3-18), modified by the Coulomb function $F(Z, W)$, is convenient. We then have for the probability of electron momentum between η and $\eta + d\eta$ (with η in units of m_0c)

$$P(\eta)\,d\eta \propto F(Z, W)\eta^2(W_0 - W)^2\,d\eta, \tag{3-21}$$

provided that the transition matrix element M_{if} is independent of the energy partition between electron and neutrino. As pointed out by F. N. D. Kurie et al. (K1), it follows from (3-21) that a plot of $[P(\eta)/\eta^2F(Z, W)]^{1/2}$ against W should be a straight line for allowed transitions, with the intercept on the energy axis at W_0. Such plots with $P(\eta)$ from spectrometer data are known as Kurie plots, Fermi plots, or F-K plots, and have proved exceedingly useful for the analysis of β spectra. Even when β spectra are measured in terms of energy (e.g., with semiconductor detectors), the results are often transformed by means of tables to momentum spectra for the purpose of constructing Kurie plots. Extrapolation of Kurie plots to the energy axis is the only reliable means for the determination of β-ray endpoint energies.[18] The theoretical predictions of allowed spectrum shapes have now been extremely well verified, a major triumph for the Fermi theory. However, for many years much confusion and misinterpretation resulted from the effects of electron scattering in β-spectrometer sources and their mountings, which can cause sizable shifts of β spectra to lower energies. Extremely thin sources and source backings are required for careful measurements of spectral shapes.

Comparative Half Lives. Returning now to (3-13) and inserting the spectrum shape from (3-19) and the Coulomb correction $F(Z, W)$, we obtain for the probability per unit time that an electron in the energy interval $W \rightarrow W + dW$ is emitted:

$$P(W)\,dW = \frac{64\pi^4 m_0^5 c^4 g^2}{h^7} |M_{if}|^2 F(Z, W) W(W^2 - 1)^{1/2}(W_0 - W)^2\, dW. \tag{3-22}$$

[Note that the electron and neutrino wave functions $\psi_e(0)$ and $\psi_\nu(0)$ no longer appear because of the normalization used; see footnote 17.]

Integrating (3-22) over all values of W from 1 to W_0, we obtain the total probability per unit time that a β particle is emitted, which is just the decay constant λ:

$$\lambda = \frac{\ln 2}{t_{1/2}} = K\,|M_{if}|^2 f_0, \tag{3-23}$$

where

$$K = 64\pi^4 m_0^5 c^4 g^2/h^7 \quad \text{and} \quad f_0 = \int_1^{W_0} F(Z, W) W(W^2 - 1)^{1/2}(W_0 - W)^2\, dW.$$

We have here again assumed that the nuclear matrix element is independent of β energy, an assumption valid for allowed transitions only. It follows from (3-23) that the product $f_0 t_{1/2}$, usually denoted as the $f_0 t$ (or, more loosely, ft) value and called the comparative half life of a transition, should be approximately the same for all transitions with similar matrix elements. The ft value may be thought of as the half life corrected for differences in Z and W_0.

The integral f_0 can be evaluated in closed form only when $F(Z, W) = 1$, that is, for very small Z. For the more general case in which $F(Z, W)$ cannot be neglected, extensive tabulations are available (G6). For rapid estimation it is convenient to use the nomograms connecting $f_0 t$, half life,

[18] It should be mentioned that Kurie plots, even for allowed transitions, will be straight all the way to the energy axis only if the neutrino rest mass m_ν is zero. For finite values of m_ν the spectrum shape near the endpoint would be modified through the necessary modification of (3-13). Careful measurements of the shape of the β spectrum of ^3H near the endpoint have set an upper limit of 60 eV, lower than the limit based on energy balance, quoted on p. 77.

and energy, devised by S. A. Moszkowski (M2) and reproduced in several standard works (e.g., L1 and W1). Approximate values (errors less than 0.3 in log f for $0 < Z \leq 100$ and for $0.1\,\text{MeV} < E_0 < 10\,\text{MeV}$) can be obtained with the following purely empirical expressions:

$$\log f_{\beta^-} = 4.0 \log E_0 + 0.78 + 0.02Z - 0.005(Z-1)\log E_0, \tag{3-24}$$

$$\log f_{\beta^+} = 4.0 \log E_0 + 0.79 - 0.007Z - 0.009(Z+1)\left(\log \frac{E_0}{3}\right)^2. \tag{3-25}$$

Note that in these equations Z, as before, is the atomic number of the *product* nuclide and E_0 is the kinetic energy, in millions of electron volts, of the upper limit of the spectrum.

Electron Capture. Before proceeding with the discussion of ft values and their significance, we note that we have talked so far entirely about β^- and β^+ emission and have ignored the third type of β transition, orbital electron capture (EC). In EC an electron bound with energy E_B MeV is captured and a neutrino of energy E_0 MeV is emitted, where E_0 is the difference in atomic masses of parent and daughter.

Although EC is a very common mode of decay, it escaped discovery until 1938 (L. Alvarez) because it is not accompanied by the emission of detectable *nuclear* radiation except when the product nuclei are left in excited states so that γ rays are emitted. The most characteristic radiations accompanying EC are the X rays emitted as a consequence of the vacancy created in the electron shell from which the capture took place. The atomic rearrangements following EC are discussed below.

A continuous spectrum of electromagnetic radiation of very low intensity is found to be emitted in EC processes and, in fact, in all β-decay processes. The quanta of this so-called **inner bremsstrahlung** (see chapter 6, section B for a discussion of ordinary or external bremsstrahlung) have part of the energy ordinarily carried away by the neutrino. The total number of quanta per EC disintegration is approximately $7.4 \times 10^{-4} E_0^2$, where E_0 is in MeV. When nuclear γ rays are emitted the inner bremsstrahlung usually escapes detection because of its low intensity. However, for EC transitions not accompanied by γ emission, measurement of the upper energy limit of the inner-bremsstrahlung spectrum is a very useful method for the determination of the transition energy. In fact, it is the only *direct* way of measuring decay energy in EC. Note that, to obtain E_0, we must add to the bremsstrahlung endpoint energy the mean atomic excitation energy of the product atom (R1). In those events in which an internal-bremsstrahlung quantum is emitted, the neutrino energy is smaller than E_0 by the energy of that quantum.

The fact that *monoenergetic* neutrinos are emitted in EC decay simplifies the calculation of the statistical factor greatly, since only the neutrino phase space needs to be considered and no integration over energy is involved. Again values of f_{EC} can be obtained from the sources quoted or they can be approximated in the spirit of (3-24) and (3-25) (but with appreciable errors for $E_0 < 0.5$ MeV at high Z):

$$\log f_{EC} = 2.0 \log E_0 - 5.6 + 3.5 \log (Z+1). \tag{3-26}$$

Capture-to-Positron Ratios. It is to be noted that whenever β^+ emission is energetically possible it competes with EC. Since initial and final nuclear states are the same for the two modes of decay, the ratio $\lambda_{EC}/\lambda_{\beta^+}$ is, at least for allowed transitions, expected to be completely independent of the nuclear matrix element and just equal to f_{EC}/f_{β^+}. Measurements of EC-to-positron branching ratios thus constitute an important test for β-decay theory. In general, these ratios increase with decreasing decay energy (going to infinity when β^+ emission becomes impossible at decay energies $\leq 2m_0c^2$) and with increasing Z. The latter trend comes about through the increase in the expectation value for finding orbital (especially K) electrons at the nucleus and through the increasing suppressive effect of the Coulomb factor $F(Z, W)$ on β^+ emission (3-20). The dependence of EC/β^+ ratios on decay energy provides a useful tool for the determination of the latter quantity.

K/L Ratios. Whenever energetically possible, capture of $K(1s)$ electrons predominates over capture of electrons with higher principal quantum numbers because, of all the electron wave functions, those of the K electrons have the largest amplitudes at the nucleus. However, at decay energies below the binding energy of the K electrons, EC is possible only from the $L(2s+2p)$, $M(3s, 3p, 3d)$, and so on, shells. The ratio of L_I capture[19] to K capture as a function of decay energy has been calculated (R2) for allowed transitions. The results for $Z \gtrsim 14$ can be represented by the approximate formula

$$\frac{L_I}{K} = (0.06 + 0.0011Z)\left[\frac{E_0^L(\nu)}{E_0^K(\nu)}\right]^2, \tag{3-27}$$

where $E_0^L(\nu)$ and $E_0^K(\nu)$ are the neutrino energies accompanying the two processes; $E_0^L(\nu)$ exceeds $E_0^K(\nu)$ by the difference between the binding energies of the two shells. At decay energies not too far in excess of the K-binding energy, $E_0^L(\nu)/E_0^K(\nu)$ differs appreciably from unity, and a measurement of the L- to K-capture ratio then permits an estimation of the decay energy by use of (3-27). Note that the electron binding energies and the Z in (3-27) are those of the parent nucleus.

Extranuclear Effects of Electron Capture. As already mentioned, the

[19] In allowed transitions most of the L captures take place from the L_I $(2s_{1/2})$ subshell; the capture probability is small for $L_{II}(p_{1/2})$ electrons and zero for $L_{III}(p_{3/2})$ electrons. The contribution of M capture can usually be neglected too. Complete expressions for capture in the L and M shells are available in reference M3.

only abundant radiations resulting from EC, other than the essentially undetectable neutrinos, are of extranuclear origin.

If a K-shell vacancy is filled by an L electron, the difference between the K- and L-binding energies may be emitted as a characteristic X ray or may be used in an internal photoelectric process in which an additional extranuclear electron from the L, or M, or other shell is emitted with a kinetic energy equal to the characteristic X-ray energy minus its own binding energy. Such electrons are called **Auger electrons**. The whole process of readjustment in a heavy atom may involve many X-ray emissions and Auger processes in successively higher shells. The fraction of vacancies in a given shell that is filled with accompanying X-ray emission is called the **fluorescence yield**, and the fraction that is filled by Auger processes is the Auger yield. The K-shell fluorescence yield ω_K increases with increasing Z as shown in figure 3-11. The L-shell fluorescence yield

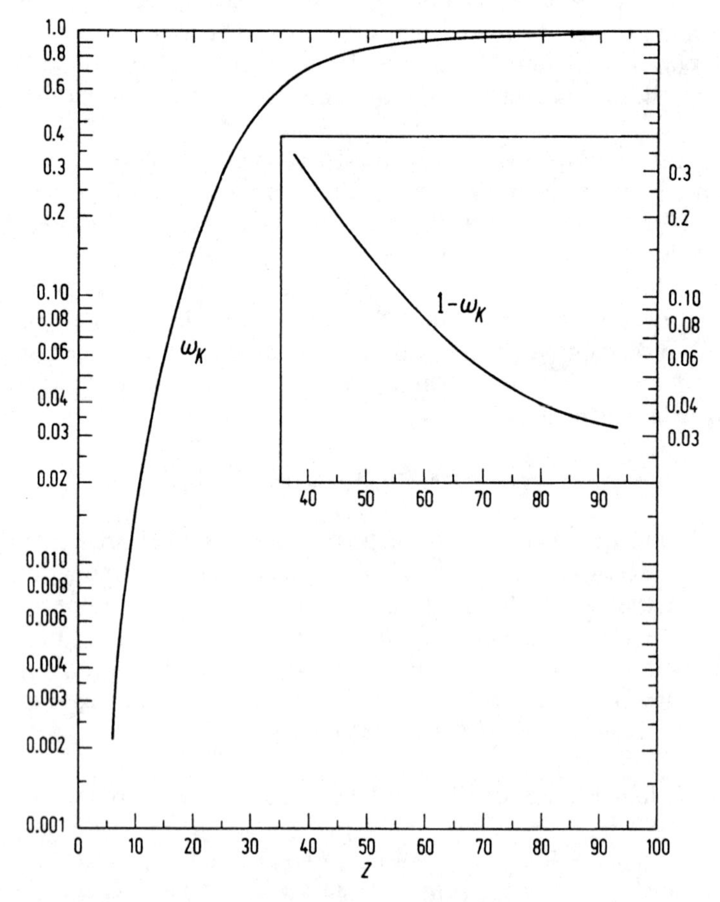

Fig. 3-11 K-shell fluorescence yield as a function of Z. At high Z a curve of $1 - \omega_K$ is shown for ease of reading. (Based on a table of "best" values given by Burhop and Asaad, reference B5.)

varies with Z in a similar manner but is several times smaller than the K yield for a given Z or about the same as the K fluorescence yield for a given electron binding energy. Knowledge of the fluorescence yield is important in the measurement of disintegration rates of EC nuclides since the radiations most frequently detected are the X rays. However, accurate experimental determination of fluorescence yields is difficult, particularly for ω_L, and theoretical calculations are often resorted to in order to supplement the experimental data. References B5 and B6 give critical reviews of the subject, including tables of "best" values.

Selection Rules. We now return to the subject of ft values and transition probabilities. We remarked (on p. 79) that transitions in which electron and neutrino carry away no orbital angular momentum are expected to have the largest transition probabilities (for a given energy release) and we have called these transitions "allowed." If electron and neutrino do not carry off angular momentum, the spins of initial and final nucleus cannot differ by more than one unit of $\hbar$ and their parities must be the same. In fact, if electron and neutrino are emitted with their *intrinsic* spins antiparallel (singlet state), the *nuclear* spin change ΔI must be strictly zero; if electron and neutrino spins are parallel (triplet state), ΔI may be $+1$, 0, or -1 (but $0 \rightarrow 0$ transitions are forbidden). The former selection rule was the one originally proposed by Fermi; the latter was subsequently suggested by Gamow and E. Teller. Which of these selection rules applies depends on the form of the interaction operator in the matrix element M_{if} in (3-13), specifically on its behavior under rotation and space inversion. If the operator is a scalar (S) or vector (V) quantity, the Fermi rules apply; if it is a tensor (T) or axial vector (A), the Gamow-Teller rules hold. Much effort went into the determination of the exact form of the interaction that applies and, without being able to go into the evidence, we merely state the result that β decay is governed by a linear combination of V and A interactions, with a ratio of coupling constants $G_A/G_V = 1.24$. Thus a mixture of Fermi and Gamow-Teller selection rules is applicable.

From what has been said so far we might expect that all allowed β transitions, that is, all transitions between states of $\Delta I = 0$ or 1 with no parity change, should have (1) the allowed spectrum shape, (2) closely similar f_0t values. Whereas the first expectation is borne out by all experiments to date, the second is not. The values of f_0t extend from $\sim 10^3$ (for example $n \rightarrow {}^1H$) to $\sim 10^9$ (e.g., ${}^{14}C \rightarrow {}^{14}N$). There is, however, a strong clustering of $\log f_0t$ values around 3–3.5 and another broader peak with $\log f_0t$ between 4 and 7. Transitions characterized by the very low f_0t values in the first of these groups are called "favored" or "superallowed." They are found mainly among β emitters of low Z and particularly between so-called **mirror nuclei**. Two nuclei constitute a mirror pair if one contains n neutrons and $n+1$ protons, the other $n+1$ neutrons and n protons; examples are ${}^3_1H$ and ${}^3_2He$, ${}^{23}_{12}Mg$ and ${}^{23}_{11}Na$. Provided neutron-

neutron and proton-proton forces are the same except for a Coulomb interaction, the wave functions characterizing two mirror nuclei are certainly expected to be very nearly the same, and therefore the square of the nuclear matrix element $|M_{if}|^2$ for a mirror transition should be $\cong 1$. It is from the decay rates of these superallowed transitions (the simplest one being the decay of the free neutron with $f_0 t \cong 1100$ s) that the magnitude of the β-decay coupling constant $g \cong 1.4 \times 10^{-49}$ erg cm^3 has been estimated. Once the value of g is known, ft values of other β transitions can be used to obtain information about nuclear matrix elements.

The rather wide range of $f_0 t$ values found for allowed transitions (other than the superallowed ones) indicates that our assumption of approximately equal $|M_{if}|^2$ values for all transitions with $\Delta I = 0, \pm 1$ without parity change was too naïve. The nuclear matrix elements are evidently sensitive to other factors. As an extreme illustration we mention the so-called l-forbidden transitions, of which $^{32}P \rightarrow {}^{32}S + \beta^- + \bar{\nu}$ is an example. Here the spins of ^{32}P and ^{32}S have been measured as 1 and 0, respectively, and both parities are unquestionably even. Yet the log $f_0 t$ value is 7.9, and this large value apparently comes about (as will be made clearer in the discussion of shell-model states in chapter 10, section D) because a $d_{3/2}$ neutron is transformed into an $s_{1/2}$ proton, so that $\Delta l = 2$ even though $\Delta I = 1$.

Forbidden Transitions. The discussion so far has been confined to allowed transitions. Let us now consider briefly what happens when the transition from initial to final nucleus cannot take place by the emission of s-wave electron and neutrino. That electron and neutrino emission with orbital angular momenta other than zero is possible at all comes about because of the finite size of nuclei. The wave functions $\psi_e(0)$ and $\psi_\nu(0)$ "at the nucleus" that appear in (3-13) thus have to be evaluated over the entire nuclear volume; therefore they do not vanish for p-, d-, and, higher-wave emission. However, the magnitudes of these electron and neutrino wave functions over the nuclear volume decrease rapidly with increasing orbital angular momentum. Hence for each unit of angular momentum l carried off by the two light particles together, the β-transition probability decreases by several orders of magnitude, and β transitions with $l = 1,2,3$, and so on, are classified as first, second, third, and so on, forbidden transitions. The l value associated with a given transition can be deduced from indirect evidence only, such as ft values or spectrum shapes (see below). The various transition orders, the ranges of log $f_0 t$ corresponding to them, and some examples are listed in table 3-2.

The selection rules for the various orders of forbiddenness are readily derived. If l is odd, initial and final nucleus must have opposite parities ($\Delta\Pi$, yes); for even l values the parities must be the same ($\Delta\Pi$, no). Furthermore, as in allowed transitions, the emission of electron and neutrino in the singlet state (Fermi selection rules) requires $\Delta I \leq l$, whereas triplet-state emission (Gamow-Teller selection rules) allows $\Delta I \leq l + 1$.

Table 3-2 Classification of Beta Transitions and Selection Rules

Type	l	ΔI	$\Delta\Pi$	Log ft^a	Log $[(W_0^2-1)^{\Delta I-1}ft]$	Examples
Superallowed	0	0 or 1	No	3		^3H, ^{23}Mg
Allowed (normal)	0	0 or 1	No	4–7		^{35}S, ^{69}Zn
Allowed (l-forbidden)	0	1	No	6–12		^{14}C, ^{32}P
First forbidden	1	0 or 1	Yes	6–15		^{111}Ag, ^{143}Ce
First forbidden (unique)	1	2	Yes	9–13	~10	^{38}Cl, ^{90}Sr
Second forbidden	2	2	No	11–15		^{36}Cl, ^{135}Cs
Second forbidden (unique)	2	3	No	13–18	~15	^{10}Be, ^{22}Na
Third forbidden	3	3	Yes	17–19		^{87}Rb
Third forbidden (unique)	3	4	Yes		~21	^{40}K
Fourth forbidden	4	4	No	~23		^{115}In
Fourth forbidden (unique)	4	5	No		~28	

[a] The log ft ranges are very approximate. Occasional examples may even fall outside the ranges shown (R3).

Assuming again a mixture of Fermi and Gamow-Teller-type interactions, the selection rules listed in table 3-2 result. Note that values of $\Delta I < l$ appear only in first forbidden transitions ($l = 1$), because in all other cases transitions with such spin changes ΔI are also possible with lower degrees of forbiddenness ($l - 2$, etc.). The ranges of $\log f_0 t$ values show a fair amount of overlap, and the determination of $\log f_0 t$ alone can rarely give unambiguous information on ΔI and $\Delta\Pi$; but with other data, and particularly in conjunction with the predictions of nuclear models (see chapter 10), $\log f_0 t$ values are an important aid in making spin and parity assignments (R3).

An illustration of these concepts may be seen in figure 3-12, which shows the decay scheme of $^{24}_{11}Na$. Almost all the β^- decays are to the second excited state of ^{24}Mg at 4.12 MeV. For this transition $t_{1/2} = 15\,h = 5.4 \times 10^4\,s$ and $E_{max} = 1.39$ MeV. From (3-24) we estimate $\log f \approx 1.6$, and thus $\log ft \approx 6.3$, which agrees with the "normal allowed" classification for $4^+ \to 4^+$, $\Delta I = 0$, no. For the three rare branches to the first, third, and fourth excited states, we find the following:

	E_{max} (MeV)	$\log f$	t (s)	$\log ft$
β_1	4.15	3.46	1.8×10^9	12.7
β_3	1.28	1.44	2.7×10^9	10.9
β_4	0.29	−1.10	6.0×10^7	6.7

All the $\log ft$ values are consistent with the level assignments, β_1 and β_3 being second forbidden transitions with $\Delta I = 2$, no, whereas β_4 corresponds to an

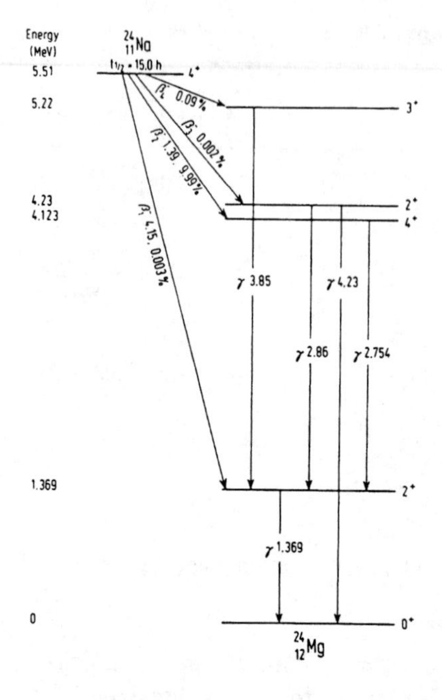

Fig. 3-12 Decay scheme of ^{24}Na. The transition energies are in MeV. Spin and parity are shown to the right of each level, energies above the ^{24}Mg ground state to the left.

allowed transition with $\Delta I = 1$, no. The transition to the ground state is not observed, which is not surprising, since it would be fourth forbidden and might thus have a $\log ft$ value of ≈ 23. With its $\log f = 4.0$ we estimate $\log t \approx 19$, or $t \approx 3 \times 10^{11}$ y. Thus only about a 5×10^{-13} percent branch would be expected to go to the ground state of ^{24}Mg, and that would be quite unobservable. Some applications of β-decay selection rules to decay-scheme determinations are discussed in chapter 8, section E.

Spectrum Shapes. Additional identification of transition types sometimes comes from spectrum shapes. As noted earlier, the assumption that M_{if} is independent of the energy partition between electron and neutrino applies in general to allowed transitions only. For other transition types (3-21) is usually not valid, and Kurie plots therefore do not give straight lines. However, for each of the various forms of basic β-decay interaction (p. 87) and for each order of forbiddenness it is possible to calculate the additional energy dependence and (in good approximation) to factor out of the matrix element an energy-dependent term (B1, p. 726ff.; K2). By multiplying the right-hand side of (3-21) by the appropriate one of these shape-correction factors, we again obtain a function which, if plotted against the β energy W, gives a straight line. The correction factors take on a particularly simple form for the transitions with $\Delta I = \ell + 1$, which are forbidden by the Fermi selection rules and therefore involve axial vector interactions only. This restriction makes the predictions of the theory much less ambiguous for these than for any other transitions—hence they are called "unique" (see table 3-2). The shape-correction factor for a unique transition of order ℓ can be written

$$\frac{(p_e + p_\nu)^{2\ell+2} - (p_e - p_\nu)^{2\ell+2}}{4p_e p_\nu},$$

which, for first forbidden unique transitions ($\ell = 1$) reduces to $2(p_e^2 + p_\nu^2)$, in the literature

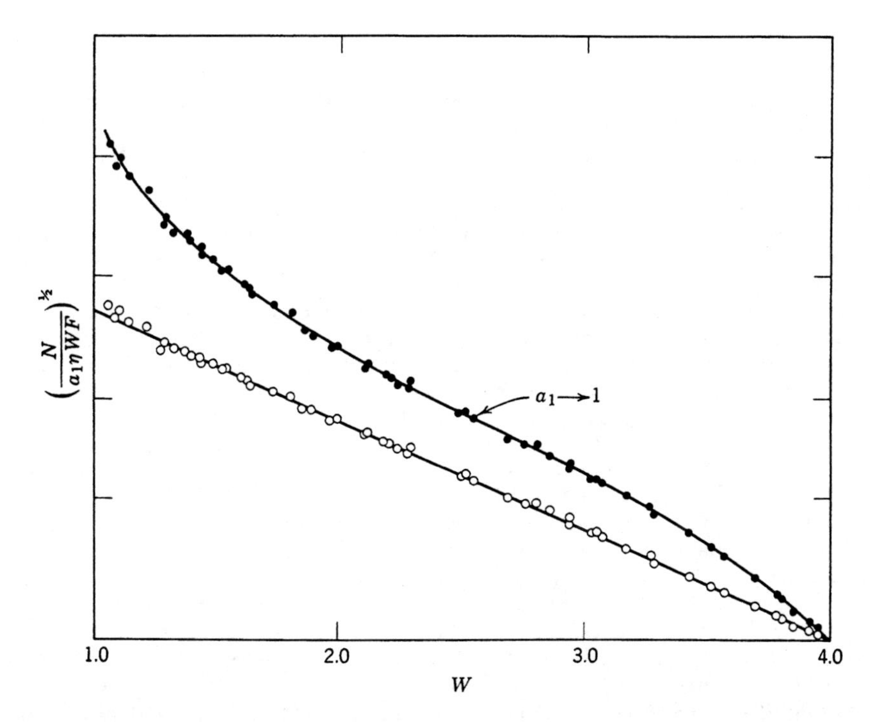

Fig. 3-13 Kurie plot of the ^{91}Y β^- spectrum. The open circles are the data points corrected by the shape factor a_1; they are seen to fall on a straight line. The closed circles represent the same data but treated as if the transition were allowed ($a_1 = 1$). [From E. J. Konopinski and L. M. Langer, *Ann. Rev. Nucl. Sci.* 2, 261 (1953).]

often called the "a_1 correction factor." As illustrated in figure 3-13 for the ^{91}Y β spectrum, the use of this correction term indeed linearizes the Kurie plots of β spectra emitted in decays between states characterized by $\Delta I = 2$, yes. The same is true for the unique transitions of higher order when the appropriate correction factors are used. Until about 1960 it appeared that spectra for most nonunique first forbidden transitions ($\Delta I = 0$ or 1, yes) had the allowed shape (3-21). More recently it has been found that small correction factors of the form $(1 + aW)$ are needed to make Kurie plots for these transitions truly linear (see the ^{198}Au example on p. 313). For higher forbidden nonunique transitions the situation can become quite complex.

If the β spectrum does not have the allowed shape, (3-23) for the decay constant is no longer strictly correct because M_{if} is then not independent of W. Thus we should not use $f_0 t$ as the "comparative half life," but a corrected ft value. However, this is not customary, and the tabulated ft values are almost always $f_0 t$ values. For the unique forbidden transitions the corrected f value can be approximated by $(W_0^2 - 1)^{\Delta I - 1} f_0$. In table 3-2, therefore we list $\log [(W_0^2 - 1)^{\Delta I - 1} f_0 t]$ for these transitions; this quantity is much more nearly constant for a given order of unique transition than is $\log f_0 t$. Note, however, that this method of correcting ft values of "unique" transitions, based as it is on considerations of spectrum shapes, should not be applied to EC transitions.

Nonconservation of Parity. We mentioned in passing (chapter 2, p. 38) that the conservation of parity, long accepted as one of the universal conservation laws, does not hold for weak interactions. This possibility was suggested in 1956 by Lee and Yang (L2) to explain what appeared to be two different decay modes of a single type of particle, a K meson—one to an even-parity (two-pion), one to an odd-parity (three-pion) final state. Lee and Yang pointed out that no then-existing experimental data proved parity conservation in weak interactions (whereas it was well established for strong and electromagnetic interactions) and suggested some experimental tests. The first experimental verification of nonconservation of parity in weak interactions came in the historic experiment by C. S. Wu et al. (W2) in which the emission of β particles from ^{60}Co nuclei whose spins were aligned by a magnetic field at very low temperatures (to suppress thermal agitation) was found to be preferentially along the direction opposite to the ^{60}Co spin vector.

To understand the implications of this experiment, we must consider the properties of different quantities under space inversion, that is, reflection through a point, or change of sign of space coordinates. So-called polar vectors, such as linear momentum, velocity, or electric field, change sign under this operation, whereas so-called axial vectors, such as angular momentum or magnetic field (which are characterized not only by direction but also by a screw sense), do not change sign. Any observed quantity that is the (scalar) product of two polar vectors or of two axial vectors will be invariant under space inversion; such quantities are called scalars. A number that is the scalar product of one polar vector and one axial vector changes sign under space inversion; the occurrence of such quantities, called pseudoscalars, is prohibited by the requirement of parity conservation.

The important point made by Lee and Yang in their 1956 paper was that none of the experimental data on weak interactions then available could throw any light on the question of parity conservation because the observed quantities were always scalars. The experiment with aligned ^{60}Co nuclei was specifically designed to look for a pseudoscalar quantity, namely a component of β-particle intensity proportional to the product of the nuclear spin (an axial vector) and the electron velocity (a polar vector). The asymmetry found established the existence of this pseudoscalar component and thus proved that parity was not conserved in β decay. Since then many other experiments have corroborated nonconservation of parity in all weak interactions. Thus we now know that nature, in this class of processes, distinguishes left from right. In fact, an ingenious experiment by M. Goldhaber, L. Grodzins, and A. W. Sunyar (G7) has shown that the neutrinos accompanying electron capture (and presumably those accompanying β^+ emission) are "left-handed," that is, they have their spins antiparallel to their direction of motion. Positrons then must be "right-handed," negatrons "left-handed," and the antineutrinos accompanying β^- decay "right-handed."

It is worthwhile to emphasize once more that everything that has been said in this section about spectrum shapes and lifetimes in β decay is unaffected by the overthrow of parity conservation because only scalar quantities are involved. Thus Fermi's basic theory is largely unaffected, except for the need for the inclusion of some additional parity-nonconserving coupling constants in the interaction. On the other hand, the discovery of parity nonconservation has stimulated whole new classes of experiments involving observations of (1) asymmetry of emission from aligned nuclei, (2) polarization of β particles, and (3) correlations between β particles and polarized γ rays. These experiments, which can shed some light on the form of the β-decay interaction, are discussed in W1 and K2.

Neutrinos and Antineutrinos. Partly as a result of the discovery of parity nonconservation and partly through difficult experimental work on the very rare neutrino interactions, there has been an important clarification and simplification of our ideas about neutrinos (see, for example, R4). As we mentioned on p. 77, F. Reines and C. Cowan experimentally established the capture of antineutrinos (from a nuclear reactor) by protons and measured a cross section of about 10^{-43} cm^2 for this process, in rough agreement with theoretical expectations. On the other hand, R. Davis obtained a null result in attempts to measure the capture of reactor *antineutrinos* ($\bar{\nu}$) in ^{37}Cl to form ^{37}Ar, presumably because reversal of the ^{37}Ar EC decay ($^{37}\text{Ar} + e^- \rightarrow {}^{37}\text{Cl} + \nu$) requires *neutrinos*: $^{37}\text{Cl} + \nu \rightarrow {}^{37}\text{Ar} + e^-$. The upper limit for the antineutrino cross section set in this experiment was about one tenth of the calculated neutrino cross section. Thus neutrinos and antineutrinos are evidently different particles. This conclusion is in accord with the expectations from parity nonconservation in β decay. All present evidence is consistent with the view that in β-decay processes there are two and only two types of neutral massless particles: left-handed neutrinos and right-handed antineutrinos.[20]

Double Beta Decay. The distinction between ν and $\bar{\nu}$ removes an ambiguity that has existed with respect to the expected half lives for double β decay. If neutrino and antineutrino were identical, this process could take place through a virtual intermediate state, with a "neutrino" produced in the first step and absorbed in the second, each step producing one β^-. With $\nu \neq \bar{\nu}$ this type of process is excluded, since the first step produces $\bar{\nu}$ and the second would require the absorption of ν. Instead, the production of $2\beta^- + 2\bar{\nu}$ is required, and the expected lifetime for this process is several orders of magnitude greater than that of the neutrinoless one. The half lives found for the two established cases of double β decay, 2×10^{21} y for ^{130}Te and 10^{20} y for ^{82}Se (see p. 48), as well as a number of lower limits for other double-β-decay half lives are consistent with the theoretical expectations for $2\bar{\nu}$ emission.

E. GAMMA TRANSITIONS

An α- or β-decay process may leave the product nucleus either in its ground state or, more frequently, in an excited state. Excited states may also arise as the result of nuclear reactions or of direct excitation from the ground state. In this section we deal with the phenomena that occur in the de-excitation of excited states.

[20] It has, however, been conclusively shown in accelerator experiments (L3) that the neutrinos emitted along with μ mesons in π-meson decay are not identical with those associated with β decay. Thus there are at least two neutrinos and two antineutrinos: ν_e, $\bar{\nu}_e$, ν_μ, $\bar{\nu}_\mu$. In μ-meson decay neutrinos of both the electron and the muon type are emitted: $\mu^+ \rightarrow e^+ + \nu_e + \bar{\nu}_\mu$ and $\mu^- \rightarrow e^- - \bar{\nu}_e + \nu_\mu$. In all known processes the muonic leptons ($\mu^\mp$, ν_μ, $\bar{\nu}_\mu$) and the electronic ones ($e^\pm$, ν_e, $\bar{\nu}_e$) are separately conserved. There is believed to be a third kind of neutrino, ν_τ, associated with the heavy lepton called τ meson.

De-excitation Processes. A nucleus in an excited state may give up its excitation energy and return to the ground state in a variety of ways. The most obvious, and the most common, transition is by the emission of electromagnetic radiation.[21] Such radiation is called γ radiation; the γ rays have a frequency determined by their energy $E = h\nu$. Frequently the transition does not proceed directly from an upper state to the ground state but may go in several steps involving intermediate excited states. Gamma rays with energies between a few thousand electron volts and about 7 MeV have been observed in radioactive processes.

Gamma-ray emission may be accompanied, or even replaced, by another process, the emission of internal-conversion electrons. **Internal conversion** comes about by the (purely electromagnetic) interaction between nucleus and extranuclear electrons leading to the emission of an electron with a kinetic energy equal to the difference between the energy of the nuclear transition involved and the binding energy of the electron in the atom.

A third process for the de-excitation of a nucleus is possible if the available energy exceeds 1.02 MeV. This energy is equivalent to the mass of two electrons. It is possible for the excited nucleus to create simultaneously one new electron and one positron and to emit them with kinetic energies that total the excitation energy minus 1.02 MeV. This is an uncommon mode of de-excitation.

All the processes just described we call γ transitions, although only in the first is a γ ray emitted by the nucleus. All are characterized by a change in energy without change in Z and A.

In a number of instances a nuclide in an excited state decays predominantly by α- or β-decay.[22] It is even possible for such a β decay to be followed by another β process leading back to the original nucleus in its ground state. One instance of such a sequence (which we would not like to designate as a γ transition) has been observed: the isomer $^{87}Sr^m$ decays partially by electron capture to ^{87}Rb, a β^- emitter that decays to ^{87}Sr.

Lifetimes of Excited States. The overwhelming majority of γ transitions take place on a time scale too short for direct measurement, that is, in less than about 10^{-12} s, as would be expected for a dipole of nuclear dimensions and unit electronic charge. As was already indicated in the discussion of α and β decay, γ-de-excitation processes are of vital im-

[21] This statement applies to bound states. As the excitation energy exceeds the binding energy of the most loosely bound nucleon (most often a neutron), emission of this nucleon rapidly becomes a more probable process than γ emission.

[22] The so-called long-range α particles of ^{212}Po (ThC′) and ^{214}Po (RaC′) arise from the decay of excited states in these nuclei fed by β decays and so unstable with respect to α emission that α decay can compete with γ emission. This phenomenon is quite analogous to the β-delayed neutron and proton emission already discussed. Usually the lifetime of an excited state for γ emission is much shorter than that for β or α decay, although this is, of course, not true for some metastable states (see discussion of isomerism below).

portance in all types of radioactivity measurements and in the establishment of nuclear level schemes, whether or not their lifetimes can be measured. However, in this section we are mostly concerned with the factors that affect the lifetimes of γ transitions and make possible the existence of metastable or isomeric nuclear states. As we remarked in chapter 2 (p. 23), the definition of a nuclear isomer in terms of a "measurable half life" has become somewhat vague, since the development of new direct and indirect techniques keeps extending the lower limit of what is measurable.[23] At the upper end of the scale there is probably no limit either; $^{210}Bi^m$ holds the record in 1980, with $t_{1/2} = 3.5 \times 10^6$ y.

Gamma decay from an isomeric state is called an **isomeric transition** (IT). In the following paragraphs we discuss the connection between transition probabilities (or half lives) for γ decay, decay energy, and spins and parities of initial and final states. The systematics of isomer lifetimes and their dependence on energy and spin changes was important in the development of the nuclear shell model. In discussing half lives for γ transitions we are concerned with *partial* half lives, since β decay and, in heavy elements, α decay often compete with ITs in the decay of metastable states.

Multipole Radiation and Selection Rules. Gamma radiation arises from purely electromagnetic effects that may be thought of as changes in the charge and current distributions in nuclei. Since charge distributions give rise to electric moments and current distributions to magnetic moments, γ-ray transitions are correspondingly classified as electric (E) and magnetic (M). In addition, it is convenient, as in β decay, to characterize transitions according to the angular momentum l (in units of $\hbar$), which the γ ray carries off. We see that, as in β decay, transition probabilities fall off rapidly with increasing angular-momentum changes. The accepted nomenclature[24] is to refer to radiations carrying off $l = 1, 2, 3, 4, 5$ units of $\hbar$ as dipole, quadrupole, octupole, 2^4-pole, and 2^5-pole radiations. The shorthand notation for electric (or magnetic) 2^l-pole radiation is El (or Ml); thus $E2$ means electric quadrupole, $M4$ magnetic 2^4 pole, and so on. The electric and magnetic multipole radiations differ in

[23] As a practical measure we designate with the superscript m only those states with $t_{1/2} \geq 10^{-6}$ s. In appendix D only isomers with $t_{1/2} \geq 1$ s are listed.

[24] The multipole nomenclature has developed because the radiation field around a system of oscillating charges can always be expressed as an expansion in spherical harmonics of orders 1, 2, 3 . . . ; for a pure dipole radiator the first nonvanishing term in this series is the first term, for a quadrupole radiator the second term, etc. Furthermore, the successive terms in this multipole expansion correspond to the photon carrying off 1, 2, 3, etc., units of angular momentum. The lth term in the expansion of the field is proportional to $(R/\lambda\!\!\!^{-})^l$, where R is the dimension of the radiator ($\cong$ nuclear radius) and $\lambda\!\!\!^{-}$ the wavelength of the emitted radiation divided by 2π. For γ rays from radioactive decay $\lambda\!\!\!^{-}$ is always large compared with nuclear dimensions (for a 1-MeV photon, $\lambda\!\!\!^{-} \cong 2 \times 10^{-11}$ cm), so that the series converges rapidly and only the first nonvanishing term usually needs to be considered.

their parity properties. If we denote even and odd parity of the radiation by $+1$ and -1, then electric 2^l-pole radiation has parity $(-1)^l$, whereas magnetic 2^l-pole radiation has parity $(-1)^{l+1}$.

We can now formulate the selection rules for γ transition between an initial state of spin I_i and a final state of spin I_f and with either equal or opposite parities. It follows immediately from what was said above about angular momenta associated with 2^l-pole radiation that $l \geq |I_i - I_f|$. However, consideration of the vector addition of the angular momenta involved leads to the further restriction that l cannot exceed $I_i + I_f$. Thus we have, for both electric and magnetic radiations,

$$I_i + I_f \geq l \geq |I_i - I_f|. \tag{3-28}$$

If initial and final state have the same parity, electric multipoles of even l and magnetic multipoles of odd l are allowed. If initial and final states have opposite parities, electric multipoles of odd l and magnetic multipoles of even l are allowed. As an example, if the transition is between a 4^+ and a 2^+ state, multipole orders l can range from 2 to 6, but because of the parity rules $E2$, $M3$, $E4$, $M5$, and $E6$ are the only radiations possible.

The actual situation is simpler because as a rule only the lowest multipole order (sometimes the lowest two) allowed by the selection rules contributes appreciably to the intensity. This comes about as follows: the transition probability is proportional to the square of the matrix element for the interaction; the contribution of each term in the power-series expansion of the field (see footnote 24) to the transition probability is therefore proportional to $(R/\lambda\!\!\!^{-})^{2l}$. Since $R/\lambda\!\!\!^{-}$ is always a small number ($\cong 10^{-2}$–10^{-3}), only the lowest allowed multipole order will generally contribute. Exceptions to this rule occur commonly when the lowest allowed radiation is magnetic dipole ($M1$); here electric quadrupole ($E2$) transitions can often compete favorably. This can be understood if we remember that the current densities in nuclei (which give rise to the magnetic multipoles) are smaller than the charge densities (which produce the electric multipoles) by $\sim v/c$, where v represents the speed of the charges (protons) in the nucleus; for a given multipole order the magnetic transitions therefore can be expected to be weaker than the electric ones by a factor of the order of $(v/c)^2 \cong 10^{-2}$. (This neglects the contributions of the intrinsic magnetic moments of nucleons.) Thus we might expect $E(l+1)$ radiation to compete with Ml; this expectation, as we remarked, is often borne out experimentally for $l = 1$.

The selection rules discussed are summarized in table 3-3. Some significant special cases not yet mentioned are noted in the table. They all stem from the restriction $l \leq I_i + I_f$. In particular, $0 \rightarrow 0$ transitions ($I_i = 0$, $I_f = 0$) cannot take place by photon emission, essentially because a photon has spin 1 and therefore must (vectorially) remove at least one unit of angular momentum (this condition can always by fulfilled for other $\Delta I = 0$ transitions by proper orientation of the vectors $\mathbf{I}_i$ and $\mathbf{I}_f$). If there is no

Table 3-3 Selection Rules for Gamma Transitions

ΔI	0^a	0^a	1	1	2	2	3	3	4	4	etc.
$\Delta\Pi$	No	Yes	No	Yes	No	Yes	No	Yes	No	Yes	
Transition type	$M1$	$E1$	$M1$	$E1$	$E2$	$M2$	$M3$	$E3$	$E4$	$M4$	
	$E2$		$E2^b$			$(E3^b)$	$(E4^b)$			$(E5^b)$	

[a] The more complete selection rule (3-28) excludes *any* single-photon transitions when $I_i = I_f = 0$ (photons have intrinsic spin 1). For alternative de-excitation processes in this situation see text. For transitions between two $I = \frac{1}{2}$ states of equal parity $E2$ is forbidden by (3-28), but $M1$ is allowed.
[b] When one of the states involved in a transition has spin 0 and the allowed transition of lowest order is a magnetic multipole, the next higher electric multipole is strictly forbidden by (3-28).

change in parity in a $0 \rightarrow 0$ transition, de-excitation may occur by emission of an internal-conversion electron (see below) or, if $\Delta E > 1.02$ MeV, by simultaneous emission of an electron-positron pair. Examples of the former mode are known for transitions to the ground states (0^+) from other 0^+ states in ^{72}Ge ($\Delta E = 0.691$ MeV; $t_{1/2} = 0.42\ \mu$s) and in ^{214}Po ($\Delta E = 1.415$ MeV; partial $t_{1/2}$ for this transition is 0.8 ns). Pair emission occurs, for example, from the first excited state (6.05 MeV) in ^{16}O ($t_{1/2} = 0.05$ ns) and from a state at 1.84 MeV in ^{42}Ca ($t_{1/2} = 0.33$ ns). Transitions between two $I = 0$ states of opposite parity cannot take place by any first-order process; it would require simultaneous emission of two γ quanta or two conversion electrons. No such transition has been established.

Isomeric Transitions. Having stated the selection rules, we are now ready to return to a more quantitative discussion of actual lifetimes for γ transitions, with the eventual aim of comparing theoretical predictions with the experimental observations on isomeric transitions. We have already stated that the transition probability for emission of 2^l-pole radiation of wavelength $2\pi\lambda\!\!\!^-$ from a nucleus of radius R should be roughly proportional to $(R/\lambda\!\!\!^-)^{2l}$. Since $R \propto A^{1/3}$ and transition energy $E \propto 1/\lambda\!\!\!^-$, we get for the transition probability or partial decay constant for γ emission

$$\lambda_\gamma \propto E^{2l} A^{2l/3}. \tag{3-29}$$

Thus for a given spin change half lives will decrease rapidly with increasing A and even more rapidly with increasing E (a more detailed analysis actually gives an E^{2l+1} dependence), and both the A and E dependence become steeper with increasing multipole order.

To go beyond these qualitative statements and to calculate absolute transition probabilities or half lives, it becomes necessary to make more specific assumptions about the charge and current distributions in nuclei, that is, to choose a nuclear model. The simplest model for this purpose is the extreme single-particle model (see chapter 10, section D). The assumption is that a γ transition can be described as the transition of a single nucleon from one angular-momentum state to another, the rest of the nucleus being represented as a potential well. On this basis V. F. Weisskopf has derived expressions for decay constants for electric and magnetic 2^l-pole transitions (B1, p. 583). These somewhat unwieldy general formulas (see for example, W3) reduce to the simple expressions given in table 3-4 for the first few multipole orders. The numerical values given for the illustrative case of $A = 125$, $E = 0.1$MeV indicate the enormous effect of multipole order on half life. Measured half lives for γ decay[25] are usually sufficiently close to the values predicted by this single-particle theory to allow determination of the spin change. Particularly for the $M4$ transitions, which are very common among isomers, the agreement is good (usually within a factor of 2 or 3), and for other transitions with $\Delta I > 2$ the calculated values are rarely wrong by more than a factor of 100 (G8).

The success of the independent-particle calculations of isomer lifetimes was a strong argument in favor of the shell model (which is a particular form of single-particle model—see chapter 10) and gave great impetus to its

Table 3-4 Partial Half Lives for Gamma Transitions Calculated on the Single-Particle Model[a]

Transition Type	Partial Half Life t_γ (s)	Illustrative t_γ Values (s) for $A = 125$, $E = 0.1$ MeV
$E1$	$5.7 \times 10^{-15}\ E^{-3}A^{-2/3}$	2×10^{-13}
$E2$	$6.7 \times 10^{-9}\ E^{-5}A^{-4/3}$	1×10^{-6}
$E3$	$1.2 \times 10^{-2}\ E^{-7}A^{-2}$	8
$E4$	$3.4 \times 10^{4}\ E^{-9}A^{-8/3}$	9×10^{7}
$E5$	$1.3 \times 10^{11}\ E^{-11}A^{-10/3}$	1×10^{15}
$M1$	$2.2 \times 10^{-14}\ E^{-3}$	2×10^{-11}
$M2$	$2.6 \times 10^{-8}\ E^{-5}A^{-2/3}$	1×10^{-4}
$M3$	$4.9 \times 10^{-2}\ E^{-7}A^{-4/3}$	8×10^{2}
$M4$	$1.3 \times 10^{5}\ E^{-9}A^{-2}$	8×10^{9}
$M5$	$5.0 \times 10^{11}\ E^{-11}A^{-8/3}$	1×10^{17}

[a] The energies E are expressed in MeV. The nuclear radius parameter r_0 has been taken as 1.3 fm. Note that t_γ is the *partial* half life for γ emission only; the occurrence of internal conversion will always shorten the measured half life.

[25] Note that the expressions in table 3-4 give the partial half life t_γ for γ emission only. If the internal-conversion coefficient is α, the half life for de-excitation (which is the measured half life if there is no other competing decay mode) is $t_{1/2} = t_\gamma/(1 + \alpha)$.

further development in the early 1950s (G8, G9). As outlined in chapter 10, section D, the shell model predicts, for a given nucleus, low-lying states of widely differing spins in certain regions of neutron and proton numbers, namely just preceding the shell closures at N or Z values of 50, 82, and 126. These regions coincide exactly with the so-called "islands of isomerism" empirically found, and the shell model indeed accounts remarkably well for the properties of these isomers, including their lifetimes.

The Weisskopf formula (table 3-4) is thought to give lower limits for γ half lives in the sense that transitions between states whose spins and parities cannot be ascribed completely to the properties of individual nucleons, but come about through interaction between several nucleons outside a closed shell, (see chapter 10, p. 386) are expected to be slower. Many deviations from the Weisskopf lifetimes are in this direction and are ascribed to various forms of multiple-particle configurations, although still in the spirit of the single-particle model.

On the other hand, there is a large group of $E2$ transitions in heavy nuclei that are of the order of 100 times faster than the single-particle model would predict. This suggests some type of collective motion involving not one but many protons. The properties of these fast $E2$ transitions, which occur mainly between the low-lying states of nuclei with neutron numbers between 90 and 120 and above ~140, are indeed best accounted for by the collective model (see chapter 10, section E) that predicts bands of "rotational" states for these spheroidally deformed nuclei; for the even-even nuclei the spins of the successive rotational states are $0, 2, 4, 6, \ldots$. Interestingly enough, in the same general region of nuclei some extremely slow (10^3 to 10^8 times single-particle lifetimes) $E1$ transitions between states of opposite parity are observed. On the basis of the collective model this phenomenon is ascribed to the occurrence of less symmetric forms of deformation (such as pear shapes) and the operation of special selection rules for transitions between such states and the normal spheroidal or ellipsoidal states (W3).

Internal Conversion Coefficients. As mentioned earlier, internal conversion is an alternative to γ-ray emission. The ratio of the rate of the internal conversion process to the rate of γ emission (or the ratio of the number of internal conversion electrons to the number of γ quanta emitted) is known as the internal conversion coefficient α; it may have any value between 0 and ∞. Separate coefficients for internal conversion in the K, L, M shell, and so on (α_K, α_L, α_M, etc.) and even in the subshells (α_{L_I}, $\alpha_{L_{II}}$, $\alpha_{L_{III}}$, etc.) may be measured as well as computed. In general, the coefficients for any shell increase with decreasing energy, increasing ΔI, and increasing Z.

The calculation of internal-conversion coefficients is a problem in atomic physics. It involves the computation of the amplitudes of electron wave

Table 3-5 K-Shell Conversion Coefficients and K/L Conversion Ratios[a,b]

		E_γ							
		100 keV		220 keV		460 keV		1000 keV	
Z	Transition Type	α_K	K/L	α_K	K/L	α_K	K/L	α_K	K/L
30	$E1$	5.37 (−2)	9.89	5.06 (−3)	9.94	6.56 (−4)	10.0	1.15 (−4)	10.1
	$E2$	5.79 (−1)	8.14	3.03 (−2)	9.27	2.33 (−3)	9.71	2.65 (−4)	9.93
	$E3$	5.16 (0)	5.23	1.54 (−1)	7.82	7.28 (−3)	9.12	5.53 (−4)	9.74
	$E4$	4.41 (1)	3.00	7.50 (−1)	6.10	2.19 (−2)	8.30	1.11 (−3)	9.41
	$M1$	5.53 (−2)	9.50	7.11 (−3)	9.73	1.21 (−3)	9.84	2.27 (−4)	10.0
	$M2$	5.70 (−1)	8.10	3.90 (−2)	9.01	4.08 (−3)	9.56	5.16 (−4)	9.89
	$M3$	5.46 (0)	6.28	2.03 (−1)	8.02	1.28 (−2)	9.08	1.08 (−3)	9.73
	$M4$	5.23 (1)	4.67	1.05 (0)	6.91	3.94 (−2)	8.47	2.18 (−3)	9.44
50	$E1$	1.57 (−1)	7.81	1.73 (−2)	8.12	2.55 (−3)	8.33	4.80 (−4)	8.53
	$E2$	1.15 (0)	3.19	8.27 (−2)	5.70	8.06 (−3)	7.21	1.12 (−3)	8.07
	$E3$	6.89 (0)	0.46	3.41 (−1)	2.94	2.28 (−2)	5.48	2.33 (−3)	7.43
	$E4$	3.96 (1)	0.27	1.35 (0)	1.43	6.24 (−2)	3.87	4.61 (−3)	6.64
	$M1$	4.95 (−1)	7.82	5.67 (−2)	7.96	8.60 (−3)	8.17	1.39 (−3)	8.38
	$M2$	4.72 (0)	5.26	2.98 (−1)	6.52	2.89 (−2)	7.39	3.36 (−3)	8.01
	$M3$	3.52 (1)	2.74	1.38 (0)	4.81	8.55 (−2)	6.38	6.95 (−3)	7.54
	$M4$	2.54 (2)	1.35	6.29 (0)	3.35	2.48 (−1)	5.36	1.38 (−2)	7.01

70	*E*1	2.83 (−1)	6.01	3.65 (−2)	6.59	6.20 (−3)	6.95	1.28 (−3)	7.28
	*E*2	1.01 (0)	1.00	1.26 (−1)	2.29	1.73 (−2)	4.47	3.11 (−3)	6.32
	*E*3	2.83 (0)	0.060	3.96 (−1)	0.64	4.49 (−2)	2.37	6.53 (−3)	5.06
	*E*4	7.72 (0)	0.010	1.25 (0)	0.23	1.15 (−1)	1.32	1.29 (−2)	3.97
	*M*1	2.93 (0)	6.53	3.18 (−1)	6.62	4.43 (−2)	6.75	6.30 (−3)	6.94
	*M*2	2.31 (1)	3.29	1.47 (0)	4.60	1.38 (−1)	5.56	1.53 (−2)	6.34
	*M*3	9.78 (1)	0.86	5.39 (0)	2.59	3.64 (−1)	4.26	3.01 (−2)	5.67
	*M*4	3.62 (2)	0.21	1.94 (1)	1.37	9.37 (−1)	3.10	5.62 (−2)	4.90
90	*E*1			6.04 (−2)	5.06	1.22 (−2)	5.60	2.88 (−3)	6.06
	*E*2			1.34 (−1)	0.54	3.20 (−2)	2.06	7.75 (−3)	4.31
	*E*3			3.16 (−1)	0.101	8.07 (−2)	0.84	1.69 (−2)	2.89
	*E*4			7.73 (−1)	0.030	1.95 (−1)	0.42	3.31 (−2)	2.04
	*M*1			1.72 (0)	5.33	2.28 (−1)	5.42	2.94 (−2)	5.57
	*M*2			5.95 (0)	2.99	5.84 (−1)	3.97	6.51 (−2)	4.75
	*M*3			1.43 (1)	1.13	1.24 (0)	2.57	1.14 (−1)	3.97
	*M*4			3.29 (1)	0.39	2.61 (0)	1.57	1.91 (−1)	2.94

[a] From reference R5.
[b] The number in parentheses is the power of 10 that multiplies the preceding number. Thus 5.37 (−2) means 5.37×10^{-2}.

functions at the nucleus and can be carried out without regard to nuclear forces. Calculations have been done in various approximations. The most sophisticated ones take into account small effects such as those of finite nuclear size and of the shape of nuclear charge distributions. Extensive tabulations of K-, L-, M-, and N-shell coefficients and various subshell coefficients for multipole orders up to $l = 5$ may be found in R5, B7, and B8. A small sample of α_K values and α_K/α_L ratios is shown in table 3-5. It can be used to obtain very approximate numbers at other values of E and Z by interpolation, best done with the aid of plots of $\log \alpha_i$ versus $\log E$ (quite linear) and of $\log \alpha_i$ versus Z for each multipole order. Even when the complete tables are used, some interpolation is almost always necessary, and H2 contains a convenient computer program for this purpose.

Internal Conversion and Nuclear Spectroscopy. Internal conversion electrons, examined in an electron spectrograph, show a line spectrum with lines corresponding to the γ-transition energy minus the binding energies of the $K, L, M, \ldots$ shells in which conversion occurs. The differences in energy between successive lines serve to identify Z and to classify groups of lines resulting from different γ transitions. To obtain the energy of a transition from the measured conversion-electron energies, we must add the appropriate electron-binding energies. Compilations of electron-binding energies in the various shells are available (H2, L1).

Experimental determination of absolute conversion coefficients is difficult, since it entails the measurement of conversion-electron and γ-ray intensities with known detection efficiencies. In practice, it is much easier to determine, in an electron spectrograph, the *relative* intensities of two or more conversion-electron lines belonging to the same transition and to compare these ratios with theoretical values. As can be seen from table 3-5, the α_K/α_L ratios, although they do not vary over ranges as wide as the individual coefficients, can be used to great advantage to characterize multipole order and thus ΔI and $\Delta\Pi$, especially at relatively high Z and low energy. However, as may also be seen from the sample in table 3-5, some ambiguities often still remain, and one then has to resort to the use of conversion coefficients in the higher shells (M, N, etc.), and particularly L- and sometimes M-subshell ratios. The ratios $\alpha_{L_I}/\alpha_{L_{II}}$ and $\alpha_{L_{II}}/\alpha_{L_{III}}$[26] usually differ strongly for electric and magnetic multipoles of the same order, where α_K/α_L ratios are often quite similar.

Thus for example, at $Z = 30$ and $E = 1000$ keV, where table 3-5 shows α_K/α_L for each of the eight multipoles listed to be in the narrow range 9.4–10.1, the L_I/L_{II} ratios are well spread out between 165 for $E1$ and 20 for $E4$, and the L_{II}/L_{III} ratios range from 0.44 for $E1$ to 2.55 for $M2$; in this particular example any multipole order could be clearly identified from L-subshell ratios except for a possible remaining confusion between $E3$ ($L_I/L_{II} = 36$; $L_{II}/L_{III} = 1.55$) and $M4$ ($L_I/L_{II} = 32$; $L_{II}/L_{III} = 1.86$).

[26] The L_I, L_{II}, L_{III} subshells are those containing $2s_{1/2}$, $2p_{1/2}$, and $2p_{3/2}$ electrons, respectively.

For mixed $M1$-$E2$ transitions the subshell ratios usually allow the mixing ratios to be deduced.

The resolution of electron lines originating from the different L subshells and from higher shells requires very thin sources (to minimize broadening of the lines by scattering) as well as spectrometers of high resolution.

An internal-conversion process leaves the atom with a vacancy in one of its shells. This leads, as in EC, to the emission of X rays and Auger electrons (see pp. 86). Note that, if the chemical identity (Z) of an X-ray-emitting species is known, determination of the X-ray energy will reveal whether it decays by EC or IT, since the characteristic X rays will, in one case, be those of $Z-1$, in the other those of Z.

Angular Correlations (F3). In discussing the various techniques for identification of multipole character of γ transitions—half lives and conversion coefficients—we have made the tacit assumption that once removed from the decaying nuclei the γ rays themselves bear no recognizable mark of the multipole interaction that gave birth to them. This is indeed correct under ordinary circumstances. However, it is true that different multipole fields give rise to different angular distributions of the emitted radiation with respect to the nuclear-spin direction of the emitting nucleus. We ordinarily deal with samples of radioactive material that contain randomly oriented nuclei and therefore the observed angular distribution of γ rays is isotropic.

If it were possible to align the nuclear spins in a γ-emitting sample in one direction, the angular distribution of emitted γ-ray intensity would depend in a definite and theoretically calculable way on the initial nuclear spin and the multipole character of the radiation. One method for obtaining alignment of nuclear spins is the application of strong external electric or magnetic fields at temperatures near 0 K. This technique, which requires costly specialized equipment, has found limited but important applications; we do not discuss it here but refer the reader to reviews (see, for example, A1, G10).

A second and more widely applicable method for obtaining partially "oriented" nuclei is to observe a γ ray in coincidence with a preceding radiation (α, β, or γ). By selecting a particular direction of emission for this first radiation, we then in effect select a preferred direction for the spin orientation of the intermediate nucleus; provided the lifetime of this intermediate state is short enough for the spin orientation to be preserved until the γ ray is emitted, the direction of the γ-ray emission will be correlated with the direction of emission of the preceding radiation. If a coincidence experiment is done, in which the angle θ between the two sample-detector axes is varied, the coincidence rate will in general vary as a function of θ.

Theoretical correlation functions $W(\theta)\,d\Omega$ have been calculated for a great variety of situations (W4, T1). Here $W(\theta)\,d\Omega$ denotes the relative probability that the second radiation will be emitted into solid angle $d\Omega$ if the angle between the two directions of emission is θ. Usually the correlation function is normalized so that $\int W(\theta)\,d\Omega = 1$. The correlation function can then always be written in the form

$$W(\theta) = 1 + a_2 \cos^2 \theta + a_4 \cos^4 \theta + \cdots, \tag{3-30}$$

where only even powers of $\cos \theta$ appear. If the angular momenta carried away by the first and second radiation are denoted by l_1 and l_2 and the spin of the intermediate state by I, the highest power of $\cos \theta$ that occurs is less than or equal to twice the smallest of these three numbers (l_1, l_2, I).[27] If something is known about two or at least one of these quantities, angular-correlation experiments can then give information on the other(s). For example, the angular correlations involving a $\Delta I = 1$ transition without parity change are different, depending on whether the transition is $M1$ or $E2$. Comparison of measured and calculated correlation functions can thus give the $M1/E2$ mixing ratio for the transition. More often, angular-correlation measurements are used to determine the spin of an excited state when the multipole orders of the γ rays feeding and deexciting it are known from other measurements. The coefficients a_2, a_4, and so on, have been tabulated for many types of cascades (W4, T1). Directional correlations in themselves cannot distinguish electric and magnetic transitions of the same multipole order; the parity change in a transition *can* be inferred if, in addition to the directional correlation, the polarization of the γ rays is measured.

Angular-correlation experiments can usually be performed only when transitions with low multipole orders are involved; for $l_2 > 2$, the lifetime of the intermediate state is usually so long that the correlation is destroyed. Thus the $\cos^4 \theta$ term in (3-30) is the highest term that appears in practical cases, and only two parameters need to be determined experimentally. Angular-correlation measurements therefore do not usually require data at many angles. A quantity often used to express experimental results is the anisotropy parameter

$$A = \frac{W(180°) - W(90°)}{W(90°)}, \tag{3-31}$$

which is, of course, related to a_2 and a_4 in (3-30): $A = a_2 + a_4$.

The interesting effects that external fields and chemical environment can have on angular correlations are discussed in chapter 12, section D.

REFERENCES

A1 E. Ambler, "Nuclear Orientation," in *Methods of Experimental Physics*, Vol. 5B, *Nuclear Physics* (L. C. L. Yuan and C. S. Wu, Eds.), Academic, New York, 1963, pp. 162–214.

[27] Angular correlations thus cannot occur if the intermediate state has spin 0 or $\frac{1}{2}$.

*B1 J. M. Blatt and V. F. Weisskopf, *Theoretical Nuclear Physics*, Wiley, New York, 1952.

B2 N. Bohr and J. A. Wheeler, "The Mechanism of Nuclear Fission," *Phys. Rev.* **56**, 426 (1939).

B3 M. Brack et al., "Funny Hills: The Shell Correction Approach to Nuclear Shell Effects and its Application to the Fission Process," *Rev. Mod. Phys.* **44**, 320 (1972).

B4 H. Behrens and J. Jänecke, "Numerical Tables for Beta Decay and Electron Capture," *Landolt Börnstein New Series*, Vol. I/4, Springer, Berlin, 1969.

B5 E. H. S. Burhop and W. N. Asaad, "The Auger Effect," *Adv. At. Mol. Phys.* **8**, 163 (1972).

B6 W. Bambynek et al., "X-Ray Fluorescence Yields, Auger, and Coster-Kronig Transition Probabilities," *Rev. Mod. Phys.* **44**, 716 (1972).

B7 I. M. Band, M. B. Trzhaskovskaya, and M. A. Listengarten, "Internal Conversion Coefficients for $Z \leq 30$," *At. Data Nucl. Data Tables* **18**, 433 (1976).

B8 I. M. Band, M. B. Trzhaskovskaya, and M. A. Listengarten, "Internal Conversion Coefficients for $E5$ and $M5$ Nuclear Transitions, $30 \leq Z \leq 104$," *At. Data Nucl. Data Tables* **21**, 1 (1978).

*E1 R. D. Evans, *The Atomic Nucleus*, McGraw-Hill, New York, 1955.

F1 J. A. Frenkel, "On the Splitting of Heavy Nuclei by Slow Neutrons," *Phys. Rev.* **55**, 987 (1939).

F2 E. Fermi, "Versuch einer Theorie der β-Strahlen," *Z. Phys.* **88**, 161 (1934).

F3 H. Frauenfelder, "Angular Correlation," in *Methods of Experimental Physics* Vol. 5B, *Nuclear Physics* (L. C. L. Yuan and C. S. Wu, Eds.), Academic, New York, 1963, pp. 129–151.

G1 H. Geiger and J. M. Nuttall, "The Ranges of the α-Particles from Various Radioactive Substances and a Relation between Range and Period of Transformation," *Phil. Mag.* **22**, 613 (1911); **23**, 439 (1912).

G2 G. Gamow, "Zur Quantentheorie des Atomkernes," *Z. Phys.* **51**, 204 (1928).

G3 R. W. Gurney and E. U. Condon, "Quantum Mechanics and Radioactive Disintegration," *Nature* **122**, 439 (1928); *Phys. Rev.* **33**, 127 (1929).

G4 V. I. Goldanskii, "On Neutron-Deficient Isotopes of Light Nuclei and the Phenomena of Proton and Two-Proton Radioactivity," *Nucl. Phys.* **19**, 482 (1960).

G5 V. I. Goldanskii, "Two-Proton Radioactivity," *Nucl. Phys.* **27**, 648 (1961).

G6 N. B. Gove and M. J. Martin, "Log-f Tables for Beta Decay," *Nucl. Data Tables* **A10**, 205 (1971).

G7 M. Goldhaber, L. Grodzins, and A. W. Sunyar, "Helicity of Neutrinos," *Phys. Rev.* **109**, 1015 (1958).

G8 M. Goldhaber and A. W. Sunyar, "Classification of Nuclear Isomers," *Phys. Rev.* **83**, 906 (1951).

G9 M. Goldhaber and R. D. Hill, "Nuclear Isomerism and Shell Structure," *Rev. Mod. Phys.* **24**, 179 (1952).

G10 S. R. de Groot, H. A. Tolhoek, and W. J. Huiskamp, "Orientation of Nuclei at Low Temperatures," in *α-, β-, and γ-Ray Spectroscopy*, Vol. 2 (K. Siegbahn, Ed.) North Holland, Amsterdam, 1966, pp. 1199–1262.

*H1 G. C. Hanna, "Alpha Radioactivity," in *Experimental Nuclear Physics*, Vol. III (E. Segre, Ed.), Wiley, New York, 1959, pp. 54–257.

H2 S. Hagström, C. Nordling, and K. Siegbahn, "Tables of Electron Binding Energies and Kinetic Energy vs. Magnetic Rigidity," in *α-, β-, and γ-Ray Spectroscopy*, Vol. 1 (K. Siegbahn, Ed.), North Holland, Amsterdam, 1966, pp. 845–862.

K1 F. N. D. Kurie, J. R. Richardson, and H. C. Paxton, "The Radiations Emitted from

Artificially Produced Radioactive Substances. I. The Upper Limit and Shapes of the β-Ray Spectra from Several Elements," *Phys. Rev.* **49**, 368 (1936).

*K2 E. J. Konopinsky, *The Theory of Beta Radioactivity*, Clarendon, Oxford University Press, London 1966.

*L1 C. M. Lederer and V. S. Shirley (Eds.), *Table of Isotopes*, 7th Ed., Wiley, New York, 1978.

L2 T. D. Lee and C. N. Yang, "Question of Parity Conservation in Weak Interactions," *Phys. Rev.* **104**, 254 (1956).

L3 L. Lederman, "The Two-Neutrino Experiment," *Sci. Am.* **208**(3), 80 (March 1963).

M1 H. J. Mang, "Alpha Decay," *Ann. Rev. Nucl. Sci.* **14**, 1 (1964).

M2 S. A. Moszkowski, "Rapid Method of Calculating Log (ft) Values," *Phys. Rev.* **82**, 35 (1951).

M3 M. J. Martin and P. H. Blichert-Toft, "Radioactive Atoms, Appendix IIIB: Electron Capture," *Nucl. Data Tables* **A8**, 157 (1970).

N1 J. R. Nix, "Calculation of Fission Barriers for Heavy and Superheavy Nuclei," *Ann. Rev. Nucl. Sci.* **22**, 65 (1972).

P1 I. Perlman, A. Ghiorso, and G. T. Seaborg, "Systematics of Alpha Radioactivity," *Phys. Rev.* **77**, 26 (1950).

P2 S. M. Polikanov et al., "Spontaneous Fission with an Anomalously Short Period. I.," *Soviet Phys. JETP* **15**, 1016 (1962).

R1 W. Rubinson, "The Correction for Atomic Excitation Energy in Measurements of Energies of Electron Capture Decay," *Nucl. Phys.* **A169**, 629 (1971).

R2 M. E. Rose and J. L. Jackson, "The Ratio of L_I to K Capture," *Phys. Rev.* **76**, 1540 (1949).

R3 S. Raman and N. B. Gove, "Rules for Spin and Parity Assignments Based on Log ft Values," *Phys. Rev.* **C7**, 1995 (1973).

R4 F. Reines, "Neutrino Interactions," *Ann. Rev. Nucl. Sci.* **10**, 1 (1960).

R5 F. Rösel et al., "Internal Conversion Coefficients for All Atomic Shells $30 \leq Z \leq 104$," *At. Data Nucl. Data Tables* **21**, 89 (1978).

S1 V. M. Strutinsky, "Shell Effects in Nuclear Masses and Deformation Energies," *Nucl. Phys.* **A95**, 420 (1967).

T1 H. W. Taylor et al., "A Tabulation of γ-γ Directional-Correlation Coefficients," *Nucl. Data Tables* **A9**, 1 (1971).

V1 R. Vanderbosch, "Spontaneous Fission Isomers," *Ann. Rev. Nucl. Sci.* **27**, 1 (1977).

*W1 C. S. Wu and S. A. Moszkowski, *Beta Decay*, Interscience, New York, 1966.

W2 C. S. Wu et al., "Experimental Test of Parity Conservation in Beta Decay," *Phys. Rev.* **105**, 1413 (1957).

W3 D. H. Wilkinson, "Analysis of Gamma Decay Data," in *Nuclear Spectroscopy*, Part B (F. Ajzenberg-Selove, Ed.), Academic, New York, 1960, pp. 852–889.

W4 K. Way and F. W. Hurley, "Directory to Tables and Reviews of Angular-Momentum and Angular-Correlation Coefficients," *Nucl. Data* **1**, 473 (1966).

EXERCISES

1. In the α decay of ^{211}Bi to ^{207}Tl two groups of α particles, with kinetic energies 6.623 and 6.279 MeV, are observed. They populate the ground state and first excited state of ^{207}Tl. What is the energy difference between these two states?

   Answer: 0.351 MeV.

2. Estimate the rate of energy deposition (in calories per hour) in a large calorimeter when each of the following samples is placed in it; (a) ^{14}C, (b) ^{146}Sm, (c) ^{254}Cf, each undergoing 10^{10} dis min^{-1}. *Answer:* (b) 0.058 cal h^{-1}.

3. The nuclide ^{256}Cf decays predominantly by spontaneous fission (SF). Estimate its α-decay energy from systematics and, from that estimate and from the known half life (appendix D), derive a rough prediction of its α/SF branching ratio. *Answer:* $\sim 10^{-9}$.

4. Verify that (3-10) can be derived from (2-7).

5. (a) Use (3-6) to verify the value listed in table 3-1 for the calculated decay constant of the α transition of ^{228}Th to the ground state of ^{224}Ra. (b) With the aid of the same equation calculate the partial decay constant for the decay of ^{228}Th to the first excited state of ^{224}Ra at 84 keV above ground. (c) Compare the calculated branching ratio for the two α transitions that you have just obtained with the experimentally determined one shown in figure 3-1 and comment on the degree of agreement.

6. ^{243}Cm decays by α emission with a 28.5-y half life, in 74 percent of the disintegrations to an excited state of ^{239}Pu at 285 keV above the ground state. The transitions to the ground and first excited states of ^{239}Pu take place in only 1.5 and 4.7 percent of the disintegrations with α-particle kinetic energies of 6.065 and 6.056 MeV, respectively. Estimate the hindrance factors for these two transitions, assuming the transition to the 285-keV state to be unhindered. *Answer:* 1.5×10^3; 4×10^2.

7. From figures 3-2, 3-4, and 3-6, predict (in order-of-magnitude fashion) the partial half lives for α decay and spontaneous fission of ^{246}Pu.

8. Show that the production of a positron-electron pair by a photon in vacuum is impossible. (*Note.* Set up the conditions for momentum and energy conservation, using relativistic expressions, and show that they lead to a contradiction, for example, to the inequality $\cos\theta > 1$, where θ is the angle between the directions of motion of positron and electron.)

9. Using (2-5), verify the statement on p. 68 that ^{67}Kr may be a candidate for decay by two-proton emission. What is the decay energy estimated for this process (admittedly not a very reliable quantitative result)?

10. Calculate the energy release for spontaneous fission of ^{252}Cf (a) into two equal fragments (^{125}In) and two neutrons, (b) into ^{140}Xe and ^{110}Ru and two neutrons. (c) Which of these two fission modes will have the lower Coulomb barrier? *Answer:* (a) 221 MeV.

11. With the aid of data in appendix D, calculate approximate values of $\log f_0 t$ for (a) the β^+ decay of ^{37}K, (b) the β^- decay of ^{45}Ca, (c) the EC decay of ^{41}Ca. Note that no γ emission accompanies any of these decays. Identify the most likely transition type for each of the transitions. *Answers:* (a) 3.6; (c) 10.7.

12. (a) Derive an expression for the average total energy of electrons in an allowed β spectrum with maximum total energy W_0, neglecting the Coulomb correction and assuming that the nuclear matrix elements are independent of the electron energy. (b) Compute, under the assumptions of part (a), the ratios of average to maximum kinetic energies for β spectra with maximum kinetic energies of 0.51 and 1.53 MeV. (c) Qualitatively, how would the results of part

(b) for β^- emission be affected by inclusion of the Coulomb correction?
Answers:

(a) $$\frac{W_0(W_0^2-1)^{1/2}(W_0^4-2W_0^2+12.25)-7.5(W_0^2+0.5)\ln[W_0+(W_0^2-1)^{1/2}]}{(W_0^2-1)^{1/2}(2W_0^4-9W_0^2-8)+15W_0\ln[W_0+(W_0^2-1)^{1/2}]},$$

(b) 0.37; 0.41.

13. Recalling that a particle of charge e emu and momentum p g cm s^{-1} in a magnetic field of H gauss (G) moves in a circle of radius $\rho = p/He$ cm, derive an expression for the magnetic rigidity $H\rho$ (in G cm) of an electron in terms of its kinetic energy T (in MeV). [*Note.* The required relativistic relations are given in appendix B.] *Answer*: $H\rho = 3.3356 \times 10^3 \, (T^2 + 1.022T)^{1/2}$.

14. The following data were obtained in a measurement of the ^{3}H β spectrum with a magnetic spectrometer:

$H\rho$ (G cm)	150	200	250	300	350	400	425	450	460
Intensity (counts/min)	4995	6812	7645	7274	5336	2434	1040	150	85

Using the expression derived in exercise 13, construct a Kurie plot from these data, taking the Coulomb correction as constant throughout the spectrum. What endpoint energy do you find for the β spectrum? Comment on the probable causes for any deviation from a straight-line Kurie plot that you may find.

15. The ground state of ^{61}Ni has spin $I = \frac{3}{2}$ and negative parity. ^{61}Co decays with a half life of 1.65 h and with the emission of a β^- spectrum with endpoint energy of 1.24 MeV to an excited state of ^{61}Ni 0.067 MeV above the ground state. The 0.067-MeV transition has a K-conversion coefficient of about 0.10 and a K/L conversion ratio of about 8. The β^- transition of ^{61}Co to the ^{61}Ni ground state takes place in less than 10^{-6} of the disintegrations. What are the most likely spin and parity assignments for (a) ^{61}Co, (b) the 0.067-MeV state of ^{61}Ni? Give all your reasoning. *Answers:* (a) $7/2^-$, (b) $5/2^-$.

16. The nuclide ^{54}Mn ($t_{1/2} = 312$ d, $I = 3$) decays by EC to the first excited state of ^{54}Cr (2^+, 0.835 MeV above ground). (a) From the masses in appendix D determine the energy of this EC transition. (b) Calculate its approximate ft value. (c) From the information you now have can you deduce the parity of ^{54}Mn? (d) What would be the endpoint energy of the β^+ spectrum corresponding to the decay of ^{54}Mn to the ^{54}Cr ground state? (e) Estimate an upper limit for the fraction of the ^{54}Mn decays that might go by this ground-state transition.

17. The 120-d isomer ^{123}Te^m decays by an 88-keV isomeric transition to the first excited state, which in turn de-excites to the ground state with emission of a 159-keV γ ray with half life 0.20 ns. The following conversion coefficient data have been measured for the two transitions:

88 keV:	$\alpha_K = \sim 450$	$\alpha_K/\alpha_L = 0.94,$
159 keV:	$\alpha_K = 0.17$	$\alpha_K/\alpha_L = 7.6.$

On the basis of these data select the most likely transition types for the two steps. Check the partial γ half lives for the two transitions against the approximate formulas in table 3-4.

18. The first excited state of ^{194}Pt is 328 keV above the ground state. The 328-keV

transition has a K-conversion coefficient of 0.050 and a K/L-conversion ratio of 2.2. (a) Deduce the transition type, hence the spin and parity of the 328-keV state. (b) The half life of this state has been measured as 4.5×10^{-11} s. Compare this result with the half life calculated from the single-particle model. (c) What is the natural width of the 0.328-keV state (in eV)?

19. Show how the special rules stated in the footnotes to table 3-3 follow from the general selection rule (3-28).

20. The nuclide ^{179}Ta ($t_{1/2} = 1.8$ y) decays by EC to the ground state of ^{179}Hf. From measurements of K and L X-ray intensities a value of 0.56 ± 0.13 has been deduced for the ratio of L_I capture to K capture. Estimate the energy difference between the ground states of ^{179}Ta and ^{179}Hf and the uncertainty in that quantity. The binding energies of the K and L_I electrons in tantalum are 67.4 and 11.7 keV, respectively. Compare your answer with the mass difference as found from appendix D. *Answer:* 123^{+19}_{-10} keV.

21. The energy levels of three neighboring isobars of mass number A are as follows (spins and parities to the left, energies in keV above the ^{A}Z ground state to the right of each level):

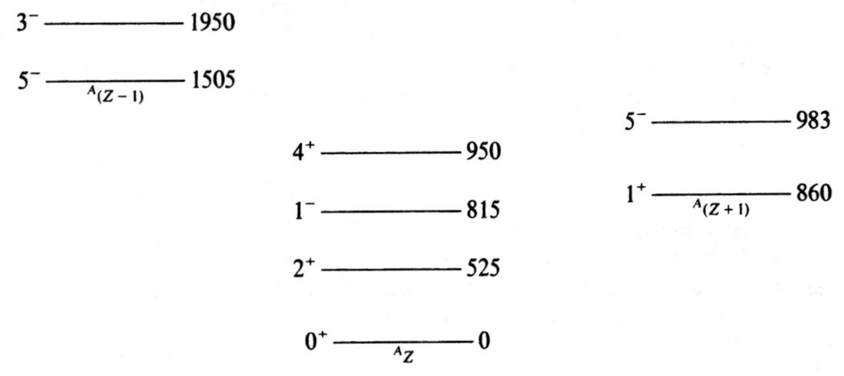

What radiations of what energies are expected to be found in the decay of the ground states (a) of $^{A}(Z-1)$, (b) of $^{A}(Z+1)$? (c) Which, if any, of the excited states shown are expected to have half lives > 1 s? (d) For the state(s) selected in part (c), estimate the order of magnitude of half life, assuming $A \approx 100$. State your reasoning in parts (a)–(c).

22. Derive the expressions for the mass balance for β^-, β^+, and EC decay processes on p. 74 by considering the masses of nuclei and electrons.

Chapter 4

Nuclear Reactions

A nuclear reaction is a process in which a nucleus reacts with another nucleus, an elementary particle, or a photon to produce, within a time of the order of 10^{-12} s or less, one or more other nuclei, and possibly other particles. The variety of reactions that have been studied is bewildering, as may be readily perceived from a consideration of the variety of bombarding particles available. They include neutrons, protons, photons, electrons, various mesons, and nuclei all the way from deuterons to uranium nuclei. The constantly advancing technology of particle accelerators (see chapter 15) has provided many of these projectiles over wide ranges of kinetic energies, for example, protons up to 500 GeV,[1] many other nuclei up to thousands of MeV per nucleon.

The phenomenon of nuclear reactions was discovered by Rutherford in 1919 when he observed that, in the bombardment of nitrogen with the 7.69-MeV α particles of RaC′ (^{214}Po), scintillations of a zinc sulfide screen persisted even when enough material to absorb all the α particles was interposed between the nitrogen and the screen. Further experiments proved the long-range particles causing the scintillations to be protons, and the results were interpreted in terms of a nuclear reaction between α particles and nitrogen to give oxygen and protons.

A. ENERGETICS

Notation. The notation used for nuclear reactions is analogous to that for chemical reactions, with the reactants on the left- and the products on the right-hand side of the equation. Thus Rutherford's first reaction may be written

$$^{14}_{7}\text{N} + ^{4}_{2}\text{He} \rightarrow ^{17}_{8}\text{O} + ^{1}_{1}\text{H}.$$

In all reactions so far observed (except those involving creation or annihilation of antinucleons) the total number of nucleons (total A) is conserved. Also conserved in nuclear reactions are charge, energy, momentum, angular momentum, statistics, and parity.

The recognition of reaction products is greatly facilitated when they are unstable, because characteristic radioactive radiations can then be obser-

[1] One GeV (gigaelectron volt) = 1000 MeV.

ved. The discovery of artificial radioactivity by Joliot and Curie thus gave enormous impetus to the field of nuclear reaction studies. The first artificially produced radionuclide observed and radiochemically characterized by them, was ^{30}P, made in the reaction

$$^{27}_{13}Al + ^{4}_{2}He \rightarrow ^{30}_{15}P + ^{1}_{0}n.$$

A short-hand notation is often used for the representation of nuclear reactions. The light bombarding particle and the light fragments (in that order) are written in parentheses between the initial and final nucleus; in this notation the two reactions mentioned above would read

$$^{14}N(\alpha, p)\,^{17}O \qquad \text{and} \qquad ^{27}Al\,(\alpha, n)\,^{30}P.$$

As indicated here, atomic numbers are commonly omitted. The symbols n, p, d, t, α, e^-, γ, π, and $\bar{p}$ are used in this notation to represent neutron, proton, deuteron, triton (^3H), alpha particle, electron, gamma ray, pi meson, and antiproton, respectively. Nuclei other than the ones mentioned are represented by their usual symbols, such as ^3He, ^{12}C, and so on, even when they are projectiles. Thus we write

$$^{139}La\,(^{12}C, 4n)\,^{147}Eu$$

for a reaction in which the bombardment of ^{139}La with ^{12}C ions results in the formation of ^{147}Eu.

Comparison of Nuclear and Chemical Reactions. Nuclear reactions, like chemical reactions, are always accompanied by a release or absorption of energy, and this is expressed by adding the term Q to the *right-hand* side of the equation. Thus a more complete statement of Rutherford's first transmutation reaction reads

$$^{14}_{7}N + ^{4}_{2}He \rightarrow ^{17}_{8}O + ^{1}_{1}H + Q.$$

The quantity Q is called the energy of the reaction or more frequently just "the Q of the reaction." Positive Q corresponds to energy release (exoergic reaction); negative Q to energy absorption (endoergic reaction).

Here an important difference between chemical and nuclear reactions must be pointed out. In treating chemical reactions we always consider macroscopic amounts of material undergoing reactions, and consequently heats of reaction are usually given per mole or occasionally per gram of one of the reactants. In the case of nuclear reactions we usually consider single processes, and the Q values are therefore given *per nucleus transformed.* If the two are calculated on the same basis, the energy release in a representative nuclear reaction is found to be many orders of magnitude larger than that in any chemical reaction. For example, the reaction $^{14}N(\alpha, p)^{17}O$ has a Q value of -1.193 MeV or -1.912×10^{-6} erg or -4.57×10^{-14} cal *per* ^{14}N *atom* transformed. To convert 1 g atom of ^{14}N to ^{17}O would thus require an energy of $6.02 \times 10^{23} \times 4.57 \times 10^{-14}$ cal =

2.75×10^{10} cal. This is about 10^5 times as large as the largest values observed for heats of chemical reactions.

Q Values and Reaction Thresholds. The energy changes in nuclear reactions are so large that the corresponding mass changes are quite significant and observable (in contrast to the situation in chemical reactions). If the masses of all the particles participating in a nuclear reaction are known from mass-spectrographic data, as is the case for the $^{14}N(\alpha, p)^{17}O$ reaction, the Q of the reaction can be calculated. The sum of the ^{14}N and ^{4}He masses is 18.0056777 mass units, and the sum of the ^{17}O and ^{1}H masses is 18.0069585 mass units; thus an amount of energy equivalent to 0.0012808 mass unit has to be supplied to make the reaction energetically possible, or $Q = -0.0012808 \times 931.5$ MeV $= -1.193$ MeV.

Conversely, when the Q value is known experimentally (from the kinetic energies of the bombarding particle and the reaction products) it is possible to calculate the unknown mass of one of the participating nuclei from the known masses of others. By this method the masses of many radioactive nuclei have been determined (see exercise 3).

Sometimes the Q value of a reaction can be calculated even if the masses of the nuclei involved are not known, provided that the product nucleus is radioactive and decays back to the initial nucleus with known decay energy.

Consider, for example, the reaction $^{106}Pd(n, p)^{106}Rh$. The product ^{106}Rh decays with a 30-s half life and the emission of β^- particles of 3.54-MeV maximum energy to the ground state of ^{106}Pd. We can write this sequence of events as follows:

$$^{106}_{46}\mathrm{Pd} + ^{1}_{0}n \rightarrow ^{106}_{45}\mathrm{Rh} + ^{1}_{1}\mathrm{H} + Q;$$

$$^{106}_{45}\mathrm{Rh} \rightarrow ^{106}_{46}\mathrm{Pd} + \beta^- + \bar{\nu} + 3.54\ \mathrm{MeV}.$$

Adding the two equations, we see that the net change is just the transformation of a neutron into a proton, an electron, and an antineutrino, with accompanying energy change, or, symbolically,

$$^{1}_{0}n \rightarrow ^{1}_{1}\mathrm{H} + \beta^- + \bar{\nu} + Q + 3.54\ \mathrm{MeV}.$$

Note that the symbol $^{1}_{1}H$ must here stand for a bare proton (evident from the charge conservation), whereas the listed "proton mass" includes the mass of one orbital electron.[2] For energy balance we therefore write

$$M_n = M_{\mathrm{H}} + Q + 3.54\ \mathrm{MeV},$$

where $M_n = 1.008665$ and $M_{\mathrm{H}} = 1.007825$ mass units. Then $Q = (1.008665 - 1.007825) \times 931.5 - 3.54 = -2.76$ MeV.

[2] In general, for β^- emission and EC the masses of electrons do not have to be included in calculations when atomic masses are used. However, whenever emission of a positron is involved two electron masses have to be taken into account: one for the positron and one for the extra electron that has to leave the electron shells to preserve electrical neutrality.

In the first example calculated we found the Q value of the reaction $^{14}N(\alpha, p)^{17}O$ to be -1.19 MeV. Does that mean that this reaction can actually be produced by α particles whose kinetic energies are just over 1.19 MeV? The answer is no, for two reasons. First, in the collision between the α particle and the ^{14}N nucleus conservation of momentum requires that at least $\frac{4}{18}$ of the kinetic energy of the α particle must be retained by the products as kinetic energy. Thus only $\frac{14}{18}$ of the α particle's kinetic energy is available for the reaction. The *threshold* energy of α particles for the $^{14}N(\alpha, p)^{17}O$ reaction, that is, the kinetic energy of α particles just capable of making the reaction energetically *possible*, is $\frac{18}{14} \times 1.19\ \text{MeV} = 1.53\ \text{MeV}$. The fraction of the bombarding particle's kinetic energy that is retained as kinetic energy of the products becomes smaller with increasing mass of the target nucleus (see exercise 5).

Barriers for Charged Particles. The second reason why the α particles must have higher energies than is evident from the Q value to produce the reaction $^{14}N(\alpha, p)^{17}O$ in *good yield* is the Coulomb repulsion between the α particle and the ^{14}N nucleus. The repulsion increases with decreasing distance of separation until the α particle comes within the range of the nuclear forces of the ^{14}N nucleus. This Coulomb repulsion gives rise to the potential barrier already discussed in connection with nuclear radii in chapter 2. The height V_c of the potential barrier around a spherical nucleus of charge Z_1e and radius R_1 for a particle of positive charge Z_2e and radius R_2 may be estimated as the energy of Coulomb repulsion when the two particles are just in contact (just as in the discussion of spontaneous fission, chapter 3 section C):

$$V_c = \frac{Z_1 Z_2 e^2}{(R_1 + R_2)} \tag{4-1}$$

If R_1 and R_2 are expressed in fermis

$$V_c = 1.44 \frac{Z_1 Z_2}{R_1 + R_2}\ \text{MeV}. \tag{4-2}$$

Setting $R = 1.5A^{1/3}$ fm, we get from (4-2) a value of about 3.4 MeV for the barrier height between ^{14}N and ^{4}He.[3] Classically an α particle must thus have at least $\frac{18}{14} \times 3.4 = 4.4$ MeV kinetic energy to enter a ^{14}N nucleus and produce the α, p reaction, even though the energetic threshold for the reaction is only 1.53 MeV. In the quantum-mechanical treatment of the problem there exists a finite probability for "tunneling through the barrier"

[3] We should keep in mind that the use of (4-1) and (4-2) is equivalent to assuming spherical nuclei and square-well potentials. Thus these equations serve to give only rough estimates of barrier heights, but they are quite useful for that purpose. For more sophisticated calculations of barriers we would use Woods-Saxon shapes (see chapter 2, section C, 2) for the nuclear potentials.

by lower-energy particles, but this probability drops rapidly as the energy of the particle decreases, as we saw in the discussion of α decay in chapter 3.

It follows from (4-1) that the Coulomb barrier around a given nucleus is about half as high for protons and for deuterons as it is for α particles.[4] The height of the barrier is roughly proportional to $Z^{2/3}$ (because the nuclear radius R increases approximately as $Z^{1/3}$). For the heaviest elements the potential barriers are about 12 MeV for protons and deuterons and about 25 MeV for α particles. In order to study nuclear reactions induced by charged particles, especially reactions involving heavy elements, it was therefore necessary to develop machines capable of accelerating charged particles to energies of many millions of electron volts.

In the context of nuclear reactions we emphasize again a point made already in connection with α decay, namely that Coulomb barriers affect particles not only on entering but also on leaving nuclei. For this reason a charged particle has to be excited to a rather high energy inside the nucleus before it can leak through the barrier with appreciable probability. Therefore charged particles emitted from nuclei have considerable kinetic energies (>1 MeV).

Neutrons. It is apparent that the entry of a neutron into a nucleus is not opposed by any Coulomb barrier, and even neutrons of very low energy react readily with even the heaviest nuclei. In fact, the so-called **thermal neutrons**, that is, neutrons whose energy distribution is approximately that of gas molecules in thermal equilibrium at ordinary temperatures,[5] have particularly high probabilities for reaction with target nuclei. This important effect was discovered at the University of Rome by Fermi and co-workers in 1934 in experiments on the neutron irradiation of silver; they found that the neutron-induced radioactivity was much greater when a bulk of hydrogen-containing material such as paraffin was present to modify the neutron beam. Fermi reasoned correctly that fast neutrons would lose energy in collisions with protons, that repeated collisions might reduce the energy to the thermal range, and that such slow neutrons could show large capture cross sections. Other workers found the effect to be sensitive to the temperature of the paraffin, thus demonstrating that the neutrons were actually slowed to approximately thermal energies.

[4] In estimating proton barriers of medium and heavy nuclei one usually considers the proton as a point charge ($R_2 = 0$).

[5] The energies of thermal neutrons are small fractions of an electron volt (at 20°C the most probable energy is 0.025 eV). Neutrons of somewhat higher energies (up to about 1 keV) are often called **epithermal** or **resonance neutrons**. Neutrons with kinetic energies of several thousand electron volts or more are called **fast neutrons**. The slowing down of fast neutrons is treated in chapter 6, section D.

B. CROSS SECTIONS

Definitions, Units, and Examples. We now turn to a more quantitative consideration of reaction probabilities. The probability of a nuclear process is generally expressed in terms of a cross section σ that has the dimensions of an area. This originates from the simple picture that the probability for the reaction between a nucleus and an impinging particle is proportional to the cross-sectional target area presented by the nucleus. Although this classical picture does not hold for reactions with charged particles that have to overcome Coulomb barriers or for slow neutrons (it does hold fairly well for the total probability of a fast neutron interacting with a nucleus), the cross section is a useful measure of the probability for any nuclear reaction. For a beam of particles striking a thin target, that is, a target in which the beam is attenuated only infinitesimally, the cross section for a particular process is defined by the equation

$$\mathbf{R}_i = \mathbf{I} n x \sigma_i, \tag{4-3}$$

where $\mathbf{R}_i$ is the number of processes of the type under consideration occurring in the target per unit time,
- $\mathbf{I}$ is the number of incident particles per unit time,
- n is the number of target nuclei per cubic centimeter of target,
- σ_i is the cross section for the specified process, expressed in square centimeters, and
- x is the target thickness in centimeters.

The target thickness is often given in terms of weight per unit area, which can be readily converted to nx, the number of target nuclei per square centimeter.

The total cross section for collision with a fast particle is never greater than twice[6] the geometrical cross-sectional area of the nucleus, and therefore fast-particle cross sections are rarely much larger than 10^{-24} cm^2 (radii of the heaviest nuclei are about 10^{-12} cm). Hence a cross section of 10^{-24} cm^2 is considered "as big as a barn," and 10^{-24} cm^2 has been named the **barn,** a unit generally used in expressing cross sections and often abbreviated b. The millibarn (mb, 10^{-3} b), microbarn (μb, 10^{-6} b), and nanobarn (nb, 10^{-9} b) are also commonly used.

As an example of the application of (4-3), consider a 1-h bombardment of a foil of metallic manganese, 10 mg/cm^2 thick, in a 1-μA beam of 35-MeV α particles. If the cross section of the $(\alpha, 2n)$ reaction on ^{55}Mn at this energy is 200 mb and if energy degradation of the beam in traversing the target can be neglected, how many ^{57}Co nuclei ($t_{1/2} = 270$ d) will be formed? First remember

[6] The reason why total cross sections may be as large as $2\pi R^2$ is briefly mentioned in footnote 12 on p. 31.

that 1 A = 6.2×10^{18} electronic charges per second so that 1 μA of (doubly charged) α particles is 3.1×10^{12} α particles per second. The number of $(\alpha, 2n)$ reactions is, from (4-3), $3.1 \times 10^{12} \times (0.01/55) \times 6.02 \times 10^{23} \times 200 \times 10^{-27} = 6.8 \times 10^7$. Neglecting decay during the 1-h irradiation, we get for the number of ^{57}Co nuclei formed $3600 \times 6.8 \times 10^7 = 2.4 \times 10^{11}$. The ^{57}Co disintegration rate at the end of the irradiation, from $dN/dt = \lambda N$, would be $[0.693/(270 \times 24 \times 60)] \times 2.4 \times 10^{11} = 4.3 \times 10^5\ \text{min}^{-1}$.

Equation 4-3 applies when there is a well-defined beam of particles incident on a target. Another important situation concerns a sample embedded in a uniform flux of particles incident on it from all directions. This is what happens, for example, in a nuclear reactor. It can be shown that, for a sample containing N nuclei in a flux of ϕ particles per square centimeter per second, the rate of reactions of type i, which have a cross section σ_i, is given by

$$\mathbf{R}_i = \phi N \sigma_i. \tag{4-4}$$

This applies, regardless of the shape of the sample, provided that the particle flux is not appreciably attenuated by sample absorption anywhere in the sample.

As an example, we calculate how long a 60-mg piece of Co wire has to be placed in a flux of 5×10^{13} thermal neutrons per square centimeter per second to make 1 mCi (1 millicurie = 3.7×10^7 dis s^{-1}; see chapter 1, section B) of 5.27-y ^{60}Co. The cross section for the reaction ^{59}Co (n, γ) ^{60}Co is 37 b. From (4-4) we have

$$\mathbf{R} = 5 \times 10^{13} \times \frac{0.060}{59} \times 6.02 \times 10^{23} \times 37 \times 10^{-24} = 1.13 \times 10^{12}\ \text{atoms s}^{-1}.$$

From $dN/dt = \lambda N$ we find that 1 mCi of ^{60}Co corresponds to 8.87×10^{15} atoms. Thus it will take $8.87 \times 10^{15}/1.13 \times 10^{12} = 7.85 \times 10^3$ s, or approximately 2.2 h, to produce 1 mCi of ^{60}Co.

Beam Attenuation Measurements. If instead of a thin target we consider a thick target, that is, one in which the intensity of the incident particle beam is attenuated, the attenuation $-d\mathbf{I}$ in the infinitesimal thickness dx is given by the equation

$$-d\mathbf{I} = \mathbf{I} n \sigma_t\, dx,$$

where σ_t is the total cross section for removal of the incident particles from the beam. Integration gives

$$\mathbf{I} = \mathbf{I}_0 e^{-n\sigma_t x}. \tag{4-5}$$

Just what processes are included in σ_t depends considerably on the particular experimental arrangement, especially on the energy selectivity of the detector used to measure the transmitted beam and on the solid angle it subtends. Thus for example, the cross section for small-angle elastic scattering may or may not be included in σ_t.

Beam attenuation measurements, of course, measure always the effect of the entire target substance, whether it is a single nuclide, an isotopic mixture, or even a compound.

Partial Cross Sections. As we emphasized in the preceding paragraph, beam attenuation or transmission experiments can be used only to determine total interaction cross sections, and (4-5) is not applicable to cross sections for specific reactions that constitute only part of the total interaction. Yet it is usually cross sections for particular processes on elementary, or even isotopically pure, substances that are of interest, such as the (n, p) reaction on ^{35}Cl or the $(\alpha, 3n)$ reaction on ^{65}Cu. Thin-target experiments are then needed so that (4-3) or (4-4) is applicable. The requirements for target thickness are particularly stringent if the cross section of interest varies rapidly with bombarding energy, as is the case for most medium-energy charged-particle reactions; the target then must be thin enough to avoid not only intensity attenuation but also appreciable energy degradation.

Sometimes the angular distribution of particles resulting from a particular process is of interest. In this case it is convenient to define a differential cross section $d\sigma/d\Omega$; this is the cross section for that part of the process in which the particles are emitted into unit solid angle at a particular angle Ω. Then the cross section for the overall process under consideration is $\sigma = \int (d\sigma/d\Omega)\, d\Omega$.

Elastic Scattering. The simplest consequence of a nuclear collision is so-called elastic scattering; this is a process that can occur at all energies and with all particles and that is not properly a reaction at all. An event is termed an elastic scattering if the particles do not change their identity during the process and if the sum of their kinetic energies (ignoring molecular and atomic excitations and bremsstrahlung) remains constant. Elastic scattering of charged particles with energies below the Coulomb barrier of the target nucleus is the Rutherford scattering described in chapter 2. As the energy of the bombarding particle is increased, the particle may penetrate the Coulomb barrier to the surface of the target nucleus, and the elastic scattering will then also have a contribution from the nuclear forces. For neutrons, of course, elastic scattering is caused by nuclear forces at all energies.

Elastic scattering may generally be considered to arise from the optical-model potential discussed in section D. With neutrons of very low energies there is also a significant contribution from so-called compound elastic scattering, since the compound nucleus formed by the amalgamation of such a neutron with the target nucleus (see section D) has a small but finite probability of emitting a neutron with all its original energy. For all other particles compound elastic scattering is negligible.

We designate the cross section for all events other than (potential) elastic scattering as the *reaction cross section.* Compound elastic scattering is formally included in the reaction cross section, although it cannot be distinguished experimentally from other elastic scattering.

Maximum Reaction Cross Sections for Neutrons. It might be expected that a nucleus that interacts with everything that hits it would have a reaction cross section of πR^2, where R is the sum of the radii of the interacting particles. As we see, this is correct at high energies only, because the wave nature of the incident particle causes the upper limit of the reaction cross section to be

$$\sigma_r = \pi (R + \lambda\!\!\!^{-})^2,$$

where $\lambda\!\!\!^{-}$ is the reduced de Broglie wavelength $(\lambda/2\pi)$ of the incident particle in the center-of-mass system and may be obtained from $\lambda\!\!\!^{-} = \hbar/p$. Here p is the *relative* momentum of the two particles computed from (C-6) in appendix C.

Although cross section limits are properly derived by quantum-mechanical methods (see, e.g., B1, chapter 8), we give a semiclassical treatment that shows the essence of the problem and points up the important role played by angular-momentum considerations. We first treat reactions with incident neutrons and then proceed to discuss the additional effects of Coulomb repulsion.

Angular Momentum in Nuclear Reactions. A collision between a neutron and a target nucleus may be characterized classically by what would be the distance of closest approach of the two particles if there were no interaction between them. This distance b, usually called the **impact parameter**, is shown in figure 4-1. The angular momentum of the system is normal to the relative momentum p and of magnitude

$$L = pb. \tag{4-6}$$

The de Broglie relation between momentum and wavelength of a particle

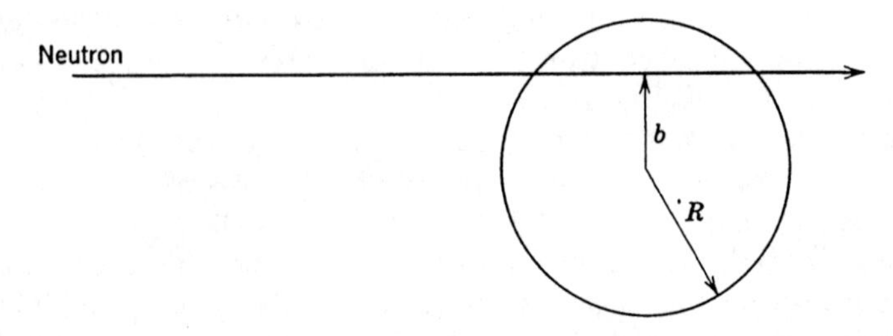

Fig. 4-1 Collision with impact parameter b between a neutron and target nucleus with interaction radius R.

allows (4-6) to be rewritten as

$$L = \frac{\hbar b}{\lambda\!\!\!^{-}}. \tag{4-7}$$

Note that the entire treatment is in the center-of-mass system and that $\lambda\!\!\!^{-}$ is thus the reduced wavelength in that system. See appendix C for transformations between laboratory and center-of-mass systems.

As b may evidently assume any value between 0 and R, the relative angular momentum will vary continuously between 0 and $\hbar R/\lambda\!\!\!^{-}$. We know, though, that this is not acceptable; quantum mechanics requires that the component of angular momentum in a particular direction be an integer when expressed in units of $\hbar$:

$$L = l\hbar, \quad \text{where} \quad l = 0, 1, 2, \ldots. \tag{4-8}$$

Combination of (4-7) and (4-8) gives

$$b = l\lambda\!\!\!^{-}. \tag{4-9}$$

Equation 4-9 is not to be interpreted as meaning that only certain values of b are possible; such control over b would violate the uncertainty principle. Rather it means that a range of values of b corresponds to the same value of the angular momentum. In particular,

$$l\lambda\!\!\!^{-} < b < (l+1)\,\lambda\!\!\!^{-} \tag{4-10}$$

corresponds to an angular momentum of $l\hbar$. This interpretation is illustrated in figure 4-2. From this figure it can be seen that the cross-sectional area that corresponds to a collision with angular momentum $l\hbar$ is

$$\begin{aligned}\sigma_l &= \pi\lambda\!\!\!^{-2}[(l+1)^2 - l^2] \\ &= \pi\lambda\!\!\!^{-2}(2l+1).\end{aligned} \tag{4-11}$$

If it is assumed that each particle hitting the nucleus causes a reaction, then

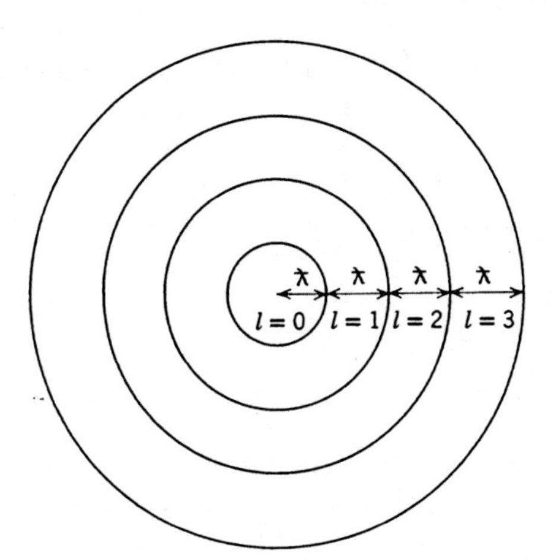

Fig. 4-2 The incident beam is perpendicular to the plane of the figure. The particles with a particular l are considered to strike within the designated ring.

(4-11) gives the partial cross section for a nuclear reaction characterized by angular momentum $l\hbar$, and the reaction cross section may be obtained by summing (4-11) over all values of l from 0 to the maximum l_m:

$$\sigma_r = \pi \lambda\!\!\!^{-2} \sum_{0}^{l_m} (2l + 1). \tag{4-12}$$

The summation in (4-12) may be easily evaluated if it is recalled that the sum of the first N integers is equal to $[N(N+1)]/2$. The expression for the reaction cross section becomes

$$\sigma_r = \pi \lambda\!\!\!^{-2} (l_m + 1)^2. \tag{4-13}$$

The maximum value of l may be estimated from (4-9) by limiting the maximum impact parameter to the interaction radius R:

$$l_m = \frac{R}{\lambda\!\!\!^{-}}. \tag{4-14}$$

Substitution of (4-14) into (4-13) yields the result already given on p. 118 for the maximum possible reaction cross section:

$$\sigma_r = \pi (R + \lambda\!\!\!^{-})^2. \tag{4-15}$$

This result suggests the possibility of nuclear-reaction cross sections that are several orders of magnitude larger than the geometrical cross section of the nucleus, a possibility that is realized in slow-neutron reactions. The largest thermal-neutron cross section known is that of ^{135}Xe, 2.65×10^6 b. (See also section E.)

In the quantum-mechanical treatment of the problem (B1) the result for the total reaction cross section is not (4-12), but

$$\sigma_r = \pi \lambda\!\!\!^{-2} \sum_{l=0}^{\infty} (2l + 1) T_l, \tag{4-16}$$

where T_l is defined as the **transmission coefficient** for the reaction of a neutron with angular momentum l and may have values between zero and one; it represents the fraction of incident particles with angular momentum l that penetrate within the range of nuclear forces. Our semiclassical treatment assigns unity to T_l for all values of l up to and including l_m, as defined in (4-14); for all higher values of l, the transmission coefficients are zero. The role of angular momentum here is analogous to the one that it plays in β and γ emission, discussed in chapter 3. It should be mentioned here that the semiclassical result is quite right for $R/\lambda\!\!\!^{-} < 1$, where the only contribution comes from $l = 0$ and the reaction cross section has $\pi \lambda\!\!\!^{-2}$ as its upper limit.

Centrifugal Barrier. Expression 4-14 for the maximum l value can be reinterpreted to mean that a particle that approaches a nucleus with relative angular momentum l must have a reduced de Broglie wavelength

$\lambdabar \leq R/l$. Since $\epsilon = p^2/2\mu = \hbar^2/2\mu\lambdabar^2$, where ϵ is the relative kinetic energy, p the relative momentum, and μ the reduced mass of the system, we have the condition

$$\epsilon \geq \frac{l^2\hbar^2}{2\mu R^2}, \tag{4-17}$$

where R may be taken as the sum of the radii of projectile particle and target nucleus. Condition 4-17 implies that, quite apart from any Coulomb barrier, there is for particles of angular momentum l an additional barrier, called the centrifugal barrier. In the quantum-mechanical treatment of the problem l^2 is replaced by $l(l+1)$, so that the proper expression for the centrifugal barrier or centrifugal potential is

$$V_l = \frac{l(l+1)\hbar^2}{2\mu R^2}. \tag{4-18}$$

Reaction Cross Sections with Charged Particles. The effect of the Coulomb repulsion on a reaction cross section may be easily estimated within the spirit of the semiclassical analysis. The Coulomb repulsion will bring the relative kinetic energy of the system from ϵ when the particles are very far apart to $\epsilon - V_c$ when the two particles are just touching, where V_c is the Coulomb barrier:

$$V_c = \frac{Z_a Z_A e^2}{R}, \tag{4-19}$$

and where Z_a and Z_A are the atomic numbers of incident particle and target nucleus, respectively. Further, the deflection of the particles causes the maximum impact parameter that leads to a reaction to be less than R, as illustrated in figure 4-3. From this figure it is seen that the trajectory of the particle is tangential to the nuclear surface when it approaches with the maximum impact parameter b_m and that the relative momentum at the

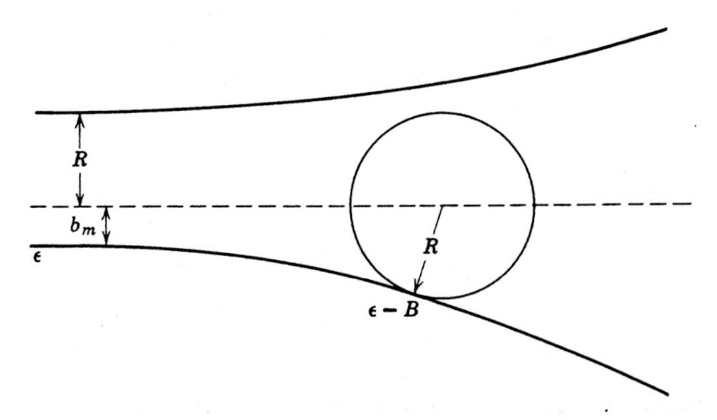

Fig. 4-3 Classical trajectories for charged particles with impact parameters R and b_m.

point of contact is

$$p = (2\mu)^{1/2}(\epsilon - V_c)^{1/2} = (2\mu\epsilon)^{1/2}\left(1 - \frac{V_c}{\epsilon}\right)^{1/2}, \qquad (4\text{-}20)$$

where μ is the reduced mass of the system. The magnitude of the maximum angular momentum is then obtained from the product of the interaction radius and the relative momentum:

$$L_m = R\,(2\mu\epsilon)^{1/2}\left(1 - \frac{V_c}{\epsilon}\right)^{1/2}. \qquad (4\text{-}21)$$

Recognizing that $(2\mu\epsilon)^{1/2}$ is the relative momentum of the two particles when they are far apart, we obtain from (4-21) in conjunction with (4-6)

$$b_m = R\left(1 - \frac{V_c}{\epsilon}\right)^{1/2} \qquad (4\text{-}22)$$

for the maximum impact parameter. Equation 4-22 has meaning only for $\epsilon \geqslant V_c$; for lower ϵ the Coulomb potential, classically, prevents nuclear reactions. Thus the Coulomb barrier diminishes the l_m of (4-14) by a factor of $(1 - V_c/\epsilon)^{1/2}$. The upper limit for the capture of charged particles can be estimated as the area of the disk of radius b_m:

$$\sigma_r = \pi R^2\left(1 - \frac{V_c}{\epsilon}\right). \qquad (4\text{-}23)$$

It is to be noted that the method of estimating the upper limit to the reaction cross section for charged particles is different from that for neutrons, where the value of l_m was found and substituted directly into (4-13). This procedure would not be appropriate for charged particles, as the Coulomb barrier has an important effect on the transmission coefficients of (4-16). In particular, it may be seen from (4-21) that $l_m \to 0$ as $\epsilon \to V_c$ for charged particles and from (4-14) that $l_m \to 0$ as $\epsilon \to 0$ for neutrons. However, the Coulomb barrier causes the transmission coefficient for charged particles to approach zero under these circumstances, whereas that for the neutron remains finite. The result is a vanishing cross section for charged particles of energies approaching that of the Coulomb barrier to be compared with the upper limit of $\pi(R + \lambda\!\!\!^{-})^2$ given in (4-15) for neutrons of very low energy. Further, since the Coulomb barrier is the most important factor in determining reaction cross sections with charged particles, (4-23) is an estimate of the reaction cross section rather than just its upper limit.

Equation 4-23, though approximate, has been useful for the estimation of reaction cross sections for charged particles, particularly when the Coulomb barrier is changed to an "effective" Coulomb barrier (see, e.g., D1) to allow for tunneling through the diffuse nuclear surface. Again, the complete analysis of the reaction cross section is properly carried out with (4-16), and the effect of the Coulomb interaction appears in the transmission coefficients (S1, H1).

C. TYPES OF EXPERIMENTS

Nuclear reactions are studied in a variety of ways. Among the important types of experimental information we usually wish to obtain are the reaction cross section, in particular its variation with incident energy, and the energy spectra and angular distributions of the reaction products. The types of experiments performed to obtain these data and the kind of information deduced from them are sketched in the following paragraphs.[7]

Excitation Functions. Frequently the variation of a particular reaction cross section with incident energy is of interest; the relation between the two is called an excitation function. Examples of excitation functions are shown in figures 4-4 and 4-5. Variable-energy beams are obtainable from various types of accelerators (see chapter 15). If only a fixed energy source is available, energy degradation by absorption is resorted to, with resulting spread in beam energy (see chapter 6). Methods for determining beam energies and beam intensities, both essential for accurate, absolute excitation function measurements, are discussed in chapter 15, section D. The determination of an absolute cross section further requires measurement of the number of reactions in the target, usually via determination of the

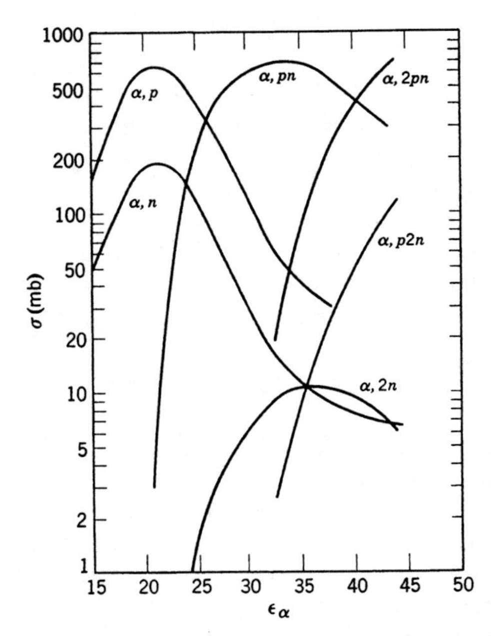

Fig. 4-4 Excitation functions for various reactions between α particles and ^{54}Fe nuclei. The abscissa is the kinetic energy of the α particle in the laboratory system. [Data from F. S. Houck and J. M. Miller, *Phys. Rev.* **123**, 231 (1961).]

[7] An important class of experiments, which is not discussed here, is aimed primarily at obtaining information on the energy states of product nuclei rather than on the mechanisms of the reactions. This field (reaction spectroscopy) is considered in chapter 8, section F.

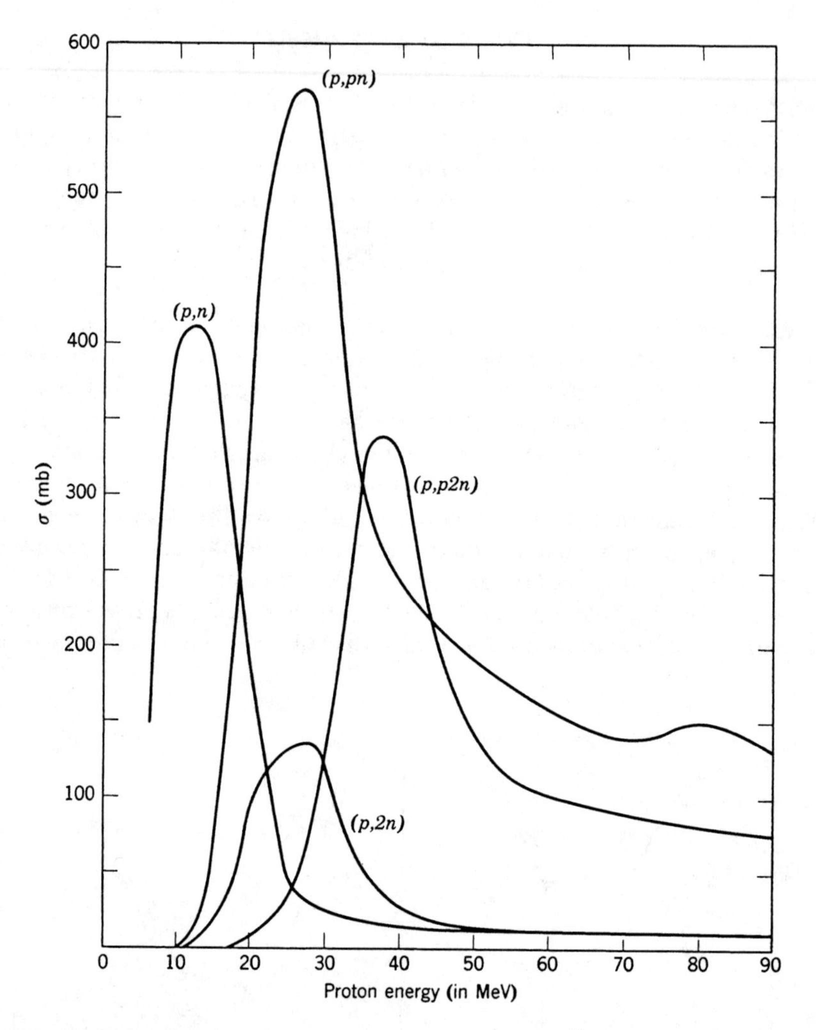

Fig. 4-5 Excitation functions for proton-induced reactions on ^{63}Cu. [From J. W. Meadows, *Phys. Rev.* **91**, 885 (1953).]

absolute disintegration rate of a radioactive product; the relevant techniques are discussed in chapter 8, section G.

The *shape* of an excitation function can often be determined in a much simpler way by the so-called stacked-foil method, that is, by exposing several target foils in the same beam, with appropriate energy-degrading foils interposed. An absolute calibration, if desired, can then be done at a single energy. In general, the higher the energy of a bombarding particle, the more complex the possible reactions. For example, with a few exceptions among the lightest nuclides, thermal neutrons can induce (n, γ) reactions only. With neutrons of several million electron volts kinetic energy, (n, p) reactions become possible and prevalent; at still higher energies $(n, 2n)$, (n, α), and (n, np) reactions set in. In the bombarding energy range up to about 50 MeV the cross section of a given reaction rises with increasing bombarding energy from threshold to some maximum value that is usually reached about 10 MeV above threshold and then drops again to some low value; the drop is accompanied by the rise of cross sections for other more complex reactions. This behavior is illustrated by the excitation functions shown in figure 4-5.

Excitation functions provide some information about the probabilities for the emission of various kinds of particles and combinations of particles in nuclear reactions because the formation of a given product implies what particles were ejected from the target nuclide. For example, figure 4-4 shows that in the reactions of α particles with ^{54}Fe the emission of a single proton is about three times as likely as that of a single neutron and the emission of a proton-neutron pair is about 50 times more probable than that of two neutrons. It is not possible from these data alone, though, to know whether the proton-neutron pair was emitted as a deuteron; in this instance it very likely was not.

It is possible to get some information about the kinetic energies of the emitted particles both from the energies at which the various excitation functions reach their maxima and from the slopes of the excitation functions, but these are at best rather crude estimates. Excitation functions will not yield any information about the angular distribution of the emitted particles.

Total Reaction Cross Sections. These quantities (which we designate as σ_r) are not as easily determined as individual activation cross sections. Summing of all experimentally measured excitation functions for individual reactions rarely yields an excitation function for σ_r since, for most target-projectile combinations, some reactions lead to stable products and thus cannot be measured by the activation technique. For example, in the $^{54}Fe + \alpha$ system illustrated in figure 4-4, the (α, γ) reaction leads to stable ^{58}Ni, the $(\alpha, 2p)$ reaction to stable ^{56}Fe, and so on. However, such unmeasured cross sections can often be estimated quite well by statistical theory (see section D) and, if the unmeasured contributions are not too large, this combination of experimental and calculated partial cross sections can lead to good data on excitation functions for σ_r.

The other general method for σ_r is to measure the attenuation of a beam, that is, to determine directly the quantity $(\mathbf{I} - \mathbf{I}_0)/\mathbf{I}_0$ [see (4-5)]. In energy regions in which σ_r varies rapidly with projectile energy, this method is difficult to apply because, on the one hand, the target must be kept thin enough to minimize energy degradation of the beam, but on the other hand, a thin target also produces a small attenuation in intensity, which is hard to measure with good accuracy.

Particle Spectra. In contrast to excitation functions the second type of experiment focuses attention on the energy and angular distributions of the emitted particles. In its simplest form this information may be collected experimentally by detection of the emitted particles in an energy-sensitive detector placed at various angles θ with respect to the incident beam. The quantity that is usually reported is $\partial^2\sigma/\partial\epsilon\partial\Omega$, which is a function of the kinetic energy ϵ of the emitted particle, and of the angle of emission θ. This quantity is the differential cross section for the emission of the particle with kinetic energy between ϵ and $\epsilon + d\epsilon$ into an element of solid angle $d\Omega$ at an angle θ with respect to the incident beam. In the laboratory system of coordinates the solid angle $d\Omega$ may be roughly approximated by dividing the area of the detector normal to the emission direction of the particle by the square of the distance between the target and the detector. Bearing in mind that the differential cross section is a function of ϵ and θ, the total cross section for the emission of the particle is obtained by integrating over all angles and energies:

$$\sigma = 2\pi \int_0^\infty \int_0^\pi \frac{\partial^2\sigma}{\partial\epsilon\,\partial\Omega} \sin\theta\, d\theta\, d\epsilon.$$

Some examples of energy and angular distributions are shown in figures 4-6 and 4-7.

An obvious limitation on the information that measurements of particle spectra provide for theoretical analysis lies in the lack of knowledge about the other particles that may be emitted in the same event with the one being detected. This difficulty may be circumvented either by using an energy so low that the probability for the emission of more than one

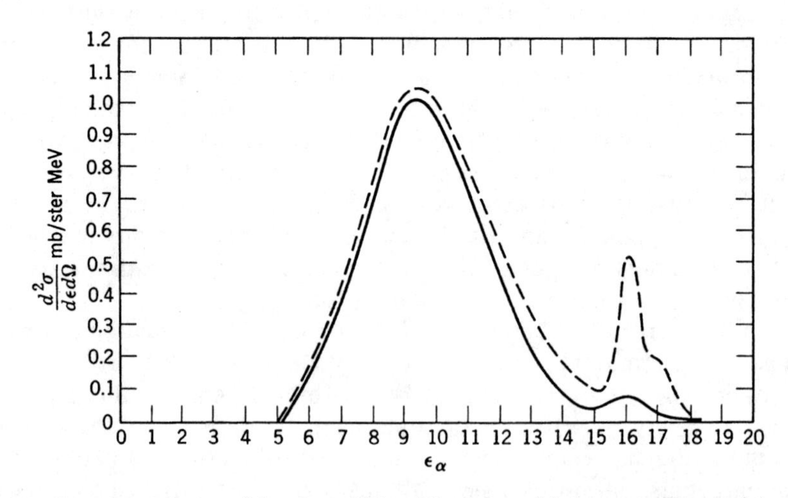

Fig. 4-6 Spectrum of α particles from Ni(p, α)Co for a nickel target of normal isotopic composition bombarded with 17.6-MeV protons. The dashed curve is the spectrum at 30° with respect to the incident beam, the solid curve is that at 120°. [Reproduced from R. Sherr and F. P. Brady, *Phys. Rev.* **124**, 1928 (1961).]

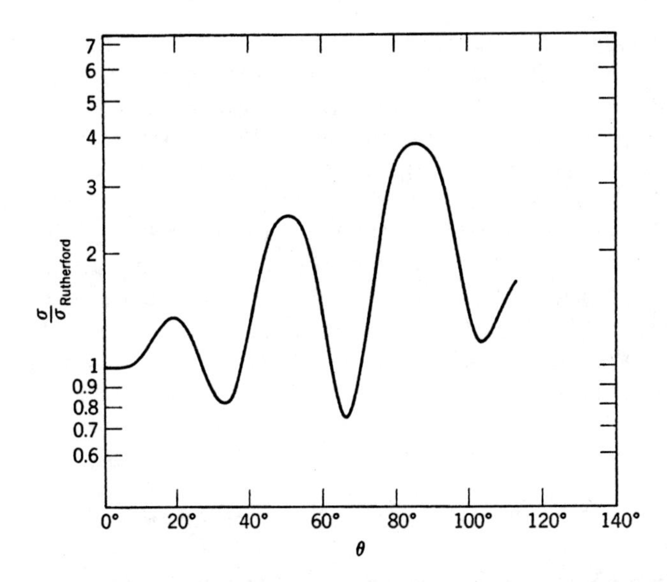

Fig. 4-7 Angular distribution of 22-MeV protons elastically scattered by nickel. The ordinate is the observed scattering cross section divided by the scattering expected from purely Coulombic interaction (Rutherford scattering). The abscissa is the scattering angle in the center-of-mass system. [Data from B. L. Cohen and R. V. Neidigh, *Phys. Rev.* **93**, 282 (1954).]

particle is negligible or by having several detectors and demanding coincidences among them before an event is recorded. The latter technique is fine in principle, but the rate of gathering information with it diminishes rapidly as the multiplicity of coincidence requirements increases. This limitation becomes less severe with the use of multiple-detector arrays.

In addition to measuring the energy and direction of emission of a particle, it is often necessary to establish its identity, that is, to determine whether it is a proton, deuteron, α particle, pion, carbon nucleus, or whatever, and to distinguish it from other particles that may be emitted from the same target. This is generally accomplished through measurement of some appropriate combination of quantities such as total kinetic energy E, specific energy loss dE/dx (cf. chapter 6), momentum p, velocity v, and mass m. The instruments and techniques used for such measurements are discussed in chapters 7 and 8.

Radiochemical Recoil Measurements. The techniques mentioned in the preceding paragraph allow unambiguous identification by A and Z for light particles (perhaps up to $A \approx 25$) emitted in nuclear reactions. For heavier fragments and product nuclei one has to resort to other methods to obtain angular distributions and kinetic-energy spectra. Specifically one can combine the activation technique with angular and energy measurements provided the product of interest is radioactive.

The experimental techniques that are employed for the measurements vary in complexity. In the simplest experiment, catcher foils placed before and behind the target determine the fraction of a given product that recoils out of the target in the forward and in the backward directions. This measurement is sensitive to the relative amounts of momentum carried away by the particles emitted in the forward and in the backward direction. In the more elegant experiment stacks of very thin (thin compared to the range of the recoil product) catcher foils are placed at various angles with respect to the incident beam, and a direct measurement of the angular and kinetic-energy distribution of the products is made. For this type of experiment the target itself must, of course, also be thin relative to the recoil ranges of interest. Reaction studies by recoil techniques are reviewed in A1.

D. REACTION MODELS AND MECHANISMS

Before proceeding to a phenomenological survey of the various types of nuclear reactions in the following sections, we give in this section a brief account of the theoretical framework in which nuclear reactions are discussed. When we remember that, even for the interaction between two individual nucleons, a complete theory in terms of nuclear forces is still lacking, it should not come as a surprise that the subject of nuclear reactions is far too complex and multifaceted to be understood in terms of a single, exact theory. Instead, it has been necessary to rely on simplified models for the description, systematization, and "understanding" of the observed phenomena as well as for predictive purposes. These models have indeed proved very useful.

1. *Optical Model*

The earliest model considered in attempts to understand cross sections for nuclear processes was one in which the interactions of the incident particle with the nucleons of the nucleus were replaced by its interaction with a potential-energy well. Although abandoned in the 1930s when it could not account for the phenomenon of slow-neutron resonances (see below), the model was revived in modified form in 1949 (F1) for the description of nuclear reactions at higher energies (≈ 100 MeV), and since that time it has been used fruitfully in the interpretation of elastic-scattering and total-reaction cross sections at energies down to a few million electron volts.

The analogy with a beam of light passing through a transparent glass ball has caused the model to be called the optical model. In its simplest form it represents the nucleus by a square-well potential V_0 MeV deep and R fm wide as illustrated in figure 4-8. The kinetic energy of a neutron entering

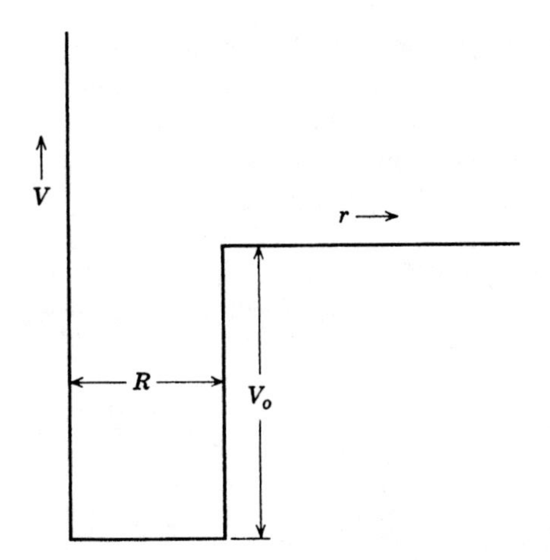

Fig. 4-8 Schematic diagram of square-well optical-model potential energy.

the nucleus will be higher (by V_0 MeV) inside the well than outside or, in terms of the wave picture, the wavelength of the neutron will be shorter inside than outside; in other words, there will be refraction at the nuclear surface, and the index of refraction, as in optics, is defined as the ratio of wavelengths. The net effect of a neutron passing through the potential well would simply be two successive refractions resulting in a change of direction: elastic scattering. No other process could occur.

Complex Potential. To account for interactions other than elastic scattering the model must be modified to allow for absorption of the incident particle. Again in analogy to optics, this modification is achieved by changing the index of refraction from a real to a complex number, which has the effect of damping the wave inside the potential well or, in other words, of making the medium somewhat absorbent rather than completely transparent: the ball of glass has become a "cloudy crystal ball."

The introduction of a complex index of refraction is equivalent to replacing the real potential V_0 of figure 4-8 by a complex optical-model potential:

$$\begin{aligned} V &= -(V_0 + iW_0) \quad \text{for } r < R, \\ V &= 0 \quad \text{for } r > R. \end{aligned} \tag{4-24}$$

For a neutron interacting with such a potential the Schrödinger equation is

$$\left(\frac{p^2}{2m} + V - E\right)\psi = 0.$$

This equation is solved both inside and outside the well, with the boundary condition that the two parts of the wave function must join smoothly (have the same value and the same slope) at $r = R$. We confine ourselves here to a brief sketch of some salient features of the results.[8] The solution of the wave equation is a wave function that, for each value of the angular momentum l, is proportional to the sum of two exponentials, one representing the incoming, the other the outgoing wave. The ratio of the coefficients of these two terms is the amplitude η_l of the outgoing partial wave of angular momentum l.

The cross sections for elastic scattering and reaction (or absorption) for a given l can be expressed in terms of this quantity η_l, which is a complex number [with real part Re (η_l)]:

$$\sigma_{sc,l} = \pi \lambda\!\!\!^{-2}(2l+1)|1-\eta_l|^2, \tag{4-25}$$

$$\sigma_{r,l} = \pi \lambda\!\!\!^{-2}(2l+1)(1-|\eta_l|^2), \tag{4-26}$$

$$\sigma_{total,l} = \sigma_{sc,l} + \sigma_{r,l} = 2\pi \lambda\!\!\!^{-2}(2l+1)[1-\mathrm{Re}(\eta_l)]. \tag{4-27}$$

Comparison between (4-16) and (4-26) shows the relationship between the transmission coefficient and the amplitude of the outgoing wave: $T_l = (1-|\eta_l|^2)$.

Two important results may be deduced from (4-25) and (4-26):

1. The maximum value of the reaction cross section for a given l is $\pi \lambda\!\!\!^{-2}(2l+1)$, in agreement with our semiclassical result (4-11), and occurs for $\eta_l = 0$, which means according to (4-25) that there must be a scattering cross section of equal magnitude. Indeed, nuclear reactions must always be accompanied by nuclear scattering; this is the source of the so-called shadow scattering.
2. The maximum scattering cross section is $4\pi \lambda\!\!\!^{-2}(2l+1)$. It occurs for $\eta_l = -1$, which means that the reaction cross section vanishes.

Optical-Model Parameters. One expectation from the optical model is the appearance of resonances in the elastic-scattering cross sections at energies corresponding to single-particle states of the incident particle in the effective potential. With a purely real potential these resonances would be quite sharp, which is at variance with observations. The introduction of the complex potential (4-24) causes the resonances to broaden to a width of the same order as the depth of the imaginary part of the potential, W_0 MeV. It was, in fact, the observation of broad resonances in the scattering cross section for neutrons of several million electron volts (B2) that led to the extension of the optical model to low energies (F2) and to its wide acceptance.

Via the optical model it is then possible to use data on elastic-scattering and reaction cross sections to obtain information about nuclei in terms of the real (V_0) and imaginary (W_0) parts of the optical potential. More

[8] For a more complete treatment of the problem the reader is referred to other sources: e.g. references N1, p. 42; S2, p. 632; B1, p. 317.

detailed information about V_0 and W_0 comes from the angular distribution in elastic scattering as exemplified in figure 4-7. The structure in the angular distribution arises from interferences between waves of various l values and is thus a more sensitive measure of η_l, and so of V_0 and W_0, than are the reaction and elastic-scattering cross sections. The differential cross section for scattering into a unit solid angle at an angle θ (in the center-of-mass system) may be obtained from the wave function of the scattered wave and turns out to be

$$\frac{d\sigma_{sc}(\theta)}{d\Omega} = \pi \lambda\!\!\!^{-2} \left| \sum_{l=0}^{\infty} \sqrt{2l+1}\, Y_{l,0}(\theta)(1-\eta_l) \right|^2 \tag{4-28}$$

where $Y_{l,0}(\theta)$ are spherical harmonics.[9]

Analyses of angular distributions in elastic scattering in terms of (4-28) have shown that the square-well potential is much too simple: a potential well with rounded corners is required and the Woods-Saxon potential of (2-3) and figure 2-2*b* is most frequently used. We then have six parameters at our disposal: the depths, radii, and skin thicknesses of the real and imaginary parts of the potential. Much effort has gone into fitting optical-model parameters to the large amount of scattering data amassed for many systems. In turns out that both V_0 and W_0 vary with the energy of the incident particle: V_0 ranges from about 50 MeV at low energies (<10 MeV) to about 15 MeV at 150 MeV and even becomes negative at still higher energies (~ −10 MeV at 300 MeV), whereas W_0 varies in the opposite direction, from about 2 MeV at low energies to about 20 MeV at 100 MeV, and then stays nearly constant to 300 MeV. Although we have talked only about neutron interactions, the optical model is also quite applicable to charged particles, in which case the Coulomb potential has to be taken into account.

Reviews of optical-model analyses may be found in H2 and P1.

Mean Free Path. As we indicated earlier, it is the imaginary part W_0 of the optical potential that brings about absorption. Absorption of an incident nucleon in the nucleus may be expressed in terms of an absorption coefficient K or of its reciprocal, the mean free path Λ. It turns out that, for a square-well potential and for incident energies high compared to the depth of the potential, Λ can be related to the depth W_0 of the imaginary potential by

$$\Lambda = \frac{v\hbar}{2W_0} \tag{4-29}$$

where v is the relative velocity. Since the kinetic energy within the potential well is $\epsilon + V_0 = \frac{1}{2}\mu v^2$, where ϵ is the relative kinetic energy outside the well and μ the reduced mass of the system, we can rewrite (4-29) as

$$\Lambda = \frac{\hbar}{W_0}\sqrt{\frac{\epsilon + V_0}{2\mu}}. \tag{4-30}$$

[9] $Y_{l,0}(\theta) = [(2l+1)/4\pi]^{1/2} P_l(\cos\theta)$, where the P_l's are Legendre polynomials.

The mean free path may also be immediately expressed in terms of the density of nucleons within the nucleus and the effective average cross section $\bar{\sigma}$ for the interaction of the incident particle with the nucleons within the nucleus:

$$\Lambda = \frac{1}{\rho\bar{\sigma}}, \tag{4-31}$$

which establishes a relationship between optical-model parameters and the effective nucleon-nucleon interaction cross section within the nucleus.

Summary. Before leaving the subject of the optical model we review what it can and cannot do. It can be used to calculate, via the quantity η_l,

1. The cross section for elastic scattering by (4-25).
2. The total-reaction cross section by (4-26).
3. The angular distribution for elastic scattering by (4-28).

The optical model can give no information about the relative probabilities of the various reactions that may occur after the incident particle has been absorbed in the nucleus, nor can it account for the very pronounced resonances seen in slow-neutron reactions.

2. *Compound-Nucleus Model*

The first model for nuclear reactions that enjoyed much success in the detailed interpretation of experimental data was the compound-nucleus model introduced by Bohr (B3) in 1936.

Basic Ideas. In the compound-nucleus model it is assumed that the incident particle, upon entering the target nucleus, amalgamates with it in such a way that its kinetic energy (which has been increased by the depth of the potential well on entering the nucleus) is distributed randomly among all the nucleons. The resulting nucleus, which is in an excited quasi-stationary state, is called the compound nucleus. The state is said to be quasi-stationary because its excitation energy makes it unstable with respect to the emission of particles, although its lifetime is thought to be long (typically 10^{-14}–10^{-19} s) compared to the time for a nucleon to traverse the nucleus (10^{-20}–10^{-23} s). The nucleons in the compound nucleus presumably exchange energy with each other through many collisions, and the finite lifetime comes about because it is possible for a statistical fluctuation in the energy distribution to concentrate enough energy on a nucleon (or a cluster of nucleons) to allow it to escape. The most probable fluctuations are those that concentrate only a part of the excitation energy on the escaping particle, and so we expect that its kinetic energy will be less than the maximum possible and that the residual nucleus will still be in an excited state. Thus if the original excitation energy of the compound nucleus is great enough, there may be the sequential emission of several particles from the excited compound nucleus, each with a relatively low kinetic energy. The similarity of this model to that for the escape of molecules from a drop of hot liquid has caused the emission of particles from excited nuclei to be called "evaporation."

In the compound-nucleus model, then, a nuclear reaction is divided into two distinct and independent steps:

1. Capture of the incident particle with a random sharing of the energy among the nucleons in the compound nucleus.
2. The evaporation of particles from the excited compound nucleus.

The independence of the two steps is one of the central features of the model. It means that, if a compound nucleus can be produced in more than one way, its subsequent decay into reaction products should be quite independent of its mode of formation.

The excitation energy U of the compound nucleus is given by

$$U = \frac{M_A}{M_A + M_a} T_a + S_a, \tag{4-32}$$

where M_A and M_a are the atomic masses of the target and bombarding particles, respectively; T_a is the laboratory kinetic energy of the bombarding particle; and S_a is the binding energy of particle a in the compound nucleus.

Because beams of bombarding particles generally have a finite energy spread, the "quasi-stationary state" of the compound nucleus includes, in fact, many excited states. Lack of detailed knowledge about this composite of states causes most of the difficulties in the analysis of the compound-nucleus model. This problem, however, is not serious for thermal neutrons because only a single excited state is involved.

Slow-Neutron Reactions. Since the excitation energy of a compound nucleus formed by capture of a slow neutron is only slightly higher than the binding energy of the neutron in the compound nucleus, a very long time would be required before enough energy would, through a fluctuation, be concentrated on a neutron again to allow it to escape from the potential well. The probability for de-excitation by γ emission is therefore much higher, and the main reaction with slow neutrons is the (n, γ) reaction.

A typical excitation function for a slow-neutron reaction, that with silver as a target, is shown in figure 4-9 for the energy range from 0.01 to 100 eV. Three important characteristics of such slow-neutron excitation functions can be seen in figure 4-9:

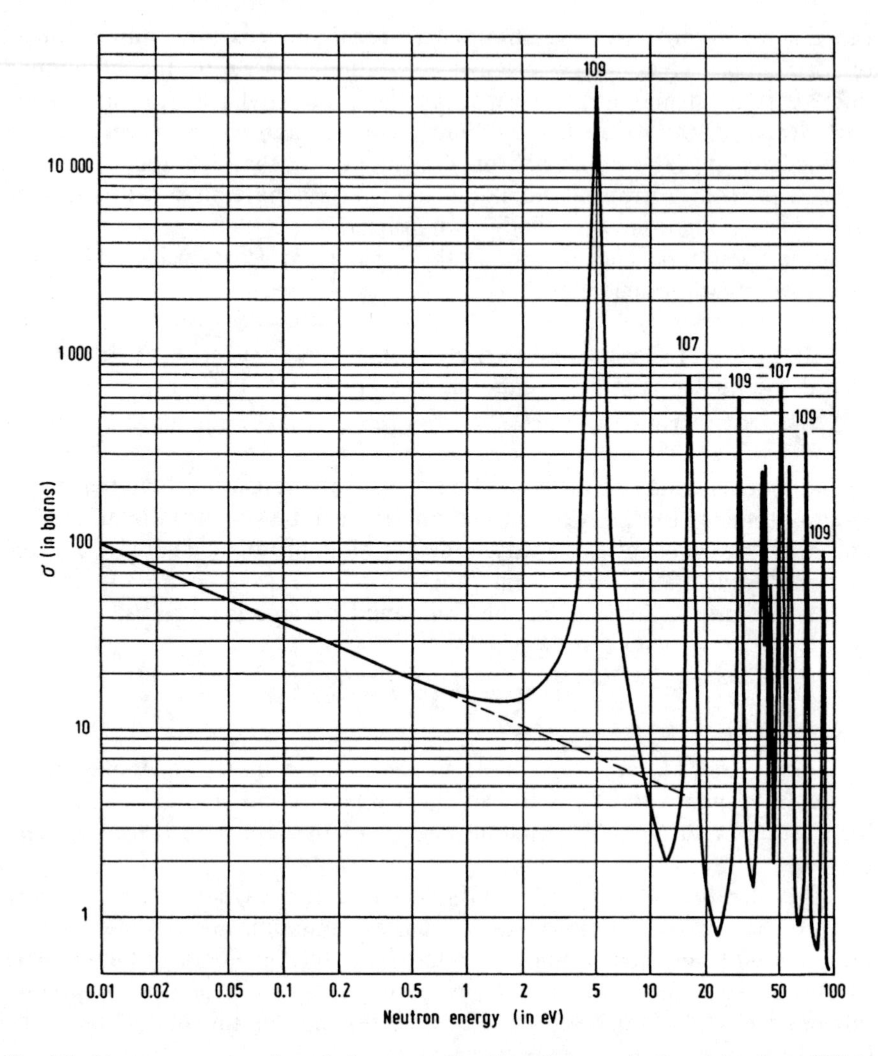

Fig. 4-9 Neutron cross section of silver as a function of energy in the region from 0.01 to 100 eV. The data as plotted are for silver of normal isotopic composition; however, each resonance peak has been assigned to one of the two silver isotopes, as indicated by mass number for a few of the peaks. (Data from reference N2.)

1. The cross sections show enormous fluctuations over a very small energy range, that is, resonances are apparent.
2. The widths of the resonances are small (~0.1 eV).
3. The spacing between the resonances is large compared to their widths; the spacings vary from the order of keV in the lightest elements to the order of eV for the heaviest).

The small widths of the resonances lead to the conclusion, by use of the

Heisenberg uncertainty principle, that the compound nucleus has a lifetime of about 10^{-14}–10^{-15} s, which is long compared to the transit time of a thermal neutron across a medium-weight nucleus, $\sim 10^{-18}$ s. This conclusion suggested the idea of the quasi-stationary state for the compound nucleus. Further, the observation that the average spacing between the resonances is 100 to 1000 times smaller than the average spacing between single-particle levels showed that the quasi-stationary excited state of the compound nucleus must involve the excitation of many particles. For these reasons the optical model is not directly applicable to slow-neutron reactions; however, it is possible to make a connection between optical-model parameters and cross sections *averaged over many resonances* (F2).

Independence Hypothesis in the Resonance Region. Although the (n, γ) reaction is by far the most likely process with low-energy neutrons, it is not the only possible one. In some light elements, where Coulomb barriers are low, (n, p) or (n, α) reactions may compete with (n, γ) if binding energies of these charged particles are sufficiently low. For the heaviest elements fission is often the most probable process. In these reactions resonances are again observed in the excitation functions.

Since the compound-nucleus model divides the reaction into two parts—formation and decay of the compound nucleus—the relative probabilities of the various possible events should be completely determined by the quantum state of the compound nucleus. In particular, if the resonances do not overlap, the behavior of the compound nucleus is essentially governed by the properties of a single quantum state (the resonant state) and should thus be independent of the manner in which the state was formed. This means, for example, that the relative amount of γ-ray emission and neutron emission will be the same when nucleus ${}^{A}_{Z}X$ is irradiated with neutrons and ${}_{Z-1}^{\ \ A}X$ is irradiated with protons as long as the energies of the particles are such that they form the same nonoverlapping resonant state. This conclusion is known as the "independence hypothesis." We return to it again in the more ambiguous situation of overlapping states.

Breit-Wigner Formula. The rapidly varying cross section illustrated in figure 4-9 shows that the amplitudes η_l of the outgoing waves (see p. 130) are very sensitive functions of the energy in this low-energy region. In the first solution of this problem by G. Breit and E. Wigner,[10] the quantities η_l were not directly calculated; rather, perturbation theory was used to solve the problem in the two steps suggested by Bohr involving the formation and decay of the compound nucleus. It is useful to give the results of their calculations for a general reaction:

$$a + A \rightarrow C \rightarrow B + b, \tag{4-33}$$

[10] See reference R1 for a complete discussion of theory and experiment with slow neutrons.

going through a compound nucleus C that is in a *single* well-defined quasi-stationary quantum state.

The cross section for the particular reaction (4-33) would be written

$$\sigma_{A\to C\to B} = \sigma_{A\to C} W_B, \tag{4-34}$$

where $\sigma_{A\to C}$ is the cross section for forming the compound nucleus C, and W_B is the probability that the compound nucleus decays in the particular manner prescribed by reaction (4-33) or goes into channel Bb.[11] Equation 4-34 explicitly presents the two-stage and the independence hypotheses. The Breit-Wigner treatment gives the expression

$$\sigma_{A\to C} = \pi\bar{\lambda}^2_{Aa} \frac{2I_C + 1}{(2I_A + 1)(2I_a + 1)} \frac{\Gamma_{Aa}\Gamma}{(\epsilon - \epsilon_0)^2 + (\Gamma/2)^2}, \tag{4-35}$$

where the I's are spins, $\bar{\lambda}_{Aa}$ is the relative wavelength in the entrance channel, ϵ_0 is the center-of-mass energy at which resonance occurs, Γ is the total width of the level, and Γ_{Aa} is the partial width of the level for decay into channel Aa. The meaning of "width-of-level" lies in the statement that $\Gamma_J/\hbar$ is the probability per unit time that the compound nucleus decays into channel J. This means that

$$\Gamma = \sum_J \Gamma_J \qquad \text{(summed over all channels)} \tag{4-36}$$

and that

$$W_B = \frac{\Gamma_{Bb}}{\Gamma}. \tag{4-37}$$

The substitution of (4-35) and (4-37) into (4-34) gives the famous Breit-Wigner one-level formula

$$\sigma_{A\to C\to B} = \pi\bar{\lambda}^2_{Aa} \frac{2I_C + 1}{(2I_A + 1)(2I_a + 1)} \frac{\Gamma_{Aa}\Gamma_{Bb}}{(\epsilon - \epsilon_0)^2 + (\Gamma/2)^2}. \tag{4-38}$$

For the (n, γ) reaction in particular

$$\sigma_{(n,\gamma)} = \pi\bar{\lambda}^2 \frac{2I_C + 1}{2(2I_A + 1)} \frac{\Gamma_n\Gamma_\gamma}{(\epsilon - \epsilon_0)^2 + (\Gamma/2)^2}, \tag{4-39}$$

where Γ_n and Γ_γ are the partial widths for neutron and γ emission, respectively. Equation 4-39 is meant to describe the cross section at any particular resonance such as those shown in figure 4-9. For example, the first resonance seen in this figure is at $\epsilon_0 = 5.19$ eV and is characterized by $\Gamma_\gamma = 136 \times 10^{-3}$ eV and $\Gamma_n = 5.5 \times \epsilon^{1/2} \times 10^{-3}$ eV.

[11] The particular manner of the formation and the decay of the compound nucleus are often referred to as **channels**; reaction (4-33) would be said to go from channel Aa to channel Bb. The definition of a channel in general requires specification of the relative energy of the particles, the total angular momentum, and the internal quantum numbers (excited states) of the particles.

The resonant state need not correspond to a positive energy for the incident neutron; the state may be at an excitation energy that is below the binding energy of a neutron in the compound nucleus. Although under these circumstances the resonance will not be directly observable with neutrons, the resonance can cause large capture cross sections for neutrons of thermal energy (of the order of 0.025 eV) if the width of the resonance is not too small when compared with the energy difference between the position of its peak and the excitation energy of the compound nucleus produced in thermal-neutron capture.

It is clear, then, that the observation of neutron-capture resonances yields information about the energies of nuclear excited states and about their widths. Except for the lightest nuclei, this type of experiment is not possible with incident charged particles because the Coulomb barrier causes Γ_A to become vanishingly small at low energies. With slow neutrons the centrifugal barrier causes Γ_n to be most important for $l = 0$; the spin of the compound-nucleus state therefore must be $I_A \pm \frac{1}{2}$.

1/v Law. It is of interest to examine the cross section of silver for neutrons (cf. figure 4-9) with energies below about 0.4 eV. In this region the dominant term in the denominator of (4-39) is clearly ϵ_0, and thus the denominator is essentially a constant.

The energy dependence of the cross section will depend on three factors:

1. $\lambda\!\!\!^{-2} \propto 1/v^2$, where v is the relative velocity of neutron and target nucleus.
2. $\Gamma_n \propto v$ because Γ_n is proportional to the density of final states for the system and thus to the relative velocity of neutron and target nucleus.
3. Γ_γ is independent of changes in neutron energy of a few eV because the energy of the γ ray is several MeV.

The result of these three factors is to make $\sigma_{n,\gamma} \propto 1/v$ in the region where $\epsilon \ll |\epsilon_0|$. The $1/v$ dependence for the neutron-capture cross section of silver is shown as the dotted line in figure 4-9.

It is seen from the preceding discussion that the thermal-neutron capture cross section of any particular nuclide will depend critically on the energies and widths of its resonant states. In particular, if there is a resonant state at an energy within about 0.01 eV (either positive or negative) of the binding energy of the neutron, the capture cross section can be enormous. On the other hand, if there are no close resonances, the capture cross section may be quite small and follow the $1/v$ law.

Cross Section Data. Thermal-neutron cross sections are listed in appendix D. A word of caution is indicated concerning the term "thermal." Measurements are sometimes made with the neutron spectrum present in a particular nuclear reactor. Other cross sections have been measured in a thermal-neutron flux characterized to good approximation by the velocity distribution at about 20°C. Still others have been determined at particular neutron velocities by the use of neutron monochromators. The usual practice is to tabulate all thermal-neutron cross sections for the discrete neutron velocity of 2.20×10^5 cm s^{-1}, which corresponds to the discrete energy 0.025 eV and is the most probable velocity in a Maxwellian distribution at 20°C. See chapter 6, section D for a discussion of neutron-velocity distributions.

Extensive tabulations and graphs of neutron cross sections as a function of energy and of resonance parameters are available (N2).

The Statistical Assumption. As the energy of the bombarding particle is increased, two effects act in concert to make (4-38) increasingly difficult to use:

1. The width Γ of each level becomes larger and larger because more outgoing channels become available.
2. As is usually true in many-particle systems, the energy spacing D between levels becomes smaller and smaller.

The net effect is that the resonances begin to overlap, and it is no longer possible, in general, even with ideal energy resolution of the incident beam, to excite but a single state of the compound nucleus. Under these circumstances the various states of the compound nucleus that enter into the reaction do not behave independently; interferences among them must be taken into consideration, and the cross section would *not* simply be given by a sum of terms, each of which has the form (4-38). These interferences could have two important effects:

1. The angular distribution of the emitted particles would not be symmetric about a plane normal to the direction of the incident beam, as it must be if the compound nucleus is in a single nonoverlapping quantum state. The lack of symmetry can arise from interferences between particles emitted with, for example, $l = 0$ and $l = 1$ (s and p waves), for s waves are an even function of θ and p waves are an odd function of θ.
2. The relative values of the various interferences, which would affect the relative probabilities for the emission of various kinds of particles, would depend on how the compound nucleus was made, and the independence hypothesis would no longer be true.

Further, aside from the interferences, if each of the overlapping states had a different width for a particular mode of decay of the compound nucleus, then again the independence hypothesis would be invalid.

These problems, caused by the interferences and by the fluctuating partial widths, are removed if two assumptions are made which together are called the statistical assumption. It is first assumed that the interference terms, which may be either positive or negative, have random signs and thus cancel out; this reinstates a symmetrical angular distribution. It is further assumed that the overlapping states all have essentially the same relative partial widths for the various possible decay channels of the compound nucleus; this reinstates the independence hypothesis. The statistical assumption, then, allows the extension of the Bohr model to the region of overlapping energy levels and may be tested by the measurement of the angular distribution of evaporated particles and by experimental tests of the independence hypothesis.

It has been observed, as exemplified in figure 4-6, that the angular distribution of most of the particles emitted by compound nuclei excited up to a few tens of MeV has the required symmetry, and thus the statistical assumption has some validity. However, some of the particles, usually of relatively high energy, tend to be preferentially emitted in the forward direction and thus represent a partial failure of the statistical assumption.

Independence Hypothesis in the Continuum Region. Whether the independence hypothesis, which states that a compound system retains no memory of the particular entrance channel by which it was formed, holds in the region of overlapping levels has been tested in a number of ways. Most of these tests involve the measurement of excitation functions for the production of two or more radioactive nuclides via a compound nucleus formed in different ways. Named after the author of the first such experiment (G1), these tests are referred to as "Ghoshal experiments."

S. N. Ghoshal investigated the behavior of an excited ^{64}Zn nucleus made in two different ways:

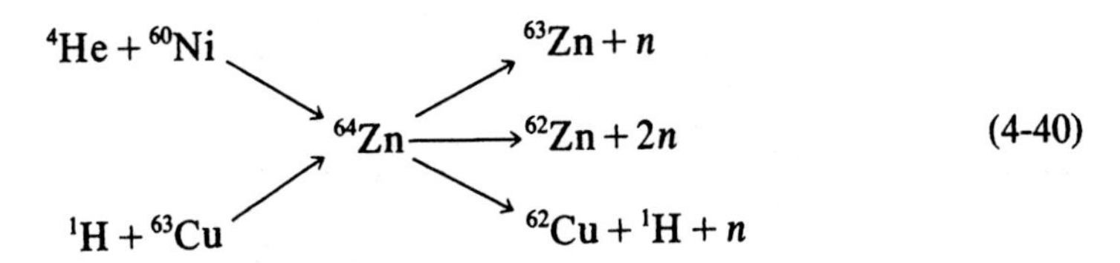

(4-40)

In light of (4-34), the independence hypothesis demands, for example, that

$$\frac{\sigma(\alpha, pn)}{\sigma(\alpha, 2n)} = \frac{W(pn)}{W(2n)} = \frac{\sigma(p, pn)}{\sigma(p, 2n)}, \tag{4-41}$$

where all of the cross sections are measured under the same conditions for the compound nucleus. Results derived from Ghoshal's data are presented in figure 4-10, in which it is seen that the independence hypothesis seems to be confirmed. The situation can be less clear-cut for ratio curves that vary more rapidly with excitation energy than do those in figure 4-10; it is often found, for example, that a ratio curve for α-particle-induced reactions has a shape similar to that of the corresponding proton-induced reactions but is

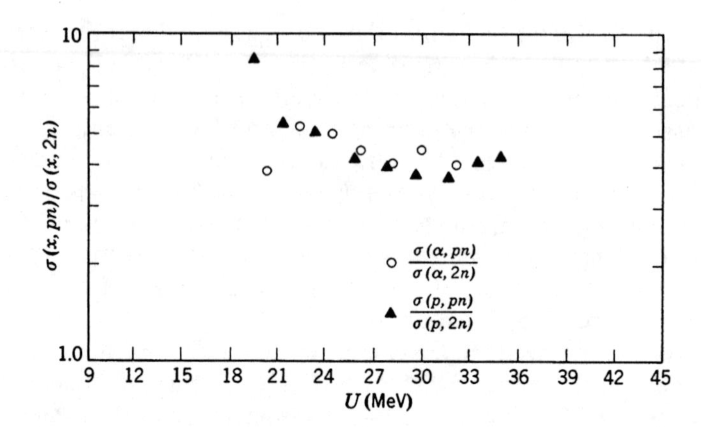

Fig. 4-10 Comparison of the behavior of an excited ^{64}Zn compound nucleus made in two different ways: ^{63}Cu + p and ^{60}Ni + α. U = excitation energy. (Data from reference G1.)

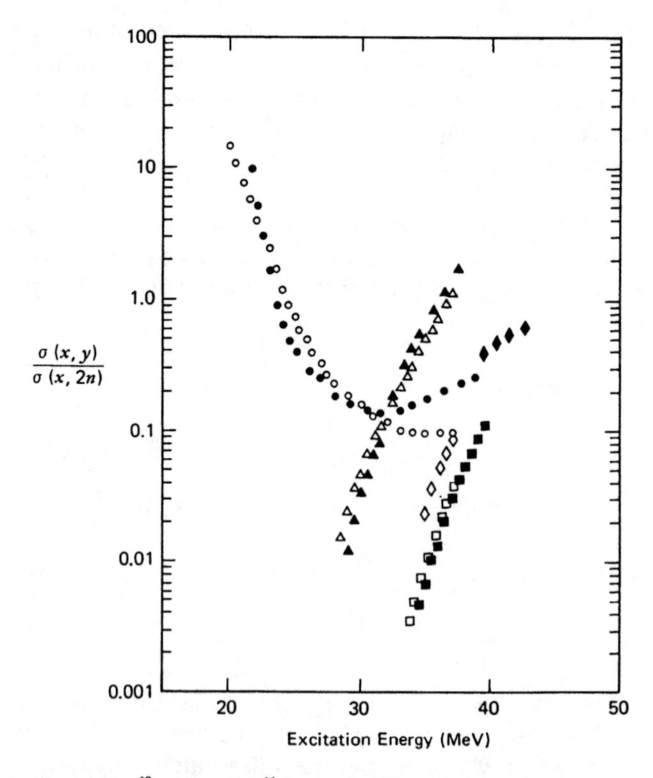

Fig. 4-11 Comparison of ^{69}Ga + p and ^{66}Zn + α reactions. ● (p, n); ▲ $(p, p2n)$; ■ (p, pn); ◆ $(p, 3n)$; ○ (α, n); △ $(\alpha, p2n)$; □ (α, pn); ◇ $(\alpha, 3n)$. The ordinate is the ratio of the cross section of each of these reactions to the corresponding $(x, 2n)$ cross section. The α-induced reaction cross sections have been plotted at 3 MeV less than the actual excitation energies (see text). [From N. T. Porile et al., *Nucl. Phys.* **43**, 500 (1963).]

displaced on the energy scale, usually to higher energies. This is illustrated in figure 4-11, which shows relative cross sections for several nuclides produced both by $^{69}Ga+p$ and $^{66}Zn+\alpha$ reactions, that is, through the compound nucleus ^{70}Ge. The agreement between the two sets of data is seen to be excellent over a large range of cross section ratios, but it was obtained by displacing all the points for α-induced reactions to lower energies by 3 MeV. Such a displacement can be rationalized when it is realized that, to reach the same excitation energy of the compound nucleus ^{70}Ge, the α particles bring in more angular momentum than the protons, and that the *rotational*-energy component of the excitation energy does not contribute to particle emission. As we indicated earlier, to be truly in the same state two compound nuclei must have not only the same energy but also the same angular momentum. Correcting angular-momentum effects by shifting energy scales as in figure 4-11 is only approximately correct, and we return to a consideration of this problem later in this section.

More detailed and stringent tests of the independence hypothesis than are provided by the integral experiments of the Ghoshal type involve the measurement of differential cross sections $d^2\sigma/d\Omega\, dE$ for the emission of particles as a function of energy and angle, when a given compound nucleus is produced in different ways. Such differential experiments, for example, on proton and α-particle spectra emitted in $^{62}Ni+p$ and $^{59}Co+\alpha$ reactions (F3), have also shown the independence hypothesis to be valid when angular-momentum effects are properly taken into account.

The statistical assumption has proved to be successful for the description of a very large body of reactions induced by nucleons and helium ions with energies up to about 40 or 50 MeV and by more complex particles with energies up to about 10 MeV per nucleon. At energies much above 100 MeV the statistical assumption is known to fail. Not many data are available at intermediate energies.

Statistical Model—Evaporation Theory. The excitation functions in figures 4-4 and 4-5 and the energy spectrum of emitted particles in figure 4-6 are typical of data that can be interpreted within the compound-nucleus model extended into the medium-energy region through the statistical assumption. Figures 4-4 and 4-5 show the competition among some of the various channels that are available for the decay of the compound nucleus. Explicit in the spectrum shown in figure 4-6 and implicit in the fairly sharp maxima shown by the excitation functions is the fact that most of the particles are emitted with considerably less than the maximum energy available. As discussed on p. 132 this would be qualitatively expected from the compound-nucleus model.

These qualitative remarks can be given quantitative expression because the statistical assumption implies that statistical equilibrium exists during a compound-nucleus reaction. Statistical equilibrium means that the relative numbers of compound nuclei and of sets of particles that correspond to the various decay channels are determined by their relative state densities (cf. footnote 17 on p. 80 for the meaning of translational state densities). These concepts are incorporated in what is known as the statistical model. Its most powerful aspect lies in its ability to predict the energy spectrum of evaporated particles as well as excitation functions for various products in terms of certain average nuclear properties, in a manner described in the following paragraphs.

The key to the problem is use of the **principle of detailed balance.** Consider the decay of compound nucleus C with excitation energy U_C into residual nucleus B at excitation U_B and particle b with kinetic energy ϵ_{Bb} relative to B[12]:

$$C(U_C) \xrightarrow{\epsilon_{Bb}} B(U_B) + b. \tag{4-42}$$

The principle of detailed balance demands that statistical equilibrium be maintained by reactions such as (4-42) proceeding forward and backward at precisely the same rate. By equating the probabilities per unit time for the forward and reverse reactions and expressing these quantities in terms of the densities of states, one can derive (see, e.g., B1, p. 365, and E1) an expression for the energy spectrum of emitted particles b:

$$I(\epsilon_{Bb})\, d\epsilon_{Bb} = \frac{\mu_{Bb}}{\pi^2 \hbar^3} \sigma_{Bb} \epsilon_{Bb} \frac{\omega_B(U_B)}{\omega_C(U_C)}, \tag{4-43}$$

where

$I(\epsilon_{Bb})\, d\epsilon_{Bb}$	is the probability per unit time for the compound nucleus to emit particle b with relative kinetic energy between ϵ_{Bb} and $\epsilon_{Bb} + d\epsilon_{Bb}$,
μ_{Bb}	is the reduced mass,
σ_{Bb}	is the cross section for the reaction between nucleus B at excitation energy U_B with particle b at relative kinetic energy ϵ_{Bb}, and
$\omega_B(U_B)$ and $\omega_C(U_C)$	are the densities of states of B and C at their respective excitation energies.

The maximum in the energy spectra at relatively low energies (see figure 4-6) occurs because, although ϵ_{Bb} obviously increases with kinetic energy, U_B simultaneously decreases and the density of states $\omega_B(U_B)$ decreases with decreasing U_B in an approximately exponential manner as discussed below.

To use (4-43) for quantitative calculations we need expressions for the inverse cross section σ_{Bb} and for the state densities. Since cross sections

[12] A more complete specification of the channel would include the spins I_C, I_B, and I_b, and orbital angular momentum ℓ between B and b, with angular-momentum conservation requiring that $I_C = I_B + I_b + \ell$. For the moment we simplify the problem by ignoring angular-momentum restrictions.

for the reactions of excited states are not available, the usual assumption is that σ_{Bb} can be approximated by the corresponding ground-state cross section. Expressions such as (4-23) are commonly used to express the energy dependence of cross sections in analytical form. However, since the use of (4-23) means that no charged particles can be emitted with kinetic energies less than the height of the Coulomb barrier, modified expressions taking approximate account of barrier penetration probabilities have been suggested (see, e.g., D1).

The total probability per unit time for the emission of particle b is obtained by integrating (4-43) over the whole spectrum,

$$P_b = \int_0^{U_C - S_b} I(\epsilon_{Bb})\, d\epsilon_{Bb} \tag{4-44}$$

where the upper limit of the integral comes about by energy conservation: the maximum value of ϵ_{Bb} is given by the excitation energy of the compound nucleus minus the separation energy S_b of particle b from the compound nucleus. To go from expression (4-44) to the *fraction* of all the compound nuclei decaying into channel Bb, we must divide the integral in (4-44) by the sum of all such integrals for all decay channels. The cross section for a reaction such as (4-33) is then obtained by multiplying that fraction by the formation cross section of the compound nucleus.

The estimation of cross sections for reactions involving the sequential emission of two or more particles becomes complicated in that it requires the evaluation of multiple integrals; it is in general best done on electronic computers, usually by Monte Carlo methods (D1).

Level Densities. So far nothing explicit has been said concerning the state densities $\omega(U)$, which are evidently of great importance to calculations in evaporation theory. We follow the usual convention of discussing this topic in terms of *level* densities rather than *state* densities, the distinction being that a level of spin J has $(2J+1)$-fold degeneracy, that is, contains $2J+1$ states. The level density $\rho(U)$ at energy U is physically observable in terms of its reciprocal, the level spacing in the vicinity of U.

Experimentally level densities within a few million electron volts of nuclear ground states are obtained from the determination of individual levels populated in radioactive decay or nuclear reactions—the subject of nuclear spectroscopy. Slow-neutron resonances give information about level spacings in the vicinity of neutron-binding energies (6–8 MeV). The data generally show an approximately exponential increase of $\rho(U)$ with U for a given nucleus. At still higher energies we have to resort to statistical-mechanical calculations, based on various nuclear models, to obtain level densities. A review of the subject may be found in H3.

From one of the simplest nuclear models, one that considers a nucleus as a mixture of noninteracting neutron and proton Fermi gases (see chapter

10, section C), comes the most widely used level density expression:

$$\rho(U) = C \exp(2a^{1/2}U^{1/2}), \tag{4-45}$$

where a and C are constants that depend on the mass number of the nucleus; in particular, the **level density parameter** a is proportional to A. In the Fermi gas model, the nucleus may be characterized by the usual thermodynamic quantities, including a **nuclear temperature** τ, which, it turns out, is related to the excitation energy U by

$$U = a\tau^2. \tag{4-46}$$

Returning now to the energy spectrum of evaporated particles as given by (4-43), and using (4-45) for the level density, we see that the spectrum has the form

$$I(\epsilon) \propto \epsilon\sigma \exp[2a^{1/2}(\epsilon_m - \epsilon)^{1/2}], \tag{4-47}$$

where ϵ_m is the maximum kinetic energy with which the particle may be emitted. A plot, then, of $\ln[I(\epsilon)/\epsilon\sigma]$ versus $(\epsilon_m - \epsilon)^{1/2}$ should give a straight line and allow the evaluation of the level density parameter a and the nuclear temperature. Deviations of such plots from straight lines and lack of proportionality between a and A may arise from

1. the inadequacy of the approximate equation (4-45);
2. neglect of the dependence of emitted-particle spectra on the angular momentum of the compound nucleus;
3. the possibility that the statistical model is inadequate for the description of some of the reactions that occur, particularly those leading to the emission of high-energy particles.

As mentioned before, much effort has gone into more adequate (but also harder-to-use) calculations of level densities (H3). Angular-momentum effects are briefly discussed in the following paragraph, and the failure of the statistical model to account for certain types of reactions has led to the development of the direct-interaction model described below.

Angular-Momentum Effects. As mentioned in footnote 12 and in the discussion of Ghoshal experiments, angular momenta as well as energies should be taken into account in doing evaporation calculations. Complete analysis of the problem involves the proper averaging over the spectra of angular momenta of the compound nuclei, consideration of the orbital and spin angular momentum carried away by the emitted particles, and an explicit expression for the spin-dependent level density $\rho(U, I)$. The treatment becomes quite complicated and the reader is referred to other sources (E1, H3, L1). Here we only sketch some important consequences of including angular-momentum effects.

The importance of angular-momentum effects comes about chiefly

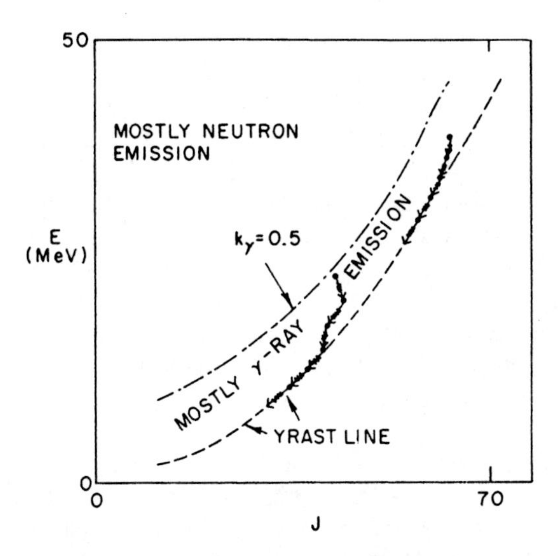

Fig. 4-12 Schematic diagram illustrating the significance of yrast levels in de-excitation processes. No levels exist below the yrast line (see text). In real nuclei the yrast line is not a smooth curve but may have many steps. The band between the yrast line and the curve marked $k_\gamma = 0.5$ denotes the region where γ emission predominates because neutrons (and other particles) would have to carry off so much angular momentum to reach any levels in the product nucleus that their emission becomes improbable. [Adapted from J. R. Grover and J. Gilat, *Phys. Rev.* **157**, 814 (1967) by permission of J. R. Grover.]

because compound nuclei can be formed in states of rather high angular momentum [see (4-21)], whereas the emitted particles, because they tend to be of low energy, do not carry away much of the angular momentum. Conservation of angular momentum demands, then, that the residual nucleus B contain the appropriate residual angular momentum and thus $\omega_B(U_B)$ in (4-43) be replaced by the spin-dependent state density $\omega_B(U_B, I_B) = (2I+1)\,\rho_B(U_B, I_B)$. The important point is that high-spin states are generally associated with high excitation (see chapter 10). In fact, for each spin I there is an excitation energy $U(I)$ below which there are on the average no states of spin I or greater (see figure 4-12). These lowest-energy states of a given spin have become of great importance, particularly in the discussion of reactions induced by heavy ions, which produce compound nuclei of extremely high spins. These lowest-energy states of a given spin have been named **yrast** levels.[13] The existence of an yrast level at excitation $U(I)$ means that the maximum kinetic energy of the emitted particle [upper limit of the integral in (4-44)] is not $U_C - S_b$, but less than that; to the approximation that b is spinless and emitted in an $l = 0$ state, the maximum kinetic energy becomes, in fact, $U_C - S_b - U(I)$. This effect will be reflected in the emission of proportionately more low-energy particles,

[13] The term was suggested by J. R. Grover (G2) and is a Swedish word meaning "dizziest."

which in turn results in an increase in the apparent value of the level density parameter a as derived from an excitation function.

It is, of course, possible for $U_C - S_b - U(I)$ to be less than zero while $U_C - S_b$ is greater than zero, in which case the particle b will not be emitted even though the excitation energy of the compound nucleus is greater than the binding energy of the particle b. The compound nucleus will instead emit some other particle (including photons) or emit particle b into a state $l > 0$ with therefore a much reduced probability. The quantity $U(I)$ effectively raises the threshold for any given reaction and can thereby contribute to the energy shifts in the Ghoshal experiment. It can also cause γ emission to compete effectively with particle emission up to a few million electron volts above the particle-emission thresholds.

Odd-Even Effects on Level Densities. We saw in the discussion of nuclear masses in chapter 2 that nuclei with even numbers of protons or neutrons are stabilized by the pairing energy. It is also well known from experimental data that, at least up to the region explored by neutron resonances, these nuclei have level spacings larger than those of their odd-Z or odd-N neighbors. These pairing effects on the level density become less marked with increasing excitation, as might be expected from the rapidly increasing number of excited-nucleon configurations that can lead to a given total excitation.

The pairing effect must be taken into account in the expressions for level density and, in view of the energy dependence noted above, this may be accomplished by introducing a fictitious ground state that the nucleus would have in the absence of enhanced stability due to pairing. This approach leads to an expression of the form

$$\rho(U) = C \exp [2a^{1/2}(U - \delta_n - \delta_p)^{1/2}] \tag{4-48}$$

in place of (4-45). The quantities δ_n and δ_p are zero for odd neutron and odd proton number, respectively; they are positive for even neutron and proton number, respectively, with a numerical value that depends on that even number. Thus the level density of an odd-odd nucleus at a given excitation energy is, in general, greater than that of an adjacent even-odd or odd-even nucleus, which, in turn, is greater than that of an adjacent even-even nucleus. The quantities δ_n and δ_p are discussed in detail in D1.

The importance of the pairing effect is illustrated in figures 4-4 and 4-5. In both sets of excitation functions a surprising result appears: the probability of the evaporation of a proton and a neutron from an excited compound nucleus is considerably greater than that of the evaporation of two neutrons, despite the fact that the Coulomb barrier to proton emission, as reflected in the inverse-cross-section term in (4-43), serves to diminish proton emission. This enhancement of proton emission occurs because the compound nucleus in both examples is an even-even nucleus that, on the evaporation of two neutrons, goes to an even-even product whose level density is low relative to that of the odd-odd isobaric product formed by the emission of a neutron and a proton. This is not an unusual situation for compound nuclei with atomic numbers up to about 30 or 40; at higher Z the Coulomb barrier becomes so high that it is usually decisive and neutron emission predominates. Considerable success in the interpretation of excitation functions has been achieved with (4-48) (D1, P2).

3. *Direct Interaction*

Types of Processes. The direct-interaction model differs from the compound-nucleus model in that it does *not* assume the energy of the incident particle to be randomly distributed among all the nucleons in the target nucleus. Rather, in one aspect of the direct-interaction model, it is assumed that the incident particle collides with only one, or at most a few, of the nucleons in the target nucleus, some of which may thereby be directly ejected. It is also possible for the incident particle to leave the target nucleus after losing but a part of its energy in these few collisions. Thus the reaction does not proceed through the formation of an intermediate excited nucleus, and it is expected that the kinetic energies of the emitted particles will usually be higher than those of particles that are evaporated from an excited compound nucleus.

In addition, direct interactions include those events in which only a part of an incident complex particle, such as a deuteron, interacts with the target nucleus. The part that does not interact will then continue on after being deflected. This kind of reaction was first characterized for incident deuterons and was dubbed a **stripping** reaction. Another direct reaction that is frequently observed is the so-called **pickup** process (which may be considered the inverse of stripping); it involves the formation of a complex particle, such as a deuteron or ^{3}He by interaction of an incident particle (e.g., a proton) with a nucleon or group of nucleons in the nucleus. Reactions of the stripping and pickup type have become particularly important with heavy-ion ($Z > 2$) projectiles. They are collectively referred to as transfer reactions—nucleons or groups of nucleons can be transferred either from target to projectile or from projectile to target.

Knock-On Reactions. When excitation functions for some nuclear reactions are extended above 30 or 40 MeV, deviations from compound-nucleus behavior often become evident. This can be seen in figures 4-4 and 4-5 where the "tails" of the excitation functions for the (α, p) and (α, n) reactions on ^{54}Fe and for the (p, pn) reaction on ^{63}Cu do not correspond to the expectations from evaporation theory: that theory would predict continuing steep drops of the cross sections as other channels for the emission of additional particles open up with increasing excitation energy of the compound nucleus. It seems likely that some direct processes involving nucleon knock-out are taking place. More direct evidence comes from angular distributions (forward-peaked) and energy spectra (flat or peaked toward high energies) of emitted particles in certain reactions, for example, (p, p') reactions.

For a direct knock-on type of interaction to take place, the mean free path Λ for the incident particle must be large compared to the average spacing between nucleons in the nucleus. Under those conditions the so-called **impulse approximation** is valid, that is, collisions with individual nucleons in the nucleus may be treated as if they occurred with free nucleons (except for such restrictions as the Pauli exclusion principle). The condition is fulfilled for high incident energies (where the de Broglie wavelength λ of the incident particle is small) and for large Λ, which corresponds, according to (4-31), to small nuclear densities. In the medium-energy range, direct knock-on reactions are therefore thought to take place only in the outer, low-density regions of nuclei, that is, at large impact parameters, whereas central collisions lead to compound-nucleus formation. At higher energies (above 100 MeV) collisions with individual nucleons are the dominant mechanism for the initial interaction, leading to the development of an **intranuclear cascade** of successive nucleon-nucleon collisions. These knock-on cascades and the subsequent phases of high-energy reactions are discussed further in connection with a survey of these reactions in section G.

The most detailed information on direct reactions comes from experiments with good energy resolution, in which emitted particles leading to discrete, low-lying levels of the product nucleus are observed. The relative populations of these states and the angular distributions of the particles leading to them provide sensitive tests for the theoretical models that have been proposed, which are essentially extensions of the optical model (see, e.g., P1).

Transfer Reactions. This class of reactions includes stripping and pickup reactions and may involve the transfer of a single nucleon, two nucleons, or clusters of three or more nucleons. All of these processes have common characteristics. The spectra of outgoing particles generally show pronounced resonances corresponding to discrete energy states being populated in the product nucleus, and the angular distributions are peaked toward the forward direction and have structure indicative of angular-momentum effects, as in elastic scattering (see figure 4-13). Transfer reactions are therefore particularly useful for the determination of energies, spins, and parities of excited states of nuclei (M1, H4). The states excited in stripping and pickup reactions are, in general, shell-model (single-particle) states, since in a (d, p) reaction a single neutron is added to the target nucleus, in a (d, t) reaction a single neutron hole is created, and so on.

We illustrate the general approach to obtaining nuclear-structure in-

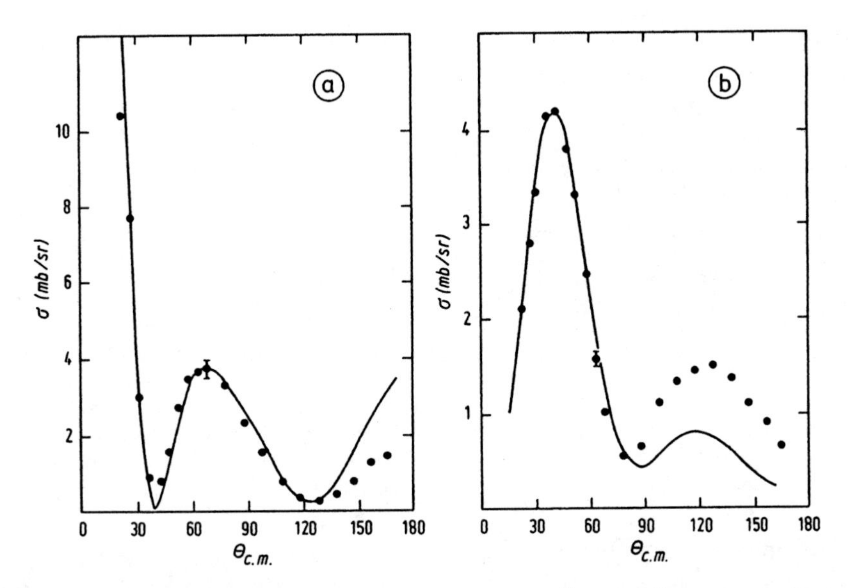

Fig. 4-13 Angular distributions of protons from the reaction $^{26}Mg(d, p)^{27}Mg$. The distributions shown are for protons going (*a*) to the ground state of ^{27}Mg and (*b*) to the first excited state of ^{27}Mg at 0.98 MeV. The points are experimental data, the curves distorted wave born approximation (DWBA) calculations with (*a*) $l = 0$, (*b*) $l = 2$. $\theta_{c.m.}$ is the center-of-mass angle. [From J. Silberstein et al., *Phys. Rev.* **136*B***, 1703 (1964)].

formation by considering as an example the simplest type of transfer, the deuteron-stripping reaction $A(d, p)B^*$, where B^* denotes a specific excited state whose energy, spin, and parity are to be determined. A measurement of the energy spectrum of the emitted protons will, by conservation of energy, give the energies of the excited states of B, provided that the binding energy of the neutron in nucleus B is known. The spin and parity of a state may be estimated in the following manner from the angular distribution of the emitted protons corresponding to formation of that state. Consider a vector (momentum) diagram for the (d, p) reaction as illustrated in figure 4-14: the deuteron approaches with momentum $\mathbf{p}_d = \mathbf{k}_d\hbar$ and the proton goes off with momentum $\mathbf{p}_p = \mathbf{k}_p\hbar$ at an angle θ with respect to the incident beam. The momentum of the captured neutron may be obtained from the conservation of momentum:

$$k_n^2 = k_d^2 + k_p^2 - 2k_d k_p \cos \theta. \tag{4-49}$$

If the neutron is captured at an impact parameter R, *orbital* angular momentum carried in by the captured neutron is classically given by [cf (4-6)] $l_n\hbar = Rk_n\hbar$, or quantum-mechanically by

$$\hbar\sqrt{l_n(l_n + 1)} = Rk_n\hbar. \tag{4-50}$$

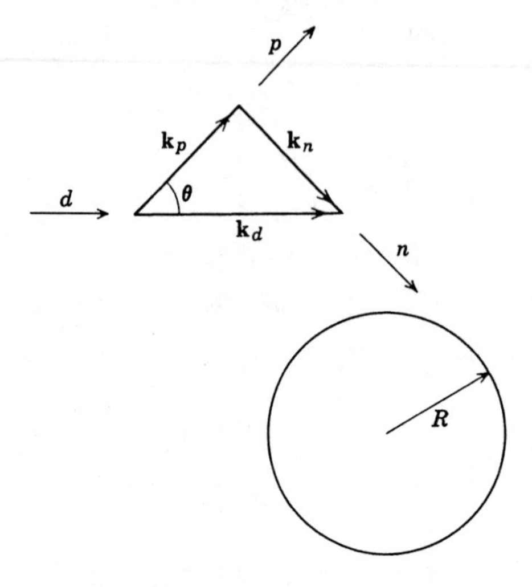

Fig. 4-14 Momentum diagram for (d, p) stripping reaction with proton emitted at an angle θ and neutron captured with impact parameter R.

Combining (4-49) and (4-50), we get

$$\frac{l_n(l_n+1)}{R^2} = k_d^2 + k_p^2 - 2k_dk_p \cos\theta. \tag{4-51}$$

Since k_d and k_p are measured quantities and R can be taken to be approximately the nuclear radius (because stripping presumably occurs mainly for peripheral collisions), there is a definite relation between l_n and θ. Although a more accurate treatment of the problem relaxes this relation somewhat, the position of the principal peak in the angular distribution usually yields an unambiguous value for l_n (see figure 4-13). The spin of the excited state of B may then be bracketed by an inequality resulting directly from the conservation of angular momentum:

$$\text{minimum of } |I_A \pm l_n \pm \tfrac{1}{2}| \leq I_B \leq I_A + l_n + \tfrac{1}{2}. \tag{4-52}$$

Conservation of parity demands that A and B have the same parity if l_n is even, and opposite parities if l_n is odd.

The analysis of transfer reactions is almost always done with the use of the so-called distorted-wave Born approximation (DWBA). In this method of quantum-mechanical analysis the wave functions representing incoming and outgoing particles are not plane waves, as in the simpler Born approximation, but are distorted by the effects of the Coulomb and nuclear potentials. The optical-model potential is generally used, and in the analysis of a particular reaction the appropriate optical-model parameters are usually obtained from elastic-scattering data for the nucleus involved or for a group of nuclei in the same region.

Transfer reactions with projectiles other than deuterons, such as (t, α), $(^3\text{He}, d)$, and (α, d), and the many possible transfer reactions with heavy ions are also often studied and can give information about excited states, but the analysis becomes considerably more complicated when more than one nucleon is transferred or when there is a large Coulomb repulsion between the reacting nuclei, as in heavy-ion reactions (see e.g., M1, K1).

4. *Preequilibrium Decay*

Although the compound-nucleus and direct-interaction models have been remarkably successful in accounting for a large body of nuclear-reaction data, there are phenomena that cannot be explained in terms of these models. This should not be surprising since the two models take rather extreme points of view: complete statistical equilibrium in one case, interaction with an individual nucleon or small cluster of nucleons in the other. In the 1960s it became clear that some intermediate model was needed to account for the frequent observation of high-energy tails on spectra of emitted particles. Following the early work of J. J. Griffin (G3), it became evident that these continuous spectra at energies too high to be accounted for by the statistical theory arise because particles can be emitted *prior* to attainment of statistical equilibrium; hence the name precompound or preequilibrium decay. The subject is reviewed in B4.

Experimental Observations. An example of the kind of data that gave impetus to the formulation of the preequilibrium model is shown in figure 4-15. The proton spectrum from the reaction ^{54}Fe (p, p') induced by 62-MeV protons clearly shows three components: an evaporation peak at low energies; some sharp resonances at 50–62 MeV corresponding to population of discrete excited states of ^{54}Fe, presumably by direct interaction; and a

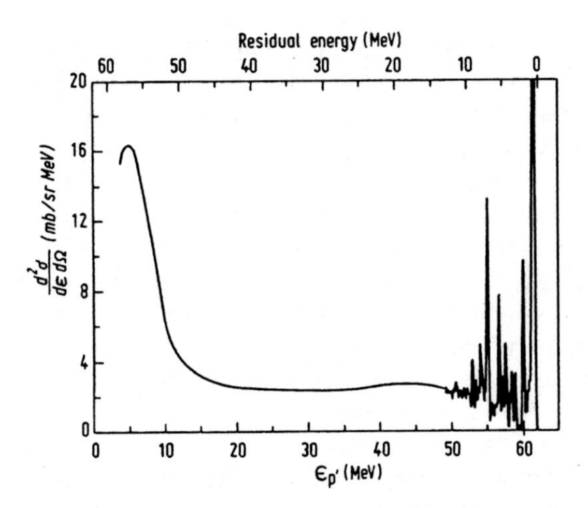

Fig. 4-15 Proton spectrum at 35° obtained in 62-MeV bombardment of ^{54}Fe. (From reference B4.) Reproduced, with permission, from the Annual Review of Nuclear Science, Volume 25 © 1975 by Annual Reviews Inc.

broad continuum in between that is not accounted for by either compound-nucleus or direct-interaction theory.

The spectra resulting from preequilibrium decay appear to be rather independent of target nucleus for a given projectile and energy, but they vary considerably with mass and energy of projectile. The angular distributions are generally forward-peaked, most strongly so for the highest-energy particles emitted.

Models for Preequilibrium Decay. Perhaps the most straightforward approach to calculating preequilibrium emission is to extend the intranuclear cascade model of high-energy reactions (cf. p. 148 and section G) to the intermediate energy range (20–100 MeV). In this model individual nucleon-nucleon collisions are followed in time and space by Monte Carlo simulations; in the course of the simulated cascades some nucleons reach the nuclear surface with sufficient energy to escape from the nuclear potential prior to complete equilibration. Comparison of the calculated energy spectra of these escaping cascade nucleons with experimental data (such as those in figure 4-15) shows quite good agreement, although the model appears to predict insufficient intensities of preequilibrium nucleons in the backward hemisphere. A shortcoming of most existing programs (although, in principle, not of the model) is that they are applicable to nucleon-induced reactions only and that they give no information about the emission of complex particles such as deuterons and α particles.

Most analyses of preequilibrium decay have been based on the exciton model of Griffin (G3) and its modifications (B4). In this model, too, successive two-body interactions are invoked, but without spatial considerations. Starting from the observation that the collision between an excited (or incoming) nucleon and a bound nucleon leads to an additional excited nucleon and a nucleon hole, the model focuses on the total number of **excitons** (excited particles plus holes) at each step. With each collision the exciton number either stays constant or increases by two, and the number of ways in which a given excitation energy can be distributed among excited particles and holes goes up rapidly with the exciton number. The further assumption is made that, for a given exciton number, every possible particle-hole configuration (including those with unbound particles) has equal a priori probability. This makes it possible to calculate the fraction of states with unbound particles as well as the energies of these unbound (and therefore escaping) nucleons at each exciton number. By summing over the different exciton numbers one can then obtain energy spectra. To calculate absolute intensities it is further necessary to specify the transition *rates* between successive states in the exciton model. In an important modification of the model these intranuclear transition rates are obtained from mean free paths of nucleons in nuclear matter based on the imaginary part of the optical potential. The modified exciton or hybrid model gives generally excellent agreement with experiment, but has the serious shortcoming of giving no information about angular distributions.

E. LOW-ENERGY REACTIONS WITH LIGHT PROJECTILES

In this and the following sections we give a survey of types of nuclear reactions, classifying them somewhat arbitrarily into low-energy reactions induced by light ($A \leq 4$) projectiles, fission, high-energy reactions, and heavy-ion reactions. In describing the phenomena we draw on the preceding mechanistic discussion. In the present section, "low energy" is to be taken as $\lesssim 50$ MeV, with the upper limit not well defined but meant to signify the energy region above which the compound-nucleus mechanism is no longer dominant.

Slow-Neutron Reactions. We have already indicated the principal features of slow-neutron reactions and recapitulate them only briefly. As expressed in (4-15), slow-neutron cross sections can be very large relative to nuclear dimensions. Slow-neutron reactions are the purest example of compound-nucleus behavior and, in fact, the narrow resonances in (n, γ) cross sections (cf. figure 4-9) led to the development of the compound-nucleus model. The $1/v$ law (p. 137) governs most neutron cross sections in the region of thermal energies.

Neutrons are available from nuclear reactions only, and they are always produced with appreciable kinetic energies. However, they are readily thermalized (that is, brought to a Maxwellian distribution of velocities corresponding to the temperature of their surroundings) by repeated collisions with light nuclei, especially protons in a hydrogenous medium.

Important slow-neutron reactions, in addition to the ubiquitous (n, γ) process, are the fission reaction discussed in section F, and a few light-element reactions such as $^{14}N(n, p)^{14}C$ and $^{10}B(n, \alpha)^{7}Li$. The latter reaction has a thermal-neutron cross section of 4×10^3 b.

Reaction Cross Sections. Except with the very lightest nuclei, the Coulomb barrier makes it impossible to study nuclear reactions with charged particles of kinetic energies below the million-electron-volt region. Reactions with charged particles or neutrons above about 1 MeV differ in two important ways from those with slow neutrons:

1. The isolated resonances are no longer observable because their spacings become small compared to their widths.
2. With increasing energy an increasing variety of reactions becomes possible.

The total reaction cross section for charged particles rises from essentially zero at energies just a little below the Coulomb barrier and asymptotically approaches πR^2 where R is the distance between centers of incident and target nuclei when they "feel" one another's nuclear forces (interaction radius). The asymptotic value of the cross section is of the order of 1 b. For neutrons the reaction cross section descends from the very high values (hundreds or thousands of barns) in the electron-volt energy region and also approaches πR^2.

Excitation Functions. As we have seen, excitation functions such as those in figures 4-4 and 4-5 provide important evidence for the compound-nucleus mechanism. The pronounced maxima are characteristic and arise from the competition that sets in as new reaction channels open up. For example, figure 4-4 shows the (α, n) and (α, p) excitation functions going through maxima at the energy at which (α, pn) and $(\alpha, 2n)$ cross sections are starting to rise, and these in turn have their maxima at the onset of the $(\alpha, 2pn)$ and $(\alpha, p2n)$ reactions. This is just the behavior expected from the compound-nucleus model, in which each emitted particle carries away only a fraction of the available excitation energy. The reactions observed may be considered as proceeding in the following manner:

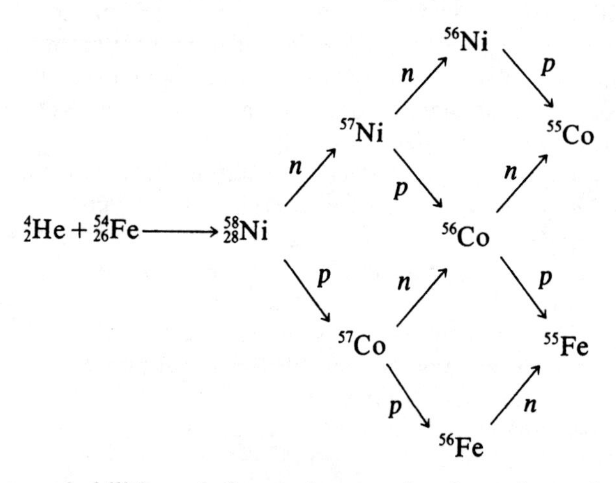

The relative probabilities of the various paths depend on the excitation energy of the ^{58}Ni compound nucleus and may be calculated by evaporation theory (cf. pp. 141–147). Agreement between calculated and observed excitation functions provides the strongest support for the interpretation of many reactions in this energy range in terms of the compound-nucleus model. Conversely, serious discrepancies between evaporation calculations and measured excitation functions are indicative of noncompound contributions to the reactions. For example, as already mentioned, the change in slope of the (α, n) and (α, p) excitation functions above about 30 MeV in figure 4-4 and the flat high-energy tails of the excitation functions in figure 4-5 are not predicted by evaporation theory and are probably due to preequilibrium emission.

Deuteron Reactions. Reactions induced by deuterons present a somewhat special case. Although compound-nucleus formation can certainly take place, it does not appear to be the dominant mechanism at any energy. Because of the large size and loose binding of the deuteron, direct reactions in which one of the nucleons is stripped off by collision with a nucleus are quite prevalent. What is more remarkable is the observation, made early in the study of nuclear reactions, that (d, p) reactions occur at deuteron energies well below the Coulomb barrier of the target nucleus and that the cross sections are considerably larger than those for the corresponding (d, n) reactions, particularly for heavy nuclei. These two observations are completely at odds with what would be expected from the compound-nucleus model. The apparent anomaly was explained by J. R. Oppenheimer and M. Phillips (O1) as the result of polarization of the deuteron by the Coulomb field of the nucleus, the deuteron being oriented with its "proton end" away from the nucleus as it approaches. Because of the relatively large neutron-proton distance in the deuteron (several fermis) the neutron comes within the range of nuclear forces while the proton is still outside most of the Coulomb barrier, and the weakly bound deuteron (binding energy 2.23 MeV) can be broken up, leaving the proton outside the barrier. An analogous mechanism appears to be responsible for the low-energy $(^3He, p)$ reaction. An interesting feature of the Oppenheimer-Phillips process is that the emergent protons have a spread of energies that includes values in excess of the incident deuteron energy, so that in a fraction of the events the excitation of the compound nucleus is that which would result from the capture of a neutron of negative kinetic energy.

Competition among Reactions. In compound-nucleus reactions the competition among the different energetically possible reactions depends on the relative probabilities for emission of various particles, such as p, n, d, α, ^3H, ^3He, and fission fragments, from the compound nucleus. The emission probability for a given particle b is determined, as expressed in (4-43) and (4-44), by the energy available $(U_C - S_b)$, the Coulomb barrier (via the inverse cross section σ_{Bb}), and the density of final states in the product nucleus (ω_B). The effect of the Coulomb barrier is to suppress charged-particle emission relative to neutron evaporation, especially at high Z. However, for a given Z neutron-binding energies increase and proton-binding energies decrease with decreasing A, and this effect can make proton emission quite competitive for light and medium elements, say $Z \leqslant 40$. The odd-even effects on level density can have an even larger influence on the relative emission probabilities, as discussed on p. 146 and illustrated by the (p, pn) and $(p, 2n)$ cross sections in figure 4-5.

To get some orientation on the reactions to be expected let us consider the bombardment of a medium-Z element $(Z \approx 30)$ with projectiles of various energies. The predominant reactions with protons of 5–15 MeV are the (p, n) and (p, p') reactions. Because α particles are not very tightly bound in most nuclei, (p, α) reactions can also be significant despite the higher Coulomb barriers. As the energy of the incident protons is increased into the 15–25 MeV interval, reactions such as $(p, 2n)$, (p, pn), $(p, 2p)$, $(p, \alpha n)$

will become dominant at the expense of the reactions in the lower-energy interval. In this energy region reactions such as $(p, ^3\mathrm{He})$, $(p, ^3\mathrm{H})$, and (p, d) are also observed, although usually with smaller cross sections and with the characteristics of a pickup rather than a compound-nucleus mechanism. As the energy of the proton is further increased, the compound-nucleus reactions in which three or more particles are emitted begin to dominate. These reactions will include, for example, $(p, 3n)$, $(p, p2n)$, $(p, \alpha pn)$, and $(p, \alpha 2n)$. The pickup reactions will continue to occur, still with rather smaller cross sections. In the vicinity of 40–50 MeV reactions with four or more emitted particles will predominate. The excitation function for the product of the (p, α) reaction may well have a second peak in this energy region, corresponding to the $(p, 2n2p)$ process, which of course requires more energy.

If the incident particle is an α particle instead of a proton, the pattern of reactions will be much the same, except that there may be some enhancement of α-particle emission because of direct reactions that may occur. Again there will be some emission of ^{3}He and ^{3}H, but now the direct contribution will be due to stripping rather than to pickup. Deuteron-induced reactions have generally similar patterns, but with important contributions from the previously mentioned pickup reactions.

When the atomic number of the target is increased, the increasing Coulomb barrier progressively suppresses the emission of charged particles until at bismuth the main processes are (p, xn) reactions where x, the number of neutrons emitted, increases with bombarding energy, reaching a value of 4 or 5 around 50 MeV. But even here there will still be some proton emission, partly from compound-nucleus reactions but largely from direct reactions. Alpha particles will still be seen (although with low cross section); the binding energy of the α particle becomes negative in the heavy elements and thus can partly compensate for the increase in Coulomb barrier. Again, a similar pattern holds for α-particle irradiation, except that the Coulomb barrier for the incident α particle is so high that, at incident energies at which the total reaction cross section is appreciable, the (α, n) reaction is already suppressed in favor of the $(\alpha, 2n)$ reaction. The emission of d, t, ^{3}He, and α by direct processes takes place with cross sections not very different from what they are with lighter targets.

Energy Spectra. The energy spectrum of the α particles emitted in the Ni (p, α) reaction shown in figure 4-6 again substantiates the expectation of the compound-nucleus model: the energy spectrum of the α particles has a maximum in the vicinity of the Coulomb barrier (about 10 MeV), and thus most of the α particles are emitted with the minimum possible energy. It is also to be noted that the energy spectra for the emitted α particles are very similar both at 30° and at 120° with respect to the incident beam, as expected from the random motion of the nucleons within the excited compound nucleus. The second peak in the 30° spectrum corresponds to the emission of α particles with all or nearly all of the available energy and is presumably due to direct interactions. A large body of experimental data on energy spectra and angular distributions of particles emitted in low-energy reactions is available, and it is on the basis of such data that fairly clear distinctions between compound-nucleus and direct reactions can be made.

In summary, the excitation function for a medium-energy reaction rises to a maximum and then diminishes because of competition from other reactions that become energetically possible. The energy spectra of most of the emitted particles have peaks in the vicinity of the lowest energy that allows the particle to escape from the nucleus. However, in addition, some high-energy particles are usually emitted preferentially in the forward direction.

Photonuclear Reactions (F4, F5). The first nuclear reaction induced by photons was discovered by Chadwick and Goldhaber in 1934 (C1): the photodisintegration of the deuteron. They used the high-energy γ's from a radiothorium source (^{208}Tl γ's of 2.61 MeV) and were able to deduce a fairly accurate value for the neutron mass from their measurement of the energy of the protons produced. The only nuclide other than deuterium with low enough threshold (neutron-binding energy) to permit photodisintegration by naturally occurring γ rays is ^{9}Be.

Reactions between nuclei and low- and medium-energy photons are dominated by what is known as a giant resonance: in all nuclei the excitation function for photon absorption (not just for a specific reaction) goes through a broad maximum a few million electron volts wide. The energy of the resonance peak varies smoothly with A, decreasing from about 24 MeV at ^{16}O to about 13 MeV at ^{209}Bi. Peak cross sections are 100–300 mb.

This giant-resonance absorption is ascribed to the excitation of dipole vibrations of all the protons against all the neutrons in the nucleus (G4), the protons and neutrons separately behaving as compressible fluids. This model makes some fairly simple predictions about the magnitude and A-dependence of the resonance that are quite well borne out by the experimental data: the integrated cross sections under the resonance peaks are given to good approximation by $0.06NZ/A$ MeV b, and the peak energies can be approximately represented by $aA^{-1/3}$; however, a is not quite constant but varies from about 60 MeV for the lightest to about 80 MeV for the heaviest nuclei.

The energy of the dipole resonance is so low that mostly rather simple processes—such as (γ, n), (γ, p), some $(\gamma, 2n)$, and (in heavy elements) photofission reactions—take place in the giant-resonance region. The competition between these processes is governed by the usual statistical considerations of compound-nucleus de-excitation, so that neutron emission usually dominates.

Above the giant resonance, from 30 to ~140 MeV, the absorption cross section remains approximately constant at roughly one tenth the peak cross section. At these and still higher energies an important mechanism appears to be the absorption of a photon by a neutron-proton pair, termed the quasi-deuteron mechanism. This comes about because a high-energy photon cannot transfer all its energy to an individual nucleon since its own low momentum would make momentum conservation impossible, whereas it can interact with a nucleon pair, with the two nucleons then flying apart in nearly opposite directions. Since pairs of like nucleons do not have dipole moments, they are not effective for photon absorption. In light nuclei the proton and neutron have a high probability of escaping from the nucleus; in heavy nuclei they have an appreciable probability of interacting with other nucleons, leading to more complex reactions. Angular distributions and energy spectra of protons give evidence for the validity of the quasi-deuteron model.

At energies above the pion production threshold (~140 MeV) photon absorption cross sections gradually rise; presumably pions produced inside the nucleus can be reabsorbed and thus distribute their energy among nucleons. The processes observed are similar to those discussed in section G.

F. FISSION

In view of our discussion of spontaneous fission in chapter 3, it should not be surprising that fission is another possible mode for the de-excitation of an excited compound nucleus and, in the region of high atomic numbers, competes with the evaporation of nucleons and small nucleon clusters. Whereas spontaneous fission requires tunneling through the Coulomb barrier, induced fission comes about when enough energy is supplied by the bombarding particle for the barrier to be surmounted. Fission by thermal neutrons has, of course, assumed enormous practical importance and is probably the most intensely studied nuclear reaction. Its unique importance results from the large energy release of close to 200 MeV accompanying the reaction and from the fact that in each neutron-produced fission process more than one neutron is emitted, which makes a divergent chain reaction possible.

In the following paragraphs we review very briefly the major experimentally observed features of fission reactions (H5, H6, V1), and then sketch qualitatively the theoretical ideas that have been developed to account for the phenomena.

1. *Experimental Observations*

Fission Cross Sections. Some nuclides, notably those with an odd number of neutrons like ^{233}U, ^{235}U, ^{239}Pu, and ^{242}Am, are fissioned by

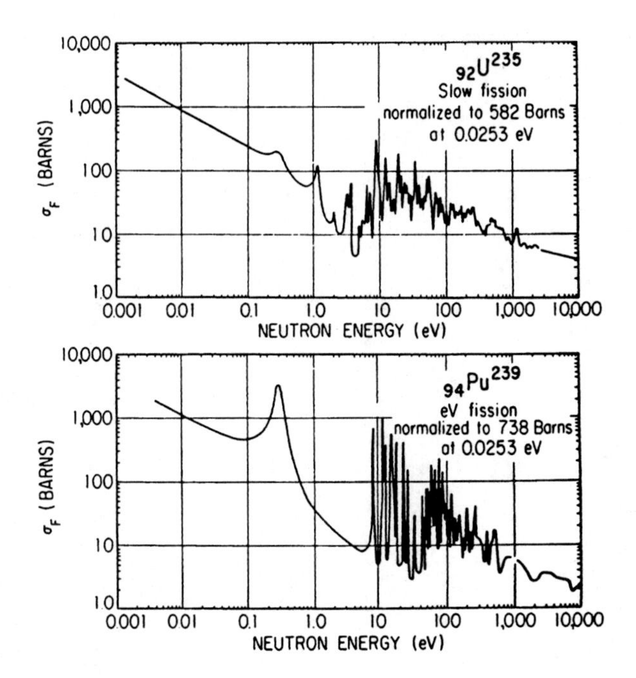

Fig. 4-16 Fission cross section of ^{235}U and ^{239}Pu as a function of neutron energy. (From reference H5.)

thermal neutrons with large cross sections (531, 580, 742, and ~2300 b for the examples given). In the energy region up to ~0.1 eV these fission cross sections follow the $1/v$ law, then at energies up to a few thousand electron volts they exhibit many sharp resonances, very much like (n, γ) reactions. This behavior is illustrated in figure 4-16. At still higher neutron energies ($\gtrsim$0.5 MeV) σ_f becomes fairly independent of energy at values of 1–2 b.

Nuclides in the high-Z region that are not fissionable with thermal neutrons, such as ^{226}Ra, ^{232}Th, ^{231}Pa, ^{238}U, and ^{242}Pu, can undergo fission by fast neutrons, with thresholds for the reaction in the range of 0.2–1.7 MeV. As shown in figure 4-17 for the case of $^{238}U + n$, σ_f tends to rise steeply from threshold to a plateau value, then a second rise occurs in the neighborhood of 6 MeV, followed by a second plateau, and sometimes, as in this instance, by further increases alternating with plateaus. This behavior comes about through so-called second- and higher-chance fissions: when the excitation energy of the compound nucleus (in our illustration ^{239}U) becomes high enough, the residual nucleus remaining after evaporation of a neutron (^{238}U) can still be sufficiently excited to undergo fission: at still higher energies ($\gtrsim$14 MeV) the evaporation of a second neutron can be followed by fission of the excited residual ^{237}U and so forth. Such processes are designated as (n, nf) and $(n, 2nf)$ reactions.

Fission can be produced by particles other than neutrons, such as

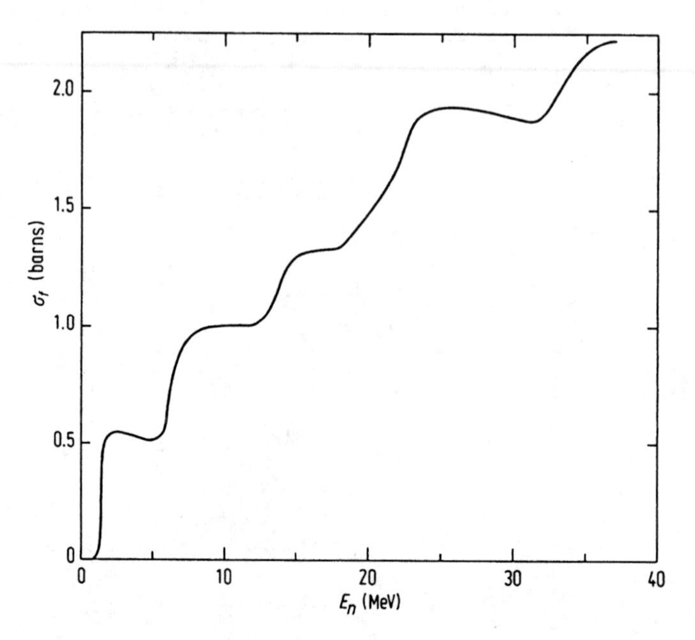

Fig. 4-17 Fission cross section of ^{238}U for neutrons up to 37 MeV. (Data from reference N2.)

protons, deuterons, and helium ions, as well as by γ rays. In charged-particle bombardments the onset of fission and the low-energy behavior of the cross section is, or course, largely determined by the Coulomb barrier. With deuterons it is possible to observe stripping accompanied by fission; with nuclei fissionable by thermal neutrons these (d, pf) processes can take place with the capture of neutrons of effectively negative kinetic energy, and this method is used for measuring the (negative) fission thresholds for such nuclei.

With increasing bombarding energies it becomes possible to induce fission in lighter and lighter elements, but in this section we are mainly concerned with fission of elements of $Z \geq 86$ at energies of ≤ 50 MeV.

Mass Distribution. The split into two fragments does not occur in a unique mode; rather, the fragments can have a wide range of mass ratios. In thermal-neutron fission of most nuclei, an asymmetric mass split, with a ratio of heavy-to-light fragment mass (M_H/M_L) of about 1.4, is much more likely than a symmetric mass split. The mass-versus-yield curves for thermal-neutron fission of ^{235}U and ^{239}Pu are shown in figure 4-18 along with that for spontaneous fission of ^{252}Cf. The curves are seen to be approximately symmetrical about the minima corresponding to equal-mass splits. The high-mass peak, especially its left-hand portion, is almost the same for the different fissioning nuclei and for others with A between 229 and 254, whereas the low-mass peak shifts more. The near-constancy of

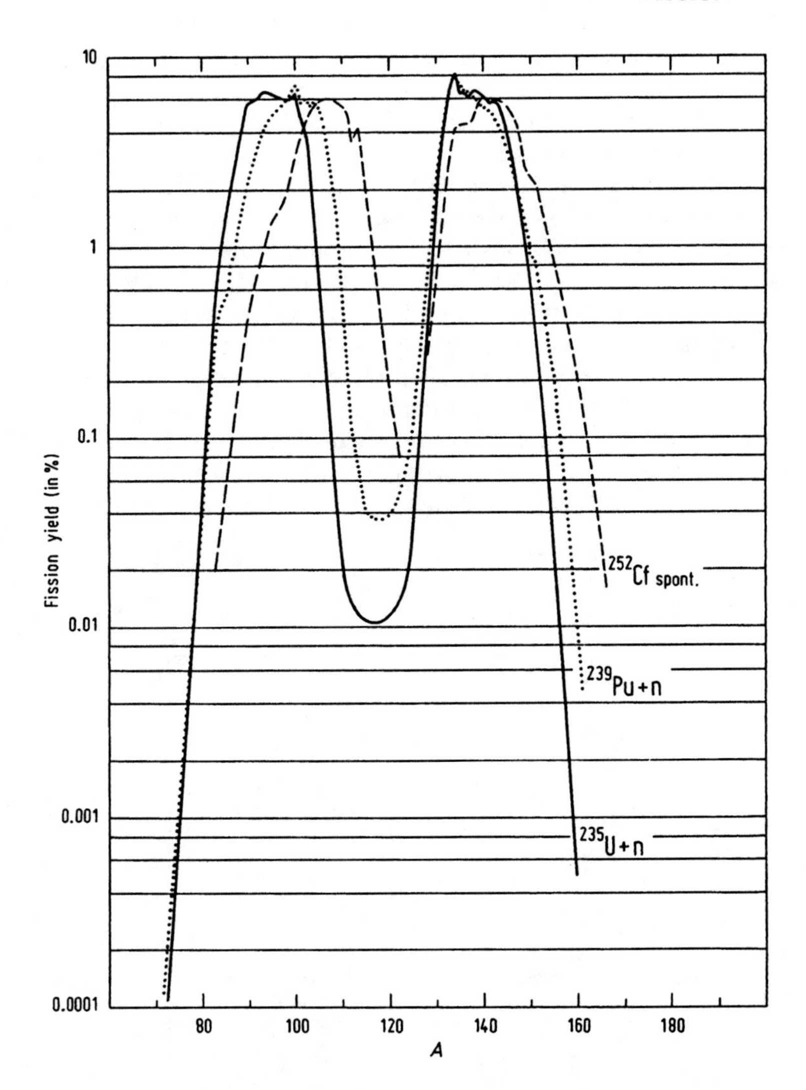

Fig. 4-18 Fission product mass distributions for the thermal-neutron-induced fission of ^{235}U and ^{239}Pu and the spontaneous fission of ^{252}Cf. (From data compiled in reference H5.)

the left-hand edge of the high-mass peak has been ascribed to the stabilizing effect of the closed shells at $Z = 50$ and $N = 82$, both of which occur in this region. Shell effects are also undoubtedly responsible for the fine structure in the fission-yield curves that can be seen near the maxima in figure 4-18. Preferential formation of primary fission fragments with closed-shell configurations, as well as post-fission neutron emission from fragments containing 51 or 83 neutrons, probably plays a part in bringing about this fine structure.

With increasing bombarding energy the valley between the two humps in the fission-yield curve gradually fills in, so that, for example, with 14-MeV neutrons on ^{235}U, the peak-to-valley ratio is only about 6 (compared to about 600 for thermal neutrons). When the bombarding energy reaches about 50 MeV the fission-yield curves for the highly fissile elements ($Z \geq 90$) exhibit single broad humps with the valley completely filled in. Much lighter elements, such as bismuth and lead, give rise to much narrower, single-peaked distributions near threshold, and these broaden with increasing bombarding energy. The fission of the intermediate elements radium and actinium near threshold exhibits a triple-humped yield distribution. Finally it has recently been established that for the very heaviest elements, that is in the region of fermium, the probability of a symmetrical mass split again increases relative to the asymmetric modes until, for the thermal-neutron fission of ^{257}Fm, symmetric fission predominates. Several of the different types of fission-yield curves are shown in figure 4-19. Exhaustive reviews of fission product-yield data may be found in C2 and D2.

Charge Distribution. Since the high-Z elements that undergo fission

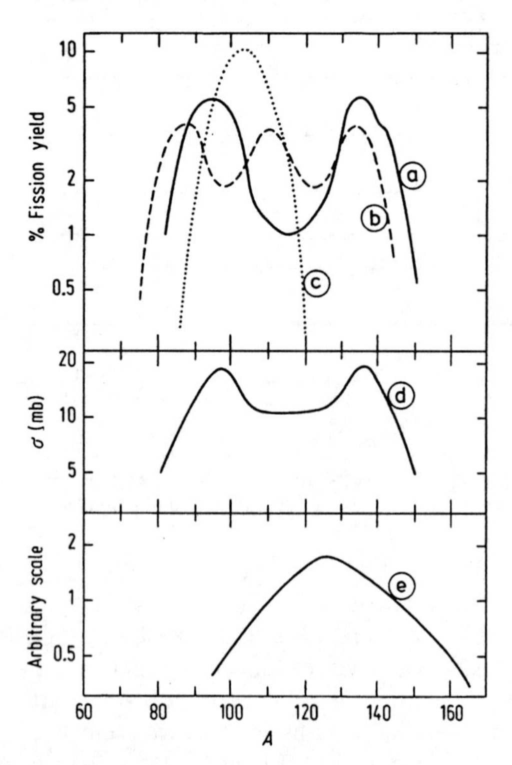

Fig. 4-19 Mass-yield curves for the fission of (*a*) ^{235}U by 14-MeV neutrons, (*b*) ^{226}Ra by 11-MeV protons, (*c*) ^{209}Bi by 22-MeV deuterons, (*d*) ^{235}U by 26.8-MeV ^{4}He, (*e*) ^{257}Fm by thermal neutrons. Note different ordinate scales. (Data from various literature sources.)

have much larger neutron-proton ratios than the stable nuclides in the fission product region, the primary fission products are always on the neutron-excess side of β stability. Each such primary product then decays by a series of successive β^- processes to its stable isobar. Beginning with the startling discovery by Hahn and Strassmann that a barium isotope resulted from the reaction of uranium with neutrons, which constituted the discovery of fission, an enormous amount of radiochemical work has been required to identify and characterize the several hundred fission products—about 90 mass chains, each with several members. As techniques are refined and shorter and shorter half lives become accessible, work on fission product characterization continues. The total or chain yield of a given A is best determined by mass-spectrometric measurement of the stable end product of the β^--decay chain or by radioassay of the last, usually fairly long-lived radioactive member. The independent yields of members along the decay chains are harder to measure because they must be rapidly separated from their radioactive precursors. A few so-called shielded nuclides—shielded from β^- decay by a stable isobar one unit lower in Z—occur in small yield (^{136}Cs, shielded by stable ^{136}Xe, is an example); they are unambiguously formed as direct, primary products.

Although, as we said, it is difficult to obtain several independent yields along a given mass chain, such data do exist now for many mass numbers and, provided they are corrected for an odd-even effect,[14] they are all consistent with a narrow Gaussian distribution of isobaric yields around a most probable charge Z_p and with a width that is approximately independent of A. In other words, the probability of a primary product having atomic number Z is given by

$$P(Z) = \frac{1}{\sqrt{c\pi}} \exp\left[-\frac{(Z - Z_p)^2}{c}\right],$$

which holds for all mass numbers, with the parameter $c = 0.79 \pm 0.14$ (for ^{235}U fission by thermal neutrons). This corresponds to a full width at half maximum of 1.50 ± 0.12 charge units (W1).

The values of Z_p obtained from the experimental data are usually discussed in terms of their displacement from the charge Z_{UCD} that would result if the postulate of unchanged charge distribution (UCD) prevailed, that is, if the primary fragments had the same charge-to-mass ratio as the fissioning nucleus. The data can be approximately represented by $Z_p - Z_{UCD} \approx 0.5$ for light fragments, and $Z_p - Z_{UCD} \approx -0.5$ for heavy fragments.[15]

[14] Recent accurate measurements have established that the yields of even-Z products are systematically higher than those of odd-Z products (the deviations from the average Gaussian being ±25% for ^{235}U).

[15] An empirical postulate that represents the observed charge division rather well is that of equal charge displacement (ECD). It states that in any given fission event the two complementary products are equally far displaced from the line of β stability, i.e., $(Z_A - Z_p)_H = (Z_A - Z_p)_L$.

In thermal-neutron fission the Z_p value for any given mass number is usually 3–4 units smaller then Z_A, the Z on the β-stability line. With increasing bombarding energy Z_p tends to move closer to Z_A, and the distribution around Z_p broadens.

Kinetic-Energy Distribution. The total energy released in a fission process is given by the difference between the mass of the fissioning system (the excited compound nucleus) and the sum of the masses of the primary fragments. Some of this energy goes into emission of prompt neutrons and into internal excitation of the fragments, but most of it appears in the form of kinetic energy of the fragments. The average total fragment kinetic energy lies between 160 and 190 MeV and varies for different fissioning nuclei approximately as $Z^2A^{-1/3}$. The actual kinetic-energy release varies rather markedly with the mass split; in thermal-neutron fission of ^{235}U and ^{239}Pu it is largest for a slightly asymmetric split ($M_H/M_L \approx 1.25$) and some 10–20 percent lower for symmetric and highly asymmetric mass splits.

In low-energy fission the two fission fragments travel in opposite directions and have momenta of equal magnitude: $M_Hv_H = M_Lv_L$. Since the ratio of kinetic energies T_H/T_L is given by $M_Hv_H^2/M_Lv_L^2$, it follows from this momentum-conservation condition that $T_H/T_L = M_L/M_H$. Thus a measurement of the kinetic-energy ratio in a fission event gives directly the mass ratio. Mass distributions can also be deduced from measurements of both T and v of single fragments or from measurement of v_L and v_H of coincident fragments by time-of-flight techniques. In comparing the mass distributions obtained from such physical measurements with the radiochemically determined ones, we must be aware that some of them, for example, those based on double-velocity measurements, give the masses of fragments *prior* to neutron emission.

Prompt Emission of Neutrons and Other Particles. The emission, on the average, of several neutrons per fission is crucial for the possibility of maintaining a chain reaction and thus for the applications of the fission process. The average number $\bar{\nu}$ of neutrons per fission is 2.41, 2.48, and 2.88, respectively, for the thermal-neutron fissions of ^{235}U, ^{233}U, and ^{239}Pu. The actual ν values for individual fission events are distributed in approximately Gaussian fashion around $\bar{\nu}$ with a width of slightly more than one neutron. The value of $\bar{\nu}$ increases with increasing bombarding energy, and higher values of $\bar{\nu}$ are found for the fissile nuclides of higher Z (e.g., 3.76 for the spontaneous fission of ^{252}Cf and 4.02 for that of ^{257}Fm).

Measurements of the angular correlations of fission neutrons with fragments have shown that 80–90 percent of them originate from the fission fragments in flight, the remainder being emitted prior to complete separation of the fragments. More detailed studies show that ν is in fact a rather strong function of the particular mass of the fragment from which the

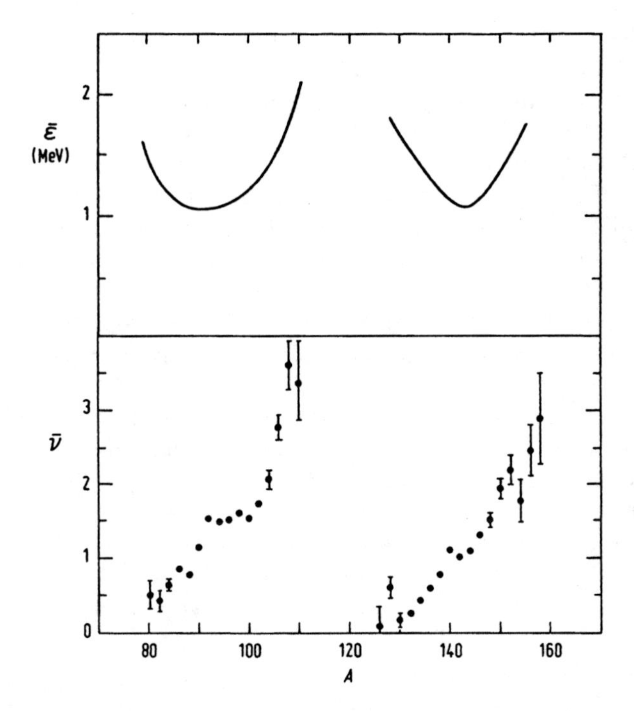

Fig. 4-20 Variation of average number $\bar{\nu}$ (lower part) and average kinetic energy $\bar{\epsilon}$ (upper part) of fission neutrons with fragment mass in the thermal-neutron fission of ^{233}U (from J. C. D. Milton and J. S. Fraser, *Physics and Chemistry of Fission*, Vol. II, International Atomic Energy Agency, Vienna, 1965, p. 39.)

neutron originates, with a plot of ν versus A always exhibiting the saw-tooth shape illustrated in figure 4-20 for thermal-neutron fission of ^{233}U. The kinetic energies of the neutrons also vary with the mass of the associated fragment as shown schematically in the upper portion of the figure.

The energy spectra of fission neutrons are very nearly Maxwellian in the frame of the moving fragment, and are well represented by $N(E) \propto E^{1/2} \exp(-E/C)$, where the constant C is approximately 1.3, but varies slightly for different systems. Average neutron energies in the laboratory system are between 1 and 2 MeV (see figure 4-20).

Neutrons are not the only particles emitted in fission, although they are by far the most abundant. One fission in every few hundred is accompanied by the emission of an α particle and, even much more rarely, protons, deuterons, tritons, and even some ^8He and Li nuclei are found. Studies of the angular correlation of the α particles with the fission fragments have been useful in elucidating some aspects of the mechanism of fission.

What we might call true ternary fission, that is, breakup into three fragments of comparable masses, has been reported as a rare event in thermal-neutron fission, but this finding remains controversial. At higher bombarding energies, and particularly in heavy-ion bombardments, such ternary fission does take place, but with cross sections never much greater than 1 percent of the binary-fission cross section.

Delayed Neutrons. In addition to the prompt neutrons, which are emitted in 10^{-14} s, a much smaller number of neutrons—of the order of 1 to 3 in 100 fissions—are emitted with time delays of between 0.08 s and nearly 1 min. These delayed neutrons, which play a very important role in the control of nuclear reactors (see chapter 14), originate from excited states of fission products that are unstable with respect to prompt neutron emission and are formed in β^--decay processes. The half lives for neutron emission are thus controlled by the preceding β^- decays. Over 60 individual delayed-neutron precursors have been identified, and for many of them the fraction of the decays leading to neutron emission as well as the neutron spectrum have been determined. Many of the delayed-neutron emitters are clustered just above the $N = 50$ and $N = 82$ closed neutron shells, as a result of the low neutron-binding energies in the β-decay daughters. Among the most prominent delayed-neutron precursors are 4.4-s ^{89}Br, 55.6-s ^{87}Br, 2.8-s ^{94}Rb, 24.5-s ^{137}I, and 1.7-s ^{135}Sb. The subject of delayed neutrons is reviewed in R2.

2. *Theoretical Framework*

Fission Barriers. In discussing spontaneous fission as a mode of radioactive decay in chapter 3, we have already made the point that fission of the heaviest elements, either spontaneous or at modest excitation energies, is made possible through the nearly perfect match between the energy release in fission and the height of the Coulomb barrier between the two fragments, both in the neighborhood of 200 MeV. As we have seen, the energy balance in the uranium region is so delicate that some nuclei, like ^{235}U, are fissile with thermal neutrons, while others, like ^{238}U, are not, although the excitation energies of the ^{236}U and ^{239}U compound nuclei formed by slow-neutron capture differ by only 1.7 MeV—they are 6.5 and 4.8 MeV, respectively. Thus we can immediately deduce that the height of the fission barrier is intermediate between these two values; it is in fact about 5.5 MeV.

Calculation of Potential-Energy Surfaces. In chapter 3 we gave a brief account of the liquid-drop approach to fission theory developed by Bohr and Wheeler. Although this theory and its various modifications were quite successful in accounting for many gross features of the fission process and formed the principal basis for "understanding" fission for a quarter century, it is clear that the calculation of absolute heights of fission barriers, and even of their variation from nucleus to nucleus, is beyond the capability of the liquid-drop model. An approach to the calculation of nuclear properties that has been remarkably successful in accounting for details of the fission process, as well as for a variety of other phenomena, was introduced by Strutinsky (S3, B5, N3). This so-called macroscopic-microscopic or shell-correction approach is based on using the liquid-drop model to represent the smooth, average properties of nuclei and correcting these by the separately evaluated single-particle or shell effects that take account of the nonuniform distribution of nucleons in phase space.

In this, as in any other theory designed to account for the details of the fission process, the first task is to map the potential-energy surfaces that connect initial and final states. This involves calculation of the potential energy as a function of nuclear deformation and, for practical computations, some functional form of the potential (such as a Woods-Saxon potential generalized to nonspherical shapes) must be chosen. The deformed shapes must also be describable in terms of a small set of parameters, typically two or three. Several parameterizations have been used; since fission must surely involve elongation along some axis and eventual formation of a neck, a convenient description is based on an elongation parameter c (defined as the ratio of the length of the deformed nucleus to the diameter of the sphere of equal volume) and a parameter h that defines the neck thickness at any given elongation. Only shapes cylindrically symmetric about the elongation axis have so far been investigated, but in order to explore the paths toward asymmetric mass splits, which are so prevalent in fission, asymmetry in the elongation direction is sometimes considered and may be expressed in terms of a third parameter α. Some typical shapes in this parameterization are shown in figure 4-21. A spherical nucleus is described by $c = 1$, $h = 0$, $\alpha = 0$.

When energies calculated for a typical actinide according to the liquid-drop model are mapped for a variety of shapes, for example, in the (c, h) representation, we obtain a surface with a rather well-developed valley along the $h = 0$ line[16] from $c = 1$ over a saddle point in the region of $c \approx 1.5$, then sloping down to $c = 1.7$ where the energy surface suddenly drops steeply towards increasing h (neck formation), eventually leading to two separate fragments. Since the deformation energy along this valley varies typically by only a few million electron volts (as a result of the near-cancellation of surface and Coulomb energies; cf. chapter 3, section C), it is not surprising that the shell corrections, although quite small relative to the *total* liquid-drop energy, can have sizable effects on the contours in the valley region. On the other hand, they will be relatively unimportant in "mountain" regions, where the liquid-drop energy is much higher than in the valley.[17] Specifically, as we see in chapter 10 (figure 10-14), single-

[16] The parameter h is in fact defined so that $h \approx 0$ corresponds to this liquid-drop valley.

[17] This is fortunate in the sense that it limits the regions of shapes for which shell-model calculations must be performed.

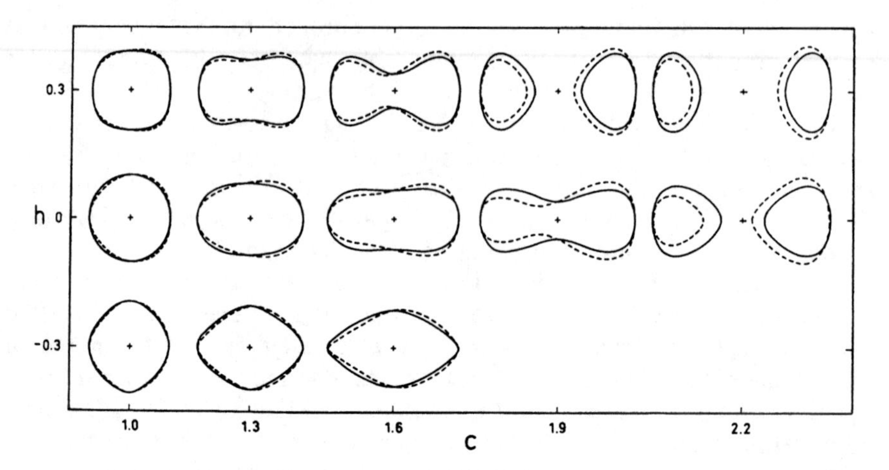

Fig. 4-21 Some nuclear shapes in the (c, h) parameterization (see text). The solid lines show symmetric shapes ($\alpha = 0$), the dotted lines represent shapes with an asymmetry parameter $\alpha = 0.2$. (From reference B5.)

particle states move up and down in energy as the deformation of a nucleus changes, and shell closures, that is, the regions of large energy gaps between adjacent levels, can occur for certain deformations at various nucleon numbers quite different from those known for spherical nuclei (20, 28, 50, 82, and so on).

The results of a typical shell-correction calculation are illustrated in figure 4-22. In the upper right a contour plot of the liquid-drop energy for ^{240}Pu is shown, with the energies normalized to that of the spherical shape as 0. The valley and the saddle point (at $c \approx 1.4$) are clearly seen. In the left-hand parts of the figure are the separate shell corrections for the 94 protons (top) and 146 neutrons (bottom). Finally in the lower right of figure 4-22 is the contour plot of the total deformation energy of ^{240}Pu, which is simply the sum of the other three maps. We see that the sphere is no longer a stable configuration; rather, the ground state—the state of lowest potential energy—is deformed ($c \approx 1.2$, $h \approx -0.15$) in accord with experimental evidence (such as ground-state quadrupole moment and rotational band structure of excited states). A very interesting and important feature is the appearance of a second potential minimum (at $c \approx 1.4$, $h \approx 0$) about 2 MeV higher than the ground state, separated from it by a barrier about 6 MeV high, and followed (at $c \approx 1.6$) by a second, slightly lower barrier. A look at the individual contributions that give rise to the final map shows that the second potential well has its origin in this instance in the neutron shell correction; $N = 146$ happens to be a magic number for an elongation of $c \approx 1.5$.

Double-Humped Fission Barriers. The potential-energy map of ^{240}Pu is

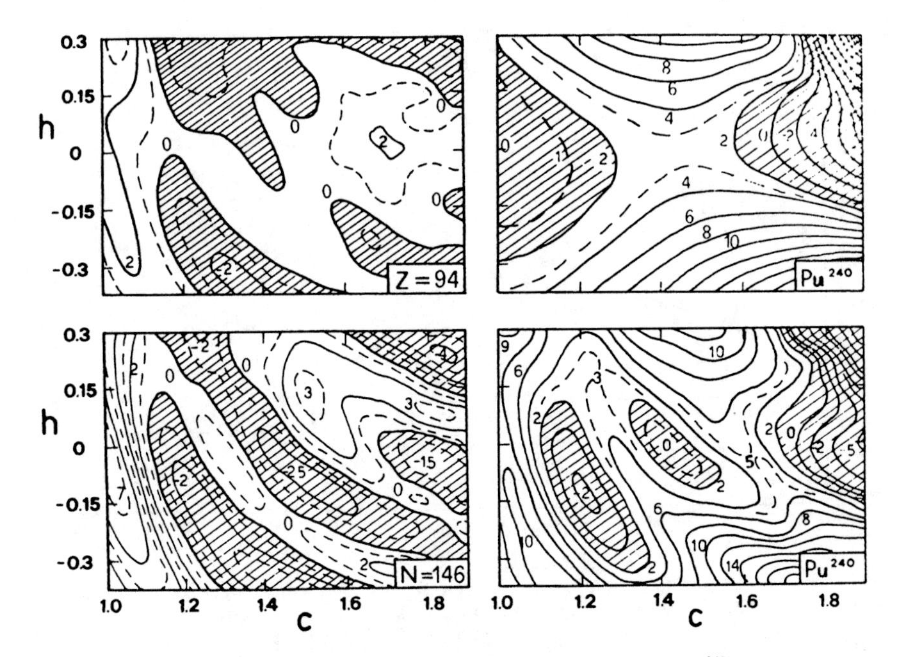

Fig. 4-22 Contour maps, in the (c, h) plane, of the potential energy of ^{240}Pu. The left-hand side shows the contour plots for the shell corrections for 94 protons (top) and 146 neutrons (bottom). The upper right-hand map represents the liquid-drop model energy of ^{240}Pu, normalized to 0 MeV for the spherical shape ($h = 0$, $c = 1$). The lower right-hand map is the sum of the other three and represents the total deformation energy of ^{240}Pu as calculated by the shell-correction method. The contour lines are drawn at 2-MeV intervals, and regions with potential energies below 2 MeV are shaded. Thus in the lower right diagram the shaded area centered around $h = -0.15$, $c = 1.2$ represents the ground state, the one around $h = 0$, $c = 1.4$ the second minimum, with the first saddle in between; the second saddle is in the vicinity of $h = 0$, $c = 1.6$. (From reference B5.)

quite representative for the lighter actinides ($228 \leq A \leq 252$). The calculations indicate for all of them a distorted ground-state shape with c between 1.12 and 1.22 and with h going from ~ -0.2 for ^{228}Ra to 0 at ^{252}Fm. For all these nuclei a second minimum is found near $c = 1.4$, $h = 0$ (corresponding to an ellipsoid of rotation with ratio of axes $\approx 2:1$). The relative heights of first and second barriers or saddle points gradually change, the second barrier decreasing with increasing A until it essentially disappears near $A = 252$. Calculated energies of first saddle, second well, and second saddle relative to the ground-state energies are diagrammed in figure 4-23 for a series of even-even nuclei. We conclude that at the lower mass numbers the second barrier is the rate-determining one, whereas this role presumably passes to the inner barrier at larger A. Another important result emerges when the asymmetry parameter α is introduced. Minimization of total deformation energy with respect to α leads, for all nuclei studied, to the conclusion that symmetric

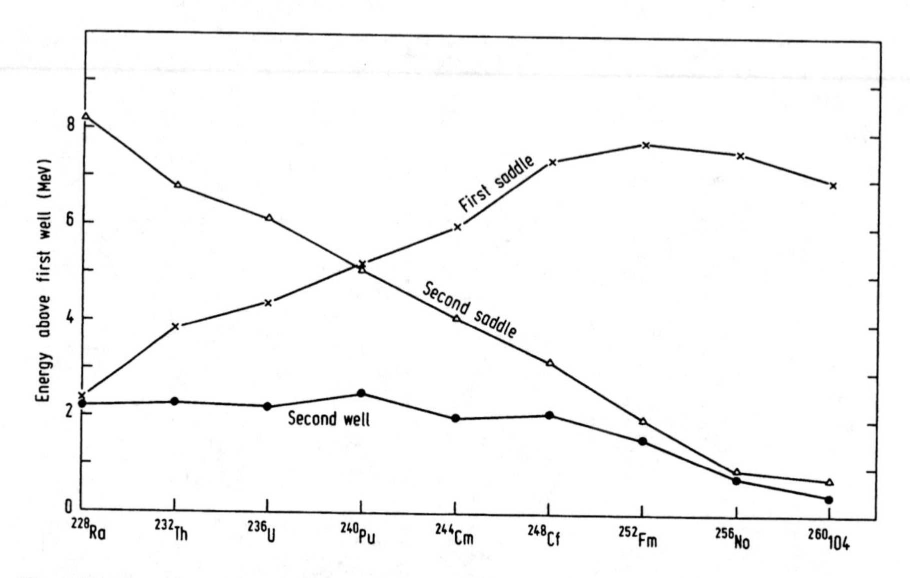

Fig. 4-23 Energies of first saddle, second well, and second saddle, relative to ground-state energy, for a series of even-even nuclides. (From data in reference B5.)

shapes are the most stable at the two potential minima and at the first saddle, but that, in the mass region we are discussing, some degree of asymmetry significantly lowers the second saddle (e.g., by about 2 MeV for ^{240}Pu). Furthermore, the introduction of asymmetry moves the second saddle toward larger *h*, that is, a thinner neck, which would lead to increased Coulomb repulsion energies for the separating fragments. The calculated asymmetries at the second saddle correspond quite well to the observed mass asymmetries which, for so long, have confounded fission theorists. The trends of the mass distributions with *A* and *Z* (see p. 160) can be, at least qualitatively, accounted for.[18] Even the observation that the total kinetic-energy release is larger for asymmetric than for symmetric mass splits follows naturally from the above mentioned increase in Coulomb repulsion that results from the narrower necks of the asymmetric saddle shapes.

Another important aspect of the shell-correction theory has already been discussed in chapter 3, that is the explanation of the spontaneously fissioning isomers in terms of the double-humped barrier.

[18] Calculations for nuclei both lighter and heavier than the region discussed above reproduce the observed trend towards symmetric mass splits. However, the triple-peaked mass distribution found in the radium region (p. 162) is not easily accounted for. The trend toward symmetric mass splits with increasing excitation energy (p. 162) is presumably associated with the decreasing importance of shell effects as the nucleons become distributed over many single-particle levels.

G. HIGH-ENERGY REACTIONS

Mass-Yield Curves. We have seen that, even in the energy region where the compound-nucleus picture accounts for most of the observed phenomena (at incident nucleon energies $\leq$50 MeV), direct reactions and preequilibrium emission also play a role. The relative importance of these mechanisms increases with increasing energy, and above 100 MeV nuclear reactions appear to proceed nearly completely by direct interactions. One reason for this remark may be seen in the contrast among the nuclear reactions induced by protons of three different energies—40, 400, and 4000 MeV—in ^{209}Bi, as illustrated schematically in figure 4-24, where the cross section for a given mass-number product is plotted against the mass number (mass-yield curve).

At 40 MeV, where most of the reactions proceed through the formation of a compound nucleus with a given excitation energy, we find that nearly all of the reactions, as expected, lead to products with mass numbers ranging from 206 to 208. At 400 MeV, on the other hand, there is a wide distribution of products that roughly divide themselves into two groups: those down to mass number 150, which are called **spallation products,**[19] and those between mass numbers 60 and 140, which we designate as fission products. At 4 GeV there is a continuous distribution of products with no evident division between fission and spallation. Raising the proton energy by another two orders of magnitude to 400 GeV causes only minor additional changes in the mass-yield curves, principally a further increase in the yields of products of $A < 30$.

It is evident from figure 4-24 that at the higher bombarding energies we do not observe the relatively few spallation products expected from the formation of a compound nucleus at a given excitation energy but rather a large array of products corresponding to various amounts of excitation energy from zero up to the maximum possible. Measurements of the energy and angular distribution of the emitted particles show some of them to be of high energy, approaching that of the incident particle, and to be preferentially emitted in the forward direction.

Cascade-Evaporation Model. The basic phenomenology of spallation reactions was discovered as soon as the first accelerators capable of accelerating protons to energies above 100 MeV became available in the late 1940s. This was before the development of the optical-model and direct-reaction theories. The ideas that we have already discussed in the context of direct interactions at lower energies—use of the impulse approximation, consideration of mean free path in nuclear matter—actually

[19] The term spallation is derived from the verb "to spall" which means to chip, to break up; the word is meant to describe reactions in which many small "chips" (nucleons, α particles, and so on) are removed from a nucleus.

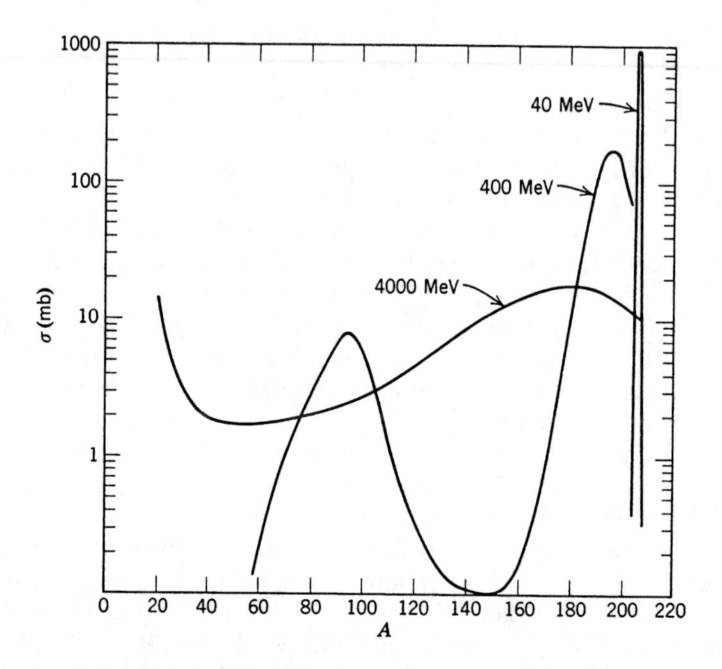

Fig. 4-24 Comparison of the approximate mass distributions of the products of the reactions of 40-, 400-, and 4000-MeV protons with ^{209}Bi.

originated in the (successful) attempts to understand spallation at high energies. In explanation of the observed phenomena R. Serber (S4) pointed out that, if the energy of the incident proton is significantly larger than the interaction energy between the nucleons in the nucleus, and its wavelength is less than the average distance between nucleons, then the incident proton will collide with one nucleon at a time within the nucleus. Further, the cross section for each collision and the angular distribution will be very nearly the same as if the collision occurred in free space rather than in the interior of a complex nucleus, that is, the impulse approximation is justified and the mean free path of the incident proton is given by (4-31). From the density of nuclear matter, $\rho \approx 10^{38}$ nucleons per cubic centimeter, and from the effective nucleon-nucleon cross section, $\bar{\sigma} \approx 30$ mb at a few hundred MeV, we obtain for the mean free path of the incident proton a value of ~3 fm, which is of the same order of magnitude as nuclear radii. Thus Serber reasoned that a high-energy proton may make only a few collisions while traversing a complex nucleus, leaving behind only a fraction of its energy and sometimes directly ejecting a nucleon with which it collides. The struck nucleons also often have considerable kinetic energy, and their passage through the nucleus can be considered in the same manner as that for the incident proton. In this fashion an intranuclear knock-on cascade of fast nucleons is generated. At energies in excess of about 350 MeV the cascade must also include the π mesons (pions) that can be created in nucleon-nucleon collisions. These pions in fact play an important part in enhancing the deposition of energy in the nucleus, because they have large cross sections for interactions with nucleons, that is, short mean free paths. Production and interactions of pions appear to be largely responsible for the rapid change in the pattern of spallation yields toward lower A above ~400-MeV bombarding energy (see figure 4-24).

This model is represented schematically in figure 4-25 for a proton incident on a complex nucleus at an impact parameter b. A cascade nucleon may either immediately escape from the nucleus, as is shown in figure 4-25 for a neutron and a proton, or it may be reduced to (or formed with) an energy so low that it is considered captured by the nucleus and gives up its energy to excitation of the whole nucleus.

It must be mentioned at this point that the other nucleons in the nucleus are not totally without effect on a collision; they occupy quantum states and so, because of the Pauli exclusion principle, make those states unavailable as final states to the two colliding nucleons. The result is a lowering of the effective collision cross section primarily through the decreased probability of very small or very large energy transfers. Collisions forbidden by the Pauli principle are shown as open circles in figure 4-25. The effect of the Pauli principle is particularly important for low-energy cascade nucleons.

At the end of an intranuclear cascade, which takes place in a time of the order of 10^{-22} s and during which several particles may be ejected, the product nucleus generally remains in an excited state. A particular target-projectile system will result in a spectrum of cascade products with a

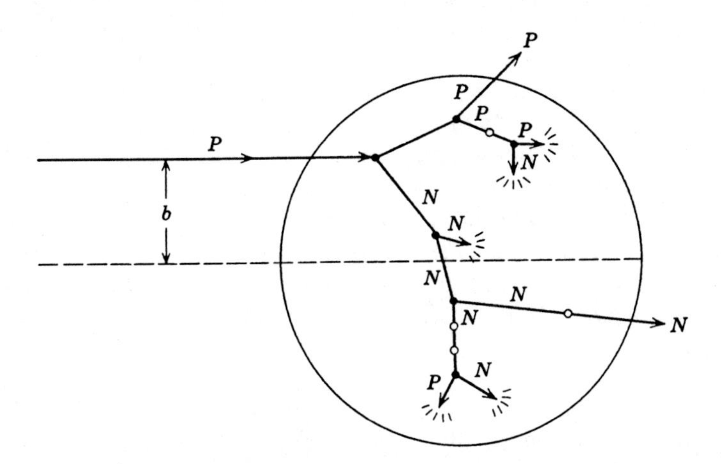

Fig. 4-25 Schematic diagram of an intranuclear cascade generated by a proton with impact parameter b. The solid circles indicate positions of collisions; the open circles represent collisions forbidden by the Pauli exclusion principle. The short arrows ending within the nucleus connote "captured" nucleons that contribute to the overall excitation.

distribution in A, Z, and excitation energy E^*. Detailed computer calculations by the Monte Carlo method have been quite successful in accounting for the distribution of cascade products as well as the energy spectra and angular distributions of cascade particles from a great variety of targets bombarded with various projectiles of energies up to about 3 GeV. (see, e.g., H7, C3, H8, B6).

The second stage of high-energy reactions, the de-excitation of the cascade products, presumably takes place by the same process as the second stage in compound-nucleus reactions: evaporation of nucleons and light nuclei, or fission. But we deal here with a *spectrum* of excited nuclei and a *spectrum* of excitation energies, rather than a well-defined compound nucleus. The products formed by evaporation are the spallation products.

Spallation Products. A large number of radiochemical studies have been made of spallation reactions with bombarding energies up to about 400 GeV. In summary of this rather complex field it is probably fair to say that essentially any spallation reaction that is energetically possible appears to occur. In general, the products in the immediate neighborhood of the target element, within perhaps 10 or 20 mass numbers on the low-mass side, are found in the highest yields. The yields for lower mass numbers then drop off rather rapidly, the rate of drop-off decreasing with increasing bombarding energy as indicated in figure 4-24. Between about 10 and 400 GeV spallation cross sections appear to remain virtually constant.

Spallation yields tend to cluster quite strongly in the region of β stability in the case of medium-weight products and increasingly more to the neutron-deficient side of stability with increasing Z of the products. This is just what is expected from evaporation theory because of the effect of the Coulomb barrier on the relative evaporation probability of neutrons and charged particles. Computer calculations of evaporation processes generally account well for the observed results (H7, B6). It is interesting to note that the ratio of yields of two isobars in spallation is nearly invariant to changes in target element and bombarding energy because it is largely determined by the final evaporation steps.

Most spallation studies have been done with protons as projectiles. Alpha-induced reactions have rather similar excitation function shapes but somewhat higher cross sections than the corresponding proton reactions.

Some typical excitation functions for spallation reactions are shown in figure 4-26.

High-Energy Fission. The characteristics of fission induced by high-energy particles differ markedly from those of thermal-neutron fission. The familiar double hump in the mass-yield curve is at these energies replaced by a single broad peak, centered around a value of A somewhat less than half the mass number of the target nuclide (cf. 400-MeV curve in figure 4-24). In contrast to low-energy fission, many neutron-deficient nuclides are

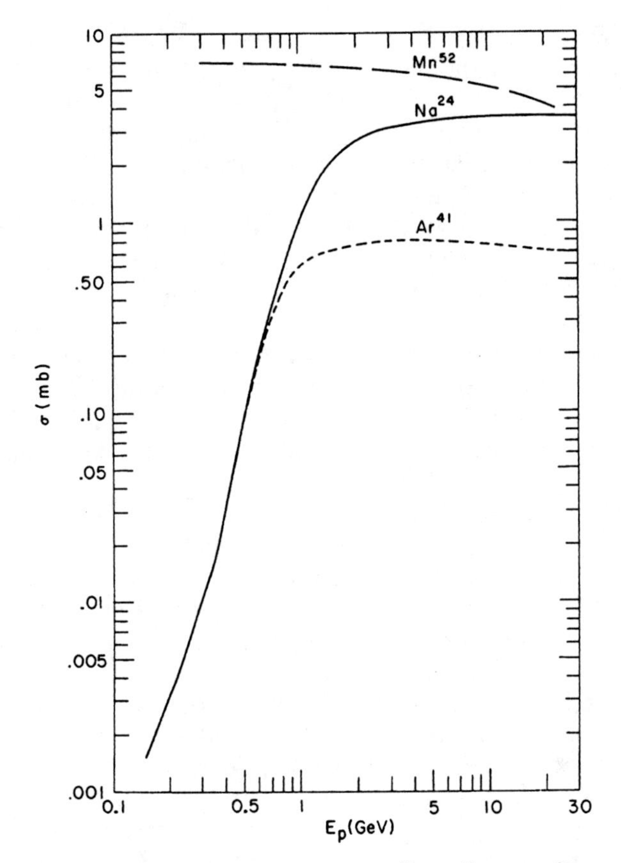

Fig. 4-26 Excitation functions for the production of ^{52}Mn, ^{41}Ar, and ^{24}Na in proton bombardments of copper. (From reference H7.)

found, especially among the heavy products. As seen in figure 4-24, in reactions of heavy ($Z \geqslant 70$) elements with projectiles of a few hundred MeV, the fission-product region is rather clearly separated from the spallation products by a region of A with very low formation cross section. In the gigaelectron volt region this separation is no longer evident in the mass-yield curves. However, measurements in which two fragments in coincidence are detected show that fission still accounts for a large fraction of the product yields in the medium-mass range. From angular-distribution and angular-correlation studies it can in fact be concluded that the bulk of these binary fission processes, even though induced by projectiles of very high energies, results from intermediate nuclei of modest excitation, that is, from cascades in which typically no more than tens of MeV have been deposited. Most of the products resulting from these true fission processes are on the neutron-rich side of or near β stability, and have kinetic-energy spectra typical of low-energy fission, that is, reflecting the Coulomb repulsion in a binary breakup. The more neutron-deficient products originate largely from processes involving higher deposition energies (hundreds of MeV), have lower kinetic energies, and do not appear to have partners of comparable mass. They are thus not believed to arise from fission processes, but rather from spallation-like or fragmentation reactions (see below). At gigaelectron volt energies and in the mass region $120 \leqslant A \leqslant 150$ these latter products, with peak yields on the neutron-deficient side of β stability, have been found to be separated from the fission products, whose peak yields are on the neutron-excess side, by a region of low yields near β stability.

At high energies fission has been reported for elements as light as copper, but the cross sections are small, and it is difficult to disentangle fission from spallation. The most convincing evidence comes from observation of binary events with correlated fragment tracks in photographic emulsions or dielectric track detectors.

Other Processes—Fragmentation. Although most phenomena observed with projectiles in the gigaelectron volt region can probably be explained within the framework of the two-step model of intranuclear cascades followed by evaporation and fission, there appear to be some exceptions. In particular, the formation, in high yield, of products with mass numbers between about 15 and 40 in the bombardment of heavy elements such as bismuth (see figure 4-24), and especially the angular and energy spectra of these fragments, cannot be completely accounted for in terms of the two-step mechanism. The term fragmentation has been coined for the process that is thought to be involved but is not yet well understood. Excitation functions for the fragments in the A range mentioned rise steeply from effective thresholds of a few hundred MeV. The observation that neutron-deficient products in the fission product mass range (see above) have very similar excitation functions has prompted the view, partially verified by coincidence experiments, that the two types of products may represent partners of some breakup process. However, more complex mechanisms, involving multiple emission of light fragments, also appear to play a role. That the neutron-deficient species in the fission product mass range arise from processes other than either fission or spallation is made evident by their recoil properties—ranges and angular distributions—which are very different from those of fission products, and which cannot be accounted for by cascade-evaporation calculations (see, e.g., B7). However, the mechanism for their production appears to change in the vicinity of 3 GeV, as indicated by rather dramatic changes in these recoil properties (B7).

Reactions with Pions (Pi Mesons). We noted in discussing proton-induced reactions that the production and subsequent interaction of pions inside nuclei appears to be an important mechanism for the deposition of excitation energy. The study of reactions induced by pions as *incident* particles is therefore of particular interest. The copious production of pions in proton-nucleus collisions (~0.5 pion per interaction with 1-GeV protons) and the development of high-current proton accelerators in the 0.5–1.0 GeV range (see chapter 15) have made fairly intense pion beams available.

Pion interactions with complex nuclei strongly reflect a prominent feature of elementary pion-nucleon interactions: the scattering of pions by nucleons exhibits a pronounced, broad resonance centered around 180 MeV, with peak cross sections of ~200 mb for π^+-p (and π^--n) and ~70 mb for π^--p (and π^+-n) scattering. This resonance is interpreted as the formation of a short-lived excited state of the nucleon, or **nucleon isobar**, called Δ, which subsequently decays again into nucleon and pion. It is the large cross section for isobar formation that accounts for the already-mentioned short mean-free paths of pions in nuclei. A second important feature of pion interactions is the possibility of pion absorption by a pair of nucleons, resulting in the *total* energy of the pion (kinetic plus rest energy) being shared by the two nucleons. In the isobar picture this pion capture is interpreted as a two-step reaction: formation of a Δ, followed by Δ-nucleon scattering leading to two ground-state nucleons.

Excitation functions for some pion-induced reactions in light- and medium-mass nuclei have peaks in the region of 150–200 MeV, a direct reflection of the pion-nucleon resonance. This is illustrated in figure 4-27 for the reactions $^{12}C(\pi^{\pm}, \pi^{\pm}n)^{11}C$. With increasing A of the target and with

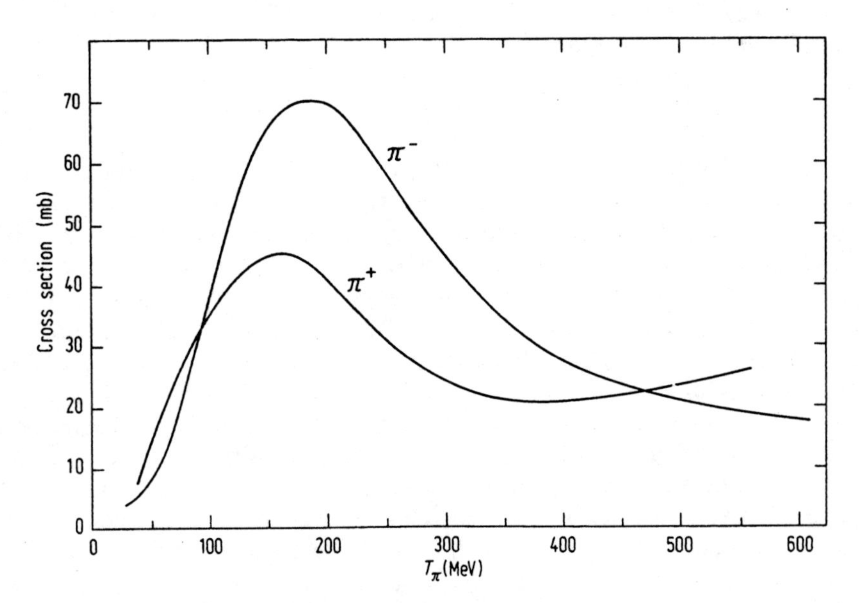

Fig. 4-27 Excitation functions for the formation of ^{11}C by π^+ and π^- interactions with ^{12}C. [Data from B. J. Dropesky et al., *Phys. Rev. Lett.* **34**, 821 (1975).]

increasing complexity of the reactions, the resonance behavior becomes less pronounced and eventually disappears. For example, with copper as a target broad peaks are seen for products as far removed from the target nuclide as ^{52}Mn, but not for ^{48}V or ^{44}Sc (O2).

Because of the effect of pion absorption, we might expect the reaction patterns of pions of a given kinetic energy to resemble those induced by protons with a kinetic energy higher by 140 MeV (the pion rest energy). This is indeed found to be approximately true in the region up to a few hundred MeV. At still higher energies proton- and pion-induced spallation patterns become very similar at equal kinetic energies. A further effect of pion absorption becomes evident in comparisons of π^+- and π^--induced reactions at energies below the resonance: neutron-rich products become more prominent in π^--induced, proton-rich products in π^+-induced reactions (02). This is expected from the change in the N/Z ratio of the target-projectile combination if the pion is absorbed.

H. HEAVY-ION REACTIONS

General Characteristics. Nuclear reactions induced by ions of $Z > 2$ deserve special consideration because, although they exhibit many of the same features observed with light-ion projectiles, they additionally have characteristics that set them apart, and that make possible the investigation of nuclear phenomena not otherwise accessible. Principal among the relevant properties of heavy ions are their large radii, the large amounts of angular momentum they can bring into a reaction, their large charges, and their small de Broglie wavelengths.[20]

The first nuclear reactions induced by heavy ions (120-MeV C^{+6} accelerated in a cyclotron) were reported in 1950, but it was not until several years later that accelerators and techniques suitable for detailed heavy-ion reaction studies became available. In the 1960s and 1970s an almost explosive growth occurred in this field, and it has become one of the most active areas of nuclear research. Accelerators capable of producing intense beams of ions up to uranium with energies sufficient to surmount the Coulomb barriers of even the heaviest elements are now in operation (see chapter 15). The variety of possible reactions is thus enormous and makes for a very rich field of study.

The reaction mechanisms invoked for light-ion reactions—elastic and inelastic scattering, compound-nucleus formation, direct interactions—all play an important role with heavy ions also. In addition, a new type of process, somewhat intermediate between transfer reaction and compound-nucleus formation and usually called **deeply inelastic reaction**, is of great

[20] The de Broglie wavelength λ for a system of two particles of relative kinetic energy ϵ and reduced mass μ is $\lambda = h/(2\mu\epsilon)^{1/2}$ (see appendix C).

importance. The major parameters that apparently determine which mechanisms predominate are the impact parameter of the collision, the kinetic energy of the projectile, and the masses of the target and projectile nuclei.

A very significant aspect of heavy-ion reactions is the applicability of classical or semiclassical considerations to the analysis of many of the phenomena in the region of Coulomb-barrier energies. Classical ideas are valid when the reduced wavelength $\lambda\!\!\!^{-}$ is small compared to the collision distance d. For a collision in which the Coulomb force is dominant, we can take d as the distance of closest approach in a head-on collision (see p. 30): $d = Z_1Z_2e^2/\epsilon$, where ϵ is the c.m. kinetic energy. Then the condition for the applicability of classical ideas becomes $Z_1Z_2e^2/\lambda\!\!\!^{-}\epsilon \gg 1$.[21] This condition is fulfilled for most heavy-ion reactions of interest. For example, for a 100-MeV ^{12}C ion incident on a Zn nucleus, $\eta \approx 10$; for a 600-MeV ^{84}Kr ion incident on Au, $\eta \approx 600$.

Elastic and Inelastic Scattering, Coulomb Excitation. In the simplest approximation, the so-called sharp cut-off model, encounters between a heavy ion and a target nucleus will not bring the nuclear forces into play if the impact parameter or, equivalently, the angular momentum, exceeds some critical value corresponding to the trajectory that lets the two nuclei just touch. Within the limitations of this model elastic-scattering measurements can thus be used to obtain information on interaction radii $R = R_1 + R_2$ between two nuclei of mass numbers A_1 and A_2, and the results are in good agreement with the formula $R = r_0(A_1^{1/3} + A_2^{1/3})$ with $r_0 \approx 1.5$–1.6. Use of more sophisticated, semiclassical models leads to information about the nuclear-skin thickness. Elastic-scattering data with heavy ions provide critical tests for the choice of parameters in the optical model.

Inelastic scattering, that is, scattering in which some of the projectile's kinetic energy is transformed into excitation of the target nucleus, is again of greatest importance at large impact parameters. What makes heavy ions particularly valuable for inelastic-scattering experiments is their ability to excite high-spin states in target nuclei by virtue of the large angular momenta they can bring in. Furthermore, because of their high charges, they can at relatively high energies (tens of MeV) still be below Coulomb barrier heights and thus are able to excite nuclei by purely electromagnetic interactions. This process is known as Coulomb excitation (S5). Much information on high-spin states of nuclei, information that has considerable impact on our understanding of nuclear structure, has come from Coulomb-excitation experiments with heavy ions. For these studies, beams of the highest possible Z are particularly desirable.

[21] The condition is more commonly stated as $\eta \gg 1$, where η, the so-called Sommerfeld parameter, is defined as $Z_1Z_2e^2/\hbar v = Z_1Z_2e^2/2\lambda\!\!\!^{-}\epsilon$, and thus differs from the above expression by a factor of 2.

Transfer Reactions. Stripping and pickup reactions are very prevalent with heavy ions. Such reactions presumably take place principally at impact parameters (orbital angular momenta) just below those at which interactions are purely Coulombic. The most thoroughly studied and best-understood reactions of this type are the single-nucleon transfer reactions, such as $^{A}Z(^{14}N, ^{13}N)^{A+1}Z$ or $^{A}Z(^{17}O, ^{18}F)^{A-1}(Z-1)$, corresponding in the first example to neutron transfer from projectile to target, in the second to proton transfer from target to projectile. In the particular examples given the projectile gets transformed into a radioactive nuclide, but this is, of course, not always the case; when it is, the radioactive product can be detected and its angular and energy distribution measured by radiochemical (catcher foil) techniques. Counter telescope techniques are applicable more generally. Sometimes the angular and energy spectra of both reaction products can be measured. The angular distributions tend to show an oscillatory, diffraction-like pattern when a transfer reaction to a single, well-defined state is observed, as is often possible with low-Z targets (figure 4-28). Optical-model and DWBA analyses have been quite successfully applied to the interpretation of such data (K1). When the transfer

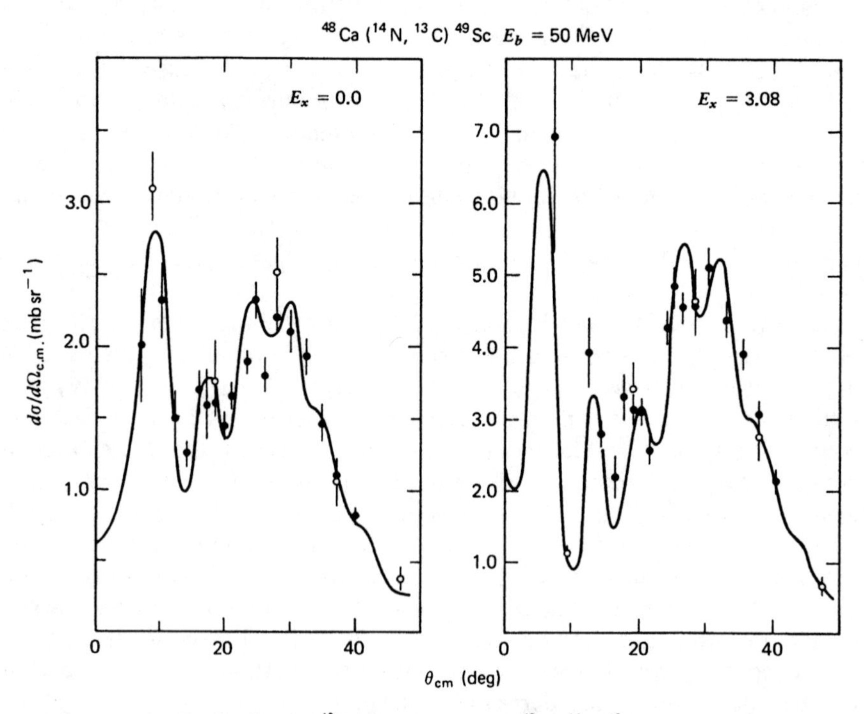

Fig. 4-28 Angular distributions of ^{13}C from the reaction $^{48}Ca(^{14}N, ^{13}C)$ going to the ground and 3.08 MeV states in ^{49}Sc. The solid lines are DWBA calculations made with an imaginary potential $W = 10$ MeV fitted to elastic-scattering data. θ_{cm} is the center-of-mass angle. (From reference K1.)

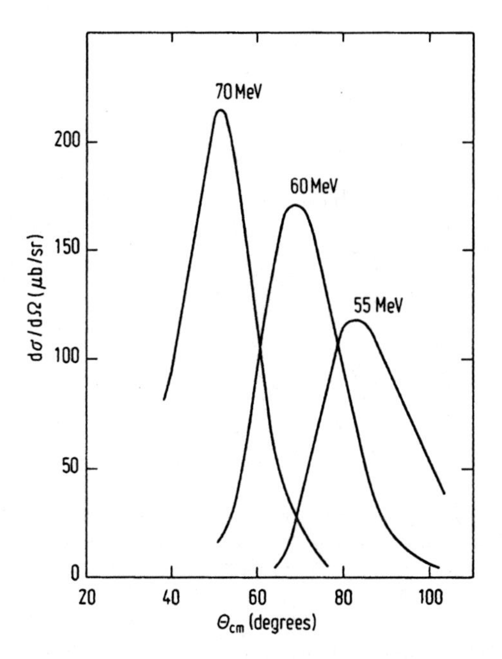

Fig. 4-29 Angular distribution of ^{16}O from the reaction $^{94}Mo(^{18}O, ^{16}O)^{96}Mo$ (ground state) at three different bombarding energies. Θ_{cm} is the center-of-mass angle. [Data from C. Chasman et al., *Phys. Rev. Lett.* **28**, 843 (1972).]

populates many overlapping states we typically find a single peak at a characteristic angle with respect to the original beam direction. This angle, called the **grazing angle**, decreases with increasing bombarding energy (see figure 4-29). The name derives from the picture of simple transfer reactions as taking place in "soft" or "grazing" collisions, that is, collisions involving little transfer of mass or energy and large impact parameters so that target and projectile overlap only in their outermost reaches; thus while nuclear forces clearly must come into play in causing the transfer of nucleons, the projectile trajectory is still essentially controlled by Coulomb forces.

One-nucleon transfer reactions have thresholds somewhat below Coulomb-barrier energies, and their cross sections rise fairly rapidly as the energy is increased, reaching typically tens of millibarns and then remaining rather constant with further increases in energy. At energies appreciably above the Coulomb barrier more complex transfer reactions come into play; a great variety of such multinucleon transfer reactions have been studied, including some very exotic ones involving simultaneous pickup and stripping of several nucleons. The excitation functions of multinucleon transfer reactions tend to rise with increasing energy. The simpler multinucleon transfers, such as two-nucleon or α transfers, still show preferential emission angles (figure 4-29) but with increasing complexity and with increasing bombarding energy the angular distributions become more and more strongly forward-peaked. Such processes presumably involve "harder" collisions, that is, somewhat smaller impact parameters and

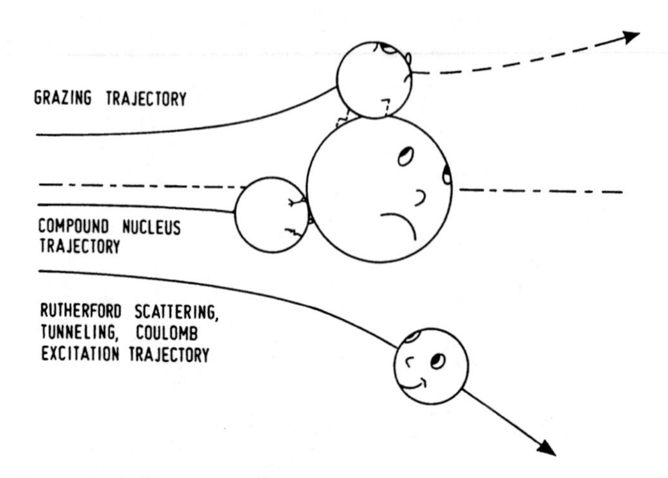

Fig. 4-30 Schematic representation of three different types of heavy-ion interaction characterized by different impact parameters. (From R. Kaufmann and R. Wolfgang, *Proc. 2nd Conf. Reactions Between Complex Nuclei*, Wiley, New York, 1960; see also reference K2.)

deeper interpenetration of the two nuclei. There is clearly no sharp dividing line between transfer reactions and the next class of processes to be considered.

Deeply Inelastic Reactions (L2, S6). Processes in which relatively large amounts of nuclear matter are transferred between target and projectile and which show strongly forward-peaked angular distributions were characterized in 1959 by R. Kaufmann and R. Wolfgang (K2) and ascribed by them to a "grazing contact mechanism" in which the projectile-target combination separates again after less than half a rotation. They illustrated the process graphically as shown in figure 4-30 as having a trajectory intermediate between a pure Coulomb and a compound-nucleus trajectory.

Over a decade elapsed before this type of process was rediscovered and aroused much interest. When ions with $A > 40$ became available it was found that what are now called deeply inelastic collisions[22] are in fact very dominant processes with heavier projectiles. The principal distinguishing feature of these reactions is found in their double differential cross sections $d^2\sigma/dE\,d\theta$: products with masses in the vicinity of the projectile mass appear at angles other than the classical grazing angle, with relatively small kinetic energies. This is illustrated in figure 4-31, which shows a contour map for the production of potassium isotopes in the bombardment of thorium with 388-MeV argon ions.[23] The large peak of potassium cross sections centered near the grazing angle of 34° and with nearly the original

[22] Other terms used by some authors are: strongly damped collisions, quasi-fission, incomplete fusion. They all have roughly the same meaning.

[23] This type of contour diagram is called a Wilczynski plot after its originator.

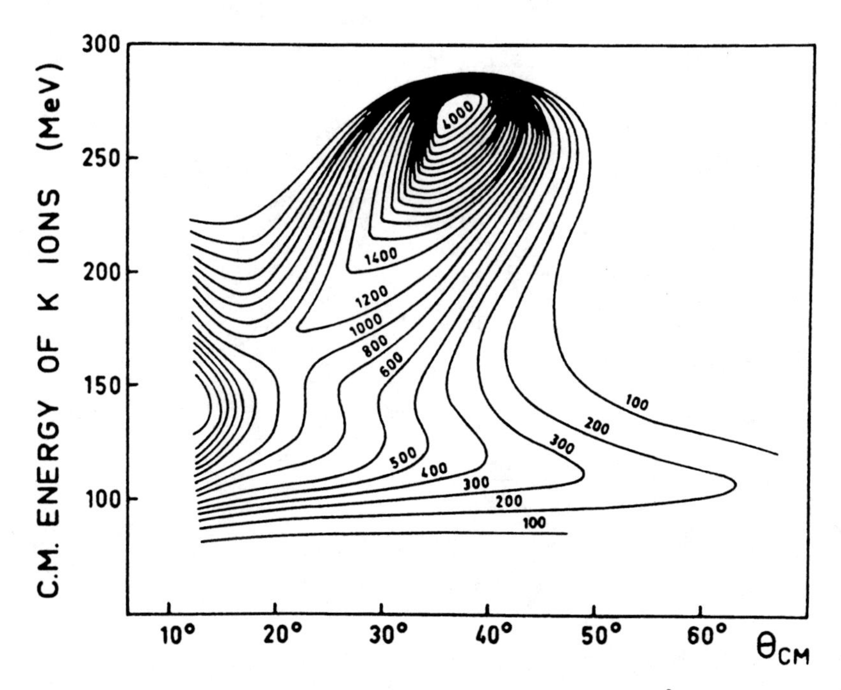

Fig. 4-31 Contour map of the triple differential cross section $d^2\sigma/d\theta\, dE\, dZ$ (in mb $rd^{-1}\, MeV^{-1}$) for the production of potassium isotopes in the bombardment of thorium with 388-MeV argon ions. The ordinate is the center-of-mass energy, the abscissa the center-of-mass angle of the potassium. [From J. Wilczynski, *Phys. Letters* **47B**, 484 (1973).]

kinetic energy is due to quasi-elastic events. Another peak is seen at small angles (<15°) and considerably reduced kinetic energies, and a ridge of relatively high cross sections extends from that peak to larger angles. These latter events represent the products of deeply inelastic collisions.

The specifics of angular and energy distributions vary considerably with the charge product Z_1Z_2 of the system studied and also with the bombarding energy. For systems with large Z_1Z_2 the distributions tend to be peaked at sideways angles, with the peaks shifting to smaller angles with increasing projectile energy (see figure 4-32). In light systems forward-peaked distributions are observed. The total kinetic energies of the products are strongly correlated with the amount of mass transfer: the more the A (or Z) of product and projectile differ in either direction, the lower the kinetic energy. The integrated cross section for deeply inelastic events is a very strong function of Z_1Z_2, going from a small fraction of the total reaction cross section for light systems to practically the entire reaction cross section for the heaviest ones. With increasing energy above the Coulomb barrier the cross section for deeply inelastic collisions tends to decrease.

Without going into the rather complicated details of the theoretical treatment of these reactions, we can visualize the mechanistic ideas that

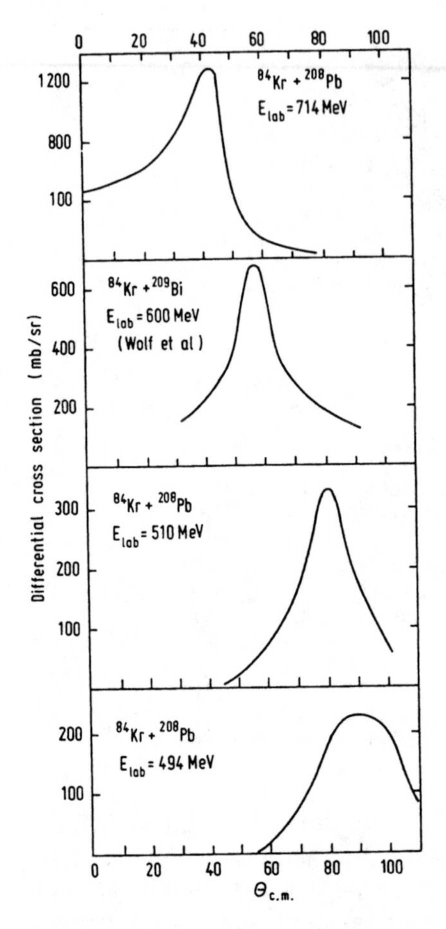

Fig. 4-32 Angular distribution of the light fragments from the reaction $^{208}Pb + ^{84}Kr$ and the similar system $^{209}Bi + ^{84}Kr$. θ_{cm} is the center-of-mass angle. (From reference S6.) Reproduced, with permission, from the Annual Review of Nuclear Science, Volume 27 © 1977 by Annual Reviews Inc.

have been developed. The pictorial representation of figure 4-30 indicates the basic point: it is thought that, at impact parameters intermediate between those for purely Coulombic interactions and those leading to compound-nucleus formation, a short-lived intermediate complex is formed that will rotate as a result of the large angular momentum brought in by the projectile. This object, being very far from spherical, will generally dissociate again into two fragments after a time corresponding to no more than about half a rotation. During the lifetime of the intermediate ($\sim 10^{-22}$ s) an appreciable fraction of the incident kinetic energy is dissipated and goes into internal excitation. The observed phenomena are quite well accounted for by theoretical treatments based on this picture. What makes the subject particularly interesting is the introduction into nuclear theory of ideas from hydrodynamics and diffusion theory, with consideration of dissipative forces such as friction.

Compound-Nucleus Reactions. From what has been said already it must be clear that compound-nucleus formation[24] can take place over a restricted range of small impact parameters only. Correspondingly we can define a critical angular momentum l_{crit} above which complete fusion cannot occur; as a result the ratio of the complete-fusion cross section σ_{cf} to the total reaction cross section σ_R decreases with increasing bombarding energy. The fraction σ_{cf}/σ_R decreases strongly with increasing Z_1Z_2 product and appears to become vanishingly small for the heaviest systems. This may be the result of deformation of the target and projectile nuclei during their relatively slow approach to each other; such dynamic (as well as any static) deformations will, of course, alter the interaction barrier. The fusion of two very heavy ions can, in some respects, be considered the inverse of fission, and the considerations of potential-energy surfaces and trajectories mentioned in connection with fission (section F2) are applicable here.

Compound-nucleus reactions with heavy ions have been very useful for producing and making available for spectroscopic study a large number of new radioactive nuclides on the neutron-deficient side of β stability. This comes about because light heavy ions such as ^{12}C, ^{16}O, and ^{20}Ne have equal numbers of neutrons and protons; as a result of the slope of the β-stability line at medium and large A (see figure 2-6) such projectiles produce compound nuclei on the neutron-deficient side. Furthermore, a heavy ion of energy above the Coulomb barrier always brings in enough excitation energy to evaporate several nucleons; reactions of the type (HI, xn), where HI stands for heavy ion and x is typically between 3 and 7, are therefore very prevalent. Heavy-ion bombardments have also been essential for the discovery and study of the heaviest known transuranium elements ($Z > 101$), since these elements could be reached from available targets only by the addition of several charge units. Finally we mention that the large range of angular momenta available with heavy ions has made possible detailed studies of the effect of angular momentum on the course of compound-nucleus reactions and the exploration of levels of high angular momenta, particularly of yrast levels (see p. 145).

Heavy-ion reactions provide the only possible means for reaching the predicted island of stability in the neighborhood of $Z = 114$ and $N = 184$ (chapter 1, section D) in the laboratory, since they can, in principle, "jump over" the region of highly unstable nuclides immediately below that island. Interest in the hypothetical superheavy elements was in fact a strong driving force toward the construction of several powerful heavy-ion accelerators. A variety of heavy-ion reactions have been tried in very intensive efforts to produce superheavy elements in the United States, USSR, France, and Germany, but none of them have as yet led to success.

[24] In heavy-ion parlance we speak more frequently of **complete fusion** rather than compound-nucleus formation. Complete fusion implies that projectile and target lose their identity and that the entrance channel is "forgotten," but it does not necessarily imply equilibration among all degrees of freedom. Preequilibrium emission is, in fact, very important, especially in reactions with light heavy ions.

It appears that either the cross sections for compound-nucleus formation become too small or fission barriers are too low. Deeply inelastic processes have perhaps the best chance of reaching the sought-after island (H10).

Relativistic Heavy Ions. This is the newest subfield in heavy-ion reaction studies and one that has aroused considerable interest. Many ions with energies up to 2 GeV per nucleon have become available in modest intensities. The interest in this field stems from theoretical speculations about the possible production of exotic states of nuclear matter due to the high compressions and temperatures that may be achieved. Pion condensates, nuclear shock waves, a superdense phase of nuclear matter (density isomers) are among the subjects discussed. Nothing so exotic has been uncovered yet, but interesting studies have begun to reveal the systematics of the reaction patterns that result from the fragmentation of both projectile and target nuclei. Projectile fragments carry large forward momenta in the laboratory system; they presumably result from fairly peripheral collisions. Target fragments carry little momentum; their mass-yield distributions are strikingly similar to those obtained with high-energy protons.

Results and prospects of relativistic-heavy-ion research are reviewed in G5.

REFERENCES

A1 J. M. Alexander, "Studies of Nuclear Reactions by Recoil Techniques," in *Nuclear Chemistry*, Vol. I (L. Yaffe, Ed.), Academic, New York, 1968 pp. 273–357.

*B1 J. Blatt and V. Weisskopf, *Theoretical Nuclear Physics*, Wiley, New York, 1952.

B2 H. H. Barschall, "Regularities in the Total Cross Sections for Fast Neutrons," *Phys. Rev.* **86**, 431 (1952).

B3 N. Bohr, "Neutron Capture and Nuclear Constitution," *Nature* **137**, 344 (1936).

B4 M. Blann, "Preequilibrium Decay," *Ann. Rev. Nucl. Sci.* **25**, 123 (1975).

B5 M. Brack et al., "Funny Hills: The Shell Correction Approach to Nuclear Shell Effects and Its Application to the Fission Process," *Rev. Mod. Phys.* **44**, 320 (1972).

B6 H. W. Bertini, "Nonelastic Interactions of Nucleons and π Mesons with Complex Nuclei at Energies Below 3 GeV," *Phys. Rev.* **C6**, 631 (1972).

B7 K. Beg and N. T. Porile, "Energy Dependence of the Recoil Properties of Products from the Interaction of ^{238}U with 0.45–11.5 GeV Protons," *Phys. Rev.* **C3**, 1631 (1971).

C1 J. Chadwick and M. Goldhaber, "A Nuclear Photoeffect: Disintegration of the Diplon by γ Rays," *Nature* **134**, 237 (1934).

C2 J. G. Cuninghame, "Status of Fission Product Yield Data," in *Fission Product Nuclear Data—1977*, IAEA Report 213, Vol. I, International Atomic Energy Agency, Vienna, 1978, p. 351.

C3 K. Chen et al. "VEGAS: A Monte Carlo Simulation of Intranuclear Cascades," *Phys. Rev.* **166**, 949 (1968).

D1 I. Dostrovsky, Z. Fraenkel, and G. Friedlander, "Monte Carlo Calculations of Nuclear Evaporation Processes. III. Applications to Low-Energy Reactions," *Phys. Rev.* **116**, 683 (1960).

D2 J. O. Denschlag, "Prediction of Unmeasured Fission Yields by Nuclear Theory or Systematics," in *Fission Product Nuclear Data—1977*, IAEA Report 213, Vol. II, International Atomic Energy Agency, Vienna, 1978, p. 421.

E1 T. Ericson, "The Statistical Model and Nuclear Level Densities," *Phil. Mag. Supp.* **9** (36), 425 (1960).

F1 S. Fernbach, R. Serber, and T. B. Taylor, "The Scattering of High-Energy Neutrons by Nuclei," *Phys. Rev.* **75**, 1352 (1949).

F2 H. Feshbach, C. E. Porter, and V. F. Weisskopf, "Model for Nuclear Reactions with Neutrons," *Phys. Rev.* **96**, 448 (1954).

F3 M. J. Fluss et al., "Investigation of the Bohr Independence Hypothesis for Nuclear Reactions in the Continuum: $\alpha + {}^{59}Co$, $p + {}^{62}Ni$, and $\alpha + {}^{56}Fe$, $p + {}^{59}Co$," *Phys. Rev.* **187**, 1449 (1969).

F4 E. G. Fuller and E. Hayward (Eds.), *Photonuclear Reactions*, Dowden, Hutchinson, and Ross, New York, 1976.

F5 F. W. K. Firk, "Low-Energy Photonuclear Reactions," *Ann. Rev. Nucl. Sci.* **20**, 39 (1970).

*F6 A. Fleury and J. M. Alexander, "Reactions Between Medium and Heavy Nuclei and Heavy Ions of Less Than 15 MeV/amu," *Ann. Rev. Nucl. Sci.* **24**, 279 (1974).

G1 S. N. Ghoshal, "An Experimental Verification of the Theory of Compound Nucleus," *Phys. Rev.* **80**, 939 (1950).

G2 J. R. Grover, "Shell Model Calculations of the Lowest-Energy Nuclear Excited States of Very High Angular Momentum," *Phys. Rev.* **157**, 832 (1967).

G3 J. J. Griffin, "Statistical Model of Intermediate Structure," *Phys. Rev. Lett.* **17**, 478 (1966).

G4 M. Goldhaber and E. Teller, "On Nuclear Dipole Vibrations," *Phys. Rev.* **74**, 1046 (1948).

G5 A. S. Goldhaber and H. H. Heckman, "High-Energy Interactions of Nuclei," *Ann. Rev. Nucl. Part. Sci.* **28**, 161 (1978).

H1 J. R. Huizenga and G. Igo, "Theoretical Reaction Cross Sections for Alpha Particles with an Optical Model," *Nucl. Phys.* **29**, 462 (1962).

*H2 P. E. Hodgson, "The Optical Model of the Nucleon-Nucleus Interaction," *Ann. Rev. Nucl. Sci.* **17**, 1 (1967).

H3 J. R. Huizenga and L. G. Moretto, "Nuclear Level Densities," *Ann. Rev. Nucl. Sci.* **22**, 427 (1972).

H4 R. Huby, "Stripping Reactions," in *Progress in Nuclear Physics*, Vol. 3 (O. Frisch, Ed.) Pergamon, New York, 1953, p. 177.

*H5 E. K. Hyde, *The Nuclear Properties of the Heavy Elements III. Fission Phenomena*, Prentice-Hall, Englewood Cliffs, N.J., 1964.

H6 D. C. Hoffman and M. M. Hoffman, "Post-Fission Phenomena," *Ann. Rev. Nucl. Sci.* **26**, 151 (1974).

*H7 J. Hudis, "High-Energy Nuclear Reactions," in *Nuclear Chemistry*, Vol. I (L. Yaffe, Ed.), Academic, New York, 1968, pp. 169–272.

H8 G. D. Harp, "Extension of the Isobar Model for Intranuclear Cascades to 1 GeV," *Phys. Rev.* **C10**, 2387 (1974).

*H9 P. E. Hodgson, *Nuclear Heavy-Ion Reactions*, Clarendon (Oxford University Press), New York, 1978.

H10 G. Herrmann, "Superheavy Element Research," *Nature* **280**, 543 (1979).

K1 S. Kahana and A. J. Baltz, "One and Two Nucleon Transfer Reactions with Heavy Ions," in *Advances in Nuclear Physics*, Vol. 9, Plenum, New York, 1977, pp. 1–122.

K2 R. Kaufmann and R. Wolfgang, "Nuclear Transfer Reactions in Grazing Collisions of Heavy Ions," *Phys. Rev.* **121**, 192 (1961).

L1 D. W. Lang, "Nuclear Correlations and Nuclear Level Densities," *Nucl. Phys.* **42**, 353 (1963).

L2 M. Lefort and C. Ngo, "Deep Inelastic Reactions with Heavy Ions. A Probe for Nuclear Macrophysics Studies," *Ann. Phys.* **3**, 5 (1978).

M1 M. H. Macfarlane and J. P. Schiffer, "Transfer Reactions," in *Nuclear Spectroscopy and Reactions*, Part B (J. Cerny, Ed.), Academic, New York, 1974, p. 169.

N1 T. D. Newton, "Nuclear Models," in *Nuclear Chemistry*, Vol. I (L. Yaffe, Ed.), Academic, New York, 1968, pp. 1–55.

N2 National Nuclear Data Center, "Neutron Cross Sections," Report BNL-325 and its Supplements, Brookhaven National Laboratory, Upton, NY.

N3 J. R. Nix, "Calculation of Fission Barriers for Heavy and Superheavy Nuclei," *Ann. Rev. Nucl. Sci.* **22**, 65 (1972).

O1 J. R. Oppenheimer and M. Phillips, "Note on the Transmutation Function for Deuterons," *Phys. Rev.* **48**, 500 (1935).

O2 C. J. Orth et al., "Pion-Induced Spallation of Copper Across the (3, 3) Resonance," *Phys. Rev.* **C18**, 1426 (1978).

P1 F. G. Perey, "Elastic and Inelastic Scattering," in *Nuclear Spectroscopy and Reactions*, Part B (J. Cerny, Ed.), Academic, New York, 1974, p. 137.

P2 N. T. Porile, "Low-Energy Nuclear Reactions," in *Nuclear Chemistry*, Vol. I (L. Yaffe, Ed.), Academic, New York, 1968, pp. 57–168.

R1 J. Rainwater, "Resonance Processes by Neutrons," in *Encyclopedia of Physics*, Vol. 40 (S. Flügge, Ed.), Springer, Berlin, 1957.

R2 G. Rudstam, "Status of Delayed Neutron Data," in *Fission Product Nuclear Data—1977*, IAEA Report 213, Vol. II, International Atomic Energy Agency, Vienna, 1978, p. 567.

S1 M. M. Shapiro, "Cross Sections for the Formation of the Compound Nucleus by Charged Particles," *Phys. Rev.* **90**, 171 (1943).

*S2 C. M. H. Smith, *A Textbook of Nuclear Physics*, Pergamon, London, 1965.

S3 V. M. Strutinsky, "Shell Effects in Masses and Deformations," *Nucl. Phys.* **A95**, 420 (1967).

S4 R. Serber, "Nuclear Reactions at High Energies," *Phys. Rev.* **72**, 1114 (1947).

S5 P. H. Stelson and F. K. McGowan, "Coulomb Excitation," *Ann. Rev. Nucl. Sci.* **23**, 163 (1963).

S6 W. U. Schroeder and J. R. Huizenga, "Damped Heavy Ion Collisions," *Ann. Rev. Nucl. Sci.* **27**, 465 (1977).

*S7 E. Segrè, *Nuclei and Particles*, 2nd ed., Benjamin, Reading, MA, 1978.

*V1 R. Vandenbosch and J. R. Huizenga, *Nuclear Fission*, Academic, New York, 1973.

W1 A. C. Wahl et al., "Nuclear-Charge Distribution in Low-Energy Fission," *Phys. Rev.* **126**, 1112 (1962).

*Y1 L. Yaffe, Ed., *Nuclear Chemistry*, 2 vols., Academic, New York, 1968.

EXERCISES

1. Compute, from masses in appendix D, the Q values for the following reactions: (a) ^{24}Mg (d, p) ^{25}Mg, (b) ^{10}B (n, α) ^{7}Li.

2. Would the ^{43}Ca (n, α) reaction proceed in good yield with thermal neutrons? Justify your answer. *Answer:* No.
3. The reaction ^{33}S$(n, p)^{33}$P is exoergic by 0.533 MeV. The mass of ^{33}S is 32.971458 amu. What is the mass of ^{33}P?
4. Estimate the Coulomb barrier and the centrifugal barrier for p-wave interaction between protons and ^{9}Be. *Answer:* 1.2 MeV, 2.2 MeV.
5. Show from the conservation of momentum that, at the threshold of an endothermic nuclear reaction A (a, b) B, the fraction of the kinetic energy of the bombarding particle a which goes into kinetic energy of the product is $M_a/(M_a + M_A)$, where M_a and M_A are the masses of a and A, respectively.
6. Calculate the approximate heights of the Coulomb barriers around ^{27}Al, ^{89}Y, ^{159}Tb, and ^{238}U for (a) protons, (b) ^{12}C ions.
7. Calculate the de Broglie wavelength of a neutron of (a) 1 eV, (b) 1 keV, (c) 1 MeV kinetic energy. *Answer:* (b) 0.9×10^{-10} cm.
8. (a) Calculate the number of ^{60}Co atoms produced in a 10-mg sample of cobalt metal exposed for 2 min to a thermal-neutron flux of 2×10^{13} cm^{-2} s^{-1} in a reactor. (b) What is the disintegration rate of the cobalt sample a few hours after the irradiation? *Answer:* (b) 2.3×10^{6} dis min^{-1}.
9. A copper foil of surface density 20 mg cm^{-2} is exposed for 5 min to a beam of 14-MeV neutrons. The beam intensity is 10^{7} s^{-1}. At the end of the irradiation the foil is found to contain 2.0×10^{5} atoms of ^{64}Cu. Assuming that they have been formed entirely by the reaction ^{65}Cu$(n, 2n)$ and neglecting any decay of the 12.8-h nuclide during the short bombardment, estimate the cross section for the $n, 2n$ reaction. Would you expect the n, γ reaction on ^{63}Cu to have an important effect on the actual result of the experiment? *Answer:* $\sigma_{n,2n} = 1.1$ b.
10. It is desired to produce 5 mCi ^{65}Zn by exposing Zn metal for 24 h in a nuclear reactor to a flux of 1×10^{14} thermal neutrons per square centimeter per second. How much Zn has to be exposed? *Answer:* ~0.19 g.
11. Give the energy in the center-of-mass system in a collision between a 10-MeV proton and a 40-MeV ^{16}O ion when (a) they are moving in the same direction, (b) they are moving in opposite directions. *Answer:* (a) 2.4 MeV.
12. What are the de Broglie wavelengths in the center-of-mass system for the collisions described in exercise 11? *Answer:* (a) 19 fm.
13. Estimate the total-reaction cross section for the interaction of 20-MeV ^{4}He ions with ^{93}Nb. What radioactive nuclides would you expect as major reaction products? Estimate the maximum angular momentum brought in by the ^{4}He ions.
14. Derive (4-27) from (4-25) and (4-26).
15. Show that (4-13) follows from (4-12).
16. Suggest two methods for producing 58.5-d ^{91}Y as free as possible of other radioactive yttrium isotopes. Assume that separated stable isotopes are available for targets.
17. Prove that the magnitude of the relative momentum of two particles in the laboratory system is the same as the magnitude of the momentum of *either* particle in the center-of-mass system.
18. Estimate the imaginary part of the potential energy felt by a 1-GeV proton in the center of a heavy nucleus. Take the central density to be about 2×10^{38}

nucleons per cubic centimeter and the effective average cross section for nucleon-nucleon collisions to be 40 mb at this energy. Assume that the real part of the potential energy is negligible compared to 1 GeV.

Answer: ~115 MeV.

19. By the use of a cyclotron that accelerates He^{2+} ions up to 40 MeV, deuterons up to 20 MeV, and protons up to 10 MeV, suggest methods (target, type, and energy of incident particle) for the synthesis of (a) ^{210}Bi, (b) ^{58}Co (with a minimal contamination from ^{56}Co), (c) ^{75}Se, (d) ^{112}Ag, (e) ^{140}Ba. Consider targets of normal isotopic composition only. Radiochemical purity and good yield are goals to be kept in mind.
20. Estimate the total energy release and (from Coulomb barrier considerations) the minimum total kinetic energy (in MeV) of the two fission fragments when a ^{235}U nucleus captures a thermal neutron and splits into (a) ^{90}Kr and ^{143}Ba, (b) ^{113}Rh and ^{121}Ag.
21. The neutron capture reaction of ^{197}Au at neutron energies up to a few hundred eV is characterized by a number of resonances. The most prominent (and the one with the lowest energy) is at 4.906 eV and has $\Gamma_\gamma = 0.124$ eV and $\Gamma_n = 0.0071\epsilon^{1/2}$ eV. The compound nucleus formed in this resonance absorption has spin 2. From these resonance parameters estimate (a) the cross section of gold for 0.025-eV neutrons and compare your result with the experimental value given in appendix D; (b) the peak cross section of the 4.906-eV resonance.

Answer: (b) 3.3×10^4 b (experimental value 3.0×10^4 b).

22. Calculate and sketch the shape of the energy spectrum of protons evaporated from a ^{67}Zn nucleus excited to an energy of 15 MeV. Take the level density parameter a as 6.6 MeV^{-1}. For the inverse cross section use the form of (4-23), with the effective barrier V_c taken as 0.7 times the full electrostatic barrier.
23. A target of pure ^{76}Ge is bombarded with protons up to 50-MeV kinetic energy. Sketch shapes and relative magnitudes of the excitation functions you would expect for the production of ^{76}As, ^{74}As, ^{73}Ga, and ^{72}Ga.
24. (a) Estimate the maximum angular momentum (in units of $\hbar$) of ^{116}Sb compound nuclei formed by the bombardment of ^{103}Rh with 50-MeV ^{13}C ions. (b) What is the excitation energy of the compound nuclei formed in (a)? (c) What kinetic energy would a ^{3}He beam have to have to produce ^{116}Sb compound nuclei at the same excitation from an ^{113}In target? (d) What is the maximum angular momentum of the compound nuclei produced in (c)?
25. Which would you expect to be more effective for producing the neutron-deficient Zr isotopes ^{84}Zr and ^{85}Zr: bombardment of ^{40}Ca with ^{48}Ca or bombardment of ^{75}As with ^{14}N? Give your reasoning and suggest approximate bombarding energies.
26. Write equations for the absorption of a pion on a pair of nucleons in the two-step model involving nucleon isobars (Δ's) described on p. 177. Consider all possible combinations of pion charge states and nucleon pairs.
27. Evaluate the Sommerfeld parameter η (see footnote 21) for (a) a 150-MeV ^{20}Ne beam incident on an aluminum target, (b) a 300-MeV ^{40}Ar beam incident on a manganese target.

Chapter 5

Equations of Radioactive Decay and Growth

A. EXPONENTIAL DECAY

Half Life. We have seen (in chapter 1) that a given radioactive species decays according to an exponential law: $N = N_0 e^{-\lambda t}$ or $\mathbf{A} = \mathbf{A}_0 e^{-\lambda t}$, where N and $\mathbf{A}$ represent the number of atoms and the measured activity, respectively, at time t, and N_0 and $\mathbf{A}_0$ the corresponding quantities when $t = 0$, and λ is the characteristic decay constant for the species. The half life $t_{1/2}$ is the time interval required for N or $\mathbf{A}$ to fall from any particular value to one half that value. The half life is conveniently determined from a plot of $\log \mathbf{A}$ versus t when the necessary data are available, and is related to the decay constant:

$$t_{1/2} = \frac{\ln 2}{\lambda} = \frac{0.69315}{\lambda}.$$

Average Life. We may determine the average life expectancy of the atoms of a radioactive species. This average life is found from the sum of the times of existence of all the atoms divided by the initial number. If we consider N to be a very large number, we may approximate this sum by an equivalent integral, finding for the average life τ

$$\tau = -\frac{1}{N_0}\int_{t=0}^{t=\infty} t\,dN = \frac{1}{N_0}\int_0^\infty t\lambda N\,dt = \lambda \int_0^\infty te^{-\lambda t}\,dt$$

$$= -\left[\frac{\lambda t + 1}{\lambda} e^{-\lambda t}\right]_0^\infty = \frac{1}{\lambda}.$$

We see that the average life is greater than the half life by the factor $1/0.693$; the difference arises because of the weight given in the averaging process to the fraction of atoms that by chance survive for a long time. It may be seen that during the time $1/\lambda$ an activity will be reduced to just $1/e$ of its initial value.

Mixtures of Independently Decaying Activities. If two radioactive species, denoted by subscripts 1 and 2, are mixed together, the observed total activity is the sum of the two separate activities: $\mathbf{A} = \mathbf{A}_1 + \mathbf{A}_2 =$

$c_1\lambda_1 N_1 + c_2\lambda_2 N_2$. The detection coefficients c_1 and c_2 are by no means necessarily the same and often are very different in magnitude. In general, $\mathbf{A} = \mathbf{A}_1 + \mathbf{A}_2 + \cdots + \mathbf{A}_n$ for mixtures of n species.

For a mixture of several *independent* activities the result of plotting $\log \mathbf{A}$ versus t is always a curve concave upward (convex toward the origin). This curvature results because the shorter-lived components become relatively less significant as time passes. In fact, after sufficient time the longest-lived activity will entirely predominate, and its half life may be read from this late portion of the decay curve. Now, if this last portion, which is a straight line, is extrapolated back to $t = 0$ and the extrapolated line subtracted from the original curve, the residual curve represents the decay of all components except the longest-lived. This curve may be treated again in the same way, and in principle any complex decay curve may be analyzed into its components. In actual practice experimental

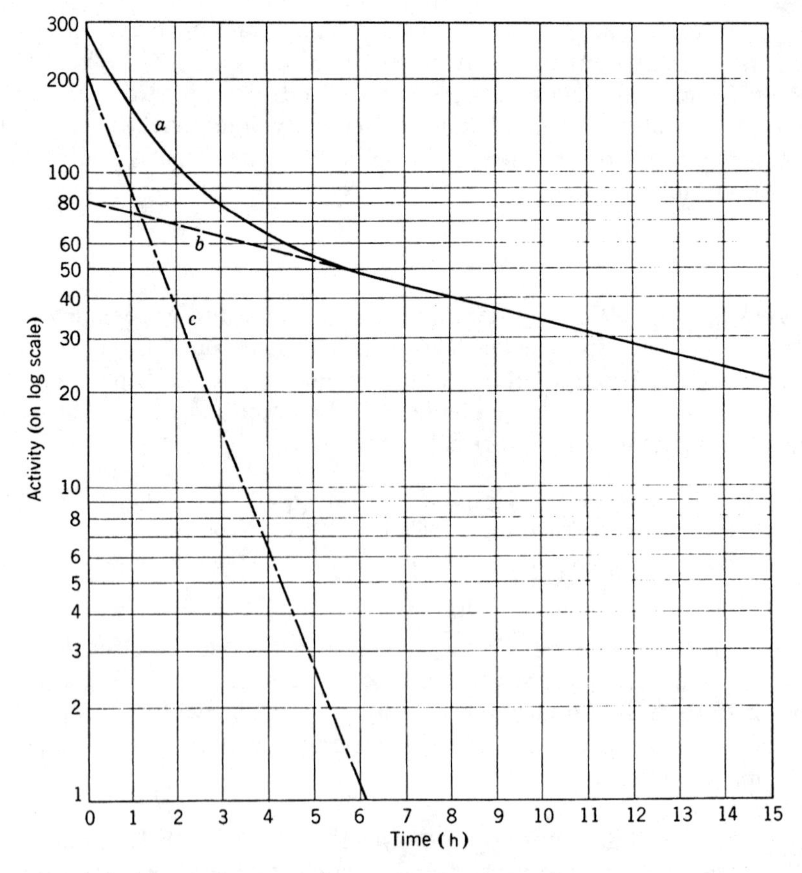

Fig. 5-1 Analysis of composite decay curve: (*a*) composite decay curve; (*b*) longer-lived component ($t_{1/2} = 8.0$ h); (*c*) shorter-lived component ($t_{1/2} = 0.8$ h).

uncertainties in the observed data may be expected to make it difficult to handle systems of more than three components, and even two-component curves may not be satisfactorily resolved if the two half lives differ by less than about a factor of two. The curve shown in figure 5-1 is for two components with half lives differing by a factor of 10.

The resolution of a decay curve consisting of two components of known but not very different half lives is greatly facilitated by the following approach. The total activity at time t is

$$\mathbf{A} = \mathbf{A}_1^0 e^{-\lambda_1 t} + \mathbf{A}_2^0 e^{-\lambda_2 t}.$$

By multiplying both sides by $e^{\lambda_1 t}$ we obtain

$$\mathbf{A} e^{\lambda_1 t} = \mathbf{A}_1^0 + \mathbf{A}_2^0 e^{(\lambda_1 - \lambda_2)t}.$$

Since λ_1 and λ_2 are known and $\mathbf{A}$ has been measured as a function of t, we can construct a plot of $\mathbf{A}e^{\lambda_1 t}$ versus $e^{(\lambda_1-\lambda_2)t}$; this will be a straight line with intercept $\mathbf{A}_1^0$ and slope $\mathbf{A}_2^0$.

Least-squares analysis is a more objective method for the resolution of complex decay curves than the graphical analysis described. Computer programs for this analysis have been developed (C1) that give values of $\mathbf{A}^0$ and its standard deviation for each of the components. Some of the programs can also be used to search for the "best values" of the decay constants.

B. GROWTH OF RADIOACTIVE PRODUCTS

General Equation. In chapter 1 we considered briefly a special case in which a radioactive daughter substance was formed in the decay of the parent. Let us take up the general case for the decay of a radioactive species, denoted by subscript 1, to produce another radioactive species, denoted by subscript 2. The behavior of N_1 is just as has been derived; that is, $-(dN_1/dt) = \lambda_1 N_1$, and $N_1 = N_1^0 e^{-\lambda_1 t}$, where we use the symbol N_1^0 to represent the value of N_1 at $t = 0$. Now the second species is formed at the rate at which the first decays, $\lambda_1 N_1$, and itself decays at the rate $\lambda_2 N_2$. Thus

$$\frac{dN_2}{dt} = \lambda_1 N_1 - \lambda_2 N_2$$

or

$$\frac{dN_2}{dt} + \lambda_2 N_2 - \lambda_1 N_1^0 e^{-\lambda_1 t} = 0. \tag{5-1}$$

The solution of this linear differential equation of the first order may be obtained by standard methods and gives

$$N_2 = \frac{\lambda_1}{\lambda_2 - \lambda_1} N_1^0 (e^{-\lambda_1 t} - e^{-\lambda_2 t}) + N_2^0 e^{-\lambda_2 t}, \tag{5-2}$$

where N_2^0 is the value of N_2 at $t = 0$. Notice that the first group of terms shows the growth of daughter from the parent and the decay of these daughter atoms; the last term gives the contribution at any time from the daughter atoms present initially.

Transient Equilibrium. In applying (5-2) to considerations of radioactive (parent and daughter) pairs, we can distinguish two general cases, depending on which of the two substances has the longer half life. If the parent is longer-lived than the daughter ($\lambda_1 < \lambda_2$), a state of so-called radioactive equilibrium is reached; that is, after a certain time the ratio of the numbers of atoms and, consequently, the ratio of the disintegration rates of parent and daughter become constant. This can be readily seen from (5-2); after t becomes sufficiently large, $e^{-\lambda_2 t}$ is negligible compared with $e^{-\lambda_1 t}$, and $N_2^0 e^{-\lambda_2 t}$ also becomes negligible; then

$$N_2 = \frac{\lambda_1}{\lambda_2 - \lambda_1} N_1^0 e^{-\lambda_1 t},$$

and, since $N_1 = N_1^0 e^{-\lambda_1 t}$,

$$\frac{N_1}{N_2} = \frac{\lambda_2 - \lambda_1}{\lambda_1}. \tag{5-3}$$

The relation of the two measured *activities* is found, from $\mathbf{A}_1 = c_1 \lambda_1 N_1$, $\mathbf{A}_2 = c_2 \lambda_2 N_2$, to be

$$\frac{\mathbf{A}_1}{\mathbf{A}_2} = \frac{c_1(\lambda_2 - \lambda_1)}{c_2 \lambda_2}. \tag{5-4}$$

In the special case of equal detection coefficients ($c_1 = c_2$) the ratio of the two activities, $\mathbf{A}_1/\mathbf{A}_2 = 1 - (\lambda_1/\lambda_2)$, may have any value between 0 and 1, depending on the ratio of λ_1 to λ_2; that is, in equilibrium the daughter activity will be greater than the parent activity by the factor $\lambda_2/(\lambda_2 - \lambda_1)$. In equilibrium both activities decay with the parent's half life.

As a consequence of the condition of transient equilibrium ($\lambda_2 > \lambda_1$), the sum of the parent and daughter disintegration rates in an initially pure parent fraction goes through a maximum before transient equilibrium is achieved. This situation is illustrated in figure 5-2. The more general condition for the total *measured activity* ($\mathbf{A}_1 + \mathbf{A}_2$) of an initially pure parent fraction to exhibit a maximum is found to be $c_2/c_1 > \lambda_1/\lambda_2$. This condition holds regardless of the relative magnitudes of λ_1 and λ_2. The condition $(\lambda_1 - \lambda_2)/\lambda_2 \leq c_2/c_1 \leq \lambda_1/\lambda_2$ will give a maximum in the total measured activity that occurs at a negative time.

Secular Equilibrium. A limiting case of radioactive equilibrium in which $\lambda_1 \ll \lambda_2$ and in which the parent activity does not decrease measurably during many daughter half lives is known as secular equilibrium. We illustrated this situation in chapter 1 and now may derive the equation

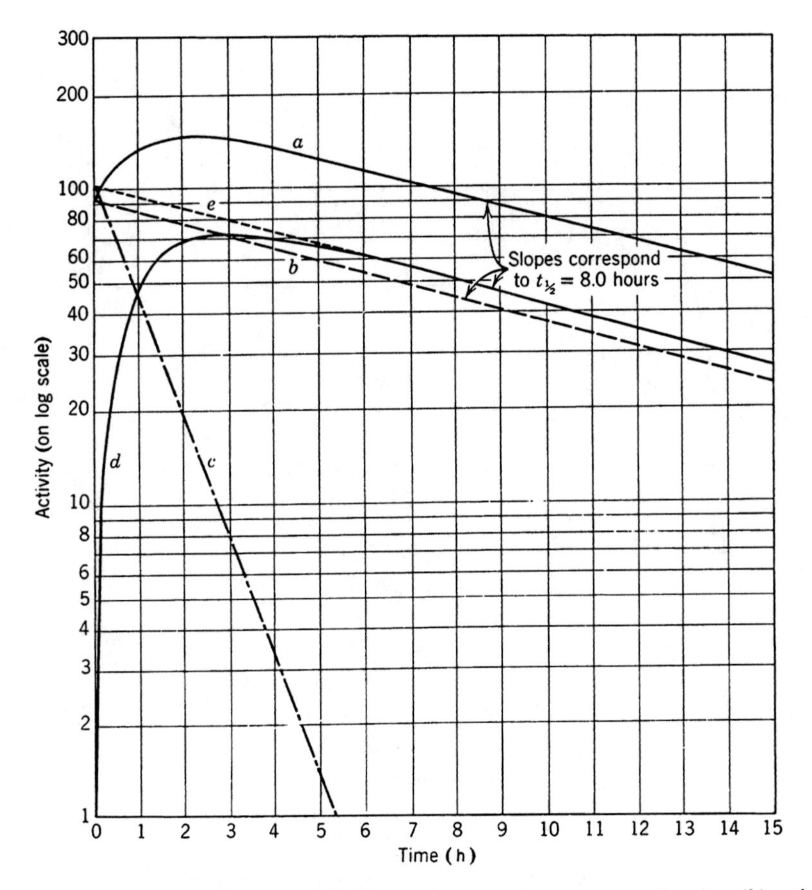

Fig. 5-2 Transient equilibrium: (*a*) total activity of an initially pure parent fraction; (*b*) activity due to parent ($t_{1/2} = 8.0$ h); (*c*) decay of freshly isolated daughter fraction ($t_{1/2} = 0.80$ h); (*d*) daughter activity growing in freshly purified parent fraction; (*e*) total daughter activity in parent-plus-daughter fractions.

presented there as a useful approximation of (5-3):

$$\frac{N_1}{N_2} = \frac{\lambda_2}{\lambda_1}, \qquad \text{or} \qquad \lambda_1 N_1 = \lambda_2 N_2.$$

In the same way (5-4) reduces to

$$\frac{\mathbf{A}_1}{\mathbf{A}_2} = \frac{c_1}{c_2},$$

and the measured activities are equal if $c_1 = c_2$.

Figure 5-2 presents an example of transient equilibrium with $\lambda_1 < \lambda_2$ (actually with $\lambda_1/\lambda_2 = 0.1$); the curves represent variations with time of the parent activity and the activity of a freshly isolated daughter fraction, the growth of daughter activity in a freshly purified parent fraction, and other

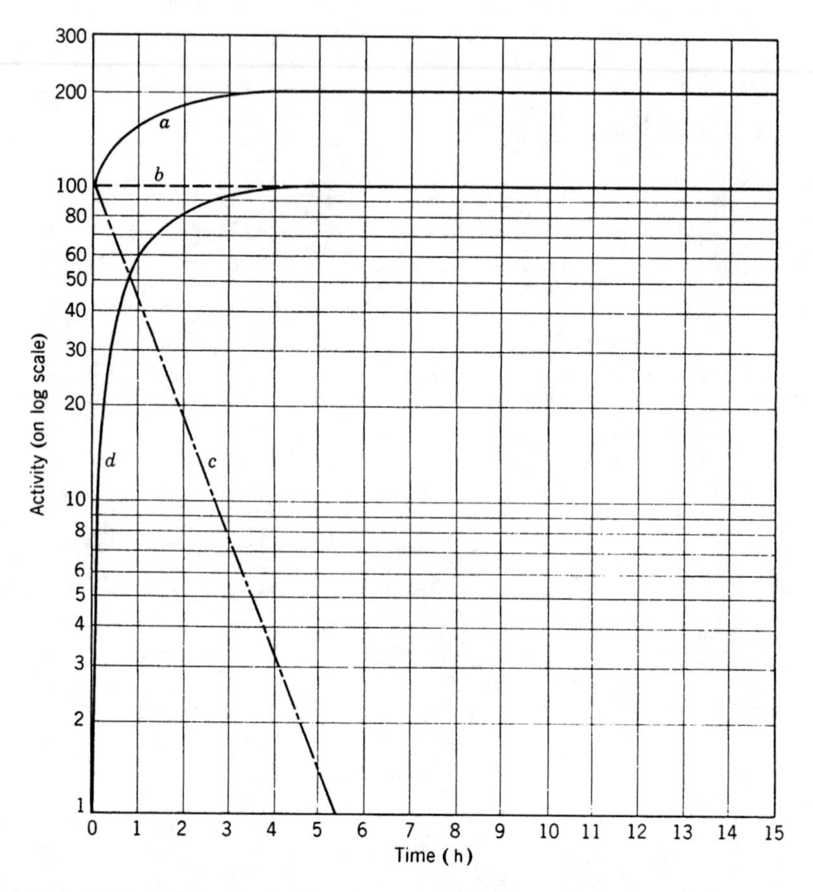

Fig. 5-3 Secular equilibrium: (*a*) total activity of an initially pure parent fraction; (*b*) activity due to parent ($t_{1/2} = \infty$); this is also the total daughter activity in parent-plus-daughter fractions; (*c*) decay of freshly isolated daughter fraction ($t_{1/2} = 0.80$ h); (*d*) daughter activity growing in freshly purified parent fraction.

relations; in preparing the figure we have taken $c_1 = c_2$. Figure 5-3 is a similar plot for secular equilibrium; it is apparent that as λ_1 becomes smaller compared to λ_2 the curves for transient equilibrium shift to approach more and more closely the limiting case shown in figure 5-3.

The Case of No Equilibrium. If the parent is shorter-lived than the daughter ($\lambda_1 > \lambda_2$), it is evident that no equilibrium is attained at any time. If the parent is made initially free of the daughter, then as the parent decays the amount of daughter will rise, pass through a maximum, and eventually decay with the characteristic half life of the daughter. This is illustrated in figure 5-4; for this plot we have taken $\lambda_1/\lambda_2 = 10$, and $c_1 = c_2$. In the figure the final exponential decay of the daughter is extrapolated

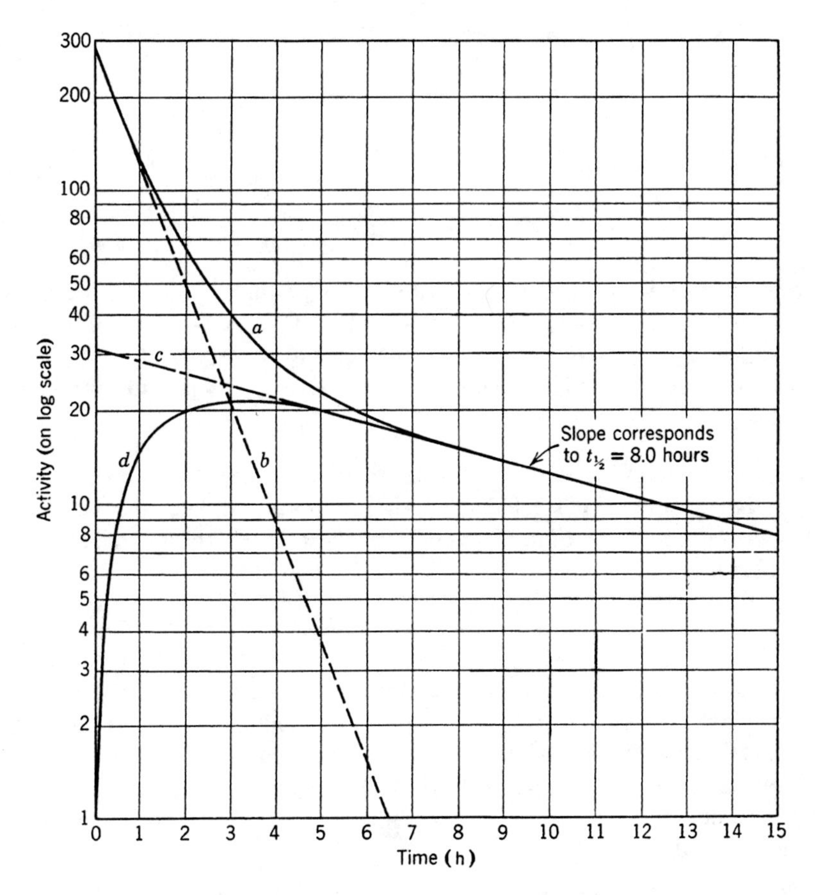

Fig. 5-4 The case of no equilibrium: (*a*) total activity; (*b*) activity due to parent ($t_{1/2} = 0.80$ h); (*c*) extrapolation of final decay curve to time zero; (*d*) daughter activity in initially pure parent.

back to $t = 0$. This method of analysis is useful if $\lambda_1 \gg \lambda_2$, for then this intercept measures the activity $c_2\lambda_2N_1^0$: the N_1^0 atoms give rise to N_2 atoms so early that N_1^0 may be set equal to the extrapolated value of N_2 at $t = 0$. The ratio of the initial activity $c_1\lambda_1N_1^0$ to this extrapolated activity gives the ratios of the half lives if the relation between c_1 and c_2 is known:

$$\frac{c_1\lambda_1N_1^0}{c_2\lambda_2N_1^0} = \frac{c_1}{c_2} \times \frac{\lambda_1}{\lambda_2} = \frac{c_1}{c_2} \times \frac{(t_{1/2})_2}{(t_{1/2})_1}.$$

If λ_2 is not negligible compared to λ_1, it can be shown that the ratio λ_1/λ_2 in this equation should be replaced by $(\lambda_1 - \lambda_2)/\lambda_2$ and the expression involving the half lives changed accordingly.

Both the transient-equilibrium and the no-equilibrium cases are sometimes analyzed in terms of the time t_m for the daughter to reach its maximum activity when growing in a freshly separated parent fraction.

This time we find from the general equation (5-2) by differentiating,

$$\frac{dN_2}{dt} = -\frac{\lambda_1^2}{\lambda_2 - \lambda_1} N_1^0 e^{-\lambda_1 t} + \frac{\lambda_1 \lambda_2}{\lambda_2 - \lambda_1} N_1^0 e^{-\lambda_2 t},$$

and setting $dN_2/dt = 0$ when $t = t_m$:

$$\frac{\lambda_2}{\lambda_1} = e^{(\lambda_2 - \lambda_1)t_m} \qquad \text{or} \qquad t_m = \frac{1}{\lambda_2 - \lambda_1} \ln \frac{\lambda_2}{\lambda_1}.$$

At this time the daughter decay rate $\lambda_2 N_2$ is just equal to the rate of formation $\lambda_1 N_1$ [this is obvious from (5-1)]; in figures 5-2 and 5-4, in which we assumed $c_1 = c_2$, we have the parent activity $\mathbf{A}_1$ intersecting the daughter growth curve *d* at the time t_m. (The time t_m is infinite for secular equilibrium.)

Many Successive Decays. If we consider a chain of three or more radioactive products, it is clear that the equations already derived for N_1 and N_2 as functions of time are valid, and N_3 may be found by solving the new differential equation:

$$\frac{dN_3}{dt} = \lambda_2 N_2 - \lambda_3 N_3. \tag{5-5}$$

This is entirely analogous to the equation for dN_2/dt, but the solution calls for more labor, since N_2 is a much more complicated function than N_1. The next solution for N_4 is still more tedious. H. Bateman (B1) has given the solution for a chain of n members with the special assumption that at $t = 0$ the parent substance alone is present, that is, that $N_2^0 = N_3^0 = \cdots = N_n^0 = 0$. This solution is

$$N_n = C_1 e^{-\lambda_1 t} + C_2 e^{-\lambda_2 t} + \cdots C_n e^{-\lambda_n t},$$

$$C_1 = \frac{\lambda_1 \lambda_2 \cdots \lambda_{n-1}}{(\lambda_2 - \lambda_1)(\lambda_3 - \lambda_1) \cdots (\lambda_n - \lambda_1)} N_1^0,$$

$$C_2 = \frac{\lambda_1 \lambda_2 \cdots \lambda_{n-1}}{(\lambda_1 - \lambda_2)(\lambda_3 - \lambda_2) \cdots (\lambda_n - \lambda_2)} N_1^0, \text{ and so on.}$$

If we do require a solution to the more general case with N_2^0, $N_3^0, \ldots, N_n^0 \neq 0$, we may construct it by adding to the Bateman solution for N_n in an n-membered chain a Bateman solution for N_n in an $(n-1)$-membered chain with substance 2 as the parent, and, therefore, $N_2 = N_2^0$ at $t = 0$, and a Bateman solution for N_n in an $(n-2)$-membered chain, and so on.

Branching Decay. The case of branching decay when a nuclide can decay by more than one mode is illustrated by

$$A \begin{array}{l} \xrightarrow{\lambda_B} B \\ \xrightarrow[\lambda_C]{} C \end{array}$$

The two **partial decay constants** λ_B and λ_C must be considered when the general relations in either branch are studied because, for example, the substance B is formed at the rate

$$\frac{dN_B}{dt} = \lambda_B N_A,$$

but A is consumed at the rate

$$\frac{dN_A}{dt} = -(\lambda_B + \lambda_C)N_A.$$

The nuclide A *has only one half life*

$$t_{1/2}(A) = \frac{0.693}{\lambda_t},$$

where $\lambda_t = \lambda_B + \lambda_C + \cdots$. By definition the half life is related to the total rate of disappearance of a substance, regardless of the mechanism by which it disappears.

If the Bateman solution is to be applied to a decay chain containing branching decays, the λ's in the *numerators* of the equations defining C_1, C_2, and so on, should be replaced by the partial decay constants; that is, λ_i in the numerators should be replaced by λ_i^*, where λ_i^* is the decay constant for the transformation of the ith chain member to the $(i+1)$th member. If a decay chain branches, and subsequently the two branches are rejoined as in the natural radioactive series, the two branches are treated by this method as separate chains; the production of a common member beyond the branch point is the sum of the numbers of atoms formed by the two paths.

C. EQUATIONS OF TRANSFORMATION DURING NUCLEAR REACTIONS

Stable Targets. When a target is irradiated by particles that induce nuclear reactions, a steady state can be reached in which radioactive products disintegrate at just the rate at which they are formed; the situation is analogous to that of secular equilibrium. If the irradiation is terminated before the steady state is achieved, then the disintegration rate of a particular active nuclide is less than its rate of formation R. The differential equation that governs the number of product atoms N present at time t during the irradiation is

$$\frac{dN}{dt} = R - \lambda N,$$

the solution to which is

$$R = \frac{N\lambda}{1 - e^{-\lambda t}}. \tag{5-6}$$

For very large irradiation times ($t \gg 1/\lambda$) the disintegration rate λN approaches the **saturation value** R. The factor $(1 - e^{-\lambda t})$ is often called the **saturation factor**. If the disintegration rate of a particular radioactive product at the end of a steady bombardment of known duration is divided by this saturation factor, the rate at which the product was formed during the bombardment is obtained.[1]

Occasionally a product is formed during irradiation both directly by a nuclear reaction and by the decay of an active parent that is produced by another reaction [e.g., the product of a (p, pn) reaction, if unstable, may decay by positron emission or EC into the product of the $(p, 2p)$ reaction on the same target]. Under these circumstances the number of atoms of the product of interest present at a time t_s after the end of a bombardment of duration t_b has three sources:

1. Those formed directly in nuclear reactions.
2. Those formed by the decay of the parent during bombardment.
3. Those formed by the decay of the parent during the interval t_s (which may, for example, be the time between the end of bombardment and the chemical separation of daughter from parent).

If R_1 and R_2 are the rates of the nuclear reactions that directly form the parent and daughter products, respectively, then the number of daughter atoms (characterized by subscript 2) arising from each of the three sources is

$$N_2' = \frac{R_2}{\lambda_2}(1 - e^{-\lambda_2 t_b})e^{-\lambda_2 t_s}, \tag{5-7}$$

$$N_2'' = \left[\frac{R_1}{\lambda_2}(1 - e^{-\lambda_2 t_b}) + \frac{R_1}{\lambda_1 - \lambda_2}(e^{-\lambda_1 t_b} - e^{-\lambda_2 t_b})\right]e^{-\lambda_2 t_s}, \tag{5-8}$$

$$N_2''' = \frac{R_1(1 - e^{-\lambda_1 t_b})(e^{-\lambda_1 t_s} - e^{-\lambda_2 t_s})}{\lambda_2 - \lambda_1}. \tag{5-9}$$

Experimentally it is, of course, only the totality of the daughter atoms

[1] If the rate of formation R is not constant during the irradiation (beam current varies during the bombardment), the bombardment may be divided into time intervals Δt_i during each of which the rate R_i is approximately steady. Under these circumstances the number of atoms present after a bombardment of total length t comes directly from an obvious modification of (5-6):

$$N = \frac{1}{\lambda}\sum_{i=1}^{n} R_i(1 - e^{-\lambda \Delta t_i})e^{-\lambda(t - t_i)},$$

where t_i is the time at the end of the ith interval. In the event that the time intervals are short compared to the half life of the product ($\lambda \Delta t_i \ll 1$), expansion of the exponential gives

$$N = \sum_{i=1}^{n} R_i \, \Delta t_i e^{-\lambda(t - t_i)}.$$

$(N_2' + N_2'' + N_2''')$ that is observed, but from a knowledge of the times t_b and t_s, of the decay constants λ_1 and λ_2, and of the rate of formation R_1 of the parent (which can be determined in a separate experiment), it is possible to calculate R_2.

Radioactive Targets in a High-Flux Reactor. When nuclear reactions are induced in a radioactive nuclide, the rate of disappearance of the substance is no longer governed by the law of radioactive transformation alone but by a modified law that takes into account the disappearance by transmutation reactions also. Under most practical bombardment conditions the rate of transformation of radioactive species by nuclear reactions is negligible compared to the rate of radioactive decay. However, in the case of long-lived nuclides, and with the large neutron fluxes available in nuclear reactors, transformations by both mechanisms sometimes have to be considered. We state the modified transformation equations for the case of a neutron flux; they are equally applicable for any other bombarding particle. The treatment given here follows that developed by W. Rubinson (R1).

Consider N atoms of a single radioactive species of decay constant λ (in reciprocal seconds) and total neutron reaction cross section σ (in square centimeters) in a constant neutron flux nv (neutrons cm^{-2} s^{-1}). The rate of radioactive transformation is λN, the rate of transformation by neutron reactions is $nv\sigma N$, and the total rate of disappearance is

$$-\frac{dN}{dt} = (\lambda + nv\sigma)N = \Lambda N, \tag{5-10}$$

where Λ may be considered as a modified decay constant. Equation 5-10 has the same form as the standard differential equation of radioactive decay and is integrated to give

$$N = N_o e^{-\Lambda t}. \tag{5-11}$$

If we consider a parent-daughter pair, the parent disappears by both transmutation and decay: $-dN_1/dt = (\lambda_1 + nv\sigma_1)N_1 = \Lambda_1 N_1$; but the daughter grows by decay of the mother only and disappears by both processes: $dN_2/dt = \lambda_1 N_1 - \Lambda_2 N_2$, or, in more general notation,

$$\frac{dN_{i+1}}{dt} = \lambda_i N_i - \Lambda_{i+1} N_{i+1}.$$

Actually we may want to consider chains in which the transformation from one member to the next may occur by nuclear reaction as well as by radioactive decay. Then λ_i must be replaced by a modified decay constant $\Lambda_i^* = \lambda_i^* + nv\sigma_i^*$, where the asterisks serve as a reminder that, if either the decay or reaction of the parent does not always lead to the next chain member, then λ_i^* must be the partial decay constant and σ_i^* must be the partial reaction cross section leading from the ith member to the $(i+1)$th member of the chain. With this notation the general solution is written, as in the Bateman equations, for $N_2^0 = N_3^0 = \cdots = N_n^0 = 0$:

$$N_n = C_1 e^{-\Lambda_1 t} + C_2 e^{-\Lambda_2 t} + \cdots + C_n e^{-\Lambda_n t}, \qquad (5\text{-}12)$$

where

$$C_1 = \frac{\Lambda_1^* \Lambda_2^* \cdots \Lambda_{n-1}^*}{(\Lambda_2 - \Lambda_1)(\Lambda_3 - \Lambda_1)\cdots(\Lambda_n - \Lambda_1)} N_1^0,$$

$$C_2 = \frac{\Lambda_1^* \Lambda_2^* \cdots \Lambda_{n-1}^*}{(\Lambda_1 - \Lambda_2)(\Lambda_3 - \Lambda_2)\cdots(\Lambda_n - \Lambda_2)} N_1^0, \text{ and so on.}$$

As an illustration, we compute the amount of 3.15-d ^{199}Au formed by two successive (n, γ) reactions when 1 g ^{197}Au is exposed for 30 h in a neutron flux of $1 \times 10^{14}\ \text{cm}^{-2}\ \text{s}^{-1}$. The chain of reactions is

$$^{197}\text{Au} \xrightarrow[n,\,\gamma]{\sigma=99\text{b}} {}^{198}\text{Au} \xrightarrow[n,\,\gamma]{\sigma=2.5\times10^4\text{b}} {}^{199}\text{Au}$$

$$^{198}\text{Au}: \beta^- \downarrow t_{1/2}=2.70\text{ d} \qquad ^{199}\text{Au}: \beta^- \downarrow t_{1/2}=3.14\text{ d}$$

We use (5-12) for this three-membered chain:

$$N_{199} = \Lambda_{197}^* \Lambda_{198}^* N_{197}^0 \left[\frac{e^{-\Lambda_{197} t}}{(\Lambda_{198} - \Lambda_{197})(\Lambda_{199} - \Lambda_{197})} + \frac{e^{-\Lambda_{198} t}}{(\Lambda_{197} - \Lambda_{198})(\Lambda_{199} - \Lambda_{198})} + \frac{e^{-\Lambda_{199} t}}{(\Lambda_{197} - \Lambda_{199})(\Lambda_{198} - \Lambda_{199})} \right].$$

The numerical values to be substituted are

$$t = 1.08 \times 10^5\ \text{s},$$

$$nv = 10^{14}\ \text{cm}^{-2}\ \text{s}^{-1},$$

$$\sigma_{197} = 9.9 \times 10^{-23}\ \text{cm}^2,$$

$$\sigma_{198} = 2.5 \times 10^{-20}\ \text{cm}^2,$$

$$N_{197}^0 = \frac{6.02 \times 10^{23}}{197} = 3.05 \times 10^{21},$$

$$\Lambda_{197}^* = \Lambda_{197} = nv\sigma_{197} = 9.9 \times 10^{-9}\ \text{s}^{-1},$$

$$\Lambda_{198} = \lambda_{198} + nv\sigma_{198} = 3.0 \times 10^{-6} + 2.5 \times 10^{-6} = 5.5 \times 10^{-6}\ \text{s}^{-1},$$

$$\Lambda_{198}^* = nv\sigma_{198} = 2.5 \times 10^{-6}\ \text{s}^{-1},$$

and

$$\Lambda_{199} = \lambda_{199} = 2.55 \times 10^{-6}\ \text{s}^{-1}.$$

Using these values, we get

$$N_{199} = 7.85 \times 10^7 \left(\frac{e^{-0.00107}}{5.5 \times 10^{-6} \times 2.55 \times 10^{-6}} + \frac{e^{-0.594}}{5.5 \times 10^{-6} \times 2.95 \times 10^{-6}} - \frac{e^{-0.275}}{2.55 \times 10^{-6} \times 2.95 \times 10^{-6}} \right)$$

$$= 7.55 \times 10^7 (7.12 \times 10^{10} + 3.40 \times 10^{10} - 1.01 \times 10^{11}) = 3.2 \times 10^{17}.$$

The disintegration rate of ^{199}Au at the end of the irradiation is $\lambda_{199} N_{199} = 0.82 \times 10^{12}\ s^{-1}$. For comparison we compute the disintegration rate of ^{198}Au in the sample [again from (5-12) for a two-membered chain]:

$$\lambda_{198} N_{198} = \lambda_{198} n v \sigma_{197} N_{197}^0 \left(\frac{e^{-\Lambda_{197} t}}{\Lambda_{198} - \Lambda_{197}} + \frac{e^{-\Lambda_{198} t}}{\Lambda_{197} - \Lambda_{198}} \right)$$

$$= 9.06 \times 10^7 \frac{0.999 - 0.552}{5.5 \times 10^{-6}} = 7.36 \times 10^{12}\ s^{-1}.$$

Thus about 10 percent of the radioactive disintegrations in the sample occur in ^{199}Au.

REFERENCES

B1 H. Bateman, "Solution of a System of Differential Equations Occurring in the Theory of Radio-active Transformations," *Proc. Cambridge Phil. Soc.* **15**, 423 (1910).

C1 J. B. Cumming, "CLSQ, The Brookhaven Decay-Curve Analysis Program," in *Application of Computers to Nuclear- and Radiochemistry* (G. D. O'Kelly, Ed.), NAS-NRC, Washington, 1963, p. 25.

R1 W. Rubinson, "The Equations of Radioactive Transformation in a Neutron Flux," *J. Chem. Phys.* **17**, 542 (1949).

*R2 E. Rutherford, J. Chadwick, and C. D. Ellis, *Radiations from Radioactive Substances*, Cambridge University Press, Cambridge, 1930.

EXERCISES

1. The following experimental data were obtained when the activity of a certain β-active sample was measured at the intervals shown.

Time (h)	Activity (counts/min)	Time (h)	Activity (counts/min)
0	7300	4.0	481
0.5	4680	5.0	371
1.0	2982	6.0	317
1.5	1958	7.0	280
2.0	1341	8.0	254
2.5	965	10.0	214
3.0	729	12.0	181
3.5	580	14.0	153

Plot the decay curve on semilog paper and analyze it into its components. What are the half lives and the initial activities of the components?

2. Compute (a) the weight of 1 Ci of ^{222}Rn, (b) the weight of 1 Ci of ^{32}P, (c) the disintegration rate of 1 cm^3 of tritium ($^3H_2$) at STP. *Answer:* (*a*) 6.5 μg.
3. What was the rate of production, in atoms per second, of ^{128}I during a constant 1-h cyclotron (neutron) irradiation of an iodine sample if the sample is found to contain 2.00 mCi of ^{128}I activity at 15 min after the end of the irradiation?
4. From data in appendix D calculate the total rate of emission of α particles from 1 mg of ordinary uranium. Calculate this answer also for the case of 1 mg of very old uranium in secular equilibrium with all its decay products. *Answer to first part:* 25.3 s^{-1}
5. A 0.100 mg sample of pure ^{239}Pu (an α-particle emitter) was found to undergo 1.38×10^7 dis min^{-1}. Calculate the half life of this nuclide. ^{239}Pu is formed by the β decay of ^{239}Np. How many curies of ^{239}Np would be required to produce a 0.100-mg sample of ^{239}Pu? *Answer to second part:* 23.2 Ci.
6. To determine the thermal-neutron capture cross section of 30.6-h ^{193}Os, a 100-mg sample of osmium metal is placed in a thermal-neutron flux of 2×10^{12} cm^{-2} s^{-1} for 30 d. The amount of 6.0-y ^{194}Os activity formed is found from subsequent decay measurements to be 5.4×10^3 dis s^{-1} at the end of irradiation. From the known activation cross section of ^{192}Os and the half lives of ^{193}Os and ^{194}Os, compute the capture cross section of ^{193}Os. Neglect neutron capture in ^{194}Os. *Answer:* See appendix D.
7. A sample of 1.00×10^{-10} g of ^{210}Bi is freshly purified at time $t = 0$. (a) If this sample is left without further treatment, when will the amount of ^{210}Po in it be a maximum? (b) At that time of maximum growth what will be the weight of ^{210}Po present, the α activity in disintegrations per second, the beta activity of the sample in disintegrations per second, the number of microcuries of ^{210}Po present? (c) Sketch on semilog paper a graph of α activity and β activity versus time.
8. (a) Show that the number of radioactive daughter atoms present in a sample at time t is given by

$$(\lambda N_1^0 t + N_2^0)e^{-\lambda t}$$

if the parent and daughter atoms have the same half life and N_1^0 and N_2^0 are the numbers of parent and daughter atoms, respectively, that were initially present. (b) Show that the sum of the disintegration rates of parent and daughter decays with the half life of the daughter if the daughter substance has twice the half life of the parent. (c) Derive the condition $c_2/c_1 > \lambda_1/\lambda_2$ (given on p. 194) for the occurrence of a maximum in the total counting rate of an initially pure parent fraction.
9. In the slow-neutron activation of a sample of separated ^{100}Mo isotope some 14.6-min ^{101}Mo is produced; this decays to 14.0-min ^{101}Tc. A sample of ^{101}Mo is chemically freed of technetium and then immediately placed under a counter. (a) Sketch the activity as a function of time, assuming the detection coefficient to be the same for the ^{101}Tc as for the ^{101}Mo radiation. (b) Sketch the activity as a function of time if radiation characteristic of ^{101}Tc decay (e.g., a 307-keV γ ray) is detected, that is, if the detection coefficient for ^{101}Mo radiations is essentially zero.

10. Carry out the solution of the differential equation 5-5. Compare your result with the Bateman solution for this case with N_2^0 and N_3^0 not equal to 0.

11. A sample of an activity whose half life is known to be 7.50 min was measured from 10:03 to 10:13. The total number of counts recorded in this 10-min interval was 34,650. What was the activity of the sample (in counts per minute) at 10:00? *Answer:* 7.01×10^3.

12. What will be the disintegration rate of ^{198}Au in a 10-mg sample of gold that has been irradiated in a nuclear reactor at a flux of (a) 10^{12} neutrons cm^{-2} s^{-1} for 1 d; (b) 10^{15} neutrons cm^{-2} s^{-1} for 1 d? *Answer:* (*a*) 6.8×10^8 s^{-1}.

13. A 1.0-mg cm^{-2} target of ^{64}Zn enriched to 100 percent is bombarded for 38 min in an α-particle beam of intensity 6×10^{12} alphas per second. The energy of the α particles is such that there is production of ^{67}Ge and ^{67}Ga by the (α, n) and (α, p) reactions, respectively, during the bombardment. From gallium and germanium samples isolated from the target it was found that the disintegration rate of the ^{67}Ge 1 h after the irradiation was 1.34×10^6 dis s^{-1}, whereas that for ^{67}Ga was 2.50×10^5 dis s^{-1}. The gallium sample from which the disintegration rate of ^{67}Ga was determined was separated from the germanium in the target 0.5 h after the end of the irradiation. From this information calculate the rates of formation of ^{67}Ge and ^{67}Ga directly by nuclear reactions during the irradiation and their production cross sections.

Answers: ^{67}Ge—1.60×10^7 s^{-1}, $\sigma = 0.28$ b.
^{67}Ga—3.19×10^7 s^{-1}, $\sigma = 0.56$ b.

Chapter 6

Interaction of Radiations with Matter

Nuclear radiations, both corpuscular and electromagnetic, are detectable only through their interactions with matter. If this interaction is not sufficiently large, as in the case of the neutrino, the radiation is nearly undetectable. For an understanding of the methods and instruments used for the detection, measurement, and characterization of nuclear radiations, it is necessary to consider the manner in which these radiations interact with matter.

The slowing down and absorption of radiation in matter are important for the reduction in energy of beams of high-energy particles, for the detection of a particular radiation in the presence of others with sufficiently different absorption characteristics, and for the application of nuclear radiations in medical therapy. Up to about 1950 absorption studies of the radiation from radioactive substances were an important technique for energy determinations. They no longer are; most serious energy determinations are now performed by the measurement of deflections in electric and magnetic fields, by the use of detectors whose output is sensitive to energy, or by the use of crystal diffraction.

The interactions of all types of radiations with matter *ultimately* have the same effect. However, the initial stages of the interactions (excitation and ionization of atoms and molecules) are sufficiently different for charged particles, electromagnetic radiation, and neutrons to merit separate treatments in the following sections of this chapter. Further, the fact that electrons are about 1800 times lighter than the lightest positive ion causes differences sufficiently large that the interactions of charged particles with matter are separated into those with positive ions (protons and heavier) and those with electrons (both positive and negative).

A. POSITIVE IONS

Processes Responsible for Energy Loss. In passing through matter positive ions lose energy chiefly by interaction with electrons. This interaction may lead to the dissociation of molecules or to the excitation or ionization of atoms and molecules. Ionization is the effect most easily measured and most often used for the detection of positive ions. Because α particles were easily available from radioactive sources, and because gas-filled ionization chambers played a key role in the early studies of radioactivity, much of the available information on the passage of positive ions through matter is for the ionization caused by α particles interacting with various gases.

A known number of α particles of known initial energy can be made to deposit their entire energy inside an ionization chamber since α particles travel only relatively short distances in matter before being reduced to thermal energies. Thus the total ionization produced per α particle is readily measured. These experiments show that on the average approximately 35 eV of energy are dissipated for each ion pair formed in air. The energy expended in the formation of an ion pair in other gases studied ranges from a minimum of 21.9 eV for xenon to a maximum of 43 eV for helium. This spread is compatible with the increase of the first ionization potentials from 12.1 eV for xenon to 25.4 eV for helium. In both of these monatomic gases the fraction of the energy expended that goes into ionization is about the same. As might be expected, the other degrees of freedom that are available for diatomic and polyatomic gases lower the fraction of the expended energy that goes into ionization. For example, NH_3 requires 39 eV for each ion pair formed although the first ionization potential is only 10.8 eV. The energy expended per ion pair is about an order of magnitude less in semiconductors than in gases, which is of great practical importance for radiation detectors (see chapter 7). The value, for example, in germanium is 2.9 eV, which reflects the fact that the energy required to raise an electron to the conduction band in germanium is of the order of 1 eV.

The energy per ion pair is insensitive to the energy and nature of the radiation; almost identical values have been obtained in experiments with α particles of a few million electron volts, with 340-MeV protons, and with β particles. Thus a very good way to measure the energy of a charged particle is to measure electrically the total number of ions produced when it is stopped in either a gas-filled ionization chamber or a semiconductor.

A large part of the energy loss of positive ions is accounted for by the kinetic energy given to the electrons removed from atoms or molecules in close collisions with the ion. It can easily be shown from conservation of momentum that the maximum velocity that an ion of velocity v can impart to an electron is about $2v$; therefore the maximum energy that an electron can receive from the impact of a 6-MeV α particle, for example, is about 3 keV. The average energy imparted to electrons by ions in their passage through matter is of the order of 100–200 eV. Many of these secondary electrons or δ rays are energetic enough to ionize other atoms. In fact about 60–80 percent of the ionization produced by positive ions is due to secondary ionization; the exact ratio of primary to secondary ionization is difficult to determine. Delta-ray tracks are often seen in cloud-chamber pictures of positive-ion tracks.

When the velocity of the ion has been reduced to a point at which it is comparable to the velocity of the valence electrons in an atom of the stopping material, a new phenomenon becomes important: the ion starts making elastic collisions with the atoms rather than exciting the atomic electrons. These ion-atom collisions give rise to what is known as **nuclear stopping**, as compared to the **electronic stopping** that occurs at the higher velocities. In addition, when the velocity of the ion becomes comparable to that of an electron in its K shell, the ion will start picking up electrons from the atom in the stopping material, and the average charge of the particles will change from Z to $Z-1$. On the average, ions passing through matter will be stripped of all orbital electrons whose orbital velocity is less than the velocity of the ion.

In sum, there are three important phenomena attendant upon the passage of a positive ion through matter:

1. At sufficiently high velocities the ion is stripped of all of its electrons and the energy loss is essentially all through electronic excitation and ionization of the stopping material.
2. At velocities comparable to the velocities of its K-shell electrons the ion starts to pick up electrons from the stopping material. The mechanism of energy loss is still essentially all electronic.
3. At velocities comparable to those of the valence electrons of the stopping material the mechanism of energy loss essentially becomes one of elastic collisions between the ion, even if it still has a charge, and the atoms of the stopping material.

There is no sharp gradation between (2) and (3); there is an energy region in which the ion makes both elastic and inelastic atomic collisions.

Range. Because of the very large mass of a positive ion compared to that of an electron, the rectilinear distances that positive ions of a given type and initial energy travel in matter before being brought to rest are the same within narrow limits. This distance is known as the range and is, of course, dependent on the type and the energy of the ion. The large mass is decisive in this situation for two related reasons: (1) the fractional energy loss per collision is very small (the maximum is about $4m/M$ where m and M are the masses of electron and ion, respectively), thereby ensuring a very large number of collisions required to stop the ion, and thus minimizing the effects of fluctuations in the average energy lost in each collision; (2) the deflection of the ion in each collision is very small, and thus the actual path length is closely the same as its projection on the initial direction of motion (the rectilinear distance). To the extent that fluctuations in average energy loss and projected path do occur, there is some dis-

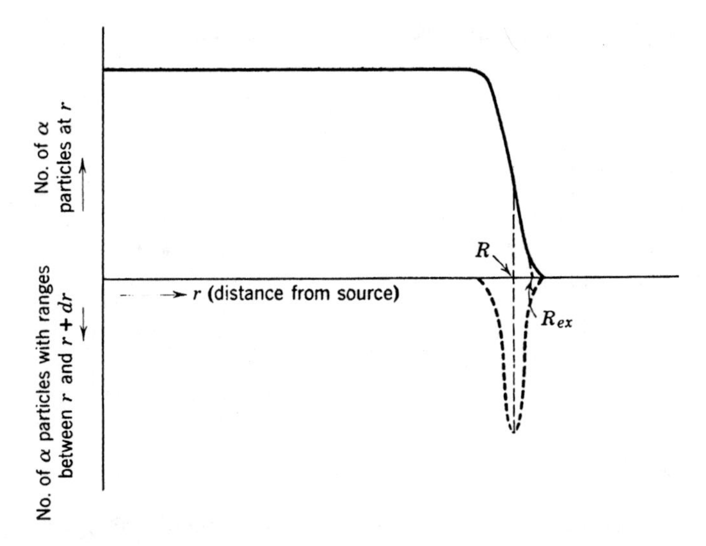

Fig. 6-1 Number of ions from a point source as a function of distance from the source (full curve). The derivative of that function is also shown (heavy dashed curve); the latter represents the distribution in ranges.

tribution in the ranges, which is known as **straggling** and is usually of the order of a few percent.

Ranges of positive ions are generally determined by absorption methods, either with solid absorbers or, more accurately, with a gaseous absorber at variable pressure. A typical absorption curve is shown in figure 6-1 where the number of ions found in a gas at a distance r from the source is plotted against r for a source that emits ions of a single energy. The heavy dashed curve in figure 6-1 is obtained by differentiation of the other (integral) curve and represents the distribution of ranges or the amount of straggling; it is approximately a Gaussian curve. The distance r corresponding to the maximum of the differential curve (point of inflection of the integral curve) is called the **mean range** R of the ions. The distance r obtained by extrapolating to the abscissa the approximately straight portion of the integral curve is the **extrapolated range** R_{ex}. The mean range is now generally used in range tables and in range-energy relations. Extrapolated ranges were often given in the older literature and are more easily determined experimentally. Relations between the two ranges are available; the difference is approximately 1.1 percent for α particles of ordinary energies.

Stopping Power. The relationship between the energy of a positive ion and its range may be more clearly seen in terms of dE/dx, the rate at which the charged particle loses energy in passing through matter. In a given medium the quantity dE/dx, often called the **stopping power** or **specific ionization**, is a function of the energy, charge, and mass of the ion. The

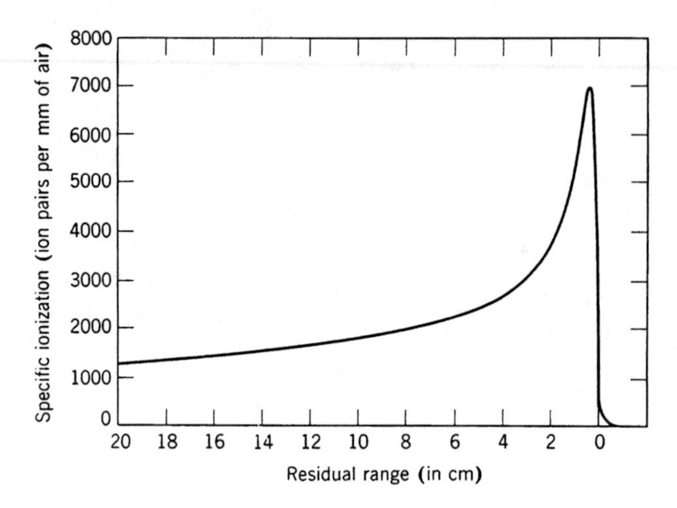

Fig. 6-2 Bragg curve for initially homogeneous α particles.

general properties of this function may be seen from figure 6-2, a classical **Bragg curve** that gives the specific ionization as a function of the distance of an α particle from the end of its path. From these data it is seen that there is a maximum rate of energy loss that occurs at rather low energies and a decrease toward higher energy that is approximately given by an inverse dependence on the energy. This behavior can be rather easily understood qualitatively.

The interaction between the charged particle and the atomic electrons is nearly completely described as a Coulomb interaction between the electrons and a positive point charge. Thus if the ion diminishes its charge by picking up electrons in its passage through matter, the Coulomb interaction and the rate of energy loss will diminish. This is what occurs on the low-energy side of the maximum and ultimately changes the mechanism of energy loss to one of elastic collisions between atoms. As mentioned on p. 208, an ion will pick up an electron whose orbital velocity is no greater than its own speed. Actually, an ion will pick up and then lose electrons many times near the end of its trajectory.

The diminution in the rate of energy loss with increasing energy on the high-energy side of the maximum is a consequence of the diminished "time of interaction" between the charged particle and the atomic electrons. If the velocity of the ion is v, the time that the ion spends within a given distance from the atom is proportional to $1/v$; hence the impulse on the electrons in the atom, or the momentum transfer, is also proportional to $1/v$. Since the energy transfer to the electrons in the atom is equal to the energy loss by the ion and is proportional to the square of the momentum transfer, it is evident that the rate of energy loss must have a term inversely proportional to v^2 or inversely proportional to E, the energy of the particle.

Derivation of Stopping-Power Formula. We now proceed to use the simple model of momentum transfer via a Coulomb interaction just outlined as the basis for a classical treatment of stopping power dE/dX, which yields many of the important features that result from a quantum-mechanical analysis of the problem. The central problem of the classical calculation is that of computing the momentum transfer to an electron from an ion of mass M and charge γze moving with a velocity v along a linear trajectory characterized by a distance of closest approach to the electron (impact parameter) b, where z is the atomic number of the *ion*[1] and γ is the average fraction of the electrons stripped off the ion. The model is illustrated in figure 6-3. The assumption that the ion moves undeflected is justified because the mass of the electron is negligible compared to that of the ion. The electron is considered to be free and stationary; this is acceptable as long as the velocity of the ion is much greater than that of the electron, or $(m/M)E \gg I$ where I is approximately the ionization potential of the electron and m is the electron mass.

The force exerted on the electron by the ion at point x (origin of coordinates taken at the point of closest approach) can be resolved into two components, normal to and parallel to the trajectory:

$$F_{\perp}(x) = \frac{\gamma ze^2}{(x^2+b^2)} \frac{b}{(x^2+b^2)^{1/2}}, \tag{6-1}$$

$$F_{\parallel}(x) = \frac{\gamma ze^2}{(x^2+b^2)} \frac{x}{(x^2+b^2)^{1/2}}. \tag{6-2}$$

From the symmetry of the problem it is evident that the average value of the parallel component will vanish. The momentum transfer to the electron will then be given by the usual impulse integral:

$$\Delta p = \int_0^{\infty} F_{\perp}(x)\, dt. \tag{6-3}$$

The integration over time may be transformed to an integration over x through the function $dx/dt = v$:

$$\Delta p = \frac{1}{v} \int_{-\infty}^{+\infty} F_{\perp}(x)\, dx \tag{6-4}$$

$$= \frac{\gamma ze^2 b}{v} \int_{-\infty}^{+\infty} \frac{dx}{(x^2+b^2)^{3/2}} \tag{6-5}$$

$$= \frac{2\gamma ze^2}{bv}. \tag{6-6}$$

The energy of the accelerated electron is $(\Delta p)^2/2m$; thus the energy transfer in a "collision" at impact parameter b is

$$\Delta E = \frac{2\gamma^2 z^2 e^4}{mb^2 v^2}. \tag{6-7}$$

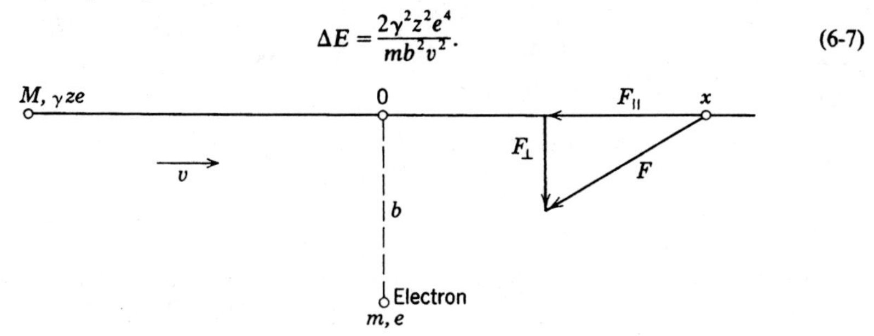

Fig. 6-3 Ion of mass M and charge γze losing energy to an electron (mass m and charge e) in an interaction with impact parameter b.

[1] The atomic number of the *absorber* is designated by Z.

The number of electrons with impact parameters between b and $b+db$ in a small thickness dx of the absorber is

$$2\pi bNZ\,db\,dx,$$

where N is the number of atoms per unit volume in the absorber of atomic number Z. Thus the energy transfer to electrons at impact parameters between b and $b+db$ is $4\pi\gamma^2 z^2 e^4 NZ\,db/mbv^2$ and the total rate of energy loss,[2] the stopping power, is obtained by integrating over all allowable values of b:

$$-\frac{dE}{dx}=\frac{4\pi\gamma^2 z^2 e^4 NZ}{mv^2}\int_{b_{\min}}^{b_{\max}}\frac{db}{b} \tag{6-8}$$

$$=\frac{4\pi\gamma^2 z^2 e^4 NZ}{mv^2}\ln\frac{b_{\max}}{b_{\min}}. \tag{6-9}$$

The negative sign of the differential arises because (6-9) represents the rate of energy *loss* by the ion. The allowable extremes of the impact parameter that appear as the limits of integration in (6-8) are determined by the assumptions made. The maximum impact parameter is limited by the assumption that the electron is both free and stationary during the time of the interaction. Although strictly speaking the range of the Coulomb interaction is infinite, the major effect, as seen from (6-1), occurs when $|x|<b$; thus the time of the interaction is approximately $2b/v$. We demand that this time be less than $1/\omega$ where ω is the classical frequency of motion of the electron in the atom. The upper limit on the impact parameter is then

$$b_{\max}=\frac{v}{2\omega}. \tag{6-10}$$

The minimum impact parameter is determined by the maximum amount of energy that can be transferred in a single collision. As stated earlier, the maximum velocity that may be imparted to an electron by collision with an ion is $2v$, which corresponds to a maximum energy transfer of $2mv^2$. From (6-7) the minimum impact parameter is

$$b_{\min}=\frac{\gamma ze^2}{mv^2}. \tag{6-11}$$

The substitution of (6-10) and (6-11) into (6-9) gives

$$-\frac{dE}{dx}=\frac{4\pi\gamma^2 z^2 e^4 NZ}{mv^2}\ln\frac{mv^3}{2\gamma ze^2\omega}, \tag{6-12}$$

a formula first derived by Bohr (B1).

A more exact analysis that proceeds through a quantum-mechanical and relativistic statement of the conditions required by the assumptions leads to the more accurate expression[3]

$$-\frac{dE}{dx}=\frac{4\pi\gamma^2 z^2 e^4 NZ}{mv^2}\left[\ln\frac{2mv^2}{I}-\ln(1-\beta^2)-\beta^2\right]. \tag{6-13}$$

[2] This ignores the special situation of a positive ion moving parallel to a crystal axis through a channel among the atoms. The rate of energy loss in this so-called **channeling** is less than that given above (D1).

[3] Equation 6-13 does not include small corrections arising from the effects of the density of the absorber and the large binding energies of the inner electrons in high-atomic-number absorbers. These effects are considered in reference F1, equations 38, 48, and 58.

The quantity I is the effective ionization potential of the atoms in the absorber, and $\beta = v/c$ where c is the velocity of light. Values of I/Z are in the range of 10–21 eV and are tabulated in F1.

For kinetic energies of ions small compared to their rest-mass energy ($\beta \ll 1$), (6-13) reduces to

$$-\frac{dE}{dx} = \frac{4\pi\gamma^2 z^2 e^4 NZ}{mv^2} \ln \frac{2mv^2}{I}, \tag{6-14}$$

which shows that, in this energy domain, the rate of energy loss decreases monotonically as the ion energy increases. Note that in (6-14) and the preceding equations m is the *electron* mass, but v is the *ion* velocity.

When the kinetic energy of the ion becomes larger than its rest-mass energy ($\beta \to 1$), the term $\ln(1-\beta^2)$ in (6-13) is the most rapidly varying one and the rate of energy loss increases with increasing energy. Thus the rate of energy loss goes through a broad minimum that occurs at approximately twice the rest-mass energy of the ion. The specific ionization of a singly charged particle at the minimum is about 1.8 MeV g^{-1} cm^2 in carbon and 1.1 MeV g^{-1} cm^2 in lead.[4] Physically the stopping power increases with energy in the relativistic region because the Lorentz contraction shortens the time of a collision and thereby allows a larger $b_{\max}$.

Stopping Power for Different Ions in Different Materials. An important feature of the equations just given is immediately evident: the rate of energy loss of all charged particles moving with the same velocity in a given absorber is proportional to the squares of their charges. Thus the rates of energy loss of protons of energy E, deuterons of energy $2E$, and tritons of energy $3E$ are all the same and are one quarter as large as those of a ^{3}He of energy $3E$ or an α particle of energy $4E$. These relatively simple relationships among the stopping powers of various ions with the same velocity require that the ion be stripped entirely of its electrons ($\gamma = 1$) and that the energy loss by nuclear stopping (elastic collisions) be negligible. Fulfillment of the former requirement ensures that of the latter. The very light ions such as hydrogen and helium are entirely stripped of their electrons at energies above approximately 1 MeV amu^{-1}. For boron through neon the energy required is of the order of 10 MeV amu^{-1}, while for uranium it approaches several hundred MeV amu^{-1}. The consequences of electron pickup and ultimately of nuclear stopping as the energy of the ion diminishes are illustrated in figure 6-4, in which the logarithm of the stopping power in aluminum divided by the square of the atomic number is plotted against the logarithm of the ion energy in MeV amu^{-1} for several ions.

[4] Equations 6-12, 6-13, and 6-14 give the rate of energy loss per unit distance traveled. Thus if cgs units were used, the rate of energy loss would be in ergs per centimeter. To convert this into the more common units of MeV mg^{-1} cm^2 it is necessary to divide by 1.6×10^{-6} ergs MeV^{-1} and by the density of the stopping material in mg cm^{-3}.

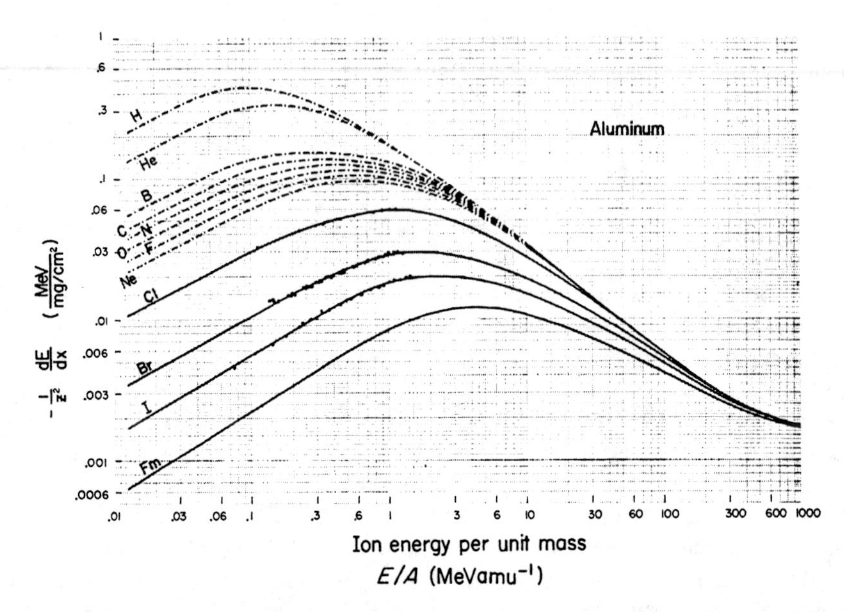

Fig. 6-4 Stopping power curves for various ions in aluminum. (From reference N1.)

As can be seen from either (6-14) or figure 6-4, the rate of energy loss is not the same for different ions with the same energy. For example, ^{16}O, ^{15}O, and ^{14}N, all at an energy of 80 MeV, lose energy in aluminum at the rate of 3.46, 3.32, and 2.49 MeV mg^{-1} cm^2, respectively. This property of stopping power makes it possible to identify atomic numbers of ions up to ~25 and individual isotopes for $Z \lesssim 20$ by means of a so-called **particle-identifier telescope**. In such a telescope the particle first passes through a thin detector in which it deposits some of its energy and then into a detector thick enough to stop the particle in which the rest of its energy is deposited. The energy of the particle is determined by the sum of the signals from the two detectors while the rate of energy loss is determined by the signal from the thin detector, thereby allowing identification of the particle as well as measurement of its energy (cf. chapter 8, section F).

The stopping power for a given ion in different solid absorbers relative to aluminum is given in figure 6-5. The use of this curve with figure 6-4 allows the determination of the stopping power of various ions in many solid absorbers. For example, the stopping power of 120-MeV ^{12}C ions in a nickel absorber is found to be 1.0 Mev mg^{-1} cm^2. This is determined from the stopping power of 120-MeV ^{12}C ions in aluminum (1.2 Mev mg^{-1} cm^2) given in figure 6-4 multiplied by the stopping power of 10-MeV amu^{-1} ions in nickel relative to aluminum (0.84) given in figure 6-5. This value agrees well with the detailed stopping-power tables given in N1.

Range-Energy Relations. The range of an ion may be immediately

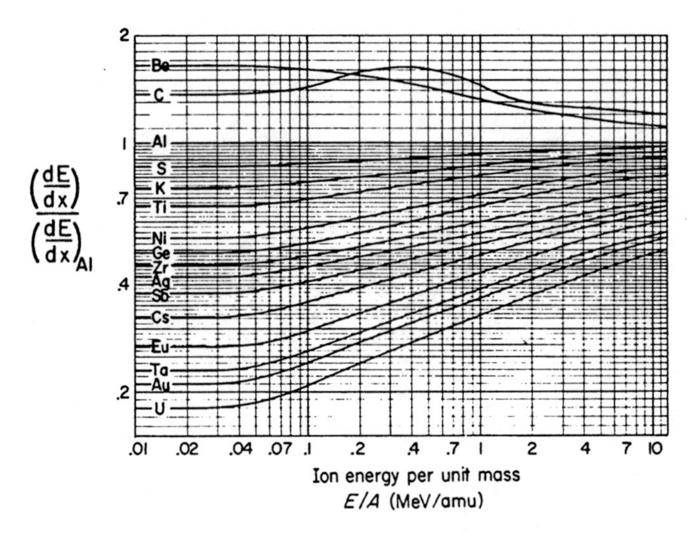

Fig. 6-5 Stopping-power curves for a given ion in different solid absorbers relative to aluminum. The curves are assumed to be the same for all ions. (From reference N1.)

computed by integration of the energy-loss expression

$$R = \int_{E_0}^{0} \frac{1}{dE/dx}\, dE. \tag{6-15}$$

Figure 6-6 gives the ranges in aluminum of protons and of helium ions (α particles) as functions of their kinetic energy. These curves are based more on theoretical calculations than on experiment because the calculated values are believed to be more accurate over most of the energy region. The only energy losses considered were those covered by (6-13). At very high energies other mechanisms may become relatively important; for example, 2-GeV protons are appreciably (approximately 15 percent) attenuated in intensity by nuclear reactions in a lead absorber 2.5 cm thick, but lose relatively little (less than 3 percent) of their energy by ionization processes in such an absorber. The ranges of some other representative ions are presented in figure 6-7.

If (6-14) for the stopping power is used in (6-15) for the range, and if the relatively slowly varying logarithmic term in (6-14) is neglected, then it would be expected that the range of a particle would be roughly proportional to the square of its energy for nonrelativistic energies. Using this qualitative idea as a guide, it has been found that a reasonable semiempirical fit to range data over a significant region of energy can be expressed as

$$R = aE_0^{\,b}, \tag{6-16}$$

where a and b are empirical constants that vary only slowly with energy,

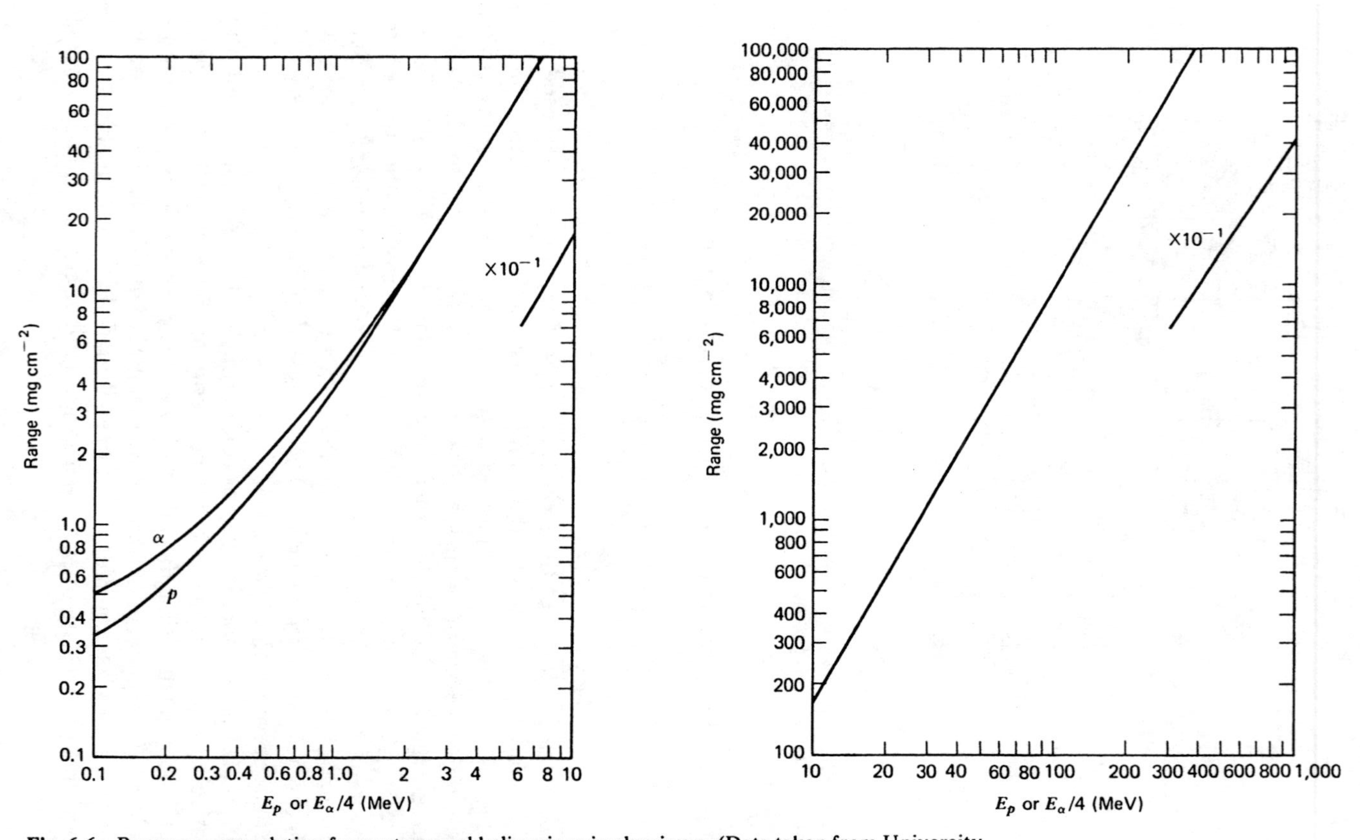

Fig. 6-6 Range-energy relation for protons and helium ions in aluminum. (Data taken from University of California Radiation Laboratory Report #2426 Rev (1966), Berkeley, CA.

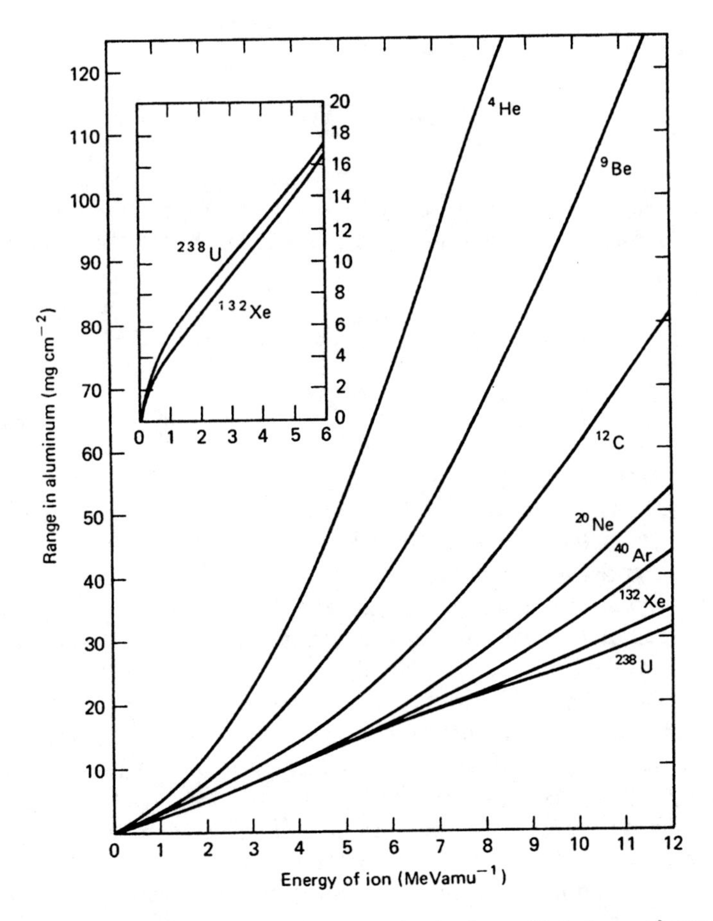

Fig. 6-7 Range-energy relation for heavy ions in aluminum. (Data from reference N1).

and E_0 is the kinetic energy of the positive ion. The constants a and b depend on particle type, but b usually has a value between 1.7 and 1.8 for ions that are stripped of all their electrons over essentially all of the range. For ions such as fission products that are not stripped of all their electrons, the range will increase more slowly with energy because increasing energy will also increase the average charge on the ion. It has been found experimentally that, for this situation, b is close to 0.5, and thus the range of the ion is proportional to its velocity. Equation 6-16 can be useful for the interpolation and extrapolation of the range tables.

Because the rate of energy loss by electronic processes for all ions at the same velocity is proportional to the square of the charge on the ion, it is possible to express the range of one ion in terms of the range of any other in a simple form for energies where both ions spend essentially all of the range completely stripped of electrons:

$$R(z_1, M_1, E_1) = \frac{M_1 z_2^2}{M_2 z_1^2} R\left(z_2, M_2, \frac{M_2}{M_1} E_1\right). \tag{6-17}$$

Here z_i, M_i, and E_i refer to the atomic number, mass, and energy of ion i. For example, the range of a 160-MeV ^{16}O ion in aluminum may be estimated from the information on protons in figure 6-6 through the use of (6-17):

$$R(8, 16, 160) = \left(\frac{16 \times 1^2}{1 \times 8^2}\right) R\left(1, 1, \frac{1}{16} \cdot 160\right)$$

$$= \left(\frac{1}{4}\right) \times 171 = 43 \text{ mg cm}^{-2}.$$

This estimate is to be compared with the experimental value of 46.8 mg cm^{-2}. The actual range is greater than the estimated one because electron pickup diminishes the rate of energy loss. The fractional error in the above expression may be approximated by the fraction of the range that is not spent in the completely ionized form.

The ranges of positive ions in absorbing materials other than aluminum are often wanted. Theoretical calculations based on (6-13) and (6-15) are normally used. These computations are time-consuming but have been carried out and tabulated for some elementary substances (N1, N2). These tables are derived from data of the kind given in figures 6-4 and 6-5.

In many actual cases, for example in air, the stopping substance is not a single element but rather a compound or mixture of elements. For practical purposes we make the further approximation that the stopping power of a molecule or of a mixture of atoms or molecules is given by the sum of the stopping powers of all the component atoms (rule due to Bragg).[5] A useful empirical expression that reflects this approximation is given by

$$\frac{1}{R_t} = \frac{w_1}{R_1} + \frac{w_2}{R_2} + \frac{w_3}{R_3} + \cdots, \tag{6-18}$$

where $R_1, R_2, R_3, \ldots$ denote the ranges (in milligrams per square centimeter) of a particular ion in each of several elements, and R_t is the range of that ion in a compound or in an essentially homogeneous mixture of these elements with respective weight fractions w_1, w_2, w_3. Because the relative stopping effects of various elements are functions of the energy of the ion, (6-18) is not to be applied to grossly heterogeneous absorbers, which contain separate phase regions large enough to produce serious changes in the particle energy.

[5] In view of the considerable fraction of the ion energy that is expended in molecular excitation and dissociation processes, this simple additivity relation is somewhat surprising. The stopping power of water vapor has been measured to be about 3 percent less than that of the equivalent mixture of hydrogen and oxygen; range measurements in a number of organic isomers show that ranges in them are the same within less than 1 percent.

Energy Dependence of Ionic Charge. It is seen in (6-13) and (6-14) that the charge γz on an ion is a decisive quantity for the stopping power. The greatest difficulty in the estimation of ranges and specific ionization is the uncertainty in the energy dependence of the average value of the ionic charge and the uncertainty of the distribution about that average value. This difficulty increases with the atomic number of the ion, as an increasing fraction of the range is spent with the ion less than completely stripped of its electrons. Fission fragments, for example, with $M = 103$, $z = 45$, and $E = 100$ MeV would be expected to have a range in aluminum of approximately 0.2 mg cm^{-2} if they remained fully stripped; the measured range is close to 4 mg cm^{-2}, which corresponds to an effective root-mean-square charge of about 10. In fact, just after entering the absorber the fragment has a charge of about 16 and gains electrons as it is slowed down until it is neutralized at about 1 MeV (approximately 0.1 mg cm^{-2} before the end of the range).

Because positive ions can lose only a small fraction of their energy in a single encounter, they make many encounters at energies within a rather narrow range. At each of these many encounters there is a probability of gaining or losing electrons that depends on the atomic number and velocity of the ion as well as on its charge at that particular encounter. Because there are many such encounters, and thus possibilities for electron exchange within a narrow range of energy, it is possible to characterize the system by an equilibrium charge distribution as a function of the energy of the particle.

The observation that three different kinds of experiments indicate three different charges for a given projectile shows the complexity of this system. For example, 100-MeV ^{103}Rh would exhibit an average charge of about 24 after passing through a thin carbon absorber, a charge of about 17 after passing through a thin gaseous nitrogen absorber, and an initial rate of energy loss in both absorbers that corresponds to a root-mean-square charge of about 16. It is thought that the charge of 16 best represents the average charge while the ion is moving in the absorber. The other two higher charges result from the loss of electrons by the Auger process *after* the ion has left the absorber but before it has entered the charge-analyzing apparatus. This loss of electrons occurs because the ion is in an excited state when travelling through and leaving the absorber. Furthermore, the ion is at a higher excitation in a dense absorber than in a gas, although at the same average charge. This happens because, in the denser medium, there is a shorter time between encounters in which to lose excitation energy.

Useful semiempirical formulas for these average charges for various ion-absorber pairs and energies are to be found in B2 and N1. An approximate general expression for *gaseous* absorbers gives for the average fractional charge γ in the stopping-power equations (6-13) and (6-14):

$$\gamma = 1 - C \exp \frac{-v}{v_0 z^{\eta}} \tag{6-19}$$

where $C = 1$, $\eta = \frac{2}{3}$, $v_0 = e^2/h = 2.188 \times 10^8$ cm s^{-1}, v is the velocity of the ion in the same units, and $v \gtrsim v_0$. The charge state of an ion emerging from a *solid* absorber is approximated by

$$\gamma = \left[1 + \left(\frac{v}{v' z^{0.45}}\right)^{-1.67}\right]^{-0.6}, \tag{6-20}$$

where $v' = 3.6 \times 10^8$ cm s^{-1}. Equations 6-19 and 6-20 give predictions, for average charge $\bar{q} = \gamma z$, good to within a charge unit or two.

The width of the charge distribution, as well as the average value, is often of interest. For partially stripped ions the distribution is found to be approximately Gaussian with standard deviation σ_q that is given (B2) by the expression

$$\sigma_q = 0.27 z^{0.5}. \tag{6-21}$$

It is of interest to note that the width of the distribution depends, to a first approximation, on neither the average charge nor the absorber. These simple generalizations about the distribution of charge fail, of course, at energies where the ion is either close to being completely stripped or, on the other hand, close to neutralization.

Straggling. The rate of energy loss, as given in (6-5), is only an average quantity; there are fluctuations in the energy lost by an ion in each collision as well as fluctuations in the number of collisions per unit path length. These fluctuations in the fractional energy loss per collision become even larger at low energies where the fluctuations in the charge of the ion occur and at even lower energies where nuclear stopping is dominent. Further, the ions, largely through nuclear stopping, will undergo scattering, and thus the distance traveled by the ion along its original direction of motion is less than the actual distance traversed. As a consequence of all of these effects, an initially monoenergetic beam of ions of a given type does not have a unique range in an absorber. As pointed out in the discussion of figure 6-1, there is a distribution of ranges. These phenomena all come under the heading of straggling.

Quantitatively the straggling S is defined as the difference between the mean and extrapolated ranges as defined in figure 6-1. For protons moving through air the straggling expressed in terms of the mean range varies from about 1.9 to 1.1 percent as the initial energy varies from about 8 to 500 MeV; the percentage of straggling diminishes by about 0.3 for each fourfold increase in energy. The straggling of any other particle of charge z and mass M may be approximated from that of protons of the *same initial velocity* if its energy is sufficiently high so that all but a negligible part of its range is spent in the completely ionized form:

$$S_{z,M} = \frac{\sqrt{M}}{z^2} S_{1,1}. \tag{6-22}$$

This expression, for example, would be useful for 40-MeV α particles but useless for fission fragments.

At the other extreme, where most of the stopping of the heavy ions is largely nuclear stopping (below about 1 MeV for ions of atomic number greater than about 35 being stopped in an absorber also of high atomic number), the mean-square deviation of the range divided by the square of the range is approximately $\frac{2}{3}[M_1M_2/(M_1+M_2)^2]$ where M_1 and M_2 are the atomic weights of the heavy ion and of the atoms in the absorber, respectively (L1).

B. ELECTRONS

Processes Responsible for Energy Loss. The interaction of electrons with matter is in many ways fundamentally similar to that of positive ions. The processes responsible for the energy loss are qualitatively the same in both cases. In fact, the average energy loss per ion pair formed is the same for electrons and ions (35 eV in air). The primary ionization by electrons accounts for only about 20–30 percent of the total ionization; the remainder is due to secondary ionization.

There are a number of differences between the interactions of the two types of particles with matter. First, for a given energy the velocity of an electron is much larger than that of a positive ion, and therefore the specific ionization is less for electrons. In table 6-1 the specific ionization in air is given for electrons of different energies. The largest specific ionization, 5950 ion pairs per milligram per square centimeter, occurs at 146 eV (velocity $= 0.024c$), which is a much lower energy but somewhat higher velocity than corresponds to the peak in the Bragg curve for α particles. In air, ionization stops when the electron energy has been reduced to 12.5 eV (the ionization potential of oxygen molecules). On the higher-energy side of the maximum the specific ionization reaches a flat minimum at about 1.4 MeV. The increase beyond this energy is a relativistic effect, as discussed in connection with (6-13). The Lorentz contraction of lengths enables the fast electron to ionize atoms at greater distances, even at distances of several molecular diameters.[6]

An electron may lose a large fraction of its energy in one collision; therefore a statistical treatment of the energy-loss processes is less justified than for ions, and straggling is much more pronounced. In the passage of an initially homogeneous beam of electrons through matter, the apparent straggling is further increased by the pronounced scattering of the electrons

[6] This has the perhaps unexpected consequence of making the physical state of the absorber of importance. For example, in liquid rather than gaseous air the dielectric polarization of the medium probably reduces the specific ionization from the values in table 6-1 by about 10 percent at 10 MeV and about 20 percent at 100 MeV, if W remains 35 eV per ion pair in liquid air.

Table 6-1 Specific Ionization and Velocity for Electrons of Various Energies in Air

Velocity (in Units of the Velocity of Light, c)	Energy (MeV)	Ion Pairs per 1.00 mg cm^{-2}
0.001979	10^{-6}	0
0.006257	10^{-5}	0
0.0240	1.46×10^{-4}	5950 (maximum)
0.1950	10^{-2}	~850
0.4127	0.05	154
0.5483	0.10	116
0.8629	0.50	50
0.9068	0.70	47
0.9411	1.0	46
0.9791	2.0	46
0.9893	3.0	47
0.9934	4.0	48
0.9957	5.0	49
0.9988	10	53
0.99969	20	57
0.999949	50	63
0.9999871	100	66

into different directions, which makes possible widely different path lengths for electrons traversing the same thickness of absorber. Nuclear scattering is responsible for most of the large-angle deflections, although energy loss is caused almost entirely by interactions with electrons.

For electrons of high energy an additional mechanism for losing energy must be taken into account: the emission of radiation **(bremsstrahlung)** when an electron is accelerated in the electric field of a nucleus. The ratio of energy loss by this radiation to energy loss by ionization in an element of atomic number Z is approximately equal to $EZ/800$, where E is the electron energy in millions of electron volts. Thus in heavy materials such as lead the radiation loss becomes appreciable even at 1 MeV, whereas in light materials (air, aluminum) it is unimportant, at least for the energies available from β emitters. The distance over which the energy of an electron is reduced by a factor e due to bremsstrahlung is called the **radiation length.**

Finally the additional fact that β particles are emitted with a continuous energy spectrum further complicates any attempt at detailed analysis of their absorption in matter.

Absorption of Beta Particles. The combined effects of continuous spectrum and scattering lead—quite fortuitously—to an approximately

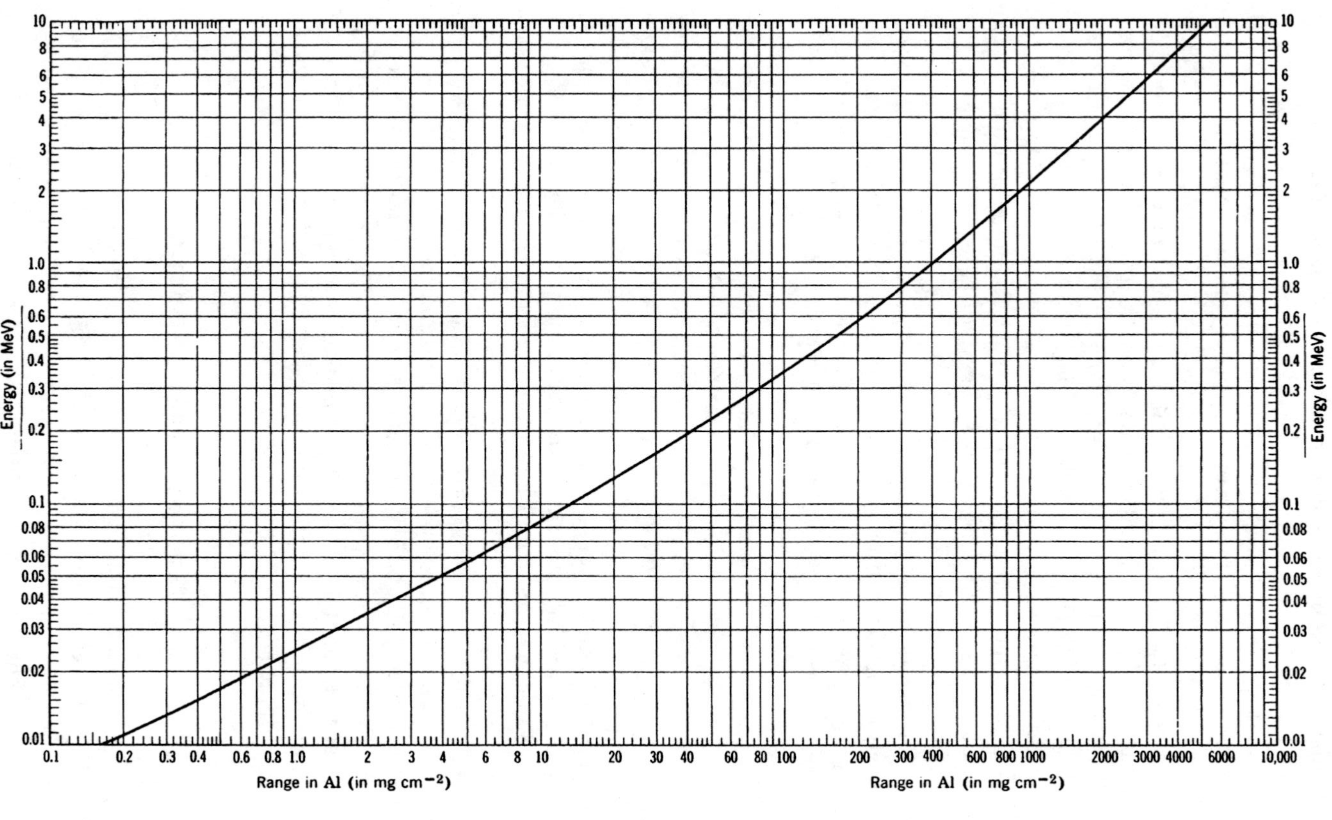

Fig. 6-8 Range-energy relation for β particles and electrons in aluminum.

exponential absorption law for β particles of a given maximum energy. Absorption curves, that is, curves of activity versus thickness of absorber traversed, are for this reason usually plotted on semilogarithmic paper. The exact shape of an absorption curve depends on the shape of the β-ray spectrum and, because of scattering effects, on the geometrical arrangement of active sample, absorber, and detector. If sample and absorber are as close as possible to the detector, the semilog absorption curve becomes most nearly a straight line; otherwise, some curvature toward the axes is generally found. When β particles belonging to two spectra of widely different maximum energies are present in a source, this is apparent from the change of slope in the absorption curve; such an absorption curve is roughly analogous to the semilog decay curve of an activity containing two different half-life periods.

Once the range of β particles or conversion electrons is known, a range-energy relation can be used to deduce the maximum energy. Many empirical relations have been proposed; however, it is best to use a range-energy curve such as that plotted in figure 6-8.

Back-Scattering. As already mentioned, scattering of electrons, both by nuclei and by electrons, is much more pronounced than scattering of heavy particles. A significant fraction of the number of electrons striking a piece of material may be reflected as a result of single and multiple scattering processes. The reflected intensity increases with increasing thickness of reflector, except that for thicknesses greater than about one third of the range of the electrons, saturation is achieved and further increase in thickness does not add to the reflected intensity. The ratio of the measured activity of a β source with reflector to that without reflector is known as the back-scattering factor. The saturation back-scattering factor is essentially independent of the maximum β-particle energy for energies above about 0.6 MeV and varies from about 1.3 for aluminum to about 1.8 for lead. These factors, though, are dependent on the particular counting arrangement used and should be determined in each configuration. There may also be a small difference between the back-scattering of positrons and negatrons of the same energy.

C. ELECTROMAGNETIC RADIATION

Processes Responsible for Energy Loss. Photons passing through matter do not lose energy continuously along their paths as do charged particles. On the contrary, in two of the three fundamental processes through which photons interact with matter the entire energy of the photon is transferred to the medium in a single interaction; in the third there is a small probability for a few energy-degrading encounters. Thus the absorption of photons in matter is expected to be exponential with, as it turns out, a half-thickness that is much greater than the range of a β particle of the same energy. A consequence of this is that the average specific ionization of a γ ray is perhaps one tenth to one hundredth of that caused by an electron of the same energy, and practical ranges are much greater. The ionization observed for γ rays is almost entirely secondary in nature, as we see from a discussion of the three processes by which γ rays (and X rays) lose their energy. The average energy loss per ion pair formed is the same as for β rays, namely, 35 eV in air.

Photoelectric Effect. At low energies the most important process is the photoelectric effect. In this process the electromagnetic quantum of energy $h\nu$ ejects a bound electron from an atom or molecule and imparts to it an energy $h\nu - \epsilon_b$, where ϵ_b is the energy with which the electron was bound. The quantum of radiation completely disappears in this process, and momentum conservation is possible only because the remainder of the atom can receive some momentum. For any photon energy greater than the K-binding energy of the absorber, photoelectric absorption takes place primarily in the K shell, with the L shell contributing only of the order of 20 percent and outer shells even less. For this reason the probability for photoelectric absorption has sharp discontinuities at energies equal to the binding energies of the K, L, M, . . . , electrons. For photon energies well above the K-binding energy of the absorber, the photoelectric absorption first falls off rapidly (about as $E_\gamma^{-7/2}$), then more slowly (eventually as E_γ^{-1}) with increasing energy. It is also approximately proportional to Z^5. The γ-ray energy at which the photoelectric contribution to the total γ-ray absorption is about 5 percent is 0.15 MeV for aluminum, 0.4 MeV for copper, 1.2 MeV for tin, and 4.7 MeV for lead. Except in the heaviest elements, photoelectric absorption is relatively unimportant for energies above 1 MeV.

The ionization produced by photoelectrons accounts largely for the ionization effect of low-energy photons. The photoelectric effect is frequently used to determine γ-ray energies. This may be accomplished by measurement of the total ionization due to the photoelectrons in a semiconductor or scintillation counter.

Compton Effect. Instead of giving up its entire energy to a bound electron, a photon may transfer only a part of its energy to an electron, which in this case may be either bound or free; the photon is not only degraded in energy but also deflected from its original path. This process is called the Compton effect or Compton scattering. The relation between energy loss and scattering angle can be derived from the relativistic conditions for conservation of momentum and energy. The important relativistic expression required relates the *total* energy E of a particle to its momentum p (cf. appendix B):

$$E = (E_0^2 + c^2p^2)^{1/2}. \quad (6\text{-}23)$$

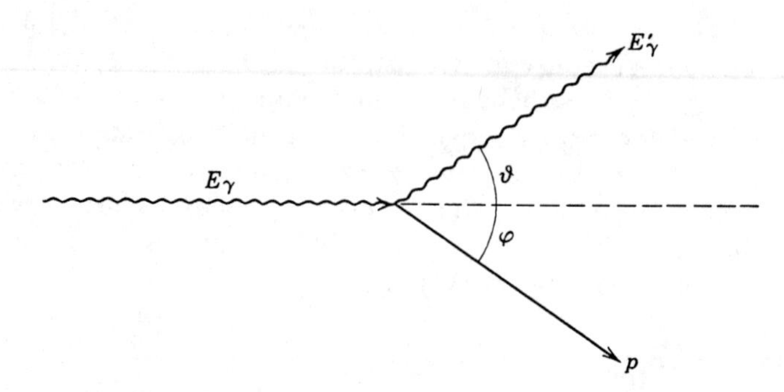

Fig. 6-9 Schematic diagram of Compton scattering of a γ ray by an electron.

The quantity E_0 is the total energy of the particle when it is at rest and is given by mc^2 where m is the rest mass of the particle. It must also be recalled that the rest mass of a photon is zero. The scattering event is depicted in figure 6-9. We define E_γ as the energy of the initial gamma ray, E'_γ as its energy after being scattered through an angle ϑ, E_0 as the rest energy of the electron (511 keV), and p as the magnitude of the momentum of the originally stationary electron after being struck by the incident γ ray and projected at an angle φ with respect to the incident direction. The conditions for conservation of total energy and of momentum components parallel and perpendicular to the incident γ-ray direction are:

$$E_\gamma + E_0 = E'_\gamma + (E_0^2 + c^2p^2)^{1/2}; \tag{6-24}$$

$$\frac{E_\gamma}{c} = \frac{E'_\gamma}{c}\cos\vartheta + p\cos\varphi; \tag{6-25}$$

$$\frac{E'_\gamma}{c}\sin\vartheta = p\sin\varphi. \tag{6-26}$$

The angle φ may be eliminated between (6-25) and (6-26) through the relation $\sin^2\varphi + \cos^2\varphi = 1$; this gives

$$E_\gamma^2 - 2E_\gamma E'_\gamma\cos\vartheta + (E'_\gamma)^2 = c^2p^2. \tag{6-27}$$

Equation 6-27 may be substituted into (6-24), which yields after some simple manipulations

$$\frac{1}{E'_\gamma} - \frac{1}{E_\gamma} = \frac{1-\cos\vartheta}{E_0}. \tag{6-28}$$

Through the relation between the energy of a photon and its wavelength, $E = hc/\lambda$, (6-28) takes the more familiar form

$$\lambda' - \lambda = \frac{h}{m_e c}(1-\cos\vartheta), \tag{6-29}$$

where m_e is the rest mass of the electron. The quantity $h/m_e c = 2.42631 \times 10^{-10}$ cm is called the **Compton wavelength** of the electron.

Equation 6-29 shows that for a given incident energy there is a minimum energy (maximum wavelength) for the scattered γ ray and that this occurs for scattering in the backward direction ($\cos\vartheta = -1$). This minimum energy is readily obtained from (6-28):

$$(E'_\gamma)_{\min} = \frac{E_0}{2}\frac{1}{1+E_0/2E_\gamma}. \tag{6-30}$$

For large incident γ-ray energies ($E_\gamma \gg \frac{1}{2}E_0$) the minimum energy of the scattered γ rays approaches $\frac{1}{2}E_0 = 250$ keV. For this reason scintillation spectra of high-energy γ rays always show a **back-scattering peak** at ≤ 250 keV, which is caused by Compton scattering in *surrounding material* (cf. chapter 8, section G) and a valley between photopeak and Compton continuum whose width corresponds to the minimum energy (≤ 250 keV) carried off by γ rays Compton-scattered in the *crystal*.

The Compton scattering per electron is independent of Z, and therefore the scattering coefficient per atom is proportional to Z. For energies in excess of 0.5 MeV it is also approximately proportional to E_γ^{-1}. Thus Compton scattering falls off much more slowly with increasing energy than photoelectric absorption, at least at moderate energies (up to 1 or 2 MeV), and even in lead it is the predominant process in the energy region from about 0.6 to 4 MeV.

Pair Production. The third mechanism by which electromagnetic radiation can be absorbed is the pair-production process (discussed in chapter 3, p. 78). Pair production cannot occur when $E_\gamma < 1.02$ MeV. Above this energy the atomic cross section for pair production first increases slowly with increasing energy and above about 4 MeV becomes approximately proportional to $\log E_\gamma$. It is also proportional to Z^2. At high energies, where pair production is the predominant process, γ-ray energies can best be determined by measurements of the total energies of positron-electron pairs. Pair production is always followed by annihilation of the positron, usually with the simultaneous emission of two 0.51-MeV photons. The absorption of quanta by the pair-production process is therefore always complicated by the appearance of this low-energy secondary radiation.

Energy and Z Dependance. The atomic cross sections for all three processes discussed increase with increasing Z, except for the photoelectric effect at very low energies. For this reason heavy elements, atom for atom, are much more effective absorbers for electromagnetic radiation than light elements, and lead is most commonly used as an absorber. Because photoelectric effect and Compton effect decrease and pair production increases with increasing energy, the total absorption in any one element has a minimum at some energy. For lead this minimum absorption, or maximum transparency, occurs at about 3 MeV; for copper at about 10 MeV; and for

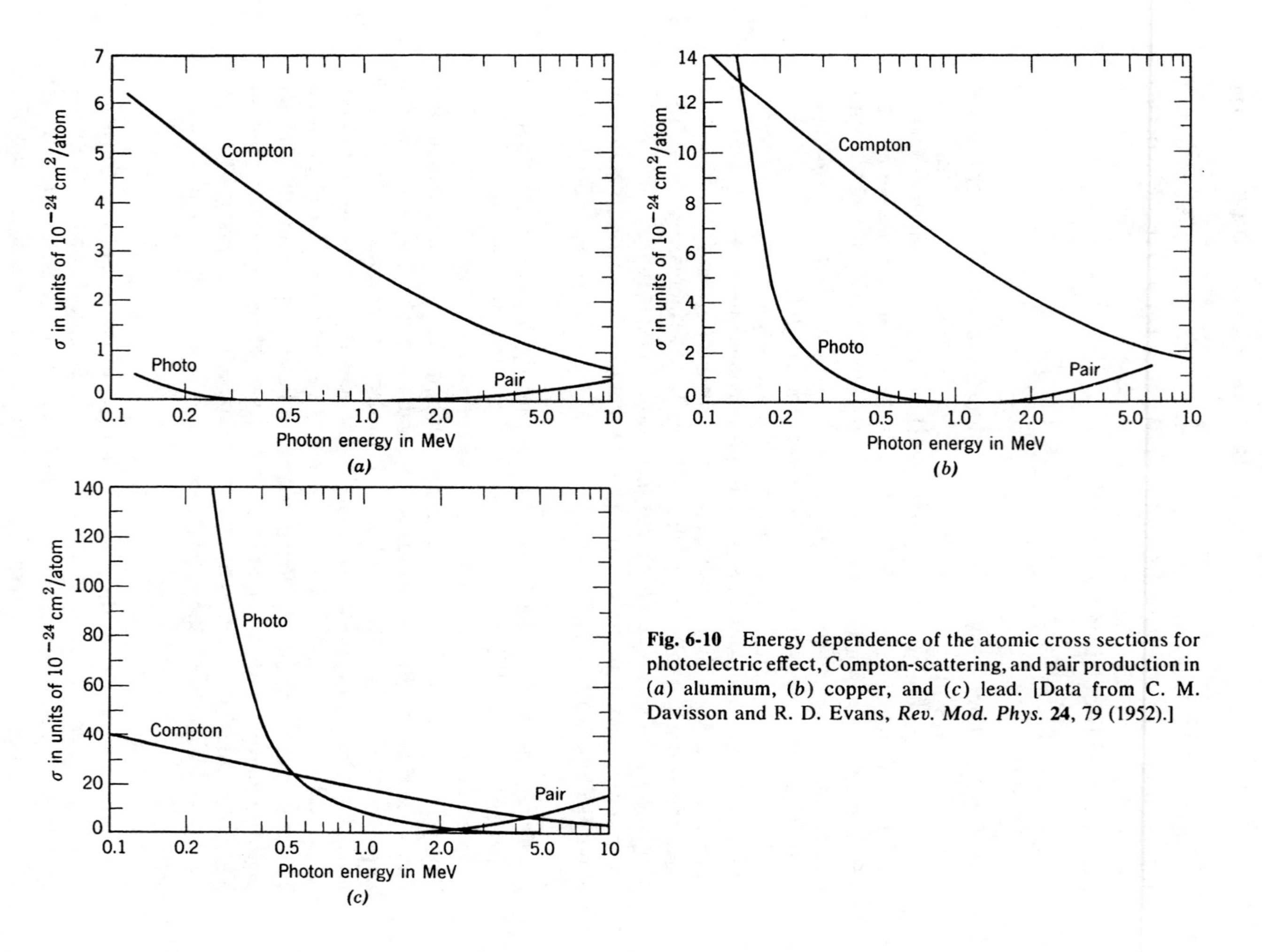

Fig. 6-10 Energy dependence of the atomic cross sections for photoelectric effect, Compton-scattering, and pair production in (*a*) aluminum, (*b*) copper, and (*c*) lead. [Data from C. M. Davisson and R. D. Evans, *Rev. Mod. Phys.* **24**, 79 (1952).]

aluminum at about 22 MeV. The energy dependence of the three processes is shown for aluminum, copper, and lead in figure 6-10.

Absorption Coefficient. If only photons of the incident energy are considered, all the processes by which γ rays interact with matter lead to exponential attenuation; that is, the intensity $\mathbf{I}_d$ transmitted through a thickness d is given by $\mathbf{I}_d = \mathbf{I}_0 e^{-\mu d}$, where $\mathbf{I}_0$ is the incident intensity and μ is called the **absorption coefficient**. Separate absorption coefficients for photoelectric effect, Compton scattering, and pair production are sometimes quoted, and the total absorption coefficient μ is the sum of the three. The **half-thickness** $d_{1/2}$ is defined as the thickness which makes $\mathbf{I}_d = \frac{1}{2}\mathbf{I}_0$; $d_{1/2} = 0.693/\mu$. Absorber thicknesses are frequently given in terms of surface density (ρd, expressed in grams per square centimeter). Then $\mathbf{I}_d = \mathbf{I}_0 e^{-(\mu/\rho)\rho d}$, and μ/ρ is called the **mass absorption coefficient**.

Unless absorption is entirely by the photoelectric process, the condition that only photons of the incident energy be measured is not always easy to meet experimentally. It requires either a very "good" geometry (large distances between source and absorber and between absorber and detector) or a detector that responds over a narrow energy range only. Curves of calculated half-thicknesses in various absorbers versus photon energy are given in figures 6-11, 6-12 and 6-13. Note that the ordinate in figures 6-11 and 6-12 is $d_{1/2}$, whereas that in figure 6-13 is $d_{1/2}\rho$.

Critical Absorption of X Rays. We have already mentioned the discontinuities in absorption coefficients at photon energies corresponding to the electron-binding energies. These absorption edges and their variation from element to element are often useful in measuring characteristic X rays. To understand this method of critical absorption we recall that the *emission* of an X ray from an atom is due to the transition of an electron from one of the outer shells to a vacancy in a shell of higher binding energy, say, from the L to the K shell.[7] Photoelectric *absorption* in a given electron shell, on the other hand, can occur only if the photon has enough energy to promote an electron from that shell to a vacant level (which means very nearly enough energy to remove the electron from the atom). It follows that an element is a poor absorber for its own characteristic X rays. The K_α X rays of an element have an energy equal to the difference between the K and L shells and so cannot promote a K electron to one of the outer vacant shells in the same element. However, the binding energy of electrons decreases with decreasing Z; therefore the K_α emission line of

[7] In X-ray terminology, X rays due to transitions from the L to the K shell are called K_α X rays ($K_{\alpha 1}$ and $K_{\alpha 2}$ corresponding to the electron originating in different sublevels of the L shell); X rays due to transitions from the M to the K shell are called K_β, etc. Similarly, there are L_α, L_β, etc., X rays.

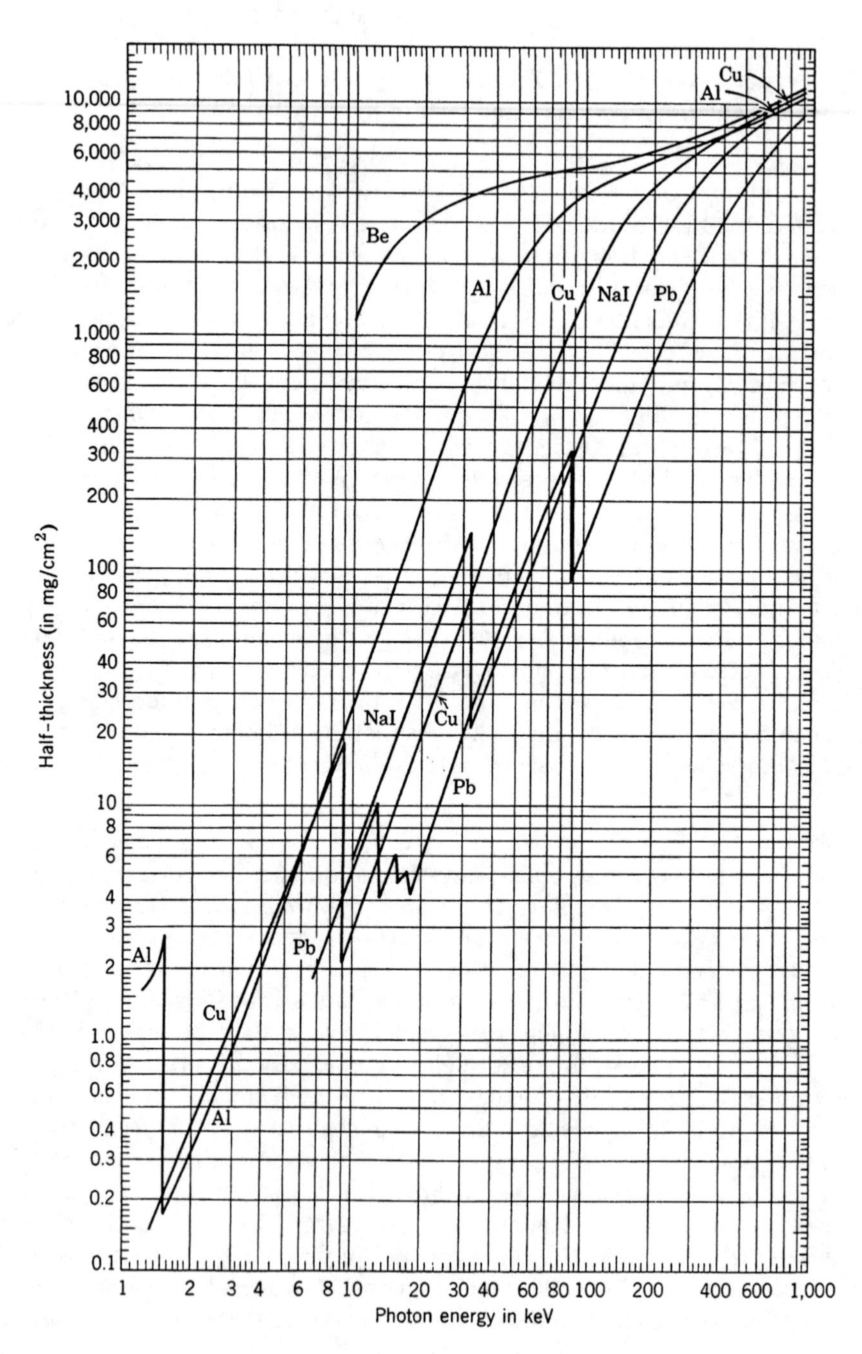

Fig. 6-11 Half-thickness values in beryllium, aluminum, copper, lead, and sodium iodide for low-energy photons. The K-absorption edges of aluminum, copper, iodine and lead, as well as the $L_{\rm I}$, $L_{\rm II}$, and $L_{\rm III}$ edges of lead are shown. (Data from G. W. Grodstein, *National Bureau of Standards Circular 583*, April 1957, and from *Handbook of Chemistry and Physics*, 44th ed., The Chemical Rubber Publishing Co., Cleveland, 1962.)

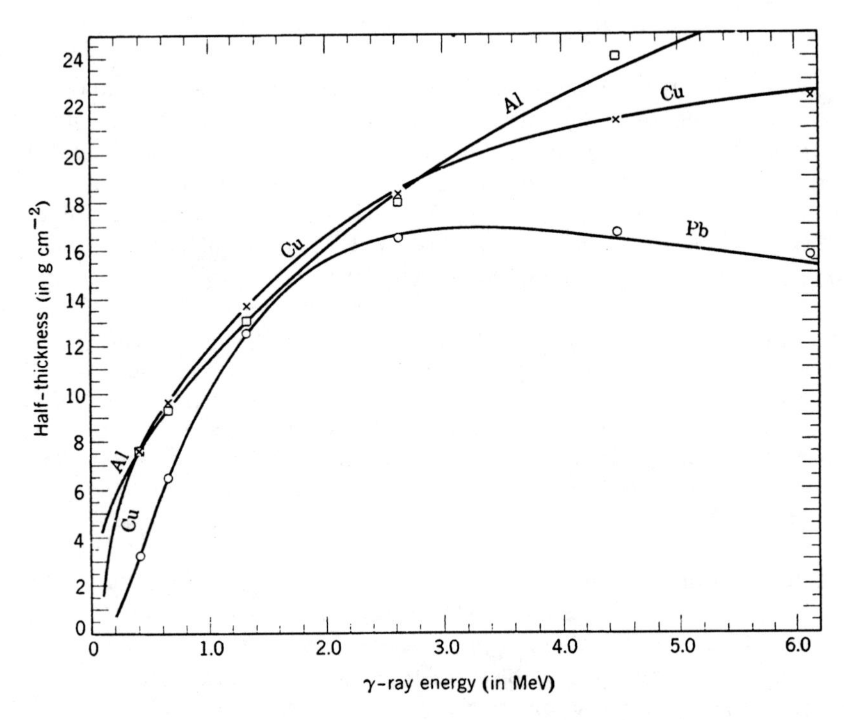

Fig. 6-12 Half-thickness values in aluminum, copper, and lead for intermediate-energy photons. The curves are based on the calculated absorption coefficients of C. M. Davisson and R. D. Evans, *Rev. Mod. Phys.* **24**, 79 (1952). Some experimental points (□ Al, × Cu, ○ Pb) are taken from S. A. Colgate, *Phys. Rev.* **87**, 592 (1952).

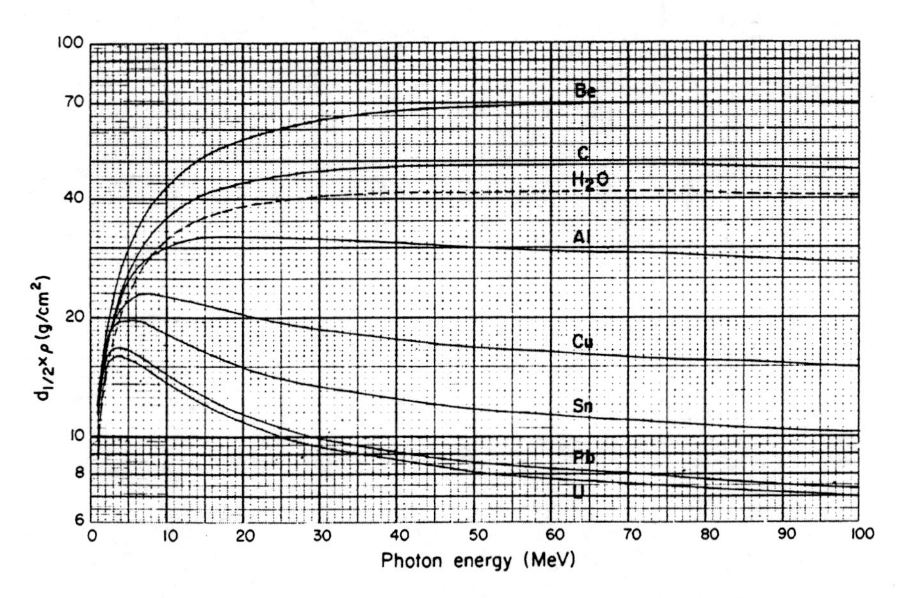

Fig. 6-13 Absorption of photons of energy up to 100 MeV in various materials. (From reference L2.)

an element Z has an energy rather close to but slightly greater than the K absorption edge of some element of slightly lower Z and is strongly absorbed by that element but not by the next higher one. These two neighboring elements will thus have very different absorption coefficients for the particular rays, and the one that absorbs more strongly is called the critical absorber for these X rays. Critical absorption can also be applied to L-emission lines, especially of heavy elements.

As an example, consider the K_α X rays of zinc ($Z = 30$) which have an energy of 8.6 keV. The K absorption edges of $_{29}$Cu and $_{28}$Ni are 9.0 and 8.3 keV, respectively. Therefore, nickel is a good absorber for zinc K_α X rays, and copper is not (figure 6-14). The K_α X rays of gallium ($Z = 31$), on the other hand, are strongly absorbed both in nickel and copper because their energy is 9.2 keV, but they are not absorbed well in zinc whose K absorption edge is 9.7 keV.

Critical absorbers can be used to advantage, for example, to suppress one X-ray line so that the spectrum of a neighboring one can be measured cleanly. Both the X-ray emission lines and the absorption edges of the elements can be found in tables (L2). An example of the use of critical absorption is discussed in chapter 11, section B,5.

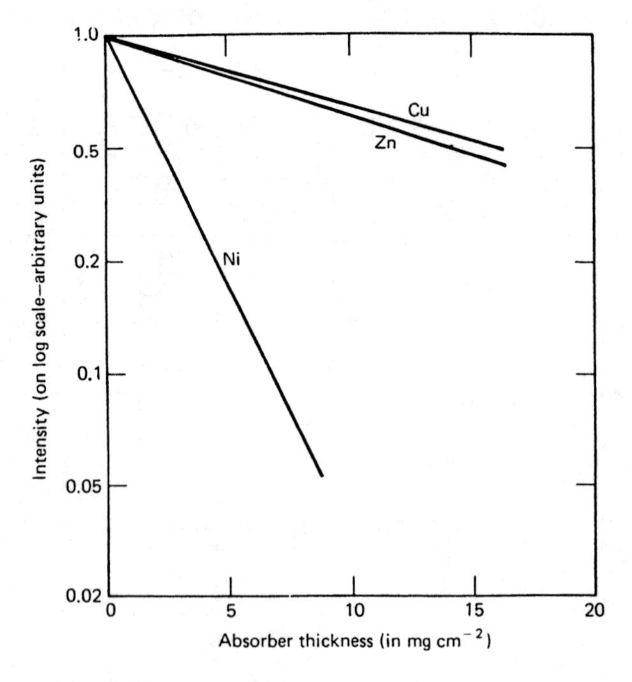

Fig. 6-14 Absorption of zinc K_α X rays in zinc, copper, and nickel. (These absorption curves were calculated from data given in reference C1.)

D. NEUTRONS

Because neutrons carry no charge, their interaction with electrons is exceedingly small, and primary ionization by neutrons is a completely negligible effect. The interaction of neutrons with matter is confined to nuclear effects, which include elastic and inelastic scattering and nuclear reactions such as (n, γ), (n, p), (n, α), $(n, 2n)$, and fission. These nuclear interactions have been discussed in chapter 4; how they are applied to detection and measurement of neutrons is discussed in chapter 7, section E.

Slowing Down of Neutrons. From the description of nuclear reactions given in chapter 4, section E, it will be recalled that thermal neutrons, neutrons whose energy distribution is approximately that of gas molecules at ordinary temperatures, are very efficient at producing nuclear reactions. Because of this fact processes for reducing the energy of high-energy neutrons produced in nuclear reactions to a thermal energy distribution have received much theoretical and experimental study.

Fast neutrons may lose large amounts of energy in inelastic collisions, especially with heavy nuclei. This process ceases to be effective after intermediate energies are reached and does not produce slow neutrons. Most slowing down is accomplished through a process of many successive *elastic* collisions with nuclei. Because of the conservation of momentum a neutron of energy E_0 making an elastic collision with a heavy nucleus bounces off with most of its original energy, giving up no more energy than $4AE_0/(A+1)^2$ to the recoil nucleus, where A is the mass number of the target nucleus. The lighter the nucleus with which a neutron collides, the greater the fraction of the neutron's kinetic energy that can be transferred in the elastic collision. For this reason hydrogen-containing substances such as paraffin or water are the most effective slowing-down media for neutrons.

In the elastic scattering of neutrons with energies below about 10 MeV all energy transfers between zero and the upper limit, $4AE_0/(A+1)^2$, are equally probable. Thus the probability that a neutron of energy E_0 has a residual energy between E and $E + dE$ is

$$P(E)\,dE = \frac{dE}{4AE_0/(A+1)^2},$$

and the average energy retained by the neutron is

$$\bar{E} = \int_{E_0[1-4A/(A+1)^2]}^{E_0} P(E)E\,dE = \frac{(A+1)^2}{4AE_0}\int_{E_0[1-4A/(A+1)^2]}^{E_0} E\,dE$$

$$= E_0\left[1 - \frac{2A}{(A+1)^2}\right]. \qquad (6\text{-}31)$$

From this result it is seen that the average value of E/E_0 is independent of E_0; therefore the average value of E/E_0 after n collisions is simply

$$\overline{\frac{E_n}{E_0}} = \left[1 - \frac{2A}{(A+1)^2}\right]^n. \tag{6-32}$$

The average value after n collisions is a rather misleading quantity, as the distribution of energies is strongly skewed. The probability that a neutron of initial energy E_0 has an energy between E_n and $E_n + dE_n$ after n elastic collisions with hydrogen nuclei may be obtained from the recursion relation[8]

$$P_n(E_n)\, dE_n = \int_{E_n}^{E_0} [dE_{n-1}P_{n-1}(E_{n-1})]\left[\frac{dE_n}{E_{n-1}}\right], \tag{6-33}$$

where the first bracketed term is the probability of obtaining energy between E_{n-1} and $E_{n-1} + dE_{n-1}$ in $n-1$ collisions, and the second bracketed term is the probability of going from the interval $E_{n-1} \rightarrow E_{n-1} + dE_{n-1}$ to the interval $E_n \rightarrow E_n + dE_n$ in the nth collision. The integration is performed over the variable E_{n-1}. Equation 6-33 has the solution (ignoring thermal motion):

$$P_n(E_n) = \frac{1}{(n-1)!E_0}\left(\ln \frac{E_0}{E_n}\right)^{n-1}. \tag{6-34}$$

Another question of interest is the average number of collisions required to slow a neutron of energy E_0 down to an energy E. We may write for the energy after n collisions

$$E = E_0 f_1 f_2 \cdots f_i \cdots f_n, \tag{6-35}$$

where

$$f_i = \frac{E_i}{E_{i-1}}. \tag{6-36}$$

As we stated before, f_i has equal probability for all values between 1 and $1 - 4A/(A+1)^2$. It is clear that (6-35) has an infinite number of possible solutions for any value of n above a certain minimum value determined by the mass of the scattering nucleus. It is tempting to estimate the average value of n by putting the average value of f from (6-32) into (6-35), but this would be wrong. It would be wrong for the same reason that the average square of a set of random numbers is, in general, not equal to the square of the average. The solution to the problem, however, can be immediately obtained by a simple transformation of (6-35), which turns the problem into a more familiar one whose answer is well known.

Take the logarithm of both sides of (6-35):

$$\ln\left(\frac{E}{E_0}\right) = \ln(f_1 f_2 \cdots f_i \cdots f_n) = \sum_{i=1}^{n} \ln f_i;$$

define $x_i = -\ln f_i$ so that

$$\ln\left(\frac{E_0}{E}\right) = \sum_{i=1}^{n} x_i. \tag{6-37}$$

[8] The recursion relations for heavier nuclei are more complicated than (6-33), since it is not possible for neutrons of all values of E_{n-1} to go to E_n in a single collision.

Again, an infinite number of values of n will satisfy (6-37), but now the average value of n is equivalent to the average number of collision-free segments (the number of collisions + 1) traversed by a gas molecule when it travels a "distance" $\ln(E_0/E)$. The answer to this problem is well known; the average number of collisions is just the distance traveled divided by the mean-free path.[9] The number of collision-free segments is then

$$\bar{n} = \frac{\ln(E_0/E)}{\bar{x}} + 1 = \frac{\ln(E_0/E)}{\overline{\ln(E_{i-1}/E_i)}} + 1, \tag{6-38}$$

where the quantities with bars over them denote mean values. The mean value of $\ln(E_{i-1}/E_i)$ may be obtained in the same manner as $\bar{E}$ in (6-31) [cf. (9-6)]; the result is

$$\overline{\ln\left(\frac{E_{i-1}}{E_i}\right)} = 1 - \frac{(A-1)^2}{2A}\ln\left(\frac{A+1}{A-1}\right). \tag{6-39}$$

Substituting (6-39) into (6-38) gives

$$\bar{n} = \frac{\ln(E_0/E)}{1 - [(A-1)^2/2A]\ln[(A+1)/(A-1)]} + 1. \tag{6-40}$$

Equation 6-40 just derived gives the average number of collisions required to slow a neutron of energy E_0 down to an energy E. For collisions with protons ($A = 1$) the denominator in (6-40) becomes unity, hence $E_n = E_0 e^{1-\bar{n}}$; approximately 20 collisions are therefore necessary to reduce neutrons from a few million electron volts to thermal energies (about 0.04 eV at ordinary room temperature). Paraffin about 20 cm thick surrounding a neutron source is adequate for reducing most neutrons to the thermal energy distribution. The whole slowing-down process requires less than 10^{-3} s.

The probable eventual fate of a thermal neutron in a hydrogenous medium like water of paraffin is capture by a proton to form a deuteron; but, since the cross section for this reaction is quite small compared with the cross section for scattering, a neutron after reaching thermal energies makes about 150 further collisions before being captured. Paraffin and water are good substances to use for the slowing down of neutrons because the capture cross sections of oxygen and carbon are even much smaller than the hydrogen capture cross section. Heavy water is better than ordinary water because of the low probability of neutron capture by deuterium. Carbon (graphite) is also useful as a slowing-down medium; many more (about 120) collisions are necessary to reduce neutrons to thermal energies in carbon than in hydrogen, but after reaching thermal energies the neutrons can exist longer in carbon. In either substance the lifetime of a neutron before capture is only a fraction of a second.

Even if neutrons could be kept in a medium in which they would not eventually be captured, they would not exist very long. The systematics of β radioactivity predict that free neutrons are unstable and should decay

[9] The probabilities of the various values of n are given by the Poisson distribution, which is discussed in chapter 9, section D.

rather quickly into protons and electrons. This decay was observed in 1950 by A. H. Snell and by J. M. Robson with reactor neutrons that were in free flight in vacuum. The energy released in the disintegration is 0.78 MeV, and the half life is 10.6 min.

Thermal Distribution. It should be apparent that not all thermal neutrons have the same energy. After neutrons are slowed to energies comparable to thermal agitation energies they may either lose or gain energy in collisions, and the result is a Maxwellian distribution of velocities in which the fraction of the total number of neutrons with velocity between v and $v + dv$ is given by

$$F(v)\,dv = 4\pi^{-1/2}\left(\frac{M}{2kT}\right)^{3/2} v^2 e^{-Mv^2/2kT}\,dv. \tag{6-41}$$

Here M is the neutron mass, T is the absolute temperature, and k is the Boltzmann constant. Some properties of this distribution, usually derived in books on the kinetic theory of gases, are that the **most probable velocity** is

$$v_m = \left(\frac{2kT}{M}\right)^{1/2},$$

the **average velocity** is

$$\bar{v} = \left(\frac{8kT}{\pi M}\right)^{1/2} = \frac{2v_m}{\pi^{1/2}},$$

and the average kinetic energy is $\bar{E} = \frac{3}{2}kT$. The average energy of the neutrons depends on the temperature of the slowing-down medium. At very low temperatures the Maxwellian distribution function becomes a poor approximation because of the discrete energy levels of the bound atoms of the medium. At all temperatures the approximation can be poor if the neutron path in the medium is too short or if the distribution is seriously altered by neutron absorption or leakage from the surface.

A significant point is the distinction between the velocity distribution present in a medium and that felt by a sample placed in the medium. The two distributions are different because the probability that a particular neutron will strike the sample in a given time is proportional to v. It is the altered or weighted distribution, denoted here by $F'(v)\,dv$, that is significant in any transmutation or cross section computation:

$$F'(v)\,dv = 2\left(\frac{M}{2kT}\right)^2 v^3 e^{-Mv^2/2kT}\,dv. \tag{6-42}$$

E. RADIATION PROTECTION

The biological effects of radiation are brought about through chemical changes in the cells caused by ionizations, excitations, dissociations, and atom displacements. In determining radiation effects on living organisms, whether from external radiation or from ingested or inhaled radioactive material, we must take into consideration not only the total dosages of ionization produced in the organism but also such factors as the density of the ionization, the dosage rate, the localization of the effect, and the rates of administration and elimination of radioactive material.

Apart from various medical procedures there is no evidence that the net direct effect of radiation on man is anything but harmful. In the absence of other clinical indications it is probable that even some diagnostic procedures that entail radiation have a greater chance of inducing rather than revealing a morbid condition. Thus except for the unavoidable background radiation, exposure to radiation is acceptable only on the basis of a risk/benefit analysis. It is, unfortunately, not yet possible to measure in a persuasive fashion the risk involved through exposure to small quantities of radiation. It is this fact that is at the root of much of the controversy on this subject.

Dosimetry in Radiation Protection (M1, M2). The unit of radiation dosage that is used in radiation protection is the **roentgen equivalent man (rem)**. The dosage in rems is equal to that in rads (defined on p. 7) multiplied by the relative damage caused by various kinds of radiation. The latter quantity depends on several factors, the most important of which is the density of ionization that in biological studies is often measured by the **linear energy transfer (LET)**, the energy that is deposited per unit path length. Note that the physical principles of LET have been outlined in the discussion of stopping power (dE/dx). Thus the dose equivalent in rems is given by the dose in rads multiplied by the **quality factor (QF)**.[10] Approximate values of QF are given in table 6-2. The ranges of values in that table reflect the energy dependence of LET and thus of QF; it is prudent to use the upper limit in the absence of persuasive evidence to the contrary.

As an example of the practical application of some of the concepts discussed, we estimate the dosage rate in rads per hour to be expected at a distance of 50 cm from a 100-mCi ^{60}Co source. Each disintegration of ^{60}Co is accompanied by two γ quanta with energies 1.17 and 1.33 MeV; for simplicity we use for each an average energy of 1.25 MeV. The source emits $2 \times 100 \times 3.7 \times 10^7 = 7.4 \times 10^9$ quanta per second. At a distance of 50 cm the γ flux is $7.4 \times 10^9/(4\pi \times 2500) = 2.3 \times 10^5$ photons cm^{-2} s^{-1} or $2.3 \times 10^5 \times 1.25 \times 10^6 = 2.9 \times 10^{11}$ eV cm^{-2} s^{-1}. Since at an energy of 1.25 MeV the mass absorption

[10] In the older literature the relative damage caused by various kinds of radiation was measured by the **relative biological effectiveness (RBE)**. This quantity is now reserved for more precise studies in radiobiology and is not used in the transformation from dose to dose equivalent in radiation protection. In the SI system the unit of radiation dose equivalent is the **sievert**, which is defined as the dose in gray (Gy) multiplied by the QF. See chapter 1, p. 7, for the definition of the gray.

Table 6-2 QF Values for Various Types of Radiation (N3)

Radiation	QF
X and γ rays	1
Electrons and positrons	1
Neutrons, energy < 10 keV	3
Neutrons, energy > 10 keV	10
Protons	1–10
Alpha particles	1–20
Heavy ions	20

coefficients in air and aluminum are about the same, the half-thickness from figure 6-10 is $12.5\,\text{g cm}^{-2}$. Then the fractional energy loss for the γ rays per g cm^{-2} of air is given by $\mu/\rho = 0.693/12.5 = 0.055$, and the energy lost by the γ rays in going through $1\,\text{g cm}^{-2}$ of air is $0.055 \times 2.9 \times 10^{11} \times 3600 = 5.7 \times 10^{13}\,\text{eV h}^{-1}$ or $92\,\text{erg h}^{-1}$. Setting the energy absorbed per gram of air equal to this energy loss,[11] we get $92/100 = 0.92\,\text{rad h}^{-1}$.

Radiation Protection Guide (N3, M2). The upper limit of the dose equivalent of radiation to which an individual should be exposed is not well defined. As mentioned earlier, this issue might be resolved on the basis of a risk/benefit analysis. While fine in principle, such an analysis is all but impossible to carry out except in some therapeutic procedures where large doses of radiation are used against neoplasms. The question might also be resolved if it could be demonstrated that there is a threshold dose below which there are no somatic or genetic effects. The difficulty of measuring the effects of vanishingly small doses of radiation in the presence of the background of "spontaneous" somatic and genetic changes leaves this, in general, an unresolved question. Under these circumstances it is usually assumed that the deleterious effects of radiation are roughly proportional to the dose equivalent that is absorbed and that there is no threshold.

In light of these difficulties, a Radiation Protection Guide (N3, S1) has evolved that has necessarily been based on the knowledge of the biological effects of radiation that has been gained over the past three quarters of a century. This information includes controlled laboratory tests with animals, effects of radiation therapy on both patients and therapists, industrial use of radiation, and damage to the populations of Hiroshima and Nagasaki

[11] This procedure leads to an overestimate (in the present case by about a factor 2) of the energy absorption in air because a fraction of the energy loss occurs by Compton scattering, and some of the secondary quanta leave the local region of interest. The method used applies when the primary radiation is in equilibrium with secondaries; this is more nearly the case inside a mass of tissue than in air.

from atomic bomb blasts. Two sets of guidelines have evolved: one for individuals whose occupation entails exposure to radiation and another for the population at large. For those who are occupationally exposed it is recommended that the whole body radiation not exceed 5 rems y^{-1}; localized doses may exceed this, reaching a maximum of 75 rem y^{-1} for hand-only exposure, but the long-term accumulation to age N should not exceed $(N-18) \times 5$ rems. For an individual within the population at large, it is recommended that the exposure in addition to background and medical procedures not exceed 0.50 rem y^{-1}. To put these quantities in some perspective it should be noted that the background dose from cosmic rays, radioactivity in the surroundings, and radionuclides deposited internally, is about 0.12 rem y^{-1} in the United States at sea level, rising to about 0.25 rem y^{-1} at 1500 m above sea level, while the dose that is 50 percent lethal in man is about 500 rems delivered over a short period of time. It is difficult to ascertain the average dose equivalent received by the population at large from medical diagnostic procedures. It appears to be a few hundred millirems (mrem) per year, and unfortunately it can be much higher in individual cases. There is not universal agreement that the doses in the recommended guidelines are prudent, and there are those who suggest that they should be lowered; some even suggest that they be lowered by an order of magnitude.

Internal Radiation Sources. The body may receive excessive irradiations from internal as well as external sources. Many radioactive nuclides when ingested or inhaled become fixed in the body for varying lengths of time. Care must therefore be taken to avoid intake of radioactive materials. Table 6-3 lists the maximum allowable concentrations of a few nuclides in inhaled air and in ingested liquids and also the maximum permissible amounts in the body.

Hazards Encountered with Radioactive Materials. It should be clear from the foregoing discussion that, whenever one is working with radioactive materials or other sources of radiation, one must endeavor to keep one's radiation exposure to a minimum, certainly below the maximum allowable levels. Dose reduction is achieved by shielding, distance, or some judicious combination of the two. Another important consideration in handling radioactive materials, even at activity levels so low that health hazards from external radiation exposure are minimal, is the prevention of the spread of radioactive contamination. Such contamination can seriously raise counter backgrounds and interfere with low-activity experiments. The degree of precaution needed, both to contain contamination and to prevent excessive radiation exposure, depends on many factors, including the amount of activity handled, the nature and energy of the radiation involved, the half life of the active substance, and possibly its chemical properties.

Table 6-3 Biologically Permissible Levels of Selected Radionuclides[a] Above Natural Background

	Occupational Exposure[b,c] (Restricted Area)		General Public[b] (Unrestricted Area)		
Nuclide	In Air (μCi ml^{-1})	In Water (μCi ml^{-1})	In Air (μCi ml^{-1})	In Water (μCi ml^{-1})	In Critical Organ[d] (μCi)
^{3}H(H_2O)	5×10^{-6}	1×10^{-1}	2×10^{-7}	3×10^{-3}	1000
^{14}C(CO_2)	4×10^{-6}	2×10^{-2}	1×10^{-7}	8×10^{-4}	300 (fat)
^{24}Na	1×10^{-6}	6×10^{-3}	4×10^{-8}	2×10^{-4}	7.0 (G.I. tract)
^{32}P	7×10^{-8}	5×10^{-4}	2×10^{-9}	2×10^{-5}	6.0 (bone)
^{35}S	3×10^{-7}	2×10^{-3}	9×10^{-9}	6×10^{-5}	90.0 (testes)
^{60}Co	3×10^{-7}	1×10^{-3}	1×10^{-8}	5×10^{-5}	10.0 (G.I. tract)
^{90}Sr	1×10^{-9}	1×10^{-5}	3×10^{-11}	3×10^{-7}	2.0 (bone)
^{131}I	9×10^{-9}	6×10^{-5}	1×10^{-10}	3×10^{-7}	0.7 (thyroid)
^{137}Cs	6×10^{-8}	4×10^{-4}	2×10^{-9}	2×10^{-5}	30 (whole body)
^{210}Po	5×10^{-10}	2×10^{-5}	2×10^{-11}	7×10^{-7}	0.03 (spleen)
^{226}Ra	3×10^{-11}	4×10^{-7}	3×10^{-12}	3×10^{-8}	0.1 (bone)
^{238}U[e]	7×10^{-11}	1×10^{-3}	3×10^{-12}	4×10^{-5}	0.005
^{239}Pu	2×10^{-12}	1×10^{-4}	6×10^{-14}	5×10^{-6}	0.04 (bone)

[a] In soluble form.
[b] From reference S1, Appendix B, p. 209 (1980).
[c] Assumes 40 hours per week for 50 years.
[d] "Maximum Permissible Body Burdens and Maximum Permissible Concentrations of Radionuclides in Air and in Water for Occupational Exposure," *Nat. Bur. Std. (U.S.) Handbook* **69**, 1959.
[e] For mixtures of ^{238}U, ^{234}U, and ^{235}U in air, chemical toxicity may be the limiting factor.

Some general precautions should be observed in all work with radioactive sources. Some survey instrument (see chapter 7, section G) should always be used to determine the actual radiation levels present, hence the type of protection required. As many of the manipulations as possible should be carried out in hoods with adequate air flow or in dry boxes. To contain possible spills it is well to work in trays or on surfaces covered with absorbent paper. Pipetting by mouth is to be avoided. Even at the microcurie level, radioactive materials should never be handled with bare hands, but with gloves or tongs or in containers. At somewhat higher levels of γ emitters (typically in the millicurie range) it becomes necessary to carry out separations behind lead shields, which are usually assembled from lead bricks to suit the particular purpose. Operations are then performed with the use of tongs and other tools. For very high activity levels (say in excess of about 10^{12} γ quanta per minute) more elaborate remote-control methods are necessary. It is obvious that chemical procedures are more difficult under these conditions and in many cases have to be modified considerably to adapt them for remote-control operation.

More detailed discussions of the safe handling of radioactive materials and of appropriate health-protection measures may be found in M1 and M2.

REFERENCES

B1 N. Bohr, "On the Theory of the Decrease of Velocity of Moving Electrified Particles on Passing through Matter," *Phil. Mag.* **25**, 10 (1913).

B2 H.-D. Betz, "Charge States and Charge-Changing Cross Sections of Fast Heavy Ions Penetrating Through Gaseous and Solid Media," *Rev. Mod. Phys.* **44**, 465 (1972).

*B3 H. A. Bethe and J. Ashkin, "Passage of Radiations through Matter," *Experimental Nuclear Physics*, Vol. 1 (E. Segre, Ed.), Wiley, New York, 1953, pp. 166–357.

C1 A. H. Compton and S. K. Allison, *X-rays in Theory and Experiment*, Van Nostrand, Princeton, N.J., 1935.

D1 S. Datz et al., "Motion of Energetic Particles in Crystals," *Ann. Rev. Nucl. Sci.* **17**, 129 (1967).

F1 U. Fano, "Penetration of Protons, Alpha Particles, and Mesons," *Ann. Rev. Nucl. Sci.* **13**, 1 (1963).

L1 J. Lindhard, M. Scharff, and H. E. Schiøtt, "Range Concepts and Heavy Ion Ranges," *Kgl. Danske Videnskab. Selskab. Mat.-Fys. Medd.* **33**, No. 14 (1963).

L2 C. M. Lederer and V. S. Shirley (Eds.), *Table of Isotopes*, 7th ed., Wiley, New York, 1978.

M1 K. Z. Morgan, "Techniques of Personnel Monitoring and Radiation Surveying," in *Nuclear Instruments and Their Uses* (A. H. Snell, Ed.), Wiley, New York, 1962, pp. 391–469.

M2 K. Z. Morgan and J. E. Turner (Eds.), *Principles of Radiation Protection*, Wiley, New York, 1967.

N1 L. C. Northcliffe and R. F. Schilling, "Ranges and Stopping-Power Tables for Heavy Ions," *Nucl. Data Tables* **7A**, 233–463 (1970).

N2 L. Northcliffe, "Passage of Heavy Ions through Matter," *Ann. Rev. Nucl. Sci.* **13**, 67 (1963).

N3 "Basic Radiation Protection Criteria," *National Council on Radiation Protection Report* Vol. 39, NCRP Publications, Washington, 1971.

*S1 Standards for Protection against Radiation, Title 10, *Code of Federal Regulations*, Part 20 (updated and published annually).

U1 A. C. Upton, "Effects of Radiation on Man," *Ann. Rev. Nucl. Sci.* **18**, 495 (1968).

W1 W. Whaling, "The Energy Loss of Charged Particles in Matter" in *Encyclopedia of Physics*, Vol. 34 (E. Flügge, Ed.), Springer-Verlag, Berlin, 1958.

EXERCISES

1. Show that the maximum velocity an electron can receive in an impact with an α particle of velocity v is approximately $2v$.
2. Estimate the range of 120 MeV ^{14}N ions in aluminum.
3. Use (6-14) to calculate the stopping power of 12-MeV α particles in lead. Take $I = 788$ eV. Compare your result with the value interpolated from figures 6-4 and 6-5.

4. (a) What thickness of aluminum foil will reduce the energy of 40-MeV ^4He ions to 32 MeV? (b) What energy loss will a 20-MeV deuteron beam suffer in passing through the same foil? *Answer:* (*a*) 55 mg cm^{-2}; (b) $\Delta E \approx 2$ MeV.
5. An absorption curve of a sample emitting β and γ rays was taken, using a gas-flow proportional counter, with aluminum absorbers. The data obtained were:

Absorber Thickness (g cm^{-2})	Activity (counts/min)	Absorber Thickness (g cm^{-2})	Activity (counts/min)
0	5800	0.700	101
0.070	3500	0.800	100
0.130	2200	1.00	98
0.200	1300	2.00	92
0.300	600	4.00	80
0.400	280	7.00	65
0.500	120	10.00	53
0.600	103	14.00	40

(a) Estimate the maximum energy of the β spectrum (in MeV). (b) Find the energy of the γ ray. (c) What would be the absorption coefficient of that γ ray in lead? *Answers:* (b) 0.8 MeV; (c) 1.0 cm^{-1}.
6. Estimate the straggling of 32-MeV α particles in air. *Answer:* ~1 mg cm^{-2}.
7. At 1.00 m from 1.00 g ^{226}Ra (in equilibrium with its decay products and enclosed in 0.5 mm of platinum) the γ-ray dosage rate is 0.84 rad h^{-1}. What is the minimum safe working distance from a 20-mg ^{226}Ra source, assuming the worker is present 40 h per week for 1 y? *Answer:* 2.6 m.
8. From data on X-ray spectra and X-ray absorption coefficients (in the Chemical Rubber Company Handbook, for example) locate the critical absorbers for the identification of the X rays following K capture (a) in ^{37}Ar, (b) in ^{101}Pd.
9. At a distance of 60 cm from a ^{137}Cs source the dosage rate due to the γ rays from this source is found to be 127 mrad h^{-1}. The decay of ^{137}Cs is accompanied by the emission of a 0.66-Mev γ ray in about 85 percent of the disintegrations; no other γ rays are emitted. (a) Estimate the strength of the ^{137}Cs source in millicuries. (b) What thickness of lead shielding (in mm) is required around the source to reduce the dosage rate at 60 cm to 3 mrad h^{-1}? *Answer:* (a) 0.07 Ci; (b) ~30 mm.
10. A 100-mCi point source of ^{137}Cs is placed 20 cm from the palm of your hand. (a) Calculate the flux of γ rays per cm^2 per second that is incident on your hand. (b) Calculate the dose rate in rad per second that your open palm is receiving. Assume an average composition of water for your hand. [Take μ/ρ for 0.66-MeV γ rays as 0.03 cm^2 g^{-1}].
11. Derive (6-33) for $n = 2$.
12. Calculate the average number of elastic collisions required to slow a neutron from 10^6 to 10^{-2} eV in (a) ^{238}U, (b) ^{12}C, and (c) ^1H. *Answer:* (a) 2.2×10^3.

Chapter 7

Radiation Detection and Measurement

All methods for detection of radioactivity are based on interactions of the charged particles or electromagnetic rays with matter traversed. The uncharged neutron is detected only indirectly, through recoil protons (from fast neutrons) or through nuclear transmutations or induced radioactivities (from fast or slow neutrons), as discussed in section E. Neutrinos have neither charge nor rest mass and therefore do not interact measurably with matter to produce either ions or recoil particles. As mentioned on p. 77, neutrinos are expected to be capable of causing nuclear transmutations that are the inverse of β-decay processes; observation of such reactions has been reported, but the cross sections are extremely small—of the order of 10^{-40} cm^2 or less (R1).

A. GASEOUS ION COLLECTION METHODS

1. Saturation Collection

Current-Voltage Characteristics. Many common radiation detectors make use of the electric conductivity of a gas resulting from the ionization produced in it. This conductivity is somewhat analogous to the electric conductivity of solutions caused by the presence of electrolyte ions. In gas conduction, as produced by radiation, the ion current first increases with applied voltage; with increasing voltage the current eventually reaches a constant value that is a direct measure of the rate of production of charged ions in the gas volume. This constant value of the current is called the **saturation current**. A schematic representation of an ionization chamber circuit is shown in figure 7-1, along with a plot of I versus V that might be obtained.

In the region of applied voltage below that necessary for the saturation current, recombination of positive and negative ions reduces the current collected. As the applied voltage is increased beyond the upper limit for saturation collection, the current increases again, and finally the gap breaks down into a glowing discharge or arc, with a very sharp rise in the current. In the measurement of gas ionization it is obviously of some advantage to

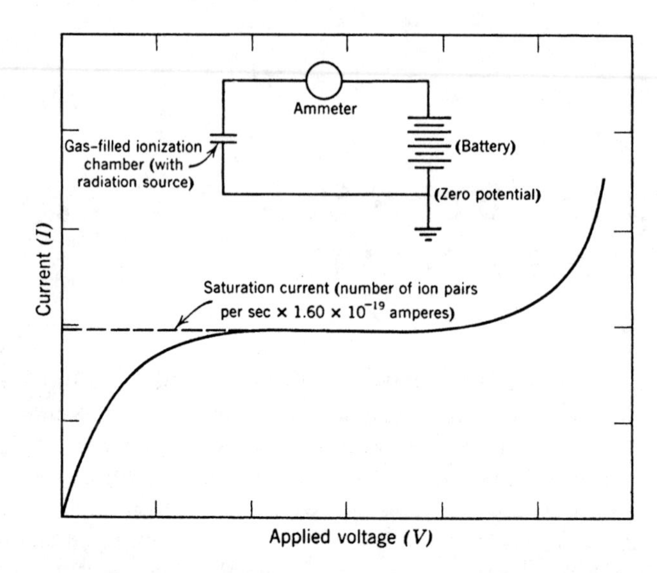

Fig. 7-1 Ionization current as a function of applied voltage as obtained in an ionization chamber. Above the curve is a schematic diagram of a simple ionization chamber circuit.

measure the saturation current: the current is easily interpreted in terms of the rate of gas ionization, and the measured current does not depend critically on the applied voltage. The range of voltage over which the saturation current is obtained depends on the geometry of the electrodes and their spacing, the nature and pressure of the gas, and the general and local density and spatial distribution of the ionization produced in the gas. In air, for many practical cases, this range may extend from $\sim 10^2$ to $\sim 10^4$ V per centimeter of distance between the electrodes.

Time Constants. The gas-filled electrode system designed for saturation collection is called an **ionization chamber**. In a complete system this chamber must be connected to a device for measurement of the very small currents obtained. Either steady-state currents or pulses resulting from individual ionizing events may be measured, depending on the time constant of the device. The time constant RC of a circuit is the time required for an initial charge on a capacitor of capacitance C to be reduced to $1/e$ of its value when the capacitor is short-circuited with a resistance R. If RC is long compared to the time between ionizing events, a steady state is reached, and a direct current (or a voltage developed across a known resistor through which this current flows) may be measured. On the other hand, if RC is small compared to the time between ionizing events, the charges collected during individual events (or the corresponding voltages) may be measured by means of appropriate ac circuitry.

Current Collection. The simplest current collection instruments are the once widely used, but now obsolete, electroscopes; here a quartz fiber or gold leaf electrode system is initially charged up to a voltage V and its rate of discharge $\Delta V/\Delta t$, resulting from ion collection, is measured visually. More versatile, more sensitive, and having a wider dynamic range are ion chambers connected to electronic dc amplifiers. However, dc amplification is notoriously tricky, and an important improvement came with the development of the vibrating-reed electrometer. In this instrument the IR voltage developed by the ion current I across a high resistance R is converted to an alternating potential by means of a reed that vibrates continuously and thus has an oscillating capacity with respect to a fixed electrode. The ac signal is then readily amplified in an ac circuit. The instrument is very stable and is sensitive to currents as small as 10^{-15} A.

Pulse Amplification. An ionization chamber may be directly connected to an ac amplifier for measurements of individual ionization pulses. A short burst of intense ionization, such as results from the passage of an α particle through the chamber, will give a sudden change of voltage on the grid of the first amplifier tube or at the control element of the first transistor. The voltage will return to normal in a time of the order of the time constant RC of the input circuit and collecting-electrode system. This voltage pulse is amplified by the circuit, usually in such a manner that the height of the output pulse is proportional to the input pulse—hence the name **linear amplifier**.

Ion chambers with linear pulse amplifiers are particularly useful for the measurement of α particles and fission fragments. Energies of such particles with discrete spectra can be accurately measured, provided the chamber is arranged such that all the particles spend their entire ranges within the chamber and that the pulse height is independent of the particles' trajectories; the latter result is usually achievable through placement of a negatively charged grid between the ionization region and the positive collecting electrode. This shields the collector from the positive ions that would otherwise lead to induced charge effects whose magnitude would depend on the distance of the charge cloud from the collector. Extremely low background rates (of the order of 0.1 α per minute and less than 1 fission per day) are easily attained in ion chambers. Further, these instruments are ideally suited to the measurement of low α rates in the presence of large β-particle fluxes and of low fission rates in the presence of large α-particle fluxes since the pulse heights for the different types of particles are so different that they can be electronically sorted out.

2. *Multiplicative Ion Collection*

As shown in figure 7-1 the ion current or pulse height in an ion chamber device eventually increases above the saturation value at some sufficiently

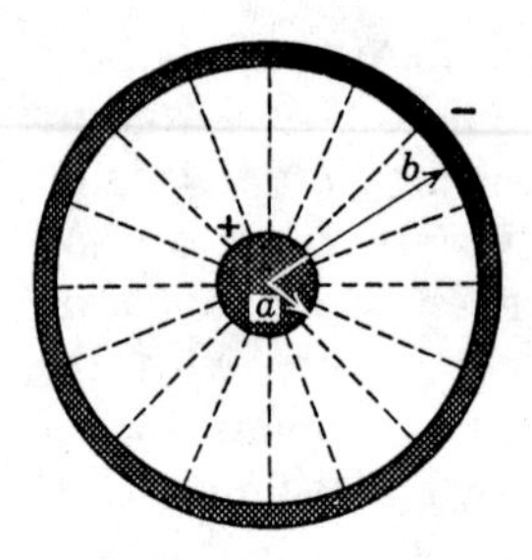

Fig. 7-2 Electrostatic lines of force between coaxial cylindrical electrodes whose radii are a and b, respectively.

high value of the applied voltage. This comes about because electrons moving in these high fields acquire enough energy to cause secondary ionization. In practice, multiplicative collection is always coupled with pulsed operation (small RC values), and the devices employing this scheme are referred to as "counters."

Voltage Gradients. In counters the cathode (negative electrode) is most often a cylinder, the anode (positive electrode) an axial wire. Note that electrons and negative ions thus move to the wire. Figure 7-2 shows a cross-sectional view with the wire radius exaggerated; the lines of force are sketched in. It is readily seen that the density of these lines, which is a measure of the voltage gradient (field), is inversely proportional to the radial distance; that is,

$$E = \frac{k}{r}. \tag{7-1}$$

By definition, $E = dV/dr$, and we may represent the voltage difference between the electrodes of radii a and b:

$$\Delta V = \int_{r=a}^{r=b} dV = \int_a^b E\,dr = \int_a^b \frac{k}{r}\,dr = k \ln\left(\frac{b}{a}\right). \tag{7-2}$$

In a practical case we might have $b = 1$ cm, $a = 4 \times 10^{-3}$ cm, and $\Delta V =$ 1000 V. Then

$$1000 = k \ln\left(\frac{1}{4 \times 10^{-3}}\right) = 5.5k; \qquad k = 180.$$

The voltage gradients at wall and wire are

$$E_b = 180 \text{ V cm}^{-1};$$

$$E_a = \frac{180}{4 \times 10^{-3}} = 4.5 \times 10^4 \text{ V cm}^{-1}.$$

The field at the wire and for a small space around it is above the maximum value for saturation collection (say $\sim 10^3$ V cm^{-1} in a practical counter gas).

Regions of Multiplicative Operation. With an electrode system of the kind just described (cylindrical cathode with central wire anode), filled with

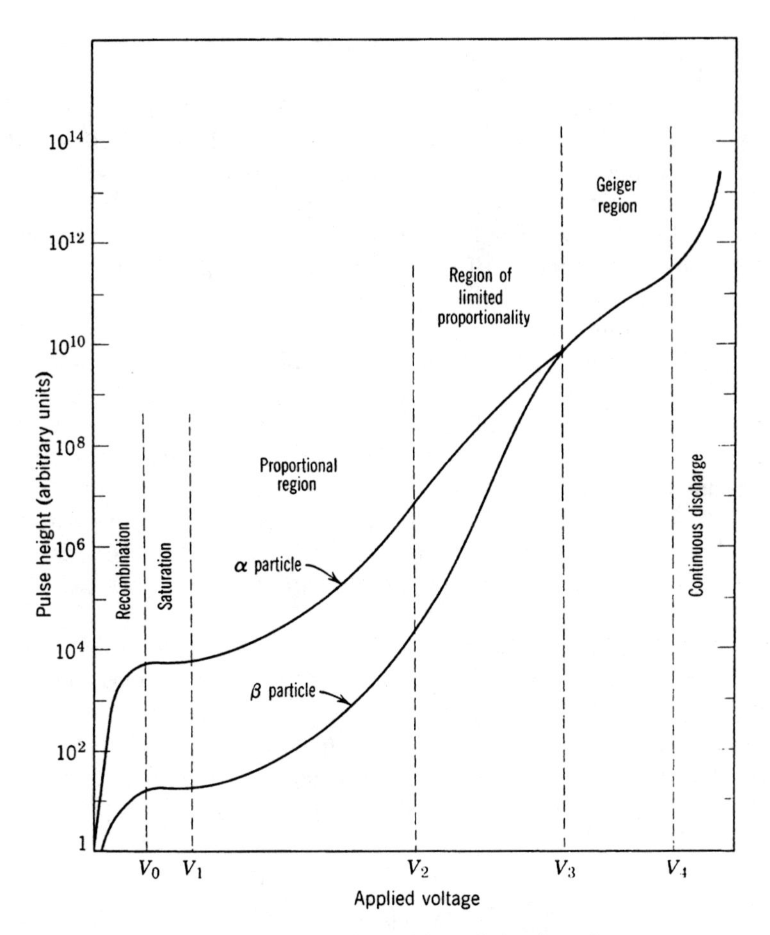

Fig. 7-3 Variation of pulse height with applied voltage in a counter.

a suitable gas and connected to a high-gain amplifier and oscilloscope, the pulse heights obtained as a function of applied voltage would be approximately as shown in figure 7-3. Curves are drawn for two types of ionizing particles, one losing several hundred times as much energy in the chamber volume as the other (they might be typical α and β particles, respectively).

In the region of saturation collection, the voltage pulses caused by β particles are, as already discussed, generally too small to be detected with practical amplifiers, whereas the α pulses are measurable with a sensitive pulse amplifier. Once the voltage is raised above the limit of the saturation region the pulse heights increase as a result of secondary ionization by the electrons accelerated in the high field gradient near the wire. For a considerable voltage range (V_1 to V_2 in figure 7-3, typically of the order of several hundred to a thousand volts) the *ratio* of pulse heights for different ionizing events remains independent of applied voltage, or, in other words, the pulse height remains proportional to the amount of energy lost in the chamber by the primary ionizing particle. In this voltage region the apparatus operates as a **proportional counter. Multiplication factors** (number of electrons collected per initial ion pair) in actual proportional counters may vary from ~ 10 to $\sim 10^4$. The gain required in the external amplifier depends on the multiplication factor used and on the radiation to be detected.

If the voltage is increased further (above V_2), pulse heights continue to increase. From V_2 to V_3 pulse heights are still related, but no longer proportional, to initial ionization intensity. This is sometimes referred to as the region of limited proportionality. Finally, at V_3 the pulse height becomes independent of initial ionization—a pulse caused by a single ion pair becomes indistinguishable from one due to a fission fragment depositing all its energy. A device operated in this region is called a **Geiger-Müller (GM) counter**. Pulse heights are typically of the order of volts and very little if any additional amplification is needed. At a still higher voltage (V_4) the Geiger-Müller action is terminated by the onset of self-excitation, and eventually the counter goes into continuous discharge.

Gas Multiplication. The gas multiplication factor obtained in a given counter tube depends on the nature and pressure of the gas, on the tube dimensions, particularly the wire diameter (cf. equation 7-1), and on the applied voltage. In general, the critical field gradient is higher for polyatomic than for rare gases. In a given gas the functional dependence of the multiplication factor M on wire radius a, cathode radius b, pressure P, and voltage V is of the form

$$M = f\left[\frac{V}{\ln (b/a)}, (Pa)\right]. \tag{7-3}$$

As an illustration, figure 7-4 (based on data in R2) shows the variation of M with voltage in argon and methane at two different pressures. As these data suggest, it is found that relatively small admixtures of argon lower the threshold and operating voltages of methane-filled proportional counters considerably. On the other hand, the presence of methane or other polyatomic gases in argon-filled counters decreases the dependence of M on applied voltage and thus improves the stability of operation with respect to voltage variations.

Proportional Counters. True proportionality between pulse height and primary ionization requires that the avalanches produced by individual primary electrons in an ionization track be essentially independent of one another; thus each avalanche must be confined to a very small region of the central wire. In the course of an avalanche, excitation of molecules can lead to the emission of ultraviolet photons, which in turn are capable of producing photoelectrons at the cathode or in some constituent of the gas.

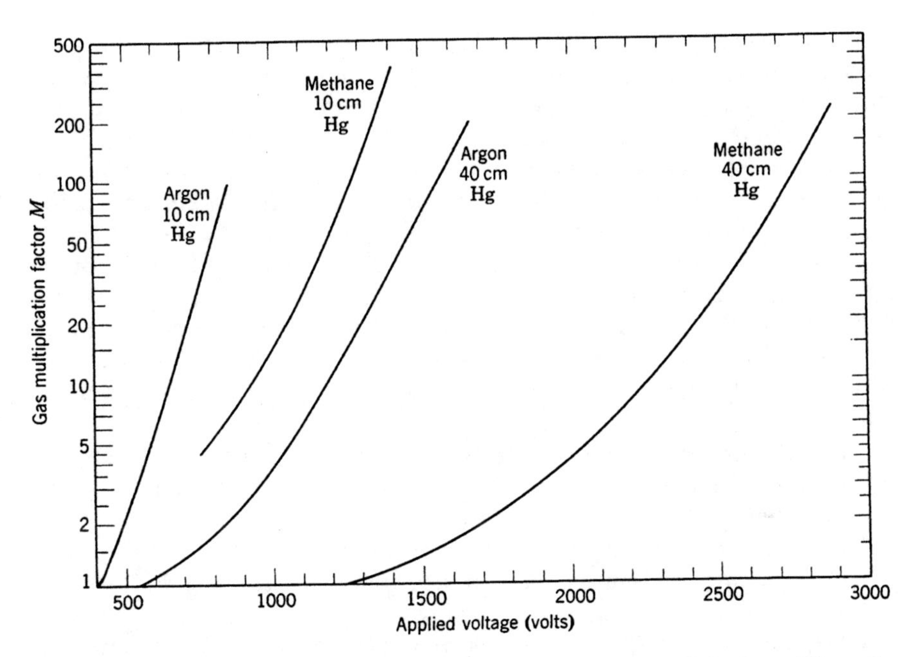

Fig. 7-4 Multiplication factors in argon and methane as a function of applied voltage. Wire radius $a = 0.13$ mm, cathode radius $b = 11$ mm (Data from reference R2.)

At sufficiently high voltages the number of photons per primary electron becomes so great that a given avalanche is likely to spread along the entire tube by these photoionization phenomena; the final pulse size is then no longer dependent on primary ionization and the Geiger region has been reached.

With any reasonable value of M, the pulse size in a proportional counter is almost entirely determined by the amount of avalanche ionization and therefore essentially independent of the location of the primary track. Because of the small extent of the multiplicative region the avalanche electrons travel through only a small part of the potential difference applied to the counter, and most of the pulse height is therefore contributed by the positive ions moving *away* from the central wire. However, although the total time for the positive ions to reach the cathode is typically of the order of 10^{-3} s, the variation of field gradient with radius is such that initially the pulse rises very rapidly and then approaches its final height slowly. If the total collection time is T, the time at which half the final pulse height is reached is about $(a/b)T$, or typically of the order of 10^{-6} s. Amplifier circuits with time constants of this order are therefore used to "clip" the pulses from proportional counters. Even with sharp clipping, proportionality is preserved because the pulse *shape* is independent of pulse *height*. With a clipping circuit a proportional counter is ready to accept a new ionizing event within $\leq 1\ \mu s$ after a count, and proportional counters can thus be used at much higher counting rates than GM counters (see below). They are also generally more stable and have better voltage plateau characteristics.

A **voltage plateau** is a region in which the counting rate (not the pulse height) caused by a given radiation source is independent of applied voltage. Proportional counters typically have plateaus of several hundred volts with slopes of ≤ 1 percent per 100 V. The onset and length of the plateau depends on the setting of the **discriminator**, an electronic device that prevents pulses below a certain size from being registered and that is needed to cut out pulses due to electronic noise. Figure 7-5 shows the response of a proportional counter to a source of both α and β particles as a function of voltage. The counter exhibits two plateaus, one in a voltage range in which only the α particles (which produce about 100 times as much primary ionization as the β particles) are registered, and a second one on which both types of radiation are recorded.

The most widely used proportional counters are of the flow type; the counting gas, usually methane or an argon-methane mixture from a compressed-gas tank, flows slowly at atmospheric pressure through the counter. Deterioration of the gas is thus avoided and very thin windows, for example, made of metal-coated Mylar, can be used to allow very soft radiations to enter the counter. In another design, especially well suited for measurement of α particles and very-low-energy electrons, the mounted sample is introduced into the counter-gas volume itself.

An extension of the proportional counter is the **multiwire proportional chamber**, now widely used in particle physics (C1). It consists of a plane of parallel wires, spaced one to several millimeters apart and mounted be-

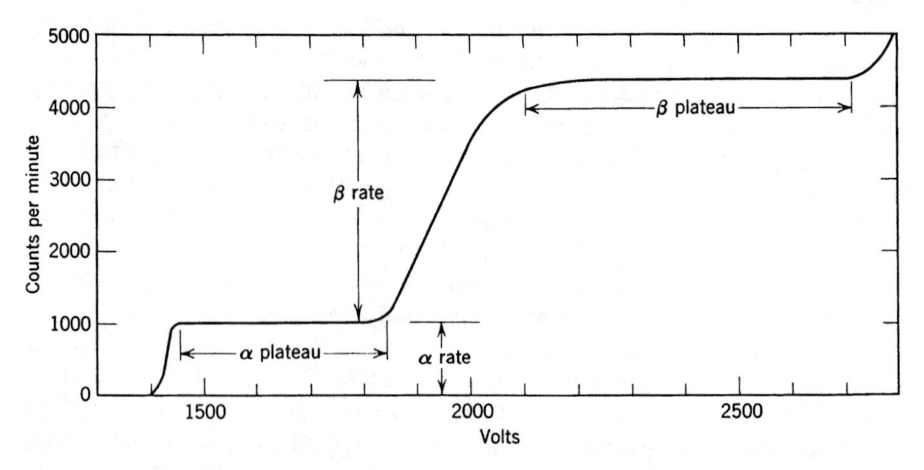

Fig. 7-5 Counting rate as a function of applied voltage for a proportional counter exposed to a source emitting both α and β particles.

tween two parallel electrodes in a gas such as argon-pentane. Excellent spatial and time resolution has been obtained with such proportional chambers.

Geiger-Müller (GM) Counters. As we have already mentioned, the proportional region of counter operation is limited at the upper voltage end by the onset of photoionization; this process spreads the intense ionization produced by a single primary electron along the anode, thus causing interaction with the avalanches produced by other primaries, and destroying proportionality of response. As the voltage is further increased, a condition is finally reached (V_3 in figure 7-3) in which each ionizing event is spread along the entire length of the wire and the final pulse size becomes independent of primary ionization—the tube now operates as a GM counter.

In a GM counter, as in a proportional counter, the negative ions formed (mostly free electrons) reach the wire very quickly, typically in about 5×10^{-7} s; however, the sheath of positive ions that now surrounds the wire reduces the voltage gradient below the value necessary for ion multiplication, and the counter cannot record another event until the positive ions reach the cathode, in about 100–500 μs. This inherent dead-time limits the use of GM counters to counting rates below a few tens of thousands per minute.

When the positive ions reach the cathode they can cause secondary electron emission from the surface, thus triggering a new counter discharge, and so forth. This secondary electron emission is usually suppressed by the addition of a **quench gas** to the main filling gas, which may be argon. Polyatomic vapors, such as alcohol, ether, or methane, are the most common quench gases; in their presence the positive ions are, by electron transfer, converted to polyatomic organic ions, and these dissipate energy by predissociation. The quench gas is thus gradually consumed, and GM counters deteriorate after 10^8 or 10^9 counts. Geiger counter plateaus are shorter and have more of a slope than do those of proportional counters. Until reliable high-gain amplifiers became available the large pulse sizes from GM counters made them preferable to proportional counters; now the advantages are strongly in the other direction.

Counter Backgrounds. Since both GM and proportional counters register individual ionizing events, their sensitivities are limited purely by background counting rates. Even in a laboratory not contaminated by radiochemical work small amounts of activity are present as impurities in construction materials. Also the air contains an appreciable and variable amount of ^{222}Rn and ^{220}Rn with their decay products. In free air at the earth's surface most of the ionization comes from these two causes, with the cosmic radiation contributing a smaller part. However, because the counter is itself closed and enclosed in a building, it is not accessible to most of the radioactive α, β, and even γ radiation, and the cosmic-ray effect is often the most significant. A β-sensitive counter with a diameter of 2.5 cm and length 6 cm may have a background rate of about 30 counts per minute (cpm); this may be reduced to 6–8 cpm by the usual lead shield of a few centimeters thickness. By use of special shielding and anti-coincidence circuits that reject those counts occurring simultaneously with counts in nearby auxiliary counters, backgrounds can be reduced by at least another order of magnitude. Further reduction of proportional-counter backgrounds is possible by the use of circuits that reject pulses of sizes that do not correspond to the radiation being measured (see section F). Proportional counters operating on their α plateaus require no special shielding and may have background rates of about 0.1 cpm if constructed of selected materials. The techniques for achieving the lowest possible background rates and their importance in various low-level radioactivity measurements are reviewed in O1.

B. SEMICONDUCTOR DETECTORS

Solid Ion Chambers. Ionization-chamber operation is not limited to gas-filled devices. The use of denser ionizing media has obvious advantages for the stopping of higher-energy particles and for the detection of radiations of low specific ionization. Numerous attempts have therefore been made to use liquid and solid dielectrics in ion-chamber devices. The only really successful and widely used detectors of this type are those using semiconductors, specifically silicon and germanium.

Principles of Operation. The process in a semiconductor or an insulator that is analogous to ionization in a gas is the lifting of an electron from the highest filled band, the valence band, to the conduction band. The energy difference between these two bands is called the **band gap** E_g. In semiconductors E_g is of the order of 1 eV, small enough for thermal excitation to lead to some conduction; in insulators E_g is several times larger. When an electron is lifted into the conduction band, thus acquiring high mobility, a positive hole is created in the valence band, and it too can travel under the influence of an electric field, not by bodily movement of the ion through the crystal but by successive electron exchanges between neighboring lattice sites. The energy ϵ required to produce an electron-hole pair always exceeds E_g because some energy goes into coupling electrons to lattice vibrations. This is analogous to the situation in gases where the energy required for ion-pair formation always exceeds the ionization potential because some energy is used in excitation and dissociation processes. Despite this effect ϵ is remarkably independent of the energy of the ionizing radiation, only slightly dependent on temperature, and practically the same for different ionizing radiations except for heavy ions. For germanium $\epsilon = 2.96$ eV (at 77 K and for γ rays and electrons); for silicon at 300 K, $\epsilon = 3.76$ eV. The corresponding values of E_g are 0.67 and 1.10 eV, respectively. The low value of E_g for germanium makes it necessary to use germanium detectors at low temperatures, typically liquid nitrogen, to avoid excessive thermal noise. Despite this disadvantage, germanium detectors have come into very wide use as γ-ray detectors because the relatively high Z of germanium gives them good sensitivity to γ rays.

As indicated above in the discussion of band gaps, it is in the very nature of semiconductors that they will pass some leakage current when an electric field is applied. To make them useful as particle detectors we have to be able to apply appreciable electric fields without excessive leakage currents. Various techniques for achieving this goal, as well as the characteristics of the resulting detector types, are discussed in the following paragraphs.

Reverse-Bias *p*-*n* Junction Detectors. This type of detector makes use of a diode structure that incorporates regions with excess negative (electron) and excess positive (hole) charge carriers, referred to as n-type and p-type semiconductors, respectively. Small impurity concentrations can be used to produce the excess charge carriers. For example, phosphorus and arsenic serve as electron donors in silicon and germanium and produce n-type, whereas boron and gallium are electron acceptors and produce p-type. The impurity may be diffused into the semiconductor by thermal treatment, or it can be introduced by ion implantation into a well-defined thin layer. *Heavily* doped p-type and n-type semiconductors are designated as p^+ and n^+.

Figure 7-6 shows a schematic diagram of a p-n junction detector. In

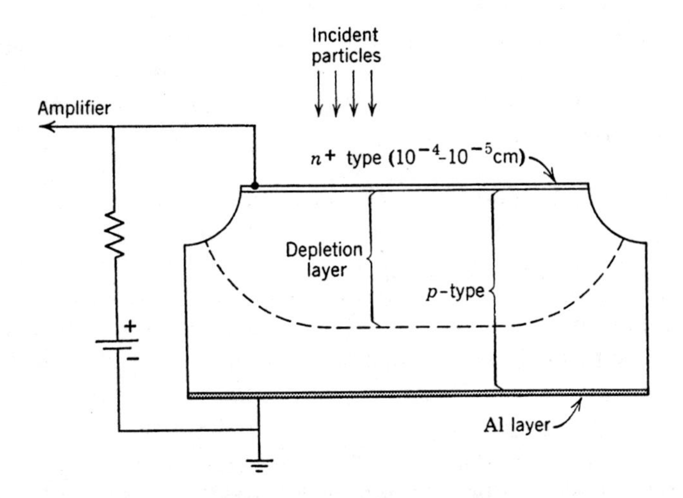

Fig. 7-6 Schematic diagram of a p-n junction detector.

this case the base material is p-type silicon. A thin layer (10^{-4}–10^{-5} cm) of n^+-type silicon has been produced at one surface of the slab. On the opposite face a thin layer of aluminum is evaporated to facilitate electrical contact, the edges of the device are etched to reduce surface leakage, and a reverse bias (+charge on n-type, −charge on p-type side) is applied, providing a field of the order of 10^3 V cm^{-1}. In the presence of this field the positive holes in the p-type silicon are pulled toward the negative electrode, the electrons toward the p-n junction, and thus a **depletion layer** with a very small concentration of free charge carriers is created in the p-type material. The result is an extremely low leakage current. The thickness of the depletion layer depends on the magnitude of the applied field. When ionizing particles enter the device (usually through the p-n-junction side), the depletion layer serves as the sensitive volume, and electron-hole pairs created there will be quickly collected (carrier velocities are 10^6–10^7 cm s^{-1}). One interesting feature of solid-state detectors is that hole mobilities are only about three times smaller than electron mobilities (in contrast to positive-ion mobilities in gases), so that all charge carriers are usually collected and the current produced is therefore independent of the location of the ionizing event within the sensitive volume.

The depth w of the depletion layer is given approximately by

$$w \approx c(\rho V)^{1/2}, \tag{7-4}$$

where ρ is the resistivity of the material in the main body of the detector in Ω cm, V is applied potential in volts, and the constant c is about 3×10^{-5} cm for p-type and about 5×10^{-5} cm for n-type silicon. The highest-resistivity silicon available has $\rho \approx 10^4$ Ω cm, and the bias voltage that can be applied is usually limited by surface leakage to a maximum of a few hundred volts. A typical p-n junction detector might thus have $\rho = 10^4$, $V = 200$, and therefore $w = 0.04$ cm, sufficient to stop 350-keV electrons, 7-MeV protons, or 30-MeV α particles. Satisfactory α-particle and fission fragment detectors can clearly be made from silicon of much lower resistivity.[1] Silicon detectors are widely used for β-ray and conversion electron spectroscopy.

Surface Barrier Detectors. Making p-n junction detectors by controlled diffusion at elevated temperatures or by ion implantation is not entirely simple. A much simpler process that leads to detectors operating in essentially the same manner is the evaporation of a thin layer of metal, usually gold, onto one face of a slab of n-type silicon. The exact mechanism responsible for the operation of such surface barrier detectors

[1] Maximum thickness is by no means always what is wanted. For the measurement of $\Delta E/\Delta x$ of heavy ions one requires very thin detectors of uniform thickness that allow the particles to pass through with minimal energy loss. Such transmission detectors are available in thicknesses down to a few micrometers.

is not fully understood; however, they behave very much like true *p-n* junctions, perhaps because a thin oxidized film under the metal surface acts as the *p*-type layer. Depletion depths up to 2 or 3 mm have been achieved in surface barrier detectors through the application of high bias voltages. Surface barrier detectors are primarily used for charged-particle detection.

Lithium-Drifted Germanium [Ge(Li)] Detectors. For some applications, particularly for γ-ray detectors, sensitive volumes larger than can be achieved with diffused-junction detectors are desirable. Also, because of the Z^5 dependence of the photoelectric effect, germanium rather than silicon is the material of choice; in fact, if a semiconductor of even higher Z were available in sufficient purity, it would probably replace germanium. The technique perfected in the 1960s that has really revolutionized γ-ray spectroscopy is the production of thick layers of high-resistivity germanium by diffusion of Li^+ ions into ordinary *p*-type material (typically 10–100 Ω cm). First a thin n^+ layer is produced by diffusing lithium into the surface at 350–450°C. Then the device is subjected to a reverse bias at a controlled temperature somewhere between room temperature and 60°C. Under these conditions Li^+ ions have quite high mobility (1.6×10^{-9} cm^2 V^{-1} s^{-1} at 60°C), and under the influence of the bias voltage they drift toward the negative electrode on the side opposite the n^+ layer. As the Li^+ ions drift across, they almost perfectly compensate the excess acceptor concentration, producing a depletion layer essentially equivalent to intrinsic germanium, that is, with resistivity of the order of 10^5 Ω cm. The resulting devices are sometimes referred as *p-i-n* detectors.

Lithium-drifted germanium [Ge(Li)] detectors are available in many sizes and shapes. In addition to cylindrical slabs in which the lithium is drifted in from one face, coaxial types in which the lithium is drifted from the cylindrical or other surface toward a central, axial electrode are also popular. Detectors with volumes up to many tens of cubic centimeters are commercially available, but the cost goes up with size because the manufacturing techniques are tedious—drift times of weeks and months under carefully controlled conditions are involved.

As already mentioned, one drawback of Ge(Li) detectors is the necessity of operating them at low temperatures. A Ge(Li) detector must at all times be kept at liquid-nitrogen temperature; permanent damage may result if such a detector is warmed to room temperature. A schematic diagram of a typical detector arrangement with associated cryostat is shown in figure 7-7. Integral detector assemblies of this kind can be bought commercially. Care must be taken to insure that the detector is mounted appropriately for the particular applications of interest. For example, if the radiations enter the detector through the n^+ layer, low-energy γ rays and X rays may get absorbed, or at least severely attenuated, in that (insensitive) layer. The

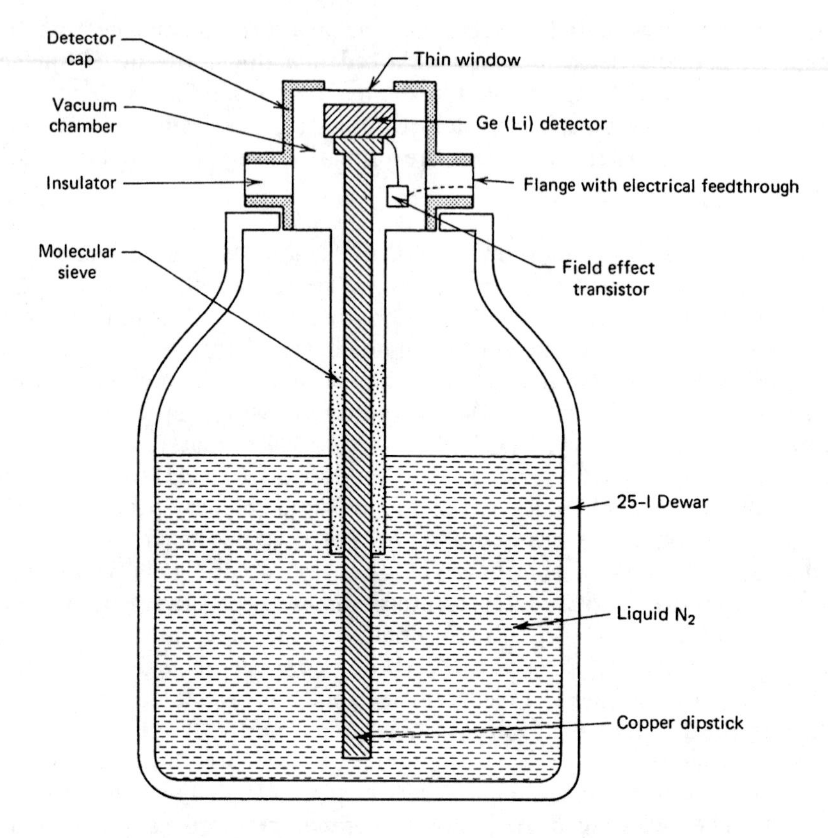

Fig. 7-7 Schematic drawing of a Ge(Li) detector system.

more recently developed **intrinsic germanium detectors** (made of germanium sufficiently pure so as not to require lithium drifting to achieve the requisite purity) avoid this problem of an insensitive layer and, furthermore, are not adversely affected by being warmed to room temperature.

Resolution, Linearity, and Timing Characteristics. The chief virtue of semiconductor detectors is the excellent energy resolution that can be attained with them. This results from the small energy required for the formation of an electron-hole pair, approximately one tenth that needed for producing an ion pair in a gas, and one hundredth the energy that gives rise to one photoelectron at the photocathode of a scintillation counter (see section C). The best resolution [full peak width at half maximum (FWHM)] attainable with large Ge(Li) detectors is in the region of 1.7 keV for 1-MeV γ rays and less than half that for 0.1-MeV γ rays. For still lower energies silicon detectors (some of them also lithium-drifted) with FWHM down to <150 eV (for very small detector diameters) are available. Typical resolutions of silicon detectors for α particles in the 5–8 MeV range are ~20 keV.

The energy resolution for heavy ions and fission fragments tends to be considerably poorer because it is limited by the statistics of ionization near the end of the particle range.

Another asset of semiconductor detectors already mentioned is their linearity over wide energy ranges.[2] The combination of good linearity and high resolution makes these detectors into excellent spectrometers when they are combined with appropriate electronic instrumentation (see section F). Practically all modern γ-ray spectroscopy is done with Ge(Li) detectors. An illustrative example of a γ spectrum taken with a 65-cm^3 Ge(Li) detector is shown in figure 7-8. The extraordinarily good energy resolution obtainable with small Si(Li) detectors makes them particularly useful as X-ray spectrometers, since it permits resolution of characteristic X rays of neighboring elements down to quite low Z as well as resolution of individual transitions in a given element (figure 7-9).

A further property of semiconductor detectors that is important for their use in nuclear spectroscopy is the large drift velocity for electrons and holes—of the order of 10^7 cm s^{-1} for modest field gradients ($\sim$1000 V cm^{-1}). Pulse rise times can therefore be quite short (in the nanosecond range), which makes semiconductor detectors particularly well suited for fast-coincidence applications.

Pulse Height Spectra. Although the analysis of γ-ray spectra is always based on areas under photopeaks, it is important for the interpretation of the spectra to understand the spectral features arising from all the energy loss processes (cf. chapter 6, section C).

A pulse height spectrum obtained with a Ge(Li) detector (and pulse height analyzer, see section F) exposed to the single γ ray emitted by ^{137}Cs (0.662 MeV) is shown in figure 7-10. For comparison a scintillation spectrum of the same γ ray is also shown; scintillation detectors are discussed in section C. The basic features—photoelectric peak and broad Compton distribution—are the same in both spectra, but the much better resolution of the Ge(Li) detector is immediately evident, both in the much narrower peak width and in the much larger peak-to-valley ratio. For any given detector and for any particular geometrical arrangement of source and detector, the efficiency (area under the photopeak) as a function of γ-ray energy must be calibrated by means of several standard sources (see section H).[3]

At energies above the pair-production threshold (1.02 MeV) additional

[2] For heavy ions and fission fragments the linearity of response to different particles is not perfect. These nonlinearities are usually described in terms of **pulse height defects**. Careful calibrations are necessary for each type of ion.

[3] It is worth noting that the "photopeak efficiency" is usually higher than corresponds to the photoelectric absorption cross section of the incident γ ray, because multiple processes can lead to full-energy deposition.

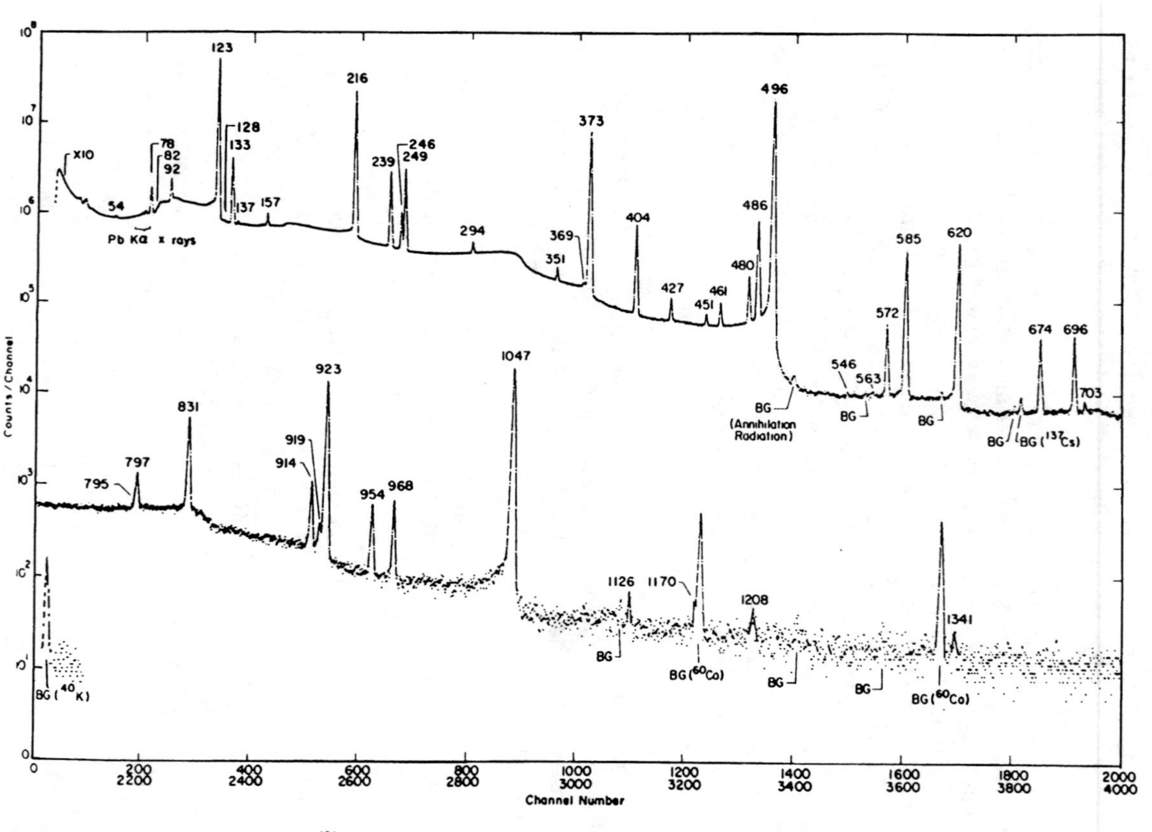

Fig. 7-8 Gamma-ray spectrum of ^{131}Ba obtained with a 65-cm^3 Ge(Li) detector. Peaks are labeled with the γ-ray energies in keV; those marked BG are background lines. The upper (lower-energy) curve is displaced upward by a factor 10. [From R.J. Gehrke et al., *Phys. Rev.* **C14**, 1896 (1976).]

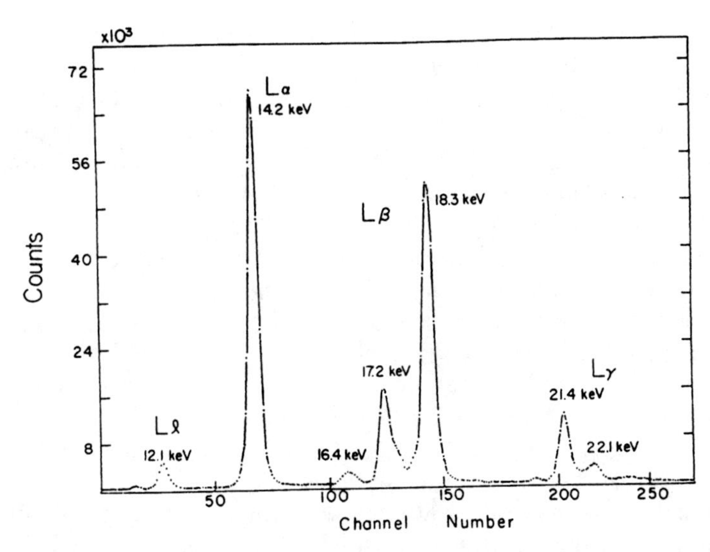

Fig. 7-9 Spectrum of plutonium L-X rays taken with a Si(Li) detector of 4-mm diameter and 3-mm sensitive depth. [From E. S. Macias and M. R. Zalutsky, *Phys. Rev.* **A9**, 2356 (1974).]

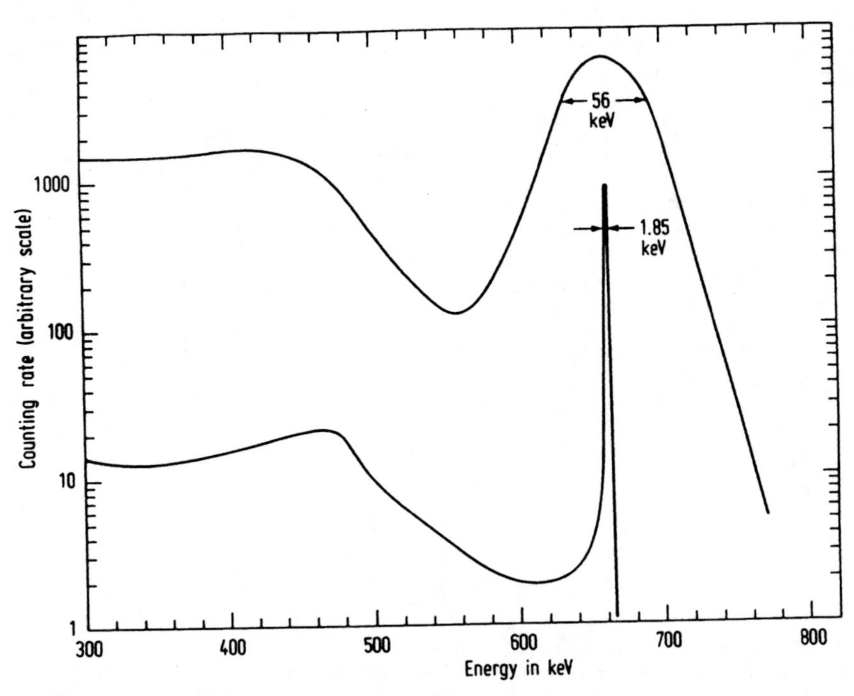

Fig. 7-10 Gamma-ray spectrum of ^{137}Cs taken with a 7.5 cm × 7.5 cm NaI(Tl) scintillation detector (top curve) and with a 50 cm^3 Ge(Li) semiconductor detector (bottom curve). The FWHM is shown for each photopeak.

complications set in. The positron formed in pair production is generally annihilated within the detector and one or both of the 511-keV annihilation quanta may escape from the detector without interaction, the probability of such escape again depending on detector size. Thus the spectrum of a high-energy γ ray will contain, in addition to the full-energy photopeak at energy E, a **single-escape peak** at $E - 0.511$ MeV and a **double-escape peak** at $E - 1.022$ MeV. This is illustrated in figure 7-11.

Both the Compton continuum and the annihilation escape peaks can be significantly suppressed relative to the full-energy peak if the detector is placed inside a second, larger detector (usually a scintillator) connected in anticoincidence, so that only those pulses in the central detector that are not in coincidence with pulses in the outer detector are recorded. Such **anti-Compton spectrometers** are particularly useful for measurement of γ rays of very high energy. The effect of such an anticoincidence arrangement on a ^{60}Co spectrum is shown in figure 7-11.

Additional peaks may occur at energies *above* the photopeaks. If two γ rays or a γ ray and an X ray are emitted in cascade, a small peak may be found at the sum of their energies. Such spurious sum peaks can be distinguished from true photopeaks by varying the sample-detector distance: the intensity of a sum peak varies with the square of the solid angle subtended by the detector, whereas the individual photopeak intensities vary linearly with the solid angle. Summing effects are more fully discussed in chapter 8, p. 330.

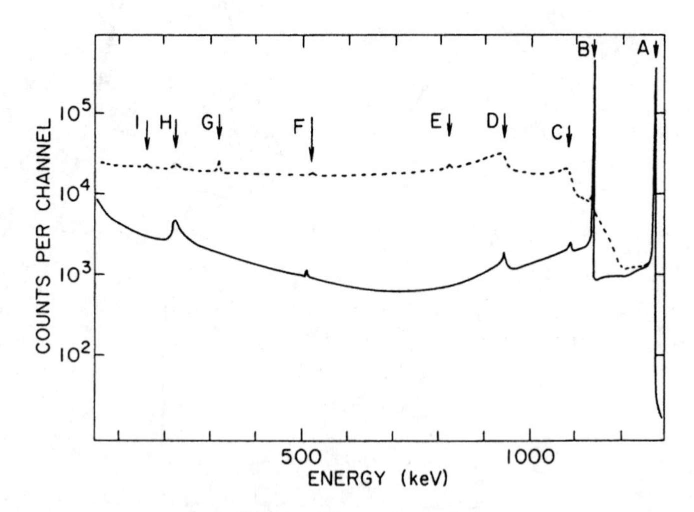

Fig. 7-11 Gamma-ray spectrum of ^{60}Co taken with a 10-cm^3 Ge(Li) detector. The solid curve was taken with, the dashed curve without a NaI(Tl) anti-Compton shield. The peaks may be identified as follows: *A*, 1.33-MeV photopeak; *B*, 1.17-MeV photopeak; *C*, 1.33-MeV Compton edge; *D*, 1.17-MeV Compton edge; *E*, 1.33-MeV single escape peak; *F*, 511 keV from e^+ annihilation following pair production; *G*, 1.33-MeV double escape peak; *H*, backscattering; *I*, 1.17-MeV double escape peak. (Courtesy D. Camp.)

The operation and applications of semiconductor detectors are reviewed in G1, A1, and E1.

Microchannel Plates. This recently developed class of detectors (L1, T1) makes use of semiconductors, but in a way that is completely different from the operating principles of the semiconductor detectors hitherto discussed.

Microchannel plates are an outgrowth of the continuous-channel electron multiplier or channeltron that, in its simplest form, is nothing more than a glass tube of 0.1–5 mm inside diameter whose inside is coated with a semiconductor and that has a potential applied between its ends. A photon or charged particle striking the inner wall of the tube near the cathode (negative) end may release an electron that then gets accelerated down the tube. If it strikes the wall with sufficient energy further down the tube, secondary electrons will be emitted, which again get accelerated and release more electrons, and so on. With typical length-to-diameter ratios of 50 to 100, amplification factors of 10^6–10^8 are obtained. By miniaturizing the tubes so that their diameters are of the order of 0.01 mm, and packing many of them into an array, one obtains a microchannel plate, typically about 1 mm thick and with an area that may be many square centimeters; more than 50 percent of the surface area can be open, so that high detection efficiencies are attained. Microchannel plates have become quite widely used in on-line accelerator experiments. They find particular application in image intensifiers.

C. DETECTORS BASED ON LIGHT EMISSION

Scintillation Counting (B1, A1). In the earliest studies of radioactivity the scintillations produced when α particles strike a fluorescent screen of zinc sulfide were of great value. The method fell into disuse for several decades, but in the 1940s a modern adaptation of scintillation counting was introduced, especially for β- and γ-ray measurements. The rays produce light in a suitable scintillator mounted on a photomultiplier tube, the light causes the ejection of photoelectrons from the first, photosensitive electrode of the tube, and the output pulse from the multiplier may be recorded. A variety of scintillators, each with particular advantages, was developed, and several of these were and are still widely used, although the advent of semiconductor detectors has diminished the importance of scintillation counters, particularly for applications requiring good energy resolution.

Organic Scintillators. Any material that luminesces in a suitable wavelength region when ionizing radiation passes through it can serve as a scintillator. Organic crystals, such as anthracene, stilbene, and terphenyl, were used extensively as electron detectors. More important nowadays are liquid solutions containing scintillators as solutes and plastics with scintillating substances incorporated in them. Such liquid and plastic scintillators can be made in any desired shape and in very large volumes for special applications.

In liquid scintillators the solvent is the main stopping medium for the radiation and must be chosen to give efficient energy transfer to the scintillating solute and to have little light absorption. Toluene and *p*-xylene are suitable solvents. When aqueous solutions are to be added to the liquid scintillator (e.g., for measuring low-energy β emitters such as ^{14}C or ^{3}H in aqueous media), dioxane-toluene or dioxane-naphthalene mixtures are often used as solvents. Among the most efficient scintillating solutes are *p*-terphenyl, 2,5-diphenyloxazol (PPO), and tetraphenylbutadiene. However, the emission spectra of some of these scintillators are at wavelengths too short for efficient absorption at the photocathodes. Therefore substances that absorb at the emission wavelengths of the primary scintillators and emit light at longer wavelengths are added in small concentrations. Among these **wavelength shifters** are 1,4-bis-[2-(5-phenyl-oxazolyl)]-benzene (called POPOP) and dimethyl-POPOP.

Liquid scintillators are used for the efficient, routine measurement of β emitters, particularly those of low energy. They are especially well suited for the measurement of large samples with high sensitivity. The sample must be dissolved or at least uniformly dispersed in the scintillation liquid, and this may require the addition of complexing agents for inorganic ions. A variety of liquid-scintillation systems are commercially available.

Plastic scintillators are also commercially available. They are produced by mixing scintillator (e.g., PPO or *p*-terphenyl), wavelength shifter (e.g., POPOP), and a monomer such as styrene, and then polymerizing the mixture. Large plastic scintillators are often used as anticoincidence counters surrounding low-level detectors.

NaI(Tl) Scintillation Counters. Among inorganic scintillators NaI activated with 0.1–0.2 percent thallium is by far the most widely used. The high density (3.7 g cm^{-3}) of NaI and the high Z of iodine make this a very efficient γ-ray detector. Crystals of many shapes and sizes up to thousands of cubic centimeters are commercially available and can be bought hermetically sealed,[4] provided with appropriate light reflectors, and optically coupled to photomultipliers. A popular size for routine γ-ray measurements is a 7.5-cm diameter, 7.5-cm high cylinder. Another useful type has a re-entrant well in the center to allow measurement of liquid (or solid) samples in nearly 4π geometry.

Approximately 30 eV of energy deposition in a NaI(Tl) crystal is required to produce one light photon, and it takes on the average about 10

[4] NaI crystals are very hygroscopic.

photons to release one photoelectron at the photocathode of the multiplier. These photoelectrons are then accelerated by a potential of the order of 100 V to the first dynode where each one produces n secondary electrons; these secondary electrons are similarly accelerated to and multiplied n-fold at the second dynode, and so on. With 10 dynodes and with n typically about 3 or 4, the total multiplication factor is n^{10} or of the order of 10^5 or 10^6. Thus a 0.3-MeV γ ray absorbed in a NaI(Tl) crystal might produce 10^4 light photons, giving 10^3 photoelectrons and leading eventually to an output pulse of about 10^8 electrons or 1.6×10^{-11} coulomb (C). In an output circuit of 10^{-10}-F capacity this would be a pulse of about 0.16 V requiring further modest amplification in a pulse amplifier. With careful regulation of dynode voltages there is good proportionality between the energy absorbed in the scintillator and the size of the output pulse.

Pulse height spectra taken with NaI(Tl) detectors have the same basic characteristics that were discussed in the context of semiconductor detectors: photopeaks, Compton distributions, annihilation radiation escape peaks. The resolution is much poorer for scintillation counters (see figure 7-10). An additional feature in NaI(Tl) spectra is so-called **iodine escape peak** about 28 keV below the photopeak; it results from absorption of a γ ray near the surface of the detector and subsequent escape of a K-X ray of iodine. The effect, which is illustrated in figure 7-12, becomes less pronounced with increasing γ-ray energy because fewer of the initial interactions take place near the surface. It is also generally unimportant for germanium detectors.

Background rates in NaI(Tl) counters are high—of the order of 50 cpm cm^{-3} of scintillator in an unshielded room. Massive shielding can be effective in reducing the background effects due to cosmic rays and to γ rays from surrounding material, and further background reduction can be achieved with anticoincidence arrangements. Pulse height selection helps, of course, to obtain improved ratios of sample to background rates. In certain low-level activity measurements it is worthwhile to use phototubes with quartz rather than glass envelopes in order to avoid the background contribution of the ^{40}K γ rays originating in the glass.

Čerenkov Counters (L2). These devices, which are now among the major tools of the high-energy physicist, are based on the then rather surprising discovery (C2) by P. A. Čerenkov (1934) that a beam of γ rays in water was accompanied by the emission of light at a definite angle to the beam direction (~40°). The effect was subsequently explained by I. M. Frank and I. E. Tamm in terms of the electromagnetic shock wave produced when a charged particle travels through a transparent medium at a speed exceeding the velocity of light in the same medium. Thus if n is the refractive index of the medium and βc the velocity of the particle, the condition for emission of Čerenkov light is

$$n\beta > 1. \qquad (7\text{-}5)$$

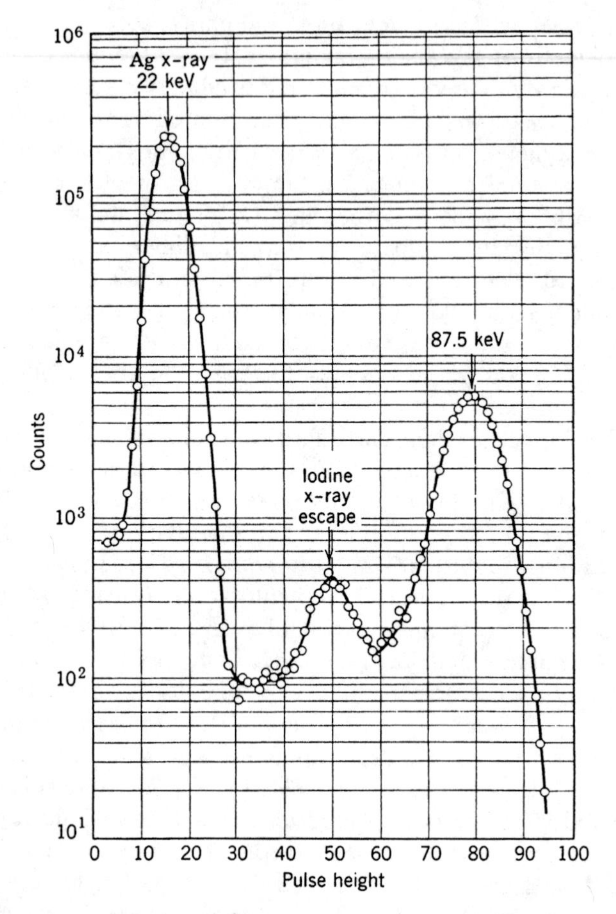

Fig. 7-12 Spectrum of the 87.5-keV γ rays and 22-keV X rays from ^{109}Cd decay, observed with a 7.5 cm × 7.5 cm NaI(Tl) detector. In addition to the two photopeaks, the peak due to escape of iodine K-X rays is prominent. (Reproduced from N. F. Johnson, E. Eichler, and G. D. O'Kelley, *Nuclear Chemistry*, Interscience, New York, 1963.)

The angle θ between the particle trajectory and the direction of light emission is given by the relation

$$\cos\theta = \frac{1}{n\beta}. \tag{7-6}$$

Although the theory of Čerenkov radiation, including intensity and frequency relationships, was worked out in 1937, a practical Čerenkov counter was not built until 10 years later when the development of photomultiplier tubes made efficient light collection feasible. The intensity of Čerenkov light is weak compared with the light output of a scintillator, but the directional properties can be used to advantage to improve light collection, and many types of highly efficient Čerenkov counters for relativistic particles have been constructed.

The most important applications of Čerenkov counters are based on their velocity-selecting properties. From condition (7-5) it is evident that a Čerenkov counter will always act as a threshold detector, recording only those particles with $\beta > 1/n$. For example, with water ($n = 1.332$) the threshold is at $\beta = 0.751$, which corresponds to 500-MeV protons, 73-MeV π mesons, or 265-keV electrons. Liquid nitrogen ($n = 1.205$, $\beta_{min} = 0.830$) can be used (by direct coupling of a Dewar flask to a photomultiplier) to detect protons above 760 MeV or pions above 112 MeV, and so on. For still lower refractive indices (down to about 1.01) various compressed gases have been used.

The angular definition of Čerenkov light (7-6) is used in many ingenious ways to achieve selectivity in detecting particles in certain velocity intervals. An upper limit to β can be set by taking advantage of internal reflections at the exit face of the Čerenkov medium for all light arriving at less than a critical angle; blackening of other faces aids in the absorption of the internally reflected photons. In other types of arrangements focusing of light emitted in a certain angular interval is accomplished by systems of mirrors and lenses.

Čerenkov counters, because of their velocity selectivity, are particularly useful for experiments with particle beams at high-energy accelerators. They may serve as elements in counter telescopes in which, together with momentum selection by magnetic deflection, they can be used to obtain good energy resolution for specific particles. They are also ideally suited as triggering devices for cloud and spark chambers that are to be sensitive for incident particles of specified properties only.

D. TRACK DETECTORS

Photographic Film (B2). The first method for the detection of radioactivity was the general blackening or fogging of photographic negatives, apparent on chemical development (see chapter 1). This method was soon supplanted by ionization measurements but has reappeared more recently in the "film badge" for personnel exposure control (see section G) and in the γ raying (analogous to X raying) of castings and other heavy metal parts of hidden flaws. Also, in the **radioautograph technique** the distribution of a radioactive tracer (preferably an α or soft-β emitter) is revealed when a thin section, perhaps of biological material, is kept in contact with a photographic plate.

Special photographic emulsions known as **nuclear emulsions**, when exposed to ionizing radiations, such as α rays, protons, mesons, and electrons, and then developed, show blackened grains along the path of each particle. These tracks, some of which may be quite short, are observed and their characteristics (length, point of origin, direction, ionization density, and scattering) are measured under a microscope. The number of developed grains per unit track length is called the grain density and is proportional to specific ionization, but is orders of magnitude smaller than the number of ion pairs produced in the same distance.

Nuclear emulsions differ from ordinary light-sensitive plates in that they have considerably higher silver halide content and smaller grain sizes (0.1–0.6 μm). The smaller the grain size, the less sensitive the emulsion to anything but the most densely ionizing particles. Thus different commercially available emulsions, differing chiefly in grain size, can be used to discriminate between different particles. Further differentiation can be obtained by variations in development. The least sensitive emulsions will show only fission fragment tracks, whereas the most sensitive ones reveal the tracks of singly charged minimum-ionizing particles.

Nuclear-emulsion techniques have been particularly useful for the recording of very rare events, for example those of interest in cosmic-ray studies. A number of elementary particles were discovered by means of emulsions, among them the positron and the π meson.

Cloud Chamber (F1). A pictorial representation of the paths of ionizing particles similar to the photographic track but capable of finer detail is given by the cloud chamber (Wilson chamber). In this instrument, invented in 1911 by C. T. R. Wilson, the particle track through a gas is made visible by the condensation of liquid droplets on the ions produced. To accomplish this an enclosed gas saturated with vapor (water, alcohol, and the like) is suddenly cooled by adiabatic expansion to produce supersaturation. If the gas is sufficiently free of dust and other potential condensation centers, condensation will occur only along the ion tracks. The piston or diaphragm causing the expansion is operated in a cyclic way, and a small electrostatic gradient is provided to sweep out ions between expansions. There is usually an arrangement of lights, camera, and mirrors to make stereoscopic photographs of the fresh tracks at each expansion. The supersaturated vapor for cloud chamber operation can be achieved in other ways, notably by the diffusion of a saturated organic vapor into a colder region. In the **diffusion cloud chamber** the working volume is continuously sensitive rather than intermittently so, and the whole instrument is considerably less complicated than the conventional Wilson chamber. Expansion chambers were therefore largely replaced by diffusion chambers in the early 1950s until the latter were themselves made obsolete by the advent of the bubble chamber.

Bubble Chamber (B3). Some advantages of photographic emulsions and of cloud chambers are combined in the bubble chamber. This device, invented in 1952 by D. A. Glaser, makes use of the well-known fact that liquids can be heated for finite though short times above their boiling points without actually boiling. In the bubble chamber charged particles traveling through a superheated liquid cause vapor bubbles to form along their tracks, probably because of local heating; these bubble tracks can be photographed in strong illumination. Since superheated liquids are not stable for long periods of time, bubble chambers are always operated in pulsed fashion, with the liquid normally under such a pressure that at the operating temperature it is below its boiling point; it is made sensitive by sudden reduction of the pressure and remains sensitive for about 10^{-3}–10^{-2} s. The best operating temperature appears to be roughly two thirds of the way from the normal boiling point to the critical temperature, and the corresponding pressure is in the neighborhood of half the critical pressure.

Bubble chambers are among the important research tools at high-energy accelarators. The greater densities of liquids compared to gases make bubble chambers superior to cloud chambers for the study of energetic particles. Even so, the dimensions of bubble chambers used at the biggest accelerators are gigantic, some chambers having volumes of many thousands of liters. Various liquids have been used successfully as bubble chamber fillings, but hydrogen-filled chambers are probably the most important (despite the great safety and cryogenics problems they pose), because they give unambiguous identification of interactions with protons. Both cloud and bubble chambers are commonly operated in magnetic fields, which makes possible the determination of particle momenta. The complex task of analyzing bubble chamber photographs has been greatly aided by sophisticated computer programs. The achievement of cycling rates as high as $60\,s^{-1}$ has been an important development and the usefulness of bubble chambers has been enormously enhanced by incorporating them in so-called hybrid systems (B4) with other types of detectors; such detectors, e.g. wire chambers or Čerenkov counters, are placed upstream and downstream for identification and momentum determination of incoming and outgoing particles.

Spark Chamber (C3). One of the chief limitations of the bubble chamber lies in the limited number of tracks that can be tolerated in the chamber during one expansion cycle. This makes detection and study of extremely rare events in the presence of many unwanted interactions excessively time-consuming. The spark chamber, developed since about 1957 but based on previously known ideas, is suitable for this type of application. It consists in its original form of a series of parallel plates separated by gaps that may be from a few millimeters to many centimeters wide and that are filled with a noble gas at or near atmospheric pressure. If a positive high voltage is applied to alternate plates, the intermediate plates being left at ground potential, an ionizing particle crossing a gap between adjacent plates will cause a spark to develop very nearly along its trajectory. The high voltage is applied in short pulses only when actuated by some triggering counters (often scintillation or Čerenkov counters) surrounding the chamber. Thus the chamber is sensitive only for preselected types of events, even though the total flux of ionizing particles through it may be very large. This represents an enormous advantage over bubble chambers. After each event the ions are swept out by a sweeping field. Detection efficiencies, even for minimum ionizing particles, are very high, and excellent spacial resolution is obtainable.

In the original spark-chamber designs the particle trajectories were photographed. Direct conversion of the optical image into electrical signals by means of a television camera (the "Vidicon" system) has also been widely used. Further advances came with the replacement of plates by wires and particularly the introduction of crossed-wire construction, which makes possible the direct readout of the coordinates of each spark into an electronic computer. Wire spark chambers with dimensions of many meters, containing of the order of 10^5 wires, and capable of recording many simultaneous tracks have been constructed.

Other devices such as multiwire proportional chambers (see p. 250) are often used in conjunction with spark chambers as triggering elements.

Dielectric Track Detectors (F2, G2). Whereas cloud, bubble, and spark chambers are of interest primarily to the particle physicist, another technique for visualizing particle tracks that has come to the fore in recent years is widely used in nuclear physics and chemistry as well as in many applied fields such as the space sciences and geochemistry. It is based on the observation, first made by E. C. H. Silk and R. S. Barnes in 1959, that heavily ionizing radiations produce tracks of radiation damage in insulating or semiconducting solids. Originally these tracks could only be observed under an electron microscope, but a crucial advance came when R. L. Fleischer, P. B. Price, and R. M. Walker found that the tracks could be developed by suitable chemical treatment—they used HF to develop tracks in mica—to the point where they were visible in an ordinary microscope. Etching proceeds much faster along the damaged track than on the undamaged material. Track formation has since been established and development methods worked out for many materials including minerals such as olivine, zircon, quartz; glasses; various plastics such as Mylar and Lexan; and any number of synthetic inorganic crystals. For each material there is a critical value of specific ionization, $(dE/dx)_c$, below which tracks are not registered. For example, $(dE/dx)_c$ is $\sim 13\ \mathrm{MeV\ mg^{-1}\ cm^2}$ for mica and $\sim 4\ \mathrm{MeV\ mg^{-1}\ cm^2}$ for Lexan, two of the most widely used detectors. As a result mica does not register ions with $A \lesssim 30$, Lexan is insensitive to ions with $A \lesssim 12$. It is thus possible, for example, to use mica foils to detect fission fragments in the presence of much larger fluxes of lighter ions such as protons, α particles, and so on. Mica detectors have therefore been used for the determination of many fission cross sections. Also, since radiation damage tracks are very stable, many minerals both on earth and on the moon contain a "fossil record" of exposure to heavily ionizing radiations, either from fission or from cosmic rays, throughout their history.

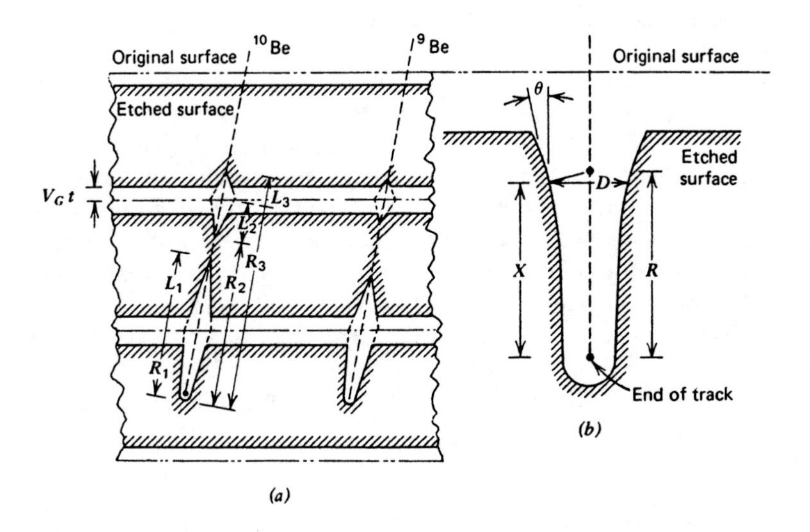

Fig. 7-13 Particle identification with dielectric track detectors by means of etching rate measurements. Ionization rates at various residual ranges R can be determined by measurements of either (a) the etched cone length L, or (b) the taper angle θ or diameter D. Reproduced, with permission, from P. B. Price and R. L. Fleischer, *Ann. Rev. Nucl. Sci.* **21**, 295 (1971) © 1971 by Annual Reviews Inc.]

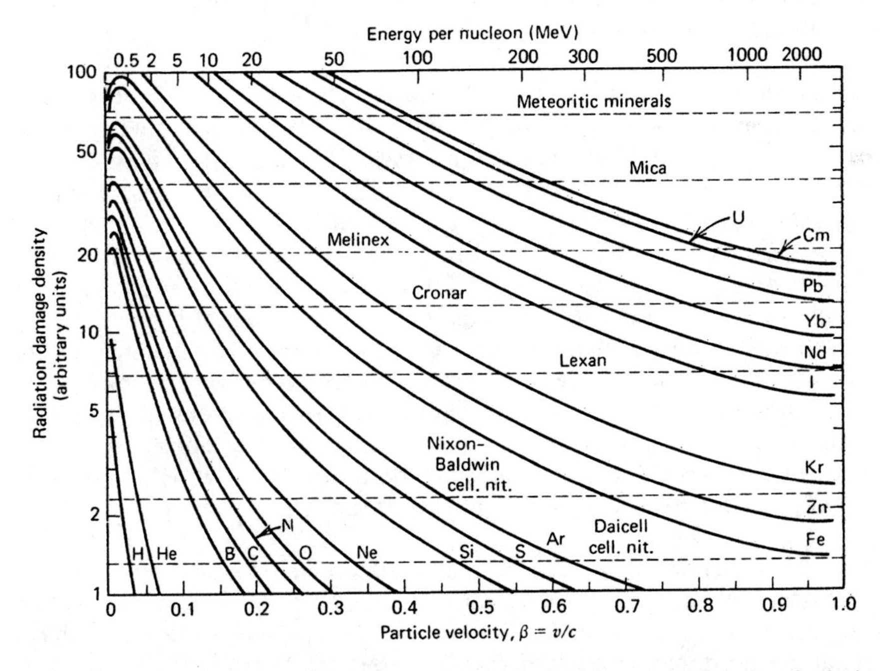

Fig. 7-14 Density of radiation damage (proportional to ionization rate) as a function of velocity (and energy per nucleon) for various bombarding nuclei. Approximate thresholds for track recording by several solids are indicated by dashed lines. [Reproduced, with permission, from P. B. Price and R. L. Fleischer, *Ann. Rev. Nucl. Sci.* **21**, 295 (1971) © 1971 by Annual Reviews Inc.]

The process responsible for the formation of an etchable track is thought to be the mutual repulsion of the ions produced, which propels them into interstitial positions, at the same time creating a large number of vacancies in the lattice. The resulting damage region is known as an ion explosion spike. It turns out that the density of ions and excited atoms along the track is directly related to the specific ionization of the radiation, and that the rate at which the damage track can be etched, in turn depends on the density of ionization. Thus if the etching rate can be determined at two (or more) points along a particle track, for example by using sandwiches of detectors (figure 7-13), particles can be identified by Z and, for light elements, even by Z and A. Figure 7-14 shows curves of radiation damage density as a function of particle energy for various ions in several materials.

E. NEUTRON DETECTORS

As mentioned in the introduction to this chapter, methods for neutron detection and measurement are based largely on the detection of secondary, ionizing radiations. The instruments used are variants of the ones already discussed, and in this section we therefore confine ourselves to a brief survey of the methods and principles used (S1).

Activation Methods. Activation by (n, γ) reaction and subsequent measurement of the induced radioactivity is a widely used technique for the determination of thermal-neutron fluxes. The relevant activation cross section must, of course, be known, and an absolute flux determination requires the absolute measurement of the induced activity (cf. chapter 8, section F). Gold, indium, and cobalt are useful for such activation measurements and are used in the form of foils or wires. Activation by epithermal (resonance) neutrons must be corrected for and is measured by activating a second, identical flux monitor wrapped in gadolinium or cadmium, both of which are excellent absorbers for thermal neutrons.

To measure fast-neutron fluxes activation by reactions with energy thresholds, such as (n, p) or (n, α) reactions, is sometimes used. By employing several detectors with different thresholds it is possible to obtain approximate information on neutron spectra.

Ionization Chambers. The charged particles emitted in neutron-induced reactions such as (n, p), (n, α), or (n, f) can be detected in ionization chambers operated either in the dc or the pulsed mode. The most frequently used reactions for thermal neutrons are $^{10}B(n, \alpha)^{7}Li$, $^{3}He(n, p)^{3}H$, and $^{235}U(n, f)$. The chamber may be lined with a thin layer of ^{10}B- or ^{235}U-containing material or filled with an appropriate gas such as $^{10}BF_3$ or ^{3}He. For fast-neutron detection a hydrogen-containing filling gas may be used, and the recoil protons produced in n-p collisions measured. Ion chamber detectors are useful for integral flux measurements at high neutron fluxes and over large ranges of intensity and can be operated in the presence of high γ backgrounds (important in reactor applications). The γ sensitivity can be further reduced by shielding with γ-absorbing materials of low neutron cross section such as lead or bismuth.

Gridded ion chambers filled with ^{3}He and used in a pulsed mode have found important application as neutron spectrometers in the energy range from ~10 keV to ~2 MeV, a range of particular interest for the measurement of neutron spectra of delayed-neutron emitters (F3). For this purpose the chambers are filled to high pressure (~10 atm) with a mixture of ^{3}He, Ar, and CH_4, surrounded with cadmium and a boron compound to shield them from thermal and epithermal neutrons. They can be calibrated with monoenergetic neutrons of various energies, for example from the ^{7}Li(p, n)^{7}Be reaction.

Proportional Counters. Most of the types of counters discussed in sections A–C are applicable to neutron detection and measurement. The classes of nuclear reactions used are those already mentioned in the preceding paragraph.

Proportional counters filled with $^{10}BF_3$ or ^{3}He are used for integral measurement of thermal and epithermal neutrons. Fluxes and spectra of intermediate-energy and fast neutrons can be measured with ^{3}He- or H_2-filled proportional counters via the ^{3}He(n, p) reaction or proton recoils, respectively. Instead of H_2 in the gas phase, the source of recoil protons can be solid hydrogenous material, such as polyethylene, introduced into the counter in the form of a liner, coated with a conducting layer.

Scintillation Counters. The efficiency for neutron detection is generally higher in scintillators than in gas-filled counters. On the other hand, scintillation counters have the disadvantage of poor discrimination against γ rays. For detection of slow neutrons (or of fast neutrons after moderation by a hydrogenous medium) the ^{6}Li(n, α) and ^{10}B(n, α) reactions are used, with ^{6}Li or ^{10}B incorporated in ZnS or in liquid or glass scintillators. Crystals of ^{6}LiI (analogous to NaI scintillators) are also used. Fast-neutron spectra can be determined via proton recoil measurements in large organic scintillators, solid or liquid. For very low neutron fluxes the high efficiency attainable with gadolinium-loaded scintillators is advantageous; what is detected here is the total γ-ray energy (~8 MeV) from the Gd(n, γ) reaction.

Semiconductor Detectors. If an appropriate "converter" material such as ^{235}U, ^{6}Li, or ^{3}He is deposited on the surface of a semiconductor detector, one obtains a neutron counter. For this purpose a lithium or uranium compound may be evaporated onto the surface, or ^{3}He may be sealed into the vacuum-tight detector housing. The small size and high efficiency (up to ~30 percent for thermal neutrons with ^{6}Li converter) of such detectors makes them useful for applications requiring good spatial resolution. They cannot, however, be used in high neutron fluxes, because they deteriorate (develop large leakage currents) after exposure to neutron fluences of 10^{14}–10^{16} cm^{-2}.

Track Detectors. Boron- or lithium-loaded photographic emulsions are used for the measurement of small fluxes of slow neutrons, particularly for health physics purposes (see section G). The ranges of recoil protons originating in the hydrogen atoms of emulsions give information on fast-neutron spectra.

High sensitivity for neutron detection can be obtained with dielectric track detectors coated with a fissile material such as ^{235}U. Recoil proton tracks due to neutron interactions in plastics such as polycarbonates or cellulose acetate have been employed for fast-neutron detection.

Other Devices. A variety of chemical and physical effects of radiation have been successfully applied to neutron dosimetry. Among chemical effects used are the radiolysis of water in ^{10}B-containing solutions and hydrogen evolution from organic compounds such as cyclohexane. Thermoluminescence—the formation of metastable states in a crystal that, on later heating, are de-excited with the emission of light—can be used for neutron dosimetry. LiF crystals and ^{10}B-coated CaF_2 crystals are good thermoluminescence detectors. Another solid-state phenomenon used for neutron dosimetry is the formation of color centers, for example in boron-containing glasses.

F. AUXILIARY INSTRUMENTATION

The most widely used detectors of nuclear radiations discussed in the preceding sections—GM, proportional, scintillation, and semiconductor counters as well as pulsed ion chambers—put out voltage or charge pulses of various sizes and durations. In general, these pulses need to be amplified, sorted according to pulse height (and possibly shape), recorded, and manipulated in various ways. The electronic instrumentation for these various tasks has developed and continues to develop very rapidly. Only the sketchiest idea of some of the principles and some of the types of instruments used can be given here. For more thorough coverage the reader is referred to books such as C4 and H1 and to the voluminous literature in journals such as *Nuclear Instruments and Methods* and *Review of Scientific Instruments.*

Amplifiers and Scalers. The pulses originating from pulsed ion chambers and counters (other than GM counters) require amplification by factors of 10^2–10^5. A preamplifier is usually placed as close as possible to the detector and its output is sent through a cable to a linear amplifier. In ordinary counting practice the amplified pulses then go into a scaling circuit that reduces the pulse rate electronically, usually by some power of ten. Finally the scaled pulses drive a mechanical register, printer, paper tape punch, or other recording device. Automatic turnoff of the counter after a preset time or preset number of counts is often provided. Automatic sample-changing devices are also available.

Pulse Height Analysis. Whenever the output pulse from a counter or ionization chamber is proportional to the energy dissipation in the detector, the measurement of pulse heights is a useful tool for energy determinations. Some pulse height selection is used even in the simplest scalers in the form of a **discriminator**, which allows only pulses above a certain minimum size to be recorded. In a **single-channel analyzer** there are two discriminators, and usually an anticoincidence arrangement is used to pass only pulses of such a height that they fall between the two discriminator settings. The two discriminators are moved up and down the voltage scale together, with a constant "channel" or "window" width between them. The pulse height range might typically be 0–10 V, and different parts of a pulse height spectrum can be brought into this range by the choice of suitable amplifier gains.

Pulse height analysis is made much more versatile and rapid if, instead of a single-channel analyzer, a **multichannel analyzer** is used in which the pulses are sorted according to size and simultaneously recorded in many consecutive channels. A variety of such instruments has been developed, the most popular ones having between 200 and 4000 channels. Data are usually stored in magnetic-core memories like those used in high-speed computers, and provisions must, of course, be made for obtaining in useful form the information stored in the memory. This may include display of the spectrum on an oscilloscope, conversion from binary to decimal numbers, and the facility to drive an automatic printer, plotter, or magnetic tape unit so that the content of each channel can be appropriately recorded. In some commercial units it is also possible to add the counts in any selected region of the spectrum (e.g., in a peak), subtract background, and perform simple calculations with the data. An alternative to the use of "hard-wired" pulse height analyzers is to feed the output pulses from the amplifier through a digitizing circuit [**analog-to-digital converter** (ADC)] directly to a programmable computer or to store them on magnetic tape for later computer analysis.

Coincidence Techniques. Studies of the time relations between various radiations emitted from one nucleus may be made by means of coincidence techniques and are very useful in decay scheme studies.

Whether a β ray goes to the ground state of the product nucleus or is followed by γ emission can be established by a coincidence experiment in which the sample is placed between a β and a γ counter and time-coincident pulses in the two counters are recorded. Similarly, with appropriately chosen detectors, and possibly with the aid of absorbers or pulse height discrimination to make detection more selective, γ-γ, α-γ, X-γ, β-e^-, e^--γ, and other types of coincidences may be studied. Coincidence measurements with pulse height analysis at one or both detectors offer a particularly powerful tool for detailed decay scheme studies. A number of the available multichannel analyzers are of the so-called two-dimensional or two-parameter variety; that is, they can be used to record simultaneously the coincidences between each of n pulse height groups from one detector with each of m groups from the other. Thus a 4096-channel machine might be used to give arrays of 64×64 or of 32×128. The display of the output from these two-parameter analyzers can take various forms, and the analysis of the data obtained with these devices often requires computer methods. It is, in fact, often preferable to store all the data from each event on magnetic tape and perform the data analysis subsequently with an appropriate computer program. However, even if the ultimate data analysis is done in this manner, it is of great value to have an instantaneous display of sample results by means of a pulse height analyzer or minicomputer while the experiment is in progress.

In most coincidence measurements rather strong samples are used. This is because the number of coincidence counts recorded is proportional to the product of the solid angles subtended by the two counters at the sample, and frequently the sample-to-counter distances have to be rather large (at least several centimeters) to minimize scattered radiation from one counter entering the other. Since the coincidence rates are often quite low, background rates are a problem. Apart from a very small true coincidence background (e.g., due to a cosmic ray striking both detectors), there is always a certain chance or accidental background that comes about because sometimes two rays not originating from the same nucleus happen to arrive at the two counters within the resolving time of the coincidence circuit. If the single counting rates in the two counters are $\mathbf{R}_1$ and $\mathbf{R}_2$ per second and if the coincidence resolving time (the time within which the two counters have to be tripped for coincidence to be recorded) is τ seconds, then the accidental coincidence rate is $2\mathbf{R}_1\mathbf{R}_2\tau$ per second. To reduce the chance rate it is desirable to make the resolving time as short as possible. Coincidence resolving times of 10^{-6}–10^{-9} s are common, and for delayed-coincidence measurements of very short half lives resolving times of less than 10^{-10} s have been achieved. To be used with coincidence circuits the detectors must have pulse-rise times not much longer than the coincidence resolving time, and for this reason GM tubes are not very useful for coincidence work. Scintillation counters and semiconductor detectors are most commonly used.

Multiparameter Experiments. In many experiments, particularly those performed on-line on accelerators, it is desirable to measure a number of different parameters for each event observed. We may, for example, wish to record, for each of two particles emitted in a reaction, the energy loss in a thin dE/dx detector, the total energy deposited in a thick detector, and the time of flight between two detectors. Furthermore, in such an experiment we are usually interested in the angular correlation between the two emitted particles. Therefore in order to conserve accelerator time we would place several detector telescopes at different angles to the beam and record the coincidences between different telescope pairs. This is only one example of the many types of complex nuclear experiments that have become possible in recent years. The data from such multiparameter experiments are usually stored, event by event, in a temporary "buffer" device, transferred in batches to magnetic tape or disc storage, and later subjected to various kinds of computer analysis off-line. However, as mentioned earlier, it is always desirable to have some on-line monitoring capability as well.

The variety of multidetector arrangements and large-area arrays that have been developed in recent years precludes any detailed discussion here. Multiwire proportional counters, position-sensitive scintillation counters, spark chambers, and channel plate arrays are among the devices used. Some further mention of multidetector techniques is made in chapter 8, p. 323.

Spectrographs and Spectrometers (S2). We have discussed various types of counters that, in conjunction with pulse height analyzers, serve as energy-measuring instruments or spectrometers for nuclear radiations. In addition there are a variety of other devices for energy measurements on α, β, γ, and X rays and conversion electrons. We can make only the briefest mention of these.

Most of the methods for charged-particle spectrometry, other than the pulse height methods, involve deflection in magnetic or electric fields. The simplest type of magnetic spectrograph or spectrometer makes use of the fact that identical charged particles emerging from a point source with equal momenta but at slightly divergent angles (say within 20°) are brought to an approximate focus after traveling about 180° in a plane perpendicular to a uniform magnetic field (figure 7-15). If a constant magnetic field is used, electrons or α particles of different momenta are detected in different positions, either on a photographic film or by a movable counter. Alternatively the field may be varied to bring particles of different energies into focus at the detector. Efficiencies are very low because only particles emitted in the plane perpendicular to the magnetic field are focused. Considerable improvement in efficiency is obtained in the double-focusing spectrograph, in which the pole faces are shaped so as to achieve some focusing for out-of-plane trajectories. The focus in these devices is at $\pi\sqrt{2}$ rather than at 180° to the source. Instruments of the types mentioned are particularly suited for line spectra such as conversion-electron or α spectra.

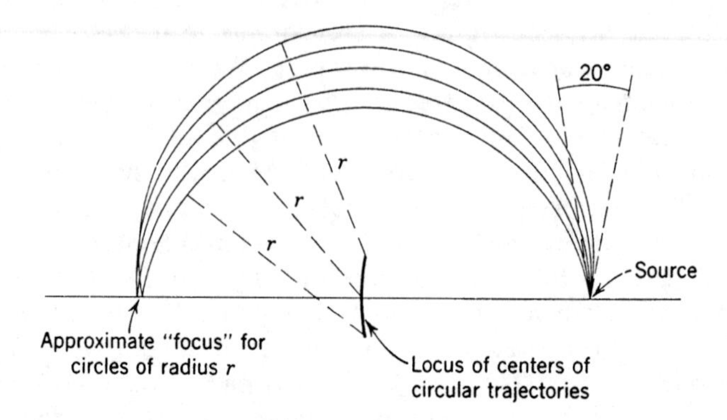

Fig. 7-15 Principle of 180° focusing. The magnetic field is perpendicular to the plane of the paper. The angular divergence of the trajectories shown is 20° at the source.

Another type of β-ray spectrometer is the lens spectrometer. Here the source and the detector are located on the axis of an axially symmetric magnetic field. It is a property of such a system that all electrons emitted with a large spread of angles but with a given momentum will, after traveling along spiral paths, be focused at some other point on the axis. Very high geometric efficiencies (several percent of 4π) and good momentum resolution (~1 percent) can be attained, although in the latter respect the 180° and double-focusing instruments are superior.

In all electron spectrometers source preparation is a problem because sources must be extremely thin (often $<0.1 \text{ mg cm}^{-2}$) and mounted on equally thin backings to minimize self-absorption and back-scattering. Also the counters used must have very thin windows.

For complex decay scheme studies it may be desirable to operate two electron spectrometers in coincidence, for example to study β-e^- or e^--e^- coincidences. More commonly a scintillation or semiconductor detector may be used in coincidence with an electron spectrometer.

An instrument for the precision measurement of X rays and low-energy γ rays is the curved-crystal spectrograph (D1). This is analogous to an optical spectrograph of the grating type, with the atomic planes of a bent crystal replacing the ruled lines of the curved grating. The detector may be a photographic plate, or a counter in the curved-crystal spectrometer.

G. HEALTH PHYSICS INSTRUMENTATION

By the term health physics instruments we refer to detection and measuring instruments designed for the monitoring of personnel radiation exposures and for the surveying of laboratories, equipment, clothing, hands, and the like for biologically harmful radioactive contaminations. Many types are in use (C5). Those that are generally available commercially may be divided into a few categories, and are all derived from the instrumentation principles already discussed in this chapter.

Film Badges. Personnel potentially exposed to nuclear radiation or handling any appreciable amounts of radioactive material should routinely wear badge-type holders containing photographic film, which gives a permanent record of general body exposure to radiation integrated over a period of time—usually one week. Ordinary dental X-ray film is generally used for β and γ dosimetry. To obtain information about the type and energy of radiation, various filters (plastic, aluminum, and cadmium foils) are placed over certain areas of the film. Development, calibration, and dosage evaluation for the badges can be obtained on a subscription basis. Film holders can be worn on the wrist or as finger rings to monitor hand exposure. Thermal-neutron exposures are monitored with boron-loaded films. Fast neutrons and radiations of very high energies (from synchrocyclotrons, proton synchrotrons, etc.) may be monitored with nuclear emulsions; track counting must then be resorted to.

Pocket Ion Chambers. Another widely used radiation monitor is the pocket ionization chamber. This is an ordinary ionization chamber in most respects, made small enough to be worn clipped in the pocket like a fountain pen. The charging potential is applied through a temporary connection, and the residual charge is read on a built-in electrometer and scale at any time without auxiliary apparatus and without effect on the indication.

These pocket meters are calibrated in roentgen units, with full scale corresponding most commonly to 0.1 to 0.2 R, so that they easily detect general radiation dosage below tolerance levels. They may not give a measure of local exposure (say, of the hands, while other parts of the body are shielded by a lead screen) and, of course, are not sensitive to soft radiations that do not penetrate the chamber wall. Some pocket ion chambers produce audible signals whose frequency is proportional to dose rate.

Portable Counters and Survey Meters. More sensitive detection instruments are used to determine the rate at which exposure is being received in a given radiation field. These may be larger ionization chambers with compact dc amplifiers operated from self-contained batteries, so that a readily portable survey meter weighing not much over 1 kg is achieved. Models of this type ordinarily have several calibrated scale ranges, from about 0–20 to about 0–3000 mR h^{-1} or have a logarithmic response that compresses a wide range into a single scale. Battery-operated portable Geiger counter sets of about the same size and weight can be used for the same purpose. They are usually arranged as counting-rate meters, with full-scale readings calibrated at about 0.2–20 mR h^{-1}. Although the counter-type of meter is much more sensitive than the ionization chamber, the chamber is usually sensitive enough and may be expected to give a response more nearly proportional to the biological effects of the radiation. Both types are ordinarily provided with a movable shield to permit a distinction between hard and soft radiations. These two types of instruments are also useful for surveying the laboratory and its apparatus for radioactive contamination. The GM counter instrument with its higher sensitivity, especially when arranged to give an audible signal, is more convenient in rapid surveys for small amounts of activity, but in its usual form it is not useful for very soft β rays such as those from ^{14}C. The ionization chamber instrument is more easily fitted with a window thin enough for this purpose (not more than a few milligrams per square centimeter), and some available models have very thin windows (or simply open screens) that will pass even α particles. The most sensitive γ-ray monitor is the portable scintillation counter.

Thermoluminescence Dosimeters. When certain crystals are exposed to radiation the electrons and holes produced are trapped on impurities, so that energy is stored in the crystal and can subsequently be released in the form of light by heating the crystal. This thermoluminescence property can be used for radiation dosimetry. The most widely used crystals are LiF and CaF_2 containing Mn impurity. The former is advantageous for health physics purposes because its response is nearly energy-independent from 30 keV to 3 MeV and its effective Z is almost the same as that of soft tissue. However, it is useful only for doses of ≥ 10 mrad, whereas CaF_2(Mn) is sensitive down to about 0.1 mrad, but becomes very energy-dependent in its response below 400 keV. Small capsules containing of the order of 50–100 mg of phosphor are used and their exposure is determined by measuring the light output with a photomultiplier while they are heated electrically.

Other Procedures. A number of other more specialized instruments have been devised. Geiger counters and atmospheric-pressure proportional counters may be arranged particularly to detect β and α contaminations on the hands. The monitoring of air-borne contamination requires special instruments and may be particularly important in laboratories handling long-lived α activities. One method for air-borne dusts is to filter a large volume of the air and assay the activity left on the filter paper with a standard detector. (The radon decay products ordinarily present in air can be detected in this way.) A very simple and widely applicable semiquantitative method of contamination monitoring that requires no special instrumentation is the so-called swipe method. A small piece of clean filter paper of a standard size is wiped over a roughly uniform path length on the suspected desk top, floor, wall, laboratory ware, or almost anywhere, and then measured for α, β, or γ activity on a standard instrument. Even air-borne contamination may be checked in a rough way by swipe samples of accumulated dust from an electric-light fixture or some such place exposed only to contamination from the air.

H. CALIBRATION OF INSTRUMENTS

General Counting-Room Practices. A typical counting room may contain a variety of instruments. It is convenient to have a standard arrangement for holding standard-size samples at various reproducible distances from the detectors. This is usually a carefully machined lucite stand with slots for sample cards as shown in figure 7-16; identical stands should be used for the various instruments.

To ensure constancy of response with time all the measuring instruments should be checked routinely—preferably daily—with standard samples. Ideally a standard should have radiations similar to those of the activity to be measured. With multipurpose counters this is not practical, and, in any case, other criteria for the choice of a standard source, such as long half life and rugged physical form, may be of overriding importance. A very useful standard sample for routine checks of β counters can be prepared from ^{36}Cl ($t_{1/2} = 3 \times 10^5$ y, $E_{max} = 0.7$ MeV), fused into a metal backing and

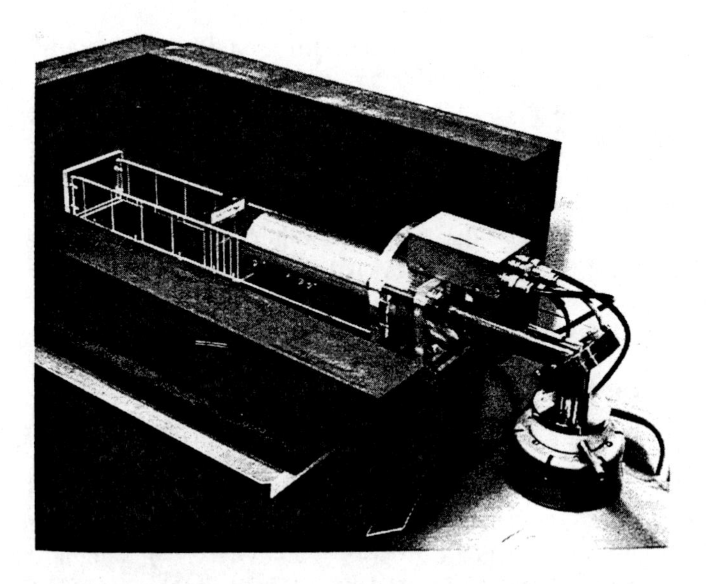

Fig. 7-16 A Ge(Li) detector assembly. The top of the cryostat is seen on the right, the horizontal, cylindrical detector housing in the center. Lucite shelves permit samples to be measured at various distances from the detector. An aluminum sample card is seen in one of the slots. The iron block shielding has been partially removed. (Courtesy Brookhaven National Laboratory.)

covered with a thin evaporated layer of gold. Background rates should be measured at least daily. To reduce backgrounds due to cosmic rays and strong samples in the laboratory, most counters, including their stands, are enclosed in lead shields 2–5 cm thick. Voltage plateaus should be checked occasionally. Scintillation and semiconductor detectors for γ rays should have their energy resolution for a specific photopeak (e.g., that of the 0.66-MeV γ ray of ^{137}Cs) determined from time to time. An intercalibration of various instruments for activities of interest can be useful but ordinarily should be depended on only for semiquantitative results. A knowledge of at least the relative geometry factors for samples placed on various shelves is often useful.

Each instrument must have its response to samples of different activity levels determined; outside the linear-response region it should be used cautiously, with calibrated corrections. This calibration can be made in several ways: (1) with samples of different activity levels carefully prepared from aliquots of an active solution; (2) by comparison of the decay curve of a very pure short-lived activity of known half life with the exponential decay to be expected; and (3) by measurements of the separate and combined effects of samples located in reproducible assigned positions. With counters the failure of linearity at high counting rates is attributed to dead-time losses; the correction is known as a dead-time correction (see chapter 9, p. 361). Ordinarily the necessity for corrections amounting to more than a few percent should be avoided.

Energy Calibrations. Any instrument that is to be used as a spectrometer requires calibration from time to time with sources that emit radiations of known energies. Linearity of response with energy (or momentum, in the case of magnetic spectrometers) is not to be taken for granted, although highly desirable, and calibration with several sources over the entire energy range of interest is therefore advisable. For γ-ray measurements with NaI(Tl) or Ge(Li) counters it is, of course, the position of the photopeak (full-energy peak) that is relevant for energy measurements.

Lists of nuclides suitable as γ-ray or α-particle energy standards, that is, emitting γ rays or α particles of accurately known energies and having convenient half lives, may be found in L3. The γ-ray energies range from 14.413 keV (^{57}Co) to 3547.9 keV (^{56}Co), the α-particle energies from 2.470 MeV (^{146}Sm) to 8.7844 MeV (^{212}Po, a member of the ^{232}Th decay chain). The most commonly used γ-ray energy standards are listed in table 7-1. Many of the nuclides may be purchased from commercial companies, from the International Atomic Energy Agency (IAEA), Vienna, Austria, or from the National Bureau of Standards (NBS), Washington, D.C. A particularly useful mixed standard containing a range of γ emitters is also available from NBS.

As X-ray standards, one can use appropriate nuclides decaying by EC or

Table 7-1 Frequently Used γ-Ray Energy Standards[a]

Nuclide	Gamma-Ray Energy (keV)	Abundance (Number of Quanta per 100 Disintegrations)	Half Life
^{241}Am	26.345	2.4	433 y
	59.537	35.7	
^{57}Co	14.413	9.8	271 d
	122.061	85.6	
	136.474	11.1	
^{203}Hg	279.197	81.5	46.8 d
^{51}Cr	320.084	10.0	27.7 d
^{137}Cs	661.66	85.0	30.17 y
^{54}Mn	834.85	100	312 d
^{60}Co[b]	1173.24	100	5.271 y
	1332.51	100	
^{22}Na	1274.5	99.9	2.602 y
	511.0[c]	181	
^{88}Y[b]	898.05	91.3	106.6 d
	1836.06	99.3	

[a] Data from reference L3. Additional standards are listed in Appendix II of that reference.
[b] The peak corresponding to the sum of the two γ rays in cascade (2505.75 keV for ^{60}Co, 2734.11 keV for ^{88}Y) may also be used if a calibration point at a higher energy is desired.
[c] Annihilation radiation. The intensity given is based on the assumption that all positrons are annihilated in close proximity to the ^{22}Na sample.

IT, or one can produce fluorescent X rays by means of a pure β emitter (such as ^{14}C or ^{32}P) mixed with or placed adjacent to a fluorescent radiator of the desired material. Electron spectrometers are calibrated with conversion electron lines from γ transitions of well-determined energies; carefully prepared thin sources are necessary (see chapter 8, section B).

Efficiency Calibrations. Measurements of absolute (and even of relative) intensities of radiations are appreciably more difficult than energy determinations, and the matter of intensity or efficiency calibrations deserves careful attention.

What is required for γ-ray efficiency calibrations of scintillation or semiconductor detectors is a series of sources emitting γ rays of different energies and having known γ-ray emission rates. Suitable sets of such calibrated standards are available, as already mentioned above, from the IAEA and the NBS. The nuclides listed in table 7-1 are among the most

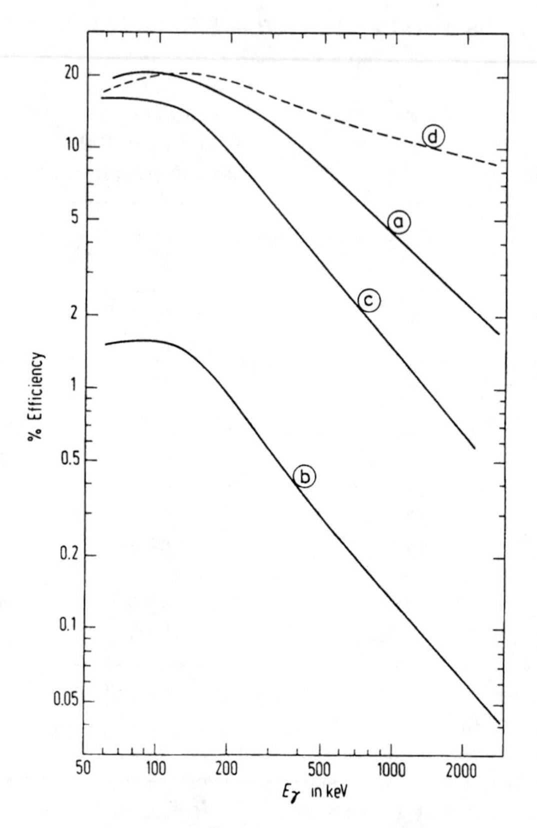

Fig. 7-17 Efficiency-versus-energy curves for various γ-ray detectors. (*a*) Photopeak efficiency of a 7.5 cm × 7.5 cm NaI(Tl) scintillator for sources at 2-cm distance; (*b*) photopeak efficiency of a coaxial 36-cm^3 Ge(Li) detector for a source-detector distance of 5 cm; (*c*) and (*d*) are photopeak and total efficiencies, respectively, for sources mounted directly adjacent to a 67-cm^3 coaxial Ge(Li) detector.

widely used for efficiency as well as energy standards. The standards must, of course, be measured in the same geometrical arrangement as the samples whose γ-emission rates are to be determined.

Typical curves of photopeak efficiency versus energy are shown in figure 7-17 for a NaI(Tl) crystal and for two different coaxial Ge(Li) detectors. The evaluation of the pulse height distributions obtained both for the standards and for any samples measured involves determination of the areas under the photopeaks. From any photopeak the background arising from Compton distributions of higher-energy γ rays must be subtracted. Various procedures may be used, such as successive "peeling" of spectral distributions of individual γ rays, starting with the γ ray of highest energy and using spectra taken with single-γ-ray sources to infer the shapes. For complex spectra (such as the one shown in figure 7-8) this becomes a very tedious procedure unless done by computer; a number of computer programs have been developed for the analysis of complex γ spectra (e.g., the program SAMPO described in R3). The summing effects mentioned on p. 260 and discussed in chapter 8, p. 330, must be taken into account or avoided. Absolute intensities of γ rays in moderately complex spectra can usually be determined to about ± 5 percent.

The efficiencies of β counters can also be calibrated with samples of known disintegration rate. However, the problem is complicated by the rather pronounced effects of scattering and absorption of β particles in the samples themselves, in counter windows, and in sample backings. These effects are discussed in chapter 8, and standardization of β-emitting samples is taken up there.

REFERENCES

*A1 F. Adams and R. Dams, *Applied Gamma-Ray Spectrometry*, 2nd ed., Pergamon, Oxford, 1970.

B1 J. B. Birks, *The Theory and Practice of Scintillation Counting*, Pergamon, New York, 1964.

B2 M. Blau, "Photographic Emulsions," in *Methods of Experimental Physics*, Vol. 5A, *Nuclear Physics* (L. C. L. Yuan and C. S. Wu, Eds.), Academic, New York, 1961, pp. 208–264.

B3 H. Bradner, "Bubble Chambers," *Ann. Rev. Nucl. Sci.* **10**, 109 (1960).

B4 J. Ballam and R. D. Watt, "Hybrid Bubble Chamber Systems," *Ann. Rev. Nucl. Sci.* **27**, 75 (1977).

*B5 K. Bächmann, *Messung Radioaktiver Nuklide*, Verlag Chemie, Weinheim, West Germany, 1970.

C1 G. Charpak et al., "The Use of Multiwire Proportional Counters to Select and Localize Charged Particles," *Nucl. Instr. Methods* **62**, 262 (1968).

C2 P. A. Čerenkov, "Visible Glow of Pure Liquids under the Influence of γ Rays," *Dokl. Akad. Nauk SSSR* **2**, 451 (1934).

C3 G. Charpak, "Evolution of the Automatic Spark Chambers," *Ann. Rev. Nucl. Sci.* **20**, 195 (1970).

C4 H. H. Chiang, *Basic Nuclear Electronics*, Wiley-Interscience, New York, 1969.

*C5 H. Cember, *Introduction to Health Physics*, Pergamon, Oxford, 1969.

D1 J. W. M. DuMond, "Gamma-Ray Spectrometry by Direct Crystal Diffraction," *Ann. Rev. Nucl. Sci.* **8**, 163 (1958).

E1 G. T. Ewan, "Semiconductor Spectrometers," in *Progress in Nuclear Techniques and Instrumentation*, Vol. III (F. J. M. Farley, Ed.), North Holland, Amsterdam, 1968.

F1 W. B. Fretter, "Nuclear Particle Detection (Cloud Chambers and Bubble Chambers)," *Ann. Rev. Nucl. Sci.* **5**, 145 (1955).

*F2 R. L. Fleischer, P. B. Price, and R. M. Walker, *Nuclear Tracks in Solids*, University of California Press, Berkeley, 1975.

F3 H. Franz et al., "Delayed Neutron Spectroscopy with ^{3}He Spectrometers," *Nucl. Instr. Methods* **144**, 253 (1977).

F4 W. Franzen and L. W. Cochran, "Pulse Ionization Chambers and Proportional Counters," in *Nuclear Instruments and Their Uses*, Vol. I (A. H. Snell, Ed.), Wiley, New York, 1962, pp. 3–81.

*G1 F. S. Goulding and A. H. Pehl, "Semiconductor Radiation Detectors," in *Nuclear Spectroscopy and Reactions*, Vol. A (J. Cerny, Ed.), Academic, New York, 1974, pp. 289–343.

G2 F. Granzer, H. Paretzke, and E. Schopper, Eds., *Proc. 9th Int. Conf. Solid State Nuclear Track Detectors*, Pergamon, Oxford, 1978.

H1 L. J. Herbst, *Electronics for Nuclear Particle Analysis*, Oxford, London, 1970.

L1 P. Lecomte and V. Perez-Mendez, "Channel Electron Multipliers: Properties, Development, and Applications," *IEEE Trans. Nucl. Sci.* **NS-25**, 964 (1978).

L2 J. Litt and R. Meunier, "Čerenkov Counter Technique in High-Energy Physics," *Ann. Rev. Nucl. Sci.* **23**, 1 (1973).

L3 C. M. Lederer and V. S. Shirley, Eds., *Table of Isotopes*, 7th ed., Wiley-Interscience, New York, 1978.

O1 H. Oeschger and M. Wahlen, "Low-Level Counting Techniques," *Ann. Rev. Nucl. Sci.* **25**, 423 (1975).

R1 F. Reines, "Neutrino Interactions," *Ann. Rev. Nucl. Sci.* **10**, 1 (1960).

R2 B. B. Rossi and H. H. Staub, *Ionization Chambers and Counters*, Nat. Nucl. Energy Series Div. V, Vol. 2, McGraw-Hill, New York, 1949.

R3 J. T. Routti and S. G. Prussin, "Photopeak Method for the Computer Analysis of Gamma-Ray Spectra from Semiconductor Detectors," *Nucl. Instr. Methods* **72**, 125 (1969).

S1 W. Schneider, *Neutronenmesstechnik und ihre Anwendung an Kernreaktoren*, Walter de Gruyter, Berlin, 1973.

*S2 K. Siegbahn, Ed., *Alpha-, Beta, and Gamma-Ray Spectroscopy*, North Holland, Amsterdam, 1966.

T1 J. G. Timothy and R. L. Bybee, "Preliminary Results with Microchannel Array Plates Employing Curved Microchannels to Inhibit Ion Feedback," *Rev. Sci. Instr.* **48**, 292 (1977).

*Y1 L. C. L. Yuan and C. S. Wu, Eds., *Methods of Experimental Physics*, Vol. 5A, *Nuclear Physics*, Academic, New York, 1961.

EXERCISES

1. Estimate roughly the voltage (IR) applied to the grid of a dc amplifier tube; use these assumptions: $R = 10^{11}\,\Omega$; the sample emits 1-MeV γ rays at the rate of $10^6\,\text{min}^{-1}$; the geometry is such that 30 percent of the γ's spend an average 8-cm path length in the ionization chamber, which is filled with CF_2Cl_2 at 2 bar total pressure.
2. Calculate the time required for a positive ion to move from the wire to the wall of a Geiger counter; take 0.13 mm for the wire diameter, 28 mm for the cathode diameter, 1000 V as the applied voltage, 130 mbar as the gas pressure, and $1.5\,\text{cm s}^{-1}$ for the mobility of the ion at $1\,\text{V cm}^{-1}$ gradient and at 1 bar pressure. *Answer:* 0.46 ms.
3. What type of instrument would you use for each of the following: (a) Detection of 0.1 μCi of ^{32}P? (b) Measurement of ^{3}H samples in the range of 10^{-8}–10^{-10} Ci? (c) Measurement of the growth of ^{233}Pa in 1 μg of freshly purified ^{237}Np? (d) Following the decay of a sample of ^{64}Cu (initially $3\times 10^5\,\text{dis s}^{-1}$) over a period of 8 days? (e) Determination of the relative amounts

of ^{239}Pu and ^{242}Pu in a sample containing both? (f) Determination of the relative amounts of ^{57}Co and ^{60}Co in a sample (without decay measurements over a long period of time)? State briefly the reasons for each choice.

4. The nuclide ^{7}Be emits a single γ ray of 478 keV. Sketch what the γ spectrum of a ^{7}Be sample would look like if taken with the Ge(Li) detector whose efficiency-versus-energy characteristics are depicted in figure 7-17*b*.

5. A certain nuclide decays predominantly by emission of β^- particles of 1.25-MeV maximum energy to a 25-s isomeric state, which in turn decays to the product ground state by emission of 0.45-MeV γ rays. A rare β^- branch (0.1 percent abundance, 0.81-MeV maximum energy) leads to a state that decays by γ emission to the 25-s isomeric level. To establish this branch by a β-γ coincidence measurement with scintillation detectors, a true coincidence rate at least three times the accidental coincidence rate is desired. The sample strength available is 1000 dis s^{-1}, and background effects in the counters may be neglected. Assume equal counting efficiency of the β counter for the two β groups. What coincidence resolving time is required? *Answer:* $\leq 0.17\ \mu$s.

6. Consider the relative merits of a proportional counter, a NaI(T1) scintillation counter, and a lithium-drifted silicon detector for the measurement of 12-keV X rays in the presence of a 1-MeV β spectrum and of 0.5-MeV γ rays. Discuss details such as counter dimensions and, in the case of the proportional counter, the gas filling conducive to optimizing the measurement of the X rays. Also comment on the energy resolution attainable with each instrument (assuming that a pulse height analyzer is available).

7. A beam of 210-MeV π^- mesons has been momentum-analyzed by magnetic deflection but is contaminated with μ^- mesons of the same momentum. How could two Čerenkov counters and an anticoincidence circuit be used to detect the π mesons only? What are the requirements for the refractive indices of the substances used in the two Čerenkov counters? Could the same type of system be used to discriminate against electron contamination in the π beam?

8. An n^+p-type diffused-junction detector made from p-type silicon of $5 \times 10^3\ \Omega$ cm resistivity is to be used to measure kinetic-energy spectra of protons incident on its face. Estimate the maximum proton energy for which the detector will be useful if a bias voltage of 200 V is applied. (Take 2.4 g cm^{-3} for the density of silicon.) *Answer:* 6 MeV.

9. A proportional counter of 2 cm radius and with a center wire of 4×10^{-3} cm diameter is filled to 1 bar with methane. It is operated with an applied voltage of 4000 V to achieve a certain gas multiplication. Under what conditions of pressure and voltage would the same gas multiplication be obtained in methane-filled counters of (a) 1 cm radius and 4×10^{-3} cm wire diameter, (b) 2 cm radius and 8×10^{-3} cm wire diameter?

10. How would you propose to measure (a) the ratio of K- to L-conversion coefficients for a 150-keV transition in a germanium isotope, (b) the relative intensities of 560-keV and 820-keV γ rays in a sample, (c) the relative amounts of ^{236}U and ^{233}U present in approximately 1 μg of a mixture of the two isotopes?

11. What gas multiplication is required in a methane-flow proportional counter if a minimum-ionizing electron that spends a 2 cm path length in the active counter volume is to result in a 5×10^{-4} V pulse? Assume ~5 pF for the capacity of the counter. Take the ionization loss of minimum-ionizing electrons (MeV mg^{-1} cm^2) in methane the same as in air. *Answer:* Approximately 240.

12. In a typical liquid scintillator six photons are produced for every thousand electron volts deposited. If such a scintillator is coupled to a 10-stage photomultiplier tube with an output circuit that has a capacity of 100 pF, what will be the height of the output pulse produced by a 5-keV electron? Assume 1 photoelectron produced at the photocathode for every 10 incident photons and a multiplication factor of 4 per stage in the multiplier. *Answer:* 5 mV.

Chapter 8

Techniques in Nuclear Chemistry

A. TARGET PREPARATION

The problems encountered in the preparation of samples that are to serve as targets in nuclear bombardments vary widely, depending on the purpose and degree of sophistication of the experiment and on the nature of the particular irradiation.

Reactor Irradiations. Although target or sample preparation for reactor irradiations is generally quite straightforward, some special considerations do enter. For example, containers for samples to be exposed in high-flux reactors have to be carefully chosen, with due regard to neutron flux, ambient temperature, and length of irradiation. Pyrex vessels should be avoided because of their high boron content (boron has a very high neutron-capture cross section). For irradiations of the order of minutes in the modest fluxes of many research reactors (10^{12}–10^{13} cm^{-2} s^{-1}), plastic vials are often satisfactory, and they give rise to rather low activity levels. Aluminum-foil wrappers made of highest-purity aluminum are often convenient, if time for the decay of the 2.3-min ^{28}Al can be allowed. For longer irradiations samples are often sealed in evacuated quartz vials. However, these vials must generally be allowed to "cool" for some time after irradiation to let the intense ^{31}Si activity ($t_{1/2} = 2.6$ h) decay. Some thought must also be given to arrangements for breaking the seal without undue personnel exposure and contamination hazard. The thermal stability of the substance to be irradiated is, of course, a problem to be considered. The ambient temperatures in different types of reactors differ widely; water-cooled and water-moderated swimming-pool reactors are generally much more suitable for irradiation of organic materials than, for example, graphite reactors. Some reactors have special water-cooled or even liquid-nitrogen-cooled irradiation facilities. The irradiation of aqueous solutions creates special problems. Even if cooling is adequate to keep them below the boiling point, the radiation decomposition of water can lead to the buildup of dangerous pressures unless provisions are made for venting or catalytically recombining the gases. Another problem encountered occasionally in reactor irradiations is the **self-shielding** of materials having high neutron cross sections. For example, a 0.1-mm layer of gold (whose absorption cross section for thermal neutrons is almost 100 b) reduces a thermal-neutron flux by about 6 percent, so that the interior of a cube of gold 1 mm on an edge would receive only a fraction of the flux incident on its surface.

Thick-Target Accelerator Experiments. In accelerator bombardments the variety of possible targets and targeting problems is so large that only a few generalities can be mentioned. The simplest situation arises when production of a radionuclide is the goal, without the need for quantitative information about the reaction involved. Generally it is adequate or even desirable to use a thick target, that is, a target in which the incident bombarding particles are appreciably degraded in energy. For example, if we wished to produce a radionuclide by (α, n) reaction and had 40-MeV ^{4}He ions available, we would probably use a target thick enough to degrade the ^{4}He ions to just a few million electron volts—approximately the (α, n) threshold—to maximize the product yield. On the other hand, even in a simple production problem there could easily be complicating circumstances that would dictate a different choice of bombarding conditions. For example, if it were desirable to produce ^{84}Rb with minimal contamination by ^{83}Rb, we would not wish to use 40-MeV ^{4}He ions on bromine but would first degrade them with absorber foils below the threshold of the reaction ^{81}Br $(\alpha, 2n)$ ^{83}Rb, even though this would lower the yield of the desired reaction ^{81}Br (α, n) ^{84}Rb also. Conversely, by use of the 40-MeV beam and proper choice of target thickness we could maximize the ratio of $(\alpha, 2n)/(\alpha, n)$ yield.

The principal problem in cyclotron irradiations for radionuclide production is one of cooling, since the energy dissipation in the target can become quite large—of the order of a kilowatt over an area of a square centimeter or two. Metal targets, bolted or soldered to water-cooled backing plates, are most satisfactory. However, it is frequently necessary to resort to the bombardment of nonmetallic elements or compounds. Satisfactory targets can then often be made by pressing powders into grooves on a cooled target plate or wrapping them in metal-foil packages, which are clamped to a cooled plate. Fairly effective cooling can be achieved by flowing helium gas over the target surface. Beam currents may have to be adjusted to the particular problem at hand, but it is usually possible to use many microamperes of particles with energies of tens of MeV.

Liquids may be used as accelerator targets under special circumstances, for example in the production of ^{18}F by helium irradiation of water. Target cooling can be accomplished by means of a continuously recirculating flow system that passes the liquid through a cold bath outside the irradiation chamber. Gas targets, either stationary or flowing, are useful in some applications.

Requirements for Thin Targets. In a great variety of accelerator experiments thin targets are needed. What constitutes a "thin" target depends very much on the particular information sought. In any experiment designed for the measurement of a reaction cross section the target must be so thin that the energy degradation of the bombarding particle in its passage through the target will not cause a significant change in the cross section. However, the implications of this general requirement may differ widely for different situations. For example, a target that is "thin" for the study of (p, xn) reactions with 30–50 MeV protons may be thick indeed for the investigation of a narrow resonance in a (p, γ) reaction at 2 MeV. If the spectra of particles produced in a reaction are to be measured, the criterion for maximum target thickness will likely be set by the interactions, not of the primaries, but of these secondaries in the target. For example, if we wished to investigate the low-energy end of the α spectra produced in (p, α) reactions, the targets would have to be much thinner than those required for a study of total (p, α) cross sections measured via the activity of the reaction product (tens of micrograms versus milligrams per square centimeter). Similarly, in any experiment designed for the determination of momenta and angular distributions of the recoil nuclei, the targets need to be so thin that these recoiling reaction products will not undergo appreciable scattering or degradation on their way out of the target. This criterion may require targets of no more than a few micrograms per square centimeter.

Still another limitation on target thickness arises sometimes from the need to suppress secondary reactions caused by particles produced in the primary interactions, if the products of such secondary reactions interfere with the measurement at hand. For example, the product of a $(p, p\pi^+)$ reaction is the same as that of an (n, p) reaction on the same target. Thus in an attempt to measure the very low cross section ($\sim 10^{-4}$ b) of a $(p, p\pi^+)$ reaction with high-energy protons, the targets used must be thin enough so that (n, p) reactions caused by low-energy neutrons originating in the target will not swamp the sought-for effect. The severity of this type of problem depends on the number of secondaries per primary interaction, on the ratio of primary to secondary cross section, and also on such factors as the angular distribution of the secondaries. In practice, when the effect is likely to be of any significance, it is necessary to irradiate targets of several different thicknesses and to extrapolate the results to zero target thickness.

Techniques for Preparation of Thin Targets. It is impossible to give, in a brief space, anything like a complete summary of the methods that have been used to prepare thin targets. In a sense, each element presents a separate problem; different thickness ranges require different approaches; a method suitable when an abundant supply of target material is available may not be adaptable to a situation in which a few milligrams of a costly enriched isotope must be made into a target; whether or not a target backing can be tolerated and how big an area is required will strongly affect the method of preparation. Thus we can make only rather sketchy comments and refer the reader to reviews of thin-target preparation methods such as P1 and Y1.

Whenever suitable foils are commercially available they, of course, offer the simplest solution to targeting problems. However, not many metals can be purchased in thicknesses below a few milligrams per square centimeter; among those most readily available in the form of thin foils are aluminum, nickel, and gold. Vacuum evaporation has been used to prepare targets of a large variety of metals, some nonmetallic elements, and some compounds, over a large range of thicknesses. The method is generally wasteful of material, but has occasionally been used to make separated-isotope targets. The evaporated films may be deposited on a variety of backing materials (metal foils, plastic films). Plastic films for target backings are usually prepared by letting a few drops of a suitable solution of the plastic spread on distilled water and, after evaporation of the solvent, picking up the film on a metal frame. Among the most useful plastics are: Formvar (soluble in chloroform), collodion (in amyl acetate), and especially a very tough resin called VYNS, a polyvinyl acetate-polyvinyl chloride copolymer (soluble in cyclohexanone). Film thicknesses down to 1 $\mu g\ cm^{-2}$ can be used, although 5–10 $\mu g\ cm^{-2}$ is much easier to achieve. Plastic films down to $\sim 250\ \mu g\ cm^{-2}$ are commercially available. If an unsupported target is required, techniques are available for stripping off or dissolving the backing. Probably the most useful is evaporation onto a layer of water-soluble material (such as $BaCl_2$, NaCl, or glycerol) on glass, followed by gentle dissolution of this intermediate layer in a trough of water, whereupon the desired film is floated off. Self-supporting foils of various materials with thicknesses down to about 0.01 $mg\ cm^{-2}$ have been prepared in this way. For the deposition of small amounts of material with high efficiency cathodic sputtering is sometimes superior to vacuum evaporation.

Another important method for target preparation is electrodeposition. This is not restricted to deposition of metals, but can be used, for example, for cathodic or anodic deposition of oxides and other compounds. In most cases electrodeposition can be made nearly quantitative and is therefore suitable for use with enriched isotopes. Removal of backing materials is generally more difficult than with vacuum-evaporated targets, but if a very thin evaporated metal film on a plastic foil is used as the plating electrode, it may be possible to dissolve off the plastic. Various forms of electrophoretic deposition of finely divided materials from suspensions have been successfully used for target preparation. So-called molecular plating, which is essentially electrodeposition of molecular species from organic solvents, is becoming more and more widely used.

Many other specialized techniques have been described. Thermal decomposition of gases on hot surfaces is sometimes useful, for example, for the preparation of boron films from B_2H_6, nickel films from $Ni(CO)_4$, and carbon films from CH_3I. The preparation of separated-isotope targets often presents special problems; ideally the targets can be prepared directly in the isotope separator if the isotope of interest is collected on the target backing, but this technique can be used in a rather limited number of laboratories only.

If uniformity criteria are not too stringent, targets can often be successfully prepared by sedimentation from a slurry, perhaps with the use of some binder. A useful though tedious technique (D1), especially for use with enriched isotopes or other precious target materials, involves painting onto the target backing many successive portions of an alcohol solution of metal nitrate containing a small amount of Zapon lacquer. After each application the deposit is ignited to remove most of the organic material and rubbed with tissue paper to improve uniformity and adhesion. Very satisfactory targets of such materials as lanthanide and actinide oxides have been produced in this manner.

Measurement of Target Thickness. Whenever a thin target is required it is usually necessary to know what its thickness is. Furthermore, there are generally some requirements for uniformity. Measurements with mechanical thickness gauges are rarely applicable. Weighing an accurately measured area is often the method of choice for self-supporting targets; it can be applied to backed targets as well if the backing material is weighed separately before target deposition and if the ratio of target weight to backing weight is not too small. For very thin targets a microbalance may be required. Uniformity within the target area to be used cannot be established by this gravimetric method, but measurements on several neighboring areas can help establish the degree of uniformity on a slightly larger scale. X-ray fluorescence spectrometry can be useful for thickness measurements on thin films.

Methods based on the absorption of α and β particles in matter have found widespread use in the determination of foil thicknesses and foil uniformity (Y1). Collimated beams of monoenergetic α particles or low-energy β particles are used, and the foil to be measured is interposed between source and detector. Alpha gauges are most sensitive if the α particles reaching the detector are near the end of their range; then slight changes in interposed thickness cause large changes in counting rate. Alpha and β gauges are particularly simple to use for relative measurements and uniformity checks. If calibrated carefully, they can also be used for absolute thickness measurements with accuracies of 1–2 $\mu g\,cm^{-2}$. In a particularly useful variant of the α gauge the well-collimated monoenergetic α beam is detected by a high-resolution spectrometer, such as a semiconductor detector with pulse height analyzer. The shift of the spectral line to lower energy when a foil is interposed is a measure of the average foil thickness over the area of the beam; the line broadening can give information on nonuniformities on a microscale. A monoenergetic accelerator beam can be substituted for the α source.

In experiments performed in external beams with particle detectors it is sometimes practical to determine target thicknesses by means of Rutherford scattering [see (2-1)]. This requires measurement of both the primary-beam and scattered-beam intensities and fairly accurate knowledge of the beam energy and scattering angle; the latter also has to be small enough to ensure that one deals with Rutherford scattering only.

Occasionally it may be most practical to determine target thickness after, rather than before, an irradiation. This may be done by dissolving an accurately measured area of the target and analyzing the solution or an aliquot of it for the target material.

B. TARGET CHEMISTRY

We now turn to the problems of identification, isolation, and purification of nuclides produced in nuclear reactions. Historically this has been one of the major preoccupations of nuclear chemists, and it is still an important field. We are primarily concerned with radioactive products. When stable nuclides are of interest they generally have to be isolated by means of a mass spectrometer or isotope separator.

In dealing with an irradiated target nuclear chemists or radiochemists may be confronted with one of two tasks. They may need to prepare a known reaction product free from other radioactive contaminants and sometimes free from certain inactive impurities and in a specified chemical form for use in subsequent experimentation or for determination of its yield in a nuclear reaction; or they may wish to identify a hitherto unknown or unidentified radioactive species by its atomic number, mass number, half life, and radiation characteristics. In both cases chemical separations are usually required, although for the determination of reaction yields of known products the need for chemical isolation has in many cases been eliminated by the advent of high-resolution Ge(Li) γ-ray spectroscopy, which makes possible the analysis of even very complex mixtures.

Comparison with Ordinary Analytical Practice. In many respects the chemical separations that radiochemists carry out on irradiated targets are similar to ordinary analytical procedures. However, there are a number of important differences. One is the time factor that is often introduced by the short half lives of the species involved. An otherwise simple procedure such as the separation of two common cations may become quite difficult when it is to be performed, and the final precipitates are to be dried and mounted, in a few minutes. When the usual procedures involve long digestions, slow filtrations, or other slow steps, completely different separation procedures must be worked out for use with short-lived activities. Ingenious chemical isolation procedures, taking as little as a few seconds, have been developed for many elements (H1, T1, T2); they usually involve automation and computer control of standard operations.

In radiochemical separations, at least those subsequent to bombardments with projectiles of moderate energies, we are usually concerned with several elements of neighboring atomic numbers. Thus the procedures given in complete schemes of qualitative analysis can often be modified and shortened. On the other hand, the separation of neighboring elements sometimes presents considerable difficulties, as can readily be seen by considering such groups as ruthenium, rhodium, palladium, or any sequence of neighboring rare earths. In very-high-energy reactions and in fission the products are spread over a wide range of atomic numbers. In these cases the separation procedures either become more akin to general schemes of analysis or, more frequently, are designed for the isolation of one or a few elements free from all the others. The latter type of procedure is required particularly when a short-lived substance is to be isolated, and for such cases many specialized techniques have been developed.

High yields in radiochemical separations are not always of great importance, provided that the yields can be evaluated. It may be more valuable to get 50 percent (or perhaps even 10 percent) yield of a radioactive element separated in 10 min than to get 99 percent yield in 1 h (this is certainly so if the activity has a half life of 10 or 20 min). High *chemical* purity may or may not be required for radioactive preparations, depending on their use. For identification and study of radioactive species and for many chemical tracer applications it is not important; for most biological work it is. On the other hand *radioactive* purity is usually required and often has to be extremely good.

Hazards Encountered with Radioactive Materials. Some specific effects of the radiations from radioactive substances on the separation procedures may be noted. At very high activity levels (say 10^{12} β dis min^{-1} per ml of solution) chemical effects of the radiations, such as decomposition of water and other solvents, and heat effects may affect the procedures. However, this is generally not so important as the fact that even at much lower activity levels, especially in the case of γ-ray emitters, the person carrying out the separation receives dangerous doses of radiation unless protected by shielding or distance. At even lower activity levels, say in the handling of microcurie amounts, where the health hazards from radiation are minimal, special care is still required to prevent spread of radioactive contamination that could seriously raise counter backgrounds and interfere with low-activity experiments. The degree of precaution needed, both to contain contamination and to prevent excessive radiation exposure, depends on many factors, including the amount of activity handled, the nature and energy of the radiation involved, the half life of the active substance, and possibly its chemical properties.

Radiation protection is discussed in chapter 6, section E.

Carriers. The amount of radioactive material produced in a nuclear reaction is generally very small. Notice, for example, that a sample of

37-min ^{38}Cl undergoing 10^8 dis s^{-1} weighs about 2×10^{-11} g; a sample of 51-d ^{89}Sr of the same disintegration rate weighs 1×10^{-7} g. Thus the substance to be isolated in a radiochemical separation may often be present in a completely impalpable quantity.[1] It is clear that ordinary analytical procedures involving precipitation and filtration or centrifugation may fail for such minute quantities. In fact, solutions containing the very minute concentrations of solutes that can be investigated with radioactive tracers behave in many ways quite differently from solutions in ordinarily accessible concentration ranges. Adsorption on container surfaces, dust particles, and other suspended impurities can be important at these "tracer" concentrations.

Usually some inactive material isotopic with the radioactive transmutation product is deliberately added to act as a carrier for the active material in all subsequent chemical reactions. Furthermore, it is often necessary, particularly when precipitations are to be used, to add so-called **hold-back carriers** for radionuclides we do not wish to carry along with the product of interest. Certain precipitates such as $BaSO_4$ and $Fe(OH)_3$ have a notorious tendency to occlude or coprecipitate foreign ions.[2]

As mentioned before, extreme radioactive purity is often very important. Frequently the desired product has an activity that constitutes only a small fraction of the total target activity, yet this product may be required completely free of the other activities. Such extreme purification is usually attainable by repeated removal of the impurities with successive fresh portions of carrier until the fractions removed are sufficiently inactive. For example, radioactive iron impurity might be removed by repeated extraction of ferric chloride from 9*M* HCl into isopropyl ether, with fresh portions of $FeCl_3$ carrier added after each extraction. In applying this "washing-out" method one must, of course, make sure that the desired product is not partially removed along with the impurity in each cycle. If the washing out works properly, the activities of successive impurity fractions should decrease by large and approximately constant factors, provided that the conditions in each step are about the same.

In order that an added inactive material serve as a carrier for an active substance, the two must generally be in the same chemical form. For example, inactive iodide can hardly be expected to be a carrier for active

[1] Actually the amount of an element formed in a nuclear reaction is usually exceeded by that of the inactive isotopes of the same element present as an impurity in the target and in the reagents used in the separation procedure.

[2] In the early decades of radiochemistry there was a great deal of interest in the laws governing coprecipitation and adsorption and in the classification of various "carrying" phenomena (H2). These are no longer very active fields of research and not much more than broad general guidelines are available for the prediction of coprecipitation behavior. A useful general rule formulated in 1913 by K. Fajans may be paraphrased as follows: Conditions that favor the precipitation of a substance in macroamounts also tend to favor the coprecipitation of the same material from tracer concentrations with a foreign substance.

iodine in the form of iodate ion; sodium phosphate would not carry radioactive phorphorus in elementary form. The chemical form in which a transmutation product emerges from a nuclear reaction is usually hard to predict and has been investigated in only a few cases. However, it is often possible to treat a target in such a way that the active material of interest is transformed to a certain chemical form. For example, if a zinc target is dissolved in a strongly oxidizing medium (say HNO_3, or $HCl + H_2O_2$), any copper present as a transmutation product is found afterward in the Cu^{2+} form. If there is any uncertainty about the chemical form of the transmutation product—its oxidation state or presence in some complex or undissociated compound, for example—the only method that can be relied on to avoid difficulties is the addition of carrier in the various possible forms and a subsequent procedure for the conversion of all of these into one form. To go through such a procedure *prior* to the addition of carrier may not be adequate. In fact, it appears that it may not always be sufficient to add the carrier element (say iodine) in its highest oxidation state (IO_4^-) and carry through a reduction to a low oxidation state (I_2). In the case of the iodine compounds this procedure does not seem to reduce all the active atoms originally present in intermediate oxidation states. Repeated oxidation-reduction cycles may be necessary.

Specific Activity. The amount of carrier used depends on circumstances, but amounts between 0.1 and 10 mg are common.[3] Since the chemical yield in a procedure is usually determined by measurement of the amount of carrier in the final sample, the analytical technique to be used in that determination must be taken into account in choosing the quantity of carrier. The desired specific activity (activity per unit weight) is, however, often the deciding criterion. High specific activities are particularly essential in many biological and medical applications of radioactive isotopes and are often desirable in samples to be used in physical measurements or chemical tracer studies to ensure small absorption of the radiations in the sample itself or to permit high dilution factors.

It is often possible to prepare samples of very high specific activities by the use of a **nonisotopic carrier** in the first stages of the separation. This carrier may later be separated from the active material. In the isolation of 107-d ^{88}Y) from deuteron-bombarded strontium targets, ferric ion can be used as a carrier for the active Y^{3+}. Ferric hydroxide is then precipitated, centrifuged, washed, redissolved, and, after the addition of more strontium as hold-back carrier, it is precipitated several more times to free it of strontium activity. Finally the ferric hydroxide that carries the yttrium activity is dissolved in 9*M* HCl, and ferric chloride is extracted into

[3] In a radiochemical laboratory it is convenient to have carrier solutions for a large number of elements on hand. These may, for example, be made up to contain 1 or 10 mg of carrier element per milliliter.

isopropyl ether, leaving the active yttrium in the aqueous phase almost carrier-free. The use of nonisotopic carriers became essential in early work with the artificially produced elements that do not occur in nature (chapter 11, section E). For example, most of the chemical processes that were used during World War II for the large-scale isolation of plutonium from irradiated uranium were worked out on a tracer scale before any weighable amounts of plutonium were available. A very rough rule governing coprecipitation of tracers with nonisotopic carriers is mentioned in footnote 2.

Not all chemical procedures require the use of carriers. In particular, procedures that do not involve solid phases may sometimes be carried out at tracer concentrations without the addition of carriers. Because of the great importance of high specific activities, considerable work has been done on the preparation of carrier-free sources of many radioactive species (see for example, G1, G2). In the course of the following brief discussion of the various types of separation techniques we therefore point out those that lend themselves to the production of carrier-free preparations. A rather complete collection of radiochemical procedures for the separation and isolation of every element (except hydrogen, helium, lithium, and boron) and including carrier-free procedures is available in a series of monographs (S1).

Precipitation. In many radiochemical separations, as in conventional analytical schemes, precipitation reactions play a dominant role. The chief difficulties with precipitations arise from the carrying down of other materials. Some precipitates such as manganese dioxide and ferric hydroxide are so effective as "scavengers" that they are sometimes used deliberately to carry down foreign substances in trace amounts. Other precipitates, such as rare-earth fluorides or CuS precipitated in acid solution, have little tendency to carry substances not actually insoluble under the same conditions and therefore can sometimes be brought down without the addition of hold-back carriers for activities that are to be left in solution. Most precipitates have an intermediate behavior in this regard.

A radionuclide capable of existence in two oxidation states can be effectively purified by precipitation in one oxidation state followed by scavenging precipitations for impurities while the element of interest is in another oxidation state. For example, a useful procedure for cerium decontamination from other activities uses repeated cycles of ceric iodate precipitation, reduction to Ce(III), zirconium iodate precipitation [with Ce(III) staying in solution], and reoxidation to Ce(IV).

Adsorption on the walls of glass vessels and on filter paper, which is sometimes bothersome, has been put to successful use in special cases. Carrier-free yttrium activity can be quantitatively adsorbed on filter paper from an alkaline strontium solution at yttrium concentrations at which the solubility product of yttrium hydroxide could not have been exceeded. Adsorption of carrier-free niobium from $10N$ HNO_3 on glass fiber filters has been used for very fast and specific niobium separations (T1).

Ion Exchange. This has become one of the most useful techniques for radiochemical separations both with and without carriers. Synthetic organic resins are extensively used as both cation and anion exchangers (K1, M1, S1).

By far the most popular ion-exchange resins are crosslinked polystyrenes, produced by polymerizing styrene in the presence of divinylbenzene (DVB), the percentage of DVB controlling the degree of crosslinking.[4] Most cation exchangers (such as Dowex-50 and Amberlite IR-100) contain free sulfonic acid groups, whereas anion exchangers (such as Amberlite IRA-400 and Dowex-1) have quaternary amine groups with replaceable hydroxyl ions. Particle diameters of 0.08–0.16 mm (100–200 mesh) are commonly used, but larger particles give higher flow rates. The exchange capacity of resins is typically 3–5 meq per gram of dry resin.

The distribution of any given element between a solution and the resin depends strongly on the particular ionic forms of the element present (either hydrated ion or various cation or anion complexes) and on their concentrations and therefore on the composition of the solution. For almost any pair of ions conditions can be found under which they will show some difference in distribution.

In practice a solution containing the ions to be separated is run through a column of the finely divided resin, and conditions (solution composition, column dimensions, and flow rate) are chosen so that the ions to be adsorbed will appear in a narrow band near the top of the column. In the simplest kind of separation some ionic species will run through the column while others are adsorbed. For example, Ni(II) and Co(II) may be separated very readily by passing a 12*M* HCl solution of the two elements through a Dowex-1 column; the Co(II) forms negatively charged chloride complexes and is held on the column, whereas Ni(II) apparently does not form such complexes and appears in the effluent.

More commonly a number of ionic species may be adsorbed together on the column and separated subsequently by the use of eluting solutions differing in composition from the original input solution. Frequently complexing agents that form complexes of different stability with the various ions are used as eluants. There exists then a competition between the resin and the complexing agent for each ion, and, if the column is run close to equilibrium conditions, each ion will be exchanged between resin and complex form many times as it moves down the column.[5] The number of times an ion is adsorbed and desorbed on the resin in such a column is analogous to the number of theoretical plates in a distillation column. The

[4] Increased crosslinking reduces solubility, swelling, and porosity of the resin and tends to increase selectivity. The degree of crosslinking is indicated by the manufacturers by a number preceded by an *X*, the number giving the percentage of DVB used.

[5] Slow flow rate, high resin-to-ion ratio, and fine resin particle size favor close approach to equilibrium. In practice a compromise has to be made between high separation efficiency and speed.

rates with which different ionic species move down the column under identical conditions are different because the stabilities of both the resin compounds and the complexes vary from ion to ion. Separations are particularly efficient if both these factors work in the same direction, that is, if the complex stability increases as the metal-resin bond strength decreases. As the various adsorption bands move down the column, their spatial separations increase, until finally the ion from the lowest band appears in the effluent. The various ions can then be collected separately in successive fractions of the effluent. Transition metal ions form colored bands, which allows visual observation of their movement down a column.

The most striking application of cation-exchange columns is in the separation of rare earths from one another, both on a tracer scale and in gram or 100-gram lots. Elution with α-hydroxy isobutyric acid gives efficient, clean, and relatively fast ($\sim\frac{1}{2}$ h) rare-earth separations (K1). The rare earths are eluted in reverse order of their atomic numbers, and, if yttrium is present, it is eluted between dysprosium and holmium. Similar cation-exchange procedures serve successfully for the separation of the actinide elements from each other (K1; see also chapter 11, section E).

For rather rapid target chemistry anion exchange is often more useful than cation exchange because larger flow rates can be used with anion columns. A large number of elements form anionic complexes under some conditions, and a general scheme of analysis based entirely on ion-exchange-column separations could be worked out for these elements. If all the transition elements from manganese to zinc are present in a 12*M* HCl solution, all but Ni(II) are adsorbed on Dowex-1. Then they may be successively eluted, Mn(II) with 6*M* HCl, Co(II) with 4*M* HCl, Cu(II) with 2.5*M* HCl, Fe(III) with 0.5*M* HCl, and Zn with 0.005*M* HCl. With milligram quantities of the elements and a column a few millimeters in diameter and about 10 cm long, this entire separation can be carried out in about half an hour.

Ion-exchange separations generally work as well with carrier-free tracers as with weighable amounts of ionic species. A remarkable example was the original isolation of mendelevium at the level of a few atoms (see chapter 11, section E).

In addition to the organic ion-exchange resins, some inorganic ion exchangers have come into use. For example, excellent separations of alkali elements from one another have been obtained by elution with NH_4Cl solutions from columns of microcrystalline zirconium phosphate or zirconium molybdate. Inorganic exchangers are of interest mainly in those applications where their superior resistance to heat and radiation is an asset.

Chromatographic Methods. Techniques other than ion exchange, but also based on differential migration of different substances through a porous medium, have found increasing application in radiochemical separations (K1). In **paper chromatography** the sample is deposited near one end of a strip of filter paper and that end is dipped into the developing solvent. As the solvent moves through the pores of the paper, it leaches components of the sample and different solutes migrate at different rates, thus ending up in different zones. **Thin-layer chromatography** is an analogous technique, but the paper is replaced by an adsorbent such as silica gel applied in a thin layer to a glass plate. Again the sample is deposited at one end and the plate is placed vertically in a small amount of solvent that then migrates upward through the thin layer. In several variants of **electrochromatography** dc electric potentials are used to produce migration of ions through a medium such as paper or a gel containing an aqueous solution; again different ions move at different rates. In **extraction chromatography** an aqueous solution passes through a column packed with an inert substrate such as a halogenated polyethylene on which an organic solvent has been fixed. All these chromatographic techniques find their principal use in separations of relatively small amounts (micrograms) of materials.

Solvent Extraction (K1, F1, W1). Under certain conditions compounds of some elements can be quite selectively extracted from an aqueous solution into an organic solvent, and often the partition coefficients are approximately independent of concentration down to tracer concentrations (say 10^{-12} or $10^{-15} M$). In other cases, particularly if dimerization occurs in the organic phase (as in the ethyl ether extraction of ferric chloride), carrier-free substances are not extracted. Solvent extractions often lend themselves particularly well to rapid and specific separations. In most cases the extraction can be followed by a back-extraction into an aqueous phase of altered composition.

Extractions of the chlorides of Fe(III), Ga(III), and Tl(III) into various ethers are frequently used by radiochemists. The partition coefficients vary quite rapidly with HCl concentration. Extraction from 6*M* HCl into ethyl ether or from 8–9*M* HCl into isopropyl ether gives very good separations from nearly all other metal chlorides. The separation of gallium from iron and thallium can be achieved by ether extraction of $GaCl_3$ in the presence of reducing agents so that the reduced ions Fe(II) and Tl(I) are present.

Gold nitrate and mercuric nitrate can be extracted into ethyl acetate from nitric acid solutions. The extraction of uranyl nitrate by ethyl ether from a nitric acid solution of high nitrate concentration is sufficiently specific to serve as an excellent first step in the isolation of carrier-free fission products from the bulk of irradiated uranium. The extraction into ethyl ether of the blue peroxychromic acid formed when H_2O_2 is added to a dichromate solution is a good radiochemical decontamination step for chromium, although it tends to give low yields. Extraction of copper dithizonate into carbon tetrachloride, of cadmium thiocyanate into chloroform, of beryllium acetylacetonate into benzene, and many other examples could be cited. Judicious use of complexing agents such as ethylenediamine tetraacetate (EDTA) can often help to make extractions more specific for a particular element. The addition of EDTA is recommended, for example, in the extraction of beryllium acetylacetonate just mentioned because its complexing action prevents the extraction of some other ions that would otherwise accompany beryllium.

Various organic phosphorus compounds have become extremely important for many metal ion extractions. Among the reagents so used are tri-*n*-butylphosphate (TBP), bis-(2-ethylhexyl)*o*-phosphoric acid (HDEHP), and tri-*n*-octylphosphine oxide (TOPO), all frequently used in kerosene solution. Under various conditions the extractions can be made rather specific for particular elements.

Compounds that form chelate complexes with inorganic ions are of great importance in facilitating solvent extraction, because the chelates are usually quite soluble in nonpolar solvents. Since the dissociation constants of different metal compounds with a given chelating agent have different pH dependence, specific separation procedures can sometimes be devised with several extraction steps at different pH values. Among useful chelating agents are cupferron, dithizone, β-diketones, and theonyltrifluoroacetone (TTA).

Occasionally it may be possible to leach an active product out of a solid target material. This has been done successfully in the case of neutron- and deuteron-bombarded magnesium oxide targets; radioactive sodium is separated rather efficiently from the bulk of such a target by leaching with hot water.

Volatilization (K1). Differences in vapor pressure can be exploited in radiochemical separations. The most straightforward application is the removal of radioactive rare gases from aqueous solutions or melts by sweeping with an inert gas such as helium. The volatility of such compounds as $GeCl_4$, $AsCl_3$, and $SeCl_4$ can be used to effect separations from other chlorides by distillation from HCl solutions. Similarly, osmium, ruthenium, rhenium, and technetium can be separated from other elements and from one another by procedures involving distillations of their oxides OsO_4, RuO_4, Re_2O_7, and Tc_2O_7. Carrier-free palladium (^{103}Pd) has been prepared from a rhodium target by a method involving coprecipitation of palladium with selenium (by reduction of H_2SeO_3 with SO_2), followed by removal of selenium by a perchloric acid distillation.

Distillation and volatilization methods often give very clean separations, provided that proper precautions are taken to avoid contamination of the distillate by spray or mechanical entrapment. Most volatilization methods can be done without specific carriers, but some nonisotopic carrier gas may be required. Precautions are sometimes necessary to avoid loss of volatile radioactive substances during the dissolving of irradiated targets or during the irradiation itself.

Electrochemical Methods. Electrolysis or electrochemical deposition may be used either to plate out the active material of interest or to plate out other substances, leaving the active material in solution. For example, it is possible to separate radioactive copper from a dissolved zinc target by an electroplating process. Carrier-free radioactive zinc may be obtained from a deuteron-bombarded copper target by solution of the target and electrolysis to remove all the copper.

In attempting to use electrode processes at tracer concentrations we must keep in mind that the measured potential E for a reaction can deviate appreciably from the standard potential E^0, according to the Nernst equation:

$$E = E^0 - \frac{RT}{nF} \ln Q,$$

where R is the gas constant, T is the absolute temperature, F is the Faraday, n is the number of electrons transferred in the reaction as written, and Q is the appropriate activity ratio for the reaction (product activities divided by reactant activities, each raised to proper power, as in an equilibrium constant). If the activity of tracer deposited on the electrode is taken as unity (which is by no means always a good assumption), Q can take on very large values. Measurements of the potentials needed to deposit a tracer (relative to some suitable reference electrode) have been used to estimate standard electrode potentials for some of the artificially produced elements before they were available in macroconcentrations.

Chemical displacement may sometimes be used for the separation of carrier-free substances from bulk impurities. The separation of polonium from lead by deposition on silver is a classic example. Similarly, bismuth activity obtained in lead bombardments may be separated almost quantitatively from the lead by plating on nickel powder from hot 0.5M HCl solution. This method for lead-bismuth separations is sufficiently rapid to permit isolation of the 0.8-s $^{207}Pb^m$ isomer from its bismuth parent.

Transport Techniques. Although not strictly or at least not entirely a chemical problem, the rapid and efficient transport of reaction products from an accelerator or reactor to a measuring instrument or to an apparatus for chemical separations is of vital importance, especially for short-lived products. The simplest technique for moving targets rapidly uses a pneumatic transfer; such a system consists essentially of a tube or hose and a carrier for the target (often called a **rabbit**) that is moved through the tube by application of vacuum or pressure. Depending on the distance that needs to be traversed, rabbit systems can have transit times down to the order of a second.

The recoil energy imparted by a nuclear reaction or radioactive (particularly α) decay can be used to separate reaction products physically from the target and to transport them to a nearby rotating wheel or moving belt, which can then carry them in fractions of a second in front of detectors. A more versatile technique that also makes use of recoil separations is the **helium-jet** method (M2). Here the reaction products recoiling out of a thin target are slowed to thermal energies in helium gas at atmospheric or higher pressure, and the thermalized atoms are transported together with the helium carrier gas by differential pumping through an orifice or through a long capillary. An essential ingredient appears to be the presence of small concentrations of impurities (such as H_2O or hydrocarbons) in the helium; these are believed to form high-molecular-weight clusters under the influence of radiation, and the recoil products are transported by attaching to these clusters. Transfer capillaries as long as 200 m have been used effectively, and they can have fairly sharp bends in them without serious reduction in transfer efficiency. From the helium jet the reaction products can be collected on rotating drums, moving tapes, and so on, or, after pumping away the helium in a nozzle-skimmer arrangement, they can be introduced into a high-vacuum system (such as the source of an isotope separator). Deposition on surfaces can be used to achieve separations from volatile products. A helium-jet system with tape transport arrangement to carry reaction products to a series of detectors is shown schematically in figure 8-1.

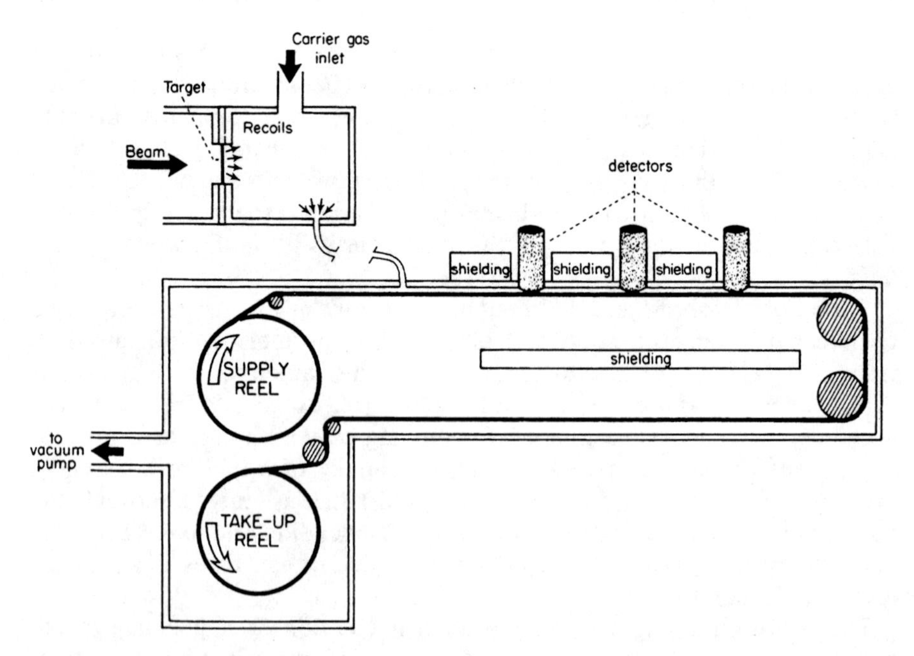

Fig. 8-1 Schematic diagram of a helium-jet system with moving-tape transport for measuring short half lives.

C. PREPARATION OF SAMPLES FOR ACTIVITY MEASUREMENTS

Many points of experimental technique arise in the preparation of samples for activity measurements. Most of them have to do with the attainment of a suitable and reproducible geometrical arrangement and with the scattering and absorption of radiations in the sample and in its support. The difficulties encountered in sample preparation are greatest when absolute disintegration rates or energies are wanted, less when samples of different radiation characteristics are compared, and least when the relative strengths of several samples of the same kind, or of the same sample at several times, are to be determined. Fortunately, the last-mentioned problem is probably the one most often met in radiochemical work. However, even here adequate reproducibility may not always be easy to achieve.

Choice of Counting Arrangement. Careful consideration must be given to the chemical and physical form in which samples are to be measured. The radiations emitted by the substance and the available measuring equipment are among the determining factors. Alpha emitters are usually counted in the form of thin deposits, preferably prepared by electrodeposition or by distillation and placed inside a proportional counter or ionization chamber or near a solid-state detector. Nuclides that emit primarily soft radiations (low-energy β rays, X rays, conversion electrons, or Auger electrons) may be very efficiently assayed for activity if they can be prepared in the form of a gas suitable as a component of a counter-filling mixture. For example, ^{14}C-labeled compounds may be burned to CO_2, which is then introduced into a proportional counter along with an appropriate amount of argon, methane, or argon-methane mixture. Essentially 100 percent counting efficiency and good counter behavior can be obtained over a fair range of CO_2 partial pressures (0.5–5 torr). This technique requires the use of a good gas-handling and purification system.

Widely used for routine measurement of β emitters, particularly emitters of low-energy β particles such as ^3H and ^{14}C, are **liquid scintillation counters**. They are especially popular for tracer applications in organic chemistry and biochemistry. A variety of ready-to-use scintillator solutions are commercially available, based on such solvents as toluene, xylene, polyglycols, or carbitols. They will readily dissolve a wide spectrum of substances and can hold large amounts of water (sometimes up to 20 percent) in solution without any appreciable effect on their scintillation efficiency. If a sample is not soluble, it can be dispersed in the scintillator solution by grinding it to a fine powder, stirring it in, and adding a gelling agent. Counting efficiencies in liquid scintillators are very high, and different nuclides such as ^3H and ^{14}C can be determined in the presence of each other by means of several counts at different discriminator settings.

In nuclear research β-active nuclides are usually prepared in the form of thin solid samples and measured with thin-window counters. However, lack of reproducibility of absorption and self-absorption effects can be troublesome in this technique. Therefore if a sample emits both β and γ rays, γ assay is generally the method of choice, since absorption effects are much smaller for γ than for β rays and since γ-ray measurements with scintillation and especially with Ge(Li) counters allow the determination of a specific γ ray in the presence of others. Samples may be prepared as solids or in solutions that are placed near the detector in some standard arrangement. A convenient device for efficient γ counting is the well-type scintillation counter. Since absorption effects are so small for γ rays of all but the lowest energies, no great precautions are usually required in the mounting of samples for γ assay, as long as reproducible positioning relative to the detector is assured.

By contrast, various problems arise with solid samples for β measurements, and since, despite the advantages of γ assays, nuclear chemists are inevitably confronted with the need for some β measurements on solid samples, we devote a few paragraphs to these problems.

Backscattering, Self-Scattering, and Self-Absorption. The phenomenon of back-scattering of electrons has been described on p. 224. To achieve reproducibility in the measurement of β activities, all samples are usually mounted on thick supports of low-Z material (plastic or aluminum) and assayed in the same geometry. This is adequate for relative measurements, except for accurate comparisons of different β emitters, which may require some correction for the energy dependence of backscattering at low energies.

In addition to backscattering, electrons also undergo scattering and absorption in the sample itself. These effects become negligible for samples $\ll 1\ \mathrm{mg\ cm^{-2}}$ thick, but it is not always practical to make samples that thin.

Whenever it becomes necessary to do β measurements on thicker samples, it is advisable either to standardize the thickness at a fixed value—this is often adequate for relative measurements, for example in tracer applications—or to prepare an empirical calibration curve for different thicknesses. In either case careful attention must be given to a reproducible mechanical form for the sample, and reproducibility should be tested by experiment. The calibration curves obtained normally include the effects of backscattering.

Self-absorption and self-scattering depend not only on β-particle energy, but also on the chemical form of the sample and on the geometrical arrangement of sample and detector. With increasing sample thickness the counting rate from a given amount of activity at first usually increases due to scattering of electrons out of the sample plane into the counter. After reaching some maximum, which may be as much as 1.3 times the counting rate for a "weightless" sample, the counting rate decreases as the absorption effects become dominant.

For work of the highest precision nearly weightless samples should be mounted on essentially weightless plastic films (<0.1 mg cm^{-2})[6] and assayed in a 4π counter (see section G).

If the specific activity of a sample rather than the total activity is of interest, as is frequently the case in tracer applications, "infinitely thick" samples, that is, samples at least as thick as the β-particle range, may be used, provided all samples to be compared have the same chemical composition and uniformly cover the same area.

Useful Sample-Mounting Techniques. A large variety of methods is available for the preparation of solid samples for radioactivity measurements. The choice will depend on the type of measurement to be performed, the total as well as the specific activity available, the physical and chemical properties of the radioelement to be measured, the thickness and degree of uniformity desired, the need for quantitative or semiquantitative transfer, and so on.

One of the simplest techniques is the evaporation of a solution to dryness in a shallow cup or, in small portions, onto a flat disk. This procedure, best carried out under an ordinary infrared lamp, always leaves a very nonuniform deposit, with most of the residue in a ring around the edge. Various tricks can be used to improve the uniformity of the deposits, for example the addition of a wetting agent such as tetraethylene glycol, or precipitation and settling of the active material prior to evaporation.

Precipitation followed by filtration and drying generally gives more uniform deposits. Figure 8-2 shows a convenient arrangement for sample preparation by filtration (similar to an arrangement ascribed to Hahn). The filter paper is supported on a sintered glass disk with a fire-polished rim, which is clamped between the thickened and ground ends of two glass tubes; the top tube serves as the area-defining chimney and the bottom tube is fitted into a rubber stopper on a filter flask. The precipitate is usually washed with alcohol or acetone, which helps both to dry it and to wash down any precipitate particles from the walls of the top chimney.

Centrifugation into the demountable bottoms of specially constructed centrifuge tubes could be an alternative way of preparing precipitated samples for measurement.

Samples prepared in any of these ways should be thoroughly dry before measurement, otherwise the self-absorption and self-scattering will change with time as water evaporates. Precipitated samples must be handled carefully to avoid shifting of the precipitate and, whenever possible, they should be covered with a thin film of plastic such as Mylar or Formvar to avoid losses and, most importantly, contamination of the measuring equipment.

Other sample preparation techniques may be appropriate in specific cases. Some metals (such as copper and iron) can be deposited electroly-

[6] The films and techniques described on p. 290 for target backings are suitable here also.

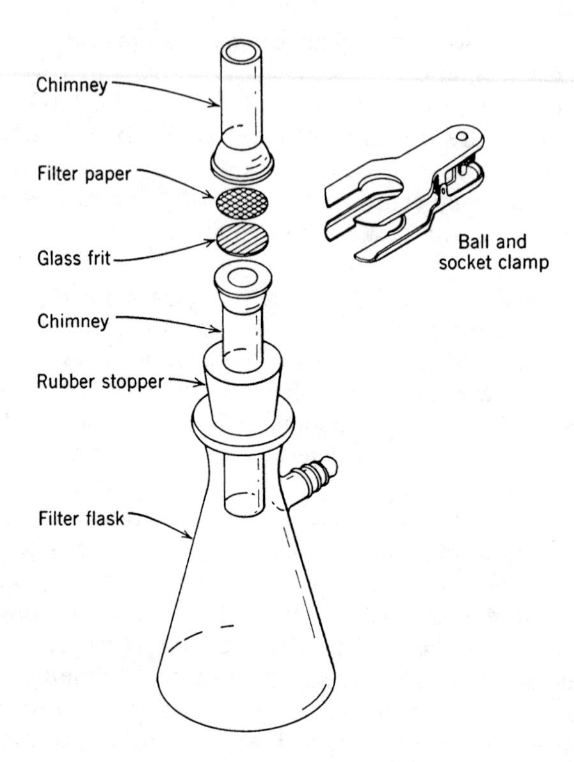

Fig. 8-2 Convenient filter apparatus for the preparation of radioactive samples for measurement. (Courtesy Brookhaven National Laboratory.)

tically, as can be certain insoluble compounds. For example, an adherent coat of UF_4 can be deposited on a cathode by reduction of a uranyl salt in the presence of F^-, and may subsequently be ignited to U_3O_8.

"Weightless" Sources. The preparation of the extremely thin (sometimes loosely called weightless) sources required for α and β spectrometry and for 4π counting presents special problems. In order to prevent broadening of lines in α-particle or conversion-electron spectra, to minimize distortions of β spectra, and to ensure virtually 100 percent efficiency in 4π measurements, such sources may have to be as thin as 1–10 $\mu g\,cm^{-2}$. Uniformity is also important, insofar as the specification of maximum surface density, set by a given experimental situation, applies not only to the source as a whole but to any small portion of it. Samples for 4π counting and for investigations of β-spectral shapes must not only be thin themselves but they must be mounted on equally thin backings. The preparation of thin plastic films for this purpose has already been mentioned (p. 290) and is discussed in review articles (Y1, P1). An insulating film with a radioactive source deposited on it can become highly charged as a result of the emission of charged particles from the source, and the source potential built up in this manner can seriously distort the spectrum of emitted particles. For this reason films used for β-spectrometer or 4π sources should always be rendered conducting, usually by evaporation of a thin ($\sim 5\ \mu g\ cm^{-2}$) metal coating, and grounded. A noble metal has obvious advantages since sources are often deposited from acid solutions. Gold coatings have been used most frequently but palladium is even more advantageous because its smaller infrared absorption (compared with gold) lowers the probability of film breakage when the source is evaporated under a heat lamp.

When quantitative deposition of a given amount of source material on a thin backing is required, as in absolute disintegration rate measurements by 4π counting, evaporation of a solution is the method of choice. Uniform spreading is usually ensured by use of a wetting agent such as insulin. An aqueous insulin solution (concentration ≈ 5 percent) is pipetted onto the spot to be covered by the source, then removed with the pipette. The residue may be dried, and the sample is then pipetted onto the spot and dried under a heat lamp. Successive portions of sample as well as washings may be added and evaporated.

When quantitative transfer is not essential, thin uniform samples may be prepared by such techniques as volatilization, electrodeposition, electrophoresis, and electrospraying. All of these methods have already been discussed in connection with target preparation (section A). Volatilization from a hot filament can be applied to most elements. Occasionally it can even be carried out in air, for example, for transferring such volatile elements as polonium and astatine from a metal holder to a counting disk placed above it. More often a simple vacuum system is used. By careful design of the filament and receiver assembly the evaporation can be made reasonably directional so that losses are not excessive. The catcher can even be a thin plastic film if heating by radiation from the filament can be kept from destroying the film. Whenever a source is prepared by volatilization, it is advisable to get rid of volatile impurities by heating the sample filament to a temperature just below that required for the evaporation of the desired material, and then bringing the source mount into position and raising the temperature to the required range.

A special technique is available for the preparation of thin samples of radionuclides, which are themselves formed by radioactive decay, especially α decay. The recoil energy imparted by the α decay is used to carry the daughter atoms out of a deposit of the parent material and onto a nearby catcher plate. Similarly, the recoil energy imparted by a nuclear reaction can be used to transfer reaction products directly from a thin-target deposit to a catcher foil placed downstream from the target in the ion beam. These techniques have been particularly useful in the investigation of short-lived transuranium nuclides produced in accelerator bombardments.

D. DETERMINATION OF HALF LIVES

Methods for the determination of half lives vary with the half life to be measured. We have divided up the enormous range of experimentally accessible half lives (10^{22} s $\geq t_{1/2} \geq 10^{-18}$ s) into three groups for the discussion of measurement techniques. The boundaries between these groups are, of course, not sharp.

Long Half Lives. If the half life, or disintegration constant, is to be determined for a substance of very long half life (very small λ), the activity $\mathbf{A} = c\lambda N$ may not change measurably in the time available for observation. In that case λ may be found from the relation $\lambda N = -dN/dt = \mathbf{A}/c$, provided $-dN/dt$ may be determined in an absolute way (through knowledge of the detection coefficient c) and N is known or can be determined (e.g., mass-spectrometrically, by the isotope dilution technique—see chapter 11, section B). This method, which is essentially a measurement of specific activity, is probably most accurate for α emitters because their absolute disintegration rates are relatively easily measured (see section G), but it has been used also for many β-active nuclides such as ^{137}Cs ($t_{1/2} = 30.17$ y), ^{99}Tc (2.14×10^5 y), and ^{205}Pb (1.4×10^7 y). The absolute rates of emission of α particles from uranium samples have been investigated with great care to measure the half life of ^{238}U. In an accurate determination of the half life of ^{239}Pu the value of $-dN/dt$ was established in a calorimetric measurement of the heating effect, with the α-particle energy known from separate measurements.

In some instances the disintegration rate is better obtained from a measurement of the equal disintegration rate of a daughter in secular equilibrium. Early determinations of the half life of ^{235}U were based on the α-particle counting rate of ^{231}Pa obtained in known yield from old uranium ores; the ^{235}U α particles were not measurable in a direct way because of the much larger number of α disintegrations occurring in ^{238}U and ^{234}U.

To determine half lives in the range of years to hundreds of years it is convenient to use differential measurements, that is, to compare, as a function of time, the activity of a sample having the half life to be determined with that of a sample with sufficiently long half life to be practically nondecaying. This may be done by using two balanced ion chambers and measuring the difference in ion currents. More generally useful is the technique of measuring the ratio R of the two activities with a single counter as a function of time (H3). Great care must be taken to ensure that the samples are always measured under exactly the same conditions (reproducible placement, equal air path from sample to detector, etc.). Then, if the decay constant of the reference source is negligible relative to the decay constant λ of the unknown, $R = ce^{-\lambda t}$ (where c is a constant). With 10^7–10^8 counts accumulated per measurement, a half life can be determined to an accuracy of 5–10 percent with measurements extending over about $0.01 t_{1/2}$.

Intermediate Half Lives. Half lives in the range from several seconds to several years are usually determined experimentally by measurements of the activity with an appropriate instrument at a number of suitable successive times. After counting, log **A** is plotted versus time and the half life may be found by inspection, provided that the activity is sufficiently free of other radioactivities that a straight line (exponential decay) is found, preferably extending over several half-life intervals. As discussed in chapter 5, the decay curve resulting from a mixture of independent activities may often be analyzed to yield the half lives of the various components. This can be accomplished by the use of computer programs that fit multiple components of different half lives to the data by a least-squares procedure (C1). It may be advantageous to use energy-selective instruments such as semiconductor detectors for γ-ray or α-particle counting to measure separately the radiations from several activities in the sample. Alternatively it is sometimes adequate to measure decay curves separately through several thicknesses of absorbing material to obtain data with some components relatively suppressed. Our treatments of the more general equations in chapter 5 have already suggested methods of finding half lives from more complicated growth and decay curves.

For half lives at the short end of the range discussed here, say a few minutes or less, it is often useful to transport the radioactive sample by means of a rabbit system (see p. 301) from the site of production to the location where chemistry and activity measurements take place.

If the number of atoms of a short-lived species produced in a single irradiation is small, it is convenient to do repetitive experiments, with identical timing between irradiation and start of counting, and to accumulate counts in corresponding time intervals by storing them in different memory locations of a computer. Many multichannel pulse height analyzers are equipped for this purpose with a so-called multiscalar mode.

Short Half Lives (F2). More sophisticated techniques and procedures are required as the half life to be determined grows shorter. There are two general types of experiments that are employed. In the first the time dependence of the decay rate of an active sample is still the observed quantity. The lower limit to the half life that can be measured in this manner is ultimately determined by the recovery time of the detector that is employed, but more practically by the time required to transport the sample from its site of formation into the detection system (M2). In the second type of experiment it is not the decay rate of a collection of radioactive atoms that is observed; rather, it is the distribution of the time intervals between the formation and the decay of an active atom that is observed experimentally. This distribution is again described by the exponential decay law.

In experiments of the first type the short-lived species is usually produced in a nuclear reaction and advantage is taken of the fact that the reaction, particularly if it is of the compound-nucleus type, imparts momentum to the products and can cause a fraction of them to recoil out of the target foil. These radioactive recoils are then caught on some sort of rapidly moving conveyor and transported from the target area to a detector. In one such system, already discussed on p. 302 and shown schematically in figure 8-1, the radioactive recoil is stopped in fast-flowing helium and then is carried with a helium jet through a small-bore tube to the detector or series of detectors (M2). Half lives down to about 10^{-3} s have been measured in this manner. It is also possible to combine the helium jet with a mass separator to transport the radioactive species from the target and to separate it according to the charge-to-mass ratio.

In experiments of the second type it is necessary to have a signal at the time that the decaying state is formed (the start signal) and at the time that the state decays (the stop signal). These two signals are sent to an electronic circuit that, after many such events, gives the distribution in elapsed time between these two signals. The result is an exponential decay, quite analogous to a conventional decay curve (see figure 8-3). If the short-lived activity results from a radioactive decay with moderate or long half life such as, for example, a γ ray that follows a β decay, the detection of a ray from the parent can supply the start signal while that from the daughter can supply the stop signal. If the short-lived activity is produced in a nuclear reaction in an accelerator, it is often possible to modulate the

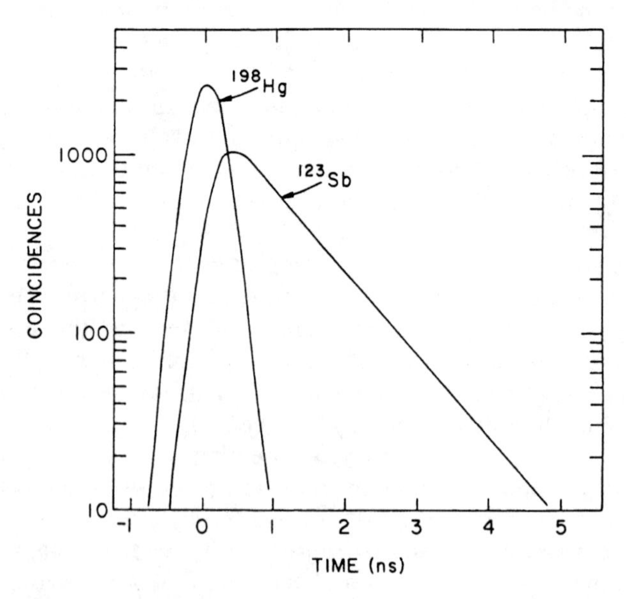

Fig. 8-3 The 0.64 ± 0.05 ns decay of the first excited state of ^{123}Sb as measured by delayed coincidences between the β^- particles of ^{123}Sn feeding this state and the conversion electrons of the 160-keV transition de-exciting it. For comparison the prompt decay of the 412-keV transition in ^{198}Hg in the decay of ^{198}Au is also shown. [Data from M. Schmorak, A. C. Li, and A. Schwarzschild, *Phys. Rev.* **130**, 727 (1963).]

beam in time so that it arrives in sequential and narrow time intervals and the start signal is provided by the beam pulse. It is possible to measure half lives down to about 10^{-11} s with this general technique.

Another way of determining the lifetime of a short-lived γ-ray emitter formed in a nuclear reaction is through the Doppler shift of the γ-ray energy. As mentioned previously, momentum will be imparted to the reaction product; this will cause a Doppler shift in the energy of a γ ray depending on the velocity of the nucleus at the time of decay. By suitable experimental arrangements it is possible to know the velocity of the recoiling nucleus as a function of time and thus the time of decay from the Doppler shift of the energy. Lifetimes down to about 10^{-15} s have been measured in this manner.

Other experiments are designed to measure half lives shorter than about 10^{-16} s from the energy width ΔE (FWHM) of an excited state; the mean life Δt is then deduced from application of the uncertainty principle: $\Delta E \cdot \Delta t = h/2\pi$.

E. DECAY SCHEME STUDIES

Considerable effort by nuclear chemists and physicists has been directed toward the collection of data on decay schemes (disintegration schemes) of radioactive nuclides. A complete decay scheme includes all the modes of decay of the nuclide, the energies and transition rates of the radiations, the sequence in which the radiations are emitted, the measurable half lives of any intermediate states, and all quantum numbers, particularly spins and parities, of all the energy levels involved in the decay.

The study of decay schemes of radioactive nuclides is only *one* branch of nuclear spectroscopy. First of all, radioactive decay populates levels only up to some energy, determined by the Q value of the decay; secondly, even in that energy range not all levels may be populated because of selection rules; thirdly, there are many nuclides whose level schemes are not conveniently accessible through decay studies (e.g., because of half life), but that can be investigated by other techniques. Among the other techniques widely used are Coulomb excitation, inelastic scattering, various nuclear reactions, and in-beam γ-ray spectroscopy. Detailed coverage of these topics is beyond the scope of this book; some have been touched on in chapter 4, and some in-beam techniques are discussed in section F.

Survey of Techniques. The amount of detail known about any given decay scheme depends very strongly on the refinements in instrumentation and technique used in its investigation. Frequently, when a previously well-studied decay scheme is reinvestigated with instruments of improved resolution or sensitivity, new features such as decay branches of low abundance are discovered. It now appears that except among the lightest nuclei there are indeed few decay schemes that are truly simple (for example, consisting of a single β transition to a ground or first excited state).

The starting point of a decay scheme investigation depends on the information already available about the nuclide under study. For a previously unknown nuclide the half life needs to be established.[7] The decay mode or modes are identified by use of appropriately selective detectors for α and β particles, conversion electrons, γ and X rays, and fission fragments (cf. chapter 7). Absorption measurements may be helpful, particularly in the initial characterization of β emission. Positron emission can be sensitively and uniquely established by detection of the two 511-keV annihilation quanta in coincidence at 180°, for example in two NaI(Tl) or two Ge(Li) counters.

Determination of the energy spectra of the various radiations emitted involves the use of the energy-sensitive detection devices discussed in chapter 7. Gamma-ray spectroscopy with Ge(Li) detectors is probably the most widely used technique in decay scheme studies. Alpha-particle and conversion-electron spectra are also relatively easily measured with semiconductor detectors. To determine end points, and particularly shapes, of β spectra is far more tedious. The particular measurements that can be done and the choice of instruments to be used may depend strongly on available source strengths and specific activities as well as on the half life.

Questions about the sequence in which various radiations are emitted and about the existence of alternative decay paths are usually answered by coincidence measurements. As already indicated in chapter 7, the more selectively each of the two detectors used in a coincidence study records one particular radiation, the more readily even very complex decay schemes may be disentangled. Since increased selectivity is almost always accompanied by decreased detection efficiency, some compromise usually needs to be made in practice. The development of multiparameter, multichannel analyzers (see chapter 7) has enormously increased the scope of coincidence measurements that can be undertaken without excessive expenditures of time. Coincidence techniques are often very helpful in energy determinations also. For example, a low-intensity, low-energy β branch in the presence of an intense high-energy component may well escape detection in a β-spectrographic measurement. If, however, the high-energy β spectrum is not coincident with γ rays, but the low-energy branch is, then a β spectrum taken in coincidence with γ rays will show the low-energy component only, and precise measurement of its end point

[7] Even when the decay scheme of a nuclide of well-known half life is under investigation, the nuclide may not be available free from other radioactive isotopes. In that case measurements have to be made as a function of time to sort out those radiations associated with the nuclide of interest.

and spectrum shape thus becomes possible. With a β spectrometer (such as a plastic scintillator) and a Ge(Li) detector in coincidence, the β spectrum coincident with each of several γ radiations may be observed.

The following two examples may serve to illustrate many of the techniques of decay scheme studies. They are relatively simple, yet encompass the same essential features that would be encountered in more complex schemes. Further details of these decay schemes and references to the original literature may be found in L1.

Gold-198. This nuclide of 2.696 d half life was for a long time thought to have a simple disintegration scheme, decaying by the emission of a single β^- group of allowed spectrum shape and upper energy limit of 0.96 MeV to the lowest excited state of ^{198}Hg at 0.412 MeV above the ground state. This scheme was verified by numerous spectrometer and coincidence measurements.[8] The nuclide has, in fact, been frequently used as a standard source for the calibration of spectrometers and coincidence circuits. The energy of the γ ray has been determined with great precision in crystal spectrometers and by magnetic spectrometer measurements of the conversion electrons and is given as 411.80441 ± 0.00015 keV. The internal-conversion coefficients have been measured in magnetic spectrometers by comparison of the areas under the conversion electron peaks with the area under the entire β spectrum as well as by other techniques. The best values appear to be $\alpha_K = 0.0300 \pm 0.0003$, $\alpha_K/\alpha_L = 2.79 \pm 0.05$, and $\alpha_{L_I}/\alpha_{L_{II}}/\alpha_{L_{III}} = 2.2/2.2/1.0$. These data establish an $E2$ assignment for the 412-keV transition. Since the ground state of even-even ^{198}Hg is presumably 0^+, it thus appears that the 412-keV level is 2^+, in accord with the general rule for first excited states of even-even nuclei. The 0.961-MeV β spectrum was long believed to have the "statistical shape" given by (3-19), but more careful recent measurements have shown that a correction factor of the form $1 + aW$, with $a = -0.05 \pm 0.02$, is required to linearize the Kurie plot (see p. 91). The transition is thus identified as nonunique first forbidden, with $\Delta I = 0$ or 1 and parity change. The $\log ft$ value is 7.4,[9] in accord with this assignment for the transition. The spin-parity assignment of ^{198}Au could thus be 1^-, 2^-, or 3^-, with 1^- immediately excluded because it would make the β transition to the ^{198}Hg ground state of the same order as that to the 412-keV state; this ground-state transition, however, is certainly not prominent and must therefore have a much higher $\log ft$ value than the observed 961-keV β transition.

[8] When reactor-produced ^{198}Au sources became available there was some confusion about the decay scheme because several workers found additional lower-energy γ rays in such sources. Subsequently the relative intensity of these γ rays was found to depend on the neutron flux in which the gold had been irradiated, and they turned out to be associated with 3.14-d ^{199}Au formed with very large cross section by neutron capture in ^{198}Au.

[9] The approximate expression in (3-24) gives $\log ft = 7.7$.

Since ^{198}Pt is stable, ^{198}Au might be expected to decay by β^+ emission or EC in addition to β^- emission. Searches for annihilation radiation by means of 180° coincidences set an upper limit of 0.003 percent for β^+ emission before mass measurements showed that the decay energy between the ^{198}Au and ^{198}Pt ground states is only 0.303 MeV, thus excluding β^+ decay completely. Searches for platinum K X rays have set an upper limit of 0.01 percent for K-EC, which [according to (3-26)] corresponds to $\log f_0 t \geq 9.4$, a reasonable result for this first forbidden unique transition.

With scintillation counters two additional γ rays of low abundance were found in ^{198}Au decay in 1950. The energies have since been determined with great accuracy (through conversion-electron measurements) to be 0.67588 and 1.08767 MeV, and their intensities relative to the 0.412-MeV γ ray are 1.1×10^{-2} and 2.4×10^{-3}, respectively. The energies of the three γ rays suggest strongly that there is a state 1.088 MeV above the ^{198}Hg ground state and that it decays both directly to ground and also to the 0.412-MeV state. Coincidence measurements indeed showed the 0.676-MeV γ rays to be in coincidence with the 0.412-MeV radiation and the 1.088-MeV γ rays not to be coincident with any other γ rays. Measured internal-conversion coefficients for the K shell and K/L conversion ratios indicate that the 0.676-MeV transition ($\alpha_K = 0.022 \pm 0.002$, $K/L = 5.7 \pm 0.5$) is an $M1$-$E2$ mixture, and the 1.088-MeV transition ($\alpha_K = 0.0045 \pm 0.0003$, $K/L = 6.3 \pm 0.5$) is $E2$. The 1.088-MeV state is then clearly 2^+ (as is the first excited state at 0.412 MeV). More detailed information about the $M1$-$E2$ mixing ratio of the 0.676-MeV transition has come from numerous angular-correlation measurements for the $\gamma\gamma$ cascade.

Coincidence experiments, with a lens spectrometer and a NaI scintillation spectrometer as the two detectors, showed the 0.676-MeV γ ray to be in coincidence not only with the conversion electrons of the 0.412-MeV γ ray but also with a β^- spectrum of upper limit 0.290 ± 0.015 MeV, which has the allowed shape and an intensity about 1.3 percent that of the main β spectrum. A third β^- transition, with an intensity 2.5×10^{-4} of the main spectrum and an upper limit of 1.37 MeV, was found in a magnetic-spectrometer study of a strong source and evidently represents the ground-state transition. The spectrum shape identifies this transition as $\Delta I = 2$, yes; thus the spin and parity assignment of ^{198}Au becomes 2^-.[10] The $\log ft$ value of ~ 12 for the ground-state transition [computed from (3-24)] is consistent with the assignment.

Finally, we mention that the half life of the 0.412-MeV excited state has been measured by delayed coincidences. It was found to be about 23 ps, to be compared with the single particle estimates (from table 3-4) of 0.5 ns and 0.3 ps for $E2$ and $M1$ transitions, respectively. The magnetic moment of the

[10] The $I = 2$ assignment for ^{198}Au has been independently established by an atomic-beam measurement.

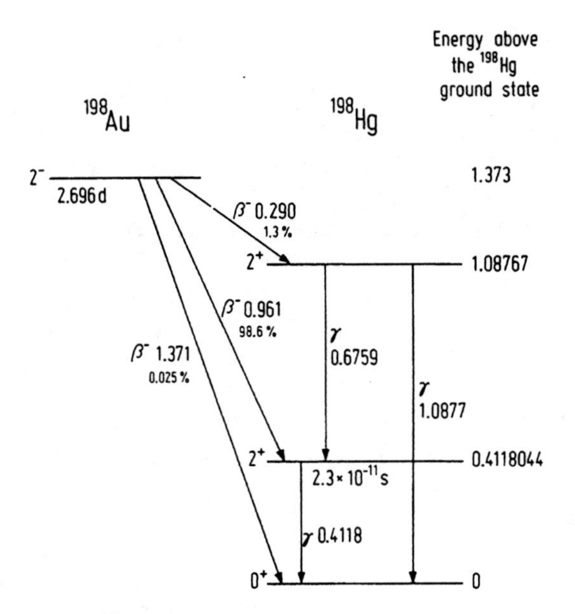

Fig. 8-4 Decay scheme of ^{198}Au. All energies are in MeV. Spin-parity assignments are shown to the left of the energy levels.

0.412-MeV state has been determined from measurements of the precession of the angular correlation in a magnetic field.

The decay scheme of ^{198}Au is shown in figure 8-4. We note in passing that additional low-lying levels of ^{198}Hg, not populated in ^{198}Au decay, have been found in ^{198}Tl decay and in nuclear-reaction studies (see L1).

Lead-204*m*. An interesting case of isomerism occurs in the even-even nuclide ^{204}Pb. A 67-min isomer decaying with the emission of γ rays of about 1 MeV has been known for some time. It is formed in the EC decay of ^{204}Bi but not in the β^- decay of ^{204}Tl. In 1950 an investigation of the electron spectrum of the lead isomer with a lens spectrometer showed *K*- and *L*-conversion lines of two γ rays, of energies 374 keV and 905 keV, with *K*/*L* conversion ratios of about 2.1 and 1.5, respectively. These *K*/*L* ratios suggest (cf. table 3-5) an $E2$ assignment for the 374-keV transition and multipole order $>2^4$ (actually $E5$, not included in table 3-5) for the 905-keV transition. Approximate values for total conversion coefficients obtained by absorption measurements were also compatible with the $E2$ and $E5$ assignments. According to table 3-4, the 67-min half life is compatible with a 905-keV $E5$ transition but not with the $E2$ transition of 374 keV. Thus it was concluded that the 905-keV step is the isomeric transition and is followed by the 374-keV transition. Delayed coincidences between the two γ rays were found with scintillation counters as detectors.

By variation of the delay time the half life of the 374-keV transition was determined to be 0.3 μs.

Later measurements with Ge(Li) detectors showed the "905-keV" γ ray to be resolved into two γ rays of 899.3 and 911.7 keV and of approximately equal intensity, and gave 374.7 ± 0.4 keV for the energy of the third γ ray. The 899- and 375-keV radiations were found to be in prompt coincidence with each other and delayed relative to the 912-keV transition with a 0.27-μs half life. The order of emission of the 375- and 899-keV γ rays shown in figure 8-5 was originally inferred from indirect evidence, such as comparison of the 0.27 μs half life with theoretical predictions (table 3-4) and absence of 375-keV γ rays in the β^- decay of ^{204}Tl (a 2^- state) and in the α decay of ^{208}Po (0^+). A 2^+ state in ^{204}Pb at 375 keV would be expected to be appreciably populated in both decay modes. More direct evidence for the γ-ray sequence 912-375-899 keV and unambiguous assignment of spins

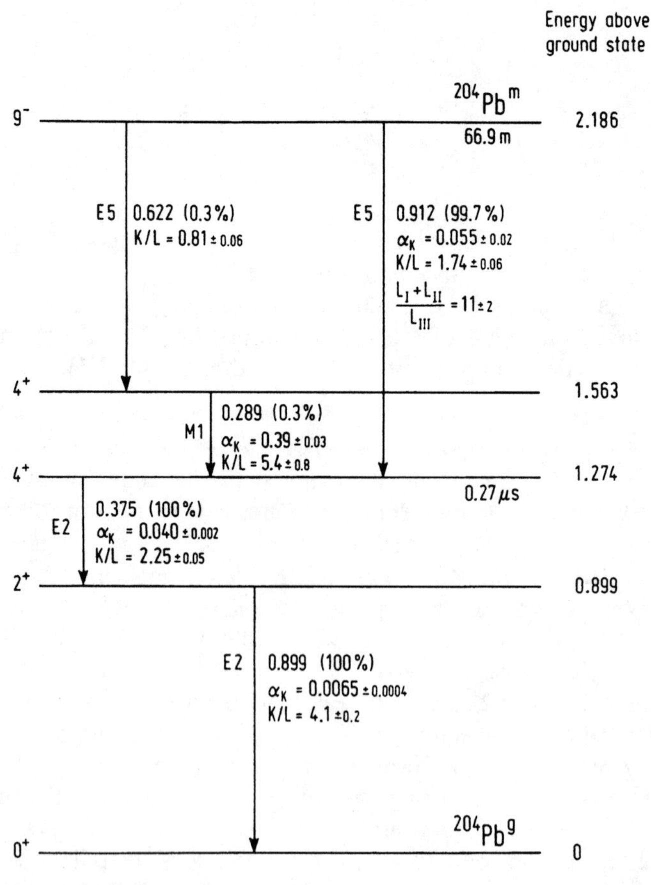

Fig. 8-5 Decay scheme of ^{204}Pb^m. All energies are in MeV; absolute transition probabilities (in percent of total transitions) are given in parentheses after the transition energies.

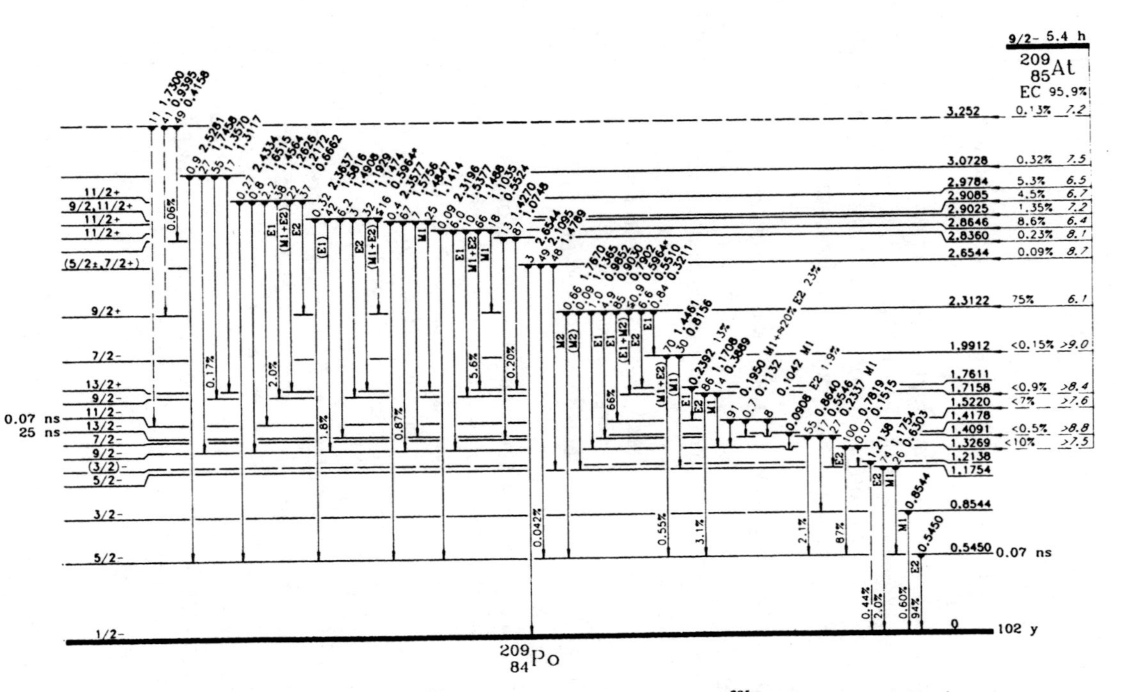

Fig. 8-6 Decay scheme for the EC decay of ^{209}At (the 4.1 percent α branch to ^{205}Bi is not shown). To the right of each level are shown: the energy above the ground state of ^{209}Po in MeV, the percentage of EC transitions to the level, and the log ft value for the transition. Spin-parity assignments are shown on the left. Above the transitions de-exciting a given level are first the percentage of the de-excitations proceeding by each transition, then the transition energy in MeV. Multipole orders are indicated where known. (From reference L1.)

to the states involved has subsequently come through angular-correlation measurements for the pairs 375-899, 912-375, and 912-899. The results of these measurements proved that the spins 9, 4, and 2 previously assigned to the levels at 2.186, 1.274, and 0.899 MeV are correct (assuming that the ground state has 0 spin). The parity designations shown in figure 8-5 for these states then follow from the multipole character assigned to each of the three transitions, which, in turn, is solidly based on rather detailed conversion-coefficient information, some of which is shown in the figure. All of these deductions are based on principles discussed in chapter 3, sections D and E.

Two additional transitions of low intensity (each about 0.3 percent of the main cascade intensity) were found at 289 and 622 keV by conversion electron spectroscopy with magnetic spectrometers. Their probable placement and multipole orders are shown in figure 8-5, which thus includes a second 4^+ level, at 1.563 MeV. Other levels in the energy region covered are not shown, although several such levels are populated in the EC decay of ^{204}Bi and in various inelastic-scattering and pickup reactions.

Complex Decay Schemes. As we emphasized before, the particular decay schemes discussed above in detail are unusually simple. They were chosen in order to illustrate some general principles while at the same time avoiding excessively lengthy and confusing discussion. To indicate what a more typical, though by no means unusually complex, decay scheme looks like, we show in figure 8-6, without discussing it, the level scheme of ^{209}Po as deduced from the EC decay of ^{209}At. The general approaches to the unravelling of such a scheme are the same ones as for the simpler cases. Often it is helpful, if not essential, to check the deductions from experimental data against the predictions of particular nuclear models or, in fact, to use the models for sorting out possible level sequences, energy spacings, and so on. This approach will become clearer in the light of the discussion of nuclear models in chapter 10.

F. IN-BEAM NUCLEAR-REACTION STUDIES

As was briefly discussed in chapter 4, section C, the important observables of a nuclear reaction include the angular and energy spectra of emitted particles as well as spatial and temporal correlations among them. "Particles" in this context can include anything from γ quanta to fission fragments and recoil nuclei, but we are here exclusively concerned with so-called on-line or in-beam studies, that is, measurements of what occurs within a very short time span, say $\lesssim 10^{-1}$ s, of the reaction. In the preceding sections we dealt with the equally important problems of identifying the reaction products "off-line."

The various instruments used in detection and identification of ions, nucleons, mesons, electrons, quanta, and so on are described in chapter 7, and some of the relevant considerations concerning interactions of radiations with matter are discussed in chapter 6. We therefore restrict ourselves here to rather brief accounts of how these instruments and interactions are applied to various problems in on-line reaction studies.

Particle Identification (G3). The unambiguous identification of particles emitted in a reaction usually requires the simultaneous measurement of their specific ionization and at least two of the following quantities: kinetic energy, momentum, and velocity.

Kinetic energy E ($=\frac{1}{2}Mv^2$ nonrelativistically) can usually, that is, for modest energies, be determined by stopping the particles completely in a detector (most frequently a gas ion chamber or semiconductor detector) in which a pulse proportional to the particle's kinetic energy is developed. The energy resolution obtainable with scintillators is typically a few percent, with semiconductor detectors an order of magnitude better.

Momentum p ($=Mv$ nonrelativistically) is most directly measured by magnetic deflection, since the radius of curvature ρ of a particle of momentum p and charge z in a magnetic field of strength B is p/Bz. Note that z is the net charge of the ion, not necessarily the atomic number.

Specific ionization dE/dx is, of course, best measured by allowing the particles to pass through a detector (semiconductor, proportional counter, or gas ionization chamber) thin compared to their range, and recording the energy deposited in that detector. (Much more crudely a measure of specific ionization can be deduced from the grain density in a nuclear emulsion or the track width in a dielectric track detector.) As discussed in chapter 6 (p. 214), ions of different charge and mass but equal kinetic energy can be distinguished by their specific ionization, and a variety of particle-identifying schemes are based on this fact. They all involve a counter telescope consisting of one or more[11] thin ("transmission") detectors to measure dE/dx (or, more properly, $\Delta E/\Delta x$) and a total-absorption detector. The ΔE detector may be a thin semiconductor wafer or a proportional counter. Particle identification was first done by using an electronic circuit that forms the product of E and dE/dx, which, according to (6-14), is approximately proportional to Mz^2, and then plotting number of events against this product; a **particle identifier spectrum** is thus obtained, with separate peaks for ions of different values of Mz^2. More commonly all pairs of values of ΔE and E are stored on magnetic tape or in a computer for subsequent analysis. On plots of ΔE versus E, points belonging to different z's (and, for light elements, to different isotopes) fall on separate curves, as shown in figure 8-7. Particle identification spectra may be constructed from the stored data, usually by algorithms somewhat

[11] With two or more ΔE detectors we sample dE/dx at more than one energy and thus gain better discrimination among particles.

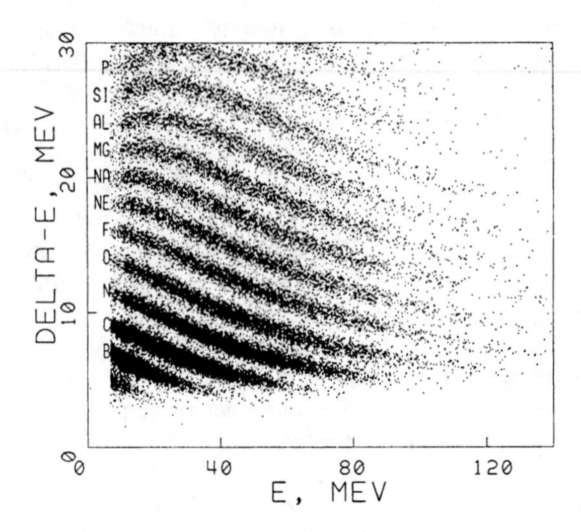

Fig. 8-7 Plot of ΔE versus E for fragments produced in the bombardment of uranium by 28-GeV protons. The data were obtained at an angle of 20° with respect to the beam direction with a silicon counter telescope consisting of a 9-μm ΔE and a 250-μm E detector. Each dot represents an event. Satisfactory element separation up to phosphorus ($Z = 15$) is seen. (Courtesy L. P. Remsberg and Brookhaven National Laboratory.)

more complicated than $E \times (dE/dx)$. An example of such a spectrum is shown in figure 8-8. Individual isotopes can be resolved up to $z \approx 10$ and individual elements up to z of 25 to 50, the limit depending on ion energy.

Velocity is most directly obtained from a **time-of-flight** (TOF) measurement: two detectors some distance l apart provide start and stop signals and a circuit called a time-to-amplitude converter (TAC) translates the time interval Δt between these two pulses into a pulse height. Then $v = l/\Delta t$. The start signal is often provided by the secondary electrons emitted from a very thin foil in the ion path, the electrons being detected, for example, in a channel plate (see p. 261). The accuracy of the TOF method depends on the length of the flight path and the resolving time of the circuitry. A particle traveling at a velocity of $0.1c$ requires about 30 ns to traverse a distance of 1 m, and that time can be measured to better than 1 percent.[12] Time-of-flight measurements can, to great advantage, be combined with E and dE/dx measurements (e.g., by using the distance between the dE/dx and E detectors as the flight path), since the combination of velocity and kinetic-energy information gives, at least in principle, unambiguous mass identification. In practice instrumental limitations make clean isotopic resolution, even with this technique, difficult for $z \gtrsim 15$, except for

[12] The TOF method becomes impractical for velocities approaching c. In the region $0.6 < \beta < 0.999$, Čerenkov counters are widely used for velocity measurements (see p. 265).

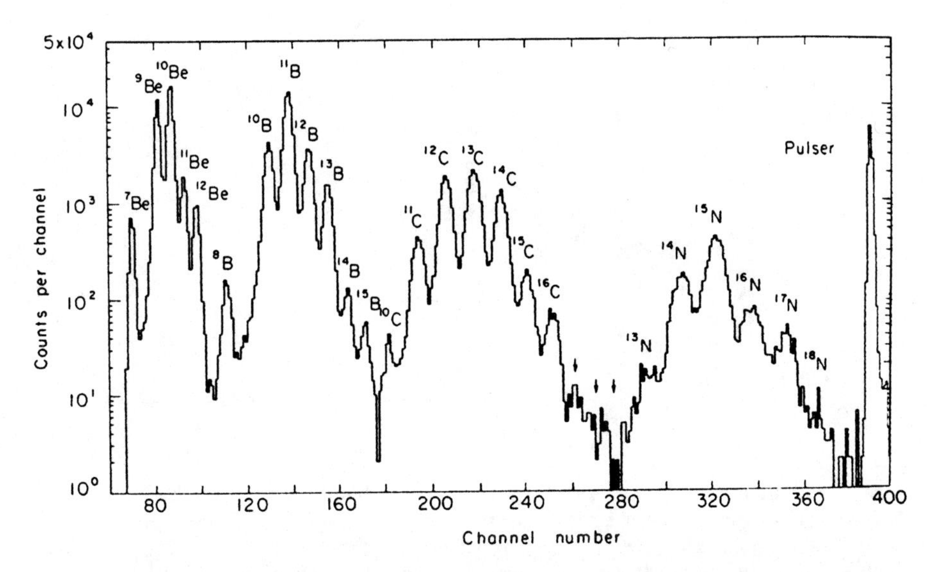

Fig. 8-8 Example of a particle spectrum derived from energy-loss measurements in semiconductor detectors. The fragments in this case were produced by the interaction of 5.5-GeV protons with uranium. [From A. M. Poskanzer et al., *Phys. Lett.* **27B**, 414 (1968).]

high kinetic energies (e.g., in certain heavy-ion reactions) when resolution can be achieved up to $z \approx 40$.

Energy, dE/dx, and TOF measurements can be used in a great variety of experiments in combination with each other and with magnetic deflection. They are employed in the study of energy and angular distributions of emitted particles and, through measurements of coincidences between two or more counter telescopes, in establishing angular correlations between emitted particles. The combination of E, dE/dx, and TOF measurements has also been used for the identification of new nuclides as illustrated in figure 8-9.

On-Line Mass Separation. Nearly instantaneous mass analysis of reaction products has become an important tool in studies of fission, spallation, and heavy-ion reactions (K2). A variety of approaches are possible. One that is used at some reactors (e.g., in the separator called LOHENGRIN at the high-flux reactor at Grenoble) is to separate *unslowed* fission fragments according to their charge-to-mass ratios in a focusing mass spectrograph of moderately high resolution. This allows, for example, the determination of kinetic-energy spectra of mass-separated fission fragments and the investigation of such details as the dependence of fission yields along a mass chain on kinetic energy. In some studies, for example of heavy-ion reactions, direct mass analysis of unslowed reaction products by

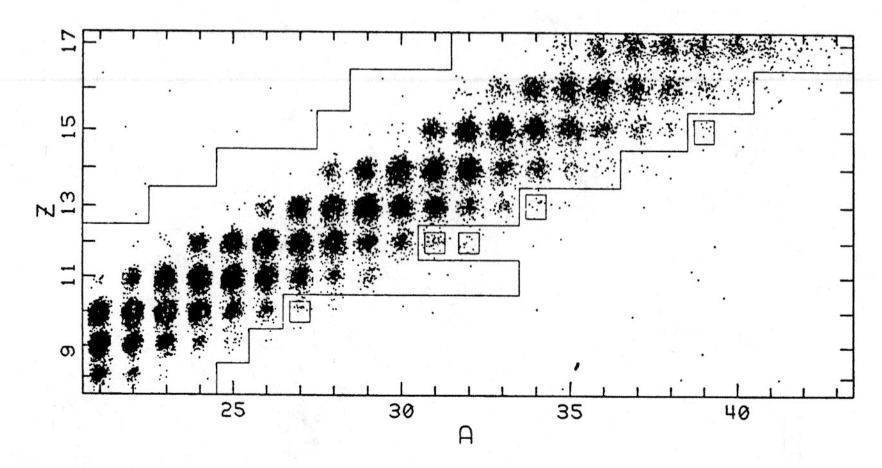

Fig. 8-9 Distribution in Z and A of fragments produced in the bombardment of uranium by 800-MeV protons and identified by a combination of $\Delta E/\Delta x$, E, and TOF measurements. Each dot represents one event. The boxes indicate the nuclides ^{27}Ne, ^{31}Mg, ^{32}Mg, ^{34}Al, and ^{39}P first identified in these measurements. The solid lines enclose the region of previously known nuclei. [From G. W. Butler et al., *Phys. Rev. Lett.* **38**, 1380 (1977).]

magnetic deflection has been combined with TOF, dE/dx, or energy measurements, or even some combination of these.

More widely used has been mass analysis of *stopped* reaction products. Both **mass spectrometers** (instruments in which mass spectra are determined in terms of *ion currents* or *numbers of ions* at different A/q values) and **isotope separators** (in which the object is to *collect samples* of different A/q values, e.g. for nuclear spectroscopy) are in use.

In a technique developed at Orsay (K2) the ion source of a mass spectrometer is placed directly in the beam of bombarding particles; thin target foils are interspersed with graphite slabs, and the entire assembly is heated to some 1800°C. Reaction products recoil out of the target foils into the graphite, and at the high temperature some elements, notably the alkali metals, diffuse out of the graphite very rapidly—in milliseconds—and are ionized on hitting a hot metal wall. The technique has been used at several types of accelerators for cross section determinations, identification of new isotopes, half-life measurements (down to a few milliseconds), and mass determinations (to an accuracy of ~0.1 amu). So far it has been applied mostly to alkali elements, but extension to halogens and possibly other elements seems likely.

On-line isotope separators have been installed at accelerators and reactors. A particularly prolific one has been ISOLDE at the CERN 600-MeV synchrocyclotron. The trickiest part of these systems is the target design, which must achieve rapid and specific removal, usually by volatilization, of one or a few product elements. Emanation of rare gases was, of course, easiest to achieve, but a surprising variety of elements have been successfully studied, including mercury, zinc, cadmium, lead, bismuth, astatine, rubidium, and cesium. Much nuclear spectroscopy of isotopes far from the line of β stability has resulted from the work with on-line isotope separators. The use of a helium-jet system in conjunction with an isotope separator provides a particularly powerful technique.

In-Beam Gamma-Ray Spectroscopy. Since the products of nuclear reactions are generally formed in excited states, in-beam measurements of γ rays can contribute importantly to nuclear spectroscopy. The levels reached in nuclear reactions are by no means necessarily the same as those populated in radioactive decay (although there is usually some overlap), so that in-beam spectroscopy and radioactive-decay spectroscopy complement each other. The detection devices are basically the same, with Ge(Li) detectors playing a dominant role, but the background problems in the vicinity of an accelerator target or at a reactor present special problems. With pulsed accelerators, coincidences between beam pulse and γ-ray pulse are used to advantage to cut down backgrounds. Similarly, coincidences with an outgoing particle, for example, with the emitted proton in a (d, p) reaction, are very useful for background suppression and also help establish the nuclide in which the γ emission occurs. As in radioactive decay, $\gamma\gamma$ coincidences and conversion electron measurements play an important part in establishing level schemes. The use of isotopically enriched targets is of great help in the assignment of γ transitions to specific product nuclides. As a simple example of in-beam γ-ray spectroscopy, we show in figure 8-10 the γ-ray spectra following $(^{40}\text{Ar}, 4n)$ reactions on separated tin isotopes, corresponding to the de-excitation of the members of the so-called ground-state rotational band (see chapter 10, section E) of even-even erbium isotopes.

For the study of complex reactions in which several or many particles and γ rays are emitted, multidetector arrays are very useful. As an illustration, we show in figure 8-11 a multidetector coincidence spectrometer used to study the formation of nuclei with very large angular momenta in heavy-ion reactions. The instrument shown has seven NaI detectors and a Ge(Li) detector with an anti-Compton shield arranged around an on-line target. Arrays have in fact been built with hundreds of detectors. The more sophisticated instruments can simultaneously measure (1) γ-ray multiplicity (the number of coincident γ rays in a cascade), (2) individual γ-ray energies, (3) total pulse height and associated γ-ray multiplicity, (4) neutron multiplicity, (5) γ-ray angular correlations, and (6) delay times between various groups of γ rays in each cascade. Experiments with such instruments generate huge amounts of multiparameter data and require complicated computer programs for data collection, reduction, and analysis.

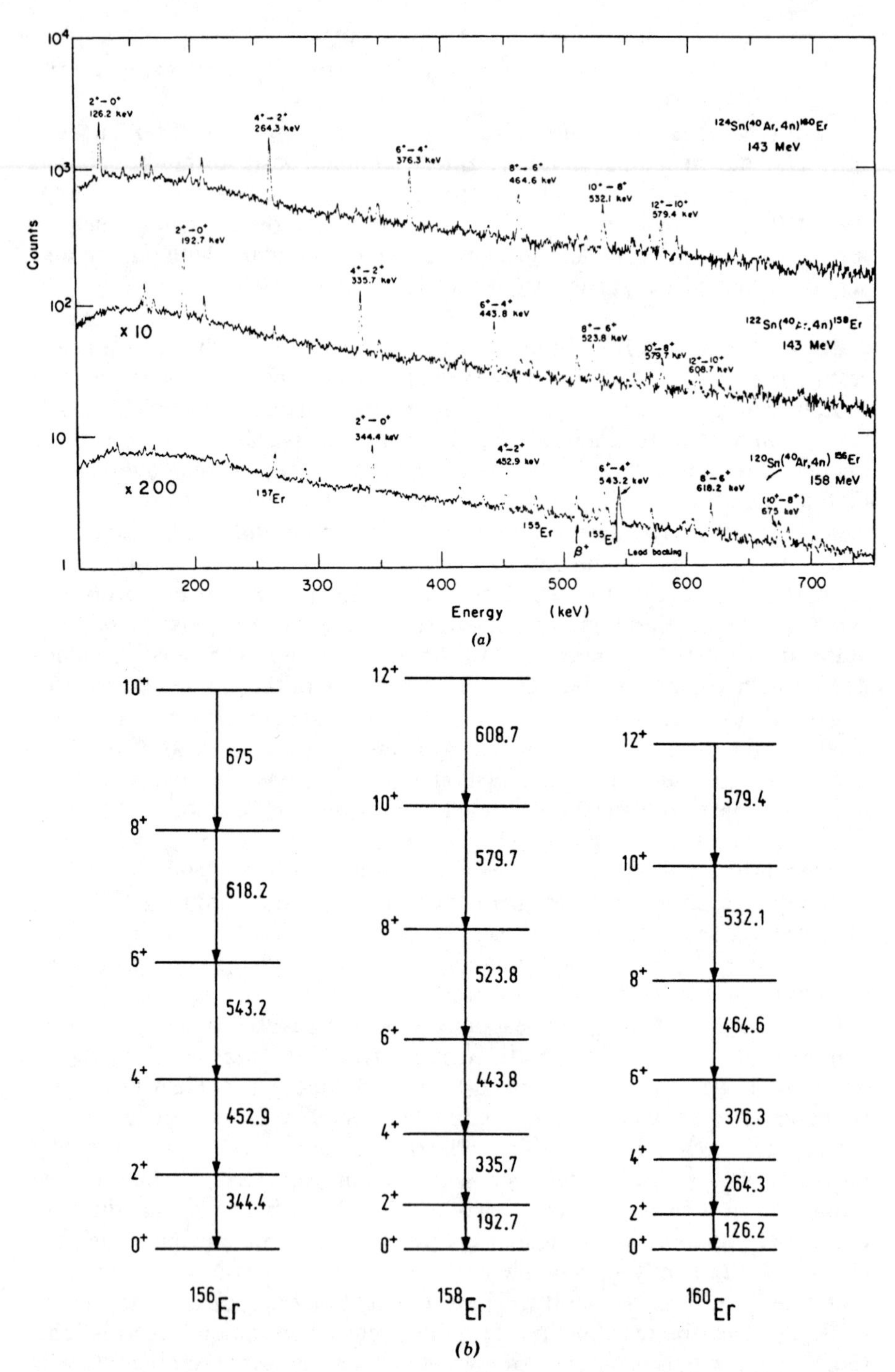

Fig. 8-10 (a) Gamma-ray spectra taken in-beam during bombardment of separated tin isotopes with ^{40}Ar and showing the de-excitation of levels in the erbium isotopes produced by (^{40}Ar, 4n) reactions. [From D. Ward, F. S. Stephens, and J. O. Newton, *Phys. Rev. Lett.* **19**, 1247 (1967).] (b) Gamma-ray cascades of the ground-state rotational bands in ^{156}Er, ^{158}Er, and ^{160}Er derived from the spectra.

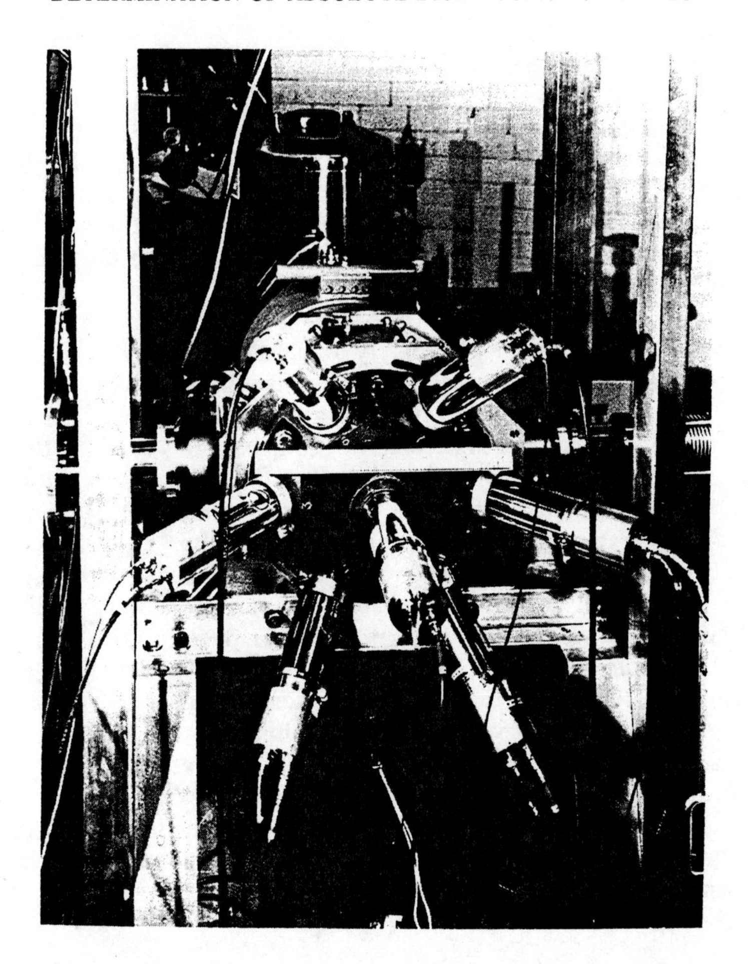

Fig. 8-11 Multidetector array with seven NaI detectors and a Ge(Li) detector used in coincidence studies of γ rays and particles emitted in heavy-ion interactions. (Courtesy D. Sarantites.)

G. DETERMINATION OF ABSOLUTE DISINTEGRATION RATES

As mentioned before, the determination of absolute disintegration rates presents special problems in addition to those encountered in relative activity measurements. Yet a knowledge of absolute disintegration rates is often required. Whenever a reaction cross section is to be determined, the number of product nuclei formed must be found, and for radioactive products this is best accomplished through knowledge of the decay constant and measurement of the disintegration rate. If the disintegration rates of a given nuclide are to be determined in many samples (e.g., in the study of an excitation function), relative measurements are quite adequate as long as the instrument used is calibrated with one sample of the nuclide whose absolute disintegration rate is known. Even in nuclear spectroscopy there is frequent need for what amounts to absolute measurements. In general, when the branching ratio of two decay modes is to be determined measurements of two different types of radiation with different instruments are involved, and the absolute efficiency of each measurement must therefore be known. This applies to EC/β^+ ratios, to absolute conversion coefficients, and often even to measurements of the number of γ quanta per β decay.

Alpha Emitters. Given adequately thin, uniform samples, the determination of absolute α-disintegration rates is relatively simple. Either a proportional counter or an ionization chamber with linear amplifier can be readily arranged to count 100 percent of the α particles entering the active volume, and 2π geometry is easily achieved if the sample is introduced into the counter or chamber volume. A correction must be applied for α particles backscattered from the sample mount, but in contrast to the β-particle case (p. 224), this correction is small (4 percent for platinum, less for lower Z). With care, accuracies of ± 1 percent can be achieved in measurements of α-disintegration rates by this method. The limitation is usually in sample preparation. Calorimetry is capable of at least comparable accuracy. It requires a larger sample activity but makes no demands on the sample's geometrical arrangement, thinness, and the like. It does require a knowledge of the α-particle energy; this is conveniently obtained from magnetic-spectrometer or semiconductor measurements. Another option for absolute α determinations is the use of a semiconductor detector in conjunction with a near-point source and a defining aperture between source and detectors. This method of known solid angles is described more fully below.

4π Counting. For the determination of absolute disintegration rates there are obvious advantages to the use of 4π geometry, particularly if the counting efficiency is 100 percent. Under these conditions every disintegration gives rise to one count, regardless of the decay scheme (provided that there is included no state with lifetime comparable to or greater than the resolving time of the equipment). The observed counting rate then equals the disintegration rate.

A number of arrangements for 4π counting have been used. The introduction of a β-active sample in gaseous form inside a counter (usually a proportional counter) provides almost a 4π geometry. The end effects and wall effects can be made small and can be evaluated by experiments with different counters in which the ratio of sensitive to insensitive volume is deliberately varied. Gas counting is particularly useful for soft-β emitters (such as ^{3}H, ^{14}C, ^{35}S, and ^{63}Ni) and for low-Z EC nuclides (such as ^{37}Ar); in the latter case even the very soft Auger electrons may be counted quantitatively. In an analogous technique the sample may be dissolved in a liquid scintillator whose dimensions are large compared to the range of the radiations. Beta particles and low-energy photons may be counted with 100 percent efficiency.

With solid samples 4π-β counting requires the use of extremely thin deposits on very thin supports. In the usual 4π-β counter such a sample is mounted between two identical proportional counters connected in parallel. Each of the two counters may be in the shape of a hemisphere, a half cylinder, or a flat cylinder, with a straight or looped wire anode. One type of 4π-β counter is shown in figure 8-12. It has a simple slide between O-rings for introducing the sample. With careful source preparation on properly conducting films, accuracies of ± 1 percent can be achieved in 4π determinations of most β emitters. It is advisable to take a voltage plateau with each sample, or at least for each nuclide to be determined. The parameters affecting the performance of 4π counters have been carefully investigated (P2). The use of 4π-β counting is not advisable for the determination of disintegration rates of nuclides that decay partially or entirely by EC, because very soft Auger electrons may be absorbed in sample and backing, and X rays may traverse the counter volume without making an ion pair.

For X-ray and γ emitters 4π geometry can be closely approximated by sandwiching a sample between two flat-faced NaI scintillators or placing it in a well scintillator, with a second scintillator covering the well; the pulse heights from the two photomultipliers are added together for each event. The method is useful for the determination of the EC disintegration rates

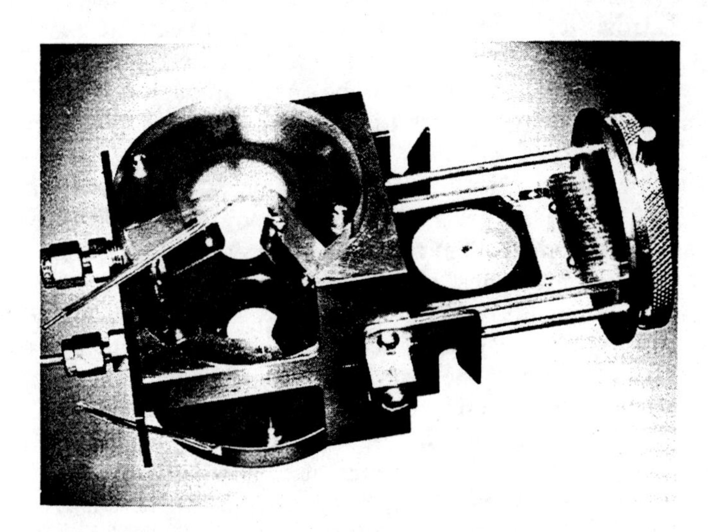

Fig. 8-12 A 4π counter with a section cut away. The two wire loops of the upper and lower counter are visible. The electrical leads and the tube connections for the counter gas are at the left. The sample slide with a source mounted on a plastic film is in the "out" position. The type of 4π counter shown is described by R. Withnell in *Nucl. Inst. Methods* **14**, 279 (1961). (Courtesy R. Withnell and Brookhaven National Laboratory.)

of high-Z nuclides. If we here denote by ω_K the K-shell fluorescence yield, then in a fraction ω_K of all the K-capture transitions K X rays are emitted, and these are detected whether or not γ rays are emitted also, so long as no X-ray-emitting delayed states are involved. Since ω_K is large and rather well known (see figure 3-11), the K-EC rate can be fairly accurately deduced. If pulse height analysis is used, summing effects—X rays adding to coincident X or γ rays—must be taken into account (see following discussion of coincidence experiments).

Coincidence Method. For nuclides with relatively simple decay schemes the absolute disintegration rates may be determined by coincidence measurements. The method is easily understood for the simple case in which the emission of one γ quantum follows each β decay and the spectrum is simple. Consider two counters arranged to count β rays and γ rays, respectively, with measured counting rates $\mathbf{R}_\beta$ and $\mathbf{R}_\gamma$ and with β-γ coincidences also measured, with $\mathbf{R}_{\beta\gamma}$. Then $\mathbf{R}_\beta = \mathbf{R}_0 c_\beta$, $\mathbf{R}_\gamma = \mathbf{R}_0 c_\gamma$ (where the coefficients c_β and c_γ may be thought of as defined by these equations and include all effects of solid angles, counting efficiencies, and absorption corrections), and $\mathbf{R}_{\beta\gamma} = \mathbf{R}_0 c_\beta c_\gamma$. Now $\mathbf{R}_\beta \mathbf{R}_\gamma / \mathbf{R}_{\beta\gamma} = \mathbf{R}_0$, and the absolute disintegration rate is given very simply in terms of this ratio of three measured counting rates. The contribution of γ rays to the counting rate in the β counter (and possibly to coincidence counts) must be measured in a separate experiment with an absorber that prevents the β rays from entering the β counter; this is essentially a background that must be subtracted from $\mathbf{R}_\beta$ (and from $\mathbf{R}_{\beta\gamma}$). No complications result from a complex γ spectrum (two or more γ rays in cascade, possibly with cross-over transitions) as long as only one β transition is involved. The coefficient c_γ then merely refers to the average overall efficiency of the γ detector for the γ rays.

Many subtle effects arise in coincidence measurements. If an extended rather than a point source is used, the response of at least one of the detectors must be independent of the location in the source from which the detected radiation originates; otherwise the simple equations given do not hold. This condition is usually more easily satisfied with the γ detector. However, the use of a 4π counter as the β detector of virtually 100 percent efficiency (for all parts of the sample and for different β branches if such are present) has also found extensive use in accurate standardization of samples (C2). The validity of the equations given in the preceding paragraph also depends on the absence of angular correlations between the directions of emission of the coincident radiations. If there is any suspicion that angular correlations do exist, measurements should be made at more than one detector-source-detector angle.

Great care is required in applying the coincidence method to more complex decay schemes. The use of energy discrimination in one or both detectors is often necessary to avoid spurious effects, the more so the better the energy resolution. A detailed discussion of the method and many of its ramifications is given in R1.

In the following we illustrate some of the problems encountered with a relatively simple example. We consider $\gamma\gamma$- rather than $\beta\gamma$-coincidence measurements. In principle the two techniques are very similar, but some particular complications arising in $\gamma\gamma$ measurements are worth noting. We discuss this problem in terms of Ge(Li) detectors, but the same considerations would apply with NaI scintillation detectors.

Consider a nuclide with the decay scheme shown in figure 8-13*a*. The disintegration rate of a sample of this nuclide is to be determined by $\gamma\gamma$-coincidence measurements. The two γ rays have energies E_1 and E_2, and the measurement is to be performed with two Ge(Li) detectors, *A* and *B*, arranged as schematically shown in figure 8-13*b*. The pulses from the detectors are sent to pulse height analyzers, with a channel on detector *A* set to encompass the photopeak of γ_1 and a channel on detector *B* to encompass

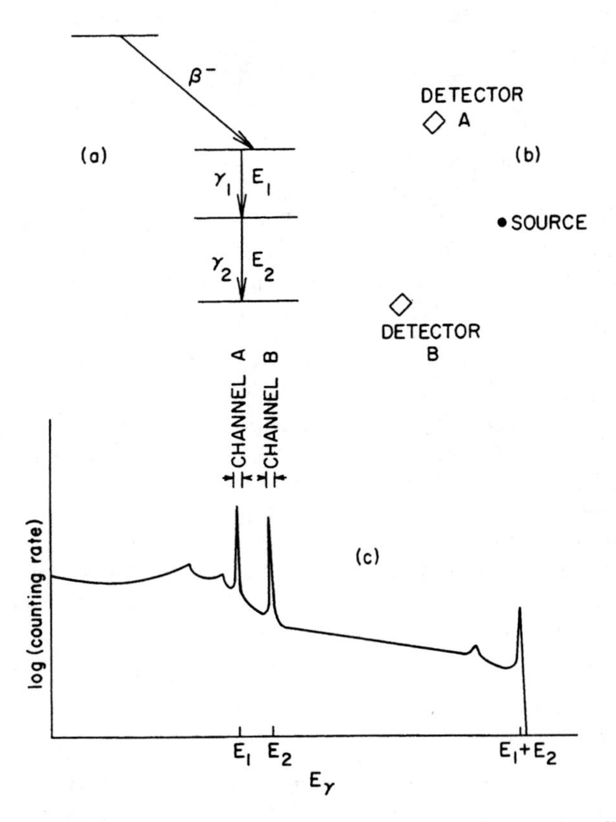

Fig. 8-13 Illustration of the use of $\gamma\gamma$-coincidence measurements for absolute disintegration-rate determination. (*a*) Decay scheme involving two γ rays in cascade, following β^- decay. (*b*) Schematic arrangement of Ge(Li) detectors *A* and *B* for $\gamma\gamma$ measurement. (*c*) "Singles" spectrum observed with either detector.

the photopeak of γ_2. Figure 8-13*c* shows what the γ spectrum obtained with either detector might look like; we have taken $E_1 < E_2$ and have indicated on the graph where the energy channels on A and B might be set.

One of the problems encountered in $\gamma\gamma$-coincidence work (and also in many other attempts to use γ spectroscopy for quantitative intensity measurements—see later) is illustrated by the small peak at energy $E_1 + E_2$ shown in figure 8-13*c*. This peak arises because there is a finite probability that a γ_1 quantum and a γ_2 quantum from the same disintegration undergo photoelectric absorption in the same detector.[13]

Similarly, there is a long Compton tail extending out to the energy $E_1 + E_2$ and including pulse addition events of three kinds: photoelectric absorption of γ_1 plus Compton effect of γ_2; Compton effect of γ_1 plus photoelectric absorption of γ_2; sum of two Compton events. From the detection efficiencies and spectral distributions for the two individual γ rays the intensity and spectral distribution of the sum spectrum can be computed in principle. However, since the intensity of the sum spectrum depends on the square of the solid angle subtended by the counter, whereas the main spectra have intensities proportional to the first power of the solid angle, it is in practice often advisable to work at solid angles small enough to make the summing effects negligible. Even when that is not possible, data taken in different geometries can be extrapolated to zero solid angle. The particular reason why pulse addition effects must be guarded against in coincidence measurements for absolute disintegration rate determinations is that they always alter the "singles" counting rates but leave the coincidence rate unaffected. In our example (although not necessarily in other situations) pulses are thrown out of the area under either of the two photopeaks in the "singles" spectrum by pulse addition with the pulse resulting from photo or Compton absorption of the other γ rays. This throw-out correction is only partly offset by the pulses thrown into the peaks by the addition of two Compton events (or, in the case of the peak at E_2, a photoelectric absorption of γ_1 coupled with an appropriate Compton scattering of γ_2). On the other hand, a coincidence count between counters A and B, with channels set on the two photopeaks, can result only when γ_1 undergoes photoelectric absorption in A, and γ_2 undergoes photoelectric absorption in B, since photoelectric absorption of a given γ ray in one detector makes it impossible for the same γ ray to deposit any energy in the other detector.

From here on we shall assume that our illustrative measurement (figure 8-13) is made in geometry low enough to justify neglect of pulse addition effects. We now define the following quantities:

ϵ_{1A} is the efficiency for detection of γ_1 in detector A,

ϵ_{2A} is the efficiency for detection of γ_2 in detector A,

ϵ_{2B} is the efficiency for detection of γ_2 in detector B,

$\mathbf{R}_A$ is the measured counting rate (in channel) of detector A,

[13] We neglect here the pulse addition that can result from "accidental" coincidences between pulses from two different decay events arriving within the resolving time of the coincidence circuit. Such effects can be minimized by proper choice of sample strength, since they depend on the square of the disintegration rate. Their magnitude can usually be determined.

$\mathbf{R}_B$ is the measured counting rate (in channel) of detector B,

$\mathbf{R}_{AB}$ is the coincidence counting rate.

The efficiency ϵ_{1B} for detection of γ_1 in detector B is zero, since the channel of detector B is set at energy E_2 and we are neglecting pulse addition phenomena. If we denote the disintegration rate as $\mathbf{R}_0$, we can write

$$\mathbf{R}_A = (\epsilon_{1A} + \epsilon_{2A})\mathbf{R}_0,$$

$$\mathbf{R}_B = \epsilon_{2B}\mathbf{R}_0,$$

$$\mathbf{R}_{AB} = \epsilon_{1A}\epsilon_{2B}\mathbf{R}_0.$$

Therefore

$$\frac{\mathbf{R}_A\mathbf{R}_B}{\mathbf{R}_{AB}} = \left(1 + \frac{\epsilon_{2A}}{\epsilon_{1A}}\right)\mathbf{R}_0. \tag{8-1}$$

Equation 8-1 differs from the corresponding expression in simple $\beta\gamma$-coincidence measurements by the additive term $\epsilon_{2A}/\epsilon_{1A}$ and therefore does not immediately give the disintegration rate in terms of three measured counting rates. Although ϵ_{1A} is readily deduced from the measurements ($\epsilon_{1A} = \mathbf{R}_{AB}/\mathbf{R}_B$), the efficiency ϵ_{2A} for detection of the higher-energy γ ray (γ_2) in the detector set at the lower energy E_1 has to be obtained separately. If a nuclide emitting a single γ ray with energy near E_2 is available, the shape of its Compton spectrum in detector A in the region of energy E_1 can be experimentally determined. This measurement, together with the "singles" spectrum of the original source in detector A (figure 8-13*c*) gives essentially the ratio $\epsilon_{2A}/\epsilon_{1A}$ needed to evaluate (8-1). A cruder approximation can be obtained by assuming that the Compton distribution of γ_2 in the region of E_1 is flat.

The detailed discussion of the rather simple example given may serve to illustrate the care that must be taken in setting up the equations relating counting rates and efficiencies for any particular case under consideration. Additional problems arise when angular correlations exist. Without discussing them in detail, we merely call attention to the extreme case of angular correlation represented by the emission of the two 511-keV quanta emitted in opposite directions when a positron is annihilated. Measurement of coincidences between two annihilation quanta is a sensitive and selective method for the detection of positron emission (see p. 312) just because the extreme angular correlation is so characteristic, but it cannot be used to yield a disintegration rate. However, if the positrons are followed by a nuclear γ ray, coincidences between that γ ray and the annihilation radiation can be used to obtain the disintegration rate in the usual way. It may be noted that, if EC as well as β^+ emission takes place, the γ-511 keV coincidence measurement will give the *total* disintegration rate, provided the γ ray measured is involved in all disintegrations. The EC branch can be considered as merely lowering the detection efficiency for positrons via the 511-keV radiation.

Measurements at Known Solid Angle. When a detector of known (preferably 100 percent) intrinsic efficiency is available, it is sometimes possible to obtain the absolute disintegration rate of a source by use of a defining aperture that makes it feasible to calculate the solid angle subtended by the detector at the source. Low-geometry arrangements of this type are often used in absolute α determinations, especially with silicon n-p junction detectors. The space between sample and detector must be evacuated, to avoid scattering and absorption of α particles. For β-disintegration measurements of good accuracy the method is not suitable because of the problems arising from self-absorption, self-scattering, and backscattering. On the other hand, measurements at known solid angles are very useful for absolute determination of X-ray intensities. For X rays of very low energy (say <10 keV) proportional counters may be used; at higher energies, NaI(Tl) detectors of a few millimeters thickness or semiconductor detectors are suitable. One must, of course, be sure that 100 percent of the rays entering the detector are registered, whatever the particular detector. For example, silicon 2 mm thick absorbs photons up to $\sim$14 keV, 5 mm-thick germanium absorbs photons up to 60 keV. The material surrounding the defining apertures must be of such a thickness that the X rays of interest are absorbed. With the use of pulse height analysis the desired X-ray counting rate can be determined even in the presence of other radiations. To convert an absolute X-ray emission rate into an EC rate, the fluorescence yield must be known (cf. figure 3-11).

Calibration of Beta Counters. Once a source of a radionuclide of known disintegration rate is available, any detector may be calibrated in terms of this standard. This calibration will be valid for other samples of the same nuclide, provided they are measured under precisely the same conditions. The calibration may also be adequate for other radionuclides emitting radiations similar to those of the standard.

Standardized sources of a number of β and γ emitters are available in various forms (solutions and mounted solid samples) from the National Bureau of Standards (Washington, D.C.), from the International Atomic Energy Agency (Vienna, Austria), and from some commerical companies.

The efficiency of an end-window β counter for detection of a particular β emitter is usually best determined via a 4π-counter standardization. The disintegration rate of a "weightless" sample of the nuclide in question prepared on a thin film is determined with a 4π counter. Then aliquots of the same activity, mixed with appropriate amounts of carrier, are prepared and mounted in the desired manner for end-window counting. The amounts of activity in the aliquots relative to the amount in the 4π sample may be determined by accurate pipetting and quantitative transfers or, more conveniently, by comparison of the activities in a device whose response is not sensitive to sample thickness, backing, and the like, such as a γ detector at a large distance. This technique allows determination of the end-window counter efficiency directly for the sample thicknesses, sample backings, and geometrical arrangements of interest, without any need for separate corrections for self-absorption, self-scattering, backscattering, air absorption, window absorption, and so on. Without too much difficulty, this method can usually be made to yield disintegration rates accurate to ± 5 percent or less. It is to be greatly preferred to any attempts at quantitative evaluation of and correction for the scattering and absorption effects.

Absolute Gamma Measurements. Absolute determination of γ-emission rates with NaI(Tl) or Ge(Li) counters in its simplest form involves merely the standardization of the counter with a source of known disintegration rate and use of the same detector for the assay of other samples of the same nuclide mounted in the same manner as the calibration standard. Pulse height analysis is not required in this application, and the method can be used with well-type scintillation crystals as well as with external source arrangements. The need for reproducible geometry cannot be overstressed; for example, the height to which the samples extend in a scintillator well must be carefully controlled. With proper precautions the accuracy of the method is limited essentially by the accuracy with which the disintegration rate of the calibration standard is known.

Greater versatility in absolute γ-ray intensity measurements can be achieved with pulse height analysis. The emission rate of a particular γ ray is then usually inferred from the total counting rate in the photopeak. The photopeak efficiency of a given detector for sources mounted in a particular geometrical arrangement can be determined as a function of γ-ray energy by means of standard sources emitting γ rays of various energies. This method is discussed in chapter 7, p. 281 ff, and typical curves of photopeak efficiency versus energy are shown for NaI(Tl) and Ge(Li) detectors in figure 7-17. It should be noted that the lower efficiency of Ge(Li) detectors relative to NaI(Tl) is more than offset in most applications by their much better energy resolution (see figure 7-10).

REFERENCES

C1 J. B. Cumming, "CLSQ, The Brookhaven Decay Curve Analysis Program," in *Applications of Computers to Nuclear and Radiochemistry* (G. D. O'Kelley, Ed.), NAS-NRC, Washington, 1963, p. 25.

C2 P. J. Campion, "The Standardization of Radioisotopes by the Beta-Gamma Coincidence Method Using High-Efficiency Detectors," *Int. J. Appl. Radiat. Isot.* **4**, 232 (1959).

D1 R. W. Dodson et al., "Preparation of Foils," in *Miscellaneous Physical and Chemical Techniques of the Los Alamos Project*, National Nuclear Energy Series Div. V, Vol. 3 (A. C. Graves and D. K. Froman, Eds.), McGraw-Hill, New York, 1952.

F1 H. Freiser and G. H. Morrison, "Solvent Extraction in Radiochemical Separations," *Ann. Rev. Nucl. Sci.* **9**, 221 (1959).

F2 D. B. Fossan and E. K. Warburton, "Lifetime Measurements," in *Nuclear Spectroscopy and Reactions*, Vol. C (J. Cerny, Ed.), Academic, New York, 1974, pp. 307–374.

G1 W. M. Garrison and J. G. Hamilton, "Production and Isolation of Carrier-Fr Radioisotopes," *Chem. Rev.* **49**, 237 (1951).

G2 I. J. Gruverman and P. Kruger, "Cyclotron-Produced Carrier-Free Radioisc Thick-Target Yield Data and Carrier-Free Separation Procedures," *Int. J. Appl. Radiat. Isot.* **5**, 21 (1959).

G3 F. S. Goulding and B. G. Harvey, "Identification of Nuclear Particles," *Ann. Rev. Nucl. Sci.* **25**, 167 (1975).

H1 G. Herrmann and H. O. Denschlag, "Rapid Chemical Separations," *Ann. Rev. Nucl. Sci.* **19**, 1 (1969).

H2 O. Hahn, *Applied Radiochemistry*, Cornell University Press, Ithaca, NY, 1936.

H3 G. Harbottle et al., "A Differential Counter for the Determination of Small Differences in Decay Rates," *Rev. Sci. Instr.* **44**, 55 (1973).

*K1 J. Korkisch, *Modern Methods for the Separation of Rarer Metal Ions*, Pergamon, Oxford, 1969.

K2 R. Klapisch, "On-Line Mass Separation," in *Nuclear Spectroscopy and Reactions*, Vol. A (J. Cerny, Ed.), Academic, New York, 1974, pp. 213–242.

L1 C. M. Lederer and V. S. Shirley, Eds., *Table of Isotopes*, 7th ed., Wiley-Interscience, New York, 1978.

M1 Y. Marcus and A. S. Kertes, *Ion Exchange and Solvent Extraction of Metal Complexes*, Wiley-Interscience, New York, 1969.

M2 R. D. MacFarlane and W. C. McHarris, "Techniques for the Study of Short-Lived Nuclei," in *Nuclear Spectroscopy and Reactions*, Vol. A (J. Cerny, Ed.), Academic, New York, 1974, pp. 243–286.

M3 M. J. Martin and P. H. Blichert-Toft, "Radioactive Atoms," *Nucl. Data Tables* **A8**, 1 (1970).

M4 H. Morinaga and T. Yamazaki, *In-Beam Gamma-Ray Spectroscopy*, North Holland, Amsterdam, 1976.

P1 W. C. Parker and H. Slätis, "Sample and Window Techniques," in *Alpha-, Beta-, and Gamma-Ray Spectroscopy*, Vol. 1 (K. Siegbahn, Ed.), North Holland, Amsterdam, 1965, pp. 379–408.

P2 B. D. Pate and L. Yaffe, "Disintegration-Rate Determination by 4π-Counting," *Can. J. Chem.* **33**, 610, 929, 1656 (1955); **34**, 265 (1956).

R1 L. P. Remsberg, "Determination of Absolute Disintegration Rates by Coincidence Methods," *Ann. Rev. Nucl. Sci.* **17**, 347 (1967).

*S1 Subcommittee on Radiochemistry, NAS-NRC, *Monographs on the Radiochemistry of the Elements*, NAS-NS 3001–3058. Available from the Office of Technical Services, Department of Commerce, Washington, DC.

*S2 K. Siegbahn, Ed., *Alpha-, Beta-, and Gamma-Ray Spectroscopy*, 2 vols., North Holland, Amsterdam, 1965.

T1 N. Trautmann and G. Herrmann, "Rapid Chemical Separation Procedures," *J. Radioanal. Chem.* **32**, 533 (1976).

T2 N. Trautmann, "Rapid Chemical Separations," in *Proc. 3rd Int. Conf. Nuclei Far From Stability* (Cargese, Corsica, May 1976), CERN Report 76-13, Geneva, 1976.

W1 B. Weaver, "Solvent Extraction in the Separation of Rare Earths and Trivalent Actinides," in *Ion Exchange and Solvent Extraction*, Vol. 6 (J. A. Marinsky and Y. Marcus, Eds.), Marcel Dekker, New York, 1974.

W2 A. H. Wapstra, "The Coincidence Method," in *Alpha-, Beta-, and Gamma-Ray Spectroscopy*, Vol. 1, (K. Siegbahn, Ed.), North Holland, Amsterdam, 1965, pp. 539–555.

*Y1 L. Yaffe, "Preparation of Thin Films, Sources, and Targets," *Ann. Rev. Nucl. Sci.* **12**, 153 (1962).

EXERCISES

1. Some trace impurities in a silver metal sample are to be determined by neutron activation analysis in a reactor. What should be the maximum diameter of the sample (a sphere), if the variation of the thermal-neutron flux within the sample is to be held below 10 percent? *Answer:* ~0.6 mm.
2. A 1-μA beam of 120-MeV ^{12}C ions is incident on an aluminum foil 20 mg cm^{-2} thick. (a) Estimate the power dissipated in the foil. (b) If the foil is 20 cm^2 in area and mounted on an insulating frame in a high vacuum, how long an irradiation would raise its temperature to the melting point of aluminum (660°C)? Neglect heat losses by radiation and take the average specific heat of aluminum between room temperature and the melting point as 0.25 cal g^{-1} deg^{-1}. Assume the ^{12}C ions are fully stripped. *Answer:* (b) ~10 s.
3. Cobalt foils are required for the following two types of experiments: (a) The excitation function of α-induced reactions are to be measured by a stacked-foil experiment for ^4He energies down to the threshold of the reaction ^{59}Co (α, n). The energy spread within a foil is not to exceed ±5 percent of the mean energy in the foil. (b) The proton spectra resulting from the (α, p) reaction are to be measured down to 2 MeV with an energy definition of ±3 percent and with protons being measured at angles as large as 30° with respect to the normal to the foil. What is the maximum thickness of cobalt foil you would use in each of the two experiments? (Assume that all the foils in a given experiment are to have the same thickness.) Suggest methods for preparing these foils and for measuring their overall thicknesses as well as their uniformity. *Answer:* (a) ~1.4 mg cm^{-2}.
4. To determine the thickness of a gold foil, a well-collimated beam of 10.50-MeV ^4He ions from a tandem Van de Graaff accelerator is passed through the foil; the transmitted energy is 8.32 MeV. What is the foil thickness in mg cm^{-2}?
5. A sample of 10 mCi ^{57}Ni is to be prepared by (α, n) reaction on ^{54}Fe. With a 10-μA beam of 25-MeV $^4He^{+2}$ ions available, estimate (from fig. 4-4) the thickness of a highly enriched ^{54}Fe target and the length of bombardment necessary. Suggest a chemical procedure for isolating the ^{57}Ni in high specific activity. If the ^{54}Fe target contains 1 ppm nickel impurity and the beam diameter is 1 cm, what is the maximum specific activity attainable?
6. Outline a procedure for each of the following tasks: (a) Separation of ^{140}La from a ^{140}La-^{140}Ba mixture; (b) preparation of ^{123}I in high specific activity and as free as possible from other radioactive iodine isotopes; (c) rapid (<10 min) isolation of selenium fission products from uranium irradiated with thermal neutrons; (d) preparation of a source for conversion electron spectroscopy of $^{191-196}Au$ from deuteron-irradiated platinum.
7. Suggest methods for the chemical identification of (a) ^{52}V produced in the fast-neutron bombardment of a chromate solution, (b) ^{52}Mn produced in the deuteron bombardment of iron, (c) ^{14}O produced in the proton bombardment of nitrogen gas.
8. A sample of sodium iodide is irradiated with fast neutrons to produce 109-d $^{127}Te^m$. Suggest a chemical procedure for the isolation of the tellurium. How would you modify this procedure if you knew that the sodium iodide contained some sodium bromide impurity?

9. A point source of ^{210}Po is placed exactly 2.0 cm from a 12.0-mm defining opening in front of a silicon surface barrier detector of larger diameter. The space between sample and detector is evacuated. In a 20-min count 38,569 counts are accumulated. What is the disintegration rate of the sample? Neglect backscattering. *Answer:* 9.14×10^4 min^{-1}.
10. A sample of neutron-irradiated silicon is found by isotope dilution to contain 1.13 ± 0.06 ng ^{32}Si. The β-disintegration rate of a 1.00 percent aliquot of this sample is determined by 4π counting to be $(2.75 \pm 0.06) \times 10^3$ dis min^{-1}. What is the half-life value for ^{32}Si, based on these data, and what is its standard deviation? [Note that the result does not agree with the half life shown in Appendix D, but with more recent data. See W. Kutschera et al., *Phys. Rev. Lett.* **45**, 592 (1980)].
11. A 1-cm^2 area of 0.025 mm aluminum foil was exposed to a fast-neutron flux to produce ^{24}Na. It was desired to determine the ^{24}Na disintegration rate in this sample by a $\beta\gamma$-coincidence measurement. For the decay scheme of ^{24}Na refer to figure 3-12. The measurement was carried out with two scintillation counters, a plastic scintillator 6 mm thick being used as the β detector and a 7.5 cm $\times$ 7.5 cm NaI(Tl) scintillator as the γ counter. The γ detector had a covering thick enough to prevent its response to the β particles; it was operated with a discriminator set to cut out all pulses corresponding to deposition of <1 MeV in the scintillator, thus making its response to bremsstrahlung negligible. Measurements were made (a) with sample in place, without additional absorbers, and without delay; (b) with sample in place, with a 0.7-g cm^{-2} aluminum absorber between sample and β detector, and without delay; (c) with sample in place, without absorber, and with the pulses from one of the detectors reaching the coincidence circuit after a 0.5 μs delay; (d) with the sample removed and with no delay. The data taken over a period of several hours were as follows:

Experiment	Time at Midpoint of Counting Interval	Length of Count (min)	Total Counts Observed in		
			β Counter	γ Counter	Coincidence
(a)	11:00	10	1.830×10^6	3.63×10^5	9158
(b)	11:40	60	1.221×10^5	2.110×10^6	307
(c)	12:45	60	1.006×10^7	1.996×10^6	557
(d)	15:00	200	5800	6.00×10^4	0

What was the disintegration rate of the ^{24}Na sample at 11:00? The response of the detectors may be considered to be the same over the entire sample area. *Answer:* 7.2×10^6 min^{-1}.
12. The aluminum foil containing ^{24}Na whose disintegration rate was determined in exercise 11 was used to calibrate an end-window proportional counter for ^{24}Na radiations. A measurement taken with that counter at 14:00 on the same day as the measurements in exercise 11 gave 227,520 net cpm with the sample in a certain shelf position. Additional measurements in the same arrangement, but taken exactly 15.0 and 30.0 h later, gave net rates of 114,880 and 57,720 cpm. Considering the statistical errors in these results as negligible and

also neglecting the possibility of any activity other than ^{24}Na in the sample, estimate (a) the dead-time of the counting device, (b) the overall efficiency of the counter for ^{24}Na radiations in the particular geometrical arrangement used. For a discussion of counter dead times see chapter 9, p. 361.

Answer: (b) 0.037.

13. An end-window proportional counter is to be calibrated for the measurement of ^{46}Sc in the form of $Sc_2(C_2O_4)_3 \cdot 5H_2O$ deposits of various thicknesses but all of 2-cm^2 area. The calibration samples are prepared by addition of various amounts of scandium carrier to aliquots of a ^{46}Sc solution of high specific activity, followed by precipitation, filtration, drying, and mounting of the oxalate. A "weightless" source of ^{46}Sc on a thin film is also prepared. It is found, by means of a 4π proportional counter, to have a disintegration rate of 63,800 min^{-1}. After this disintegration rate determination the 4π source is mounted on an aluminum card in the same manner as the oxalate deposits. The relative ^{46}Sc contents of all the samples are assayed with a NaI scintillation detector. The total scandium contents of the oxalate samples are determined analytically after completion of the activity measurements. From the following summary of data construct a curve of counting efficiency versus sample thickness (in milligrams per square centimeter) for the end-window counter measurements. All counting rates are net rates and have already been corrected for any decay during the course of the measurements.

Sample No.	Scandium Content (mg)	Net Counting Rate (cpm)	
		On γ Counter	On End-Window Counter
"4π"	—	8140	—
1	0.42	6870	3257
2	0.91	7240	3415
3	1.38	7510	3530
4	1.90	7680	3558
5	2.95	7960	3560
6	3.84	7875	3295
7	4.77	7690	2982
8	5.86	7820	2645
9	7.72	7750	2110
10	9.81	7910	1705

Answer: For sample No. 10, efficiency is 0.0275.

14. Verify the statement on p. 319 that $E \cdot (dE/dx) \approx Mz^2$.

15. The nuclide ^AZ decays by β^- emission, largely to the first excited state of $^A(Z+1)$; a small β^- branch (0.9 MeV maximum energy) goes directly to the ground state of $^A(Z+1)$. The decay of $^A(Z+2)$ proceeds entirely by K-EC to the first excited state. One sample of each of these two radionuclides is used in the following coincidence measurements with an anthracene (C_1) and a NaI (C_2) scintillation detector. The samples are placed in a standard position between the two counters, and a 0.5-g cm^{-2} copper absorber, sufficient to absorb the ^AZ β particles and the K X rays of $(Z+1)$, is placed between the

sample and C_2 during all the measurements. The following data are obtained:

Sample	Delay between C_1 and C_2 (μs)	0.5-g cm^{-2} Cu between C_1 and Sample	cpm in C_1	cpm in C_2	Coincidence cpm
^{A}Z	0	No	188,600	125,100	2928
^{A}Z	2	No	188,300	125,600	237
^{A}Z	0	Yes	2,753	125,800	3.5
$^{A}(Z+2)$	0	No	40,930	55,340	655
$^{A}(Z+2)$	2	No	41,070	55,090	23
$^{A}(Z+2)$	0	Yes	1,216	55,510	0.7

(a) What is the disintegration rate of the $^{A}(Z+2)$ sample? (b) What is the disintegration rate of the ^{A}Z sample? (c) What fraction of the ^{A}Z decays go to the $^{A}(Z+1)$ ground state directly? (d) What is the coincidence resolving time of the circuit used? (e) If the K X rays of $(Z+1)$ and the β particles emitted by ^{A}Z are counted with the same efficiency in C_1, what is the K-fluorescence yield of $(Z+1)$? Assume the two β^- groups of ^{A}Z to be counted in C_1 with equal efficiency. Sample decay during the course of the measurements may be neglected. *Answers:* (b) 8.67×10^6 min^{-1}; (c) 0.09; (d) 0.3 μs.

16. In an investigation of the decay scheme of 2.4-min ^{108}Ag the β^- spectrum was measured in an anthracene scintillation spectrometer and found to have an upper energy limit of 1.77 ± 0.06 MeV and a simple, allowed shape within the accuracy of the measurements. Measurements with a proportional counter and pulse height analyzer showed that ^{108}Ag emits palladium K X rays and that the ratio of the number of these X rays to the number of β particles emitted is 0.013 ± 0.001. The γ spectrum obtained with a NaI scintillation spectrometer showed weak γ rays of 435, 510, and 616 keV with relative intensities 1.0, 0.27, and 0.27. In a $\beta\gamma$-coincidence experiment 616-keV γ rays were found to be in coincidence with β rays; however, these coincidences could be eliminated with an aluminum absorber of 480 mg cm^{-2} placed between sample and β counter. Gamma-gamma coincidences were found between 435- and 602-keV γ rays, and between 510- and 510-keV γ rays, the latter, however, only when the two counters were 180° apart with respect to the sample. The spectrum of γ rays in coincidence with X rays showed 435- and 602-keV γ rays in the intensity ratio 1.0:0.79. The 602-keV peak in these coincidence spectra was definitely at a lower energy than the 616-keV peak found in the singles spectrum. Additional experiments proved that 85 percent of all the EC transitions lead to the ^{108}Pd ground state. Derive as much information as you can about the ^{108}Ag decay scheme, including the intensities of the various β and γ transitions, and as many of the $\log ft$ values as possible. Discuss spin and parity assignments.

Most of the information in this exercise is based on a paper by M. L. Perlman, W. Bernstein, and R. B. Schwartz, *Phys. Rev.* **92**, 1236 (1953).

Chapter 9

Statistical Considerations in Radioactivity Measurements

The radioactive-decay law discussed in chapter 5 describes the average behavior of a sample of radioactive atoms. In measurements of radioactive decay we are concerned with observations that show fluctuations about the average behavior predicted by the decay law. Therefore in this chapter we discuss the applications of statistical methods to the treatment of radioactivity measurements.

A. DATA WITH RANDOM FLUCTUATIONS

Consider the set of data actually obtained with a Geiger counter measuring a long-lived ("steady") radioactive source, as given in table 9-1. The number of counts recorded per minute (the counting rate) is clearly not uniform. What is the most accurate value of the counting rate? The most straight-

Table 9-1 Count Rate Data from a Radioactive Source

Minute	Counts	Δ_i [a]	Δ_i^2
1	89	−10	100
2	120	+21	441
3	94	−5	25
4	110	+11	121
5	105	+6	36
6	108	+9	81
7	85	−14	196
8	83	−16	256
9	101	+2	4
10	95	−4	16
Totals	990	0	1276
	Average $\bar{x} = 99$		

[a] The symbol Δ_i denotes the difference of an individual measurement from the average: $\Delta_i = x_i - \bar{x}$.

forward approach is to calculate the arithmetic mean (the average value) and consider it as representing the true counting rate. The problem is that from a small number of actual observations we are trying to estimate the results of an infinite number of measurements called the **parent population.** Furthermore, we are assuming that our data reflect the probability distribution that would be obtained if we were to make an infinite number of measurements. In particular, we wish to estimate the average value that we would find and the distribution of the observed values about that average.

Average Value. If the determinations, minute by minute, are denoted by $x_1, x_2, \ldots, x_i$ for the first, second, ..., ith minute, then the arithmetic mean value $\bar{x}$ is, by definition,

$$\bar{x} = \frac{1}{N_0} \sum_{i=1}^{i=N_0} x_i, \tag{9-1}$$

where N_0 is the number of values of x to be averaged. For the counting rates in the table $\bar{x} = 990/10 = 99.0$. This average value is the best estimate that we can make of the "true" average $\bar{x}_t$, which is the average we would find for an infinite number of observations.[1] We could also calculate the **median,** which is defined as the middle value when the observations are arranged in order of magnitude. The median and mean are equal if the distribution of an infinite number of measurements is symmetric.

Standard Deviation. The distribution of the observed results about $\bar{x}_t$ is a measure of the precision of the data and can be described by giving all of the **moments** of the distribution; that is, the quantities

$$\frac{1}{N_0} \sum_{i=1}^{N_0} (x_i - \bar{x}_t)^n \tag{9-2}$$

for all values of n. The first moment ($n = 1$) will always vanish because of the definition of $\bar{x}_t$; the other odd moments [expression (9-2) with n an odd number] will vanish only if the distribution is symmetrical about $\bar{x}_t$, and $\bar{x}_t$ is then the most probable value of x. Usually just the second moment [expression (9-2) with $n = 2$, called the **variance** and denoted by σ_x^2] is given in practice. The square root of the variance is called the **standard deviation** σ_x. This quantity is particularly significant because of the form of the so-called **normal distribution law,** which is expected to describe the distribution of experimental results with random errors:

$$P(x)\,dx = \frac{1}{\sqrt{2\pi\sigma_x^2}} \exp\left[\frac{-(x - \bar{x}_t)^2}{2\sigma_x^2}\right] dx, \tag{9-3}$$

[1] By the use of the method of maximum likelihood we can obtain an estimate of a parent population parameter from our measurements. Using this technique, it can be shown that, for a normally distributed sample, the sample mean is the best estimate of the mean of the parent population. (See section E and references B1, p. 209, and B2, p. 67)

where $P(x)\,dx$ is the probability of observing a value of x in the interval $x \to x + dx$.

In our example, which contains a finite number of observations, we do not know $\bar{x}_t$; we have only an estimate of it: $\bar{x}$. Under these circumstances the best possible estimate of the variance is

$$\sigma_x^2 = \frac{1}{N_0 - 1} \sum_{i=1}^{N_0} (x_i - \bar{x})^2. \qquad (9\text{-}4)$$

For the data in table 9-1 we compute $\sigma_x^2 = 1276/9 = 141.8$; $\sigma_x = 11.9$. The difference between (9-2) and (9-4) is noteworthy. The division by $(N_0 - 1)$ in (9-4) instead of by N_0 is a consequence of estimating the unknown quantity $\bar{x}_t$ from N_0 observations; this estimation uses up one of the observations and leaves only $(N_0 - 1)$ independent quantities for the estimation of the variance. The validity of this reasoning becomes clear when we consider the extreme case of only a single observation. Evidently, from a single observation we can have no idea of the precision of the measurement, unless special assumptions are made. This problem is a fundamental one in statistical analysis and is discussed in standard texts on the subject (for example, B3 and F1).

If the number of observations is fairly large (say ≥ 50) and if the observations follow the normal distribution, then the interval $\bar{x} \pm \sigma$ will include $\sim\frac{2}{3}$ of the observations, $\bar{x} \pm 2\sigma$ will include $\sim\frac{19}{20}$ of the observations, and $\bar{x} \pm 3\sigma$ will include $\sim\frac{997}{1000}$ of the observations.

Occasionally the standard deviation is expressed as a percentage of the average of the data ($\bar{x}$) and is then called the **coefficient of variability**. This measure of the precision of the data is of limited use because of the difficulty in determining its statistical significance.

Precision of Average Value. In the preceding discussion we have been concerned with estimating, from N_0 observations, the results that would be obtained from a very large number of observations. It is now necessary to discuss the precision of our estimation, which is not to be confused with the precision of the data, although the two quantities are related. We are here concerned with two things:

1. The distribution of the values of $\bar{x}$ given by (9-1) from many sets of experiments, each with a finite N_0.
2. The distribution of the quantities σ_x^2 obtained from the same sets of observations by (9-4).

The formal statistical analysis of these two problems, as discussed in standard texts (F1), is contained in the χ^2-test of the randomness of the data, the t-test of the reliability of $\bar{x}$ as an estimate of $\bar{x}_t$, and the F-test of the reliability of σ_x^2 as an estimate of the true variance of the sample.

Our main interest is in the first question, the reliability of $\bar{x}$; a measure of this reliability is the **variance of a mean**, which is estimated by the variance of the set of observations divided by N_0:

$$\sigma_{\bar{x}}^2 = \frac{\sigma_x^2}{N_0} = \frac{1}{N_0(N_0 - 1)} \sum_{i=1}^{N_0} (x_i - \bar{x})^2. \tag{9-5}$$

The quantity $\sigma_{\bar{x}}^2$ is our best estimate of the second moment of the distribution of average values that would be found from an infinite number of *sets* of experiments, each containing N_0 observations of which table 9-1 is an example. The value of $\sigma_{\bar{x}}$ from table 9-1 is $\sqrt{141.8/10} = 3.76$.

The significance of this quantity, for a normal distribution, is found in the statement that the probability of observing a value of $\bar{x}$ between $\bar{x}$ and $\bar{x} + d\bar{x}$ is

$$P(\bar{x})\, d\bar{x} = \frac{1}{\sqrt{2\pi\sigma_{\bar{x}}^2}} \exp\left[\frac{-(\bar{x} - \bar{x}_t)^2}{2\sigma_{\bar{x}}^2}\right] d\bar{x}.$$

Rejection of Data. The question often arises whether a particular datum should be rejected because of its relatively large deviation from the mean. In table 9-1 the observation of 120 counts during the second minute is suspect, as perhaps, though to a lesser degree, is the observation of 83 counts during the eighth minute. This is not necessarily to say that these observations are wrong (that the error is systematic rather than random), but that deviations of this magnitude among a small number of observations may have an undue influence on the mean value that is computed. Thus the criteria for rejection should consider not only the magnitude of the deviation but also the number of observations made. Chauvenet's criterion includes both factors (the magnitude of the deviation and the number of observations) and allows rejection of an observation if deviations from the mean that are equal to or greater than the one in question have a probability of occurrence that is less than $1/(2N_0)$. In our example the counting rate during the second minute may be rejected only if the probability of observing counting rates that deviate by at least 21 counts from the mean of 99 counts is less than 0.05. We compute this probability by using (9-3) to obtain the probability P of observing a count between 78 and 120:

$$P = \int_{78}^{120} \frac{1}{\sqrt{2\pi \cdot 141.8}} \exp\left[\frac{-(x - 99)^2}{2 \cdot 141.8}\right] dx.$$

The value of the integral, as found in the *Handbook of Chemistry and Physics*, is 0.92; thus $1 - P$ is 0.08 and the datum must be retained. If N_0 had been six, or less, then the datum would have been rejected. When a datum is rejected a new $\bar{x}$ must be computed, and Chauvenet's criterion may be applied to the remaining suspect data, but with N_0 being decreased by one each time that an observation is excluded.

B. PROBABILITY AND THE BINOMIAL DISTRIBUTION

The ideas and definitions just presented may be applied, with varying degrees of usefulness, to any set of data, whether or not strictly random phenomena are involved. However, for the case of radioactivity measurements we can use our *sample* of a limited number of observations to make inferences concerning the *parent population* of an infinite number of measurements if we know something about the expected behavior of the sample from a given population. This can be done if we assume that parent population is represented by a distribution function and that our measurements reflect the distribution. By using the ideas of probability we can determine the statistical significance of our "random" sample of the parent population. Before proceeding we must consider the concept of probability in greater detail. As illustrations we investigate the answers to questions such as these:

1. What is the probability that a card drawn from a deck will be an ace?
2. If a coin is flipped twice, what is the probability that it will fall "heads up" both times?
3. Given a sample of a radioactive material, what is the probability that exactly 100 disintegrations will occur during the next minute?

We define probability in this way: given a set of N_0 objects (or events, or results, etc.) containing n_1 objects of the first kind, n_2 objects of the second kind, and n_i objects of the ith kind, the probability p_i that an object specified only as belonging to the set is of the ith kind is given by $p_i = n_i/N_0$. By applying this definition we find that the probability that one card drawn from a full deck will be an ace is just $\frac{4}{52}$.

We may now rewrite the definition of the average value $\bar{x}$ of a set of quantities x_i, taking into account the possibility that any particular value may appear several, say n_i, times. Then

$$\bar{x} = \frac{1}{N_0} \sum n_i x_i = \sum p_i x_i.$$

This may be generalized, and the expression for the average value of any function of x is

$$\overline{f(x)} = \sum p_i f(x_i). \tag{9-6}$$

In particular,

$$\sigma_x^2 = \sum p_i (x_i - \bar{x})^2 = \overline{x^2} - \bar{x}^2, \tag{9-7}$$

a result that will be useful to us later.

In experimental measurements we may make a large number K of observations and find the ith result k_i times. Now the ratio k_i/K is not the probability p_i of the ith result as we have defined it, but for our purposes we assume that k_i/K approaches arbitrarily closely to p_i as K becomes very large:

$$\lim_{K \to \infty} \frac{k_i}{K} = p_i.$$

This assumption is not subject to mathematical proof because a limit may not be evaluated for a series with no law of sequences of terms.

Addition Theorem. We turn now to the compounding of several probabilities and consider first the addition theorem. Given a set of N_0 objects (or events, or results, etc.) containing n_i objects of the kind a_i and given that the kinds $a_1, a_2, \ldots, a_i$ have no members in common, the probability that one of the N_0 objects belongs to a combined group $a_1 + a_2 + \cdots a_j$ is just $\sum_{i=1}^{i=j} p_i$. Thus for two mutually exclusive events with probabilities p_1 and p_2 the probability of one or the other occurring is just $p_1 + p_2$. When one card is drawn from a full deck, the chance of its being either a five or a ten is $\frac{4}{52} + \frac{4}{52} = \frac{2}{13}$. (When one draws one card while already holding, say, four cards, none of which is a five or ten, the probability then of getting either a five or a ten is slightly greater, $\frac{4}{48} + \frac{4}{48} = \frac{1}{6}$, provided that there is available no information regarding the identity of other cards that may already have been withdrawn.) When a coin is tossed, the probability of either "heads" or "tails" is $\frac{1}{2} + \frac{1}{2} = 1$.

Multiplication Theorem. Another type of compounding of probabilities is described by the multiplication theorem. If the probabilitiy of an event i is p_i and if after i has happened the probability of another event j is p_j, then the probability that first i and then j will happen is $p_i \times p_j$. If a coin is tossed twice, the probability of getting "heads" twice is $\frac{1}{2} \times \frac{1}{2} = \frac{1}{4}$. If two cards are drawn from an initially full deck, the probability of two aces is $\frac{4}{52} \times \frac{3}{51}$. The probability of four aces in four cards drawn is $\frac{4}{52} \times \frac{3}{51} \times \frac{2}{50} \times \frac{1}{49}$. (The probability of drawing five aces in five cards is $\frac{4}{52} \times \frac{3}{51} \times \frac{2}{50} \times \frac{1}{49} \times \frac{0}{48} = 0$.)

Binomial Distribution. The binomial distribution law treats one fairly general case of compounding probabilities and can be derived by the application of the addition and multiplication theorems. Given a very large set of objects in which the probability of occurrence of an object of a particular kind w is p, then, if n objects are withdrawn from the set, the probability $W(r)$ that exactly r of the objects are of the kind w is given by

$$W(r) = \frac{n!}{(n-r)!r!} p^r (1-p)^{n-r}. \tag{9-8}$$

To see how this combination of terms actually represents the probability in question, think for a moment of just r of the n objects. That the first of these is of the kind w has the probability p; that the first and second are of the kind w has the probability p^2, and so on; and the probability that all r objects are of the kind w is p^r. But, if exactly r of the n objects are to be of this kind, the remaining $n-r$ objects must be of some other kind; this probability is $(1-p)^{n-r}$. Thus we see that for a particular choice of r objects out of the n objects the probability of exactly r of kind w is $p^r(1-p)^{n-r}$; this particular choice is not the only one. The first of the r objects might be chosen (from the n objects) in n different ways, the second in $n-1$ ways, the third in $n-2$ ways and the rth in $n-r+1$ ways. The product of these terms, $n(n-1)(n-2)\cdots(n-r+1)$, is $n!/(n-r)!$, and this coefficient must be used to multiply the probability just found. But this coefficient is actually too large in that it not only gives the total number of possible arrangements of the objects in the way required but also includes the number of arrangements that differ only in the order of selection of the r objects. So we must divide by the number of permutations of r objects, which is $r!$. Thus the final coefficient is $n!/(n-r)!r!$, which is that in (9-8). The law (9-8) is known as the binomial distribution law because this coefficient is just the coefficient of $x^r y^{n-r}$ in the binomial expansion of $(x+y)^n$. Since in (9-8)

$$x+y=p+(1-p),$$

we have

$$\sum_{r=0}^{n} W(r)=1,$$

and the binomial distribution is seen to be normalized.

C. RADIOACTIVITY AS A STATISTICAL PHENOMENON

Binomial Distribution for Radioactive Disintegrations. As we discussed earlier in chapters 1 and 5, radioactive nuclei decay independently of each other. A decay constant can be defined for a large sample of a given nuclide but each individual nucleus decays according to its own "clock." This problem is statistically similar to the case of flipping coins, and thus we may apply the binomial distribution law to find the probability $W(m)$ of obtaining just m disintegrations in time t from N_0 original radioactive atoms. We think of N_0 as the number n of objects chosen for observation (in our derivation of the binomial law), and we think of m as the number r that is to have a certain property (namely, that of disintegrating in time t), so that for this case the binomial law becomes

$$W(m)=\frac{N_0!}{(N_0-m)!m!}p^m(1-p)^{N_0-m}. \tag{9-9}$$

Now the probability of an atom not decaying in time t, $1-p$ in (9-9), is given by the ratio of the number N that survive the time interval t to the initial number N_0,

$$\frac{N}{N_0} = e^{-\lambda t};$$

p is then $1 - e^{-\lambda t}$. We now have

$$W(m) = \frac{N_0!}{(N_0 - m)!m!}(1 - e^{-\lambda t})^m (e^{-\lambda t})^{N_0 - m}. \quad (9\text{-}10)$$

Time Intervals between Disintegrations. Since the time of von Schweidler's derivation of the exponential decay law from probability considerations, the applicability of these statistical laws to the phenomena of radioactivity has been tested in a number of experiments. As an example of the positive evidence obtained, we consider the distribution of time intervals between disintegrations. The probability of this time interval having a value between t and $t + dt$, which we write as $P(t)\,dt$, is given by the product of the probability of no disintegration between 0 and t and the probability of a disintegration between t and $t + dt$. The first of these two probabilities is given by (9-10) with $m = 0$:

$$W(0) = \frac{N_0!}{N_0!0!}(1 - e^{-\lambda t})^0 (e^{-\lambda t})^{N_0} = e^{-N_0 \lambda t}.$$

(Notice that $0! = 1$.) The probability of any one of the N_0 atoms disintegrating in the time dt is clearly, from the addition theorem, $N_0\lambda\,dt$. [See chapter 1, p. 5, or obtain this result as $W(1)$ from (9-10) with $m = 1$, t replaced by dt, and all terms in $(dt)^2$ and higher powers of dt neglected.] Then

$$P(t)\,dt = N_0 \lambda e^{-N_0 \lambda t}\,dt. \quad (9\text{-}11)$$

Experiments designed to test this result usually measure a large number s of time intervals between disintegrations and classify them into intervals differing by the short but finite length Δt. Then the probability for intervals between t and $t + \Delta t$ should be $N_0 \lambda e^{-N_0 \lambda t}\,\Delta t$, and the number of measured intervals between t and $t + \Delta t$ should be $sN_0 \lambda e^{-N_0 \lambda t}\,\Delta t$. For example, N. Feather found experimentally that the logarithm of the number of intervals between t and $t + \Delta t$ is proportional to t, as required by this formula.

Average Disintegration Rate. Another application of the binomial law to radioactive disintegrations may be seen if we calculate the average value of a set of numbers obeying the binomial distribution law. For the moment we shall revert to the notation of (9-8) and for further convenience represent $1 - p$ by q:

$$W(r) = \frac{n!}{(n - r)!r!} p^r q^{n-r}. \quad (9\text{-}12)$$

The average value to be expected for r is obtained from (9-6):

$$\bar{r} = \sum_{r=0}^{r=n} rW(r) = \sum_{r=0}^{r=n} r\frac{n!}{(n-r)!r!}p^r q^{n-r}.$$

To evaluate this awkward-appearing summation consider the binomial expansion of $(px+q)^n$:

$$(px+q)^n = \sum_{r=0}^{r=n} \frac{n!}{(n-r)!r!}p^r x^r q^{n-r} = \sum_{r=0}^{r=n} x^r\, W(r).$$

Differentiating with respect to x, we obtain

$$np(px+q)^{n-1} = \sum_{r=0}^{r=n} rx^{r-1}\, W(r). \qquad (9\text{-}13)$$

Now letting $x = 1$ and using $q = 1 - p$, we have the desired expression

$$np = \sum_{r=0}^{r=n} r\, W(r) = \bar{r}.$$

This result should not be surprising; it means that the average number $\bar{r}$ of the n objects that are of the kind w is just n times the probability for any given one of the objects to be of the kind w.

The foregoing result may be interpreted for radioactive disintegration if n is set equal to N_0 and $p = 1 - e^{-\lambda t}$, as before. Then the average number M of atoms disintegrating in the time t is $M = N_0(1 - e^{-\lambda t})$. For small values of λt, that is, for times of observation short compared to the half-life, we may use the approximation $e^{-\lambda t} = 1 - \lambda t$ and then $M = N_0\lambda t$. The disintegration rate $\mathbf{R}$ to be expected is $\mathbf{R} = M/t = N_0\lambda$. (This corresponds to the familiar equation $-dN/dt = \lambda N$.)

Expected Standard Deviation. What may we expect for the standard deviation of a binomial distribution? If we differentiate (9-13) again with respect to x, we obtain

$$n(n-1)p^2(px+q)^{n-2} = \sum_{r=0}^{r=n} r(r-1)x^{r-2}\, W(r).$$

Again letting $x = 1$ and using $p + q = 1$, we have

$$n(n-1)p^2 = \sum_{r=0}^{r=n} r(r-1)\, W(r) = \sum_{r=0}^{r=n} r^2\, W(r) - \sum_{r=0}^{r=n} r\, W(r),$$

$$n(n-1)p^2 = \overline{r^2} - \bar{r}.$$

Recall from (9-7) that the variance σ_r^2 is given by

$$\sigma_r^2 = \overline{r^2} - \bar{r}^2.$$

Now, combining, we have

$$\sigma_r^2 = n(n-1)p^2 + \bar{r} - \bar{r}^2,$$

and with $\bar{r} = np$

$$\sigma_r^2 = n^2p^2 - np^2 + np - n^2p^2 = np(1-p) = npq,$$

$$\sigma_r = \sqrt{npq}.$$

For radioactive disintegration this becomes

$$\sigma = \sqrt{N_0(1-e^{-\lambda t})e^{-\lambda t}} = \sqrt{Me^{-\lambda t}}. \tag{9-14}$$

In counting practice λt is usually small; that is, the observation time t is short compared to the half life, and when this is so,

$$\sigma = \sqrt{M}. \tag{9-14a}$$

We see here a particular example of a very important property of the binomial distribution that, as presently shown, is true for the Poisson distribution also; that is, there is a simple relationship between the true mean and the true variance of the distribution. As a consequence, a single observation from a distribution that is expected to be binomial, as is true of radioactive disintegration rates, gives both an estimate of the mean and an estimate of the variance of the distribution. Further, for a single observation the estimate of the variance of the distribution is also an estimate of the variance of the mean. It must be immediately emphasized that these remarks are not true in general; the variance of a thermometer reading, of a length measured by a meter stick, or of the reading of a voltmeter cannot be estimated from a single observation and is not in general expected to be equal to the value observed.

If a reasonably large number m of counts has been obtained, that number m may be used in the place of M for the purpose of evaluating σ. Thus if 100 counts are recorded in 1 min, the expected standard deviation is $\sigma \approx \sqrt{100} = 10$, and the counting rate might be written 100 ± 10 cpm. If 1000 counts are recorded in 10 min, the standard deviation of this number is $\sigma \approx \sqrt{1000} = 32$. The counting rate is $(1000 \pm 32)/10 = 100 \pm 3.2$ cpm. Thus we see that for a given counting rate $\mathbf{R}$ the σ for the rate is inversely proportional to the square root of the time of measurement:

$$\mathbf{R} = \frac{m}{t};$$

$$\sigma_{\mathbf{R}} = \frac{\sqrt{m}}{t} = \frac{\sqrt{\mathbf{R}t}}{t} = \sqrt{\frac{\mathbf{R}}{t}}. \tag{9-15}$$

What is the result in an experiment in which the counting time is long compared to the half life? As $\lambda t \to \infty$, $e^{-\lambda t} \to 0$, and in this limit $\sigma = \sqrt{Me^{-\lambda t}} = 0$. The explanation is clear; if we start with N_0 atoms and wait for all to disintegrate, then the number of disintegrations is exactly N_0. However, in actual practice we observe not the number of disintegrations but that number times a coefficient c that denotes the probability of a disintegration resulting in an observed count. Taking this into account, we see that in this limiting case the proper representation of $\sigma = \sqrt{npq}$ is $\sigma = \sqrt{N_0 c(1-c)}$. If $c \ll 1$, then $\sigma = \sqrt{N_0 c} = \sqrt{\text{number of counts}}$ as before. When $\lambda t \approx 1$ and c is neither unity nor very small, a more exact analysis based on $\sigma = \sqrt{npq}$ should be made, with the result that $\sigma = \sqrt{Mc(1-c+ce^{-\lambda t})}$.

The introduction of the detection coefficient c in the preceding paragraph may raise the question why it is not necessary to take account of this coefficient in the more familiar case with λt small, where we have written $\sigma = \sqrt{m}$. If we do consider c in this case, we have for the probability of one atom producing a count in time t, $p = (1 - e^{-\lambda t})c$ and $q = 1 - p = 1 - c + ce^{-\lambda t}$. Then

$$\sigma = \sqrt{N_0(1-e^{-\lambda t})c(1-c+ce^{-\lambda t})},$$

and for λt small and the same approximations as before

$$\sigma = \sqrt{N_0 \lambda t c} = \sqrt{Mc} = \sqrt{\text{number of counts recorded}}.$$

This is just the conclusion we had reached without bothering about the detection efficiency. It should be emphasized, however, that actual counts and not scaled counts from a scaling circuit must be used in these equations.

D. POISSON AND GAUSSIAN DISTRIBUTIONS

Poisson Distribution. The binomial distribution law (9-10) can be put into a more convenient form if we impose the restrictions $\lambda t \ll 1$, $N_0 \gg 1$, $m \ll N_0$, that is, if we consider a large number of active atoms observed for a time short compared to their half lives. The derivation of this more convenient form requires the well-known mathematical approximation:

$$\ln(1+x) \approx x - \frac{x^2}{2} \cdots \qquad \text{if } x \ll 1. \tag{9-16}$$

Let us first define the average value of the distribution (9-10):

$$M = N_0(1 - e^{-\lambda t}).$$

The binomial distribution may then be written as

$$W(m) = \frac{N_0!}{(N_0-m)!\,m!}\left(\frac{M}{N_0}\right)^m \left(1 - \frac{M}{N_0}\right)^{N_0} \left(1 - \frac{M}{N_0}\right)^{-m}.$$

Consider the term

$$\frac{N_0!}{(N_0-m)!} = N_0(N_0-1)\cdots(N_0-m+1) = N_0^m\left(1-\frac{1}{N_0}\right)\cdots\left(1-\frac{m-1}{N_0}\right).$$

For $m \ll N_0$ this term may be estimated by taking its logarithm and using the first term of the approximation (9-16). The result is

$$\frac{N_0!}{(N_0 - m)!} \approx N_0^m \exp\left[-\frac{m(m-1)}{2N_0}\right]. \tag{9-17a}$$

The term $[1 - (M/N_0)]^{N_0}$ may also be estimated by the use of (9-16), since $M/N_0 \ll 1$, a condition that is equivalent to $\lambda t \ll 1$:

$$\ln\left(1 - \frac{M}{N_0}\right)^{N_0} = N_0 \ln\left(1 - \frac{M}{N_0}\right)$$

$$\approx -M - \frac{M^2}{2N_0},$$

$$\therefore \left(1 - \frac{M}{N_0}\right)^{N_0} \approx e^{-M} e^{-M^2/2N_0}. \tag{9-17b}$$

Note that this time we use two terms of the expansion, since $M^2/2N_0$ is not necessarily small, even for $M/N_0 \ll 1$.

Again, for $M/N_0 \ll 1$ we immediately have from (9-16)

$$\ln\left(1 - \frac{M}{N_0}\right)^{-m} \approx \frac{mM}{N_0},$$

$$\left(1 - \frac{M}{N_0}\right)^{-m} \approx e^{mM/N_0}. \tag{9-17c}$$

When the three approximate results (9-17a, b, c) are put into the binomial distribution, the result is

$$W(m) \approx \frac{M^m e^{-M}}{m!}\left[e^{-(M-m)^2/2N_0} e^{m/2N_0}\right], \tag{9-18}$$

where $W(m)$ is the probability of obtaining the particular number of counts m when M is the average number to be expected. The term outside the brackets in (9-18):

$$W(m) \approx \frac{M^m e^{-M}}{m!} \tag{9-19}$$

is the famous Poisson distribution. The term within the brackets may be considered as a correction factor and is a measure of how well the binomial distribution is approximated by the Poisson. It is to be emphasized that *the validity of* (9-19) *as an approximation to* (9-10) *requires not only that a large number of atoms be observed for a time short compared to their half lives, but also that the absolute value of* $(M - m)$ *be substantially smaller than* $\sqrt{N_0}$. For example, if $N_0 = 100$ and $M = 1$, both the Poisson and binomial distributions give $W(0) = 0.37$, but the binomial distribution gives $W(10) = 0.7 \times 10^{-7}$, whereas the Poisson distribution (9-19) gives $W(10) = 1.0 \times 10^{-7}$. The corrected Poisson distribution (9-18) gives $W(10) = 0.7 \times 10^{-7}$.

Two features of the Poisson distribution (9-19) might be noticed in particular. The probability of obtaining $m = M - 1$ is equal to the probability of obtaining $m = M$, or $W(M) = W(M-1)$. For large M the distribution is very nearly symmetrical about $m = M$ if values of m very far from M are excluded.

Gaussian Distribution. A further approximation of the distribution law may be made for large m (say > 100) and for $|M - m| \ll M$. With these additional restrictions, with the approximate expansion,

$$\ln\left(1 + \frac{M-m}{m}\right) = \frac{M-m}{m} - \frac{(M-m)^2}{2m^2},$$

neglecting subsequent terms, and with the use of Stirling's approximation

$$x! = \sqrt{2\pi x}\, x^x e^{-x},$$

we may modify the Poisson distribution to obtain the Gaussian distribution:

$$W(m) = \frac{1}{\sqrt{2\pi M}} e^{-(M-m)^2/2M}. \tag{9-20}$$

It will be noticed that this distribution is symmetrical about $m = M$. For both the Poisson and Gaussian distributions[2] we may derive $\sigma = \sqrt{M}$, or, for large m, $\sigma \approx \sqrt{m}$.

E. STATISTICAL INFERENCE

So far we have largely discussed **a priori** or prior probability, that is, the probability that a given event or set of events will occur, as calculated prior to any experimental observation. In practice we are more often concerned with a somewhat different concept of probability, sometimes called inverse probability: we may wish to deduce, from a (necessarily limited) set of observations, the probability that some particular distribution of events gave rise to these observations, or to determine which of several possible hypotheses best accounts for the observed results. This is the subject of statistical inference and we briefly discuss two aspects of it.

The Method of Maximum Likelihood (B2, M1). The **a posteriori**

[2] The functional dependence $\sigma = \sqrt{M}$ is a necessary condition of the Poisson but not of the Gaussian distribution. The general form of the Gaussian is

$$W(m) = \frac{1}{\sqrt{2\pi\sigma^2}} \exp\left[\frac{(M-m)^2}{2\sigma^2}\right],$$

where there is as a rule no relationship between M and σ. The relationship between σ and M for the Gaussian distribution of *counting rates* is a consequence of the particular source of random error: the fluctuation in the decay rate consistent with a decay probability per unit time that is independent of time.

probability of a given result is the probability of that result as deduced *after* the result has been obtained. In a series of coin tosses with an actual coin, the observed (a posteriori) probability of "heads" might, for example, be somewhat different from the a priori probability of 0.5 expected for the ideal coin. To distinguish a posteriori from a priori probability, the term **likelihood** is often substituted for the former.

Suppose a sequence of results x_i has been obtained, and we wish to determine which of several hypotheses $A, B, C, \ldots$ best accounts for these results. For each of the hypotheses a likelihood function $L(x_i \mid H)$ can be defined, where H stands for hypothesis $A, B, C, \ldots$ and $L(x_i \mid A) = \prod_{i=1}^{n} P(x_i \mid A)$ with $P(x_i \mid A)$ being the a posteriori probabilities for the individual values x_i, given hypothesis A. The method of maximum likelihood then consists in choosing, from the likelihood functions for all the different hypotheses, the one that has the maximum value. Usually this maximization can be done by expressing the different hypotheses in terms of some parameter choice and taking the first and second derivatives of the likelihood function with respect to that parameter. If the value of a parameter is to be deduced from a series of experimental measurements, the "best" value is the one that gives the maximum value of the likelihood function for the set of measurements.

Chi Square Test and Method of Least Squares. As an example we show how the method of maximum likelihood can be used to determine the "goodness" of fit to experimental data. Consider data that can be described by the linear relationship

$$y(x) = a_0 + b_0 x. \tag{9-21}$$

In general, we expect the distribution of measurements to be Gaussian or Poisson and, since these distributions are indistinguishable in most experiments, we assume that the data follow a Gaussian distribution. Then for any value of $x = x_i$ we can calculate the a posteriori probability P_i for making the observed measurement y_i with a standard deviation σ_i of observations around the actual value $y(x_i)$ as follows:

$$P_i = \frac{1}{\sigma_i \sqrt{2\pi}} \exp\left\{-\frac{1}{2}\left(\frac{y_i - y(x_i)}{\sigma_i}\right)^2\right\}. \tag{9-22}$$

For any values of a and b the likelihood function of the observed set of measurements is given by

$$L(y|a, b) = \prod P_i = \prod \left(\frac{1}{\sigma_i \sqrt{2\pi}}\right) \exp\left[-\frac{1}{2}\sum \left(\frac{\Delta y_i}{\sigma_i}\right)^2\right], \tag{9-23}$$

where the product Π is taken for i from 1 to N. The terms $\Delta y_i = y_i - a - bx_i$ are the deviations between each of the observed values y_i and the corresponding calculated values. If we assume that the observed measurements are more likely to have come from the parent distribution of (9-21) than any other distributions, then the method of maximum likelihood states that the best estimates for a and b are those that maximize the probability given in (9-23). This can be accomplished by minimizing the sum in the exponential, often called χ^2:

$$\chi^2 \equiv \sum \left(\frac{\Delta y_i}{\sigma_i}\right)^2 = \sum \frac{1}{\sigma^2}(y_i - a - bx_i)^2. \tag{9-24}$$

The parameter χ^2 therefore is a measure of goodness of fit of the parameters a and b, the optimum fit to the data being that which minimizes this weighted sum of squares of deviations. This minimization procedure is often called the method of **least squares**.

Bayes' Theorem. As we mentioned at the outset of this chapter, the primary problem of statistical inference is to estimate, from information available after only a finite number of observations, the average value that would be obtained after an infinite number of experimental observations of a given physical quantity. In terms of (9-19) what we really wish to know, for example, is the probability that the number of detected disintegrations of a radioactive sample (counts) is characterized by a mean value M when we have observed a value m [we may denote this probability by $W'(M|m)$]. Equation 9-19 gives us the inverse of what we wish to know: the probability of observing m counts when the sample is characterized by a mean value M [this inverse probability we may denote as $W(m|M)$]. These two conditional probabilities are related to each other:

$$P'(m)W'(M|m) = P(M)W(m|M), \tag{9-25}$$

where $P'(m)$ is the *prior probability* that the sample will give m counts before any observations have been made on the sample and $P(M)$ is the *prior probability* that the sample is characterized by a *mean number of counts* M before any observations have been made on the sample. The reader will readily perceive that these so-called prior probabilities are troublesome quantities. The two sides of (9-25) are equal to each other because each of them is equal to the joint probability that a sample will be characterized by a mean of M counts *and* will exhibit experimentally m counts. The quantity of interest $W'(M|m)$ may be readily obtained from (9-25):

$$W'(M|m) = \frac{P(M)W(m|M)}{P'(m)}, \tag{9-26}$$

an expression that was first discussed by the Reverend T. Bayes some two centuries ago.[3] The *prior probabilities* $P'(m)$ and $P(M)$ are related:

$$P'(m) = \sum_{M=0}^{\infty} P(M)W(m|M),$$

[3] For a discussion of conditional probability, see reference F1, chapter 5.

which states that if, in some manner, $P(M)$ is known, then through a combination of the addition theorem and the multiplication theorem the *prior probability* $P'(m)$ must also be known. The final expression, then, is

$$W'(M|m) = \frac{P(M)W(m|M)}{\sum_{M=0}^{\infty} P(M)W(m|M)}. \tag{9-27}$$

It is of interest to note the implications of (9-27) for a sample that complies with the restrictions required by the Poisson distribution, that is, a sample containing a large number of atoms that is observed for a time short compared to their half life. Taking (9-19) for $W(m|M)$, we obtain from (9-27)

$$W'(M|m) = \frac{P(M)(M^m e^{-M}/m!)}{\sum_{M=0}^{\infty} P(M)(M^m e^{-M}/m!)}. \tag{9-28}$$

It is impossible to proceed without an explicit expression for $P(M)$; it is at this point that the analysis can become metaphysical. We proceed by taking all values of M as being equally probable:

$$P(M)\, dM = K\, dM. \tag{9-29}$$

With this assumption, the summation in the denominator of (9-28) becomes an integral and we obtain

$$W'(M|m)\, dM = \frac{KM^m e^{-M}/m!}{\int_0^\infty K(M^m e^{-M}/m!)\, dM}\, dM = \frac{M^m e^{-M}}{m!}\, dM, \tag{9-30}$$

since

$$\int_0^\infty M^m e^{-M}\, dM = m!. \tag{9-31}$$

It is to be carefully noted that although the right side of (9-30) is similar to that of (9-19), it has a different meaning. Equation 9-30 gives the probability, under our choice of $P(M)$, that the sample has a *mean* between M and $M + dM$ counts when m counts have been observed. From (9-30) it is easily found that the *most probable* value of M is m; through the use of (9-6), (9-7), and (9-31), it is found that the average value of M is $m + 1$ and that the standard deviation of the distribution law (9-30) is $\sqrt{m+1}$. The difference between the average and the most probable value of M is unimportant for values of m that are not too small. For small values of m, for example $m = 0$, there is the question whether to estimate M by the average or by the most probable value.

To answer this question we must be clear about the meaning of the average value of M. It is the value that would be obtained in the following experiment: take a very large collection of samples, each of which had given m counts in a given time interval. Then the mean number of counts expected from *each* sample is determined from the average of a very large number of observations on *each* of the very large number of samples. It is then the average value of this very large number of mean values that is given by $m+1$, and m is the mean value that is most frequently observed.

Now the observation of m counts was made on one of this large number of samples; the question is: which one? The best answer is the most probable one, that is, the sample for which $M = m$. This answer becomes more familiar if we consider the estimate of the mean counts expected from a sample after n observations which gave results $m_1, m_2, \ldots, m_n$ have been made upon it. An expression for $W(M|m_i, m_2, \ldots, m_n)$, the probability that the sample is characterized by a mean value M when n observations give the results $m_1, m_2, \ldots, m_n$, can be derived in the same way as (9-30):

$$W(M|m_1, m_2, \ldots, m_n) = \frac{n(nM)^{m_1+m_2+\cdots+m_n} e^{-nM}}{(m_i + m_2 + \cdots + m_n)!}. \tag{9-32}$$

The maximum of this distribution function occurs for

$$M = \frac{m_1 + m_2 + \cdots + m_n}{n}, \tag{9-33}$$

which is the average value of the set of observations just as is expected from (9-1).

Information on the precision of the estimate for M is contained in the expressions for the distribution function: (9-30) or (9-32). The precision of the estimate of M may be characterized by the variance of its distribution function: $m+1$ for a single observation and $(m_1 + m_2 \cdots + m_n + 1)/n^2$ for n observations.

Variance, as computed above, may be used in the normal distribution law (9-3). For small values of m, though, it is probably best to discuss the data directly in terms of the distribution function (9-32). For example, if there is a single observation that gives $m = 0$, (9-32) says that there is a probability of 0.99 that M will be less than 4.6. If the value of zero is obtained in 10 independent observations, then there is a probability of 0.99 that M will be less than 0.46.

In summary, then, for an observed number of counts in excess of about 100, the best statement that can be made is the customary one (9-14a) that the mean value is $m \pm \sqrt{m}$ (taking $m+1 \approx m$). For a small number of counts the statement would be that the mean value is m and the confidence in the statement can be obtained from (9-32).

F. EXPERIMENTAL APPLICATIONS

Propagation of Errors. Whenever experimental data are used in the computation of a derived quantity, there is the question of the relationship between the precision of the computed values and the precision of the input information. For example, a background counting rate is to be subtracted from an observed counting rate, or the ratio of the counting rates of two samples is used as a measure of the relative numbers of atoms in the samples. The errors in the computed values may be more readily estimated from those of the input data if the error attached to each input datum is independent of that attached to any other.

Consider the independent measurements of two quantities x and y, which lead to the result that the probability of observing a value of x between x and $x+dx$ is $X(x)\,dx$, and similarly for y; then the independence of the measurements means that the probability of having a result with x between x and $x+dx$ while y is between y and $y+dy$ is

$$P(x, y)\,dx\,dy = X(x)\,Y(y)\,dx\,dy.$$

We now ask what is our best estimate of some quantity f, which is a function $f(x, y)$ of the variables x and y, and what is the precision of our estimate of f? The answer to this question is suggested by (9-6). Our best estimate of $f(x, y)$ is its average value[4]:

$$\overline{f(x, y)} = \int\int X(x)\,Y(y)\,f(x, y)\,dx\,dy. \tag{9-34}$$

Since the quantity that is sought is $f(\bar{x}_t, \bar{y}_t)$, it is instructive to examine the properties of (9-34) by making a Taylor expansion of $f(x, y)$ about the point $\bar{x}, \bar{y}$ that is our best estimate of $\bar{x}_t, \bar{y}_t$:

$$\begin{aligned}\overline{f(x, y)} = \int\int X(x)\,Y(y)\Big[&f(\bar{x}, \bar{y}) + (x-\bar{x})\,f_x(\bar{x}, \bar{y}) + (y-\bar{y})\,f_y(\bar{x}, \bar{y})\\ &+\frac{(x-\bar{x})^2}{2}\,f_{xx}(\bar{x}, \bar{y}) + \frac{(y-\bar{y})^2}{2}\,f_{yy}(\bar{x}, \bar{y})\\ &+(x-\bar{x})(y-\bar{y})\,f_{xy}(\bar{x}, \bar{y}) + \cdots\Big]\,dx\,dy,\end{aligned} \tag{9-35}$$

where $f_x(\bar{x}, \bar{y})$, $f_{xx}(\bar{x}, \bar{y})$, $f_{xy}(\bar{x}, \bar{y})$, and so on, mean the partial derivatives $\partial f/\partial x$, $\partial^2 f/\partial x^2$, $\partial^2 f/\partial x\,\partial y$, and so on, evaluated at the point $\bar{x}, \bar{y}$.

If $f(x, y)$ is a sufficiently slowly varying function in the region of $\bar{x}, \bar{y}$ so that the higher derivatives are negligible, then

$$\overline{f(x, y)} \approx f(\bar{x}, \bar{y}), \tag{9-36}$$

since

$$\int\int X(x)\,Y(y)(x-\bar{x}) = \bar{x} - \bar{x} = 0,$$

[4] See discussion in references B1, p. 51, and B3, p. 54.

and

$$\iint X(x)\,Y(y)(y-\bar{y}) = \bar{y} - \bar{y} = 0.$$

For the three elementary arithmetic operations, addition, subtraction, and multiplication, the Taylor series terminates after a finite number of terms, and the exact results

$$\overline{x+y} = \bar{x} + \bar{y}, \quad (9\text{-}37)$$

$$\overline{x-y} = \bar{x} - \bar{y}, \quad (9\text{-}38)$$

$$\overline{xy} = \bar{x}\bar{y} \quad (9\text{-}39)$$

are obtained. This is not the result, however, for the elementary operation of division.[5]

The estimate of the variance is given by

$$\sigma_{\bar{f}}^2 = \overline{[f(x,y) - \overline{f(x,y)}]^2} = \iint X(x)\,Y(y)[f(x,y) - \overline{f(x,y)}]^2\,dx\,dy. \quad (9\text{-}40)$$

If again a Taylor expansion is used and the higher order terms are neglected, then

$$\sigma_{\bar{f}}^2 = f_x^2(\bar{x},\bar{y})\sigma_{\bar{x}}^2 + f_y^2(\bar{x},\bar{y})\sigma_{\bar{y}}^2 + \cdots. \quad (9\text{-}41)$$

Exact expressions again result for the variance of three of the elementary arithmetic operations:

$$\sigma_{\overline{x+y}}^2 = \sigma_{\bar{x}}^2 + \sigma_{\bar{y}}^2, \quad (9\text{-}42)$$

$$\sigma_{\overline{x-y}}^2 = \sigma_{\bar{x}}^2 + \sigma_{\bar{y}}^2, \quad (9\text{-}43)$$

$$\frac{\sigma_{\overline{xy}}^2}{\bar{x}^2\bar{y}^2} = \frac{\sigma_{\bar{x}}^2}{\bar{x}^2} + \frac{\sigma_{\bar{y}}^2}{\bar{y}^2} + \frac{\sigma_{\bar{x}}^2\sigma_{\bar{y}}^2}{\bar{x}^2\bar{y}^2}. \quad (9\text{-}44)$$

The third term in expression (9-44) is usually small compared to the first two and may be neglected. Similarly, the first two terms of (9-41) are usually a good approximation for the variance of other functions x and y.

As an example, suppose that the background counting rate of a counter is measured and 600 counts are recorded in 15 min. Then with a sample in place the total counting rate is measured, and 1000 counts are recorded in 10 min. We wish to know the net counting rate due to the sample and the standard deviation of this net rate. First the background rate $\mathbf{R}_b$ is

$$\mathbf{R}_b = \frac{600 \pm \sqrt{600}}{15} = 40 \pm 1.6\ \text{cpm}.$$

[5] The quantity x/y as evaluated by (9-34) will be infinite unless $Y(y)$ approaches zero more rapidly than does y ($\lim_{y\to 0}[Y(y)/y] \neq \infty$). This infinity catastrophe is usually avoided by restricting the values of y to those that have a relatively large likelihood— that is, are close to $\bar{y}$. When this is done (9-36) gives the estimate of f.

The total rate $\mathbf{R}_t$ is

$$\mathbf{R}_t = \frac{1000 \pm \sqrt{1000}}{10} = 100 \pm 3.2 \text{ cpm}.$$

The net rate $\mathbf{R}_n = 100 - 40 = 60$ cpm, its standard deviation is $\sigma_n = \sqrt{1.6^2 + 3.2^2} = 3.6$, and $\mathbf{R}_n = 60 \pm 3.6$ cpm.

Gaussian Error Curve. Knowledge of the distribution law permits a quantitative evaluation of the probability of a given deviation of a measured result m from the proper average M to be expected. With the absolute error $|M - m| = \epsilon$ and with the assumption that the integral numbers are so large that the distribution may be treated as continuous, the probability $W(\epsilon)\, d\epsilon$ of an error between ϵ and $\epsilon + d\epsilon$ for the normal distribution is given by

$$W(\epsilon)\, d\epsilon = \frac{2}{\sqrt{2\pi M}} e^{-\epsilon^2/2M}\, d\epsilon. \tag{9-45}$$

The factor 2 arises from the existence of positive and negative errors with equal probability within the limits of validity of this approximation. Recalling that $\sigma = \sqrt{M}$, we have

$$W(\epsilon)\, d\epsilon = \frac{1}{\sigma}\sqrt{\frac{2}{\pi}} e^{-\epsilon^2/2\sigma^2}\, d\epsilon. \tag{9-46}$$

The probability of an error greater than $k\sigma$ is obtained by integration from $\epsilon = k\sigma$ to $\epsilon = \infty$. Numerical values of this integral as a function of k may be found in handbooks. For example, we have taken for table 9-2 some representative values from the table, "Probability of Occurrence of Deviations" in the Chemical Rubber Publishing Company's *Handbook of Chemistry and Physics.*

Notice that errors greater than and smaller than 0.674σ are equally probable; 0.674σ is called the **probable error** and is sometimes given rather than the standard deviation when counting data are reported. In plots of experimental curves it can be convenient to indicate the probable error of each point (by a mark of the proper length). Then on the average the smooth curve drawn should be expected to pass through as many error bars as it misses. It is unfortunately not strictly correct to use (9-46) with (9-41) in the estimation of the probability of an error of a function of random variables. For example, the distribution of the differences of two random variables that have Gaussian distributions is not

Table 9-2

k	0	0.674	1	2	3	4
Probability of $\epsilon > k\sigma$	1.00	0.50	0.32	0.046	0.0027	0.00006

itself Gaussian. Nevertheless, if the function does not vary too rapidly in the vicinity of its average, the distribution of values about the average is essentially Gaussian with a variance as given in (9-41).

Comparison with Experiment. We now return to a consideration of the typical counting data in table 9-1. We have already found from the deviations among the 10 measurements $\sigma = \sqrt{(N_0 - 1)^{-1}\Sigma(x_i - \bar{x})^2} = 11.9$. If the counting rate measured there represents a random phenomenon, as we expect it should, we may evaluate the expected σ for the result in any minute as the square root of the number of counts. For a typical minute, the ninth, we find $\sigma = \sqrt{101} = 10$, and for other minutes other values not much different. Because these values agree reasonably with the 11.9 there is evidence for the random nature of the observed counting rate. This test should occasionally be made on the data from a counting instrument.

In addition to estimating the σ for each entry in table 9-1, we may also estimate the $\sigma_{\bar{x}}$ for the average of the 10 observations. This estimate can be performed in three different ways, and it is instructive to compare them:

1. Since the 10 data are observations of a radioactive decay, we expect from (9-14*a*) that each datum has a standard deviation given by the square root of the number of counts. The mean is calculated by summing the data and dividing by the number of observations (10). The standard deviation of the mean, then, can be obtained from (9-41) for the propagation of fluctuations for a function of random variables (the number 10 has zero standard deviation). The result is

$$\sigma_{\bar{x}} = 99\sqrt{\frac{990}{990^2}} = \frac{1}{10}\sqrt{990} = 3.1.$$

2. The individual counting rates can be summed, which is equivalent to an observation of 990 counts in 10 min. Again, since we are dealing with radioactive decay, the standard deviation of the mean is given by (9-14*a*):

$$\sigma_{\bar{x}} = \frac{1}{10}\sqrt{990} = 3.1.$$

3. If the fact that these data are from radioactive decay is ignored and no special relation such as (9-14*a*) is assumed to exist between each observation and its standard deviation, then the standard deviation of the mean is computed from (9-5):

$$\sigma_{\bar{x}} = \sqrt{\frac{1276}{9 \times 10}} = 3.8.$$

It is important to note that methods 1 and 2 give the same answer, as they must; it is not possible to gain more information about the standard deviation of the mean by breaking a 10-min observation into 10 1-min observations. The $\sigma_{\bar{x}} = 3.1$ given by methods 1 and 2 is the correct answer. It is also of interest to see that relinquishing the information contained in (9-14a), as in method 3, diminishes the precision of the estimate of the mean.

The average counting rate with its standard deviation is $\bar{x} = (990 \pm \sqrt{990})/10 = 99.0 \pm 3.1$ cpm. This means that the probability that the true average is between 95.9 and 102.1 is, from table 9-2, just $1 - 0.32 = 0.68$. Actually, when the counting data given in table 9-1 were obtained, the average rate was measured much more accurately in a 100-min count, and the result was $(10{,}042 \pm \sqrt{10{,}042})/100 = 100.4 \pm 1.0$ cpm.

Counter Efficiencies. As another application of the methods of this chapter to counting techniques, we may estimate the efficiency of a Geiger counter for rays of a given ionizing power, with the assumptions that any ray that produces at least one ion pair in the counter gas is counted and that effects at the counter walls are negligible. Knowledge of the nature of the radiation and the information given in chapter 6 permit an estimate of the average number of ion pairs a to be expected within the path length of the radiation in the counter filling gas. The problem then is to find the probability that a ray will pass through the counter, leave no ion pairs, and thus will not be counted. We think of the path of the ray in the counter as divided into n segments of equal length. If n is very large, each segment will be so small that we may neglect the possibility of having two ion pairs in any segment. Then just a of the n segments will contain ion pairs, and by definition the probability of having an ion pair in a given segment is $p = a/n$. Now by (9-8) for the binomial distribution we have the probability of no ion pairs in n segments, that is, of $r = 0$:

$$W(0) = \frac{n!}{n!0!} p^0(1-p)^n = (1-p)^n = \left(1 - \frac{a}{n}\right)^n.$$

Since the probability[6] is evaluated correctly only as n becomes very large,

$$W(0) = \lim_{n\to\infty} \left(1 - \frac{a}{n}\right)^n = e^{-a}.$$

The probability of counting the ray, which is the efficiency to be determined, is then $1 - W(0) = 1 - e^{-a}$. As a particular example, consider a fast β particle with the relatively low specific ionization of 5 ion pairs per millimeter in air and a path length of 10 mm in a counter gas that is almost pure argon at 10 torr pressure. We estimate a from these assumptions, correcting for the relative densities of air and the argon:

$$a = 5 \times 10 \times \frac{7.6}{76} \times \frac{40}{29} = 7.$$

[6] We might have evaluated this probability more easily from the Poisson distribution expression: $W(0) = a^0 e^{-a}/0! = e^{-a}$.

The corresponding estimated counter efficiency is $1 - e^{-7} = 99.9$ percent. It should not be expected that an efficiency calculated in this way is very precise. Wall effects may be important, and the assumption of random distribution of ion pairs along the β-ray path is not entirely consistent with the mechanism of energy loss by ionization presented in chapter 6.

Dead-Time Correction[7]. If a counter has a recovery time (or dead time or resolving time) τ after each recorded count during which it is completely insensitive, the total insensitive time per unit time is $\mathbf{R}\tau$, where $\mathbf{R}$ is the observed counting rate. If $\mathbf{R}^*$ is the rate that would be recorded if there were no dead-time losses, the number of lost counts per unit time is $\mathbf{R}^* - \mathbf{R}$ and is given by the product of the rate $\mathbf{R}^*$ and the fraction of insensitive time $\mathbf{R}\tau$:

$$\mathbf{R}^* - \mathbf{R} = \mathbf{R}^*\mathbf{R}\tau,$$

$$\mathbf{R}^* = \frac{\mathbf{R}}{1 - \mathbf{R}\tau}. \tag{9-47}$$

A number of variants of this formula are also in use. One expression (the Schiff formula is $\mathbf{R}^* = \mathbf{R}e^{\mathbf{R}^*\tau}$. This is derived from a calculation of the probability $W(0)$ of having had no event during the time τ immediately preceding any event. An event, whether recorded or not, is here considered to prevent the recording of a second event occurring within the time τ.[8] Another approximate expression is derived from the first two terms in the binomial expansion of $(1 - \mathbf{R}\tau)^{-1}$ appearing in (9-47):

$$\mathbf{R}^* = \mathbf{R}(1 + \mathbf{R}\tau) = \mathbf{R} + \mathbf{R}^2\tau.$$

This form is especially convenient for the interpretation of an experiment designed to measure τ by measuring the rates $\mathbf{R}_1$ and $\mathbf{R}_2$ produced by two separate sources and the rate $\mathbf{R}_t$ produced by the two sources together, each of these rates including the background effect $\mathbf{R}_b$. Obviously,

$$\mathbf{R}_1^* + \mathbf{R}_2^* = \mathbf{R}_t^* + \mathbf{R}_b,$$

where we have neglected the dead-time loss in the measurement of the low background rate. Replacing by $\mathbf{R}_1^* = \mathbf{R}_1 + \mathbf{R}_1^2\tau$, and so on, and rearranging, we have

$$\tau = \frac{\mathbf{R}_1 + \mathbf{R}_2 - \mathbf{R}_t - \mathbf{R}_b}{\mathbf{R}_t^2 - \mathbf{R}_1^2 - \mathbf{R}_2^2}.$$

Statistics of Pulse Height Distributions. When a monoenergetic source of radiation is measured with a proportional, scintillation, or semi-

[7] The term coincidence correction is also used.

[8] It may be noticed that the Schiff formula might be expected to correspond more closely to the conditions of dead-time loss in a mechanical register, in which a new pulse within a dead time could initiate a new dead-time period, although it would not be recorded. There exists also the opportunity for dead-time losses in the electronic circuits.

conductor spectrometer, the observed pulse heights have a Gaussian distribution around the most probable value. The energy resolution of such an instrument is usually expressed in terms of the **full width at half maximum** (FWHM) of the pulse height distribution curve, stated as a fraction or percentage of the most probable pulse height H. The pulse height $h_{1/2}$ at the half maximum of the distribution curve may be obtained from the ratio of probabilities

$$\frac{W(h_{1/2})}{W(H)} = \exp\left[\frac{-(H-h_{1/2})^2}{2\sigma_h^2}\right] = 0.5.$$

Then $(H-h_{1/2})^2/\sigma_h^2 = \ln 2$, and the FWHM is

$$\frac{2|H-h_{1/2}|}{H} = 2\sqrt{2\ln 2}\frac{\sigma_h}{H} = \frac{2.35\sigma_h}{H},$$

where σ_h is the standard deviation of the pulse height distribution.

In a proportional counter the spread in pulse heights for monoenergetic rays absorbed in the counter volume arises from statistical fluctuations in the number of ion pairs formed and statistical fluctuations in the gas amplification factor. The pulse height is proportional to the product of the gas amplification and the number of ion pairs, and therefore the fractional standard deviation of the pulse height equals the square root of the sum of the squares of the fractional standard deviations of these two quantities.

As an example, consider the pulse height spectrum produced by the absorption of manganese K X rays in a proportional counter filled with 90 percent argon and 10 percent methane and operating with a gas gain of 1000. The energy per ion pair is estimated to be ~27 eV, and therefore the number of ion pairs formed by a 5.95-keV X ray is $5950/27 = 220 \pm \sqrt{220}$. If the numbers of ions collected per initial ion pair have a Poisson distribution, the fractional standard deviation in the gas gain is $\sqrt{1000}/1000$. Thus

$$\frac{\sigma_h}{H} = \sqrt{\frac{220}{220^2} + \frac{1000}{1000^2}} = \sqrt{0.00455 + 0.00100} = 0.0745,$$

and the FWHM is $2.35 \times 0.0745 = 0.715$ or 17.5 percent.

If the gas gain is made sufficiently large, the fluctuations in the number of ion pairs determine the resolution, and in this case the resolution of a given counter is seen to be inversely proportional to the square root of the energy of the ionizing radiation absorbed.

In a scintillation counter the statistical fluctuations in output pulse heights arise from several sources (B4). The conversion of energy of ionizing radiation into photons in the scintillator, the electron emission at the photocathode, and the electron multiplication at each dynode are all subject to statistical variations. Although the photocathode emission has been shown to have somewhat larger fluctuations than correspond to the Poisson law, the observed pulse height distributions are for most practical purposes in sufficiently close agreement with those calculated on the assumption of Poisson distributions for all the statistical processes involved. With this assumption the standard deviation of a pulse height distribution for a single energy of ionizing radiation absorbed in the phosphor turns out to be approximately

$$\sigma_h \approx H\sqrt{\frac{\bar{n}}{E\bar{q}\bar{f}\bar{p}(\bar{n}-1)}}, \tag{9-48}$$

where H is the most probable pulse height for an incident energy E keV, $\bar{q}$ is the mean value of the phosphor efficiency (number of light quanta emitted per 1000 eV of incident energy), $\bar{f}$ is the mean value of the light collection efficiency at the photocathode, $\bar{p}$ is the mean value of the photocathode efficiency (number of photoelectrons arriving at the first dynode for each photon incident on the photocathode), and $\bar{n}$ is the average electron multiplication per dynode.

In practice $\bar{f}$ can be made almost unity, $\bar{p}$ is of the order of 0.1, $\bar{n}$ is usually about 3 to 5, and $\bar{q}$ is approximately 30 for NaI (T1), 15 for anthracene, and 7 for stilbene and for the best liquid scintillators.

As an example, we estimate the resolution attainable for the 662-keV photopeak of the ^{137}Cs γ rays with a sodium iodide scintillation counter. Taking $\bar{f}\bar{p} = 0.1$ and $\bar{n} = 4$, we obtain

$$\frac{\sigma_h}{H} \approx \sqrt{\frac{4}{662 \times 30 \times 0.1 \times 3}} = 0.026.$$

The corresponding FWHM is 2.35 $\sigma_h/H = 0.061$ or 6.1 percent, which is indeed not far from the best resolution obtained experimentally. (See the experimental pulse height distribution with 8.5 percent width at half maximum shown in figure 7-10.)

In a semiconductor detector fluctuations in output pulse height result from the sharing of energy between ionization processes and lattice excitation (G1). For the case of a fixed energy E absorbed for each event in the detector, the pulse height standard deviation is given by

$$\sigma_h = H\sqrt{\frac{F\epsilon}{E}}, \tag{9-49}$$

where F is the Fano factor determined by the charge production processes in the detector[9] and ϵ is the average energy required to produce an electron-hole pair in the detector material. Because several poorly understood factors degrade resolution in a semiconductor detector, an empirical value of F must be used. The Fano factor is 0.12 for silicon and large

[9]The Fano factor can be defined as the ratio of the variance of the number of electron-hole pairs to the average number. It is essentially the ratio of the energy that goes into phonons to the total energy absorbed in the semiconductor.

germanium detectors and 0.08 for the best small-volume germanium detectors. The value of ϵ at 90 K is 3.76 eV for silicon and 2.96 eV for germanium (see chapter 7, p. 253).

The FWHM for 1-MeV γ rays in germanium at 90 K is $2.35 \times \sqrt{0.12 \times 2.96/10^6} = 1.4 \times 10^{-3}$ or 1.4×10^3 eV. The absolute value of the FWHM increases as E increases, but the percent resolution decreases as E increases. For example, the FWHM for 10-MeV γ rays in germanium is 4.4×10^3 eV or 0.04 percent, while for 0.1-MeV γ rays it is 4.4×10^2 eV or 0.4 percent.

REFERENCES

B1 C. A. Bennett and N. L. Franklin, *Statistical Analysis in Chemistry and Chemical Industry*, Wiley, New York, 1954.

*B2 P. R. Bevington, *Data Reduction and Error Analysis for the Physical Sciences*, McGraw-Hill, New York, 1969.

B3 K. A. Brownlee, *Statistical Theory and Methodology in Science and Engineering*, Wiley New York, 1960.

B4 E. Breitenberger, "Scintillation-Spectrometer Statistics," in *Progress in Nuclear Physics*, Vol. 4 (O. R. Frisch, Ed.), Pergamon, London, 1955, pp. 56–94.

*E1 R. D. Evans, *The Atomic Nucleus*, McGraw-Hill, New York, 1955, chapters 26–28.

F1 W. Feller, *Probability Theory and Its Applications*, Wiley, New York, 1950.

G1 F. S. Goulding and D. A. Landis, "Semiconductor Detector Spectrometer Electronics". in *Nuclear Spectroscopy and Reactions, Part A* (J. Cerny, Ed.), Academic, New York, 1974.

*M1 S. L. Meyer, *Data Analysis for Scientists and Engineers*, Wiley, New York, 1975.

EXERCISES

1. Mr. Jones's automobile license carriers a six-digit number. What is the probability that it has (a) exactly one 4, (b) at least one 4? Make the assumption that the numbers 0 to 9 are equally probable for each of the six digits.

   Answer: (b) 0.46856.

2. Consider the following set of observations:

Minute	Counts
1	203
2	194
3	201
4	217
5	195
6	189
7	210
8	207
9	230
10	188

(a) Calculate the average value. (b) What is the standard deviation of the set? (c) What is the standard deviation of the mean? (d) What is the probability that an eleventh observation would have a value greater than 230? (e) What is the probability that a subsequent set of 10 1-min observations will have an *average* value that is greater than 212? (f) Should any of the data be rejected? If so, what is the new average value? *Answers:* (c) 4.16. (e) 0.019.

3. Given an atom of a radioactive substance with decay constant λ, what is (a) the probability of its decaying between 0 and dt, (b) the probability of its decaying between 0 and t?

4. A sample contains 4 atoms of Lw. What is the probability that exactly 2 of the atoms will have decayed in (a) one half life, (b) two half lives?

5. A given proportional counter has a measured background rate of 900 counts in 30 min. With a sample of a long-lived activity in place, the total measured rate was 1100 counts in 20 min. What is the net sample counting rate and its standard deviation? *Answer:* 25.0 ± 1.9 cpm.

6. Denote by $\mathbf{R}_t$ and $\mathbf{R}_b$ the total and background counting rates for a long-lived sample and calculate the optimum division of available counting time between sample and background for minimum σ on the net counting rate. *Answer:* $t_t/t_b = \sqrt{\mathbf{R}_t/\mathbf{R}_b}$.

7. (a) Sample A, sample B, and background alone were each counted for 10 min; the observed total rates were 110, 205, and 44 cpm, respectively. Find the ratio of the activity of sample A to that of sample B and the standard deviation of this ratio. (b) Sample C was counted on the same counter for 2 min and the observed total rate was 155 cpm. Find the ratio and its standard deviation, of the activity of C to that of A. *Answer:* (a) 0.41 ± 0.027.

8. Derive (9-32) for the probability of a value M when given a set of observations $m_1, m_2, \ldots, m_n$.

9. Derive an equation for the FWHM for a semiconductor detector from the variance σ_n^2 in the number of electron-hole pairs

$$\sigma_n^2 = \frac{FE}{\epsilon}.$$

10. Commercially available small-volume germanium detectors have energy resolution as low as 150 eV (FWHM) for 5.6-keV X rays. How close is this to the "theoretical" limit of resolution?

Chapter 10

Nuclear Models

The central theoretical problem of nuclear physics is the derivation of the properties of nuclei from the laws that govern the interactions among nucleons. The central problem in theoretical chemistry is entirely analogous: the derivation of the properties of chemical compounds from the laws (electromagnetic and quantum-mechanical) that determine the interactions among electrons and nuclei. The chemical problem is complicated by the lack of mathematical techniques, other than approximate ones, for analyzing the properties of systems that contain more than two particles. The nuclear problem also suffers from this difficulty, but in addition it has two others:

1. The law that describes the force between two free nucleons is not completely known.
2. There is reason to believe that the force exerted by one nucleon on another when they are both also interacting with other nucleons is not identical to that which they exert on each other when they are free; in other words, there apparently are many-body forces.

Under these circumstances there is no alternative but to make simplifying assumptions that provide approximate solutions of the fundamental problem. These assumptions lead to the various models employed; or, more usually, a model for a nucleus or an atom is suggested by experimental results, and subsequently the assumptions consistent with the model are worked out. Consequently, several different models may exist for the description of the same physical situation; each model is used to describe a different aspect of the problem. For example, the Fermi-Thomas model of the atom is particularly useful for calculating quantities such as atomic form factors, which depend mainly on the spatial distribution of electron charge within the atom, but is less good than Hartree's self-consistent field approximation when questions of chemical binding are under analysis.

In the following sections we describe the models that have been found useful in codifying a large array of nuclear data, in particular, the energies, spins, and parities of nuclear states, as discussed in chapter 3, as well as nuclear magnetic and quadrupole moments. First we sketch what is known about nuclear forces and their implications for the properties of complex nuclei.

A. NUCLEAR FORCES

Information about the forces that exist between two free nucleons may be obtained most directly from observations on the scattering of one nucleon by another and from the properties of the deuteron. The quantity that is immediately useful for calculation is not the force between two nucleons, but rather the potential energy as a function of the coordinates (space, spin, and nucleon type) of the system. The quantity that we seek, therefore, plays a role similar to that of the Coulomb potential in the analysis of atomic and molecular properties and of the gravitational potential in the analysis of the motion of planets and satellites. The nuclear potential, though, seems to be considerably more complex than either the Coulomb or the gravitational potential. Although it is not yet possible to write down a unique expression for the nuclear potential, several of its properties are well known.

Characteristics of Nuclear Potential (B1, B2, C1). The potential energy of two nucleons shows great similarity to the potential-energy function that describes the stretching of a chemical bond.

1. *It is not spherically symmetrical.* For the chemical system this is simply a statement of the directional character of the chemical bond, the direction being determined by the other atoms in the molecule. For the nuclear interaction the direction is determined by the angles between the spin axis of each nucleon and the vector that connects the two nucleons. The quadrupole moment of the deuteron gives unambiguous evidence that the ground state of the deuteron lacks spherical symmetry, hence the potential cannot be a purely central one. The spherically symmetric part of the potential is called a **central potential**; the asymmetric part is the **tensor interaction.**

2. *It has a finite range and becomes large and repulsive at small distances.* The potential energy involved in the stretching of a chemical bond is adequately described by the well-known Morse potential, which is large and repulsive for the small distances at which electron clouds start to overlap, goes through a minimum several electron volts deep at distances of a few angstroms, and then essentially vanishes at distances of several angstroms. The nuclear potential behaves in much the same way, except that the distances are about 10^5 times smaller and the energies about 10^7 times larger. The nuclear potential becomes repulsive at distances smaller than about 0.5 fm and has essentially vanished when the internucleon separation is between 2 and 3 fm.

The detailed knowledge of the potential energy of the chemical bond comes mainly from information about excited vibrational states and from the determination of bond lengths from either diffraction studies or rotational spectra. The range and depth of the nuclear potential are derived from the binding energy of the only bound state of the deuteron (there are no excited states of the deuteron that are stable with respect to decomposition) and from studies of the collisions between nucleons. The size and binding energy of the deuteron are reasonably consistent with an attractive square-well potential about 25 MeV deep with a range of about 2.4 fm. More detailed information on the nuclear potential at smaller distances comes from the angular distributions of nucleon-nucleon scattering at several hundred MeV. These data require a repulsive core at a distance of about 0.5 fm and an attractive potential of about 200 MeV just before the repulsive potential sets in. At larger distances, rather than resembling a square well, the potential approaches zero in an approximately exponential fashion. The potential energy diagram for these two cases is given in figure 10-1.

The factor of 10^7 in the relative strengths of the nuclear and chemical forces is the source of the usual remark that nuclear forces are very strong; nevertheless, in view of their short range, nuclear forces behave, in point of fact, as if they were very weak. This apparently paradoxical statement can be easily understood when it is recalled that, if two particles are to be confined within a distance R of each other, they must have a de Broglie wavelength in the center-of-mass system that is no larger than $2R$. If

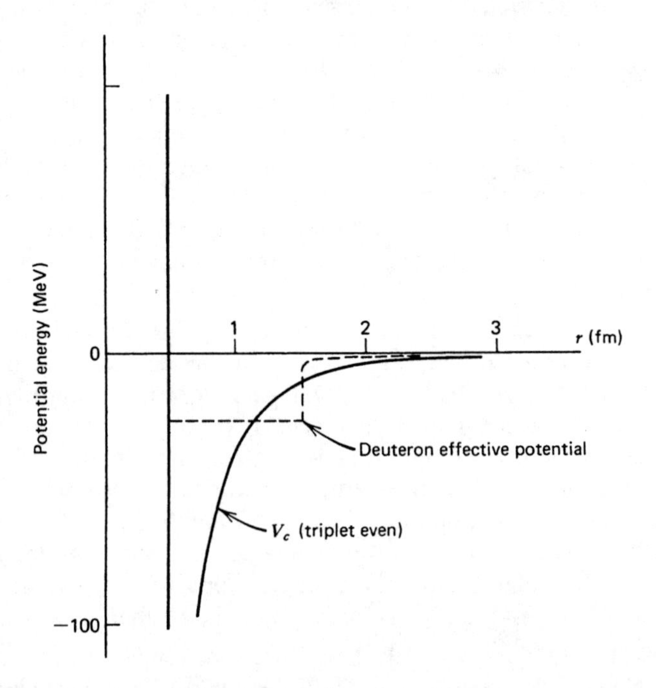

Fig. 10-1 Schematic diagram of nucleon–nucleon potential energy as a function of separation. The solid curve is the central potential for parallel spins and even relative angular momentum from reference H1. The dashed curve is the effective potential that can describe the deuteron.

$\mu = m_1 m_2/(m_1 + m_2)$ is the reduced mass of the two-particle system and v is the relative velocity, this condition can be written as

$$\lambda = \frac{h}{\mu v} \leq 2R, \tag{10-1}$$

or

$$\mu v \geq \frac{h}{2R}; \tag{10-2}$$

hence the kinetic energy of the particles is

$$\frac{1}{2}\mu v^2 \geq \frac{h^2}{8\mu R^2}. \tag{10-3}$$

The kinetic energy of two nucleons that are to remain within the range of nuclear forces (2.4 fm) must be at least

$$\frac{(6.6 \times 10^{-27})^2}{8 \times \frac{1}{2}(1.66 \times 10^{-24}) \times (2.4 \times 10^{-13})^2 \times 1.6 \times 10^{-6}} = 71 \text{ MeV},$$

which is greater than the depth of the potential well that is meant to hold them together. Thus the absence of excited states of the deuteron, its low binding energy (~2.2 MeV), and its large size (the proton and neutron spend about one half the time outside the range of the nuclear force) result from the weakness of the nuclear force when viewed in the context of its small range.

The chemical bond, on the other hand, has a range of about 10^5 times that of a nuclear force, and so the kinetic energy requirement is 10^{10} times smaller, or only 10^{-2} eV, which is but a small fraction of the depth of the potential. This large difference between the "real" strengths of the interatomic and internucleon forces is of great importance to our understanding of the properties of nuclear matter.

3. *It depends on the quantum state of the system.* The potential-energy curve that describes the stretching of a chemical bond depends on the electronic state of the molecule. For example, the stable H_2 molecule is one in which the two electrons have opposed spin (singlet state); when the electrons have parallel spin (triplet state) the molecule is unstable with respect to dissociation into two atoms.

The stable state of the deuteron is the one in which neutron and proton have parallel spins (triplet state); the potential energy of the singlet state is sufficiently different from that of the triplet so that there are no bound states of the isolated system consisting of one neutron and one proton with opposed spins. In addition to this spin dependence of the nucleon-nucleon potential, scattering experiments show that the potential also depends on the relative angular momentum of the two particles as well as on the orientation of this relative angular momentum with respect to the intrinsic spins of the nucleons. This latter term represents spin-orbit coupling which can lead to a partly polarized beam of scattered nucleons arising from an initially unpolarized beam.

4. *It has exchange character.* Our understanding of the chemical bond entails the exchange of electrons between the bonded atoms. If, for example, a beam of hydrogen ions were incident on a target of hydrogen atoms and many hydrogen atoms were observed to be ejected in the same direction as the incident beam, any analysis of the problem would have to include the process in which a hydrogen atom in the target merely handed an electron over to a passing hydrogen ion. The formal result would be that a hydrogen ion and a hydrogen atom would have exchanged coordinates.

It has been observed that the interaction between a beam of high-energy neutrons and a target of protons leads to many events (more than can be explained by head-on collisions), in which a high-energy proton is emitted in the direction of the incident neutron beam. The analysis of the observation entails the idea that the neutron and proton, when within the range of nuclear forces, may exchange roles. The observation is an excellent example of what is meant by the exchange character of the nuclear potential. The exchange character of the potential in conjunction with the requirement that the wave function describing the two-nucleon system be antisymmetric can give rise to the type of force described in (3) (B1, C1).

5. *It can be described by semiempirical formulas.* Despite the complexity of the potential between two nucleons it has been possible to construct semiempirical formulas for it that do reasonably well at describing the scattering of one nucleon by another up to energies of several hundred MeV. These potentials are rather complex as they must contain at least a purely central part with four components to account for the effect of parallel or antiparallel spins as well as the evenness or oddness of the relative angular momentum. In the most general form these potentials must also contain two components each of a tensor force and spin-orbit force that occur only for parallel spins but with either odd or even relative angular momentum, and four components of a second-order spin-orbit force that can occur for both parallel and antiparallel spins. An example of the central force for parallel spins and even relative angular momentum is given in figure 10-1 where it is compared with the effective potential that can describe the deuteron. More details are given in B2, C1, and W1.

Charge Symmetry and Charge Independence. So far we have not distinguished among neutron-neutron forces, proton-proton forces, and neutron-proton forces. The first evident difference is the Coulomb repulsion that must exist between two protons. At distances of the order of 1 fm this is much smaller than the attractive nuclear potential. Second, since the neutron and proton have differing magnetic moments, there will be different potential energies because of the magnetic interaction; this effect is even smaller than the Coulomb repulsion and is generally neglected.

From the observation that the difference in properties of a pair of **mirror nuclei** (nuclei in which the number of neutrons and the number of protons is interchanged, for example, $^{41}_{20}Ca$ and $^{41}_{21}Sc$) can be accounted for by the differing Coulomb interactions in the two nuclei, the purely nuclear part of the proton-proton interaction in a given quantum state has been taken to be identical to that of two neutrons in the same quantum state as the protons. This identity is known as charge symmetry. A more powerful generalization arises from the similarity between neutron-proton scattering and proton-proton scattering when the two systems are in the same spin state and have equal momenta and angular momenta.[1] This similarity leads to the assumption of charge independence, which asserts that the interaction of two nucleons depends only on their quantum state and not at all on their type, except, of course, for the Coulomb repulsion between two protons. So far there is no sizable divergence between this assertion and experimental results, but the search for small deviations continues to be actively pursued.

Isospin. The charge independence of nuclear forces leads to the idea that the proton and neutron can be considered as two different quantum states of a single particle, the **nucleon**. Since only two states occur, the situation is analogous to that of the two spin states an electron may exhibit, and thus the whole quantum-mechanical formalism developed for a system of electron spins has been taken over for the description of the charge state of a group of nucleons. The physical property involved is called variously isospin, isotopic spin, or isobaric spin (T). Each nucleon has a total isospin of $\frac{1}{2}$ just as the electron has a total spin of $\frac{1}{2}$. The z component of the isospin (T_z) may be either $+\frac{1}{2}$ or $-\frac{1}{2}$; in nuclear physics the $+\frac{1}{2}$ state is taken to correspond to a neutron and the $-\frac{1}{2}$ state to a proton.[2] For example, ^{9}Be with 5 neutrons and 4 protons has $T_z = +\frac{1}{2}$. The concept of isospin for individual nucleons approximately carries over to complex nuclei, where the corresponding quantity is the vector sum of the isospins of the constituent nucleons, which is *nearly* a good quantum number and thus *nearly* a conserved quantity.[3]

Two nucleons, for example, may have a total isospin of either 1 or 0. For

[1] This does not mean that the scattering of neutrons by protons is identical to that of protons by protons. The Pauli exclusion principle makes certain states inaccessible to the two protons that may be quite important in the neutron-proton scattering. For example, in low-energy scattering that takes place in states without orbital angular momentum (s states), the two protons must have opposite spins ($^1s_0$), whereas the neutron and proton may have either opposite spins ($^1s_0$) or parallel spins ($^3s_1$).

[2] In elementary-particle physics, the opposite convention is used.

[3] The concept of *nearly* good quantum numbers is well known in quantum mechanics. Small deviations from rigorously conserved quantities are treated by perturbation theory in terms of small parameters. The mixing of isospin states results from the force that makes nucleon-nucleon interactions not really independent of nucleon type: the Coulomb force.

$T = 1$, T_z may be -1 (2 protons), 0 (a proton and a neutron), or $+1$ (2 neutrons). For a total isospin of 0 the z component can only be 0 (a proton and a neutron). Thus a system containing a proton and a neutron must have $T_z = 0$ but may have $T = 1$ or $T = 0$; a system of two neutrons or of two protons must have $T = 1$. The demands of the Pauli principle for proton-proton and neutron-neutron pairs are satisfied within this formalism by requiring antisymmetry of the wave function describing the system, which is now, however, a function of three classes of variables: space, spin, and isospin:

$$\psi(\text{system}) = \psi(\text{space})\, \psi(\text{spin})\, \psi(\text{isospin}).$$

In the ground state of the deuteron, for example, ψ(space) is symmetric (it is a mixture of an s state and a d state), ψ(spin) is symmetric (the two spins are parallel), so that the ψ(isospin) must be antisymmetric and thus $T = 0$ (the two isospins are oppositely oriented). The lowest state of the deuteron in which the two nucleon spins are opposed [ψ(spin) is then antisymmetric] is the lowest one in which $T = 1$. The concept of isospin and its applications are described in detail in W1, B2, and C1. The implications of isospin for complex nuclei are discussed below.

Isobaric Analog States. The z-component of the isospin (T_z) defines the charge state of the nucleus. Thus for a nucleus containing N neutrons and Z protons

$$T_z = \frac{N - Z}{2}. \tag{10-4}$$

Accordingly,

$$\frac{|N - Z|}{2} \leq T \leq \frac{A}{2}, \tag{10-5}$$

for a nucleus of mass number A. Except for a few odd-odd nuclei with $Z = N$, all nuclei have $T = T_z$ in the ground state. As an example, $^{235}_{92}U$ in the ground state has $T = T_z = 51/2$. The other possible values of T as given in (10-5) are to be found in the excited states of the nucleus.

Let us briefly consider the nucleus (N, Z), which is characterized by a set of quantum numbers including the quantum numbers T and T_z. Suppose now that the state of the nucleus is changed by changing only T_z to $T_z - 1$ and leaving all other quantum numbers, including T, the same. Because of the lack of dependence of nuclear forces on charge state, we must again have a nucleus whose space and spin quantum states are exactly as before except that it now contains $Z + 1$ protons and $N - 1$ neutrons; a neutron has been changed into a proton. These two states, that of the original and that of the new nucleus, are called isobaric analog states for the obvious reasons that the two nuclei are isobars and the two quantum states are corresponding ones. The z-component of isospin of the new nucleus T'_z is

$$T'_z = T_z - 1. \tag{10-6}$$

The *ground state* of the new nucleus would therefore be expected to have an isospin T', where

$$T' = T'_z = T_z - 1, \tag{10-7}$$

whereas the ground-state isospin of the original nucleus is $T = T_z$. Thus the isobaric analog state in the nucleus $(N - 1, Z + 1)$ is an *excited* state with isospin one unit greater than that of the *ground* state of the isobaric analog nucleus (N, Z). In general, each state of A nucleons that is characterized by isospin T_0 will have $2T_0 + 1$ isobaric analog states with T_z going from $+T_0$ to $-T_0$ in integral units. The situation is illustrated schematically in figure 10-2.

Transition rates between isobaric analog states are strongly enhanced because of the nearly complete overlap of the space and spin parts of the wave function. Beta-decay transitions between mirror nuclei as described on p. 87 are a special case of this phenomenon.

Energies of Isobaric Analog States. The energy difference between isobaric analog states results from the change in Coulomb energy and the neutron-proton mass difference when a neutron is effectively transformed into a proton or vice versa. If the Coulomb force were somehow switched off, the energies of isobaric analog states would be precisely the same because there would then be no Coulomb repulsion among the protons in the nucleus, and the neutron-proton mass difference would also vanish. Thus the energy difference between isobaric analog states of N, Z and $N - 1$, $Z + 1$ may be expressed as

$$E_{IA}(Z + 1) = E_{IA}(Z) + \Delta E_c - (m_n - m_H)c^2, \tag{10-8}$$

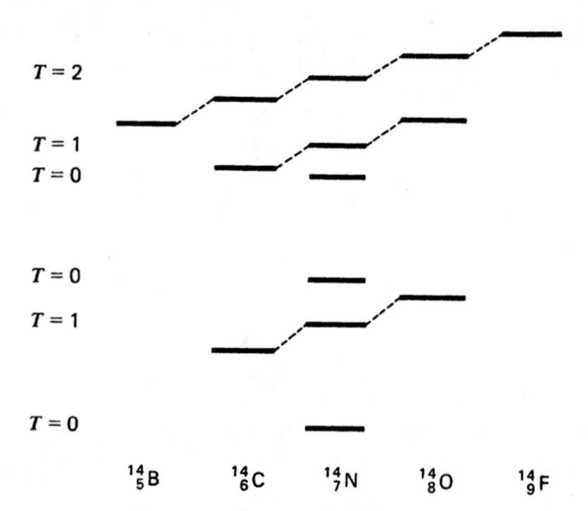

Fig. 10-2 Isobaric analog states in $A = 14$ nuclei. States are classified according to the T quantum numbers. [Adapted from *Concepts of Nuclear Physics* by B. L. Cohen. Copyright © 1971 by McGraw Hill, Inc. Used with the permission of McGraw Hill Book Company.]

where m_n and m_H are the masses of neutron and H atom, respectively The change in Coulomb energy, ΔE_c, between isobaric analog states may be estimated from the third and fourth terms of (2-5).

Meson Theories of Nuclear Forces. The qualitative similarity between the properties of chemical forces and nuclear forces led early investigators, notably H. Yukawa (Y1), to explore the possibility that nuclear forces resulted from the exchange of a particle between two nucleons in a manner analogous to the chemical force (which depends on the exchange of an electron between two atoms). This is not to say that the nucleon was now to be considered a composite particle, as is the atom, but rather that the particle to be exchanged, so to speak, was created at the instant of emission from one nucleon and vanished at the instant of absorption by the other nucleon. Processes of this type in which **virtual particles** are exchanged are important in all aspects of modern field theory that go beyond the classical idea of action-at-a-distance. For example, the Coulomb interaction between two charged particles is now analyzed in terms of the exchange of virtual photons between the two charges. The creation of the virtual particle immediately brings up the question of energy conservation. It takes energy to create particles; where does this energy come from? The answer is "nowhere," and that is why the particle is "virtual"; energy conservation is accounted for by making sure that the virtual particle does not live too long. From the Heisenberg uncertainty principle we know that

$$\Delta E \, \Delta t \gtrsim \hbar,$$

where Δt is the time available for measuring the energy of a system and ΔE is the accuracy within which the energy may be determined in the time Δt. All that is required, then, for energy conservation is that the lifetime Δt of the state produced by the creation of the virtual particle be such that

$$\Delta t \gtrsim \frac{\hbar}{\Delta E}.$$

Since the energy required to create a particle of mass m is given by the Einstein equation

$$\Delta E = mc^2,$$

$$\Delta t \gtrsim \frac{\hbar}{mc^2}.$$

If the virtual particle moves with the velocity of light, then the range of the force is about

$$R \approx c \, \Delta t \gtrsim \frac{\hbar}{mc}. \qquad (10\text{-}9)$$

A range of about 2 fm requires a virtual particle with a mass about 200 times that of an electron. Further, just as the quantum of the electromagnetic field (the virtual photon) may become a real particle in the physical world by absorbing some of the energy available in the collision between two charged particles, so the quantum of the nuclear field should become a physical particle in a collision between nucleons in which sufficient energy is available to supply the rest-mass energy of the quantum. This process does indeed occur, and the π-meson, a particle of 273 electron masses, is observed and is taken to be the quantum of the nuclear field.[4] Unfortunately the picture is not quite so simple: as the available energy is increased other particles are also created whose role in the nuclear force field is not fully understood. So far no complete field theory of nuclear forces in terms of meson exchange exists, but the approximate theories provide a valuable guide.

B. NUCLEAR MATTER

We first consider the properties of an infinite chunk of nuclear matter that contains essentially equal numbers of neutrons and protons. This hypothetical infinite nucleus is probably a good description of the central region of heavy nuclei. It is a good starting point in a discussion of nuclei because the complexities caused by boundary conditions at the surface of the nucleus may be ignored.

There are two immediately evident and important characteristics of nuclear matter exhibited by nuclei of mass number larger than about 20:

1. The binding energies per nucleon are essentially independent of mass number as reflected by the first term in the binding-energy formula (2-5). This means that all nucleons in a nucleus *do not* interact with all other nucleons (if they did, the binding energy *per nucleon* would be proportional to the mass number).
2. The densities are also essentially independent of mass number, which means that all nuclei *do not* simply collapse until the diameter is about equal to the range of nuclear forces so that all nucleons may be within one another's force field. Although the density of the nucleus is quite high, the nucleons are by no means densely packed.

These two general characteristics of nuclear matter are related and should have a common explanation. Two different possible causes of these

[4] The first particle with approximately the right mass that was discovered was the μ meson. The discovery, though, was quite a blow to the theory, since the μ meson interacted only very weakly with nuclei—hardly an acceptable behavior for the quantum of the nuclear field. Several years later it was found that the μ meson was the decay product of another meson, the π meson, which does interact strongly with nuclei.

characteristics that have immediate analogies in the domain of chemical forces come to mind.

1. A drop of liquid argon, for example, has a density and a binding energy per atom independent of the size of the drop as long as it is not too small. These characteristics result from the Van der Waals forces, which are attractive and large only for nearest neighbors. As an approximation, each argon atom interacts strongly with at most 12 other argon atoms. The key to the situation here is the Van der Waals *repulsion*, which sets in when the atoms touch. The corresponding repulsion that exists in nuclear forces at small distances would lead to the same effect. The observed density of nuclear matter, though, is much smaller than this effect by itself would give. Thus there must be an additional factor.

2. A piece of diamond also exhibits a density and a binding energy per atom independent of size, but the reason is different from that for a drop of liquid argon. In diamond each carbon atom is covalently bonded to four other carbon atoms and thus interacts strongly with only these four. It pays little attention to a fifth that may be brought near to it because the chemical bond has *saturation properties* and the first four carbon atoms have saturated the valency of the central carbon atom. The saturation property of the chemical bond arises from the limited number of valence electrons available for exchange between bonded atoms. The exchange character of nuclear forces also causes the interaction between nucleons to be strong only if the nucleons are in the proper states of relative motion.

Many-Body Calculations. Unfortunately, it is not simple to show that the repulsive core, in conjunction with the exchange character of nuclear forces, results in the approximate constancy of the density of nuclear matter and of the binding energy per nucleon. This result is difficult to obtain because it involves the many-body aspects of a quantum system in an essential manner that is further complicated by the repulsive core. Nevertheless, the problem has been successfully analyzed by an approach developed by K. Brueckner and collaborators utilizing nucleon-nucleon potentials described in section A and neglecting the Coulomb repulsion (see B3 and G1 for a review and description of this analysis). The results of this calculation, illustrated in figure 10-3, yield a binding energy per nucleon in "infinite nuclear matter" (before corrections for surface, Coulomb, and symmetry effects) that is in rough agreement with the volume term in (2-5) and with the central density of heavy nuclei.

These calculations also provide information about the motion of the neutrons and protons in nuclear matter, or, in quantum-mechanical language, the wave function that describes nuclear matter. The *effective* weakness of nuclear forces, discussed in section A, and the Pauli exclusion principle result in the nucleons moving about much as free particles in nuclear matter, with little perturbation of their motions by collisions with

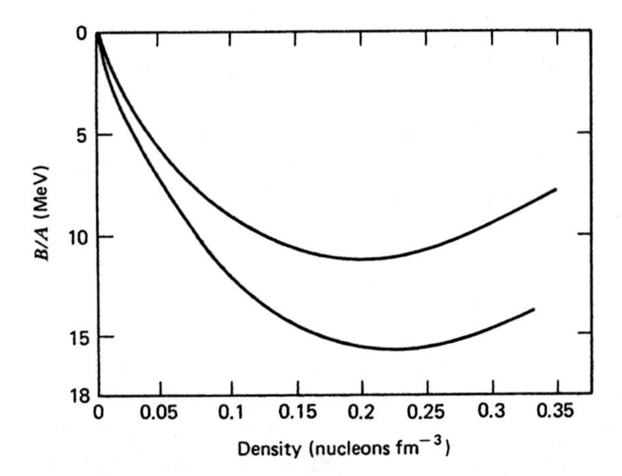

Fig. 10-3 Binding energy per nucleon versus density of nuclear matter calculated from infinite-nuclear-matter theory. The upper curve is the result from theory. The lower curve gives a result somewhat improved by arbitrarily increasing the potential energy by 12 percent. (From reference B3, reproduced, with permission, from the Annual Review of Nuclear Science, Volume 21 © 1971 by Annual Reviews Inc.)

other nucleons. Equivalently, a good first approximation to the wave function for nuclear matter is simply a properly antisymmetrized product of the free-particle wave functions for each of the nearly free nucleons in the nucleus. The collisions between nucleons are essentially quenched because for a collision to be effective the colliding particles must transfer some momentum to one another. But all states of lower momentum are already occupied by other nucleons, and the Pauli principle therefore forbids the occurrence of any momentum transfers. The effectiveness of the Pauli principle could be diminished if the internucleon forces were strong enough. For example, deuterium atoms, despite the fact that they must also obey the Pauli principle, will not move as essentially free particles at low temperatures but will couple together to form D_2 molecules. This is a manifestation of the *effective* strength of chemical forces compared to nuclear forces.

It is expected that nucleons in the nucleus would form clusters in regions where the nuclear density is so low that the average distance between nucleons, were they randomly distributed, would exceed the range of the nucleon-nucleon potential. The diffuse surfaces of finite nuclei must have such regions of very low density, and there is some evidence that α-particle clusters may have a transient existence in these diffuse edges.

Summary. The characteristics of nucleon-nucleon forces in conjunction with the Pauli exclusion principle cause nuclear matter to exhibit apparently contradictory behavior: the *macroscopic* properties, such as density and binding energy, resemble those of a drop of liquid; the *microscopic* properties, such as nuclear wave functions and particle motions, resemble those of a weakly interacting gas. The resemblance to a drop of liquid has already been exploited in the development of the binding-energy equation in chapter 2 and appears again in the treatment of the collective model; the resemblance to a weakly interacting gas serves as the basis for the Fermi gas model and for the shell model.

C. FERMI GAS MODEL

The simplest nuclear model that emphasizes the free-particle character of the motion of nucleons within the nucleus is the Fermi gas model. In this model the nucleus is taken to be composed of a degenerate Fermi gas of neutrons and protons contained within a volume defined by the nuclear surface. The gas is considered degenerate because all the particles are crowded into the lowest possible states in a manner consistent with the requirements of the Pauli principle. The gas, for each type of particle, may be characterized by the kinetic energy of the highest filled state, the **Fermi energy**. The Fermi energy is easily determined through the condition that there be enough states up to and including the highest filled state to accommodate the particles in the nucleus. Recalling that there may be two identical nucleons with opposed spins in each quantum state, this condition gives, for neutrons [refer to footnote 17, p. 80],

$$\frac{N}{2} = \frac{(4\pi/3)p_f^3 V}{h^3}, \tag{10-10}$$

where N is the neutron number of the nucleus, V is the volume of the nucleus, and p_f is the momentum of the neutron in the highest filled state, the Fermi momentum. By rearrangement of (10-10) and substitution of the classical relation between kinetic energy ϵ and momentum $p = (2M\epsilon)^{1/2}$, the expression for the Fermi energy of the neutron gas is

$$\epsilon_f = \frac{1}{8}\left(\frac{3}{\pi}\right)^{2/3}\left(\frac{N}{V}\right)^{2/3}\frac{h^2}{M}, \tag{10-11}$$

where M is the mass of the neutron. For the center of complex nuclei, where the density is about 2×10^{38} nucleons cm^{-3}, the Fermi energy for $N = A/2$ is about 43 MeV. To give the proper value for the binding energy of the neutron, approximately 8 MeV, this value of the Fermi energy implies that the neutron gas is contained in a potential-energy well that, at the center of complex nuclei, is about 50 MeV deep.

The Fermi gas model is not useful for the prediction of the detailed properties of low-lying states of nuclei observed in the radioactive decay processes described in chapter 3. It is useful, though, for the estimation of the momentum distribution of nucleons within the nucleus and for the approximate thermodynamic treatment of the properties of nuclei that are excited up into the continuum. These two aspects of the model are of importance in the study of nuclear reactions and were mentioned briefly in chapter 4.

D. SHELL MODEL

The shell model of the nucleus is similar to the Fermi gas model in that the interactions among the nucleons in the nucleus are again replaced by a potential-energy well within which each particle moves freely. In the Fermi gas model, as we have just seen, the nucleus is characterized by the energy of the highest filled level, the Fermi energy. In the shell model we are concerned with the detailed properties of the quantum states; these properties are determined by the shape of the potential-energy well. Before discussing the shell model in detail, it would be well to review briefly the experimental evidence that forced the shell model into nuclear theory. It did not develop from first principles; indeed, it appeared despite first principles, and much of the theoretical work sketched in section B was motivated by a desire to make the successes of the shell model respectable.

Experimental Evidence. In addition to the evidence for "magic numbers" (closed shells) cited in chapter 2, information about energies, spins, parities, and magnetic moments of nuclear states, partly gathered by the methods discussed in chapter 3, gave decisive impetus to a serious consideration of the shell model. Briefly, the observations were the following:

1. Ground-state spin of 0 for all nuclei with even neutron and proton number.
2. The systematics of the ground-state spins (half-integral) of odd-mass-number nuclei.
3. The form of the dependence of magnetic moments of nuclei upon their spins.

These observations suggested that the properties of the ground states of odd-mass-number nuclei to a first approximation could be considered to be those of the odd nucleon alone. The important point is the implication that all the other nucleons play no role except that of providing a potential-energy field that determines the single-particle quantum states and of filling those quantum states up to the one in which the odd particle moves. For details on the experimental evidence for the shell model, as well as the calculations of wave functions and energy levels, the reader is referred to Mayer and Jensen (M1).

Effective Potential. The important simplification in the shell model is to replace the nucleon-nucleon interactions inside the nucleus with an effective potential energy that acts on each nucleon and may be a function of its coordinates but not of the coordinates of the other nucleons. Thus it is the nuclear counterpart of the Hartree method for the many-electron atom. The problem then reduces to solving the Schrödinger equation for a particle moving in the chosen potential-energy field.

Two potentials are usually discussed. Both are taken to be spherically symmetric, but they differ in their radial dependence. The first is the **harmonic-oscillator** potential,

$$V(r) = -V_0\left[1 - \left(\frac{r}{R}\right)^2\right], \tag{10-12}$$

and the other is the **square-well** potential,

$$\begin{aligned} V(r) &= -V_0 \qquad r < R, \\ V(r) &= \infty \qquad r > R, \end{aligned} \tag{10-13}$$

where $V(r)$ is the potential at a distance r from the center of the nucleus and R is the nuclear radius. To simplify the mathematical solution of the problem both potentials, unrealistically, go to infinity rather than to zero outside of the nucleus. This further simplification has only a small effect on the energies of the states and on their relative stabilities. The Woods-Saxon potential (2-3) discussed in the context of scattering problems in chapter 2 has a shape intermediate between the square-well and harmonic-oscillator potentials. It is more difficult to manipulate for shell-model calculations than either of the other two.

Shell-Model States. The solution to the problem of the three-dimensional isotropic harmonic oscillator is well known to give energy levels

$$\begin{aligned} \epsilon &= (n_x + n_y + n_z + \tfrac{3}{2})\hbar\omega_0 \\ &= (N + \tfrac{3}{2})\hbar\omega_0, \end{aligned} \tag{10-14}$$

$$\omega_0 = \left(\frac{2V_0}{MR^2}\right)^{1/2}, \tag{10-15}$$

where M is the nucleon mass and n_x, n_y, n_z, and N are 0 or positive integers. The orbital angular momentum of the system may take on the usual values for a spherically symmetric potential,

$$L\hbar = [l(l+1)]^{1/2}\hbar, \tag{10-16}$$

where $l = N, N-2, \ldots, 0$ or 1. As in atomic spectroscopy, the states with $l = 0, 1, 2, 3, \ldots$ are designated as $s, p, d, f, \ldots$, respectively.

Because of the spherical symmetry of the problem, it is more convenient to use quantum numbers that describe the solution in spherical coordinates. For the harmonic oscillator this means that

$$N = 2(n-1) + l, \tag{10-17}$$

$$\epsilon = [2(n-1) + l]\hbar\omega_0 + \tfrac{3}{2}\hbar\omega_0, \tag{10-18}$$

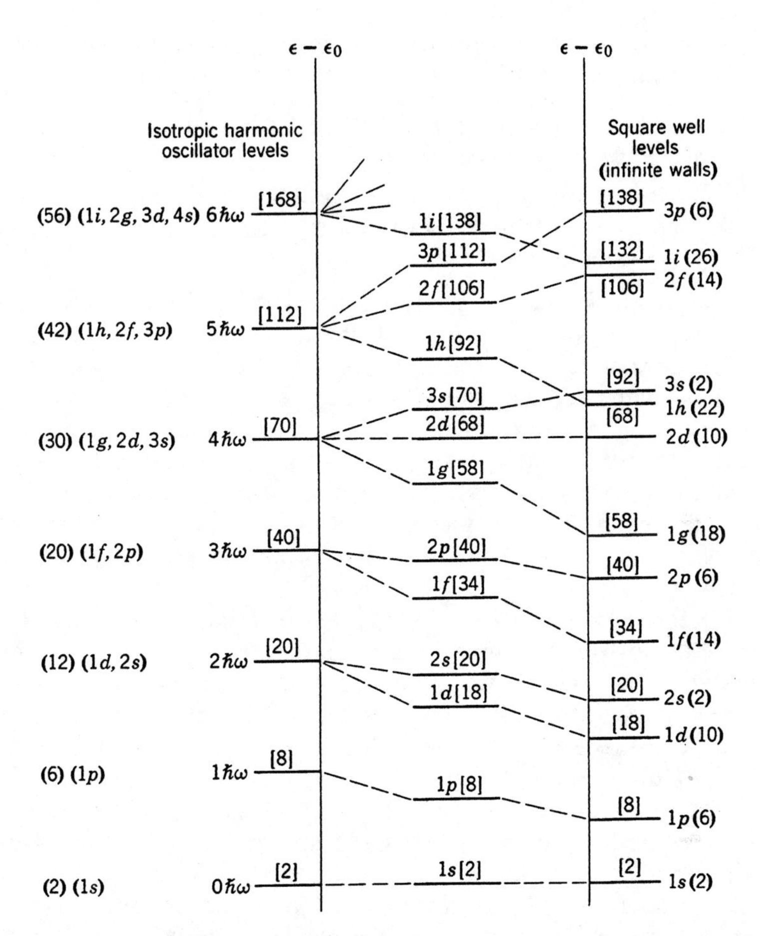

Fig. 10-4 Energy levels of the three-dimensional isotropic harmonic oscillator and of the square well with infinitely high walls. The numbers in parentheses are the numbers of nucleons of one kind required to fill the various levels; the numbers in brackets are the numbers of nucleons of one kind that are required to fill the levels up to and including a given level. (From reference M1.)

where $n = 1, 2, 3, \ldots$. Thus the states in the three-dimensional spherical potential are defined by the quantum numbers n and l, and will be identified as $3s$, $1d$, $2f$, and so on.[5] In addition, of course, there is the usual quantum number m, which takes on integral values from $-l$ to $+l$ such that $m\hbar$ is the projection of the angular momentum on a space-fixed axis. These energy levels are shown schematically at the left in figure 10-4 with respect to the

[5] The existence of states such as $1d$ and $2f$, which apparently violates the terms in atomic spectroscopy (where $n \geq l + 1$), is merely a consequence of the definition of n. For the usual hydrogenic wave functions n is defined so that each state has $n - l - 1$ radial nodes; the definition of n used above gives each state $n - 1$ radial nodes.

lowest level (zero-point energy of $\frac{3}{2}\hbar\omega_0$). Two important properties of these levels should be noted:

1. All states with the same value of $2n + l$ have the same energy and are therefore accidentally degenerate.
2. Since the energy goes as $2(n-1)+l$, the states of a given energy must either all have even or all have odd values of l. Hence all degenerate states have the same parity.

The pattern of eigenstates for the square-well potential shown on the right of figure 10-4 is similar to that of the harmonic oscillator except that all of the accidential degeneracies are removed. The change from a harmonic oscillator to a square well lowers the potential energy (a negative quantity) near the edge of the nucleus and thus enhances the stability of those states that concentrate particles near the edge of the nucleus. This means that states with largest angular momentum are most stabilized. The sequence of levels in real nuclei might be expected to be someplace between these two extremes and is indicated by the levels in the center of figure 10-4.

As has been mentioned in chapter 2, the experimental evidence for the shell structure of nuclei points to the magic numbers of 2, 8, 20, 28, 50, 82, and 126 as the numbers of neutrons or protons that occur at closed shells and correspond to the atomic numbers of the rare gases in chemistry. The sequence of levels in figure 10-4 shows the possibility of predicting the first three of these numbers, but the others are certainly not evident. The same situation occurs with the chemical elements; hydrogenic wave functions predict closed shells at atomic numbers 2, 10, 28, 60, 110, . . . , only the first two of which correspond to experimental fact. The atomic problem is now thoroughly understood in terms of the removal of degeneracies by the interactions of the electrons with one another. This suggests the search for an interaction in nuclear matter that would split the levels in figure 10-4 even further and perhaps reveal the magic numbers.

Spin-Orbit Interaction. This important interaction, which had not yet been included, was pointed out independently by M. G. Mayer (M2) and by O. Haxel, J. H. D. Jensen, and H. E. Suess (H2); it was the interaction between the orbital angular momentum and the intrinsic angular momentum (spin) of a particle. This interaction was already well known in the atomic problem, where, however, it plays a relatively minor role. It is also seen (see section A) in the polarization of scattered particles.

Consider, for example, a nucleon in a $1p$ state; it has an orbital angular momentum of $1\hbar$ and a spin of $\frac{1}{2}\hbar$. By the rules of quantum mechanics the total angular momentum of the particle may be either $J = \frac{3}{2}\hbar$ or $J = \frac{1}{2}\hbar$, states that we designate as $1p_{3/2}$ and $1p_{1/2}$, respectively. Spin-orbit interaction means that the energies of the $1p_{3/2}$ and $1p_{1/2}$ states are not the same; the sixfold degenerate $1p$ state is split into the fourfold degenerate $1p_{3/2}$

and the twofold degenerate $1p_{1/2}$ states (the remaining degeneracy is simply that of the orientation of the total angular momentum vector, **J**, in space).

If, in particular, the energy difference of states split by spin-orbit interaction is taken to be of the same order as the spacing between shell-model states and if the states with the higher j ($j = l + \frac{1}{2}$) are made more stable as those with the lower j ($j = l - \frac{1}{2}$) are made less stable, then the sequence of levels becomes something like that illustrated in figure 10-5. There it is seen that the closed shells at 28, 50, 82, and 126 nucleons appear because of the splitting of the $1f$, $1g$, $1h$, and $1i$ levels, respectively, and these shell closures occur at exactly the experimentally determined nuclear magic numbers.

Level Order. There are several important features of this energy level diagram. First, the level order given is to be applied independently to neutrons and to protons. Thus the nucleus $^4_2$He contains two protons and two neutrons all in the $1s_{1/2}$ level; $^9_4$Be contains four protons, two in the $1s_{1/2}$ and two in $2p_{3/2}$ (indicated more briefly by $1s^2_{1/2}2p^2_{3/2}$), and five neutrons, $1s^2_{1/2}2p^3_{3/2}$. On an absolute energy scale the proton levels are increasingly higher than neutron levels as Z increases. This is the familiar Coulomb repulsion effect, and in first approximation it does not change the order of the levels for a particular kind of nucleon. But there is a small tendency for the proton levels in nuclei of large Z to shift in relative stability, those levels with maximum orbital angular momentum (1f, 1g, 1h, 1i) appearing at relatively lower energies, apparently because the proton suffers less from Coulombic repulsion when traveling in the outermost region of the nucleus.

Second, the order given within each shell is essentially schematic and may not represent the exact order of filling. Indeed this order may differ slightly in different nuclides, depending on the number of nucleons in the outermost shell. (Similar level shifts are quite familiar in the atomic structure of the heavier elements.)

Ground States of Nuclei. If a nucleus contains 2, 8, 20, 28, 50, 82, or 126 neutrons, the level scheme just described permits a good prediction of the quantum states occupied by the neutrons. Thus $^{88}_{38}$Sr has its 50 neutrons filling the five shells: ($1s^2$), ($1p^6$), ($1d^{10}2s^2$), ($1f^8_{7/2}$), ($1f^6_{5/2}2p^6 1g^{10}_{9/2}$). Similarly, the proton structure is obvious for nuclides with magic atomic numbers: He, O, Ca, Ni, Sn, and Pb. It is a well-known theorem in atomic structure that filled shells are spherically symmetric and have no spin or orbital angular momentum and no magnetic moment. In the extreme single-particle model for nuclei there is the added assumption that not only filled nucleon shells but any even number of either neutrons or protons has no net angular momentum in the ground state. This is consistent with the observation that the *ground states of all even-even nuclei have zero spin and even parity.* This pairing of like nucleons also results in the increased binding

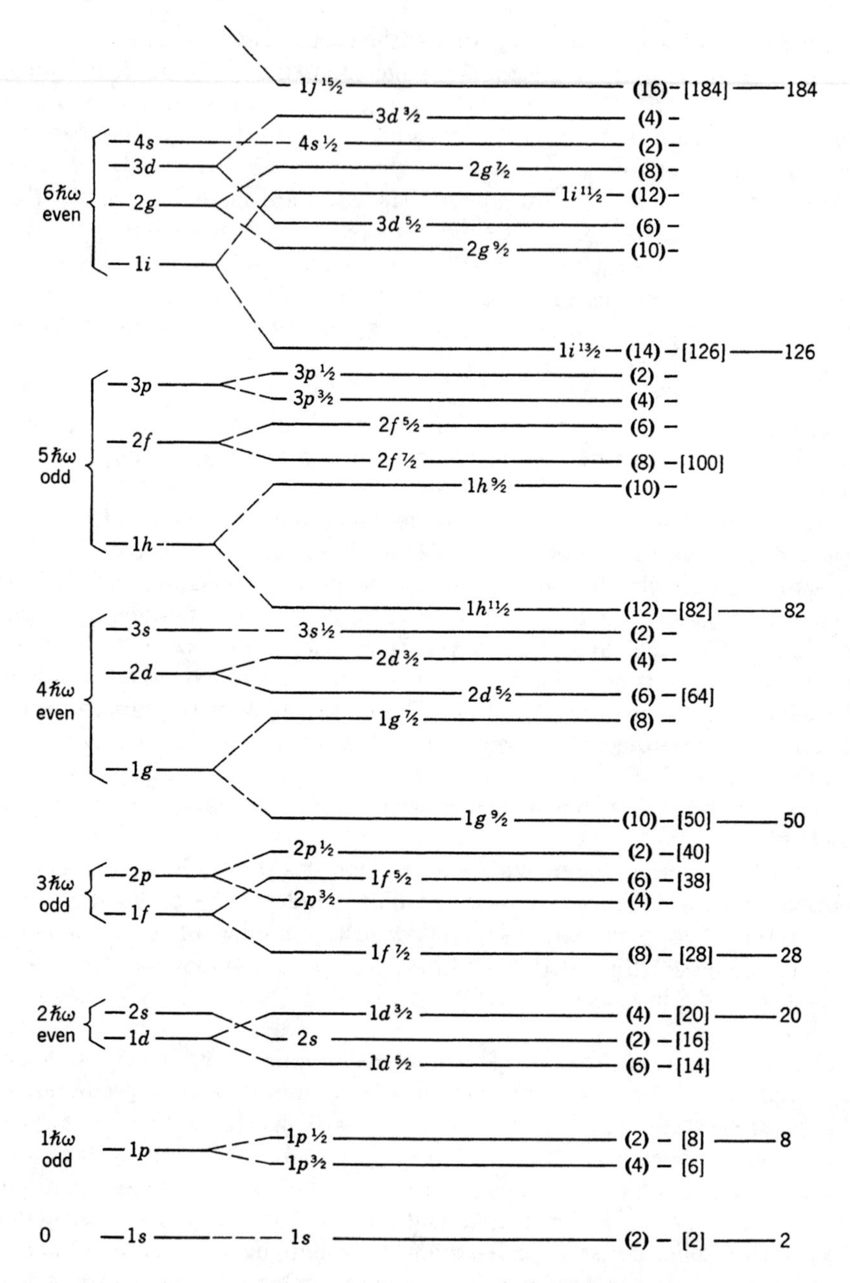

Fig. 10-5 Splitting of the energy levels of the three-dimensional isotropic harmonic oscillator by spin-orbit coupling. The numbers in parentheses and brackets have the same meaning as in figure 10-4, the "magic numbers" are given on the far right. (From reference M1.)

energy of a nucleon in nuclei with an even number of like nucleons as discussed in chapter 2, section D. Such an enhancement in the binding energy of paired nucleons suggests that, beyond the average potential felt by all nucleons, there exists a residual attractive interaction between two paired nucleons when their angular momenta couple to zero. This effect is not surprising in view of the properties of nuclear forces and has important consequences for the ground-state properties of even-even nuclei, which will be discussed in section E.

Odd-*A* Nuclei. For our present purposes, then, in any nucleus of odd *A* all but one of the nucleons are considered to have their angular momenta paired off, forming an even-even **core**. The single odd nucleon is thought to move essentially independently in (or outside) this core, and the net angular momentum of the entire nucleus is determined by the quantum state of this nucleon. For example, consider the even-odd nucleus ${}^{13}_{6}C$. The six protons and six of the seven neutrons are paired up (in the configuration $1s^2 1p^4_{3/2}$); the odd neutron is in the $1p_{1/2}$ level, and the entire nucleus in its ground state is characterized by the $p_{1/2}$ designation. The nuclear spin of ${}^{13}C$ has been measured, and the value $I = \frac{1}{2}$ corresponds to the resultant angular momentum indicated as the subscript in $p_{1/2}$. As a second example consider ${}^{51}_{23}V$. The odd nucleon in this case is the twenty-third proton and belongs in the $1f_{7/2}$ level; the ground state of the nucleus is expected to be $f_{7/2}$. The measured spin of this nucleus is $\frac{7}{2}$.

Without going into details we may say that the measured magnetic moments for ${}^{13}C$ and ${}^{51}V$ lend some support to the spin evidence for the correct assignment of these ground states. For a given spin the magnitude of the magnetic moment of a nucleus depends on whether the spin and orbital angular momenta of the odd nucleon are parallel or antiparallel. An $s_{1/2}$ and a $p_{1/2}$ nucleus will, for example, have quite different magnetic moments, and the differences can be at least qualitatively predicted.

For nuclei in general the situation is not nearly so simple as is indicated in the examples just cited. The order of the levels within each shell may often be different from that in figure 10-5, especially for two or three adjacent levels. In such a case we conclude from the single-particle model only that several particular states of the nucleus are close together in energy without knowing which is the lowest, or ground state. (Sometimes this information alone is useful.) As an example, the odd nucleon in ${}^{137}_{56}Ba$ is the eighty-first neutron, and figure 10-5 indicates that the ground state is probably $h_{11/2}$, $s_{1/2}$, or $d_{3/2}$, depending on the order of filling of these three levels; the measured spin is $I = \frac{3}{2}$. In general the high spin states such as $h_{11/2}$ and $i_{13/2}$ do not appear as ground states of odd nuclei.

The extreme single-particle model is useful in the characterization of excited states for nuclei that are very near closed shells. The low-lying states of ${}^{207}Pb$ (filled shell of 82 protons, a single hole in the 126-neutron shell) provide an excellent example. In figure 10-6 it is seen that the first

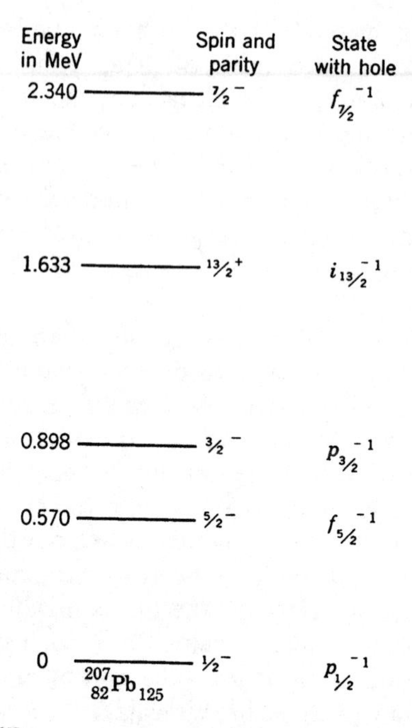

Fig. 10-6 Energy levels of ^{207}Pb with energies given on the left side and spins and parities on the right. The superscript −1 on any spectroscopic term indicates a "hole" in that state.

four excited states in ^{207}Pb correspond to transitions of the neutron hole among the various available single-particle states: different intrinsic states. It should be noted that the relative stabilities of these states emphasize that the order given in figure 10-5 for states within a given shell is not to be taken seriously.

Configuration Interaction. The prediction of the properties of odd-A nuclei in this fashion is most reliable when neither the neutron number nor the proton number is far removed from a magic number. For nuclides with either neutron or proton number near to half way between magic numbers, the situation becomes more complex. For these nuclei the single-particle model is certainly an oversimplification. As an illustration of one kind of evidence for this statement, $^{23}_{11}$Na would be expected to have a $d_{5/2}$ ground state, and the only other reasonable single-particle model possibility is $s_{1/2}$. The measured spin is $\frac{3}{2}$. This is not to be attributed to an odd proton in the $1d_{3/2}$ level because $1d_{3/2}$ certainly should lie higher than $1d_{5/2}$. Moreover, the magnetic moment is definitely in disagreement with the $d_{3/2}$ interpretation. Another such case is $^{55}_{25}$Mn, in which the odd nucleon clearly should be $1f_{7/2}$, yet the measured spin is $\frac{5}{2}$. Anomalies such as these can be caused by the interactions among all of the nucleons outside the closed shells that have not already been included in the effective potential that determines the shell-model states. Thus, evidently, the interactions among the five $f_{7/2}$ protons and the two neutrons beyond the closed shell of 28 cause the ground state of ^{55}Mn to have a spin of $\frac{5}{2}$ instead of $\frac{7}{2}$. It is interesting to note that ^{53}Mn, which does not have the two neutrons beyond the closed shell of 28, has the expected ground-state spin of $\frac{7}{2}$. We return to "anomalous" ground-state spins in section E and discuss them in terms of nuclear deformation.

Very powerful techniques have been developed for performing shell-model calculations on quite complex nuclei, that is, nuclei with several nucleons outside closed shells. Such calculations, made possible through advances in computer technology, have had remarkable successes in accounting for the level of many nuclei (B4, M5).

Odd-Odd Nuclei. What can be said about the states of odd-odd nuclides? Most of these nuclides are radioactive (the stable ones known are $^2_1$H, $^6_3$Li, $^{10}_5$B, and $^{14}_7$N), and there are fewer directly measured data on spins and magnetic moments. The single-particle model assumption of pairing leaves, in every case, one odd proton and one odd neutron, each producing an effect on the nuclear moments. No universal rule can be given to predict the resultant ground state; however, the following rules are very helpful for ground states and long-lived low-lying isomeric states with mass numbers in the range $20 < A < 120$. They were first proposed by M. H. Brennan and A. M. Bernstein (B5) as improvements to an earlier set of rules proposed by L. W. Nordheim (N1). For configurations in which the odd proton and odd neutron are *both* particles (or *both* holes) in their respective unfilled subshells, the coupling rules are: (1) if the so-called Nordheim number N $(= j_1 + j_2 + l_1 + l_2)$ is even, then $I = |j_1 - j_2|$; and (2) if N is odd, then $I = |j_1 \pm j_2|$. The prediction for configurations in which there is a combination of particles and holes is: (3) $I = j_1 + j_2 - 1$, which is less certain than the other rules.

An example of rule (1) is $^{38}_{17}$Cl, for which the shell model predicts that the odd proton is in a $d_{3/2}$ orbital and the odd neutron is in an $f_{7/2}$ state. Since N is even, rule (1) predicts $I = |j_1 - j_2| = 2$. Odd (−) parity is predicted for this nucleus in the ground state since the odd nucleons are in states of opposite parity. This agrees with the measured ^{38}Cl ground-state spin and parity. An example of rule (2) is ^{26}Al, for which both the odd proton and neutron configurations are $d_{5/2}^{-1}$. The measured spin and parity are 5^+. Rule (3) may be illustrated by ^{56}Co ($J^\pi = 4^+$), for which the shell-model predicts for the odd proton $f_{7/2}^{-1}$ and for the odd neutron $p_{3/2}^{1}$.

Application of Single-Particle Model to Nuclear Isomerism. The concept of closed shells and the single-particle model have applications in the study of excited states of nuclei, particularly for low excitation energies. Generally, when the model predicts several possible low-lying configurations, all except the one that happens to be energetically favored are eligible for existence as excited states, particularly for odd-nucleon nuclides. We have seen in chapter 3 that γ transitions, especially where ΔI is large and E is small, can have appreciable lifetimes and are then known as isomeric transitions (ITs). An important aspect of nuclear shell structure has been the correlation of nuclear spins and isomer lifetimes.

E. Feenberg (F1) in 1949 called attention to the abundant groupings of isomers with odd Z or odd N just below the magic number values 50, 82, and 126. This phenomenon is connected with the appearance, just before shell closure at these numbers, of a new level of very high spin ($1g_{9/2}$ before 50, $1h_{11/2}$ before 82, $1i_{13/2}$ before 126). As an illustration consider $^{113}_{48}Cd$; its odd nucleon—the sixty-fifth neutron—is assigned to the $3s_{1/2}$ state to accord with the measured ground-state spin $I = \frac{1}{2}$. Other possible unfilled states within the same shell, all probably low-lying, are $2d_{3/2}$ and $1h_{11/2}$. If $1h_{11/2}$ happens to be the first excited level, which is the case for this nuclide, then the γ transition to ground is $h_{11/2} \rightarrow s_{1/2}$, $\Delta I = 5$, yes (the parity changes), an $E5$ transiton that should be very long-lived. The isomer actually observed, $^{113}Cd^m$, decays predominantly by β-particle emission with an observed half life of 14 y. The branching ratio for the IT implies a partial half life of 1.4×10^4 y for that mode of decay.

The large number of even-odd isomers from $^{111}_{48}Cd$ to $^{137}_{56}Ba$, with 63 to 81 neutrons, have similar explanations. The upper state in most of these pairs is $1h_{11/2}$, and the corresponding IT in most is $h_{11/2} \rightarrow d_{3/2}$, $\Delta I = 4$, yes, classification $M4$, often followed by the $M1$ transition $d_{3/2} \rightarrow s_{1/2}$. When the next neutron shell (82 to 126) is partly filled in the even-odd nuclei such as $^{195}_{78}Pt$, ^{197}Pt, $^{197}_{80}Hg$, ^{199}Hg, and $^{207}_{82}Pb$, the long-lived isomeric level is $1i_{13/2}$; the transition is generally $i_{13/2} \rightarrow f_{5/2}$, which is again $M4$. Another group of ITs is between $g_{9/2}$ and $p_{1/2}$ states in the region of odd nucleon numbers just below 50.

There is a sizable number of odd-odd isomers, but because of the difficulties in assignment of configurations to the two-nucleon states these isomers are not easy to classify in any organized way. There are some very interesting even-even isomers. In one, $^{72}_{32}Ge^m$, $t_{1/2} = 4 \times 10^{-7}$ s, $E = 0.69$ MeV, the ground and first excited states both have $I = 0^+$. The transition is thus of the $0 \rightarrow 0$ type, and, as required by the selection rules (chapter 3, p. 97), takes place entirely by emission of internal-conversion electrons, in spite of the rather large transition energy. Most even-even isomers have very short half lives; one of the interesting exceptions is $^{180}_{72}Hf^m$ (5.5 h) discussed below (p. 393).

E. COLLECTIVE MOTION IN NUCLEI

The shell model, as just discussed, approximates the complicated inter-nucleon forces that hold the nucleus together by an effective spherically symmetric potential that is meant to represent the average potential energy experienced by a nucleon in the nucleus. The outstanding success of the shell model attests to the usefulness of this approximation; it is to be expected, though, that this description cannot be complete. In this section we examine the effects of the interactions that are not included in the shell-model description.

We have already seen both from the pairing term in the binding-energy expression and from the coupling of an even number of like nucleons to zero total spin that there must be an attractive force between a pair of nucleons whose angular momenta cancel. This force, not included in the shell-model potential, is called the **pairing force** and has a decisive effect on the enhanced stability of the ground state of even-even nuclei.

Furthermore, it must be remembered that there is no central source in the nucleus for the spherically symmetric potential as, for example, the Coulomb field that the nucleus itself provides in the atom. Instead, each nucleon in the nucleus contributes to the nuclear potential. Thus if the nucleons are not distributed with spherical symmetry in space, then the average potential that they generate will also not be spherically symmetric. Since it is only for a completed shell that the wave function leads to a spherically symmetric distribution of particles, it is expected that effects due to a nonspherically symmetric potential will be most important for nuclei with partly filled shells.

In principle both these effects could be addressed by including the residual interactions in shell-model calculations. In practice it is easier to use a better starting point. While many effects of the residual interactions (which are not large) can be evaluated by means of perturbation theory, phenomena in which the residual interactions add coherently cannot be handled so simply. The effects of the pairing force are treated by methods that were first developed for the treatment of superconductivity, based on the pairing of conduction electrons in solids. A brief discussion of this problem will be presented later in this chapter.

The deviation from sphericity also involves the coherent effects of many nucleons in forming the common potential, and indeed Hartree-Fock calculations for nuclei with partly filled shells do show a greater stability for deformation from sphericity. Rather than describe these complex details, which are still being investigated, we present a semiphenomenological model that brings out the special dynamical consequences inherent in the hypothesis of deformation.

Nonspherical Potential. The nuclear potential energy is plotted in figure 10-7 against deformation from a spherical to a spheroidal shape. In that figure, curve *a* represents a nucleus with no or at most a few nucleons beyond a closed shell. For this nucleus the spherical shape is the most stable. As more nucleons are added, the nucleus becomes deformable, as illustrated by curve *b*, and finally reaches the point, as illustrated in curves *c* and *d*, where the stable shape of the nucleus is no longer spherical. As

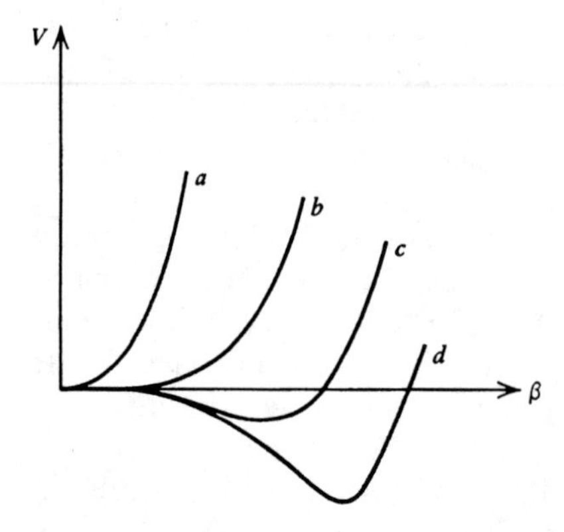

Fig. 10-7 Potential energy surfaces for even–even nuclei. The nuclear potential energy V is plotted as a function of the parameter β. The various curves illustrate the behavior of the nuclear potential as one moves away from closed shells. (From reference A1.)

still more nucleons are added, a new shell closure is approached and the potential curves will shift back again towards spherical stability. These potential-energy curves all refer to the nucleus in its lowest **intrinsic state**—the most stable distribution of nucleons among the available single-particle states. Excited intrinsic states would have different potential curves just as the potential-energy surface that governs the vibrations of a molecule depends on its electronic state (intrinsic state).

What are the consequences of these potential-energy curves? Firstly, since the energy required for deformation is finite, it is expected that nuclei can oscillate about their equilibrium shapes, and thus vibrational energy levels should be seen. If the restoring force is rather large, though, as exemplified by curve a, the spacing between vibrational energy levels may become as large as or larger than that between intrinsic states, and thus the two modes of excitation can become mixed. It is for those nuclei exemplified by curves b and c that clearly identifiable vibrational states are to be expected.

Secondly, nuclei with a stable nonspherical shape, as exemplified in curves c and d, have distinguishable orientations in space and thus are expected to exhibit rotational energy levels. Just as in molecules, each intrinsic and vibrational state of such nuclei would have a corresponding set of rotational states.

Lastly, the energies of the shell-model states will be changed in the nuclei with nonspherical shapes and, as discussed later, some degeneracies will be removed. It is just such changes in the shell-model energies that lead to the sequence of curves in figure 10-7.

These three consequences, rotational states, vibrational states, and altered shell-model states, are discussed in the following paragraphs.

Rotational States. The first experimental evidence for the deformability of nuclei away from a spherical shape came from measurements of nuclear quadrupole moments: measured quadrupole moments of odd-A nuclei are several times larger than those expected for the odd nucleon moving in the field of a spherical core. Related to this enhancement of static quadrupole moments is the observation that electric quadrupole transitions ($E2$ transitions) are often much faster than given in table 3-4, which corresponds to transitions between single-particle states. Physically, enhanced quadrupole moments and $E2$ transition rates imply that the nucleus has a spheroidal rather than a spherical charge distribution.

It was first suggested by L. J. Rainwater (R1) that these discrepancies might be overcome by considering the polarization of the even-even core by the motion (not spherically symmetric) of the odd nucleon. In this manner all of the nucleons in the nucleus could contribute collectively to static quadrupole moments and to quadrupolar transition rates. The further implications of a spheroidal even-even core for nuclear energy levels were investigated in an important series of papers mainly by A. Bohr and B. R. Mottelson (B6).

This permanent deformation away from spherical symmetry means that the orientation of the nucleus in space can in principle be determined and thus that there must be the usual conjugate angular momentum and the quantized states of rotational energy. If all of the nucleons in an even-even nucleus remain paired and thus all of the angular momentum arises from the collective rotation of the deformed spheroidal nucleus with axial symmetry, rotational energy levels of the form

$$E = \hbar^2 \frac{I(I+1)}{2\mathcal{I}}. \tag{10-19}$$

are expected. In this expression $\mathcal{I}$ is the **effective moment of inertia** about an axis that is perpendicular to the symmetry axis and $[I(I+1)]^{1/2}\hbar$ is the total angular momentum of the nucleus. Because of the symmetry of the spheroid with respect to a rotation of 180°, the allowed values of I are 0, 2, 4, 6, . . . , and all of the states are of positive parity. The effective moment of inertia $\mathcal{I}$ does not correspond to that expected for the rotation of a rigid spheroid, but rather to motion in which there is considerable slippage of the individual nucleon motions relative to the rotation of the average shape. Detailed many-body calculations that include the pairing interaction show good agreement with the experimentally observed moments of inertia, which are indeed much less than the rigid value.

Since the permanent spheroidal deformation is expected to be largest for nuclei between closed shells, the rotational bands of states should be most prominent for nuclei with A between 150 and 190 and with $A > 200$. This expectation is indeed borne out.

An example is shown in figure 10-8 where the low-lying levels of ^{242}Pu are presented. The first six levels are members of the ground-state rotational band and their energies are, within a precision of a few percent, given by (10-19) with $\hbar^2/2\mathcal{I} = 7.3$ keV, which corresponds to a moment of inertia about half that of the rigid spheroid. The existence of states from different rotational bands of similar energy can lead to drastic alterations in

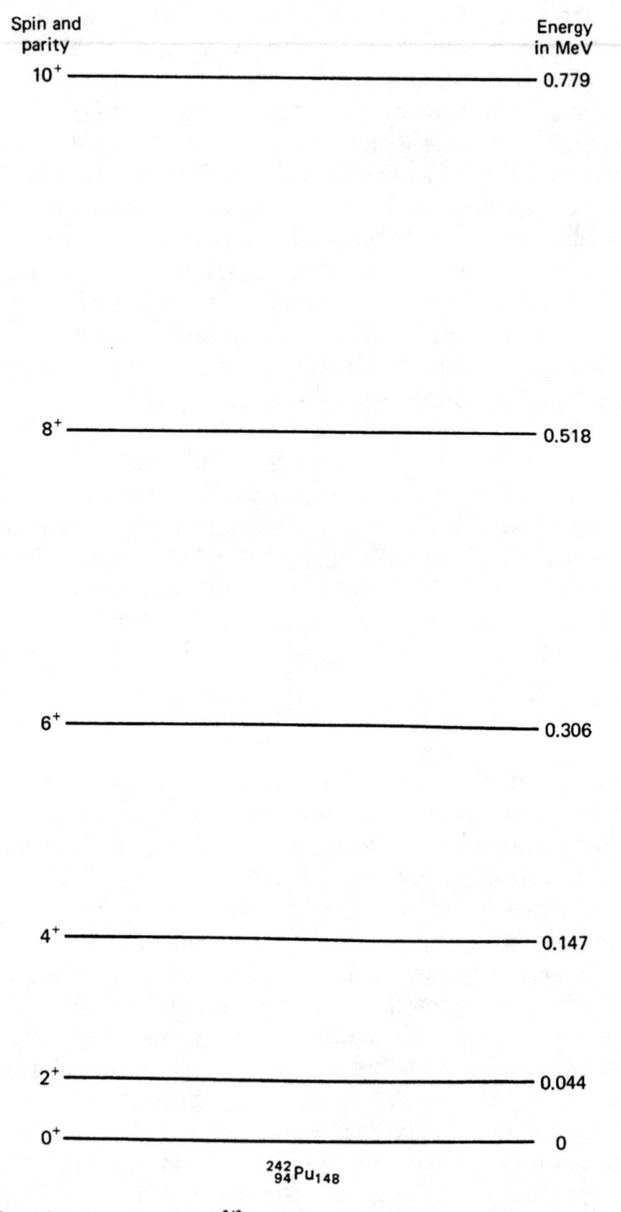

Fig. 10-8 The first six energy levels of ^{242}Pu. Spins and parities are listed on the left, energies above the ground state on the right. These levels are members of the ground-state rotational band.

transition rates. An example is the 5.5-h 8^- isomer in ^{180}Hf, the lowest member of a $K = 8$ rotational band that decays to an 8^+ state that is the fourth excited state of a $K = 0$ rotational band (the quantum number K is defined below). Although this is a spin change of 0, it corresponds to a transition with a spin change of 8 in the intrinsic state.

Rotational States in Odd-A Nuclei. The angular momentum of the odd nucleon in an odd-A nucleus implies that there can be two contributions to the total angular momentum: that from the rotation of the spheroidal even-even core and that from the intrinsic state of the odd nucleon. As stated in the previous section and illustrated in figure 10-9, the angular momentum from the rotation of the even-even spheroidal core, **R**, classically is perpendicular to the symmetry axis of that core, while that of the odd nucleon, **j**, may be in any direction. The total angular momentum, **I**, is

$$\mathbf{I} = \mathbf{R} + \mathbf{j}. \tag{10-20}$$

The rotational energy of the core is, classically,

$$E_{\text{rot}} = \frac{R^2}{2\mathcal{I}} = \frac{(\mathbf{I} - \mathbf{j})^2}{2\mathcal{I}} \tag{10-21}$$

where, as before, $\mathcal{I}$ is the moment of inertia perpendicular to the symmetry axis. If, as in figure 10-9, a coordinate system is taken in which the z axis is along the symmetry axis of the spheroidal core and $R_z = 0$, $j_z = I_z$, E_{rot} can be written as the sum of three terms:

$$E_{\text{rot}} = \frac{\mathbf{I}^2 - I_z^2}{2\mathcal{I}} - \frac{I_x j_x + I_y j_y}{\mathcal{I}} + \frac{j_x^2 + j_y^2}{2\mathcal{I}}.$$

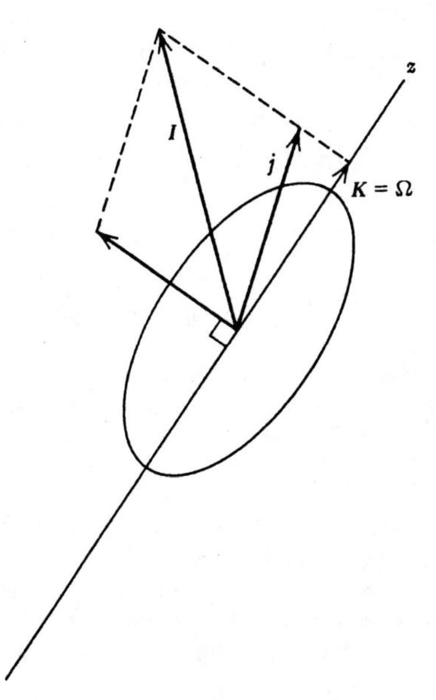

Fig. 10-9 Vector diagram of the total angular momentum of a deformed nucleus. See text for definitions of the axes and symbols.

Then the sum of the particle energy E_p and the core energy E_{rot} can be expressed as

$$E(I, K) = \frac{[I(I+1) - K^2]\hbar^2}{2\mathcal{I}} + \left(E_p + \frac{j_x^2 + j_y^2}{2\mathcal{I}}\right) - \frac{I_x j_x + I_y j_y}{\mathcal{I}}, \tag{10-22}$$

where the classical quantities have been quantized in the usual manner, I_z has been replaced by the more usual symbol $K\hbar$ for the projection of I on the symmetry axis, and $I = K+1, K+2, \ldots$. For the ground-state band with axial symmetry, K also equals j_z.

The third term in (10-22) gives rise to what is known as the Coriolis coupling.[6] Coriolis coupling similar to the nuclear effect indicated in (10-22) is responsible for the tendency of a spinning gyrocompass to align its axis with that of the rotating earth. The effects of the Coriolis term on nuclear structure have been observed in a number of nuclei and are more pronounced as j, I, and/or $\hbar^2/2\mathcal{I}$ become relatively large. A detailed discussion of Coriolis coupling in nuclei is given in S1 and B2.

Equation 10-22 indicates that odd-A nuclei with spheroidal even-even cores will exhibit bands of rotational energy levels, each of which is built upon a different state for the odd nucleon with its concomitant value of K. An example of this is shown in figure 10-10 for the well investigated energy-level scheme of ^{25}Al where levels up to an excitation of about 4 MeV are interpreted as rotational bands built on the first four intrinsic states of the odd proton.

Vibrational States. When a shell is close to half-filled, the residual interactions change the equilibrium shape of the nucleus away from spherical, as depicted in curves c and d of figure 10-7, and can be treated, as was just done, by the consideration of rotational states. With nucleon numbers nearer to those of the closed shells, however, the equilibrium shape appears to remain spherical, as illustrated in curves a and b of figure 10-7, but the residual interactions still cause formidable difficulties for a pure shell-model description of the nuclear states. In this situation it has proved useful to describe the consequences of the configuration interactions in terms of a fluctuation of the nucleus about a spherical shape. This gives rise to the concept of vibrational states for nuclei. Of course, it

[6] The Coriolis force can be understood classically as arising when a spinning particle moves in a rotating frame of reference. In an inertial system the equation of motion is simply $\mathbf{F} = m\mathbf{a}$, while in the rotating system it appears that the particle is moving under an effective force $\mathbf{F}_{\text{eff}}$:

$$\mathbf{F}_{\text{eff}} = \mathbf{F} - 2m(\boldsymbol{\omega} \times \mathbf{V}_r) - m\boldsymbol{\omega} \times (\boldsymbol{\omega} \times \mathbf{r}), \tag{10-23}$$

where m is the mass of the particle, $\boldsymbol{\omega}$ is its rotational frequency, $\mathbf{V}_r$ is the velocity of the particle relative to the rotating set of axes, and $\mathbf{r}$ is the radial coordinate of the particle in the rotating frame. The first term in (10-23) is the potential force, the second is the Coriolis force, and the third is the centrifugal force. The third term in (10-22) is conventionally called the Coriolis coupling of the odd particle to the rotating core, though it contains both the Coriolis and centrifugal energies.

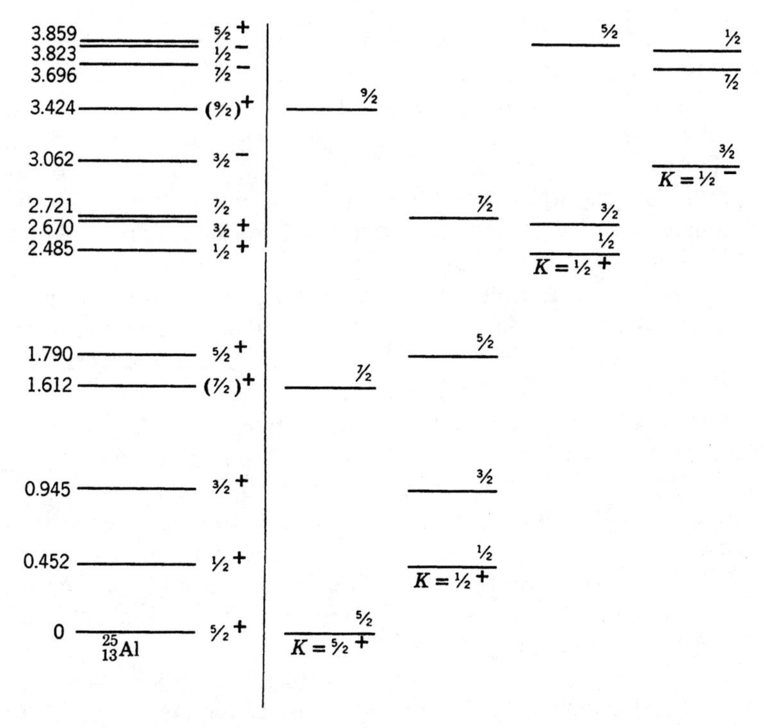

Fig. 10-10 Energy-level spectrum of ^{25}Al. On the left are drawn all of the levels of ^{25}Al that have been observed up to 4 MeV; the measured spins and parities are also shown. On the right these levels have been classified in terms of the rotational bands associated with the different intrinsic states. (Adapted from reference B7.)

is expected that the permanently deformed nuclei (curves c and d) will also oscillate about their equilibrium configurations.

If the vibrations are of very small amplitude about the equilibrium, it is reasonable to suppose that they will be harmonic. When the vibrations are quantized we get a set of independent harmonic oscillators with energy quanta

$$\hbar\omega_\lambda = \hbar\left(\frac{C_\lambda}{B_\lambda}\right)^{1/2}, \tag{10-24}$$

where B_λ is the effective mass, and C_λ is the effective spring constant of the vibration in the mode λ. Borrowing language from the analysis of vibrations in solids, each vibrational quantum is called a **phonon** and each phonon has parity $(-1)^\lambda$, angular momentum $[\lambda(\lambda+1)]^{1/2}\hbar$, and a projection of angular momentum on the polar axis that is an integral multiple of $\hbar$.

It is not surprising that the higher the order of the vibration, and thus the more complicated the shape of the nucleus, the higher is the vibrational frequency and thus the energy of that vibrational quantum. The shape of the nucleus can be expressed in terms of a sum of spherical harmonics of order λ. It is expected that the lowest-order oscillation will be the most important. However, the first term ($\lambda = 1$) corresponds to a displacement of the center of mass and is therefore of no interest in the absence of external forces. Thus the second-order term is the dominant term and corresponds to quadrupolar distortion of the nucleus. While the first 2^+ states of nondeformed nuclei show the collective characteristics expected of a $\lambda = 2$ or quadrupolar vibration, their properties are not just those of a simple harmonic mode of motion. Nor are the higher excited states simply described in terms of multiple harmonic excitations. The proper description of this situation is an active research topic. A recent attempt that shows promise is the interacting boson model briefly described later.

For spheroidal even-even nuclei rotational bands based upon vibrational states are often observed. Special variables are useful for describing these states. The nuclear shape is usually parameterized in terms of the quantities β and γ. If $\gamma = 0$, the nucleus has cylindrical symmetry and a spheroidal shape that for $\beta > 0$ is prolate (football-shaped) and for $\beta < 0$ is oblate (disc-shaped). If both β and γ differ from zero, the nucleus assumes an ellipsoidal shape. Within this model nuclear vibrations are described as either β or γ vibrations, depending on whether β or γ oscillates, as shown in figure 10-11. Even where β and γ are fixed so that a stable shape is rotating, the nucleons appear to be undergoing collective oscillations rather than massive circular movements characteristic of rigid rotation. A difference occurs for the projection of the phonon spin along the polar axis: for β vibrations the projection is zero while for γ vibrations it is nonzero.

The energy of the rotational states built on vibrations is given approximately by

$$E = \frac{\hbar^2}{2\mathcal{I}}[I(I+1) - K^2], \tag{10-25}$$

where K is the projection of the angular momentum I. For β vibrations ($\lambda = 2, K = 0$) the values of I^π are $0^+, 2^+, 4^+, \ldots$; for γ vibrations ($\lambda = 2, K = 2$) the spin parity sequence is $2^+, 3^+, 4^+, \ldots$; for octopole vibrations ($\lambda = 3, K = 1$) the sequence is $1^-, 3^-, 5^-, \ldots$. The levels of ^{232}U given in figure 10-12 display this structure. A detailed discussion of vibrational spectra is presented in N2.

Single-Particle States in Deformed Nuclei and the Unified Model. The shell-model states discussed in Section D are appropriate to a spherically symmetric potential energy. As we have seen in the preceding discussion of collective states, nuclei with partly filled shells have a spheroidal rather than spherical shape because of the residual interactions that are not included in the average effective potential. The question of what effect this equilibrium deformation has on the single-particle states

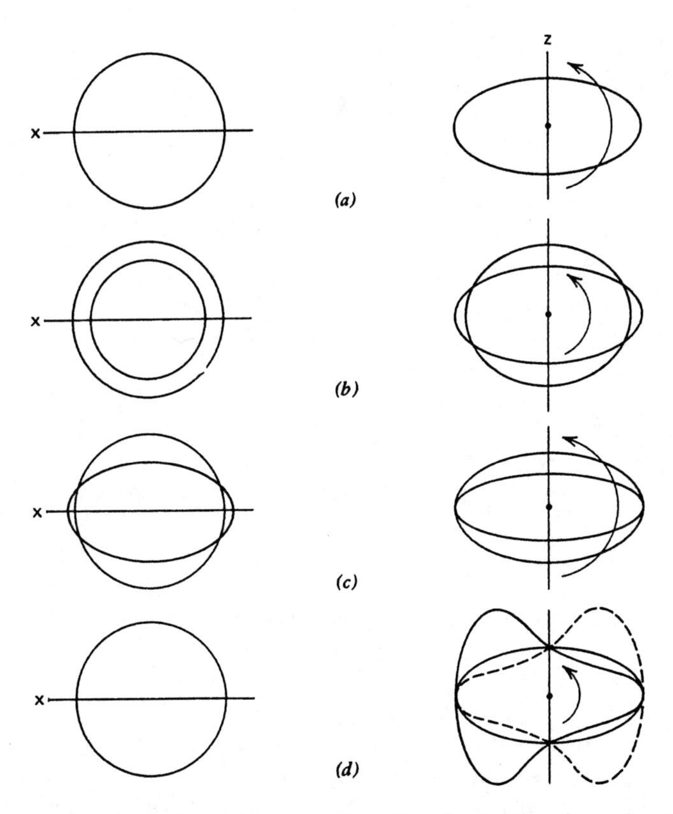

Fig. 10-11 Simple modes of collective motion of a distorted nucleus. A cross section perpendicular to the z axis is shown on the left; a cross section in the y–z plane is shown on the right. The arrows represent one possible rotation. (a) Quadrupolar rotation; (b) β vibration; (c) γ vibration; and (d) octupole vibration, $\nu = 0$. (From M. A. Preston, *Physics of the Nucleus*, © 1962, Addison-Wesley Publishing Company Inc., chapter 10, figure 6. Reprinted by permission.)

immediately arises. This question was first investigated by S. G. Nilsson (N3) and the resulting states have come to be known as **Nilsson states.**

Even without a detailed model, it is possible to say something about the new single-particle states that arise from the distortion of a potential energy with spherical symmetry to one with spheroidal symmetry. Since the potential energy would then become a function of the polar angle θ but still remain independent of the azimuthal angle ϕ, angular momentum of the single-particle state as such would no longer be a constant of the motion but its projection on the symmetry axis, usually denoted by Ω, would be. Thus for example, the $f_{7/2}$ state, which is eight-fold degenerate in a spherical nucleus because the projections of the angular momenta on the symmetry axis of $\frac{7}{2}, \frac{5}{2}, \ldots, -\frac{5}{2}, -\frac{7}{2}$ are all at the same energy, splits into four doubly degenerate states in a spheroidal nucleus because the states Ω and

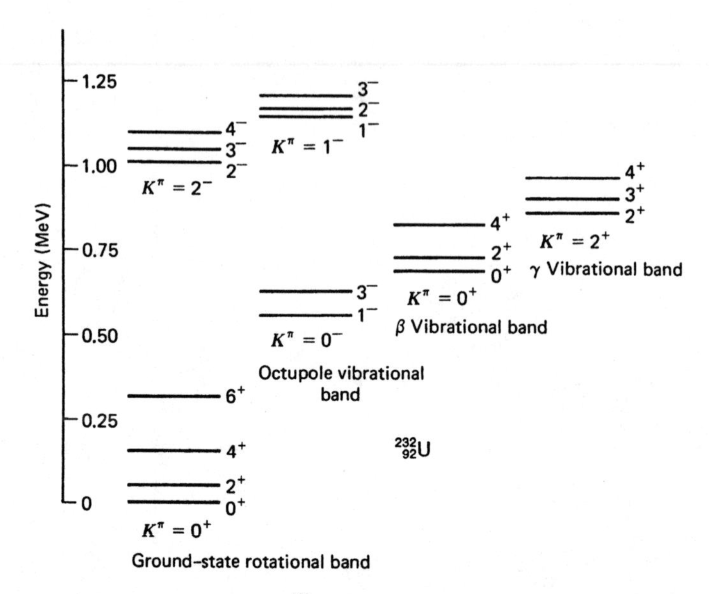

Fig. 10-12 The energy-level scheme of ^{232}U. Rotational bands, some built on vibrational states, are indicated.

$-\Omega$ are degenerate whereas states with different $|\Omega|$ are not. The $f_{7/2}$ state splits into four states characterized by $|\Omega| = \frac{7}{2}, \frac{5}{2}, \frac{3}{2}$, and $\frac{1}{2}$. This situation is illustrated schematically in figure 10-13. In general, a single-particle state with quantum number J in a spherical nucleus splits into $(2J + 1)/2$ doubly degenerate levels in a spheroidal nucleus and these levels are characterized by the quantum number Ω running from J in integral steps down to $\frac{1}{2}$. As can be seen from figure 10-13, for prolate deformations the stability of the state decreases with increasing Ω, while for oblate deformations the stability increases with increasing Ω.

These general considerations serve to explain the deviation from shell-model predictions for ^{23}Na and ^{55}Mn that are mentioned as examples on page 386. The spin of ^{23}Na is expected to be determined by the odd eleventh proton that, by the spherical shell model, would enter the $\frac{5}{2}$ state. In a spheroidal nucleus, however, the sixfold degenerate $d_{5/2}$ state splits into three doubly degenerate states with spins $\frac{1}{2}, \frac{3}{2}$, and $\frac{5}{2}$. The ninth and tenth protons fill the $\frac{1}{2}$ state and the eleventh goes into the $\frac{3}{2}$ state. The spin of $\frac{5}{2}$ for ^{55}Mn may be explained in a similar manner from the splitting of the $f_{7/2}$ level: the twenty-first and twenty-second protons enter the $\frac{1}{2}$ state, the twenty-third and twenty-fourth enter the $\frac{3}{2}$ state, and the twenty-fifth goes into the $\frac{5}{2}$ state.

This splitting of single-particle states in spheroidal nuclei was investigated quantitatively by Nilsson (N3) utilizing a three-dimensional harmonic-oscillator potential with two equal force constants perpendicular

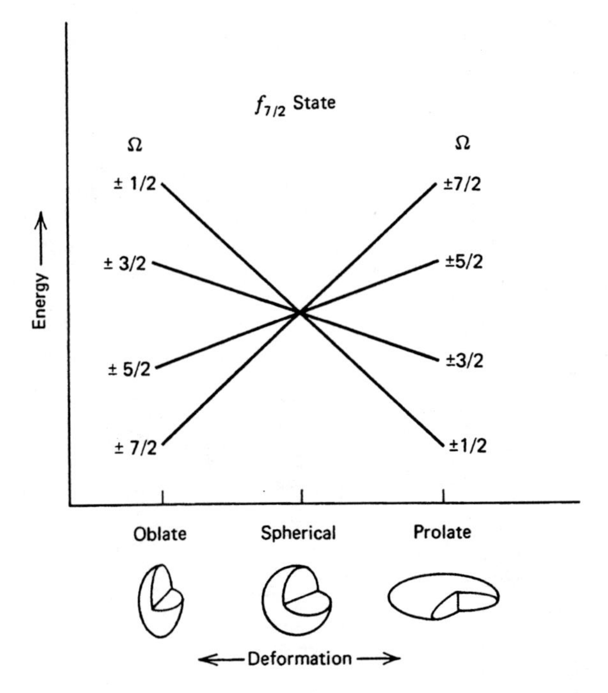

Fig. 10-13 Schematic diagram of the energy of the $f_{7/2}$ state with nuclear deformation. Cutaway views of spherical, oblate, and prolate nuclear shapes are shown below the energy diagram.

to the symmetry axis and a different one along it. An example of the results of his investigation is presented in figure 10-14 for the single-particle states that are between the closed shells of 2 and 28 in spherical nuclei. Positive and negative values of the deformation parameter correspond, respectively, to prolate and oblate spheroids. Each state in the deformed nucleus is characterized by the quantities $\Omega^{\pi}[N, n_z, \Lambda]$; Ω and π are good quantum numbers that correspond to the constants of the motion for the state, while the terms in the bracket are the so-called **asymptotic quantum numbers** and describe the state which the Nilsson state approaches for large deformations. The quantity Ω, as defined before, is the projection of the angular momentum on the symmetry axis, $\pi = (-1)^N$ is the parity of the state, N is the total number of oscillation quanta as in the shell model, $n_z < N$ is the number of oscillation quanta along the axis of symmetry, and Λ is the component of the *orbital* angular momentum along the symmetry axis.

As discussed earlier on p. 393 and illustrated by the level structure of ^{25}Al in figure 10-10, rotational states can be built upon these Nilsson intrinsic states for deformed nuclei. The lowest such state has $I = K = \Omega$; the excited rotational states have $I = K + 1$, $K + 2, \ldots$. For example, the rotational bands of ^{25}Al are built on the $\frac{5}{2}^{+}[202]$, $\frac{1}{2}^{+}[211]$, $\frac{1}{2}^{+}[200]$, and $\frac{1}{2}^{-}[200]$

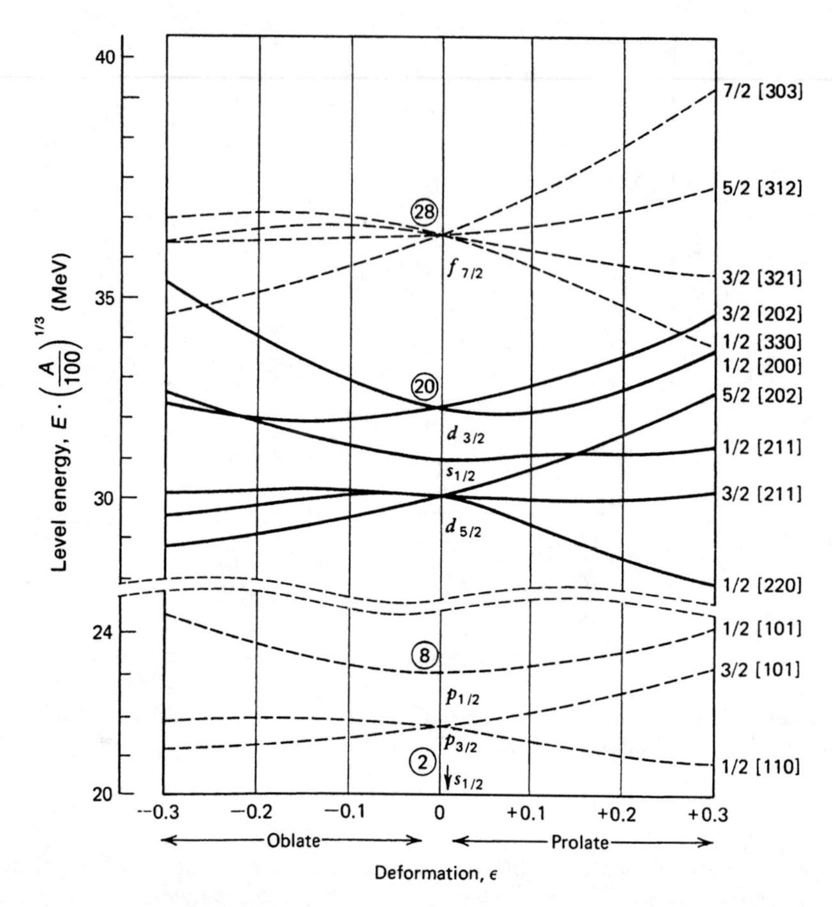

Fig. 10–14 Nilsson diagram for protons or neutrons for N or $Z \leq 28$. Levels are labeled by the asymptotic quantum numbers $K[Nn_z\Lambda]$ and at zero deformation by quantum numbers l_j. Even parity levels are given as solid lines, odd parity levels as dashed lines. (From references W2 and L1.)

single-proton intrinsic states shown in figure 10-10. The model that includes these collective motions built upon appropriately modified single-particle states is sometimes referred to as the unified model.

The Pairing Force and Quasi-Particles (C1, K1). Thus far in this section the consequences of the residual interactions among nucleons have been described semiphenomenologically in terms of the distortion of nuclei from spherical to spheroidal or even ellipsoidal shapes. From these distortions there resulted rotational states, vibrational states, and the Nilsson single-particle states. There remains, however, a systematic trend in nuclear energy levels that requires a somewhat more microscopic examination of the effects of the residual interactions among nucleons: the observation that the first excited intrinsic states of even-even nuclei are much higher in energy than those of neighboring odd-A nuclei.

It has already been mentioned that the enhanced binding energy of even-even nuclei as well as their 0^+ ground states are evidence for the enhanced attraction between pairs of like nucleons with angular momenta coupled to yield no net spin. This implies that the first excited intrinsic state of even-even nuclei would be higher than expected if only one of the nucleons in the pair were excited and thus the special effects of pairing were destroyed. However, within this context it should be possible to raise both of the particles in the pair to the next higher single-particle state, thereby maintaining the pairing and placing the state about twice as high as expected since both particles would have to be excited. The observed effect is even larger than this. For example, consider the isotopes ^{58}Ni, ^{59}Ni, and ^{60}Ni. From simple shell-model considerations the three neutrons beyond the closed shell of 28 in ^{59}Ni should occupy the $2p_{3/2}$ level (see figure 10-5), and indeed the ground state of that nucleus is $\frac{3}{2}^-$. It might be expected that the first excited state would involve raising the thirty-first neutron to either the $1f_{5/2}$ or the $2p_{1/2}$ state, whichever lies lower. It is found that the first excited state is $\frac{5}{2}^-$ at an excitation of 0.339 MeV. Accordingly, it would be expected that there would be an excited state in both ^{58}Ni and ^{60}Ni at an energy of about 0.7 MeV that corresponds to raising a pair of neutrons from the $2p_{3/2}$ state to the $2f_{5/2}$ state. Instead, ^{58}Ni and ^{60}Ni have 2^+ first excited states at 1.45 and 1.33 MeV, respectively. It is significant that there is no state at 0.7 MeV that corresponds to the excitation of a pair of neutrons. It is the absence of this state and the resulting gap in the energy spectrum that require explanation.

The pairing force may be viewed as simply the consequence of the attractive nature of nuclear forces that causes nucleons to be as close together as possible consistent with the other constraints on the system. The nuclear potential underlying the shell-model states approximately accounts for the average effect of this attractive force at distances corresponding to the average spacing between nucleons in the nucleus. In addition there is the residual attractive force between two nucleons in particular shell-model states such that they are, on the average, closer to each other than they are to the other nucleons in the nucleus. Within the constraints of the Pauli principle the two particular single-particle states are those with quantum numbers (n, l, j, m) and $(n, l, j, -m)$ which are identical except for the opposite projections of the angular momentum on a space-fixed axis.

Classically this corresponds to the two particles moving in the same orbit but in opposite directions. Thus the ground state of an even-even nucleus contains pairs of particles with each pair occupying a particular pair of states (n, l, j, m) and $(n, l, j, -m)$ in a manner that is consistent with the Pauli exclusion principle.

The question of which pairs of states are occupied then arises. The extreme single-particle model, which neglects the residual interaction,

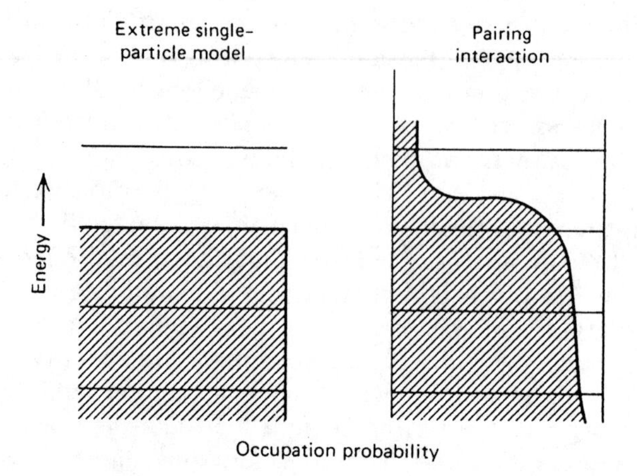

Fig. 10-15 Schematic diagram of the occupation probability of nucleons in the ground state of an even-even nucleus. The extreme single-particle model prediction is shown on the left while the result considering the pairing residual interaction is shown on the right.

would simply fill the pairs of states in order of increasing energy until all of the nucleons are accounted for as illustrated in figure 10-15, left. The residual interaction between the particles in each paired state alters this distribution (figure 10-15, right) by removing the sharp cutoff at the upper end and smearing out the distribution toward higher-energy single-particle states. At first glance it may seem strange that partly filling *higher-energy* single-particle states can result in a *lower* energy of the system. This is, however, the usual result of perturbation theory in which the perturbing potential (in this instance the pairing interaction) causes the wave function to become a linear combination of unperturbed states (in this instance the single-particle states), even if they are of higher energy. Because of this partial occupancy of states, the picture of a particle occupying a state in the extreme single-particle view becomes partly a particle and partly a hole occupying a state when pairing interactions are included. It is essentially this combination of particle and hole that is known as a **quasi-particle**.

The theory originally developed by J. Bardeen, L. N. Cooper, and J. R. Schrieffer (B8) to explain superconductivity as resulting from the pairing of electrons in metals was applied to nucleon pairing in nuclei by Bohr, Mottelson, and D. Pines (B9). With this theory a simple expression can be derived for the probability V_i^2 that a given pair of single-particle states (n_i, l_i, j_i, m_i) and $(n_i, l_i, j_i, -m_i)$ will be occupied by a pair of particles:

$$V_i^2 = \frac{1}{2}\left(1 - \frac{\epsilon_i - \lambda}{E_i}\right). \tag{10-26}$$

In this expression ϵ_i is the energy of the single-particle state and λ, often called the chemical potential of the system, is essentially the energy of the uppermost state that would be filled in the absence of the pairing force or, in terms of the Fermi-gas model, it is the Fermi energy. The important quantity E_i, which plays the same role for quasi-particles as ϵ_i does for particles in the absence of the pairing interaction, is given by the expression

$$E_i = [(\epsilon_i - \lambda)^2 + \Delta^2]^{1/2}, \tag{10-27}$$

where Δ, a measure of the strength of the pairing interaction, is often called the **gap parameter** and has a value approximately equal to that of δ in (2-5) for nuclei with odd mass number.

Just as the ground state in the single-particle picture contains no excited particles, there are no quasi-particles present in the ground state of an even-even nucleus. In the single-particle picture the lowest intrinsic excitation involves raising one particle to the first excited single-particle state, thereby generating an excited particle and a hole in the state that was previously occupied. Similarly, when the pairing interaction is included the lowest intrinsic excitation involves going from zero to *two* quasi-particles, each of which must have an excitation energy given by (10-27). The lowest possible excitation, therefore, will be approximately 2Δ and it is this quantity that is the energy gap in the spectrum of intrinsic energy levels of even-even nuclei.

Nuclei with odd mass numbers must contain at least one unpaired nucleon and thus even in the ground state contain one quasi-particle with an energy of about Δ as given by (10-27). This means that the ground states of nuclei with odd mass numbers will be less stable than those of adjacent even-even nuclei by about the quantity Δ. It is interesting to note, though, that the spacing between intrinsic quasi-particle levels is less than that between the corresponding single-particle levels.

As a result of the pairing-energy gap the level densities of even-even nuclei near the ground state are much lower than those of odd-odd nuclei; nuclei with odd mass numbers have densities of low-lying levels that are intermediate between these two extremes. The level density increases rapidly above the energy gap. As discussed in chapter 4 (p. 146), these level-density effects resulting from the pairing energy play an important role in determining relative yields of nuclear-reaction products.

F. SUMMARY AND COMPARISONS OF NUCLEAR MODELS

The various nuclear models we have discussed result in a spectrum of nuclear states that bears a strong resemblance to that for a polyatomic molecule: there are intrinsic states (single-particle for nuclei and electronic for molecules), rotational states, and vibrational states. It must be immediately stated, though, that this resemblance is much more a consequence of the interaction of scientists with the many-body problem than it is of any resemblance between the interactions in molecules and those in nuclei.

The fundamental model is the independent-particle model. The difficulty lies in the nucleon-nucleon interactions that are not included in the effective potentials exemplified in (10-12) and (10-13), the so-called residual interactions. The residual interactions cause any description of the nucleus with a particular assignment of nucleons to single-particle states (the configuration) to be inaccurate; rather the nucleus must be described by a superposition of many different configurations (configuration mixing).

The extreme single-particle model that was discussed in section D uses the residual interactions to cause an even number of identical nucleons with the same n, l, and j quantum numbers to couple to a net angular momentum of zero. Configuration mixing is neglected. This extreme assumption is found to work rather well at or near closed shells, a fact that suggests that configuration mixing is important primarily for the nucleons outside closed shells and that the mixed configurations include mainly the single-particle states within a given shell.

The pairing of nucleons in the extreme single-particle model roughly takes account of the short-range correlations in nucleon motions expected from the residual interactions; the collective model and the unified model attempt to include the long-range correlations also. They accomplish this by replacing the configuration mixing by a spheroidal deformation, which represents a time average of the spatial distribution expected for the appropriate mixture of single-particle configurations. It is assumed that the oscillations of the deformed nucleus about its equilibrium shape are slow compared to single-particle motions, and thus the single-particle states and the collective states may be treated separately. This approximation is roughly equivalent to the Born-Oppenheimer approximation in the theory of molecular structure.

The range of applicability and the successes of these various approximations to configuration mixing are most easily seen in the particular examples that follow.

Intrinsic States. The outstanding success of the extreme single-particle model lies in its ability to predict the ground-state spins and parities of nearly all odd-mass nuclei. Where it fails, such as in ^{23}Na, as mentioned on p. 386, the failure may usually be remedied by taking account of the spheroidal deformation in the region between closed shells, a deformation that splits the single-particle states. For example, the $\frac{7}{2}^+$ isomeric states of ^{107}Ag^m and ^{109}Ag^m apparently arise from the splitting of the $1g_{9/2}$ single-particle proton state by the spheroidal deformation; the $\frac{1}{2}^+$, $\frac{3}{2}^+$, and $\frac{5}{2}^+$ states so produced are filled and the forty-seventh proton is in the $\frac{7}{2}^+$ state. The extreme single-particle model is also useful in describing the excited states of nuclei, particularly near closed shells such as in ^{207}Pb (figure 10-6) discussed earlier.

Rotational States. As still more nucleons or holes are added beyond the closed shells, the configuration mixing that results from the residual interactions causes a permanent spheroidal deformation of the nucleus, and the excited states are better described as rotational states. This disagreeable metamorphosis also occurs in molecular spectroscopy: CO_2, because it is linear, has four degrees of vibrational freedom and two degrees of rotational freedom; H_2O, because it is nonlinear, has three degrees of vibrational freedom and three degrees of rotational freedom. Thus straightening out the molecule turns a rotation into a vibration. Examples of a rotational band built on the ground intrinsic state of ^{242}Pu and of rotational states built on the first four intrinsic states of ^{25}Al are shown in figures 10-8 and 10-10, respectively.

Collective Nonrotational States. When we move away from the well-deformed states, the picture becomes less clear. There are indeed collective characteristics that make it reasonable to speak of quadrupolar fluctuations about a nondeformed shape, even though a description in terms of simple harmonic vibrations is not even approximately sufficient. Whether a description in terms of interacting vibrations will be useful is under active investigation.

The specific examples that have been given for intrinsic, rotational, and collective nonrotational states illustrate the usefulness of the appropriate nuclear models. It must also be stated, though, that the unambiguous identification of the character of nuclear states is still the exception rather than the rule. This is because of the mixing of the three kinds of states for the majority of the nuclides that are intermediate cases, that is neither near enough to nor far enough from closed shells.

An interesting effect called **backbending** has been observed around spin 16 in the ground-state rotational band (yrast states) of some rare-earth nuclei (S1). This effect is manifested in a change of slope in a plot of level energy versus spin that is barely perceptible around spin 16 in figure 10-16. The insert of that figure, in which quantities proportional to the moment of inertia and the square of the rotational frequency are plotted, is the more conventional method of displaying backbending. One likely explanation of this effect is that the ground rotational band crosses another band. The nature of this crossing band has been under intensive investigation, and it appears to be due to a still different mode of nuclear motion. In this mode the motions of a few nucleons with very large angular momenta are aligned by the Coriolis interaction (which is strong for large values of j) so that their angular momentum parallels the rotation axis rather than the symmetry axis.

Interacting-Boson Model. A model for even-even nuclei that can handle intermediate nuclei has been proposed by A. Arima and F. Iachello (A1). This description treats the nucleus as if it were composed of nucleon

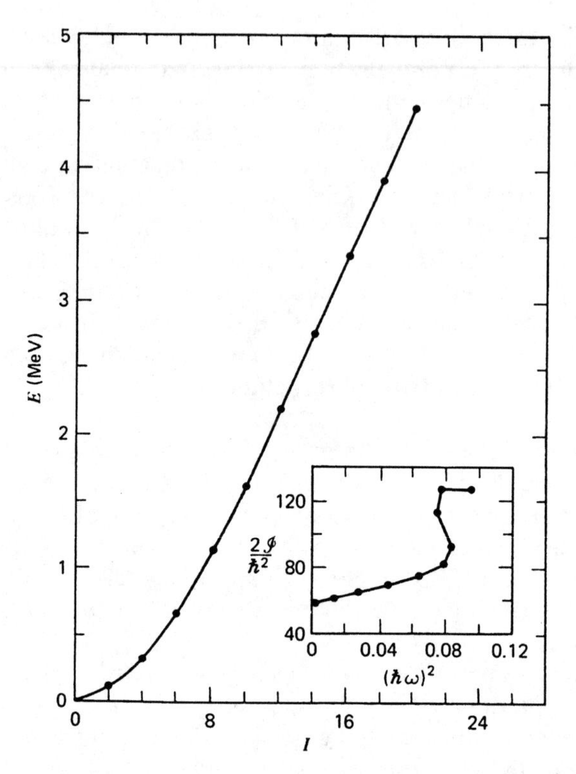

Fig. 10–16 A plot of excitation energy versus spin for the ground-state rotational band of ^{162}Er. The insert, which shows the same data plotted differently (ordinate proportional to square of moment of inertia, abscissa to square of rotation frequency) clearly shows why the phenomenon is called backbending. [From reference S1.)

pairs. Whereas both the proton and neutron are fermions (spin $\frac{1}{2}$), the nucleon pair is considered a particle with integral spin—a boson. In its simplest form the model assumes that only valence nucleons, paired to $l = 0$ or 2, contribute to the low-lying excited states. For example, in $^{118}_{54}$Xe it is assumed that only the 18 valence particles constituting 9 active nucleon pairs contribute. Shell-model calculations (M3) support the primary assumption by showing evidence for the coupling of the valence nucleons (predominantly into $l = 0$ and $l = 2$ states) to form the low-lying excited states. Work on this model is still in an early stage (I1) but it is hoped that it may eventually lead to a more complete unification of the single-particle and collective descriptions of the nucleus.

REFERENCES

A1 A. Arima and F. Iachello, "Collective Nuclear States as Representations of a SU(6) Group," *Phys. Rev. Lett.* **35**, 1069 (1975); "Interacting Boson Model of Collective States I. The Vibrational Limit," *Ann. Phys.* **99**, 253 (1976).

A2 K. Alder, A. Bohr, T. Huus, B. R. Mottelson, and A. Winther, "Study of Nuclear Structure by Electromagnetic Excitation with Accelerated Ions," *Rev. Mod. Phys.* **28**, 432 (1956).

*B1 H. A. Bethe and P. Morrison, *Elementary Nuclear Theory*, Wiley, New York, 1956.

*B2 A. Bohr and B. R. Mottelson, *Nuclear Structure*, Vol. I—*Single Particle Motion*, Vol. II—*Nuclear Deformations*, Benjamin, New York, 1969 and 1975.

B3 H. A. Bethe, "Theory of Nuclear Matter," *Ann. Rev. Nucl. Sci.* **21**, 93 (1971).

B4 P. J. Brussaard and P. W. M. Glaudemans, *Shell Model Applications in Nuclear Spectroscopy*, North Holland, Amsterdam, 1977.

B5 M. H. Brennan and A. M. Bernstein, "*jj* Coupling Model in Odd-Odd Nuclei," *Phys. Rev.* **120**, 927 (1960).

B6 A. Bohr and B. R. Mottelson, "Collective and Individual Particle Aspects of Nuclear Structure," *Dan. Mat.-Fys. Medd.* **27**(16) (1953); "Collective Nuclear Motion and the Unified Model," in *Beta and Gamma Ray Spectroscopy* (K. Siegbahn, Ed.), North Holland, Amsterdam, 1955.

B7 A. Bohr and B. R. Mottelson, "Collective Motion and Nuclear Spectra," in *Nuclear Spectroscopy*, Part B (F. Ajzenberg-Selove, Ed.), Academic, New York, 1960.

B8 J. Bardeen, L. N. Cooper, and J. R. Schrieffer, "Theory of Superconductivity," *Phys. Rev.* **108**, 1175 (1957).

B9 A. Bohr, B. R. Mottelson, and D. Pines, "Possible Analogy Between the Excitation Spectra of Nuclei and Those of the Superconducting Metallic State," *Phys. Rev.* **110**, 936 (1958).

*C1 B. L. Cohen, *Concepts of Nuclear Physics*, McGraw-Hill, New York, 1971.

F1 E. Feenberg, "Nuclear Shell Structure and Isomerism," *Phys. Rev.* **75**, 320 (1949).

G1 L. C. Gomes, J. D. Walecka, and V. F. Weisskopf, "Properties of Nuclear Matter," *Ann. Phys.* **3**, 241 (1958).

H1 T. Hamada and I. D. Johnston, "A Potential Model Representation of Two-Nucleon Data Below 315 MeV," *Nucl. Phys.* **34**, 382 (1962).

H2 O. Haxel, J. H. D. Jensen, and H. E. Suess, "Modellmässige Deutung der ausgezeichneten Nukleonen-Zahlen im Kernbau," *Z. Phys.* **128**, 295 (1950).

I1 F. Iachello, Ed., *Interacting Bosons in Nuclear Physics*, Plenum, New York, 1979.

K1 L. S. Kisslinger and R. A. Sorenson, "Pairing Plus Long Range Force for Single Closed Shell Nuclei," *Dan. Mat.-Fys. Medd.* **32**(9) (1960); "Spherical Nuclei with Simple Residual Forces," *Rev. Mod. Phys.* **35**, 853 (1963).

L1 C. M. Lederer and V. S. Shirley, Eds., *Table of Isotopes*, 7th ed., Wiley-Interscience, New York, 1978.

*M1 M. G. Mayer and J. H. D. Jensen, *Elementary Theory of Nuclear Shell Structure*, Wiley, New York, 1955.

M2 M. G. Mayer, "Nuclear Configurations in the Spin Orbit Coupling Model. I. Empirical Evidence," *Phys. Rev.* **78**, 16 (1950).

M3 J. B. McGrory, "Shell Model Tests of the Interacting-Boson Model Description of Nuclear Collective Motion," *Phys. Rev. Lett.* **41**, 533 (1978).

*M4 P. Marmier and E. Sheldon, *Physics of Nuclei and Particles*, Academic, New York, 1970.

M5 J. B. McGrory and B. H. Wildenthal, "Large-Scale Shell Model Calculations," *Ann. Rev. Nucl. Part. Sci.* **30**, 383 (1980).

N1 L. W. Nordheim, "Nuclear Shell Structure and Beta-Decay. II. Even *A* Nuclei," *Rev. Mod. Phys.* **23**, 322 (1951).

N2 O. Nathan and S. G. Nilsson, "Collective Nuclear Motion and the Unified Model," in *Alpha-, Beta- and Gamma-Ray Spectroscopy*, Vol. 1 (K. Siegbahn, Ed.), North Holland, Amsterdam, 1965, pp. 601–700.

N3 S. G. Nilsson, "Binding States of Individual Nucleons in Strongly Deformed Nuclei," *Dan. Mat.-Fys. Medd.* **29**(16) (1955).

*P1 M. A. Preston, *Physics of the Nucleus*, Addison-Wesley, Reading, MA, 1962.

R1 L. J. Rainwater, "Nuclear Energy Level Argument for a Spheroidal Nuclear Model," *Phys. Rev.* **79**, 432 (1950).

*R2 J. O. Rasmussen, "Models of Heavy Nuclei," in *Nuclear Spectroscopy and Reactions*, Part C (J. Cerny, Ed.), Academic, New York, 1974, pp. 97–178.

S1 F. S. Stephens, "Coriolis Effects and Rotational Alignment in Nuclei," *Rev. Mod. Phys.* **47**, 43 (1975).

W1 D. H. Wilkinson, Ed., *Isospin in Nuclear Physics*, North Holland, Amsterdam, 1969.

W2 A. H. Wapstra, G. J. Nijgh, and R. van Lieshout, *Nuclear Spectroscopy Tables*, North Holland, Amsterdam, 1959.

Y1 H. Yukawa, "On the Interaction of Elementary Particles. I.," *Proc. Physico-Math. Soc. Japan* **17**, 48 (1935).

EXERCISES

1. Estimate the radii of the nuclei ^{41}Sc and ^{41}Ca from the observation that the maximum energy of the β^+ spectrum emitted in the decay of ^{41}Sc to the ground state of ^{41}Ca is 5.9 MeV. Approximate both nuclei as uniformly charged spheres for which the electrostatic energy is $(\frac{3}{5})[Z(Z-1)e^2]/R$ where Ze and R are the charge and the radius of the sphere, respectively. *Answer:* 5.3 fm.

2. (a) Estimate the Fermi energies of neutrons and protons in the center of ^{238}U. Assume the density of nuclear matter in the center of the ^{238}U nucleus to be 2×10^{38} nucleons cm^{-3}. (b) Compare the difference in Fermi energies of neutrons and protons in ^{238}U with the Coulomb repulsion experienced by a proton approaching ^{237}Pa. Take the radius constant $r_0 = 1.5$ fm.
Answer: (a) 36 MeV for protons, 49 MeV for neutrons.

3. The β^+ and EC decay of ^{89}Zr lead to the 16-s ^{89}Y^m rather than to the stable ^{89}Y of spin $I = \frac{1}{2}$. The isomer is de-excited by a 909-keV transition with $\alpha_K \cong 0.01$ and $\alpha_K/\alpha_L = 7.0$. (a) Using these data and shell structure considerations, assign spins and parities to ^{89}Y^m and to the 78.4-h ^{89}Zr. (b) Estimate the partial half life for direct decay of ^{89}Zr to the ground state of ^{89}Y. (c) The 1.488-MeV β^- spectrum of 50.5-d ^{89}Sr is not accompanied by γ radiation and has the unique first-forbidden shape. What is the log ft value for this transition and the spin and parity of ^{89}Sr? (d) Estimate the fraction of ^{89}Sr decays that might lead to ^{89}Y^m.
Answer: (b) $\sim 10^7$ y.

4. What do you expect the ground-state spins and parities to be for (a) ^{39}Ar; (b) ^{196}Pt; (c) ^{89}Zr; (d) ^{55}Co; (e) ^{16}N; (f) ^{42}Sc; (g) ^{40}Cl?

5. Estimate the smallest distance within which a neutron and proton can move so as just to give a bound deuteron. (The average separation of neutron and proton is about half this distance and is the "size" of the deuteron.) *Answer:* $\sim$4.0 fm.

6. (a) What is the smallest possible value for the isospin of $^{59}_{27}$Co? (b) The isospin of a π-meson is 1 with z components +1, 0, and −1 corresponding to π^+, π°, and π^-, respectively. What total isospins are possible in the interaction of $\pi^+ + p$, $\pi^\circ + p$, $\pi^- + p$? Note that in part (b) the particle physics convention for the proton isospin should be used. *Answer:* (a) $\frac{5}{2}$.

7. The first excited state of ^{182}W is 2^+ and is 0.100 MeV above the ground state. Estimate the energies of the lowest lying 4^+ and 6^+ states of ^{182}W. (The observed values are 0.329 and 0.680 MeV.) (b) The half life of the 2^+ state is 1.4 ns. Compare this value with that calculated from table 3-4. Explain the relative values.

8. (a) Derive an expression for the *excitation* energy of the state in nucleus $Z+1$ that is the isobaric analog of a state in nucleus Z in terms of the energies of each state and the change in Coulomb energy. (b) Using information from chapter 2, derive an expression for the Coulomb energy change ΔE_C appearing in equation (10-8).

   Answer: (a) $U_{IA}(Z+1) = U_{IA}(Z) + \Delta E_C + (M_Z - M_{Z+1} + M_H - M_n)c^2$.
   (b) $\Delta E_C = (2Z+1)(0.717A^{-1/3} - 1.211A^{-1})$ MeV.

9. Using the results of exercise 8 and atomic masses given in Appendix D, determine the excitation energy of the first $T = 3$ state in ^{62}Cu.

   Answer: 4.5 MeV.

Chapter 11

Radiochemical Applications

A. TRACERS IN CHEMICAL APPLICATIONS

1. The Tracer Method

The Method and its Limitations. The isotopic tracer method is based on the fact that a mixture of isotopes of an element remains essentially invariant through the course of physical, chemical, and biological processes (exceptions are noted below). Thus a radioactive or separated stable isotope added to a mixture (usually the naturally occurring mixture) of isotopes may serve as a tracer, label, or indicator in the sense that its behavior will be the same as that of the other atoms of the same element originally present in the same chemical form. Following any chemical, physical, or biological processes, the fate of the original material of interest can then be determined by assay of the tracer—for radioactive tracers by measurement of radiations emitted in their decay, for stable tracers by isotopic analysis (usually mass spectrometry or activation analysis).

The tracer principle has some important limitations. Isotopic fractionation may be appreciable for the lightest elements (where the percentage mass difference between isotopes is greatest), and this effect must always be considered in the use of hydrogen tracer isotopes. However, apart from these isotopes and ^{7}Be, which differs in mass from stable ^{9}Be by about 25 percent, the next heaviest tracer is in carbon where the specific isotope effect may already be neglected in most tracer work of ordinary precision. The interesting and important divergences from this assumption are examined in section A, 3.

Because the isotopic tracer atoms are detected by their radioactivity, they behave normally up to the moment of detection; after that moment they are not detected, and their fate is of no consequence. Of course, if the atoms resulting from the nuclear transformation are themselves radioactive and capable of a further nuclear change, the detection method must be arranged to give a response that measures the proper (in this case the first) radioactive species only. For example, if ^{210}Bi is used as a tracer for bismuth, the α particle from its daughter ^{210}Po should not be allowed to enter the detection instrument but should be absorbed by a suitable absorber or by the counter wall. As a tracer for thorium, ^{234}Th is suitable in spite of the fact that most of the detectable radiation will be from its daughter $^{234}Pa^m$. The reason is that the short half life, 1.18 min, of $^{234}Pa^m$ ensures that it will be in transient equilibrium with ^{234}Th by the time the sample is mounted and ready for measurement, so that the total activity will be proportional to the ^{234}Th content.

Another source of interference with the tracer principle is a possible chemical (or biological) effect produced by the ionizing rays; this radiation chemistry effect is not often encountered at the usual tracer activity levels and may always be checked by experiments at varying levels of radioactivity.

A limitation of the radioactive tracer method for some applications is the absence of known active isotopes of suitable half life for a few elements, especially for some light elements such as oxygen. There are radioactive oxygen isotopes, ^{14}O, ^{15}O, and ^{19}O, but these have half lives of 71, 122, and 27 s, respectively. The 7-s ^{16}N and 4-s ^{17}N are useless as tracers, but a number of tracer studies have used the 10-min ^{13}N β^+ activity successfully. Also, there are no helium, lithium, and boron isotopes with half lives longer than 1 s.

Stable Isotope Tracers. The use of separated stable isotopes as tracers is a valuable technique. Enriched ^{18}O and ^{15}N are essential for many interesting and important purposes, and ^{13}C offers significant advantages for some carbon tracer experiments. Deuterium (^2H) has found many applications as a hydrogen tracer; the use of tritium (^3H) is not entirely equivalent because its properties are even more different from those of protium (^1H). Stable isotope tracers are most commonly assayed by mass spectrometers. Samples for introduction into a mass spectrometer are ordinarily put into gaseous form. Carbon-13 is commonly analyzed as CO_2, and the same gas is often used for ^{18}O measurements. Since CO_2 and H_2O reach isotopic equilibrium by exchange in a day or so, a convenient analysis for ^{18}O in H_2O is thus available.

For some light-element isotopes assay methods based on properties other than mass are available. In particular, nuclear magnetic resonance is commonly used for ^{13}C and ^{17}O assays. Some stable tracers (e.g., zinc isotopes) are conveniently assayed by activation analysis (see section B, 1 below).

Examples of Isotopic Tracer Applications. Isotopic tracers have come into great use in many fields. In some of these applications the tracer is necessary in principle while in others it may not be so but is a great practical convenience. The fields of study in which tracers have been successfully applied are too diverse to be thoroughly covered in this book. Some of this work is reviewed in E1. For example, much of the information on the absorption and subsequent behavior of essential trace elements in plants and animals (especially humans) has been obtained by tracer studies. Several unique and convenient applications of tracers in various fields of study are given below. Others are described in later sections of this chapter.

Chemical Reaction Mechanisms. Tracers such as ^{14}C have played an important role in determining the mechanisms and rates of reactions in chemistry. A classic example is the elucidation of the molecular rearrangement during the deamination of 1,2,2-triphenylethylamine with nitrous acid (B1):

$$\phi_2CH-\overset{*}{C}H\phi^*(NH_2) \xrightarrow[H_2O]{HONO} \phi_2C^*HC^*H\phi^*(OH)$$

where ϕ stands for a phenyl group and the asterisk indicates atoms tagged with a radioactive source. The distribution of the ^{14}C label was used in this study to distinguish between the formation of a bridged ion followed by attack of water:

$$\phi_2CH-C^*H\phi^*(N_2^{\oplus}) \rightarrow \phi CH-C^*H\phi^* \text{ (bridged, } \oplus) \xrightarrow{H_2O} \phi CH(OH)-C^*H\phi^*\phi + \phi_2CH-C^*H\phi^*(OH)$$

and the formation of a normal carbonium ion followed by rapid phenyl shifts:

$$\phi_2CH-C^*H\phi^*(N_2^{\oplus})$$

$$\downarrow$$

$$\phi_2CH-\overset{\oplus}{C^*}H\phi^* \rightleftharpoons \phi\overset{\oplus}{C}H-C^*H\phi^*\phi \rightleftharpoons \phi^*\phi CH-\overset{\oplus}{C^*}H\phi \rightleftharpoons \phi^*\overset{\oplus}{C}H-C^*H\phi_2$$

$$\downarrow H_2O \qquad \downarrow H_2O \qquad \downarrow H_2O \qquad \downarrow H_2O$$

$$\phi_2CH-C^*H\phi^*(OH) \qquad \phi CH(OH)-C^*H\phi^*\phi \qquad \phi^*\phi CH-C^*H\phi(OH) \qquad \phi^*CH(OH)-C^*H\phi_2.$$

This analysis showed that the carbonium ion mechanism is important in this reaction. Labeling studies of this kind are now routine in investigations of chemical reactions.

Self-Diffusion. To illustrate the unique tracer method we discuss an early study of self-diffusion. The rates of diffusion of various metals, including gold, silver, bismuth, thallium, and tin, in solid lead at elevated temperatures have been investigated by the use of sensitive spectroscopic analyses. The first attempt (by G. Hevesy and his collaborators) to observe the diffusion of radioactive lead into ordinary lead failed, showing that the diffusion rate must be at least 100 times smaller than that for gold in lead (which is the fastest of those just named, the others showing decreasing rates in the order listed). The method first used was a rather gross mechanical one, and the workers evolved a much more sensitive method based on the short range of the α particles from ^{212}Bi in transient equilibrium with ^{212}Pb. The lead, containing ^{212}Pb isotopic tracer, was pressed into contact with a thin foil of inactive lead, which was chosen just thick enough to stop all the α rays, and then as diffusion progressed an α activity appeared and increased as measured through this foil. A similar but even more sensitive technique than the α-range method was based on the much shorter ranges (a few millionths of a centimeter in lead) of the nuclei recoiling from α emission, with the radioactivity of the resulting ^{208}Tl as an indicator of the emergence of recoil nuclei from the thin lead foils. The diffusion of lead in lead was found to be about 10^5 times slower than that of gold in lead.

Other Migration Problems Radioactive tracers are useful in the study of numerous migration problems other than self-diffusion, particularly when movements of very small amounts of material are involved. In most applications of this kind the tracer serves only as a sensitive and relatively convenient analytical tool. Erosion and corrosion of surfaces may be measured with great sensitivity if the surface to be tested can be made intensely radioactive. Transfer of minute amounts of bearing-surface materials during friction has been studied in this way. Radioactive gases or vapors may be detected in small concentrations, and leakage, flow, and diffusion rates of gases may therefore be studied by the tracer method.

Radioimmunoassay. The radioimmunoassay (RIA) tracer technique was developed by S. Berson and R. Yalow in the 1950s for the measurement of insulin in unextracted human plasma (Y1). They showed that the binding of ^{131}I-labeled insulin to a fixed concentration of antibody depends quantitatively on the amount of insulin present. This observation provided the basis for the RIA of plasma insulin. In the general RIA technique, illustrated in figure 11-1, the concentration of an unknown unlabeled antigen is obtained by comparing its inhibitory effect on the binding of radioactively labeled antigen to specific antibody with the inhibitory effect of known standards. The method is extremely sensitive, in many instances measuring amounts less than 1 pmol.

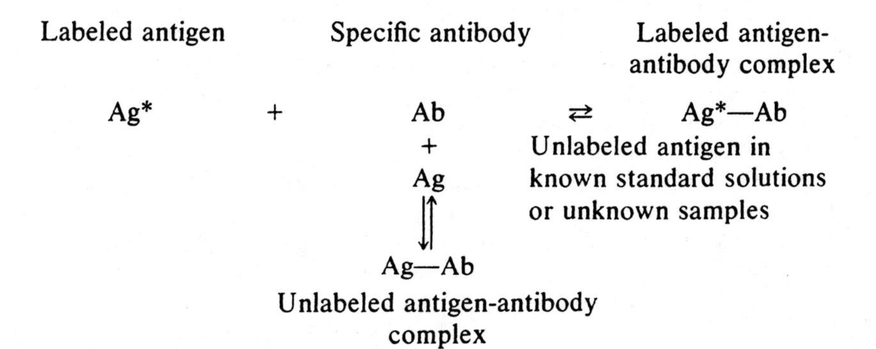

Fig. 11-1 Competing reactions that form the basis for radioimmunoassay. (From reference Y1 © The Nobel Foundation, 1978.)

The validity of the technique is dependent on identical *immunologic* behavior of antigen in unknown samples with the antigen in known standards. However, it is not necessary for standards and unknowns to be identical *chemically* or to have identical *biological* behavior. The RIA technique has been applied to many diverse areas of biomedical interest such as the measurement of peptidal and nonpeptidal hormones, drugs, vitamins, enzymes, viruses, tumor antigens, and serum proteins.

Skeleton Formation in Marine Organisms (P1). The rates and mechanisms of the formation of skeletal structures in marine organisms have been studied using tracer methods, particularly in the study of calcification (deposition of calcium carbonate on an organic matrix) in marine invertebrates and algae. These studies are of particular interest because the organisms have commercial value and the skeletal material is important in the marine calcium and carbon cycles. The utilization of dissolved calcium as a source of skeletal mineral has been studied by placing the organism in seawater containing ^{45}Ca in solution and subsequently measuring ^{45}Ca in the newly deposited skeletal material. Similarly, ^{14}C-labeled $NaHCO_3$ dissolved in seawater has been used to demonstrate the source of carbonate in skeletons of marine organisms such as bivalve molusks, corals, and algae. These studies of the uptake of dissolved materials have greatly facilitated the understanding of skeletal tissue growth and metabolism in marine biology.

Test of Separations. Radioactive tracers can be conveniently used to follow the progress and test the completeness of chemical separation procedures. If one component of a mixture is radioactive, it can often be followed satisfactorily through successive operations if beakers containing filtrates, funnels with precipitates, and so on, are merely held near a counter. Good chemical isolations have been made by these methods in the almost complete absence of knowledge of specific chemical properties. This crude qualitative procedure may be refined as far as desired, and valuable tests of analytical separation have been made with tracers. Further, it is possible to follow simultaneously the behavior of several radioactive tracers with characteristic γ-ray spectra by the use of a semiconductor or scintillation detector in conjunction with a multichannel analyzer.

Hydroxyl Concentration in Boundary Layer Air (C1). The OH radical plays a central role in a number of important reactions in atmospheric chemistry such as the conversion of CO to CO_2, SO_2 to H_2SO_4, and NO_2 to HNO_3. The desire to understand these conversion processes has provided the impetus for measuring the atmospheric concentration of the OH radical. A tracer method for measuring the rate of oxidation of CO in essentially unperturbed air has been used to determine the ambient OH concentration. The reaction

$$OH + CO \rightarrow CO_2 + H \qquad (11\text{-}1)$$

is responsible for ~90 percent of the CO oxidation. Therefore the rate expression $d[CO_2]/dt = k[CO][OH]$ should describe the oxidation well. Here k is the rate constant for (11-1), which has been determined to be $2.91 \times$

10^{-13} cm^3 s^{-1} at 760 torr of air with a pressure dependence of 2.2×10^{-16} cm^3 s^{-1} torr^{-1}.

In this method ^{14}CO is rapidly mixed into the air in a bag made of Teflon, which minimizes the absorption of incident light, particularly in the UV region between 280 and 320 nm, since photolysis of O_3 by these photons would produce additional OH. Samples of air taken as the reaction proceeds are frozen to capture the $^{14}CO_2$. These samples are later purified by sublimation and freezing to remove unreacted ^{14}CO and counted.

The OH concentration can be extracted from these data:

$$\frac{[^{14}CO_2]_t}{[^{14}CO]_0} = \frac{[^{14}CO_2]_0}{[^{14}CO]_0} + k[OH]t, \tag{11-2}$$

where the subscripts 0 and t indicate the concentration initially and at the time t, respectively. The method is extremely sensitive, capable of measuring as few as 3×10^5 OH radicals per cubic centimeter in the presence of $\sim 3 \times 10^{19}$ other molecules per cubic centimeter.

2. *Isotope Exchange and Other Tracer Reactions*

In an early exchange experiment in 1920 Hevesy demonstrated, by the use of ^{212}Pb, the rapid interchange of lead atoms between $Pb(NO_3)_2$ and $PbCl_2$ in water solution. The experiment was performed by the addition of an active $Pb(NO_3)_2$ solution to an inactive $PbCl_2$ solution and by the subsequent crystallization of $PbCl_2$ from the mixture. The result is not at all surprising because the well-known process of ionization for these salts leads to chemically identical lead ions, Pb^{2+}. This pioneering experiment opened an important field of chemical investigation, and many exchange systems have been examined since that time, particularly since the advent of artificial radioactivity (J1, S1). Some exchange studies using the tracer ^{35}S provide an interesting example. Sulfur and sulfide ions exchange in polysulfide solution. On the other hand, S^{2-} and SO_4^{2-}, SO_3^{2-} and SO_4^{2-}, H_2SO_3 and HSO_4^- do not exchange appreciably even at 100°C. If active sulfur is reacted with inactive SO_3^{2-} to form $S_2O_3^{2-}$, and then the sulfur removed with acid, the H_2SO_3 is regenerated inactive; therefore the two sulfur atoms in thiosulfate are not equivalent. The ions $S_2O_3^{2-}$ and SO_3^{2-} exchange only very slowly at room temperature, but exchange one sulfur fairly rapidly at 100°C. (Notice that this result can be found only by labeling the proper sulfur atom, the one attached directly to the oxygen atoms.)

Quantitative Exchange Law. Because exchange reactions occur at equilibrium with respect to the chemical species involved, although not with respect to the distribution of isotopes among the various chemical species, these reactions are particularly useful for the investigation of the theories of rates of reactions. This is so because, strictly speaking, the existing theories for rates of reactions assume that equilibrium conditions prevail. In this section we see how the rate of a chemical exchange reaction may be extracted from information on the rate at which a tracer atom is exchanged.

Consider a schematic exchange-producing reaction:

$$AX + BX^0 = AX^0 + BX,$$

where X^0 represents a radioactive atom of X. The radioactive decay of this species will be neglected; in practice, if the decay is appreciable, correction of all measured activities to some common time must be used to avoid error from this condition. The rate of the reaction between AX and BX in the dynamic equilibrium we call **R**, in units of moles per liter per second. Notice that **R** is quite independent of the concentration and even of the existence of the active tracer X^0, but is, in general, dependent on the total concentrations of the species $AX + AX^0$ and $BX + BX^0$. We indicate mole-per-liter concentrations as follows: $(AX)+(AX^0) = a$, $(BX)+(BX^0) = b$, $(AX^0) = x$, and $(BX^0) = y$. The rate of increase (dx/dt) of (AX^0) is given by the rate of its formation minus the rate of its destruction. The rate of formation of AX^0 is given by **R** times the factor y/b, which is the fraction of reactions that occur with an active molecule BX^0, and times the factor $(a-x)/a$, which is the fraction of reactions with the molecule AX initially inactive. The rate of destruction of AX^0 is given by **R** times the factor x/a, which is the fraction of reactions in the reverse direction that occur with an active molecule AX^0, and times the factor $(b-y)/b$, which is the fraction of reverse reactions with the molecule BX initially inactive. The differential equation is then

$$\frac{dx}{dt} = \mathbf{R}\frac{y}{b}\frac{(a-x)}{a} - \mathbf{R}\frac{x}{a}\frac{(b-y)}{b} = \mathbf{R}\left(\frac{y}{b} - \frac{x}{a}\right). \tag{11-3}$$

After a sufficiently long time, that is, at $t = \infty$, let $x = x_\infty$ and $y = y_\infty$. The conservation of radioactive atoms (after correction for any decay) demands that

$$x + y = x_\infty + y_\infty. \tag{11-4}$$

Further, at $t = \infty$ the exchange reaction is completed, which means that $dx/dt = 0$ and so, from (11-3),

$$\frac{x_\infty}{a} = \frac{y_\infty}{b}, \tag{11-5}$$

which constitutes an algebraic expresssion for the reasonable and well-known rule that when exchange is complete the specific activity (activity per mole or per gram of X) is the same in both chemical species. By the use of (11-4) and (11-5), y may be eliminated from (11-3), resulting in

$$\frac{dx}{dt} = \mathbf{R}\frac{(a+b)}{ab}(x_\infty - x). \tag{11-6}$$

This differential equation with separable variables may be integrated to give

$$\ln \frac{(x_\infty - x_0)}{(x_\infty - x)} = \mathbf{R} \frac{(a+b)}{ab} t, \tag{11-7}$$

where x_0 is the value of x at $t = 0$. Under the special, but common, condition that $x_0 = 0$, the more familiar forms emerge:

$$\ln \left(1 - \frac{x}{x_\infty}\right) = -\frac{a+b}{ab} \mathbf{R}t, \tag{11-8}$$

$$1 - \frac{x}{x_\infty} = \exp\left[-\frac{a+b}{ab} \mathbf{R}t\right]. \tag{11-9}$$

The last result shows that **R** may be evaluated from the slope of a plot of $\log [1 - (x/x_\infty)]$ versus t. Probably the most convenient procedure is to plot $[1 - (x/x_\infty)]$ on semilog paper against t, read off the half time $T_{1/2}$ at which the fraction exchanged, x/x_∞, is $\frac{1}{2}$, and find **R** from an equation derived immediately from (11-9):

$$\mathbf{R} = \frac{ab}{a+b} \cdot \frac{0.693}{T_{1/2}}. \tag{11-10}$$

It is important to notice that if a or b or both should be varied the variation in half time for the exchange would not directly reflect the variation in **R** because of the factor $ab/(a + b)$.

For a number of practical exchange studies the simple formulas AX and BX may not represent the reacting molecules; for example, AX_2 or BX_n might be involved. So long as the several atoms of X are entirely equivalent (or at least indistinguishable in exchange experiments) in each of these molecules, the equations just derived may be applied without modification, provided only that we redefine all the concentrations in gram atoms of X per liter rather than moles of AX or AX_2, or such, per liter. This is equivalent to considering (for this purpose only) one molecule of AX_2 as replaceable by two molecules of $A_{1/2}X$, and so on, in the derivation. If in a molecule like AX_2 the two X atoms are *not* equivalent and if they exchange through two different reactions with rates $\mathbf{R}_1$ and $\mathbf{R}_2$, it may be seen that the resulting semilog plot will be not a straight line but a complex curve. The differential equations for the exchanges involving the several positions may be set up and solved simultaneously, so that the curve may, at least in principle, be resolved to give values for the several **R**'s; however, this becomes very difficult for more than about two rates. A simplification may be made if $a \ll b$, with the several nonequivalent positions in the molecule AX_n. Here the value of y is very nearly a constant, and in this limit the complex semilog curve is resolvable in the same way as a radioactive decay curve into straight lines measuring $\mathbf{R}_1$, $\mathbf{R}_2$, and so on.

It must also be noted that in the analysis leading to (11-8) and (11-9) it is assumed that there are no other chemical reactions involving AX and BX

in the system. If there are other reactions, (11-8) and (11-9) will no longer be true in general. The problem will then usually entail the solution of a set of coupled differential equations.

Reaction Kinetics and Mechanisms. Radioactive tracers are finding an important place in the investigations of reaction kinetics and mechanisms. We discuss several examples to illustrate the kinds of information in this field that can be obtained with tracers but hardly in any other way. Consider the reversible reaction:

$$HAsO_2 + I_3^- + 2H_2O \rightleftharpoons H_3AsO_4 + 3I^- + 2H^+.$$

In the familiar theory of dynamic equilibrium, $K = k_f/k_r$, where K is the equilibrium constant and k_f and k_r are the rate constants of the forward and reverse reactions. Ordinarily K may be measured only at equilibrium and k_f or k_r far from equilibrium. Using radioactive arsenic to measure the rate of exchange between arsenious and arsenic acids induced by an iodine catalyst in accordance with the foregoing equilibrium reaction, it has been possible to find the rate law and rate constant at equilibrium. For the reverse direction as written it has been found that $\mathbf{R} = k_r\,(H_3AsO_4)(H^+)(I^-)$, with $k_r = 0.057$ liter^2 mol^{-2} min^{-1}, which is in satisfactory agreement with the information from ordinary rate studies made far from equilibrium, $\mathbf{R} = 0.071\,(H_3AsO_4)(H^+)(I^-)$.

Some of the applications of tracers to reaction mechanism studies are essentially qualitative. For example, when HClO labeled with ^{38}Cl oxidizes ClO_2^-, the product Cl^- contains the tracer, and the product ClO_3^- is inactive. Also, when Cl^- is oxidized by ClO_3^-, the product Cl_2 is formed from the Cl^-, and the product ClO_2 is formed from the ClO_3^-. When labeled I^- reduces IO_4^- to IO_3^- the tracer appears only in the I_2 product. Clearly any reaction intermediates containing the reactants must be unsymmetrical in that the two halogen atoms of initially different oxidation state are distinguished. This information at least rules out some of the conceivable reaction paths.

Electron-Exchange Reactions. In many oxidation-reduction reactions the net change appears to be a transfer of one or more electrons, as, for example, in the oxidation of ferrous ion by ceric ion:

$$Fe^{2+} + Ce^{4+} = Fe^{3+} + Ce^{3+}.$$

Isotopic tracers make possible the study of a relatively simple class of electron-transfer reactions, the exchange reactions between different oxidation states of the same element. For example, radioactive iron has been used by several investigators to study the rate of oxidation of ferrous ion by ferric ion,

$$Fe^{*2+} + Fe^{3+} = Fe^{*3+} + Fe^{2+}.$$

Here the asterisk indicates atoms tagged with a radioactive tracer. Such a reaction, of course, follows the quantitative exchange law already derived, and the rate is measured by the rate at which the tracer becomes randomly mixed between the two oxidation states. If the exchange is carried out in 6*M* HCl, the ferric iron is readily separated from the ferrous iron by ether extraction but the rate is too rapid for measurement. In this system chloride complexes such as $FeCl_3$ are surely present, and the exchange observed may proceed through such species. In perchloric acid the exchange rate is fast, but measurable in dilute solutions at 0°C. When fluoride ion is added so that species such as FeF^{2+}, FeF_2^+, and FeF_3 are present, the variation of the rate with F^- concentration can be used to show that the reaction proceeds through all of these forms and that the exchange rate is fastest with FeF^{2+}.

A large body of information now exists on the rates of electron-transfer reactions. For very fast reactions where rates cannot be determined with tracer techniques, other methods such as temperature jump, nuclear magnetic resonance, electron spin resonance, or stopped flow have been relied on (C2, F1, K1, S2). This work has shown that oxidation-reduction reactions can be classified into two main types: **outer sphere reactions**, where the coordination shells of the metal ions remain intact during reaction; and **inner sphere reactions**, which are often accompanied by the transfer of a bridging group. The rates of these reactions, particularly outer sphere reactions, have been calculated with some success using models that treat four factors—the Coulombic interaction energy, the inner-shell reorganization energy, the solvent reorganization energy, and the free energy change of the net reaction (M1, S3).

3. *Effects of Isotopic Substitution on Equilibria and Rates of Chemical Reactions* (W1, B2)

In the previous sections we have assumed that labeled species are physically distinguishable but chemically indistinguishable from their unlabeled counterparts. In this section we investigate the divergences from that assumption as they appear in equilibrium constants and rate constants. For example, (11-5) is not strictly true; there is some isotopic fractionation, although it is usually minute when compared to experimental error. Furthermore, the rate **R** in (11-3) is not exactly the same for all the isotopic species. But, again, the variation is usually too small to be detected experimentally.

Effect of Molecular Symmetry on Equilibrium Constants The first matter that should be considered is the value of the equilibrium constant when there is *no* isotopic fractionation and (11-5) is obeyed. We may ask, for example, what the numerical value of K is for the following reaction when the two isotopes of hydrogen are randomly distributed:

$$H_2 + D_2 = 2HD.$$

It is tempting to conclude that a random distribution of isotopes implies that $K = 1$, but this would be wrong as can rather easily be seen: consider a sample containing N hydrogen atoms of which a fraction f_H is protium and a fraction f_D is deuterium ($f_H + f_D = 1$). A random distribution of isotopes means that any particular atom in the diatomic hydrogen molecule has a probability f_H of being a protium and f_D of being a deuterium, regardless of the nature of the other atom in the molecule. This means that the numbers of H_2 molecules and D_2 molecules are $f_H^2(N/2)$ and $f_D^2(N/2)$, respectively. The number of HD + DH molecules is $2f_Df_H(N/2)$; the factor of 2 arises because of the two ways to have the same molecule: HD and DH. It is to be noted that the total number of molecules is given by

$$\frac{N}{2}(f_H^2 + 2f_Hf_D + f_D^2) = \frac{N}{2}(f_H + f_D)^2 = \frac{N}{2}.$$

If the system is contained in a volume V, the equilibrium expression is

$$K = \frac{\left[\frac{2f_Df_H(N/2)}{V}\right]^2}{\left[\frac{f_H^2(N/2)}{V}\right]\left[\frac{f_D^2(N/2)}{V}\right]} = 4. \tag{11-11}$$

Thus because there are two ways of having the HD molecule, the equilibrium constant is 4 instead of 1. The situation would be different if the two ways of having the molecule were distinguishable. For example, for the reaction

$$HCOOH + DCOOD = HCOOD + DCOOH$$

the equilibrium constant is

$$K = \frac{[HCOOD][DCOOH]}{[HCOOH][DCOOD]} = \frac{\left[\frac{f_Hf_D(N/2)}{V}\right]\left[\frac{f_Df_H(N/2)}{V}\right]}{\left[\frac{f_H^2(N/2)}{V}\right]\left[\frac{f_D^2(N/2)}{V}\right]} = 1.$$

Here the constant is 1 because HCOOD and DCOOH are distinguishable chemical species. It should be noted that if the analysis of the mixture were done in a mass spectrometer in which HCOOD and DCOOH are not necessarily distinguished from each other, the reaction would be written

$$H_2CO_2 + D_2CO_2 \rightarrow 2HDCO_2,$$

and the observed equilibrium constant would be 4.

This apparent dependence of the equilibrium constant on the state of innocence of the observer points up that the effect under consideration is contained in the entropy change in the reaction. Indeed, for exchange reactions in which there is no isotopic fractionation (there is no energy change in the reaction) it is just the entropy increase attendant on randomizing the distribution of isotopes that provides the driving force for the reaction. The equilibrium constant for the hydrogen reaction may be obtained from entropy considerations by realizing that the entropy of a particular isotopic molecular species at a temperature high enough so that the spacing between rotational energy levels is small compared to thermal energy may be written as

$$S^0 = S'^0 - R \ln \sigma, \tag{11-12}$$

where S'^0 is the entropy when no attention is paid to isotopic composition, and σ, the symmetry number, may be defined for our purposes as the number of *indistinguishable* ways that a molecule may be oriented in space under the condition that the various isotopes of an atom are considered to be *distinguishable*.[1] For example, $\sigma_{H_2} = \sigma_{D_2} = 2$ and $\sigma_{HD} = 1$; the entropy change for the hydrogen exchange reaction, then, is

$$\Delta S^0 = 2R \ln 2,$$

and from the usual thermodynamic relationships $K = 4$. The apparent contradiction with the formic acid exchange hinges on the knowledge of whether the two hydrogens are equivalent; they, of course, are not.

As another example, consider the exchange reaction

$$P^{35}Cl_3 + P^{37}Cl_3 \rightarrow P^{35}Cl_2{}^{37}Cl + P^{35}Cl^{37}Cl_2.$$

The pyramidal PCl_3 molecule has $\sigma = 3$ when all three isotopes are the same and $\sigma = 1$ when only two are the same. From (11-12) the entropy change is

$$\Delta S^0 = 2R \ln 3,$$

and the equilibrium constant is 9. The value of 9 can also be obtained from a probability argument of the type illustrated in (11-11). To use this method, though, it must be realized that the number of distinguishable ways of picking N objects when m are of one type and $N - m$ of another is[2]

$$\frac{N!}{(N-m)!m!}.$$

Isotope Effect on Equilibrium Constants. The preceding discussion, based solely on entropy effects, neglected any energy effects that may accompany isotopic substitution and so represents the high-temperature limit in which equilibria are largely governed by entropy changes. The data

[1] This represents a restricted and special use of the concept of symmetry number. For a more general discussion of the whole question, see reference D1.

[2] Refer to the discussion following (9-8).

Table 11-1 Variation with Temperature of Equilibrium Constant for Reaction $H_2 + D_2 \rightleftharpoons 2HD$

T (°K)	K
100	2.3
300	3.3
500	3.6

in table 11-1 for the hydrogen-exchange reaction show the approach to 4 at high temperatures and also illustrate the significant divergence from random isotopic distribution at lower temperatures, which occurs because the energy change in the reaction is different from zero.

The source of the energy change in exchange reactions does not lie, as it does in ordinary chemical reactions, in the change in the potential-energy field in which the atoms of the molecule find themselves. The potential-energy curve that defines the motion of two protium atoms in a hydrogen molecule, for example, is not significantly different from that for the two deuterium atoms in the isotopic molecule. What do change are the energies of the molecular translational, rotational, and vibrational quantum states. These changes arise directly from the mass differences of the isotopic molecules. The investigation of the effect of these changes on equilibrium constants (B3) has shown that the main effect stems from changes in the zero-point vibrational energies and from the spacing of vibrational states. It will be recalled that the vibrational energy states of the diatomic molecule AB are given by

$$E_{\text{vib}} = h\nu_{AB}(n + \tfrac{1}{2}), \qquad n = 0, 1, 2, \ldots,$$

where ν_{AB} is the fundamental vibration frequency of the AB molecule. This fundamental frequency depends upon the masses m_A and m_B of the atoms through the relation

$$\nu_{AB} = \frac{1}{2\pi}\left(\frac{f}{\mu_{AB}}\right)^{1/2},$$

where f is the force constant for the A—B chemical bond and

$$\mu_{AB} = \frac{m_A m_B}{m_A + m_B}$$

is the reduced mass of the system. The force constant f undergoes no significant change with isotopic substitution, but obviously μ_{AB}, and therefore ν_{AB}, do. The consequences of the change in the fundamental frequencies are most easily seen for the dissociation constants for two isotopic molecules AB and AB'. The ratio of the two dissociation constants is (B3)

$$\frac{K_{AB'}}{K_{AB}} = \frac{(1 - e^{-U'})Ue^{-U/2}}{(1 - e^{-U})U'e^{-U'/2}},$$

where $U \equiv h\nu_{AB}/kT$ and $U' = h\nu'_{AB}/kT$. It will be noted that at very high temperatures (where U and U' approach 0) the ratio approaches unity and the isotope effect vanishes as is expected. At very low temperatures (where U and U' are very large) the ratio approaches $(U/U') \exp[-(U - U')/2]$ and is governed by the difference in zero-point energies.[3] If $m_B < m_{B'}$, then $\mu_{AB} < \mu_{AB'}$ and $\nu_{AB} > \nu_{AB'}$ and the AB molecule is less stable with respect to dissociation than is the AB' molecule $[(K_{AB'}/K_{AB}) < 1]$. This implies that in an exchange reaction such as

$$AB + CB' \rightarrow AB' + CB$$

the light isotope will tend to concentrate in the compound with the smaller bond energy (smaller value of f). The isotope effect will be largest for the dissociation reaction in which there is no binding in the final state, and so the full difference in the vibrational energies of the isotopic molecules will appear.

The fact that equilibrium constants for exchange reactions differ from unity may be utilized for the separation of isotopes (I1). As an example, the exchange reaction between NO and HNO_3 (J2)

$$^{15}NO + H^{14}NO_3(aq) \rightleftharpoons {}^{14}NO + H^{15}NO_3(aq), \qquad K = 1.05 \text{ at } 25°C$$

has been used in a counter current apparatus (NO gas bubbles up and a nitric acid solution flows down) to enrich ^{15}N to an abundance of 99.5 percent from the normal 0.37 percent.

The magnitude of the isotope effect, as seen from the preceding paragraphs, depends on the difference in the reduced masses of the two isotopic molecules and thus on the fractional difference in the masses of the two isotopes. As a consequence, the larger the atomic weight, the smaller the isotope effect.

Isotope Effect on Rate Constants (J3, W1). Although we often ignore quantitative differences in the *rates* of reactions of molecules containing different isotopes, these differences usually are measurable, especially for isotopes of the light elements. For example, it has been shown that the rate of the electron-exchange reaction between Fe^{2+} and Fe^{3+} ions is diminished by a factor of 2 when the solvent is changed from H_2O to D_2O. This large effect demonstrates the important role played by the solvent molecules in the mechanism of the exchange reaction when they enter into the transition-state complex (S4).

[3] This approximation is valid only if the spacing of rotational energy states is small compared to thermal energies. This condition is not fulfilled for the various isotopic hydrogen molecules because of their small moments of inertia.

These effects, when of significant magnitude, can be inconvenient in tracer studies, since they invalidate the straightforward interpretation outlined in subsection 2. On the other hand, the magnitude of the isotope effect on reaction rates should depend on the details of the mechanism and thereby afford an opportunity for new information.

The most useful attempt to construct a theory for reaction rates of isotopic molecules is based on the transition-state approach to the problem (B4). These calculations can become very complicated (M2, W1) but, since the desired quantity is usually the ratio of rate constants for isotopic molecules rather than the actual rate constants, they probably represent the most successful application of transition-state theory. More recently rate processes have been studied with quantum-mechanical and trajectory calculations—methods that are not based on transition-state theory assumptions (W1). These calculations were carried out first for the fairly simple reaction $H + H_2$.

B. ANALYTICAL APPLICATIONS

Throughout most of the tracer work discussed, radioactive isotopes are assayed by measurement of their activities. This is actually an analytical procedure, but we have not emphasized that aspect because the samples are subject to analysis only if the tracer was provided earlier in the experiment. Of course, the naturally radioactive elements, including uranium, thorium, radium, potassium, and rubidium, may be assayed by radioactive measurements (W2). In this section we discuss a number of nuclear techniques for quantitative and qualitative analysis that are more generally applicable.

1. Activation Analysis

Neutron activation analysis is an extremely powerful trace-elemental analysis technique, in which an unknown sample is bombarded with thermal neutrons for appropriately chosen lengths of time. The chemical elements are identified and assayed after irradiation by measurement of characteristic radiation emitted from radionuclides formed in the (n, γ) reaction. In early uses of the technique it was necessary to separate the elements of interest chemically to remove the interfering activities of other elements. This is carried out in the usual manner after the addition of appropriate carriers. (For a discussion of carriers in radiochemistry see chapter 8, section B). More recently, however, the use of Ge(Li) γ-ray detectors with excellent energy resolution has allowed the determination of over 30 elements in trace quantities with no chemical separations. Standardization is provided by irradiation of a standard sample, containing known amounts of the elements to be analyzed, along with the unknown sample. It is sometimes desirable to use a standard of composition similar to the unknown, or else to use small samples, to avoid errors due to strong neutron absorption by other constituents.

The specificity of activation analysis is usually excellent since the purity of the radionuclide measured may be checked by determining both the energy and the half life of the γ rays emitted. Sensitivity depends on the flux Φ of bombarding particles, the cross section σ for the reaction involved, the decay constant λ of the nuclide measured, the duration t of the irradiation, and the efficiency ϵ of the detector. The counting rate R at the end of the irradiation of a sample that contains m grams of the isotope of atomic weight M to be assayed is given by

$$R = \frac{m}{M} N\Phi\sigma\epsilon(1 - e^{-\lambda t}), \tag{11-13}$$

where N is Avogadro's number. Note that (11-13) holds, provided the total sample is thin enough so that attenuation of the neutron flux in the sample may be ignored. With a neutron flux of $10^{12}\ \mathrm{cm^{-2}\ s^{-1}}$, limits of detectability for most elements are in the microgram to picogram range. Thus analysis by neutron activation is practical for impurities present in the parts per million or even parts per billion concentration range. One of the most important advantages of activation analysis is that, subsequent to neutron irradiation, accidental contamination of the sample with small amounts of the elements being determined is not important because the contaminants, not being radioactive, are not counted in the detector. Another important characteristic of activation analysis, and of a number of other nuclear methods, is that both the projectiles and γ rays used for analysis have long ranges in most materials, and self-absorption corrections are thus negligible.

Although slow-neutron irradiation is by far the most widely used technique in activation analysis, applications of activation by high-energy neutrons, photons, and charged particles have also been reported. These complementary techniques allow analysis for certain elements such as carbon, nitrogen, and oxygen with more sensitivity than is possible with thermal-neutron irradiation. An example of the utility of combining several instrumental activation techniques is given in G1 in which reference coal standards were analyzed for 51 elements.

The activation analysis technique has been used in such diverse media as terrestrial, lunar, and meteoritic materials, marine sediments, airborne particles, natural waters, environmental contaminants, biological materials, foods, hair, blood, drugs, semiconductors, archeological objects, and petroleum, as well as coal. A guide to the voluminous literature on activation analysis may be found in P2.

An example of the power of the method is the instrumental neutron activation of atmospheric aerosol samples collected on filters (Z1, D2). This method, using high-resolution Ge(Li) γ-ray detectors, is sensitive for the determination of over 30 elements from a single sample without chemical separations. The sample is first irradiated for about five minutes and then counted for observation of species with half lives from a few minutes to several days. The same sample, or another portion of the same filter, is then irradiated for a longer time (up to five hours). The samples are allowed to "cool" for several days before observation of species with half lives from several hours up to many years. Detection limits for elements that are typically measured in atmospheric aerosol samples are given in table 11-2.

Table 11-2 Minimum Detection Limit for Elements Measured by Instrumental Neutron Activation Analysis in Atmospheric Aerosol Samples (from D2)

Element	Product Nuclide[a]	Detection Limit[b] (10^{-9} g)	Element	Product Nuclide[a]	Detection Limit[b] (10^{-9} g)
	5-min irradiation 3-min decay			5-min irradiation 15-min decay	
Al	^{28}Al	40	Na	^{24}Na	200
Ca	^{49}Ca	1000	Mg	^{27}Mg	3000
Ti	^{51}Ti	200	Cl	^{38}Cl	500
V	^{52}V	1	Mn	^{56}Mn	3
Cu	^{66}Cu	100	Br	$^{80}Br^{c}$	20
			In	$^{116}In^{m}$	0.2
			I	^{128}I	100
	2–5-h irradiation 20–30-h decay			2–5-h irradiation 20-30-d decay	
K	^{42}K	75	Sc	^{46}Sc	3
Cu	^{64}Cu	50	Cr	^{51}Cr	20
Zn	$^{69}Zn^{m}$	200	Fe	^{59}Fe	1500
Br	^{82}Br	20	Co	^{60}Co	2
As	^{76}As	40	Zn	^{65}Zn	100
Ga	^{72}Ga	10	Se	^{75}Se	10
Sb	^{122}Sb	30	Ag	$^{110}Ag^{m}$	100
La	^{140}La	2	Sb	^{124}Sb	80
Sm	^{153}Sm	0.05	Ce	^{141}Ce	20
Eu	$^{152}Eu^{m}$	0.1	Hg	^{203}Hg	10
W	^{187}W	5	Th	$^{233}Pa^{d}$	3
Au	^{198}Au	1			

[a] Half lives and neutron-capture cross sections given in appendix D.
[b] Assumes irradiation at a flux of 2×10^{12} n cm^{-2} s^{-1}. Note that sample interferences can drastically alter detection limits.
[c] From $^{80}Br^{m}$ decay.
[d] From ^{233}Th decay.

Neutron activation has found great usefulness in criminology (G2). An example is the detection of toxic elements such as arsenic in hair. Since hair grows at a relatively fixed rate and arsenic enters the hair from the blood into the hair root, a time history of arsenic ingestion (and poisoning) can be inferred from analysis of hair sections. The sensitivity of neutron activation allows this analysis to be performed on as little as one strand of hair.

2. *Analysis with Ion Beams (Z2, Z3)*

Very sensitive methods of elemental analysis have been developed using energetic charged-particle beams. These methods are generally insensitive to the effects of outer-shell electrons and thus give little or no information on chemical binding of the elements detected. Ion beams for these analyses are most often produced in cyclotrons and Van de Graaff generators (see chapter 15). The techniques described below have in common the detection of radiation, from a target stimulated by ion bombardment, *during the bombardment itself*. This is in contrast with activation analysis techniques, which detect radiation from radionuclides *after the end of the irradiation*.

Neutrons have also been used for in-beam analysis. Of particular interest is the measurement of prompt γ rays following neutron capture (F2). This technique has been used to measure up to 17 elements and has the potential for real-time measurements of many elements in process streams such as coal or iron ore moving on a conveyor belt.

Particle-Induced X-Ray Emission (PIXE). The emission of characteristic X rays induced by charged-particle beams has been used for elemental analysis of thin samples ($\leq 1\ \mathrm{mg\,cm^{-2}}$) and small areas (a few square millimeters). The technique, shown schematically in figure 11-2, involves the observation of characteristic X rays emitted when atomic inner-shell vacancies created by particle bombardment are filled from outer shells. This method is fundamentally different from others described in this chapter in that purely atomic transitions are involved. Analysis with ion-induced X rays is quite similar to photon-induced X-ray emission (X-ray fluorescence) and electron-induced X rays (used in the electron microprobe). These techniques all take advantage of the excellent resolution of semiconductor detectors, which allow the identification of virtually all elements whose X rays are detected. The major advantage of ion-beam over photon excitation is the ability to focus the ion beam and generate much greater excitation density for near-surface elements. Typical operating conditions within thin targets are irradiation with $10\ \mu\mathrm{C}$ (microcoulombs) of 4-MeV protons or $5\ \mu\mathrm{C}$ of 16-MeV α particles and detection of X rays with a 10-mm^2 Si(Li) detector 3 mm thick fitted with a thin beryllium window ($\sim 1\times 10^{-3}$ cm). The method is generally sensitive to

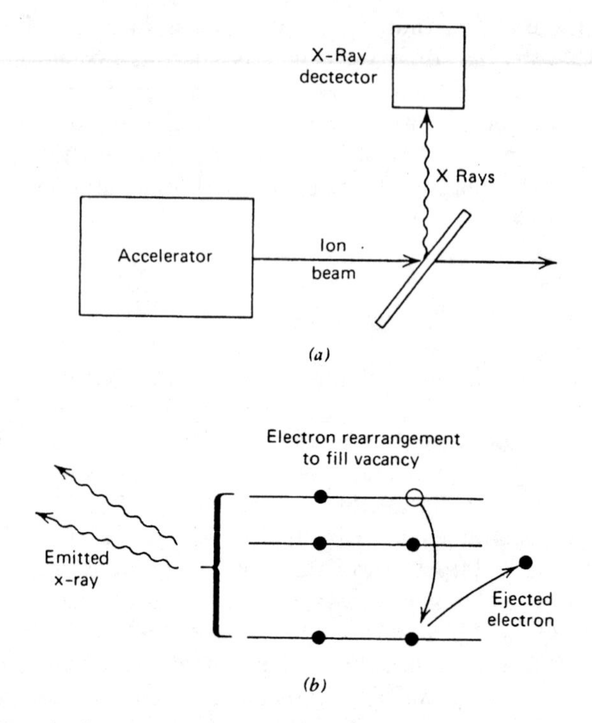

Fig. 11-2 Schematic diagrams of (*a*) the experimental arrangement for measuring particle-induced X-ray emission (PIXE); and (*b*) the physical process that leads to the production of characteristic X rays.

elements with $Z \gtrsim 11$. Details of several experimental systems are given in C3 and J4. A typical proton-induced X-ray spectrum of a multielement sample is shown in figure 11-3.

An application of the PIXE method that takes advantage of the unique features of the technique is the trace-element analysis of particles (diameter $< 20\ \mu m$). This is important in studies of air pollution, mining problems, and porous catalysts. Porous catalysts are distributed on substrates with large surfaces in order to have maximum interaction with liquids or gases forced through under pressure. When poisons deposited from the carrier stream stop catalytic activity it is interesting to carry out trace-element analysis of the catalyst to determine what reduced its effectiveness. For analysis the catalyst is pulverized into a powder and a few milligrams, deposited on thin Mylar, are analyzed by PIXE. Protons of a few million electron volts used in this analysis can penetrate $20\ \mu m$ particles and lose less than 10 percent of their energy, which minimizes cross-section changes. Therefore this analysis is nearly independent of particle size.

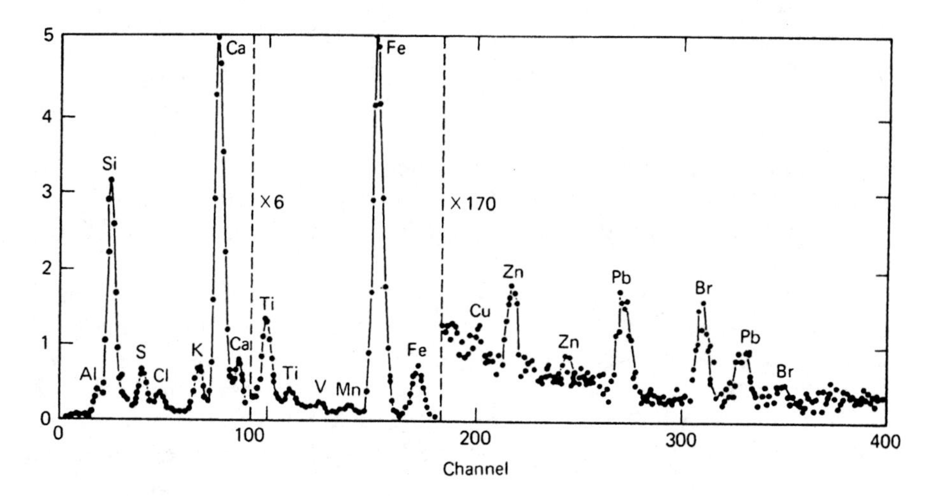

Fig. 11-3 Proton-induced X-ray spectrum of an atmospheric particulate sample (from N1). The abscissa scale corresponds to approximately 35 eV per channel. Each X-ray peak is labeled with the element it represents. The lead X rays are L_α and L_β, all others are K X rays; where there are two K peaks for an element, the left-hand one is the K_α, the right-hand one the K_β peak.

Resonant Nuclear Reactions. Quantitative depth information can be determined with sharply resonant nuclear reactions. As the incident ion energy is raised, the resonance occurs progressively deeper in the target at the point where the ions slow to the proper velocity. The detection sensitivity is determined by the size of the resonance cross section; the accuracy of the depth information is given by the uncertainty in the energy loss of the projectile. Alternatively one can pick an experimental geometry that results in one reaction product being monoenergetic and independent of the depth at which it is produced. The final energy of this product as it leaves the target surface allows the determination of the energy loss and thus the depth at which the nuclear reaction occurred. In this case a complete depth profile can be measured with the incident beam at one energy. An example of the former method is the determination of hydrogen in materials by detecting the 4.43-MeV γ ray from the $^{1}H(^{15}N, ^{12}C\gamma)^{4}He$ reaction. The detection of ^{4}He in the $^{2}H + ^{3}He \rightarrow ^{4}He + ^{1}H$ reaction has been used to determine ^{2}H or ^{3}He in metals by the latter method.

Analysis of Low-Z Elements. Multielement analysis of low-Z elements ($Z < 11$) is difficult with PIXE or neutron activation analysis. In some cases charged-particle activation can be used, for example ^{3}H activation of oxygen, producing ^{18}F. Two methods depend on prompt radiation analysis during nonresonant reactions, namely inelastic and elastic scattering. Analysis of γ rays following inelastic scattering of protons has been used for simultaneous analysis of low-Z elements such as carbon, nitrogen, oxygen, fluorine, sodium, silicon, and sulfur at microgram levels. The technique, called γ-ray analysis of light elements (GRALE), has been applied in atmospheric pollution studies by irradiating particles deposited on filters with 7-MeV protons (M3). Characteristic γ rays from the $(p, p'\gamma)$ reaction such as the 4.43- and 6.6-MeV γ rays of carbon and oxygen, respectively, are easily detected in a Ge(Li) detector without need for sample absorption corrections.

Another method that is more sensitive but that requires very thin samples ($<1 \text{ mg cm}^{-2}$) is the measurement of elastically scattered particles (generally p or α) in the forward direction (40–50° from the beam axis). The forward-scattering geometry provides increased sensitivity to low-Z elements. This can be seen from the equation for the energy loss of the scattered projectile determined from the kinematics of elastic scattering:

$$\frac{E'}{E_0} = \frac{m \cos \theta + \sqrt{M^2 + m^2 \sin^2 \theta}}{m + M}. \tag{11-14}$$

Here E_0 and E' are the projectile energies before and after scattering, m and M are the projectile and target masses in amu, and θ is the scattering angle (forward-scattering = 0°, backward-scattering = 180°). Elements with Z between 2 and 13 are usually seen as isolated peaks in thin targets, making this a complement to PIXE analysis.

Nuclear Backscattering. Nuclear backscattering was first described by Geiger and Marsden in 1909 (G3) and explained by Rutherford in 1911 (R1). The scattering is due to Coulomb repulsion, described in chapter 4, section A, which degrades the energy of the incident beam due to conservation of momentum. The energy of the backscattered projectile is used to mass analyze the elements in the target surface as given in (11-14); scattering from a light element will result in more energy transferred to the target nucleus than scattering from a heavier element. The process is summarized in figure 11-4. Because the cross section for scattering is proportional to Z^2 of the target, the method is most sensitive for heavy elements. Bombarding particles are typically protons of a few hundred keV or α particles of a few MeV. With these energies the top micrometer of the surface can be analyzed for individual elements (C4). A few of the applications of the technique are in studies of semiconductors, thin films, and corrosion. In these cases the composition variation or impurity distribution is determined as a function of depth below the surface of the sample. The method is capable of depth resolution of tens of nanometers over depths of up to hundreds of nanometers without sample erosion. Nuclear backscattering can also be used for determining bulk composition of the sample surface without the need for external standards. A very elegant example was the first elemental analysis of the moon by Surveyor 5, the first vehicle that made a soft landing on the moon (T1). In this experiment, designed by A.

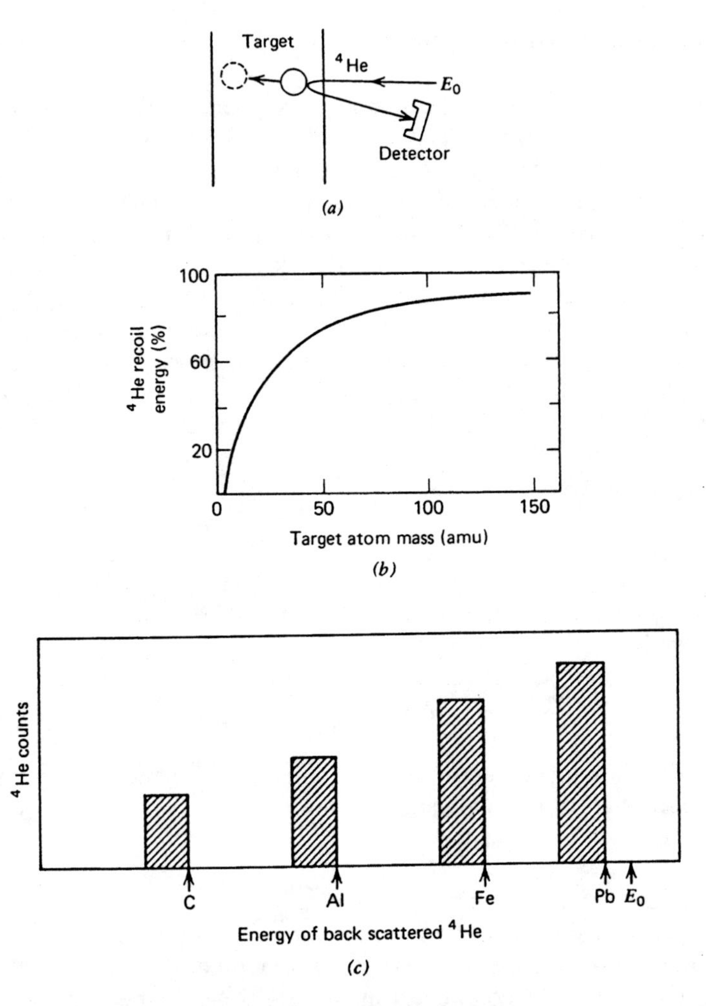

Fig. 11-4 (*a*) The arrangement for a nuclear back-scattering experiment; (*b*) the fraction of energy retained by a back-scattered ^{4}He ion plotted versus the mass of the target atom; (*c*) the spectrum of energies of back-scattered ^{4}He ions for various scatterers. Here E_0 is the original ^{4}He ion energy. Each peak gives a depth profile of target element concentration with the highest-energy ions coming from the surface. (From reference Z2.)

Turkevich and co-workers, the 5-MeV α particles from a 100 mCi ^{242}Cm source irradiated the lunar surface, and protons [from the (α, p) reaction] and backscattered α's were measured with semiconductor detectors. Pulse height analysis of the energy spectra was used to determine carbon, oxygen, sodium, magnesium, aluminum, silicon, and two groups of unresolved heavier elements. The results of this analysis agreed, within rather large uncertainties, with later analyses of moon rocks brought to earth.

3. *Analysis by Isotope Dilution (T2)*

Frequently a mixture must be quantitatively analyzed for a component, although no quantitative procedure for the isolation of this component is known. Particularly in the case of complex organic mixtures, it may be possible to isolate the desired compound with satisfactory purity but only in low and uncertain yields. In such a case the analysis may be made by the technique of isotope dilution. To a mixture containing an unknown mass M_u of the compound of interest is added a known mass M_1 of the compound containing a known activity $\mathbf{A}_1$ of radioactively tagged molecules. After the mixture is well mixed the compound of interest is purified and isolated. The mass M_2 and activity $\mathbf{A}_2$ of the isolated pure compound are determined and compared with those of the added material; the extent of dilution of the tracer shows the amount of inactive compound present in the original mixture, according to the following equality:

$$M_2 = (M_u + M_1)\frac{\mathbf{A}_2}{\mathbf{A}_1}. \tag{11-15}$$

This can be rewritten to give the mass of the compound of interest in the unknown mixture:

$$M_u = M_2\frac{\mathbf{A}_1}{\mathbf{A}_2} - M_1. \tag{11-16}$$

Equation 11-16 can be rewritten in a more convenient form in terms of specific activities[4] S as follows:

$$M_u = M_1\left(\frac{S_1}{S_2} - 1\right). \tag{11-17}$$

We may think of the tracer as serving to measure the chemical yield of the isolation procedure. However, exchange reactions that would reduce the specific activity of the compound must be absent. If the concentration of radioactive atoms is large enough to affect the molecular weight of the sample, as may be the case when long-lived radioactive tracers are used, corrections of the molecular weight must be applied for (11-15), (11-16), and (11-17) to be valid.

This powerful method may also be used with stable isotopes if mass-spectrometric analysis is employed. In this instance isotopic abundances are used instead of specific activities. Accuracy of a few percent may be

[4] The **specific activity** is defined as the ratio of the number of radioactive atoms to the total number of atoms of a given element in the sample (N^*/N). In many cases, such as the present example, where only the *ratios* of specific activities are needed, quantities proportional to N^*/N, such as activity per mole, are referred to as specific activity. In most tracer work the concentration of radioactive atoms is so small that the molecular weight of a sample is not affected. In this case specific activity can be expressed in activity per gram.

obtained for some elements that are present in concentrations as low as parts per billion to parts per trillion (I2, H1).

4. *Autoradiography (F3)*

The use of photographic emulsions and other track detectors (see chapter 7, section D) to study the occurrence and distribution of radioactive substances is called autoradiography. Exposure of photographic plates to uranium salts led Becquerel in 1896 to his discovery of radioactivity. Although various sensitive spectrometers are now used to study radioactive decay processes, autoradiography finds extensive applications in research in medicine and biology, meteorology, solid-state physics, chemical analysis, criminology, art history, and pest control. An example is its use with chromatography and electrophoresis for separating and detecting very small amounts of substances. Relative to "normal" detection by various tests such as color reactions the sensitivity of these techniques can be improved by radioactively labeling the substance and then detecting it autoradiographically. In studying the mechanism of photosynthesis radioactive starting materials were used and very small amounts of intermediates in the synthesis were detected. Autoradiography has also been employed extensively in dosimeters.

Activation of an object with a low flux of thermal neutrons followed by autoradiography using X-ray film has been used to study the distribution of pigments in oil paintings and to regenerate faded photographs (S5). If an oil painting is autoradiographed at various times after irradiation, different images are obtained, depending on the half lives of the activated nuclides in various pigments. High-quality autoradiographs of photographs that had faded due to silver oxidation have been prepared by activating silver to form either 2.4-min ^{108}Ag or 252-day $^{110}Ag^{m}$.

5. *Radiation Absorption Measurements*

The transmission of radiation passing through an absorber can be used to measure the thickness, and in some cases the elemental composition, of the absorber (chapter 8, p. 291). For example, the absorption of α radiation has been used as a method for determining film thickness. This technique is quite sensitive due to the steepness of the slope of the absorption curve near the end of the range. The use of β and γ radiation in analytical applications is described below.

Beta Attenuation Mass Monitor (M4). The attenuation of β particles from a low-energy β emitter has been used as a thickness gauge for a variety of materials. As discussed in chapter 6, β particles interact with matter through elastic and inelastic scattering with atomic electrons and through elastic nuclear scattering. For low-energy electrons ($E_\beta < 0.5$ MeV) inelastic scattering (ionization) with atomic electrons is the predominant mode of energy loss. The number of β particles passing through an absorber decreases, to a good approximation, exponentially, with absorber thickness:

$$I = I_0 e^{-\mu_m x}, \tag{11-18}$$

where I_0 is the β intensity without an absorber, I is the intensity observed through an absorber of thickness x, and μ_m is the mass absorption coefficient. The exponential form of the curve is fortuitous, since it also includes the effects of the continuous energy distribution of the β particles and the scattering of the particles by the absorber.

For low-energy β emitters the mass absorption coefficient is nearly independent of the chemical composition of the absorber. This is because the absorption of electrons depends on their initial energy and the number of electrons with which they collide in passing through the absorber. Therefore the absorption of particles depends on the ratio of atomic number to the mass number (Z/A). Although this ratio decreases from light to heavy elements, the effect of this variation on μ_m for low-energy β emitters is not large. Most chemical compounds have Z/A ratios in the range 0.44–0.53.

The β attenuation method of mass measurement has been used to determine the mass of atmospheric particulate matter collected on filters. The use of the "β gauge" is as sensitive as gravimetric analysis using a microbalance and can be easily automated.

Extended X-Ray Absorption Fine Structure. The oscillatory character of absorption probability in the vicinity of an X-ray absorption edge, called extended X-ray absorption fine structure (EXAFS), is associated with the environment of atoms surrounding the absorbing site and can be used to determine the elemental and chemical composition of the absorbing species. These oscillations result from the backscattering of the photoelectron excited by the absorbed X ray by atoms surrounding the absorbing site. The introduction of intense photon sources such as synchrotron radiation sources has made these measurements possible in a variety of materials such as crystals, amorphous materials, metalloproteins in solution, and catalysts. An example of the EXAFS phenomenon is shown in figure 11-5, where the absorption coefficient of crystalline $CuAsSe_2$ is plotted as a function of X-ray energy near the copper absorption edge. More details are given in W3, W4 and S6.

Critical Absorption. Quantitative measurements of the minor elements present in metal artifacts, such as the gold content in silver coins, provide useful information to archeologists. A nondestructive gold analysis method that does not induce radioactivity in the object has been reported for coins

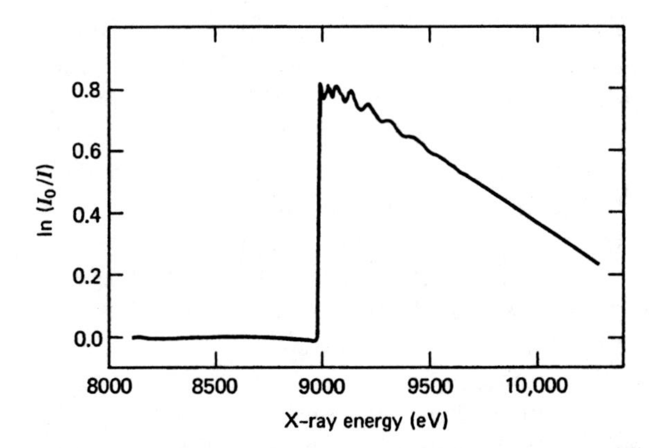

Fig. 11-5 Relative X-ray absorption coefficient of crystalline $CuAsSe_2$ as a function of X-ray energy near the copper absorption edge. The slowly varying, monotonically decreasing part of the absorption curve that would be evident in the experimental data at energies below the copper absorption edge has been subtracted. (Reproduced, with permission, from the Annual Review of Nuclear and Particle Science, Vol. 28, © 1978 by Annual Reviews, Inc., ref. W4.)

(R2). The method is analogous to critical X-ray absorption (p. 229) and is based on the differential absorption of the 79.63- and 80.88-keV γ rays of ^{133}Ba, which closely bracket the K absorption edge of gold (80.72 keV). Using a *pair* of γ rays avoids the necessity of making absolute absorption measurements. Furthermore, the transmission ratio for the γ-ray pair depends more strongly on the gold content than on the silver present; the silver absorption coefficients for such a closely separated γ-ray pair are nearly equal while the gold coefficients are quite different. This method can be used to measure samples with gold content ≥ 0.1 percent by weight.

A similar technique used in medicine, called dichromatography, improves the contrast of diagnostic X-radiographs. The difference in radiographs prepared with radiation below and above an absorption edge makes it possible to detect low concentrations of the element whose edge has been selected.

C. HOT-ATOM CHEMISTRY (H2, M5)

Szilard-Chalmers Process. The study of hot-atom chemistry, that is, the chemical reactions of atoms produced by nuclear transformations, began in 1934 when[5] L. Szilard and T. A. Chalmers (S7) showed that after

[5] Although this is the first example of the study of the *chemistry* of hot atoms, the first observation of the recoil of an atom following nuclear transformation was in 1904, when H. Brooks (B5) found that ^{218}Po atoms deposited on copper formed an activity that was transferred to the walls of an ionization chamber. Later this was shown to be due to the recoil of the daughter ^{214}Pb in the α decay of ^{218}Po.

the neutron irradiation of ethyl iodide most of the iodine activity formed could be extracted from the ethyl iodide with water; they used a small amount of iodine carrier, reduced it to I^-, and finally precipitated AgI. Evidently the iodine-carbon bond was broken when an ^{127}I nucleus was transformed by neutron capture to ^{128}I. This type of process has since been used to concentrate the products of a number of (n, γ) reactions and of some (γ, n), $(n, 2n)$, and (d, p) reactions. The chemical and physical changes following the neutron-capture reaction leading to isotope enrichment have come to be called the Szilard-Chalmers process. Three conditions have to be fulfilled to make a Szilard-Chalmers separation possible: (1) the radioactive atom in the process of its formation must be broken loose from its molecule, (2) it must neither recombine with the molecular fragment from which it separated nor rapidly interchange with inactive atoms in other target molecules; and (3) a method for the separation of the target compound from the radioactive material in its new chemical form must be available.

Most chemical bond energies are in the range of 1–5 eV (20,000–100,000 cal mol^{-1}). In any nuclear reaction involving nucleons or heavier particles entering or leaving the nucleus with energies in excess of 10 keV the kinetic energy imparted to the residual nucleus far exceeds the magnitude of bond energies.[6] In thermal-neutron capture, in which the Szilard-Chalmers method has its most important applications, the incident neutron does not impart nearly enough energy to the nucleus to cause any bond rupture. But neutron capture is almost always followed by γ-ray emission, and the nucleus receives some recoil energy in this process. A γ ray of energy E_γ has a momentum $p_\gamma = E_\gamma/c$. To conserve momentum the recoiling atom must have an identical momentum, and therefore the recoil energy $R = p_\gamma^2/2M = E_\gamma^2/2Mc^2$, where M is the mass of the atom. For M in atomic mass units and E_γ in millions of electron volts we have

$$R\ (\text{eV}) = \frac{537E_\gamma^2}{M}. \qquad (11\text{-}19)$$

Table 11-3 shows values of R for a few values of E_γ and M. Neutron capture usually excites a nucleus to about 6 or 8 MeV, and a large fraction of this excitation energy is dissipated by the emission of one or more γ rays. Unless all the successive γ rays emitted in a given capture process have low energies (say below 1 or 2 MeV), which is a relatively rare occurrence, the recoiling nucleus receives more than sufficient energy for the rupture of one or more bonds. It is not the entire recoil energy but something more like its component in the direction of a bond that should be compared with the bond energy; furthermore, the momenta of several γ

[6] For reactions other than (n, γ), particularly for (d, p) reactions, the Szilard-Chalmers technique is less useful because the energy dissipated by the incident radiation in the target is so great that many inactive molecules are also disrupted.

Table 11-3 Recoil Energies in Electron Volts Imparted to Nuclei by Gamma Rays of Various Energies

M	$E_\gamma = 2$ MeV	$E_\gamma = 4$ MeV	$E_\gamma = 6$ MeV
20	107	430	967
50	43	172	387
100	21	86	193
150	14	57	129
200	11	43	97

rays emitted in cascade and in different directions may partially cancel one another. In most (n, γ) processes the probability of rupture is certainly very high.

The second condition for the operation of the Szilard-Chalmers method requires at least that *thermal* exchange be slow between the radioactive atoms in their new chemical state and the inactive atoms in the target compound. The energetic recoil atoms may undergo exchange more readily than atoms of ordinary thermal energies. These exchange reactions and other reactions of the high-energy recoil atoms, called "hot atoms," determine to a large extent the separation efficiencies obtainable in Szilard-Chalmers processes.

A large amount of work in the field of Szilard-Chalmers separations has been done on halogen compounds. Many different organic halides (including CCL_4, $C_2H_4Cl_2$, C_2H_5Br, $C_2H_2Br_2$, C_6H_5Br, CH_3I) have been irradiated, and the products of neutron-capture reactions (^{38}Cl, ^{80}Br, ^{82}Br, ^{128}I) removed by various techniques. Many other Szilard-Chalmers processes have been studied such as the separation of halogens from chlorates, bromates, iodates, perchlorates, and periodates; separation of ^{32}P from phosphates; and separation of arsenic from arsine gas to name just a few cases. Other examples of these various methods for isotope enrichment by the Szilard-Chalmers process can be found in review articles and books (H2, W2).

Chemical Effects of Radioactive Decay (C5). Hot atoms may result from radioactive decay processes as well as from nuclear reactions. The range of recoil energies encountered in a variety of nuclear processes is given in table 11-4. The chemistry of hot atoms formed as a result of β-decay processes has been studied in a number of cases. For example, reactions such as

$$TeO_3^{2-} \rightarrow IO_3^- + \beta^- \qquad \text{and} \qquad MnO_4^- \rightarrow CrO_4^{2-} + \beta^+$$

can occur in addition to molecular disruption leading to other chemical species. Because β decay involves emission of an electron and a neutrino,

Table 11-4 Approximate Recoil Energies Expected with Various Nuclear Events (from reference C5)

Nuclear Process	Range of Recoil Energy (eV)[a]
β^- Decay	10^{-1}–10^2
β^+ Decay	10^{-1}–10^2
α Decay	$\sim 10^5$
IT	10^{-1}–1
EC	10^{-1}–10^1
n_{th}, γ	$\sim 10^2$
n, p	$\sim 10^5$
Fission	$\sim 10^8$

[a] Based on a hot-atom mass of ~100, the most probable kinetic energy for a given nuclear process, and a range of nuclear energies most frequently encountered.

the nucleus receives a spectrum of recoil energies depending on how the kinetic energy is shared and on the angular correlation between the two particles. The *maximum* recoil energy is

$$R_{\max}\,(\text{eV}) = \frac{537E_\beta(E_\beta + 1.02)}{M}, \tag{11-20}$$

where E_β is the maximum β energy in MeV. A 0.5-MeV β decay in a mass-100 nucleus produces a maximum recoil of ~4 eV. In this and all other cases of nuclear recoil not all energy is available for bond rupture; the energy is partitioned between translational, rotational, and vibrational motion of the molecule. It is the latter mode that is most effective in producing bond rupture. Thus in β decay with so little energy available for dissociation, many molecules survive the decay process with bonds intact.

The chemistry of recoil atoms following ITs has been studied more than the hot-atom chemistry of other radioactive decay processes. It is perhaps not immediately clear why ITs may lead to bond rupture. The γ-ray energies in ITs are much lower than in neutron-capture processes, often below 100 keV and rarely above 500 keV. According to (11-19), a 100-keV γ ray would give a nucleus of mass 100 a recoil energy of only about 0.05 eV, which is not sufficient to break a chemical bond. Although the recoil energy resulting from internal-conversion electron emission given by (11-20), is as much as 10 times greater than that imparted by γ emission at the same energy, even this is not sufficient for bond rupture in most cases. However, the vacancy left in an inner electron shell by the internal conversion leads to electronic rearrangements and emission of Auger electrons. The atom is therefore in a highly excited state (and positively charged), and molecular dissocation may take place if the atom is bound in a molecule.[7]

Separations of nuclear isomers analogous to Szilard-Chalmers separations have been performed in a number of cases in which the IT proceeds largely by conversion-electron emission. The 18-min ^{80}Br has been separated from its parent, the 4.4-h $^{80}Br^m$, by a number of different methods analogous to the Szilard-Chalmers methods used for bromine. The lower states of ^{121}Te (17 d), ^{127}Te (9.4 h), ^{129}Te (69 min), and ^{131}Te (25 min) have been separated as tellurite in good yield from tellurate solutions containing the corresponding upper isomeric states. Isomer separations have been useful for the assignment of isomer activities and for the elucidation of genetic relationships.

Chemistry of Recoil Atoms. Chemical effects following nuclear transitions are not only important for isotope enrichment as in the case of the Szilard-Chalmers process, but also provide an opportunity for the study of the mechanisms of chemical reactions of the energetic recoil atoms. These atoms, as already mentioned, are often referred to as hot atoms and the field of study as hot-atom chemistry. That there are such reactions is immediately seen from the observation that only a fraction of the radioactive atoms is in a chemical form different from that of the parent compound. The fraction of the active atoms that is in the same chemical form as the parent compound is often called the **retention**. The observation, for example, that the retention of pure CCl_4 is about 43 percent and diminishes to about 5 percent on the addition of 50 mol percent C_6H_{12} indicates that neither lack of bond rupture nor recombination with the fragment from which the hot atom had broken away is of much importance.

The study of hot-atom chemistry draws upon the results of many branches of chemistry such as ion-molecule studies, radiation chemistry, photochemistry, and molecular beam studies of excited atoms. In this brief discussion we do not try to cover this diverse field but rather summarize a few interesting results of recoil chemistry.

In solids and liquids the spectrum of recoil energies of atoms following thermal-neutron capture is broad and peaks at rather low energies. Thus the ranges of these (n, γ) recoils are short, not more than one or two molecular layers in solids. The calculation of ranges of ions has been discussed in chapter 6, section A. It is known that a hot zone is produced as the recoil slows down. The effect of this thermal spike on chemical reactions is not fully understood.

[7] An experimental measurement of the charges of ^{131}Xe atoms following IT of $^{131}Xe^m$ shows that on the average 8.5 electrons are lost. Similar measurements give +3.4 for the average charge of ^{37}Cl atoms following EC in ^{37}Ar. In both ITs and ECs the high charges result largely from Auger processes. In β^- decay atomic charges in the neighborhood of +1 have been reported.

The chemistry of recoil tritium is now well enough understood to be a useful prototype for hot-atom chemistry of monovalent ions. For studies of liquid and solid systems hot tritium is normally produced by the $^6Li(n, \alpha)^3H$ reaction; the $^3He(n, p)^3H$ reaction is most convenient for gas phase studies. The procedure with the ^3He simply involves irradiating a gaseous mixture of ^3He and the organic compound with neutrons from a reactor, whereas for the ^6Li reaction an intimate mixture of Li_2CO_3 and the organic compound is irradiated. It has been found that recoil tritium atoms react rapidly and with high efficiency over an energy range of ~1–50 eV. The three dominant types of simple reactions of hot recoil tritium with organic molecules are: (1) abstraction of a hydrogen to form HT; (2) substitution of hydrogen atoms, radicals, or groups by tritium; and (3) addition of tritium to unsaturated systems. These reactions are fast compared to the period of bond vibration so that the absorption of excitation energy is localized to only a few atoms in the collision area. This results in the breaking of only a few bonds. More details on the mechanisms of these reactions and their use in labeling organic molecules are given in M5 and U1.

Recoil ^{18}F atoms, produced by irradiation of ^{19}F with fast neutrons or γ rays, are extremely reactive with hydrocarbon molecules giving $H^{18}F$ as the major product. This is analogous to the abstraction reaction of recoil tritium. Many of the other reactions of recoil fluorine atoms are different from results with tritium. For example fluorine reactions involving the C—C bond are favored over reactions at the C—H site while for tritium the reverse is true. Reactions with other recoil halogens show many similarities to the fluorine results. In general the halogen systems are more difficult to understand than the recoil tritium systems.

The reactions of free carbon atoms have been studied with ^{11}C produced in $^{12}C(n, 2n)$, $^{12}C(p, pn)$, and $^{12}C(\gamma, n)$ reactions as well as with ^{14}C. The use of ^{14}C has the disadvantage of the long half life (5730 y) resulting in low specific activity. This necessitates production of a large number of atoms leading to serious radiation damage effects. Although the 20-min half life of ^{11}C limits the time available for analysis with this isotope, rapid methods of gas chromatographic analysis have allowed ^{11}C to be used in the study of organic reaction mechanisms in large numbers of organic systems and to be applied extensively for diagnostic medicine.

In typical reactions of recoil carbon with organic molecules a relatively large number of products are formed. A further complexity in these reactions is the instability of the primary products, which generally undergo further reaction. It has been found that hot and moderated carbon atoms undergo similar reactions. The difference in kinetic energy affects only the relative reaction probabilities. The main reactions are: (1) insertion of carbon atoms into C—H bonds; (2) the insertion of carbon atoms into C=C double bonds; and (3) hydrogen atom abstraction for CH_x.

A large body of data also exists on reactions of hot germanium and silicon, which have helped to elucidate the chemistry of these species.

Theoretical Models. Theoretical attempts to understand hot-atom chemistry in liquids and vapors have divided the events into two classes: those that occur before and those that occur after the hot atom has been reduced to thermal equilibrium. Analysis of the "hot" processes requires an expression for the energy spectrum of the recoil species while it is being cooled by collisions and an expression for the probabilities of the various possible reactions in each collision as a function of the energy of the recoil atom. The energy spectrum may be obtained easily under the assumption that the collisions are elastic atom-atom collisions (M6), an assumption that is probably not justified in the energy region just above thermal, where, unfortunately, most of the "hot" reactions are expected to occur. The energy dependence of the probabilities for reactions in each collision, except for a simple model proposed by W. F. Libby (L1) that is generally unreliable, has yet to be treated theoretically and is left to be experimentally determined.

In order to understand the yields observed in hot-atom reactions of gases and liquids, a phenomenological kinetic theory has been used that reduces the wide range of reaction data and expresses it in terms of a few parameters. These empirically derived parameters can be calculated theoretically from a model of the reaction mechanism and thus a framework is provided for comparing experiment and theory. This theory is analogous to the classical collision theory of reaction rates; the parameters extracted from the data are the activation energy and a steric factor (M5, G4).

A number of other approaches have been used to determine the yields of hot-atom reactions (G4). For example, trajectory calculations of reaction probabilities have been performed in order to understand aspects such as the influence of bond energies on reaction yields. Another approach is a steady-state theory of hot-atom reactions based on the Boltzmann equation. Quantum-mechanical probability calculations have also been carried out for some systems but these calculations can be rather expensive.

The reactions of the thermalized recoil atom are the usual ones expected at thermal energies, but there is, in addition, the possibility of recombination with the fragments created by the recoil atom while it was being slowed down. This process should be particularly important in liquids and in solids. Hot-atom reactions have been studied in solid inorganic compounds in connection with problems in solid-state chemistry and radiation damage (M5, H2). Particular emphasis has been placed on post-recoil annealing effects in which the increase in retention is investigated as a function of the time and temperature at which the irradiated crystal is stored before being dissolved for analysis.

D. RADIOCHEMISTRY APPLIED TO NUCLEAR MEDICINE

The visualization of organs, localization of tumors, detection of abnormalities in diagnosis, determination of metabolic pathways, and introduction of radiation sources into specific sites for therapy are among the goals of nuclear medicine. Wide use is made of radioisotope tracer techniques in this field, for example in the preparation of **radiopharmaceuticals** (radiochemicals refined to pharmaceutical purity) for clinical diagnosis of various abnormalities. The nuclides used in radiopharmaceuticals must have suitably short half lives and have a high yield of γ rays between 50 and 500 keV without causing excessive tissue irradiation from other emissions (e.g., from high-energy β particles). Radiopharmaceuticals are prepared with high specific activity to allow small administered volume with high photon flux for imaging. The chemical form of the radionuclide is chosen to yield the desired physiologic distribution. It is important to note that the effective radiopharmaceutical decay constant in the body is the sum of the physical decay constant of the radionuclide and the biological decay constant of the radiopharmaceutical for clearance from the body. Most studies employ 6-h $^{99}Tc^m$ in a variety of radiopharmaceuticals to image the thyroid, salivary glands, brain, bone, heart, kidneys, liver, spleen, and lungs. Other widely used nuclides include ^{67}Ga, ^{111}In, ^{123}I, ^{125}I, ^{131}I, and ^{201}Tl.

Two-dimensional projections of the distribution of the radioisotope are produced with a scintillation camera that consists of a thin (≤ 1 cm) large-diameter (≥ 25 cm) NaI crystal and up to 92 photomultiplier tubes (often called an Anger camera after the early developer, H. O. Anger). The phototubes nearest the interaction of the γ ray with the crystal collect the most light while those further away collect less. The scintillation photons are converted in the phototube into a voltage pulse proportional to the amount of light incident on the tube as described in chapter 7, section C. The voltage output pattern from the phototube array gives information on the two-dimensional position of the primary γ-ray interaction in the crystal. A collimator of single or multiple holes is placed in front of the detector to absorb stray radiation. Images using a technetium pyrophosphate complex are shown in figure 11-6. Details on the use of $^{99}Tc^m$-labeled radiopharmaceuticals with the Anger camera are given in N2.

In the following section we discuss two applications of great importance, namely the development of a physical detection technique for measurement of radioisotopes in vivo and the use of various labeling techniques to prepare compounds labeled with short-lived nuclides for evaluation of metabolic processes in vivo. Some of the other nuclear methods used in medicine that we do not discuss here include the Mössbauer effect (see chapter 12, section A) and the use of heavy-ion and fast-neutron beams for therapy.

Positron Imaging Devices (B6, R3). Emission tomography is a tech-

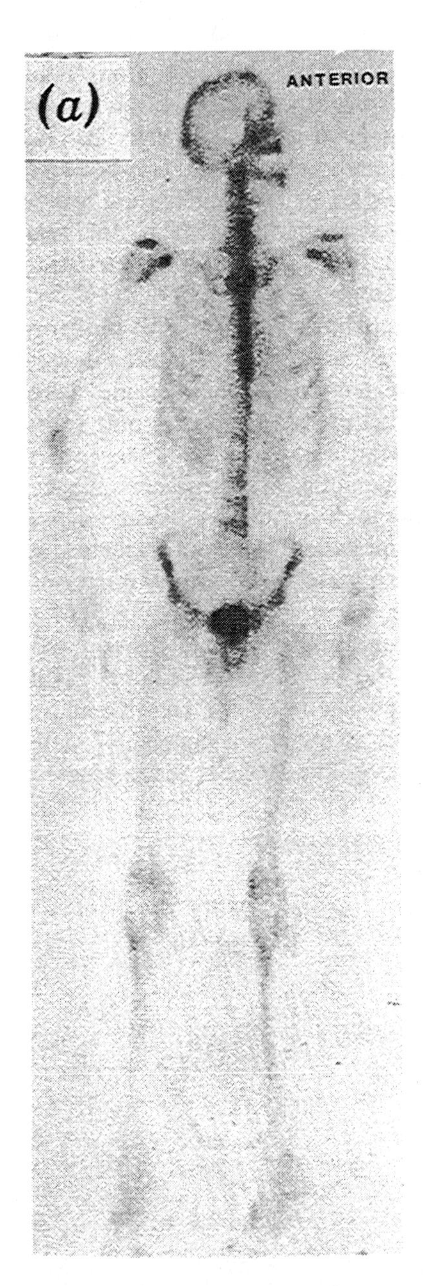

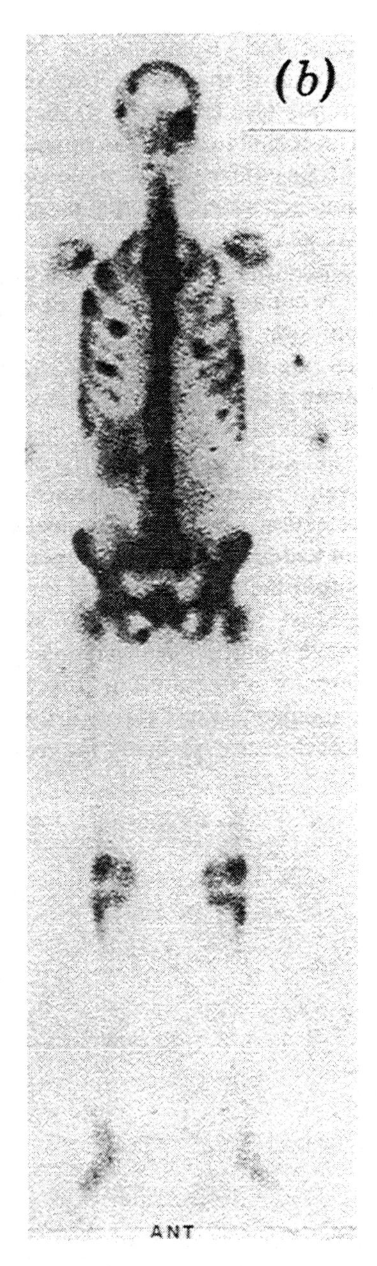

Fig. 11–6 (*a*) Normal bone scan obtained with $^{99}Tc^{m}$—pertechnetate pyrophosphate, showing a normal distribution of activity throughout the skeleton. Because of the urinary excretion of approximately 50 percent of the administered dose of this tracer, activity is also noted in the kidneys and urinary bladder. (*b*) Bone scan obtained with the same technique showing multiple skeletal metastases. Areas of increased activity are seen throughout the skeleton; they are due to reactive new bone formation in response to the presence of tumor deposits. (Courtesy Dr. B. Siegel 1980.)

nique for visualizing the distribution of a radionuclide in a transverse section of the body. This technique is a form of quantitative autoradiography (see section B, 4) that has the advantage of allowing in vivo studies. The technique is more powerful when used with positron emitters because it then utilizes the unique directional properties of the annihilation radiation generated when positrons are absorbed in matter, that is, the two 511-keV photons are emitted at an angle of $180 \pm 0.3°$. The characteristics of positron annihilation are described in chapter 12, section B.

A number of coincidence detection systems have been used for positron emission tomography (PET). In these devices annihilation radiation detected in coincidence is assumed to have originated from an event somewhere along a line joining the detector centers. This provides an electronic form of collimation and allows high sensitivity because no physical collimation is required to achieve spatial resolution. The detection of a true coincidence event requires that *neither* photon undergo a scattering event before detection. Thus the attenuation of the annihilation radiation detected in coincidence is nearly independent of the position of the source of positrons within the tissue between the two detectors.

Most positron imaging systems place multiple detectors around the imaged object, maximizing efficiency of the radiation collection. Each detector is operated in coincidence with opposing detectors, creating many coincidence lines through the imaged object as shown in figure 11-7. The detectors are normally moved about the object. The image is reconstructed

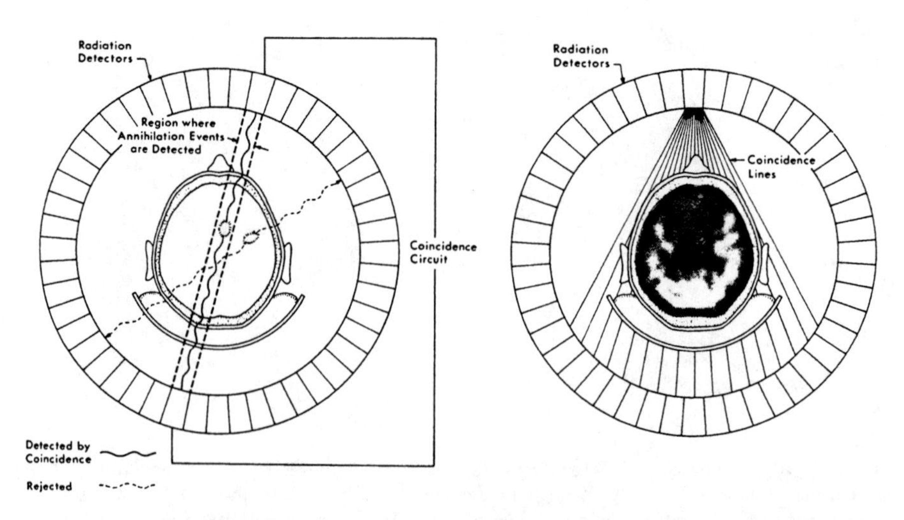

Fig. 11-7 A schematic representation of the radiation detector arrangement for positron emission tomography. The left figure shows how coincidence detection of annihilation radiation is used to localize the position of the positron-emitting nuclide. The right diagram shows the multiple coincidence arrangement used to increase the information gathered by the imaging device. (From reference R3.)

by means of computer convolution techniques giving a quantitative representation of the spatial distribution of the radionuclide in two dimensions in the object. The detector diameter is an important factor in determining spatial resolution; another factor that is independent of detector and source geometry is the distance the positron travels before annihilating with an electron (~1–6 mm in tissue). Spatial resolution in these systems is typically 1–2 cm.

A very successful positron imaging instrument is the Positron Emission Transaxial Tomograph (PETT) developed at Washington University (T3). This instrument consists of a hexagonal array of 66 NaI(Tl) detectors and uses translational and rotational motions for sampling.

In many cases it is necessary to have many tomographic slices of the organ of interest for proper visualization of the three-dimensional object. With the single-slice instrument information in the third dimension is gathered by repetitive single-slice images at different positions with respect to the organ. Several instruments have been constructed to collect multiple slices simultaneously. One of these multislice instruments, PETT IV, (shown in figure 11-8) is capable of providing images of 7 slices of an object simultaneously (T4, T5), utilizing a moving array of 48 NaI(Tl) detectors, each optically coupled to 2 photomultiplier tubes. The multislice

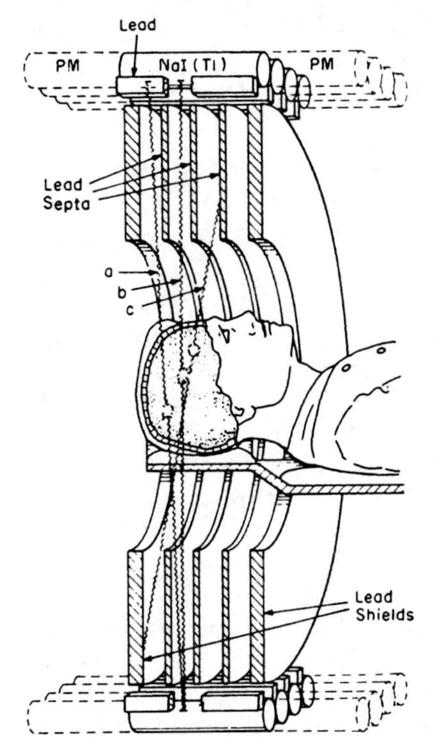

Fig. 11-8 Diagram of the PETT IV instrument. (From reference T4.) Pm stands for photomultiplier.

capability is achieved by comparing the light outputs of the two photomultiplier tubes in each detector. The ability of these instruments to obtain rapid multiple tomographic slices simultaneously is particularly advantageous for dynamic tracer studies described below.

Radiopharmaceuticals. Clinical applications of PET have been made possible with the development of ingenious techniques for rapid synthesis of radiopharmaceuticals, suitable for in vivo studies, using cyclotrons and linear accelerators within a medical complex. The short-lived positron emitters ^{15}O ($t_{1/2} = 122$ s), ^{13}N ($t_{1/2} = 10.0$ min), ^{11}C ($t_{1/2} = 20.4$ min), and ^{18}F ($t_{1/2} = 110$ min) have chemical and physical properties that make them uniquely suitable for obtaining in vivo biochemical and physiological information when used with PET. The feasibility of labeling metabolites or their analogs with these radionuclides is of particular interest. Examples of a few of the very large number of compounds that have been labeled with these nuclides are given below. The breadth of the field of radiopharmaceuticals is reviewed in W5, W6, and C6.

Oxygen-15. The $^{14}N(d, n)^{15}O$ reaction is commonly used for producing ^{15}O to label radiopharmaceuticals. For example, labeled carboxyhemoglobin can be prepared from ^{15}OO produced by irradiating N_2 with a trace (~0.01 percent) of oxygen. The labeled oxygen is converted to $C^{15}OO$ by passing it over activated charcoal at 400°C and then dissolving the gas in a sample of the patient's blood. Water labeled with ^{15}O can be produced through exchange with carbonic acid by dissolving $C^{15}OO$ in aqueous solution as follows:

$$C^{15}OO + H_2O \rightleftarrows H_2CO_2{}^{15}O \rightleftarrows H_2{}^{15}O + CO_2.$$

Nitrogen-13. Efficient reactions for producing ^{13}N to be used in radiopharmaceuticals are $^{12}C(d, n)$ and $^{16}O(p, \alpha)$. Labeled N_2 has been produced by irradiation of CO_2 containing a trace of N_2. The gaseous ^{13}NN is converted to a solution by passing the gas over cupric oxide to convert any traces of CO to CO_2 followed by absorption of the CO_2 in a soda lime trap with subsequent dissolution of the labeled ^{13}NN in saline. Ammonia labeled with ^{13}N is extensively used in clinical studies directly and as a precursor in the enzymatic synthesis of a series of amino acids. One procedure for producing $^{13}NH_3$ involves the deuteron bombardment of flowing methane followed by trapping of the activity in acid and subsequent distillation of $^{13}NH_3$.

Carbon-11. Carbon dioxide labeled with ^{11}C is a commonly used starting material for many synthetic procedures for producing labeled compounds. The $^{11}CO_2$ has been produced by a number of methods, including bombardment of boric oxide with deuterons [$^{10}B(d, n)^{11}C$] in a helium flow

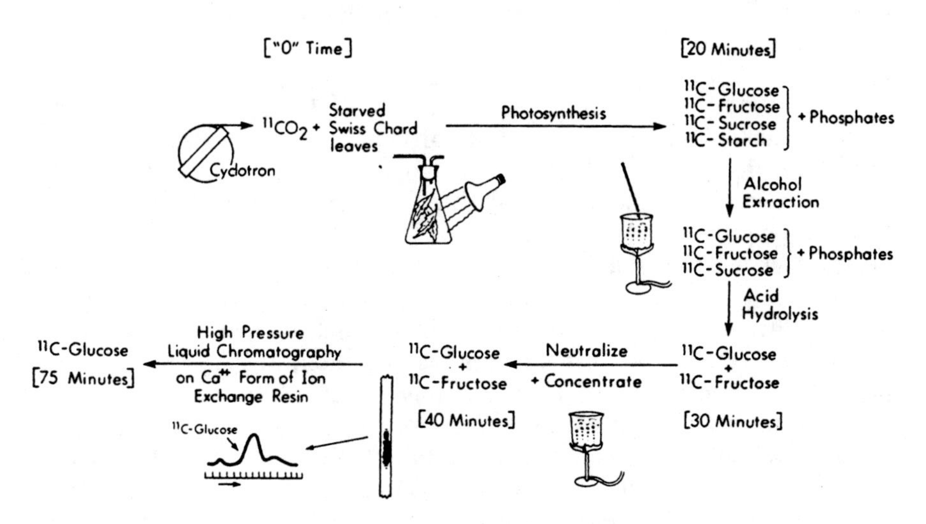

Fig. 11-9 Flow chart of the synthesis of ^{11}C-labeled glucose (From reference R3.)

system and bombardment of N_2/O_2 mixtures with protons [$^{14}N(p, \alpha)^{11}C$]. Among the compounds that have been used in clinical studies are ^{11}C-labeled carboxyhemoglobin, glucose, palmitic acid, and amino acids. As shown in figure 11-9, ^{11}C-glucose has been produced biosynthetically by passing $^{11}CO_2$ over illuminated Swiss chard leaves that had previously been light-starved. After extraction of the ^{11}C activity with ethyl alcohol and hydrolysis of the sucrose with hydrochloric acid, the mixture is neutralized and concentrated by evaporation. The mixture is then injected onto a cation-exchange column and eluted with a $Ca(OH)_2$ solution. The $Ca(OH)_2$ is removed from solution with an ion-exchange resin, and the remaining solution is concentrated, buffered to physiological pH, and filtered. This entire process takes about 75 minutes.

Generators for Short-Lived Nuclides. The use of the aforementioned short-lived nuclides requires a dedicated cyclotron in the immediate vicinity, rapid radiopharmaceutical syntheses, and metabolic processes sufficiently short to be studied. A number of parent-daughter systems can be used as generators of short-lived radionuclides extending the application of positron-emitting radiopharmaceuticals to labs without a nearby cyclotron. An example of such a generator is the ^{68}Ge/^{68}Ga system. ^{68}Ge ($t_{1/2} = 288$ d) decays by EC to ^{68}Ga ($t_{1/2} = 68$ min), which decays 88 percent by positron emission to stable ^{68}Zn. The generator consists of an alumina column as the absorbent and ethylenediaminetetraacetic acid (EDTA) as the elutant. The resulting ^{68}Ga-EDTA complex is used as a brain-scanning agent, or is decomposed before preparing other radiopharmaceuticals. The most widely used generator is $^{99}Mo/^{99}Tc^m$. Other generator systems in use or under investigation in nuclear medicine are $^{82}Rb/^{82}Sr$, $^{62}Zn/^{62}Cu$, $^{122}Xe/^{122}I$, and $^{81}Rb/^{81}Kr^m$.

Measurement of Regional Metabolism. PET can be used to monitor, in vivo and regionally, the utilization of metabolic substrates labeled with positron-emitting radioisotopes. An example is the measurement of glucose utilization in the brain with ^{11}C-labeled glucose (R4), and ^{18}F-labeled 2-deoxy-D-glucose (R5). In this work it is assumed that the tracer is transported and metabolized in the same manner and rate as the compound being traced; that the metabolized tracer is retained within the area of interest during the measurement period; and that the amount of tracer not metabolized, that is, the free tracer in blood and extracellular fluid, is either negligible or accounted for at the time of the measurement. In one investigation of this kind using ^{11}C-glucose to study the brains of rhesus monkeys a quantitative emission tomogram was begun 4 min after injection and continued for 2 min. Repeated measurements are possible during the course of one experiment, if needed to study transient effects. Experiments of this kind have demonstrated that the approach is sufficiently general to be employed with a variety of available radiopharmaceuticals utilized by brain, heart, or other organs in humans. The use of PET with positron-emitting radiopharmaceuticals has also resulted in remarkable achievements in studying regional cerebral blood volume, regional cerebral blood flow, and tissue chemical composition.

E. ARTIFICIALLY PRODUCED ELEMENTS

More than 75 years ago the methods of chemistry conventional at that time had already reached a limit in the search for new and missing elements; discoveries since that time have depended on the introduction of new physical methods. Through studies of optical spectra the elements rubidium, cesium, indium, helium, and gallium were found. The first evidence for hafnium and rhenium came from X-ray spectra. Early investigations of the natural radioelements revealed the existence (often in extremely small amounts) of polonium (84), radon (86), radium (88), actinium (89), and protactinium (91). More recently francium (87) has been found in the same way. Through studies of nuclear reactions and artificially induced radioactivities, technetium (43), promethium (61), and astatine (85) have been identified, and elements 93 to 106 have been added to the periodic chart. The artificially produced elements discussed in the remainder of this section were first studied by tracer techniques, using other elements as nonisotopic carriers or using carrier-free chemistry.

Technetium, Astatine, and Promethium. The first missing elements to be synthesized by nuclear reactions were technetium (P3) and astatine (C7). No known stable isotopes exist for either element, or for any element with $84 \leq Z \leq 106$ for that matter. The chemical properties of the two elements are, of course, those that are expected from their positions in the periodic table: technetium lying between manganese and rhenium, and astatine being the heaviest member of the halogens. A summary of the chemical properties of these two elements has been given by E. Anders (A1).

The fission of uranium produces several radioactive isotopes of promethium. The study of fission products led to the first identification of promethium (M7) by concentration of tracer activities with the ion-exchange resin adsorption and elution technique. Weighable quantities of long-lived isotopes of technetium and promethium are available commercially (e.g., ^{99}Tc, ^{145}Pm, and ^{147}Pm); however, the longest-lived astatine isotope has a half life of only 8.3 h (^{210}At), requiring the production of astatine just prior to each use.

Transuranium Elements (S8, S9, S10). When Fermi and his group in Rome first exposed uranium to slow neutrons they observed a number of activities, and in the following few years many more active species were found to be produced, most of which were at that time assigned to transuranium elements. The assignments were made because the substances were transformed by successive β^- emissions, which led to higher Z values, and because they could be shown by chemical tests to be different from all the known elements in the neighborhood of uranium in the periodic chart. This situation was resolved in the discovery by Hahn and Strassmann (H3) that these activities could be identified with known elements much lighter than uranium and that therefore the neutrons produce fission of the uranium nuclei. Further investigation of the fission process and products led to the proof by E. M. McMillan and P. Abelson (M8) that one of the activities, the one with 2.3-d half life, could not be a product of fission and was actually the daugher of the 23-min β-particle-emitting ^{239}U, which resulted from ^{238}U$(n, \gamma)^{239}$U. Also, they devised a procedure for separating chemically the element 93 tracer from all known elements through an oxidation-reduction cycle, with bromate as the oxidizing agent in acid solution and with a rare-earth fluoride precipitate as carrier for the reduced state. They gave the name neptunium, symbol Np, to the new element, taking the name from Neptune, the planet next beyond Uranus in the solar system.

The isotopes ^{239}Np and ^{238}Np [produced by $(d, 2n)$ reaction on ^{238}U] decay by β emission to element 94, named plutonium after the planet Pluto. Plutonium was discovered by G. T. Seaborg, E. M. McMillan, J. W. Kennedy, and A. C. Wahl (S11); ^{239}Pu is distinguished for its practical usefulness in slow- and fast-neutron fission.

Other transuranium elements with atomic numbers up to 106 have been synthesized since that time through nuclear reactions of various types with lighter transuranium elements. The chemical properties of each newly discovered element were first investigated on a tracer level; however, most of the transuranium elements have since been produced in weighable quantities. The first synthesis (G5) of mendelevium ($_{101}$Mv) provides an example of the remarkable tracer-level chemical manipulations that were developed for the preparation and investigation of the transuranium elements.

Element 101 was first prepared by the α-particle bombardment of a target that contained approximately 10^9 atoms of $^{253}_{99}$Es (half life 20 d) covering an area of about 0.05 cm^2 on a gold foil. Those atoms of $^{253}_{99}$Es that reacted with α particles were ejected from the target and were caught on another gold foil adjacent to the target; the atoms that had been transmuted were thereby removed from the bulk of the target. The gold catcher foil was dissolved and the gold was removed from the transmutation products by adsorption on an anion-exchange column from 6*M* HCl. The transuranium elements in the solution were then separated from one another by elution with α-hydroxy isobutyrate through a cation-exchange column. The fraction eluted just before the one identified as containing $^{256}_{100}$Fm should contain any element 101 that was produced. This fraction was found to contain a spontaneous-fission activity that was ascribed to element 101 or to one of its decay products. The production of element 101 was demonstrated by the observation of a total of 17 spontaneous-fission events in several separate experiments. Thus the various steps in the chemical separation were performed on less than 100 atoms of element 101.

Purely physical methods of identification of transuranium elements have also been successful. For example, the atomic number assignment of nobelium (102) has been made (D3) with a modification of the method used by Moseley in 1913 to identify the *Z* of an element from its characteristic X rays. In this experiment samples of ^{255}No were prepared by the ^{249}Cf (^{12}C, $\alpha 2n$) reaction and Fm *K* X rays resulting from the internal conversion process were detected in coincidence with α particles emitted from the nobelium parent. This provided an unequivocal identification of the atomic number of the α-emitting ^{255}No parent.

Actinide Series. The transuranium elements (at least through californium) and uranium and thorium all have similar precipitation properties when in the same oxidation state; they differ principally in the ease of formation and in the existence of various oxidation states. All current evidence supports the prediction by Seaborg that a new rare-earth series begins with actinium (number 89), with the 5*f* electron orbitals being filled in subsequent elements. This is analogous to the lanthanide rare-earth series beginning with lanthanum (57), with the 4*f* orbitals filling in the next 14 elements. Some of the evidence for this actinide series may be seen in these facts: (1) actinium is chemically similar to lanthanum; (2) thorium is similar to cerium in the +4 state; (3) the ease of removal of more than three electrons decreases from uranium to curium; (4) the measured II-III oxidation potentials of the heavy actinides agree very well with predicted values calculated with the actinide hypothesis. There is additional evidence for the second rare-earth series from spectroscopic and crystal-structure data, from magnetic susceptibilities, and from ion-exchange elution sequences.

It does seem evident that this new series differs from the familiar rare-earth series in that the resemblance of successive elements is less than for the lanthanide series. The lanthanide earths are for the most part separable only by multiple fractionation processes, or better by adsorption and elution from ion-exchange resins. The elements from 89 to 95 are separable by oxidation-reduction processes, but the separation of 95 to 103 is best done with an ion-exchange column as indicated in chapter 8, section B. On the actinide hypothesis curium, by analogy to gadolinium, is expected to resist oxidation or reduction in the +3 state, because the $5f^7$ and $4f^7$ structures, with one electron in each of the seven f orbitals, are particularly stable. Actually no state of curium other than +3 has been observed in solution. Americium, by analogy to europium, is reducible to a +2 state. Berkelium, with the configuration $5f^8$, might be oxidized by powerful oxidizing agents from the ordinary Bk^{3+} to Bk^{4+}; the potential of this couple is about −1.6 V.

Some of the difficulties in work with substances like ^{242}Cm may be mentioned here, difficulties in addition to those naturally associated with work on the ultramicrochemical scale. The heavy short-lived α emitters are extremely dangerous as radioactive poisons, and amounts of the order of a few micrograms taken into the body may produce harmful effects. Also, the high level of α radiation in concentrated samples can be expected to have some effect on chemical reactions; a ^{242}Cm preparation glows in the dark. In fact, the rate of energy release is so great that if cooling effects are neglected it may be estimated that a $0.1M$ ^{242}Cm solution would begin to boil in about 15 seconds and reach dryness in about 2 minutes. Longer-lived curium isotopes such as ^{247}Cm ($t_{1/2} = 1.6 \times 10^7$ y) and ^{248}Cm ($t_{1/2} = 3.5 \times 10^5$ y) make it possible to minimize these difficulties.

REFERENCES

A1 E. Anders, "Technetium and Astatine Chemistry," *Ann. Rev. Nucl. Sci* **9**, 203 (1959).

B1 W. A. Bonner and C. J. Collins, "Molecular Rearrangements: X. Rearrangement During the Deamination of 1,2,2-Triphenylethylamine with nitrous acid," *J. Am. Chem. Soc.* **78**, 5587 (1956).

B2 J. Bigeleisen, M. W. Lee, and F. Mandel, "Equilibrium Isotope Effects," *Ann. Rev. Phys. Chem* **24**, 407 (1973).

B3 J. Bigeleisen and M. G. Mayer, "Calculation of Equilibrium Constants for Isotopic Exchange Reactions," *J. Chem. Phys.* **15**, 261 (1947).

B4 J. Bigeleisen, "The Relative Reaction Velocities of Isotopic Molecules," *J. Chem. Phys.* **17**, 675 (1949).

B5 H. Brooks, "A Volatile Product from Radium," *Nature* **70**, 270 (1904).

B6 G. L. Brownell, J. A. Correia, and R. G. Zamenhof, "Positron Instrumentation," in *Recent Advances in Nuclear Medicine*, Vol. 5 (J. H. Lawrence and T. F. Budinger, Eds.), Grune and Stratton, New York, 1978.

C1 M. J. Campbell, J. C. Sheppard, and B. F. Au, "Measurement of Hydroxyl Concentration in Boundary Layer Air by Monitoring CO Oxidation," *Geophys. Res. Lett.* **6**, 175 (1979).

C2 M.-S. Chan and A. C. Wahl, "Rate of Electron Exchange between Iron, Ruthenium, and Osmium Complexes Containing 1,10-Phenanthroline, 2,2′-Bipyridyl, or Their Derivatives from Nuclear Magnetic Resonance Studies," *J. Phys. Chem.* **82**, 2542 (1978).

C3 T. A. Cahill, "Proton Microprobes and Particle-Induced X-Ray Analytical Systems," *Ann. Rev. Nucl. Part. Sci.* **30**, 211 (1980).

C4 W.-K. Chu, J. W. Mayer, and M.-A. Nicolet, *Backscattering Spectrometry*, Academic, New York, 1978.

C5 T. A. Carlson, "Primary Processes in Hot Atom Chemistry," in *Chemical Effects of Nuclear Transformations in Inorganic Systems* (G. Harbottle and A. G. Maddock, Eds.), North-Holland, Amsterdam, 1979, pp. 11–37.

C6 J. C. Clark and P. D. Buckingham, *Short-Lived Radioactive Gases for Clinical Use*, Butterworth, London (1975).

C7 D. R. Corson, K. R. MacKenzie, and E. Segrè, "Possible Production of Radioactive Isotopes of Element 85," *Phys. Rev.* **57**, 459 (1940).

D1 N. Davidson, *Statistical Mechanics*, McGraw-Hill, New York, 1962, Chapter 9.

D2 R. Dams et al., "Nondestructive Neutron Activation Analysis of Air Pollution Particulates," *Anal. Chem.* **42**, 861 (1970).

D3 P. F. Dittner et al., "Identification of the Atomic Number of Nobelium by an X-Ray Technique," *Phys. Rev. Lett.* **26**, 1037 (1971).

*E1 E. A. Evans and M. Muramatsu, Eds., *Radiotracer Techniques and Applications*, Marcel Dekker, New York, 1977.

F1 G. N. Flynn and N. Sutin, "Kinetic Studies of Very Rapid Chemical Reactions in Solution," *Chem. Biochem. Appl. Lasers* **1**, 309 (1974).

F2 M. P. Falley et al., "Neutron Capture Prompt γ-Ray Activation Analysis for Multielement Determination in Complex Samples," *Anal. Chem.* **51**, 2209 (1979).

F3 H. A. Fischer and G. Werner, *Autoradiography*, Walter De Gruyter, Berlin, 1971.

G1 M. S. Germani et al., "Concentrations of Elements in the National Bureau of Standards' Bituminous and Subbituminous Coal Standard Reference Materials," *Anal. Chem.* **52**, 240 (1980).

G2 V. P. Guinn, "Applications of Nuclear Science in Crime Investigation," *Ann. Rev. Nucl. Sci.* **24**, 561 (1974).

G3 H. Geiger and E. Marsden, "Diffuse Reflection of the α Particle," *Proc. Roy. Soc.* (London) **82**, 495 (1909).

G4 P. P. Gaspar and M. J. Welch, "Inorganic Hot-Atom Chemistry in Gaseous and One-Component Liquid Systems," in *Chemical Effects of Nuclear Transformations in Inorganic Systems* (G. Harbottle and A. G. Maddock, Eds.), North-Holland, Amsterdam, 1979, pp. 75–101.

G5 A. Ghiorso et al., "New Element Mendelevium, Atomic Number 101," *Phys. Rev.* **98**, 1518 (1955).

H1 H. Hintenberger, "High Sensitivity Mass Spectroscopy in Nuclear Studies," *Ann. Rev. Nucl. Sci.* **12**, 435 (1962).

*H2 G. Harbottle and A. G. Maddock, Eds., *Chemical Effects of Nuclear Transformations in Inorganic Systems*, North-Holland, Amsterdam, 1979.

H3 O. Hahn and F. Strassmann, "Über den Nachweis und das Verhalten der bei der Bestrahlung des Urans mittels Neutronen entstehenden Erdalkalimetalle," *Naturwissenschaften* **27**, 11 (1939).

*I1 *Isotope Effects in Chemical Processes*, Advances in Chemistry Series, Vol. 89, American Chemical Society, Washington, DC, 1969.

I2 M. Inghram, "Stable Isotope Dilution as an Analytical Tool," *Ann. Rev. Nucl. Sci.* **4**, 81 (1954).

J1 F. J. Johnston, "Isotopic Exchange Processes," in *Radiotracer Techniques and Applications*, Vol. 1 (E. A. Evans and M. Muramatsu, Eds.), Marcel Dekker, New York, 1977.

J2 M. Jeevanandam and T. I. Taylor, "Preparation of 99.5% Nitrogen-15 by Chemical Exchange between Oxides of Nitrogen in a Solvent Carrier system," in *Isotope Effects in Chemical Processes*, Advances in Chemistry Series, Vol. 89, American Chemical Society, Washington, DC, 1969, pp. 119–147.

J3 J. R. Jones, "Reaction Kinetics—Mechanisms and Isotope Effects," in *Radiotracer Techniques and Applications*, Vol. 1 (E. A. Evans and M. Muramatsu, Eds.), Marcel Dekker, New York, 1977.

J4 T. B. Johansson et al., "Elemental Trace Analysis of Small Samples by Proton Induced X Ray Emission," *Anal. Chem.* **47**, 855 (1975).

K1 M. A. Komarynsky and A. C. Wahl, "Rates of Electron Exchange between Tetracyanoethylene (TCNE) and $TCNE^-$ and between Tetracyanoquinodimethide (TCNQ) and $TCNQ^-$ and the Rate of Heisenberg Spin Exchange between $TCNE^-$ Ions in Acetonitrile," *J. Phys. Chem.* **79**, 695 (1975).

L1 W. F. Libby, "Chemistry of Energetic Atoms Produced by Nuclear Reactions," *J. Am. Chem. Soc.* **69**, 2523 (1947).

M1 R. A. Marcus, "Chemical and Electrochemical Electron-Transfer Theory," *Ann. Rev. Phys. Chem.* **15**, 155 (1954).

M2 L. Melander, *Isotope Effects on Reaction Rates*, Ronald, New York, 1960.

M3 E. S. Macias et al., "Proton Induced γ-Ray Analysis of Atmospheric Aerosols for Carbon, Nitrogen, and Sulfur Composition," *Anal. Chem.* **50**, 1120 (1978).

M4 E. S. Macias and R. B. Husar, "A Review of Atmospheric Particulate Mass Measurement via the Beta Attenuation Technique," in *Fine Particles* (B. Y. H. Liu, Ed.), Academic, New York, 1976, pp. 535–564.

*M5 A. G. Maddock and R. Wolfgang, "The Chemical Effects of Nuclear Transformations," in *Nuclear Chemistry*, Vol. II (L. Yaffe, Ed.), Academic, New York, 1968, pp. 185–249.

M6 J. M. Miller, J. W. Gryder, and R. W. Dodson, "Reactions of Recoil Atoms in Liquids," *J. Chem. Phys.* **18**, 579 (1950).

M7 J. A. Marinsky, L. E. Glendenin, and C. D. Coryell, "The Chemical Identification of Radioisotopes of Neodymium and of Element 61," *J. Am. Chem. Soc.* **69**, 278 (1947).

M8 E. M. McMillan and P. Abelson, "Radioactive Element 93," *Phys. Rev.* **57**, 1185 (1940).

N1 J. W. Nelson, "Proton Induced Aerosol Analyses: Methods and Samplers," in *X-Ray Fluorescence Analysis of Environmental Samples* (T. G. Dzubay, Ed.), Ann Arbor Science, Ann Arbor, MI, 1977, pp. 19–34.

N2 R. D. Neumann and A. Gottschalk, "Diagnostic Techniques in Nuclear Medicine," *Ann. Rev. Nucl. Part. Sci.* **29**, 283 (1979).

P1 S. R. Petrocelli, J. W. Anderson, and J. M. Neff, "Radiochemical Tracers in Marine Biology," in *Radiotracer Techniques and Applications*, Vol. 2 (E. A. Evans and M. Muramatsu, Eds.), Marcel Dekker, New York, 1977, pp. 921–968.

*P2 M. Pinta, *Modern Methods for Trace Element Analysis*, Ann Arbor Science, Ann Arbor, MI, 1978.

P3 C. Perrier and E. Segrè, "Some Chemical Properties of Element 43," *J. Chem. Phys.* **5**, 715 (1937); **7**, 155 (1939).

R1 E. Rutherford, "The Scattering of α and β Particles by Matter and the Structure of the Atom," *Phil. Mag.* **21**, 669 (1911).

R2 C. D. Radcliffe et al., "Gold Analysis by Differential Absorption of γ-Rays," *Archaeometry* **22**, 1 (1980).

R3 M. E. Raichle, "Quantitative in vivo Autoradiography with Positron Emission Tomography," *Brain Res. Rev.* **1**, 47 (1979).

R4 M. E. Raichle et al., "Measurement of Regional Substrate Utilization Rates by Emission Tomography," *Science* **199**, 986 (1978).

R5 M. Reivich et al., "The [^{18}F]fluorodeoxyglucose Method for the Measurement of Local Cerebral Glucose Metabolism in Man," *Circulat. Res.* **44**, 127 (1979).

*S1 N. Sutin, "Electron Exchange Reactions," *Ann. Rev. Nucl. Sci.* **12**, 285 (1962).

S2 N. D. Stalnaker, J. C. Solenberger, and A. C. Wahl, "Electron-Transfer between Iron, Ruthenium, and Osmium Complexes Containing 2,2′-Bipyridyl, 1,10-Phenanthroline, or their Derivatives. Effects of Electrolytes on Rates," *J. Phys. Chem.* **81**, 601 (1977).

S3 N. Sutin, "Oxidation-Reduction in Coordination Compounds", in *Inorganic Biochemistry* (G. L. Eichorn, Ed.), Elsevier, Amsterdam, 1973.

S4 N. Sutin, J. K. Rowley, and R. W. Dodson, "Chloride Complexes of Iron (III) Ions and the Kinetics of the Chloride-Catalyzed Exchange Reactions between Iron (II) and Iron (III) in Light and Heavy Water," *J. Phys. Chem.* **65**, 1248 (1961).

S5 E. V. Sayre, "Activation Analysis Applications in Art and Archaeology," in *Advances in Activation Analysis*, Vol. 2 (J. M. Lenihan and S. J. Thomson, Eds.), Academic, New York, 1972, pp. 155–184.

S6 D. R. Sandstrom and F. W. Lyth, "Developments in Extended X-ray Absorption Fine Structure Applied to Chemical Systems," *Ann. Rev. Phys. Chem.* **30**, 215 (1979).

S7 L. Szilard and T. A. Chalmers, "Chemical Separation of the Radioactive Element from its Bombarded Isotope in the Fermi Effect," *Nature* **134**, 462 (1934).

S8 G. T. Seaborg, "Elements Beyond 100, Present Status and Future Prospects," *Ann. Rev. Nucl. Sci.* **18**, 53 (1968).

S9 R. J. Silva, "Trans-Curium Elements," in *Inorganic Chemistry Series*, Vol. 8, Part One—Radiochemistry (A. G. Maddock, Ed.), University Park, Baltimore, 1972, pp. 71–105.

S10 G. T. Seaborg, *Man-Made Transplutonium Elements*, Prentice-Hall, Englewood Cliffs, NJ, 1963.

S11 G. T. Seaborg et al., "Radioactive Element 94 from Deuterons on Uranium," *Phys. Rev.* **69**, 366 (1946); G. T. Seaborg, A. C. Wahl, and J. W. Kennedy, "Radioactive Element 94 from Deuterons on Uranium," *Phys. Rev.* **69**, 367 (1946). (These letters were received for publication on January 28, 1941 and March 7, 1941, respectively, but were voluntarily withheld from publication until the end of World War II.)

T1 A. Turkevich, E. Franzgrote, and J. Patterson, "Chemical Analysis of the Moon at the Surveyor V Landing Site," *Science* **158**, 635 (1967).

T2 J. Tölgyessy, T. Braun, and M. Kyrs, *Isotope Dilution Analysis*, Pergamon, Oxford, 1972.

T3 M. M. Ter-Pogossian et al., "A Positron-Emission Transaxial Tomograph for Nuclear Imaging (PETT)," *Radiology* **114**, 89 (1975).

T4 M. M. Ter-Pogossian et al., "A Multislice Positron Emission Computed Tomograph (PETT IV) Yielding Transverse and Longitudinal Images," *Radiology* **128**, 477 (1978).

T5 M. M. Ter-Pogossian et al., "Design Considerations for a Positron Emission Transverse Tomograph (PETT V) for Imaging of the Brain," *J. Comput. Assist. Tomography* **2**, 539 (1978).

U1 D. S. Urch, "Nuclear Recoil Chemistry in Gases and Liquids," *Radiochemistry* [Specialist Rep. Chem. Soc. (London)] **2**, 1 (1975).

W1 M. Wolfsberg, "Isotope Effects," *Ann. Rev. Phys. Chem.* **20**, 449 (1969).

*W2 A. C. Wahl and N. A. Bonner, Eds., *Radioactivity Applied to Chemistry*, Wiley, New York, 1951.

W3 R. E. Watson and M. L. Perlman, "Seeing with a New Light; Synchroton Radiation," *Science*, **199**, 1295 (1978).

W4 H. Winick and A. Bienenstock, "Synchrotron Radiation Research," *Ann. Rev. Nucl. Part. Sci*, **28**, 33 (1978).

W5 M. J. Welch, Ed., "Radiopharmaceuticals and Other Compounds Labelled with Short-Lived Radionuclides," *Int. J. Appl. Radiat. Isot.* **28**, 1–234 (1977).

W6 M. J. Welch and S. J. Wagner, "Preparation of Positron-Emitting Radiopharmaceuticals," in *Recent Advances in Nuclear Medicine*, Vol. 5 (J. H. Lawrence and T. F. Budinger, Eds.), Grune and Stratton, New York, 1978.

Y1 R. S. Yalow, "Radioimmunoassay: A Probe for the Fine Structure of Biologic Systems," *Science* **200**, 1236 (1978).

Z1 W. H. Zoller and G. E. Gordon, "Instrumental Neutron Activation Analysis of Atmospheric Pollutants Utilizing Ge(Li) γ-Ray Detectors," *Anal. Chem.* **42**, 257 (1970).

*Z2 J. F. Ziegler (Ed.), *New Uses of Ion Accelerators*, Plenum, New York, 1975.

Z3 J. F. Ziegler, "Material Analysis with Ion Beams," *Phys. Today* **29**, No. 11, 52 (1976).

EXERCISES

1. A mixture is to be assayed for penicillin. You add 10.0 mg of penicillin of specific activity 0.405 μCi mg^{-1} (possibly prepared by biosynthesis). From this mixture you are able to isolate only 0.35 mg of pure crystalline penicillin, and you determine its specific activity to be 0.035 μCi mg^{-1}. What was the penicillin content of the original sample? *Answer:* 106 mg.

2. The exchange between I_2^* and IO_3^- has been studied under these conditions: $(I_2) = 0.00050M$, $(HIO_3) = 0.00100M$, $(HClO_4) = 1.00M$, at 50°C. At specified times samples were taken and measured for total (I_2 plus IO_3^-) radioactivity by γ counting. These counting rates corrected to the time $t = 0$ on the basis of the 8.0-d half life of ^{131}I are given below in the column "Corrected Total Activity." The I_2 fractions were removed by extraction with CCl_4 and the residual (IO_3^-) radioactivity measured and corrected in the same way; these rates are in the column "Corrected IO_3^- Activity."

Time (h)	Corrected Total Activity (cpm)	Corrected IO_3^- Activity (cpm)
0.9	1680	9.9 ± 3.0
19.1	1672	107 ± 4.1
47.3	1620	246 ± 6.6
92.8	1653	438 ± 9.4
169.2	1683	610 ± 13
"∞"	1640	819 ± 9.8

Find the half-time $T_{1/2}$ for the exchange and the rate **R** of the exchange reaction. *Answer:* $T_{1/2} = 89$ h; $\mathbf{R} = 3.9 \times 10^{-6}$ mole liter^{-1} h^{-1}.

3. The experiment described in exercise 2 was repeated but with this difference, $(I_2) = 0.00100M$. The results are tabulated as before.

Time (h)	Corrected Total Activity (cpm)	Corrected IO_3^- Activity (cpm)
0.9	1717	7.6 ± 2.7
19.1	1483	70.1 ± 3.6
47.3	1548	178 ± 5.2
92.8	1612	305 ± 7.3
169.2	1587	413 ± 9.8
"∞"	1592	534 ± 5.7

For these conditions find $T_{1/2}$ and **R**. What is the apparent order of the exchange reaction with respect to I_2? *Note:* Do not be surprised if the order is not an integer; according to O. E. Myers and J. W. Kennedy, *J. Am. Chem. Soc.* **72**, 897 (1950), the order is consistent with this rate law for the exchange-producing reaction: $\mathbf{R} = k(I^-)(H^+)^3(IO_3^-)^2$.

4. Bromate ion, synthesized to contain 1.13 percent ^{18}O in its oxygen, was reacted with excess sulfurous acid in ordinary water. The product sulfate was isolated and its oxygen was found to contain 0.314 percent ^{18}O. What average number of the three oxygen atoms in BrO_3^- appeared in the SO_4^{2-}? *Answer:* 1.4.

5. Neglecting isotopic fractionation, what is the equilibrium constant for the reaction

$$C^{35}Cl_4 + C^{37}Cl_4 \rightarrow 2C^{35}Cl_2{}^{37}Cl_2?$$

Answer: 36.

6. Estimate the sensitivity of the neutron activation method for the detection of (a) terbium, (b) dysprosium, and (c) lead, with a thermal neutron flux of 1×10^{11} cm^{-2} s^{-1}. Assume irradiation times no longer than 1 d and required activity levels of 10 dpm. *Answer:* (a) 2×10^{-9} g.

7. Describe a method for detecting ^{235}U in process streams based on delayed neutron emission from fission products (see p. 166).

8. Sulfur is a major constituent of atmospheric particulate matter. Which of the following would be best for detecting microgram amounts of particulate sulfur in the presence of iron, calcium, and silicon: (a) instrumental neutron activation analysis, (b) PIXE, or (c) nuclear backscattering?

9. Explain why high-energy β emitters ($E_\beta > 1$ MeV) have mass absorption coefficients that depend on the composition of the absorber and thus are not suitable for use in a beta gauge.

10. Derive an expression for the thickness of gold x_{Au} in a coin of thickness x, determined by the critical absorption of 80- and 81-keV γ rays of ^{133}Ba as described on p. 435. *Hint:* Begin with an expression for the intensity ratio of

the two γ rays as measured through the coin in terms of the intensity ratio without absorber.

11. What is the recoil energy imparted to a ^{129}Te atom by the emission of a 74-keV conversion electron? (Use the relativistic expression for the electron energy.)
Answer: 0.34 eV.

12. Suggest easily prepared compounds for use in Szilard-Chalmers processes of (a) iron, (b) mercury, (c) technetium.

13. Predict the atomic structure and some of the chemical properties of (a) element 114, (b) element 118, and (c) element 124. If you were designing experiments to discover trace quantities of each of these elements either in nature of as nuclear-reaction products, what elements would you use as carriers? How might you then separate each of the new elements from its nonisotopic carrier?

14. What is the heaviest element that can be identified following successive neutron capture and beta decay during a long irradiation (~6 months) in a high-flux reactor? Assume you are starting with pure ^{238}U and separation can not begin until 1 month after irradiation.

15. Hospitals generally purchase 66-h ^{99}Mo weekly and separate 6-h ^{99}Tc^m chromatographically in 20 ml of 0.009*M* NaCl. If 1 Ci of ^{99}Mo is received on a Monday morning to be used all week, (a) how often should the technetium be separated if the maximum *concentration* of ^{99}Tc^m (mCi/ml of solution) were required? (b) How would you design your procedure if you wished to maximize the *specific* activity of ^{99}Tc^m (mCi/mg of technetium)?

Chapter 12

Nuclear Processes as Chemical Probes

Chemical Effects on Half Lives. Nuclear processes are largely unaffected by interactions with the chemical environment in which the nucleus exists. However, those few instances in which these processes *are* affected by their environment provide probes for chemical structure. Perhaps the most obvious phenomenon is the change in decay rate for electron capture (EC) and internal conversion because both processes directly involve orbital electrons. The EC decay probability is proportional to the electron density at the nucleus, and changes in chemical structure can thus result in changes in the decay constant. One of the few EC-decay nuclides for which such a change has been observed is 53-d ^{7}Be. Decay constant changes of the order of 0.1 percent have been found between BeF_2 and beryllium metal and between several other beryllium compounds (J1).

The probability for internal conversion may be relatively large for low-energy or highly forbidden transitions as discussed in chapter 3, section E. In these cases a change in chemical structure that affects the electron density at the nucleus may produce an observable change in the decay constant. The low-energy decays of the isomeric states $^{99}Tc^{m}$, $^{235}U^{m}$, $^{90}Nb^{m}$, and $^{125}Te^{m}$ have been investigated to determine the relationship between decay constant and chemical structure. A 0.3 percent difference in $^{99}Tc^{m}$ decay rate was observed between $KTcO_4$ and Tc_2S_7. Effects of similar magnitude in $^{235}U^{m}$ have been observed between metallic uranium and sources of uranium embedded in a carbon base. In one study the variation in half life of $^{235}U^{m}$ was found to be related to the free electron concentration of the host metal into which $^{235}U^{m}$ recoils were implanted (N1).

The experimental techniques required for the investigation of chemical effects on half lives in even these rather favorable cases are quite difficult, and further study of the environmental effects on these two decay processes is not expected to yield a technique widely applicable for chemical structure studies. Reviews of these measurements are given in D1, E1, and H1.

Other, more subtle but more easily observed effects of the chemical environment on nuclear processes have been investigated. It is these processes that we discuss in this chapter, mainly from the point of view of the chemical information they can give. The topics include the Mössbauer effect, annihilation of positrons, muon studies, angular correlation of cascade radiations, and photoelectron spectroscopy. Each of these techniques has developed into a separate field of study with applications in many disciplines. In this brief chapter we have restricted the discussion to a description of each phenomenon, a few examples of some of the applications, and some references to appropriate detailed reviews. Not discussed because of space limitations are a number of related processes such as conversion electron peak intensity changes, nuclear magnetic resonance, and Auger spectroscopy.

A. MÖSSBAUER EFFECT

Recoil Effects in γ-Ray Emission and Absorption. The most thoroughly investigated nuclear process that depends critically on the chemical environment is recoilless nuclear resonance absorption or scattering (M1). To understand this phenomenon consider the energy spectrum of γ rays emitted from a nucleus $^{A}_{Z}X$ that goes from an excited state to its ground state with a transition energy E_r. The energy E_γ of the γ ray that is emitted is different from E_r for three reasons:

1. The emitting nucleus must recoil with a momentum that is equal and opposite to the momentum of the emitted γ ray; the energy associated with this nuclear recoil must come from E_r. The recoil energy, as discussed in chapter 11, section C, is given by

$$R\ (\text{eV}) = \frac{537\ E_\gamma^2}{M}, \tag{12-1}$$

where M is the atomic mass of the emitting nuclide and where E_γ is expressed in MeV. The recoil effect will generally lower the energy of low-energy photons by 10^{-2}–10^{2} eV in transitions that will be of interest to us.

2. The emitting nucleus is part of some chemical system and is in thermal equilibrium with it. The thermal motion causes the γ ray to be emitted from a moving source, and there is the consequent Doppler shift in the frequency of the emitted photon and the corresponding energy shift:

$$\Delta E = \frac{v}{c} E_\gamma \cos \vartheta, \tag{12-2}$$

where v and c are the magnitudes of the velocities of the nucleus and of light, respectively, and ϑ is the angle between the directions of motion of the emitting nucleus and the emitted γ ray. Since $\cos \vartheta$ may vary between -1 and $+1$, the Doppler shift may either increase or decrease the energy of the emitted quantum and will cause the spectrum of emitted quanta to show a distribution about the value $E_r - R$. The width of the distribution is about 0.1 eV at room temperature. It is important to realize that in Doppler broadening conservation of energy implies that either some of the energy of the chemical system goes into the γ ray or that some of the energy E_r of the transition goes not only into the recoil energy mentioned in (1), but also into phonon excitation of the solid.

3. Even if there were no Doppler broadening, the Heisenberg uncertainty principle implies that the finite half life $t_{1/2}$ of the excited state would cause a distribution in the energies of the emitted quanta. The width of that distribution would be

$$\Gamma\,(\text{eV}) = \frac{4.55 \times 10^{-16}}{t_{1/2}\,(\text{s})}. \tag{12-3}$$

[The numerical constant in (12-3) is the product of ln 2 and $\hbar$ in eV s.] It is to be noted that this natural width, as given in (12-3), will exceed the room-temperature Doppler broadening only when the half life of the excited state is less than about 10^{-15} s, which for example, means a normal $E2$ transition greater than about 7 MeV (cf. chapter 3, table 3-4).

The foregoing three effects also apply to the inverse process, resonance absorption,[1] in which the nucleus ${}^{A}_{Z}\text{X}$ in its nuclear ground state absorbs a photon and goes to the nuclear excited state E_r above the ground state. The recoil effect in this process requires that the energy of the incident photon be larger than E_r by an amount R; the Doppler broadening and the natural width will again cause a distribution about this value. An example of the two processes, emission and absorption, is shown in figure 12-1, in which the Doppler broadening is assumed to be large compared to the natural width.

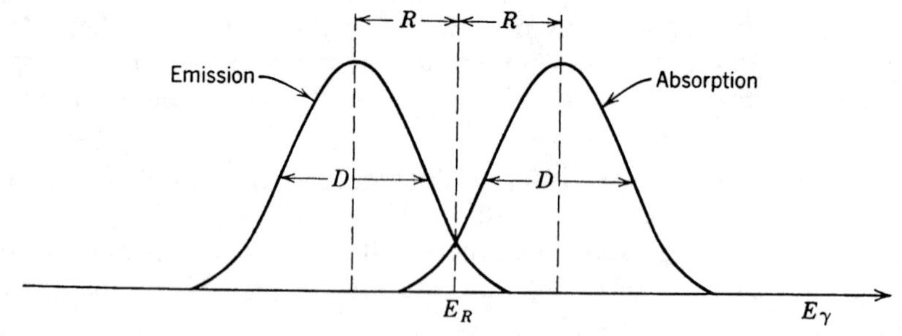

Fig. 12-1 The effect of recoil R, and Doppler broadening D on spectrum of emission and resonant absorption of γ rays.

[1] The inverse process may be investigated experimentally either by observing the change in intensity of the transmitted beam or by observing the photons that are re-emitted at some angle with respect to the incident beam. In the latter experiment it is the resonant scattering that is being studied.

Recoilless Resonance Absorption. From figure 12-1 it is seen that the recoil energy prevents the γ ray emitted by nucleus ${}^{A}_{Z}X$ in a direct transition from an excited state to ground state from being resonantly absorbed by the nucleus ${}^{A}_{Z}X$ in its ground state. This remark is not completely true because of the small overlap brought about by the Doppler effect. While investigating the temperature dependence of this overlap, R. L. Mössbauer (M2) discovered that under particular circumstances a fraction of the γ rays emitted from a solid source shows neither a measurable recoil energy loss nor any Doppler broadening; the energy of the γ ray was E_r and the line width approached the natural line width.

To understand this important observation one has to remember that, for an emitting (or absorbing) nucleus bound in a solid system, the total recoil energy R may be considered to be the sum of two parts, R_{kin} and R_{vib}. Here R_{kin} is the kinetic energy that corresponds to the linear momentum imparted to the whole crystallite in which the nucleus is bound and may be calculated from (12-1) with an effective value of M equal to the mass of the crystallite. That value of M is so large compared to the mass of a free atom that R_{kin} becomes vanishingly small. Hence $R \approx R_{\text{vib}}$ and practically the entire recoil energy goes into the lattice phonon system. Since the lattice is a quantized system (considered in a simple picture as composed of a large number of harmonic oscillators), there is a certain probability of finding the quantum-mechanical state of the phonon system unchanged after the emission (or absorption) process. It is only that fraction of the emission or absorption events that corresponds to this probability that is involved in recoilless resonance absorption. This fraction is relatively large (up to $\sim$0.9) for metallic systems, much smaller ($\lesssim$0.2) for metal-organic compounds, and increases with decreasing temperature. More generally, it is directly related to the stiffness of the crystal. This qualitative term "stiffness of the crystal" is roughly measured by the Debye temperature θ, because θ is proportional to the highest fundamental vibrational frequency in the crystal, which in turn depends on the restoring forces for the atomic vibrations. With this measure the condition for the recoilless transition is

$$R < k\theta, \tag{12-4}$$

where k is Boltzmann's constant. The fraction of the decays that occur without loss of recoil energy to the crystal increases with diminishing temperature and reaches a plateau value that depends on the relative magnitudes of the two quantities in (12-4).

When the nuclear recoil and the Doppler broadening are quenched the emission and the absorption spectra should completely overlap, as both should peak at E_r and both should be characterized by the natural width Γ. The condition given in (12-4) requires values of E_r less than about 100 keV (a condition that arises because θ generally lies between $\sim$100 and $\sim$1000 K); this, in turn, implies half lives greater than about 10^{-11} s or natural widths less than 10^{-5} eV. For a decay energy of 100 keV a Doppler shift of 10^{-5} eV, that is, equal to one line width, is brought about by a velocity of only $3\ \mathrm{cm\ s^{-1}}$ (12-2). Thus a relative velocity of only a few centimeters per second between a source and an absorber for which the Mössbauer effect holds will cause the resonance absorption to vanish. Energy shifts of the order of one part in 10^{10} may be relatively easily measured. The sensitivity of the method is great enough so that the increase in the energy of a photon that had fallen less than 30 m through the earth's gravitational field could be detected.

It would appear at this point that all effects of specific interactions between the nucleus and its environment have vanished because we now have an emission line with the energy and width determined by the characteristics of the nuclear states. Actually, it is only at this point that the specific interactions may be detected. They are of three kinds, and they cause small shifts or splittings in the energy E_r (of the order of 10^{-6} eV) but leave the width of the level unaffected and are thereby easily resolved.

Isomer Shift or Chemical Shift. The volume of a nucleus in an excited state is, in general, different from that in its ground state. As a result the probability that the orbital electrons will be found inside the nucleus will be different for the two states. This difference appears as a difference in the total binding energy of the electrons in the two states and contributes to the energy of transition:

$$E_r = \Delta E_{\mathrm{nuc}} + \Delta E_{\mathrm{elec}}, \tag{12-5}$$

where ΔE_{nuc} is the change in the nuclear binding energy and ΔE_{elec} is the change in the binding energy of the atomic electrons. Now if the emitting nucleus ($^{A}_{Z}\mathrm{X}$ in an excited state) and the absorbing nucleus ($^{A}_{Z}\mathrm{X}$ in the ground state) are in different chemical compounds, the distributions of the atomic electrons in space will be different, which will cause differences in ΔE_{elec} and therefore in E_r. This change in E_r is called the chemical shift. To a good approximation,

$$\Delta E_{\mathrm{elec}} = \tfrac{2}{5}\pi Z e^2(\overline{r^2_{\mathrm{ex}}} - \overline{r^2_{\mathrm{gr}}})[|\psi_e(0)|^2 - |\psi_a(0)|^2], \tag{12-6}$$

where $\overline{r^2_{\mathrm{ex}}}$ and $\overline{r^2_{\mathrm{gr}}}$ are the mean-square nuclear radii in the excited state and in the ground state, respectively, and $|\psi_e(0)|^2$ and $|\psi_a(0)|^2$ are the densities of electrons at the nucleus in the emitter and absorber, respectively.

Another contribution to the chemical shift, usually much smaller than the one just discussed, occurs because of the change in rest mass of the nucleus in the emission process, which in turn causes a change in the zero-point vibrational energy.

Magnetic Dipole Splitting. If either the emitting or the absorbing nucleus has a spin $\geq \frac{1}{2}$, it will also have a magnetic moment; in the presence of a magnetic field the energy of the nucleus will depend on its orientation with respect to that magnetic field. This means that, in general, another term must be added to (12-5):

$$E_r = \Delta E_{\text{nuc}} + \Delta E_{\text{elec}} + \Delta E_{\text{mag}}, \tag{12-7}$$

where ΔE_{mag} is the change in the magnetic energy of the nucleus in the transition and is determined by the change in the magnetic moment and in the projection of the spin along the magnetic field and also by the strength of the magnetic field at the nucleus. Since the projection of the spin along the magnetic field may take on the usual $(2I + 1)$ values, the effect of ΔE_{mag} is not merely to shift E_r but to split it into several components. The splitting usually corresponds to Doppler shifts caused by relative velocities of the order of $1\ \text{cm s}^{-1}$.

Electric Quadrupole Splitting. If either the emitting or the absorbing nucleus has a spin $I \geq 1$ and is in an *inhomogeneous* electric field, then, as in the magnetic interaction, E_r may be split into several lines because the interaction between the nuclear quadrupole moment and the inhomogeneous electric field causes the energy of the nucleus to depend on its orientation:

$$E_r = \Delta E_{\text{nuc}} + \Delta E_{\text{elec}} + \Delta E_{\text{mag}} + \Delta E_{\text{quad}}. \tag{12-8}$$

Again, the splitting corresponds to Doppler velocities around 1 cm/s.

Principle of Experimental Technique. The experimental observation of the three interactions may be achieved with the simple technique shown schematically in figure 12-2. The emitter contains nuclei ${}^{A}_{Z}\text{X}$ in an excited state and the absorber contains ${}^{A}_{Z}\text{X}$ in the ground state. The intensity of the γ-ray beam in the detector is then determined as a function of the relative velocity of the emitter and absorber. The output from the detector may be sent to a multichannel analyzer or, in more sophisticated systems, to a minicomputer or microprocessor. Many experimental arrangements have the capability of cooling the emitter and absorber to very low temperatures and applying an external magnetic field. Often great pains are taken to insure great stability in the relative velocities since minor fluctuations will

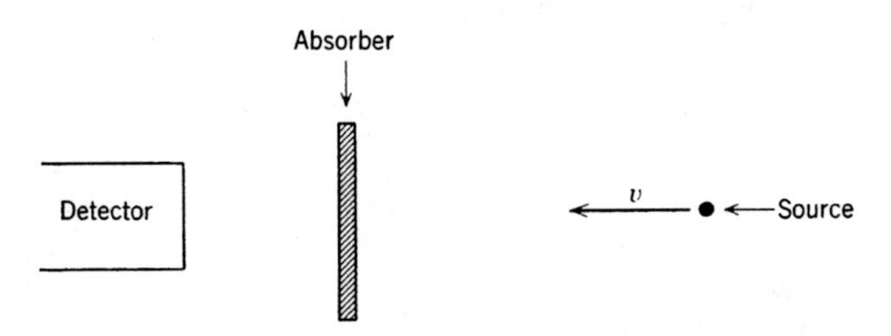

Fig. 12-2 Schematic representation of apparatus for observation of the Mössbauer effect. The source moves at velocity v with respect to the absorber.

lead to line broadening. The experimental results are more readily interpreted if the emitting nuclei are placed in a matrix in which there is no magnetic or quadrupolar splitting so that a single line is emitted; all of the splitting will then come from the absorption. Alternatively it may be interesting to study splittings from the emitting nucleus by using a single line absorber (Mössbauer emission spectroscopy).

Examples of Applications. The Mössbauer effect literature is now vast —nearly 3000 scientific articles appeared in a two-year period from 1977 to 1979. Several reviews of the various aspects of the field have appeared (e.g., M1, G1, G2, G3) and a monthly information journal is published as a bibliographic and data source (M3). By far the greatest number of applications of the Mössbauer effect deal with investigations using the 14.4-keV transition to the ground state of ^{57}Fe. The relevant nuclear information is shown in figure 12-3. The emitter is usually prepared by diffusing ^{57}Co into stainless steel in which there are evidently no electric or magnetic fields that will split the $\frac{3}{2}^-$ or $\frac{1}{2}^-$ states of ^{57}Fe produced in the EC of ^{57}Co. In the absorber, on the other hand, ^{57}Fe nuclei may be used to probe local electric and magnetic fields through the observed splitting patterns. The effects of a magnetic field and of an inhomogeneous electric field on the ground state ($\frac{1}{2}^-$) and the first excited state ($\frac{3}{2}^-$) of ^{57}Fe are shown schematically in figure 12-4. It is to be noted that the center of gravity of the four levels into which the $\frac{3}{2}^-$ state is split by the magnetic field and of the two levels in the inhomogeneous electric field does not coincide with the unsplit $\frac{3}{2}^-$ state in stainless steel; this is an example of the chemical shift.

Experimental observations made with an Fe_2O_3 absorber are shown in figure 12-5 (K1). The six lines expected from the magnetic splitting, as well as the lack of symmetry about zero velocity that is caused by the chemical

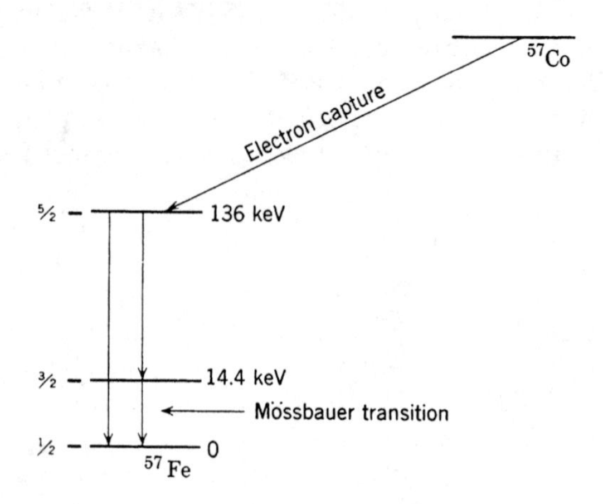

Fig. 12-3 Decay scheme of ^{57}Co showing the Mössbauer transition from the 14.4-keV level in ^{57}Fe.

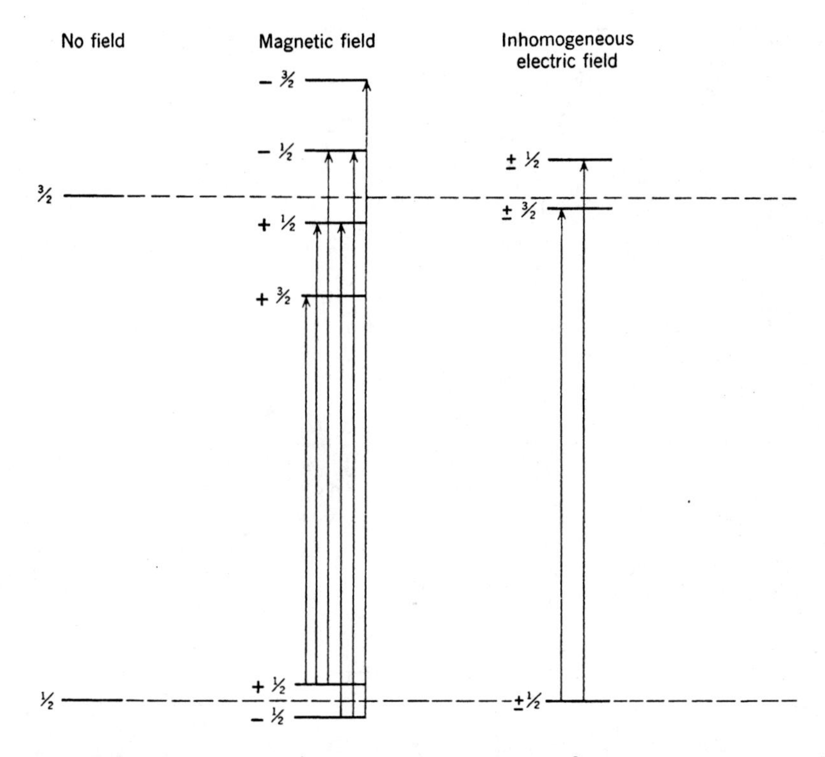

Fig. 12-4 The splitting of the $I = \frac{1}{2}$ ground state and of the $I = \frac{3}{2}$ first excited state of the ^{57}Fe nucleus in a magnetic field (as in Fe_2O_3) and in an inhomogeneous electric field (as in $FeSO_4 \cdot 7H_2O$). The lines between energy levels represent allowed transitions for γ-ray absorption; each level is characterized by its component of spin along the axis of symmetry of the field. The spacings of the energy levels are *not* drawn to scale.

shift, are seen. From this spectrum it was deduced that the splitting is 2.9×10^{-7} eV for the $\frac{1}{2}^-$ state and 1.6×10^{-7} eV for the $\frac{3}{2}^-$ state. From the known magnetic moments of the two states the splitting corresponds to a field of 5.2×10^5 oersteds at the iron nucleus in Fe_2O_3 caused by the magnetic moments of the unpaired electrons in the Fe^{3+} ions. The two lines that are expected from quadrupolar interactions are seen in figure 12-6, which shows the spectrum obtained with a $FeSO_4 \cdot 7H_2O$ absorber (D2). The splitting is a consequence of the inhomogeneous electric field produced by the sixth 3*d* electron accommodated in the lowest 3*d* orbital of the non-cubic $[Fe(H_2O)_6]^{2+}$ ion; the other five give a spherically symmetrical electric field at the nucleus. Again, the chemical shift causes a lack of symmetry about zero velocity. It is interesting to note that it was in fact the discovery of the quadrupole splitting in the Mössbauer spectrum of $FeSO_4 \cdot 7H_2O$ that led to the realization that the $[Fe(H_2O)_6]^{2+}$ ion is not cubic as had previously been thought. More precise X-ray diffraction measurements then confirmed that conclusion.

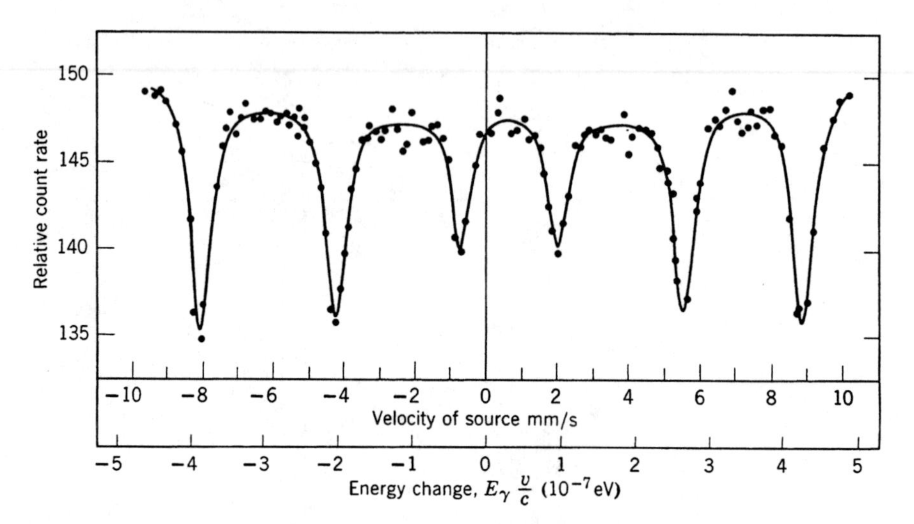

Fig. 12-5 The absorption in ^{57}Fe (bound in Fe_2O_3) of the 14.4-keV γ ray emitted in the decay of $^{57}Fe^m$ (bound in stainless steel) as a function of the relative source-absorber velocity. Positive velocity indicates motion of source toward absorber. (From reference K1.)

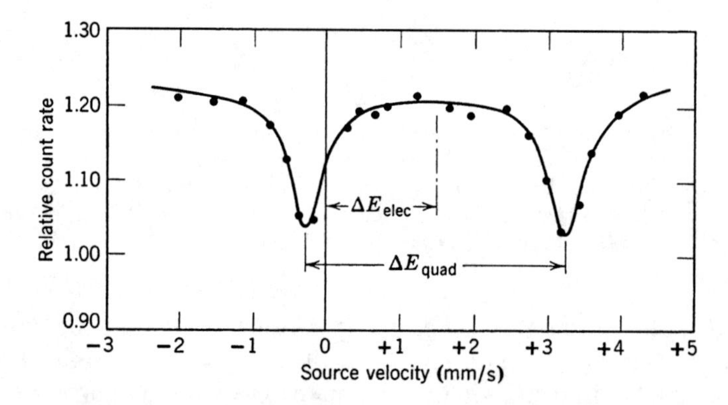

Fig. 12-6 Mössbauer spectrum of the Fe^{2+} ion in an absorber of $FeSO_4{\cdot}7H_2O$ at liquid-nitrogen temperature, taken with a room-temperature stainless-steel source. The pattern exhibits the chemical shift ΔE_{elec} and the electric quadrupole splitting ΔE_{quad} of the excited state of the ^{57}Fe nucleus. The velocity is positive for the source approaching the absorber. (From reference D2).

The Mössbauer effect, then, can serve as a sensitive probe for atomic wave functions, magnetic fields, and electric fields in the vicinity of the nuclei of atoms that are part of solid compounds. It can also yield information about the chemical consequences of any nuclear processes that immediately precede the recoilless γ ray.

Mössbauer-effect studies have found wide applicability in diverse fields such as studies of solid-state lattice dynamics of amorphous materials,

Table 12-1. Nuclides of Interest as Absorbers in Mössbauer experiments

^{40}K	^{133}Ba	^{175}Lu
^{57}Fe	$^{137,139}La$	$^{176-178,180}Hf$
^{61}Ni	^{141}Pr	^{181}Ta
^{67}Zn	^{145}Nd	$^{180,182-184,186}W$
^{73}Ge	$^{145,147}Pm$	^{187}Re
^{83}Kr	$^{147,149,151-154}Sm$	$^{186,188-190}Os$
^{99}Tc	$^{151,153}Eu$	$^{191,193}Ir$
$^{99,101}Ru$	$^{154-158}Gd$	^{195}Pt
$^{117,119}Sn$	^{159}Tb	^{197}Au
^{121}Sb	$^{160-162,164}Dy$	^{201}Hg
$^{125,127,129}Te$	^{165}Ho	^{232}Th
$^{127,129}I$	$^{164,166-168,170}Er$	^{231}Pa
$^{129,131}Xe$	^{169}Tm	$^{234,236,238}U$
^{133}Cs	$^{170-172,174,176}Yb$	^{237}Np
		$^{239,240}Pu$
		^{243}Am

catalysts and surfaces; studies of soil, coal, oil shale, lake and ocean sediments, polymeric materials; phase transitions; spin transitions; electron exchange; and biomedical problems (S1, C1). Much of this work uses the Mössbauer effect in situ to measure and/or differentiate between various chemical states of iron. For example, it has been found that EC of ^{57}Co in Co(III) acetylacetonate leads to about twice as much Fe(III) as Fe(II) and that the higher oxidation states, which are expected from the Auger effect, must be reduced in a time less than 10^{-7} s (W1).

Table 12-1 lists nuclei for which the Mössbauer effect either has been observed or is expected to occur. It is the nuclei with long-lived excited states ($t_{1/2} > 10^{-11}$ s) that emit low-energy γ rays (<200 keV) that are the likely candidates.

B. POSITRON ANNIHILATION

It has been observed that the characteristics of positron annihilation (a process described in chapter 3, p. 78) depend on the chemical composition of the medium in which the event occurs (A1, G4). In particular, the mean life of the positron as well as the fraction of the annihilations that emit three photons instead of two, quantities that we shall see are related to one another, can be grossly affected. To understand the source of the effects it is necessary to inquire further into the details of the annihilation process.

General Features. The first important point is that nearly all annihilation events involve positrons that have been slowed down to thermal energies. Evidence for this statement comes, for example, from the experimental observation that positrons have a mean life of at least 1.5×10^{-10} s in condensed media whereas the time required for slowing the positron down to thermal equilibrium is, at most 5×10^{-12} s.

The slowed positrons collide thermally with the electrons and will have a certain annihilation probability that depends on the relative orientations of the spins of the positron and of the electron: annihilation of the **singlet** state (opposed spins) is about 1115 times more probable per collision than annihilation of the **triplet** state (parallel spins). Further, because each photon must be emitted with at least $1\hbar$ of angular momentum and has only two states of polarization, annihilation in the singlet state gives two photons, whereas that in the triplet state gives three. It is this difference in multiplicity that causes the large difference between the probabilities of singlet and triplet annihilation. If the relative spins of the positron and electron were randomly oriented in each collision, the triplet state collisions would be three times as frequent as singlet state collisions and the ratio of two-photon to three-photon annihilation would be $1115/3 = 372$.

Positronium. It is also possible for the collision between a thermalized positron and an electron to result in the transient formation of a bound system, an atom of positronium (e^+e^-), before annihilation occurs. Positronium, which we denote by the symbol Ps, is a light isotope of atomic hydrogen with half the usual reduced mass and thus half the ionization potential (6.8 eV) and twice the Bohr radius. If the positronium were left undisturbed after being formed, there would be three times as many positronium atoms in the triplet (ortho) state as in the singlet (para) state, and thus there would be three times as many triple-photon as double-photon annihilations. Equivalently, since the mean life for annihilation of the para-Ps is about 10^{-10} s, whereas that for ortho-Ps is about 10^{-7} s, one quarter of the annihilation would occur with a mean life of about 10^{-10} s and three quarters with 10^{-7} s; in the absence of positronium formation the mean life would be 10^{-10} s with no long-lived component. It was through the detection of the longer mean life for annihilation (with the delayed-coincidence technique) that M. Deutsch (D3) proved the existence of positronium.

Probability of Positronium Formation. It is not to be inferred that all positrons form positronium along their way to annihilation—the formation probability depends on the stopping medium. Positronium is formed 36 percent of the time in water and 57 percent in benzene; the rest of the time the positrons are annihilated in collisions as free positrons (A1). This fact is qualitatively understood when it is realized that a positron must have an energy of at least $V - 6.8$ eV if it is to form positronium with an electron from a molecule of ionization potential V. If the energy is much in excess of the minimum value (say, about V), the collision is more likely to lead merely to ionization of the molecule without the formation of positronium. Since a previous collision is likely to leave the positron with an energy between zero and V, an upper limit to the probability of positronium formation may be estimated as $6.8/V$. Because the positronium formation rate may be inhibited by interactions between the positron and substrate molecules, chemical information can be obtained from the formation process.

Reactions of Positronium. It is from the *reactions* of positronium that the most useful chemical information may be obtained about the medium in which the annihilation occurs. After the positronium has been formed its lifetime can be quenched (shortened) by interaction with the substrate molecules. The para-Ps lifetime is so short that only interactions with the ortho-Ps need to be considered. Three main processes are responsible for the quenching of the ortho-Ps lifetime and result in rapid 2γ decay. These processes are (1) **electron pickoff**, in which a bound positron annihilates with an electron other than the one to which it is bound; (2) **spin conversion** from ortho to para states in the presence of external magnetic fields or paramagnetic species; and (3) **chemical reactions** with the substrate molecules such as oxidation, reduction, and compound formation. Much of the chemical information comes from the determination of the rate of conversion of the long-lived triplet ortho-Ps to either the short-lived singlet para-Ps or free positrons because this rate is related to the electron density in the medium.

Positron Annihilation Measurements. Positrons are normally obtained from a ^{22}Na source; ^{22}Na emits a prompt 1.28-MeV γ ray following positron decay. The positron annihilation lifetime is measured from the time delay between the 1.28-MeV γ ray and one of the 0.511-MeV annihilation γ rays as described in chapter 8, section D. The decay curve is resolved into two components: the shorter-lived component is due to the decay of free positrons, para-Ps, and "hot" ortho-Ps; the longer-lived component is attributed to the decay of thermalized ortho-Ps.

While lifetime measurements give information about the electron density in the substrate molecules, angular correlation and Doppler shift measurements give information on the momentum distribution of the electrons. In the angular correlation experiments the slight deviations from 180° emission of two 0.511-MeV γ rays due to the momentum of the electron are measured (since the positron is thermalized when it annihilates). The electron momentum also causes a Doppler shift in the energy of the annihilation γ rays that can be observed as a broadening of the 0.511-keV peak.

The directionality of γ radiation following positron annihilation in tissue has been used to make images of organs for biomedical studies, as discussed in chapter 11, section D.

Reactivity of Positronium (A1). A quantitative kinetic description of ortho-Ps conversion has been used to calculate the chemical rate constants for reaction between positronium and various substrates. The reaction of ortho-Ps (o-Ps) with a diamagnetic substrate (M) can be written

$$2\gamma \xleftarrow{\lambda_p} o\text{-Ps} + \text{M} \underset{k_2}{\overset{k_1}{\rightleftharpoons}} \text{PsM} \xrightarrow{\lambda_c} 2\gamma, \tag{12-9}$$

where k_1 and k_2 are the rate constants for formation and decomposition of the complex PsM, λ_c is the rate constant for positronium annihilation in the complex, and λ_p is the rate constant for positronium annihilation with the solvent. This assumes that, even if oxidation of positronium takes place, PsM complex formation is the rate-limiting step. A set of kinetic equations can be solved for these reactions, allowing the calculation of the time-dependent 2γ rate ($R_{2\gamma}$) from lifetime measurements:

$$R_{2\gamma} = A \exp(-\lambda_1 t) + B \exp(-\lambda_2 t). \tag{12-10}$$

Here A and B are scaling factors; λ_1 and λ_2 are the decay constants for short-and long-lived components of the experimental decay curve. λ_2 can be expressed in terms of the constants given in (12-9):

$$\lambda_2 = \lambda_p + \frac{k_1 \lambda_c}{k_2 + \lambda_c} [\text{M}]. \tag{12-11}$$

For dilute solutions λ_p approximately equals λ_2 (measured in pure solvent); thus the observed rate constant $[k_{\text{obs}} = k_1 \lambda_c/(k_2 + \lambda_c)$ in (12-11)] can be written by rearranging (12-11):

$$k_{\text{obs}} = \frac{\lambda_2 - \lambda_p}{[\text{M}]}. \tag{12-12}$$

Table 12-2 gives the reactivity of positronium with some of the large number of diamagnetic compounds that have been studied. It can be seen that compounds having high electron affinities show strong reactivity with positronium.

Metal Defect Studies (F1, H2). One area in which positron annihilation measurements are used to great advantage is in the study of imperfections in solids. Positronium is delocalized in a perfect lattice. On the other hand, the positronium atom or the free positron tends to localize at metal defects that are characterized as regions of low electron density away from the ion cores. Near these defects the repulsion between positrons and ion cores is decreased relative to the rest of the metal and thus the positrons are strongly attracted to these vacancies. The reduced electron density near the positron results in a measurable decrease in the annihilation rate and a

Table 12-2 Reactivity of Various Compounds with Thermal Positronium (M4)

Strong Interaction $k_{obs} > 10^8 M^{-1} s^{-1}$	Weak Interaction $k_{obs} < 10^8 M^{-1} s^{-1}$
Nitroaromatics	Simple aliphatic or aromatic hydrocarbons: alkanes, benzene, anthracene, etc.
Quinones	Aniline, phenol, haloalkanes
Maleic anhydride	Halobenzenes, aliphatic nitro compounds
Tetracyanoethylene	Phthalic anhydride, benzonitrile
Halogens	(Diamagnetic) inorganic ions in solution ($E_0 < -0.9$ eV)
Inorganic ions in solution ($E_0 > -0.9$ eV)[a]	
Organic ions in solution	

[a] E_0 is the standard redox potential.

change in the electron-positron momentum density as measured by the angular correlation or Doppler broadening. These changes are generally measured in a sample with defects relative to a well-annealed sample. Two successful applications of the technique are studies of the annealing behavior of radiation damage introduced in various metals and investigations of the temperature and pressure dependence of equilibrium vacancy concentrations in pure metals.

C. MUON CHEMISTRY (H3)

The fact that μ mesons interact with nuclei and electrons mainly through the electromagnetic field and the observation that their creation and decay occur through events in which parity is not conserved combine to make the μ meson useful as a chemical probe.

Muon Polarization. It was mentioned in chapter 10 that the muon (μ meson), a particle that does not interact strongly with nuclei, is formed in the decay of the pion (π meson), which does interact strongly with nuclei and is considered to be the quantum of the nuclear field. It is the pion, not the muon, that is created in high-energy nuclear collisions; the muon appears only as a secondary particle resulting from the decay of a charged pion:

$$\begin{aligned} \pi^+ &\to \mu^+ + \nu_\mu \\ \pi^- &\to \mu^- + \bar{\nu}_\mu \end{aligned} \qquad t_{1/2} = 2.6 \times 10^{-10}\ \text{s}.$$

The muon, in turn, is also unstable and decays into an electron, a neutrino, and an antineutrino:

$$\begin{aligned} \mu^+ &\rightarrow e^+ + \nu_e + \bar{\nu}_\mu \\ \mu^- &\rightarrow e^- + \bar{\nu}_e + \nu_\mu \end{aligned} \qquad t_{1/2} = 1.5 \times 10^{-6} \text{ s.}$$

The consequences of the nonconservation of parity in these two decay processes may be predicted from the discussion on p. 91 of chapter 3:

1. The muons produced in pion decay are polarized along their direction of motion: there are more muon spins (the spin of the muon is $\frac{1}{2}$) pointing in one direction than in the opposite direction.
2. In the β decay of the muon, as in the β decay of ^{60}Co, the angular distribution of the emitted electrons is *not* symmetric about a plane perpendicular to the spin of the muon.

Because of these two consequences, parity nonconservation can be detected in the following manner. Consider a beam of muons resulting from pion decay; item (1) above means that the muon spins will be aligned along their direction of motion, which we shall take as the z direction. If the muon beam is then stopped in an absorber and measurements are made on the angular distribution of decay electrons, (2) will require that the number of electrons observed at an angle ϑ with respect to the z axis be different from the number observed at the angle $\pi - \vartheta$. It was just this experiment, carried out by R. Garwin, L. Lederman, and M. Weinrich (G5), that demonstrated the violation of parity conservation in the two decay processes given above. Implicit in this experiment is the assumption that the muons are not depolarized while being stopped by the absorber and that they are not depolarized while they sit in the absorber waiting to decay. The magnetic moment of the muon will cause it to interact with any magnetic fields it may encounter in the stopping material, and the muon polarization can then be lost in the same manner as discussed for polarized nuclei in Section D. The occurrence of this depolarization was noticed in experiments in which the observed asymmetry in the β decay of the muon was found to decrease by about a factor of 2 when the stopping material was changed from graphite to a photographic emulsion (gelatin and silver bromide). It is the dependence of the depolarization on chemical environment that makes the muon useful as a chemical probe.

Muonium and Depolarization (B1, F2, P1). Positive muons lose energy in matter first by scattering with electrons in the medium (for muon energies down to about 3 keV), followed by *capture* of electrons. Thus muonium atoms (Mu), consisting of a positive muon and an electron, are formed as μ^+ mesons come to rest in nearly all materials. This seems reasonable because muonium has a higher ionization potential than most other atoms and molecules and thus can capture an electron even after it comes to rest (B1).

The μ^+ mass is only one-ninth that of the proton but the reduced mass, Bohr radius, and ionization potential of muonium are within ~0.5 percent of the hydrogen atom. As a result muonium is considered a light isotope of hydrogen and is expected to undergo similar reactions. Thermalized muonium can be of importance in various chemical studies such as kinetic isotope effects (the influence of mass differences on reaction rates) and structural isotope effects. Chemical information can be extracted from the degree of residual polarization of the muon as it is stopped in condensed matter—the depolarization is incomplete because of chemical reactions of the muonium atom.

The degree of muon spin depolarization is often measured with the transverse field muon spin rotation technique (μSR) in which a magnetic field B perpendicular to the initial muon spin direction is applied to the sample. The time delay between the stopping of a muon in the sample and the emission of a positron in the forward direction is measured with a timing circuit. A typical decay spectrum is shown in figure 12-7 in which the oscillations in the detection of the e^+ in the forward direction can be seen to be superimposed on the exponential decay of the muon ($t_{1/2} = 1.5 \times 10^{-6}$ s). These oscillations reflect the preferential emission of the e^+ along the spin direction of the muon, which is precessing at an angular frequency

$$\omega = \mu_\mu \frac{B}{\hbar}$$

where μ_μ is the muon magnetic moment (3.18 times the proton magnetic moment). The distribution of measured positrons has a time dependence that reflects the average polarization due to interactions with the medium. Information on the initial muon polarization and the spin relaxation time can be obtained from these data.

Information is normally extracted from a polarization fraction of a sample relative to the polarization in a standard material used to calibrate the experimental system. Three classes of muonic species can be differentiated by their μSR signals: bare muons or muons substituted in diamagnetic molecules (such as MuOH), muonium, and free radicals containing a muon and no other magnetic nucleus. The advent of a number of large meson factories has spurred interest in muon chemistry. In recent years there has been much activity in studies of thermal muonium kinetics in gases and liquids, and μ^+SR spectroscopy in metals, semiconductors, and insulators.

Depolarization of μ^-. The negative charge of the μ^- makes muonium formation impossible and, in general, diminishes depolarization of the μ^- by interactions with electrons in the stopping material. It does, however, cause the formation of another interesting chemical substance in which the μ^- is captured into a stable atomic or molecular orbital: the **μ-mesic atom** or molecule. Direct evidence for these new chemical species comes from

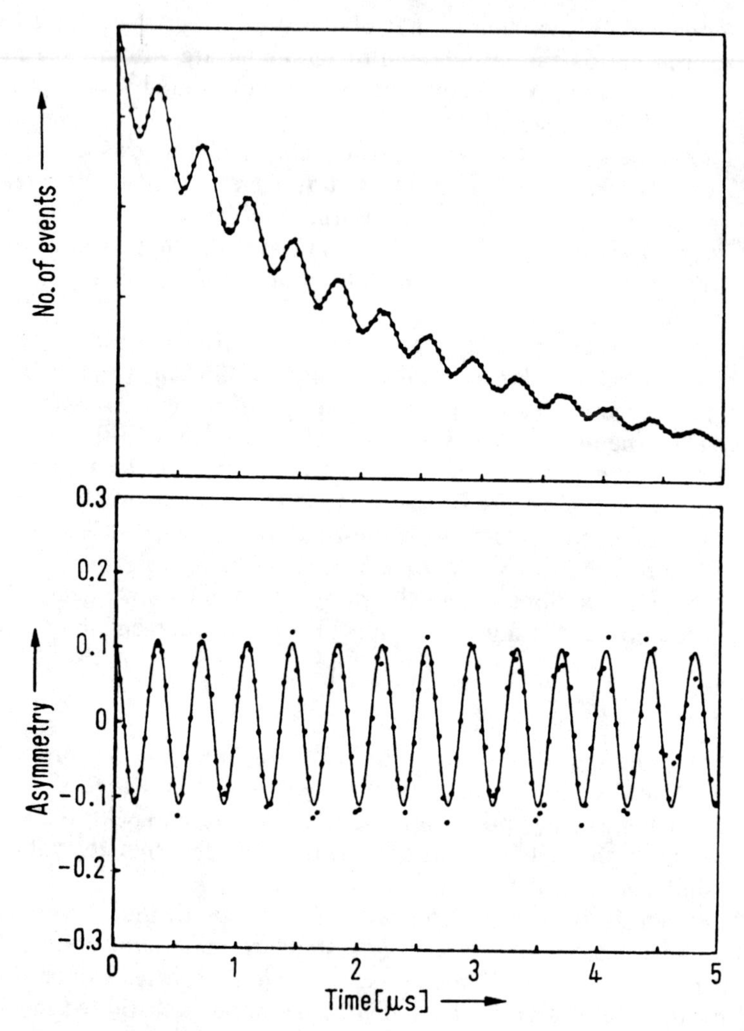

Fig. 12-7 The upper graph shows a μSR histogram obtained from water in a transverse field of 200 gauss. The lower graph shows the pure diamagnetic precession signal obtained from the histogram by subtraction of the exponential muon decay ($t_{1/2} = 1.5\ \mu s$) and of a small nondecaying background. (From reference P1.)

the characteristic X rays emitted as the μ^- cascades down to the $1s$ state.[2] Part of the μ-mesic X-ray spectrum of ^{206}Pb is shown in figure 12-8. The depolarization of the μ^- can occur not only through interaction with electrons during the capture process but also through an interaction between the μ^- meson in the atomic $1s$ state and the nuclear magnetic

[2] The fact that the muon is 207 times as heavy as the electron causes the energy of a transition to increase by a factor of 207 and the radius of an orbit to decrease by a factor of 207.

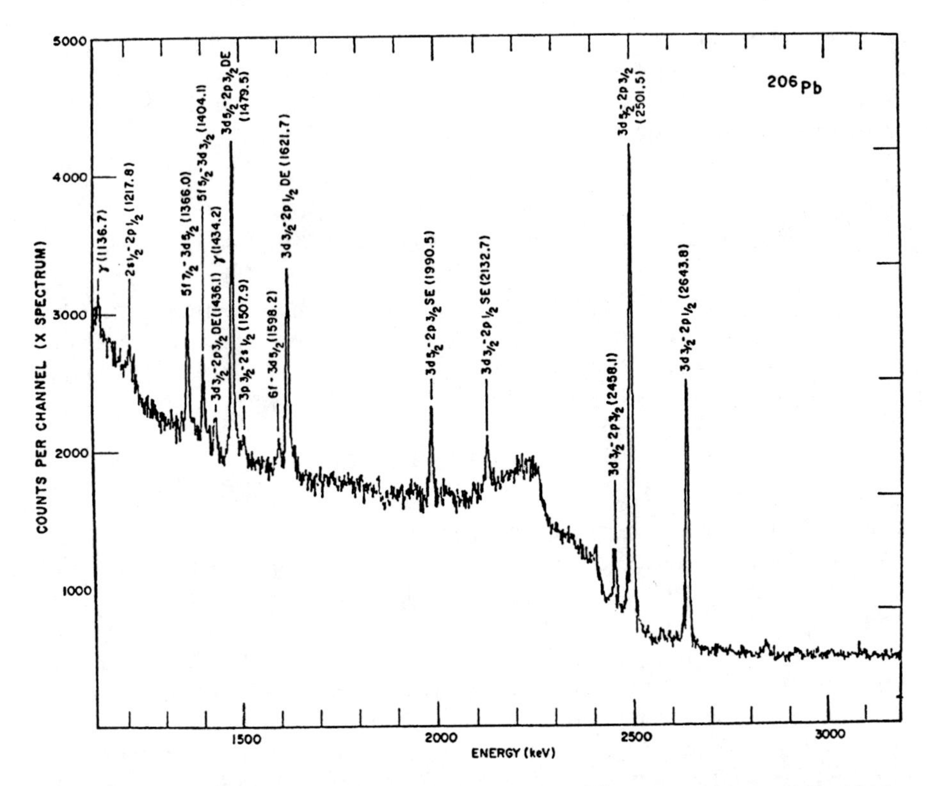

Fig. 12-8 A portion of the μ-mesic X-ray spectrum of ^{206}Pb. Peaks are labeled with the corresponding atomic transitions and energies in keV. Single and double escape peaks are denoted by SE and DE. [From H. L. Anderson et al., *Phys. Rev.* **187**, 1565 (1969).]

moment if one exists (L1). The small μ^- polarizations and high probability of μ^- capture relative to decay for high-Z targets make the μ^-SR technique difficult; thus much less work has been done on the chemical specificity in μ^- depolarization and its use as a chemical probe than has been done in μ^+ systems.

Mu-Mesic Atoms and Molecules. Aside from the depolarization process the capture of the μ^- into molecular and, finally, atomic orbitals may also provide interesting chemical information. For example, what determines the probabilities that a μ^- will be captured by the various kinds of atoms that may be present in the stopping material? A theoretical treatment leads to the prediction that the relative probabilities are proportional to the product of the atom fraction and the atomic number. This prediction, the so-called Z law, has been verified for alloys and halides but is violated for other substances such as oxides (P2). Evidently the valence electronic structure of the elements involved in the meson atomic capture process affects the capture probabilities. This effect is also seen in the nuclear absorption of π^- mesons in hydrogenous substances (P2). Further experimental and theoretical study is being actively pursued in order to evaluate mesic atoms and molecules as chemical probes.

D. PERTURBED ANGULAR CORRELATIONS OF GAMMA RAYS

It was pointed out in chapter 3 (p. 103) that when a nucleus emits two particles in sequence, such as two γ rays in succession, the angle between the two radiations is not, in general, expected to be randomly distributed. The angular correlation is measured by the correlation function

$$W(\theta) = \sum_{k_{\text{even}}} A_k P_k (\cos \theta), \tag{12-13}$$

where $W(\theta)$ is the number of events per unit solid angle that have an angle θ between the two radiations and $P_k (\cos \theta)$ is the Legendre polynomial of order k. Normally the summation is truncated at the fourth-order term. Anisotropy can be observed only if the intermediate nucleus preserves its component of angular momentum along the emission direction of the first particle: the angular momentum vector must precess about that direction. For any anisotropy to exist it is necessary for the intermediate nucleus to have a spin greater than or equal to 1 (see p. 104). It is from the perturbation of this γ-ray angular correlation that chemical information related to the electron distribution at the nucleus can be extracted.

The intermediate nucleus will, in general, have both a magnetic moment and an electric quadrupole moment. As discussed in section A, either or both of these nuclear moments will interact with the appropriate existing fields in the substance containing the intermediate nucleus and will cause the angular momentum vector to precess about the local field direction rather than the direction of emission of the first particle. This will diminish the anisotropy unless either the first particle is emitted along the local field direction or the mean life of the intermediate state is so short that the angular momentum vector will have moved only a small distance in its precession before the second particle is emitted. The latter possibility requires that, in order to measure a perturbation of the angular correlation, the condition

$$\tau > \frac{\hbar}{E} \tag{12-14}$$

must be fulfilled; here τ is the mean life of the intermediate state and E is the interaction energy of the nuclear moment with the local field (the splitting of the levels, discussed in section A). The critical lifetime for the intermediate state is of the order of 10^{-11} s.

When the angular correlation is perturbed the correlation function exhibits a time dependence given by

$$W(\theta, t) = A_k G_k(t) P_k(\cos\theta). \tag{12-15}$$

Here G_k is the **perturbation factor**, which carries all the information about interactions of the intermediate state with the extranuclear environment. At $t = 0$, $G_k(t) = 1$ and the unperturbed correlation is obtained. The complete theoretical treatment has been worked out for many years and is given in detail in F3 and S2.

Some Experimental Applications. The experimental arrangement for the determination of an angular correlation consists of the measurement of coincidence rates as a function of the angle defined by the two detectors and the source. To measure time-dependent perturbations it is also necessary to measure the coincidence rate as a function of time after the emission of the first transition. For some applications, particularly when the lifetime is very short and the perturbing field is very large, a time-integrated perturbed angular correlation is measured—that is, the weighted average correlation function between $t = 0$ and $t = \infty$:

$$\overline{W(\theta,\infty)} = \sum_k A_k \overline{G_k(\infty)}\, P_k(\cos\theta). \tag{12-16}$$

The most commonly used source in perturbed angular correlations (PAC) studies is 42.4-d ^{181}Hf. Other sources that have been widely used are 2.83-d ^{111}In, 48.6-min ^{111}Cd^m, and 7.45-d ^{111}Ag, all of which populate the 121-ns 247-keV state ($\frac{5}{2}^+$) in ^{111}Cd. Although many other nuclides are potentially useful for PAC studies, only a few sources are in general use, just as in Mössbauer spectroscopy. The experimental situation is not as well developed as might be expected given the maturity of theoretical description. This is largely due to the uncertainties resulting from the β decays preceding most γ-ray cascades of interest; the β decay disrupts the extranuclear environment in the daughter (S2). The system must recover faster than the lifetime of the *initial* state in order not to affect the otherwise well-defined angular correlation. Experimental problems notwithstanding, much progress has been made in understanding PACs in metals, insulators, solutions, and gases (S2, A2).

One area in which the PAC technique has been quite successful is the study of macromolecules in solution to determine rotational correlation times, conformation changes of macromolecules, binding constants, metal-protein interactions, and chemical structure of metal ion binding sites on proteins. The in vivo use of the method makes it possible to observe directly chemical changes in living organisms. For example, it has been shown with PAC that (^{111}Cd^m)$^{2+}$ binds at the active region (the Zn^{2+} position) of the enzyme carbonic anhydrase (M5). This provided good evidence that PAC reflects the effective molecular rotational correlation time at the metal binding site and thus ^{111}Cd^m can be used as a "rotational tracer" to label biological macromolecules.

E. PHOTOELECTRON SPECTROSCOPY

Photoelectron spectroscopy, widely used to probe the chemical environment, is not truly based on a nuclear process. However, it is an offspring of nuclear science in that it was developed by nuclear scientists and uses nuclear instrumentation. Also, it provides information complementary to that obtained from Mössbauer spectroscopy. For all these reasons we give a brief account of it here.

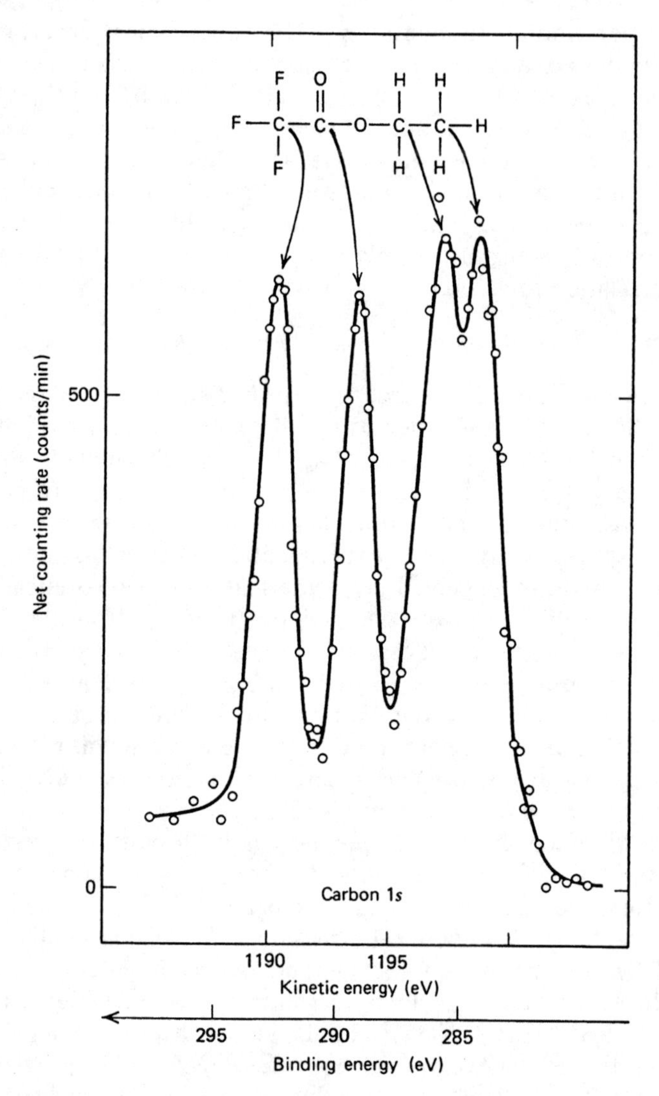

Fig. 12-9 Photoelectron spectrum of ethyl trifluoroacetate in the region corresponding to the ejection of a 1*s* electron from carbon. The four peaks correspond to the four structurally different carbon atoms shown in the formula above the spectrum. (From reference S3.)

Photoelectron spectroscopy employs the photon irradiation of materials to study the kinetic-energy distribution of emitted electrons. Chemical interactions are probed by measuring shifts in electron-binding energies. The technique was given its early impetus from the work of K. Siegbahn and co-workers in Uppsala; their development of high-resolution electron spectrometers led to the early demonstration of chemical shifts in photoelectron spectra (S3).

The method takes advantage of the photoelectric effect in which a photoelectron is ejected upon irradiation with photons[3]:

$$M + h\nu \rightarrow M^+ + e^-. \quad (12\text{-}17)$$

The binding energy E_B of the emitted electron can be calculated from the electron kinetic energy E_K, measured in an electron spectrometer:

$$E_B = E_K - h\nu, \quad (12\text{-}18)$$

where $h\nu$ is the known photon energy. Small recoil effects are neglected in (12-18), but can be taken into account. To study core electron-binding energy shifts a spectrometer with better than 0.1 percent resolution for measuring 1-keV electrons is necessary.

Core electron-binding energy shifts measured by photoelectron spectroscopy correlate well with chemical oxidation states and thus make the method a useful probe of chemical structure. An excellent illustration of observed chemical shifts is given in figure 12-9, which shows the carbon 1*s* spectrum from ethyl trifluoroacetate (S3). The four lines in the spectrum correspond to the four structurally different carbon atoms. The method has found wide use in studies of chemical bonding and in analysis (qualitative and quantitative). It also has found application in a number of other disciplines such as biology, geology, and environmental sciences. The field is thoroughly covered in C2.

REFERENCES

*A1 H. J. Ache, Ed., *Positronium and Muonium Chemistry*, Advances in Chemistry Series, Vol. 175, American Chemical Society, Washington, DC, 1979.

A2 J. P. Adloff, "Application to Chemistry of Electric Quadrupole Perturbation of γ-γ Angular Correlations," *Radiochim. Acta* **25**, 57 (1978).

B1 J. H. Brewer and K. M. Crowe, "Advances in Muon Spin Rotation," *Ann. Rev. Nucl. Sci.* **28**, 239 (1978).

C1 R. L. Cohen, Ed., *Applications of Mössbauer Spectroscopy*, Academic, New York, 1976.

*C2 T. A. Carlson, *Photoelectron and Auger Spectroscopy*, Plenum, New York, 1975.

D1 S. DeBenedetti, F. deS. Barros, and G. R. Hoy, "Chemical and Structural Effects on Nuclear Radiations," *Ann. Rev. Nucl. Sci.* **16**, 31 (1966).

[3] X rays are used for studies of inner-shell electrons while vacuum ultraviolet sources have been used to probe valence shells.

D2 S. DeBenedetti, G. Lang, and R. Ingalls, "Electric Quadrupole Splitting and the Nuclear Volume Effect in the Ions of ^{57}Fe," *Phys. Rev. Lett.* **6**, 60 (1961).

D3 M. Deutsch, "Evidence for the Formation of Positronium in Gases, "*Phys. Rev.* **82**, 455 (1951); "Three-Quantum Decay of Positronium," *Phys. Rev.* **83**, 866 (1951).

E1 G. T. Emery, "Perturbation of Nuclear Decay Rates," *Ann. Rev. Nucl. Sci.* **22**, 165 (1972).

F1 M. J. Fluss et al., "Temperature-Dependent Behavior of Positron Annihilation in Metals, " in *Positronium and Muonium Chemistry*, Advances in Chemistry Series, Vol. 175 (H. J. Ache, Ed.) Americal Chemical Society, Washington, DC, 1979, pp. 243–270.

F2 D. G. Fleming et al., "Muonium Chemistry—A Review," in *Positronium and Muonium Chemistry*, Advances in Chemistry Series, Vol. 175 (H. J. Ache, Ed.), American Chemical Society, Washington, DC, 1979, pp. 279–334.

*F3 H. Frauenfelder and R. M. Steffen, "Angular Distribution of Nuclear Radiation," in *Alpha, Beta and Gamma-Ray Spectroscopy*, Vol. II (K. Siegbahn, Ed.), North Holland, Amsterdam, 1965, pp. 997–1198.

G1 T. C. Gibb, *Principles of Mössbauer Spectroscopy*, Halsted, New York, 1976.

*G2 N. N. Greenwood and T. C. Gibb, *Mössbauer Spectroscopy*, Chapman and Hall, London, 1971.

*G3 P. Gütlich, "Mössbauer Spectroscopy in Chemistry," in *Topics in Applied Physics*, Vol. 5 (U. Gonser, Ed.) Springer, Berlin, 1975, pp. 53–96.

*G4 J. H. Green and J. Lee, *Positronium Chemistry*, Academic, New York, 1964.

G5 R. Garwin, L. Lederman, and M. Weinrich, "Observations of the Failure of Conservation of Parity and Charge Conjugation in Meson Decay: The Magnetic Moment of the Free Muon," *Phys. Rev.* **105**, 1415 (1957).

H1 H.-P. Hahn, H.-J. Born, and J. I. Kim, "Survey on the Rate Perturbation of Nuclear Decay," *Radiochim. Acta* **23**, 23 (1976).

H2 P. Hautojärvi (Ed.), *Positrons in Solids*, Topics in Current Physics Vol. 12, Springer, Berlin, 1979.

H3 V. H. Hughes and C. S. Wu (Eds.), *Muon Physics*, Volume III, Chemistry and Solids, Academic Press, New York, 1975.

J1 H. W. Johlige, D. C. Aumann, and H.-J. Born, "Determination of the Relative Electron Density at the Be Nucleus in Different Chemical Combinations, Measured as Changes in the Electron-Capture Half-Life of ^{7}Be," *Phys. Rev.* **C2**, 1616 (1970).

K1 O. C. Kistner and A. W. Sunyar, "Evidence for Quadrupole Interaction of $^{57}Fe^{m}$, and Influence of Chemical Binding on Nuclear Gamma-Ray Energy," *Phys. Rev. Lett.* **4**, 412 (1960).

L1 G. R. Lynch, J. Orear, and S. Rosendorf, "Muon Decay in Nuclear Emulsion at 25,000 Gauss," *Phys. Rev.* **118**, 284 (1960).

M1 R. L. Mössbauer, "Recoilless Nuclear Resonance Absorption," *Ann. Rev. Nucl. Sci.* **12**, 1 (1962).

M2 R. L. Mössbauer, "Kernresonanzfluoreszenz von Gammastrahlung in ^{191}Ir," *Z. Phys.* **151**, 124 (1958).

M3 *Mössbauer Effect Reference and Data Journal*, University of North Carolina, Asheville, NC, 1978–

M4 W. J. Madia, A. L. Nichols, and H. J. Ache, "Molecular Complex Formation Between Positrons and Organic Molecules in Solution," *J. Am. Chem. Soc.* **97**, 5041 (1975).

M5 C. F. Meares et al., "Study of Carbonic Anhydrase Using Perturbed Angular Correlations of Gamma Radiations," *Proc. Nat. Acad. Sci.* **64**, 1155 (1969).

N1 M. Neve de Mevergnies, "Perturbation of the $^{235}U^{m}$ Decay Rate by Implantation in Transition Metals," *Phys. Rev. Lett.* **29**, 1188 (1972).

*P1 P. W. Percival, "Muonium Chemistry," *Radiochim. Acta* **26**, 1 (1979).

P2 L. I. Ponomarev, "Molecular Structure Effects on Atomic and Nuclear Capture of Mesons," *Ann. Rev. Nucl. Sci.* **23**, 395 (1973).

S1 J. G. Stevens and L. H. Bowen, "Mössbauer Spectroscopy," *Anal. Chem.* **52**, 175R (1980).

*S2 D. A. Shirley and H. Haas, "Perturbed Angular Correlation of Gamma Rays," *Ann. Rev. Phys. Chem.* **23**, 385 (1972).

*S3 K. Siegbahn et al., *Electron Spectroscopy for Chemical Analysis—Atomic, Molecular and Solid State Structure Studied by Means of Electron Spectroscopy*, Almqvist and Wiksells, Uppsala, 1967.

W1 G. K. Wertheim, W. R. Kingston, and R. H. Herber, "Mössbauer Effect in Iron (III) Acetylacetonate and Chemical Consequences of *K* Capture in Cobalt (III) Acetylacetonate," *J. Chem. Phys.* **34**, 687 (1962).

EXERCISES

1. What relative velocity is required to compensate a shift of 10^{-6} eV between the energy levels of ^{119}Sn in an emitter and in an absorber (^{119}Sn emits a 24-keV γ ray)? *Answer:* 1.25 cm s^{-1}.
2. The half-life of the 24-keV state of ^{119}Sn is 1.9×10^{-8} s. What relative velocity will move the peak by an energy corresponding to the width of the level?
3. The Debye temperature of metallic tin is 195 K. Would a recoilless transition from the 24-keV state of ^{119}Sn be expected in metallic tin?
4. Estimate the energies of the chemical shift and of the quadrupole splitting in $FeSO_4 \cdot 7H_2O$ from the information in figure 12-6.
5. Estimate the magnitude of the interaction energy between the excited ^{60}Ni nucleus and its surroundings that would be required to destroy the angular correlation that exists between the two cascade gamma rays that follow the beta decay of ^{60}Co. The lifetime of the intermediate state in ^{60}Ni is 0.73 ps. *Answer:* $\sim 10^{-3}$ eV.
6. If the mean life of triplet positronium decreases from 1.8×10^{-9} s in pure water to 0.75×10^{-9} s in 0.10 *M* $HgCl_2$, what is the mean life for a reaction between the positronium and $HgCl_2$ in this solution? *Answer:* 1.3×10^{-9} s.
7. The 511-keV line from positron annihilation in copper is narrower in deformed copper than in annealed copper. Explain this observation in terms of the distribution of relative positron-electron momentum distributions in the two samples.
8. How is the μ^+SR technique similar to nuclear magnetic resonance?
9. Which has properties more similar to hydrogen: positronium or muonium? Explain in detail.
10. As muonium atoms slow down to thermal energies it is possible for hot-atom reactions to occur. Give examples of three different types of these reactions.

Chapter 13

Nuclear Processes in Geology and Astrophysics

A. GEO- AND COSMOCHRONOLOGY

Radioactivity and Geology. During the nineteenth century geologists amassed much information on the relative sequence of geologic ages, but a reliable absolute time scale was lacking. What was known of sedimentation rates on the one hand and of the thickness of sedimentary rocks on the other was used to derive estimates of the time required to lay down these rocks, but the estimates varied widely. A different approach taken by physicists, notably Lord Kelvin, was based on calculated cooling rates of the earth. As late as 1897 Lord Kelvin concluded from such arguments that the age of the earth must lie between 20 and 40 million years.

The discovery of radioactivity changed the picture in two profound ways. Firstly, the presence of radioactive substances in rocks provides a continuous heat source in the earth; this crucially modified the arguments based on cooling rates. Secondly, radioactive decay constitutes a "clock" provided by nature. As soon as this was realized dating by radioactive decay was turned into the first, and to this day the most important, objective method of geochronology. It radically changed our concept of the earth's history.

Radioactive Clocks. Rutherford was the first to suggest that α decay must lead to the buildup of helium in uranium minerals and that therefore the helium content of an uranium mineral could be used to determine the time elapsed since its solidification. By 1905 he had applied this method to the study of mineral ages. Shortly afterward the realization that lead is the end product of uranium decay (isotopes were not yet known!) led to a study of the lead content of uranium minerals, and by 1907 B. B. Boltwood, who pioneered these investigations, correctly concluded that geologic times had to be reckoned not in tens but in hundreds and thousands of millions of years (B1).

A variety of dating methods based on the naturally occurring radioactive nuclides has been developed. They all rest on the same basic principle: If P_0 atoms of a radioactive parent with decay constant λ and D_0 atoms of a stable daughter (or descendant) were present in a sample at time 0 (e.g., the time of solidification of a mineral), then the numbers of parent and daughter atoms at time t are

$$P_t = P_0 e^{-\lambda t} \tag{13-1}$$

and

$$D_t = D_0 + (P_0 - P_t) = D_0 + P_t(e^{\lambda t} - 1), \tag{13-2}$$

provided there has been no gain or loss of either parent or daughter other than by radioactive decay. This latter condition is often expressed by saying that the sample must be a **closed system**. Solving (13-2) for t, we obtain

$$t = \frac{1}{\lambda} \ln \frac{P_t + D_t - D_0}{P_t} = \frac{1}{\lambda} \ln \left(1 + \frac{D_t - D_0}{P_t}\right). \tag{13-3}$$

If D_0 can be assumed to be zero, as is, for example, usually the case for the helium content of a mineral at time of solidification, a measurement of the relative numbers of daughter and parent atoms (D_t/P_t) and knowledge of the decay constant are all that is necessary to determine t. If daughter atoms may have been present at $t = 0$ (i.e., $D_0 \neq 0$), additional information is needed, as we discuss in some examples.

In the following paragraphs some of the important methods of geochronology based on radioactivity are briefly discussed. For more thorough treatments the reader is referred to books and review articles such as F1, H1, and A1.

Uranium-Helium Method. The ^{238}U decay chain produces, in about 10^6 years, eight α particles per ^{238}U atom (see figure 1-1), and the resulting helium atoms will initially be trapped in the interior of the uranium-bearing rock in which they were produced. Under favorable circumstances—in impervious rocks of low uranium concentration and therefore low helium pressures—such radiogenic helium may have been retained throughout the lifetime of the rock; if so, it can now serve as an indicator of the fraction of uranium transformed since formation of the ore. The thorium in the rock also is a source of helium (six α particles per decay), and this must be taken into account. Very sensitive methods of assay for helium, uranium, and thorium are available and have permitted determinations on rocks with uranium and thorium contents below one part per million. Although the uranium-helium method was the first dating method based on radioactive decay, it is now known to be unreliable because of helium leakage over a geologic timescale. In general, U-He ages can therefore be considered as lower limits only. It was thus extremely puzzling that this method, when applied to iron meteorites, led to ages that were much longer than seemed compatible with other data. The puzzle was solved when it was realized that cosmic-ray-induced spallation reactions are an additional source of helium in meteorietes (see below). This interpretation was conclusively proved by mass spectrographic analyses of meteoritic helium that showed approximately 20 percent of it to be ^{3}He, whereas α decay can, of course, produce ^{4}He only.

Uranium-Thorium-Lead Methods. The lead isotopes ^{206}Pb, ^{207}Pb, and ^{208}Pb are the stable end products of the ^{238}U, ^{235}U, and ^{232}Th decay series (see figures 1-1, 2, 3). The amounts of these lead isotopes in uranium and thorium minerals can therefore be used as quantitative indicators of the time since the minerals have become closed systems. Modern techniques always involve isotopic analyses by mass spectrometry, and the amount of the nonradiogenic ^{204}Pb is then used to deduce the amounts of the other lead isotopes that were present at $t = 0$ [D_0 in (13-2) and (13-3)]. For the ^{238}U–^{206}Pb system (13-3) becomes

$$t = \frac{1}{\lambda_8} \ln\left(1 + \frac{^{206}\text{Pb} - {}^{206}\text{Pb}_0}{^{238}\text{U}}\right), \tag{13-3a}$$

where the subscripts t have been dropped, and λ_8 is the decay constant of ^{238}U. Dividing numerator and denominator of the right-hand fraction of (13-3a) by ^{204}Pb, we obtain

$$t = \frac{1}{\lambda_8} \ln\left(1 + \frac{^{206}\text{Pb}/^{204}\text{Pb} - (^{206}\text{Pb}/^{204}\text{Pb})_0}{^{238}\text{U}/^{204}\text{Pb}}\right). \tag{13-4}$$

Analogous equations can be written for the ^{235}U-^{207}Pb and ^{232}Th-^{208}Pb systems, denoting the respective decay constants as λ_5 and λ_2. The lead isotope ratios are measured mass-spectrometrically and the uranium, thorium, and lead concentrations are usually determined by isotope dilution (see chapter 11). For the lead isotope ratios at $t = 0$ we use either the ratios in common (i.e., nonradiogenic) modern lead or those (not very different ones) in pure lead ores (galenas) of roughly the same age as the minerals being studied.

Although gain or loss of lead, uranium, or thorium since mineral solidification is much less likely than loss of helium, such processes (e.g., through leaching) are by no means uncommon, and we therefore need objective criteria to establish whether a mineral has remained a closed system. Agreement among age determinations by different methods is considered a good indication of reliability. G. W. Wetherill (W1) introduced the following method for determining whether the ^{238}U–^{206}Pb and ^{235}U-^{207}Pb systems give concordant dates.[1] If the fraction

$$R_6 = \frac{^{206}\text{Pb}/^{204}\text{Pb} - (^{206}\text{Pb}/^{204}\text{Pb})_0}{^{238}\text{U}/^{204}\text{Pb}}$$

[1] The same scheme can be used to display concordance between dates based on the Th-^{208}Pb and one of the U-Pb systems; however, Th-^{208}Pb dates are generally less reliable than the U-Pb dates.

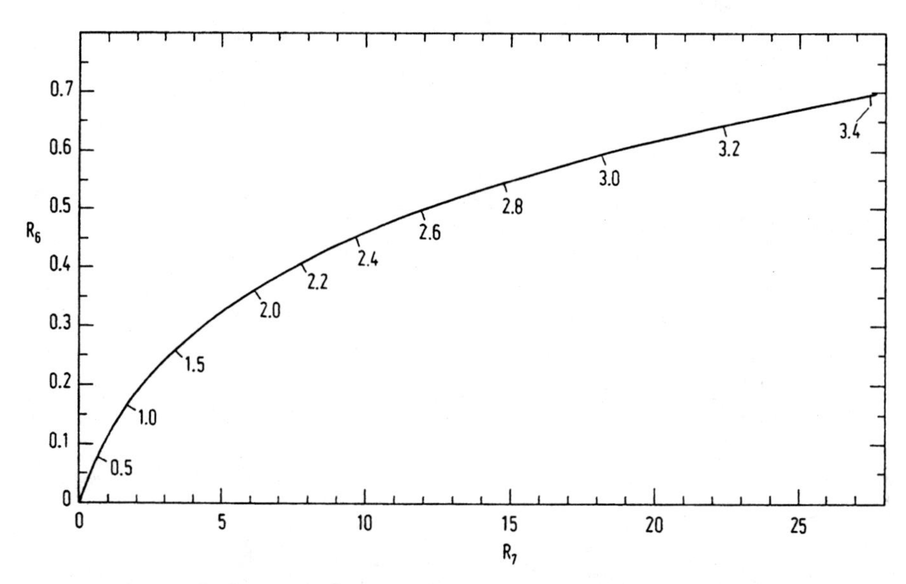

Fig. 13-1 Concordia diagram for U-Pb system. The abscissa is

$$R_7 = \frac{^{207}\text{Pb}/^{204}\text{Pb} - (^{207}\text{Pb}/^{204}\text{Pb})_0}{^{235}\text{U}/^{204}\text{Pb}},$$

the ordinate is

$$R_6 = \frac{^{206}\text{Pb}/^{204}\text{Pb} - (^{206}\text{Pb}/^{204}\text{Pb})_0}{^{238}\text{U}/^{204}\text{Pb}}.$$

in (13-4), (which equals $e^{\lambda_8 t} - 1$), and the corresponding fraction R_7 for the ^{235}U-^{207}Pb system are used as ordinate and abscissa, respectively, then the locus of all concordant dates is a universal curve that Wetherill termed **concordia**. This is shown in figure 13-1. Points will fall below concordia if loss of lead or gain of uranium has occurred, above concordia if uranium has been lost since $t = 0$. Lead addition may result in points above or below the curve, depending on the isotopic composition of the added lead.

Ratio of ^{206}Pb to ^{207}Pb. Uranium-bearing rocks that are free of nonradiogenic lead (as indicated by the absence of ^{204}Pb) may be dated straightforwardly by ^{206}Pb/^{207}Pb ratio measurements. The two equations of the form of (13-2) with $D_0 = 0$ are

$$^{206}\text{Pb} = {}^{238}\text{U}(e^{\lambda_8 t} - 1), \tag{13-5}$$

$$^{207}\text{Pb} = {}^{235}\text{U}(e^{\lambda_5 t} - 1). \tag{13-6}$$

Dividing (13-5) by (13-6) we get

$$\frac{^{206}\text{Pb}}{^{207}\text{Pb}} = \frac{^{238}\text{U}}{^{235}\text{U}} \frac{e^{\lambda_8 t} - 1}{e^{\lambda_5 t} - 1}, \tag{13-7}$$

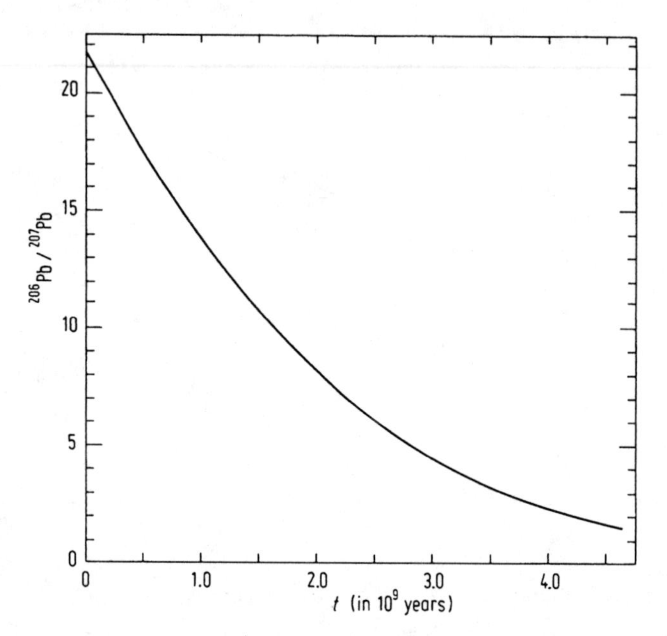

Fig. 13-2 Variation of radiogenic $^{206}Pb/^{207}Pb$ ratio with age.

where $(^{238}U/^{235}U) = 137.8$, the present ratio of the abundances. The curve of radiogenic $^{206}Pb/^{207}Pb$ ratios versus t represented by (13-7) is shown in figure 13-2. Since the lead abundance ratios can be determined very accurately, the $^{206}Pb/^{207}Pb$ method provides an excellent age scale for old ($\gg 10^8$ y) rocks containing retentive, uranium-bearing minerals. Because only isotopic abundance ratios need to be measured, the method is free from some of the experimental errors inherent in the chemical analyses required in the U-Pb and Th-Pb methods. It is also less sensitive to chemical or mechanical loss of either uranium or lead.

Rubidium-Strontium Method. Age determinations based on the decay of ^{87}Rb to ^{87}Sr are very widely used and are among the most reliable. Starting with an equation of the form of (13-2) and dividing both sides by the number of atoms of the nonradiogenic ^{86}Sr, we get

$$\frac{^{87}\mathrm{Sr}}{^{86}\mathrm{Sr}} = \left(\frac{^{87}\mathrm{Sr}}{^{86}\mathrm{Sr}}\right)_0 + \frac{^{87}\mathrm{Rb}}{^{86}\mathrm{Sr}}(e^{\lambda t} - 1). \tag{13-8}$$

The $^{87}Sr/^{86}Sr$ ratios are measured mass spectrometrically, the elemental rubidium and strontium concentrations (which are needed to evaluate $^{87}Rb/^{86}Sr)$[2] are usually determined by isotope dilution or X-ray fluorescence.

[2] Note that the isotopic composition of strontium in the particular specimen must be taken into account to convert the total strontium concentration into a ^{86}Sr concentration.

From these measurements and with the usual assumption that the system has been closed with respect to rubidium and strontium, (13-8) can be solved for t, provided we know or can otherwise obtain the initial isotope ratio $(^{87}Sr/^{86}Sr)_0$. One assumption that is usually made is that different igneous rocks crystallizing out of a magma over a time short compared to the ^{87}Rb half life of 5×10^{10} y had the same initial $^{87}Sr/^{86}Sr$ ratio, although they may have incorporated rather different relative amounts of strontium and rubidium. If we let $^{87}Sr/^{86}Sr = y$, $^{87}Rb/^{86}Sr = x$, and $(^{87}Sr/^{86}Sr)_0 = c$, we can rewrite (13-8) as

$$y = c + (e^{\lambda t} - 1)x,$$

which is the equation of a family of straight lines with common intercept c and slopes $(e^{\lambda t} - 1)$. Thus on a plot of $^{87}Sr/^{86}Sr$ versus $^{87}Rb/^{86}Sr$ all points for rock specimens of the same age t will lie on a straight line of slope $(e^{\lambda t} - 1)$, and such a line is called an **isochron.** From the slope of an isochron the age of a group of specimens is immediately obtained.

The most reliable ages come from isochrons determined for different mineral phases in a given rock formation that are cogenetic, that is, have crystallized from the same magma or lava. Figure 13-3 shows a Rb-Sr isochron for one of the oldest known terrestrial rock formations. Dates obtained for sedimentary rocks by the isochron method may often be the dates of the most recent metamorphism rather than that of the original deposition; additional geological evidence must be adduced to make such distinctions.

Potassium-Argon Dating. Since about 1950 the EC decay of ^{40}K to ^{40}Ar (10.7 percent of all ^{40}K decays[3]) has been extensively used as a basis

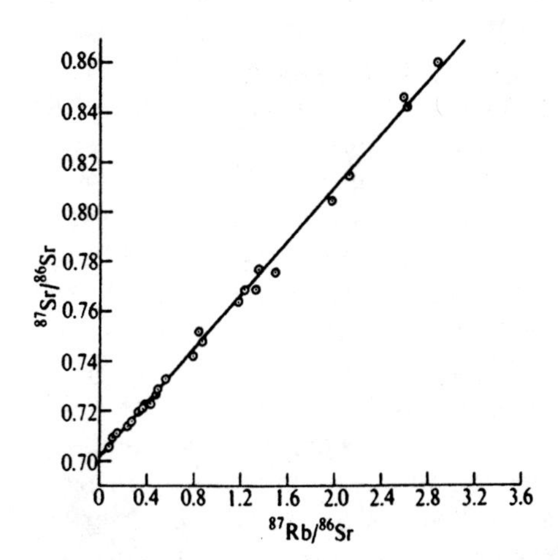

Fig. 13-3 $^{87}Rb/^{87}Sr$ isochron for a gneiss from Greenland, one of the oldest terrestrial rocks known. The slope of the isochron corresponds to $t = (3.74 \pm 0.10)\times 10^9$ y. [From S. Moorbath et al., *Nature, Phys. Sci.* **240**, 78 (1972).]

[3] The main decay branch of ^{40}K, β^- decay to ^{40}Ca, is of only very limited use for dating because calcium is a ubiquitous element and ^{40}Ca is the most abundant isotope in the normal mixture.

for dating terrestrial minerals and rocks, meteorites, and, more recently, lunar materials. Once again, the condition is that the system has been closed with respect to potassium and argon since the date of solidification or metamorphism to be determined. The ^{40}Ar is always assayed by isotope dilution, the ^{40}K by one of several methods such as atomic absorption spectroscopy, neutron activation, or isotope dilution. The specimen is usually vacuum-melted to liberate the argon, a known quantity of isotopically pure or highly enriched ^{38}Ar is added as a "spike," the argon is purified, and the isotope ratios $^{40}Ar/^{38}Ar$ and $^{38}Ar/^{36}Ar$ are determined mass-spectrometrically. The latter ratio is used to deduce the amount of atmospheric argon contamination present, the former to obtain the amount of ^{40}Ar, which must then be corrected for the atmospheric contribution. Nonradiogenic ^{40}Ar present in a sample is not necessarily all due to contamination with present-day atmospheric Ar ($^{40}Ar/^{36}Ar = 295.5$). It is therefore often advisable to construct K-Ar isochrons, analogous to the Rb-Sr isochrons discussed in the preceding paragraph, and based on the equation

$$\frac{^{40}Ar}{^{36}Ar} = \left(\frac{^{40}Ar}{^{36}Ar}\right)_0 + f\frac{^{40}K}{^{36}Ar}(e^{\lambda t} - 1), \tag{13-9}$$

where f is the fraction of ^{40}K decaying to ^{40}Ar. On a plot of $^{40}Ar/^{36}Ar$ versus $^{40}K/^{36}Ar$, an isochron will be a straight line with slope $f(e^{\lambda t} - 1)$ and intercept $(^{40}Ar/^{36}Ar)_0$. Initial $^{40}Ar/^{36}Ar$ ratios both significantly below and significantly above 295.5 have been found, which may reflect the fact that such initial argon occlusions may have originated partly from the early atmosphere (not necessarily of the same composition as the present one), partly from the deep interior of the earth (where it would be rich in ^{40}Ar due to ^{40}K decay). Essentially all atmospheric ^{40}Ar is believed to be of radiogenic origin, through outgassing of the mantle.

In a variant of K-Ar dating that has become particularly important for the study of lunar materials, fast-neutron irradiation is used to produce ^{39}Ar ($t_{1/2} = 269$ y) from ^{39}K by (n, p) reaction. The ^{39}Ar formed is then a measure of the amount of potassium present, and only isotopic $^{40}Ar/^{39}Ar$ ratios need be determined. Standards of known age are always irradiated and assayed along with the unknown, thus eliminating the need to know neutron fluxes and spectra during the irradiation. Corrections for argon isotopes produced by neutron interactions with other elements, notably calcium, must be made. The chief virtue of the $^{40}Ar/^{39}Ar$ method is that isotopic analysis of different argon fractions released during stepwise heating can give much useful additional information, since at the lowest temperatures gas tends to be released from those sites from which argon loss is most likely to have occurred (e.g., grain boundaries); thus low $^{40}Ar/^{39}Ar$ ratios will be observed from such low-temperature fractions (since ^{39}Ar represents potassium). If a plateau value in $^{40}Ar/^{39}Ar$ is reached at some temperature, it is a good indication that this value is a reliable measure of the geologic age (although the age in question may be a crystallization or a metamorphism age).

Other Dating Techniques. In principle the decay of any naturally occurring long-lived radioactive nuclide can be used for dating. Limited attempts to develop a dating method based on the β decay of ^{187}Re to ^{187}Os have been handicapped by the difficulty of measuring the ^{187}Re half life accurately, which in turn arises from the extremely low β energy ($E_{max} \approx$ 2.6 keV).

Of increasing importance is the ^{147}Sm-^{143}Nd dating method. Because of the great chemical similarity of samarium and neodymium, the Sm-Nd system is less subject to disturbance by metamorphism than any of the other parent-daughter pairs and is thus particularly reliable for establishing crystallization ages.

It is worth noting that the largest sources of uncertainty in many of the ages determined by radioactive dating, for example, those of chondritic meteorites, are the uncertainties in the half-life values, especially those of ^{87}Rb, ^{40}K, and ^{147}Sm.

Age of Earth and Solar System. The isotope methods of geochronology discussed in the preceding paragraphs have been applied to a wide variety of geological problems. Whenever possible more than one dating method is used to date a given material or group of materials to lend maximum credence to the results. Through isotope geology our knowledge of the detailed history of terrestrial, lunar, and meteoritic matter has been vastly expanded. Discussion of results in terms of geologic phenomena is, however, outside the scope of this book and the reader is referred to treatises on the subject such as F1 and H1. Suffice it to say that the oldest known ages of terrestrial rocks are in the vicinity of 3.7×10^9 y, whereas some lunar rocks appear to be as old as 4.6×10^9 y, and most meteorite ages are in the range of $(4.5\text{–}4.7) \times 10^9$ y (K1, W2).

The methods discussed can be further extended to give rather precise information on the age of the earth as a separate body. This involves the use of ^{207}Pb-^{206}Pb isochrons. Rewriting (13-4) and the corresponding equation for the ^{235}U-^{207}Pb system, we get

$$\frac{^{206}\text{Pb}}{^{204}\text{Pb}} - \left(\frac{^{206}\text{Pb}}{^{204}\text{Pb}}\right)_0 = \frac{^{238}\text{U}}{^{204}\text{Pb}}(e^{\lambda_8 t} - 1), \tag{13-10}$$

$$\frac{^{207}\text{Pb}}{^{204}\text{Pb}} - \left(\frac{^{207}\text{Pb}}{^{204}\text{Pb}}\right)_0 = \frac{^{235}\text{U}}{^{204}\text{Pb}}(e^{\lambda_5 t} - 1). \tag{13-11}$$

Dividing (13-11) by (13-10), we obtain the equation for a ^{207}Pb–^{206}Pb isochron:

$$\frac{(^{207}\text{Pb}/^{204}\text{Pb}) - (^{207}\text{Pb}/^{204}\text{Pb})_0}{(^{206}\text{Pb}/^{204}\text{Pb}) - (^{206}\text{Pb}/^{204}\text{Pb})_0} = \frac{^{235}\text{U}}{^{238}\text{U}} \frac{e^{\lambda_5 t} - 1}{e^{\lambda_8 t} - 1}. \tag{13-12}$$

C. Patterson (P1) was the first to show that, on a plot of ^{207}Pb/^{204}Pb versus ^{206}Pb/^{204}Pb, the data for several meteorites in which these ratios varied over a wide range fell on a straight line, that is, formed an isochron. The slope of this isochron [the right-hand side of (13-12)] corresponded to $t = 4.55 \pm 0.07) \times 10^9$ y. Subsequent investigations of a much larger number of meteorites of various types have fully confirmed their common age, and a value of $(4.57 \pm 0.03) \times 10^9$ y, in remarkable agreement with Patterson's early result, is generally accepted for that age (K1).

In P1 Patterson further pointed out that ordinary terrestrial lead—he used the isotopic ratios in lead of ocean sediment ($^{206}Pb/^{204}Pb = 19.0$; $^{207}Pb/^{204}Pb = 15.8$) as representing a well-mixed sample—falls on the same isochron as the meteorite lead samples. This result leads to the conclusion that the earth, as an isolated system, also originated about 4.57×10^9 y ago.

Since meteorites are believed to be fragments of several different asteroid-sized parent bodies (see p. 498), the close agreement among all their solidification ages is interpreted to mean that these parent bodies also *formed* at very nearly the same time (within tens of millions of years) and were small enough to cool rapidly. The coincidence between the time of formation of these small planets and that of the much larger planet earth is strong circumstantial evidence for the conclusion that the entire solar system originated at that time, approximately 4.6×10^9 y ago.[4]

The earliest history of the earth, from its formation about 4.6×10^9 y ago to the formation of the oldest known rocks about 0.9×10^9 y later, is a matter of conjecture. Initially most of the earth is presumed to have been molten as a result of accretional heating and of the heat from radioactive decay of uranium, thorium, and potassium. The time of formation of the earliest crust is not well known, and it is also not clear whether the crust keeps growing or is just being reworked. In any case we know that the earth remains a very active planet, as evidenced by volcanic and mountain-building activities.

Extinct Radionuclides as Chronometers. The dating methods discussed so far are all based on the decay of one of the long-lived radioactive nuclides now existing in nature. Additional information can be obtained from the existence of decay products of "extinct" radionuclides, that is, nuclides with half lives of $\ll 10^9$ y, which may have existed at the time of solar-system formation but have since decayed completely. The nuclides ^{129}I ($t_{1/2} = 1.6 \times 10^7$ y) and ^{244}Pu ($t_{1/2} = 8.1 \times 10^7$ y) are of particular interest in this connection (W2).

If the decay product of one of these extinct radionuclides is found in some system (meteorite, rock, and so on) and its origin from such decay can be established, we can conclude that the time interval between element

[4] Until recently it was generally believed that the material of the solar system was completely homogenized before formation of the sun and planets. The recent, exciting discovery of isotopic anomalies in several elements in some meteorite phases has disproved this idea and led to the conclusion that material from one or more recent nucleosynthetic events (presumably supernova explosions; see p. 514) was introduced into the solar nebula just before it began to contract and condense. See C1 for a review.

formation and isolation of the system investigated cannot have been very long compared to the half life of the extinct radionuclide. Abnormally high abundances of ^{129}Xe in some meteorites were first observed by J. H. Reynolds and attributed by him to decay of once present ^{129}I. Subsequently such excess ^{129}Xe has been found in many meteorites. Furthermore, it has been established that it is always intimately associated with the iodine in the same meteorites (P2). This is shown by neutron activation of the sample (which converts ^{127}I to ^{128}I and, by β^- decay, to ^{128}Xe, whereas neutron capture by ^{129}Xe leads to ^{130}Xe), subsequent stepwise heating, and determination of the $^{130}Xe/^{128}Xe$ ratios in the various fractions. The method is analogous to the $^{40}Ar/^{39}Ar$ method described on p. 488.

To deduce from the measured $^{129}Xe/^{127}I$ ratios the time interval between nucleosynthesis and the time meteorites could retain xenon, we must know or assume the $^{129}I/^{127}I$ ratio produced in nucleosynthesis. According to the accepted theories (see section B) this ratio is at most unity, which sets an upper limit of about 13 ^{129}I half lives, or 0.25×10^9 y, for the interval in question. A lower limit is about 0.1×10^9 y. The uncertainty arises largely because the calculated interval depends on whether nucleosynthesis took place over an extended period or in one or a few short episodes (K1). It is remarkable that the data indicate all the chondritic meteorites studied to have reached the xenon retention stage at about the same time (within 20×10^6 y).

The conclusions derived from the ^{129}Xe data are generally corroborated by measurements of effects attributable to the spontaneous fission of ^{244}Pu. Excess abundances of ^{132}Xe, ^{134}Xe, and ^{136}Xe in ratios corresponding to spontaneous-fission yields from ^{244}Pu are found in some meteorite inclusions. In some cases fossil fission fragment tracks are found in far greater abundance than can be accounted for by uranium fission and are attributed to ^{244}Pu fission.

Other Radioactive Nuclides in Nature. The so-called primary, long-lived radioactive nuclides that have survived since nucleogenesis, presumably without being replenished, are not the only radioactivities that occur in nature. In addition, there are first of all the secondary natural radioactivities, which are the short-lived descendants of the primary radionuclides ^{238}U, ^{235}U, and ^{232}Th. Although some of these, such as ^{230}Th ($t_{1/2} = 8.0 \times 10^4$ y) and ^{226}Ra ($t_{1/2} = 1.6 \times 10^3$ y), have been useful in geology, for example in dating ocean sediments (F1, L1), we do not discuss them. Also, we merely mention that man-made radioactivities that have been introduced into the earth's atmosphere since 1945, largely as a result of nuclear bomb explosions, have, in addition to their well-known deleterious effects, made some interesting scientific studies possible. Among the phenomena investigated with their aid are atmospheric mixing times between the Northern and Southern Hemispheres and residence times in the various vertical layers of the atmosphere (L1).

Of more widespread interest are the radionuclides (and stable products) formed by the interaction of cosmic rays with various objects. The following paragraphs are therefore devoted to a brief discussion of these phenomena and their application.

Cosmic Rays. Not long after the discovery of radioactivity it was known that detection instruments such as ionization chambers showed the presence of radiations even when not deliberately exposed to radioactive sources. This background effect was attributed to traces of naturally occurring radioactive substances such as uranium and thorium and their decay products, and this assumption is, of course, partly correct. Shielding the chambers with thick lead absorbers reduced but never eliminated the effect. It was reasoned that, if the radiations resulted from radioactive contamination of the ground, elevation of the ionization chambers to 1000 m or more should greatly reduce the effect because the ordinary β and γ rays would be strongly absorbed by the 100 g cm^{-2} or more of air. In the period 1910–1913 several daring experimenters carried instruments consisting of ionization chambers and electroscopes aloft in balloons to altitudes as great as 9000 m; surprisingly enough the background discharge rate at that height was about 12 times as great as on the ground. The conclusion from this and other experiments was that a radiation of extraordinary penetrating power fell continuously upon the earth from somewhere beyond. Since about 1925 this radiation has been known as cosmic radiation.

Until the advent of high-altitude rockets and artificial satellites cosmic-ray investigations were confined to the earth's surface or at least to relatively low altitudes, where the radiations observed are not the primary particles but rather are almost entirely secondary radiations produced by interactions of the primaries with the top of the atmosphere. These interactions seem to be largely very-high-energy nuclear reactions resulting in the emission of many mesons (mostly π mesons) and nucleons, many of which undergo further nuclear reactions. Mu mesons, produced in flight by π-meson decay, are found lower in the atmosphere, and they constitute most of the **hard component** of the cosmic radiation. With energies of many billions of electron volts these mesons are very penetrating and may be observed at great depths below water or ground.

The **soft component**, readily absorbed in a few inches of lead, consists largely of photons, electrons, and positrons. It accounts for about 10 percent of the cosmic-ray ionization at sea level but increases rapidly with altitude, constituting about 75 percent of all the rays at an altitude of 3 km. Most of the photons and electrons occur in showers of many particles of common origin. Initially, high-energy photons and electrons presumably result from meson decay; subsequent positron-electron pair creation by photons, and ionization and bremsstrahlung emission by electrons tend to produce large cascades of these rays. These cascades are observed in arrays of detectors in coincidence. Extensive air showers containing as

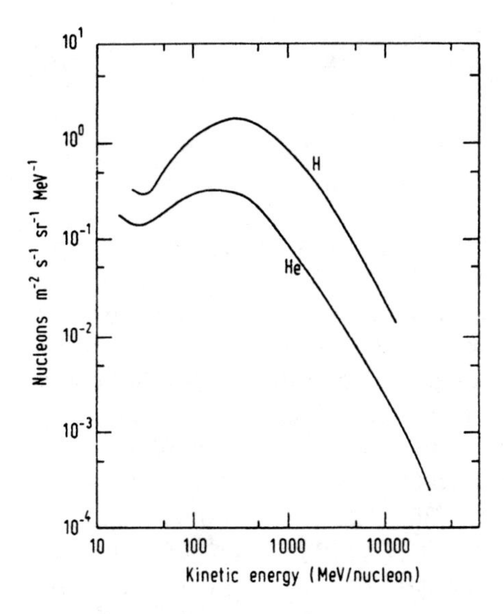

Fig. 13-4 Energy spectra of galactic cosmic-ray protons and α particles. (Data from reference M1).

many as 10^5 particles per square meter and extending over several square kilometers have been measured, corresponding to total energies up to 10^{20} eV per primary particle (M1).

The primary cosmic radiation arriving at the top of the earth's atmosphere (N1) consists predominantly of nuclei, mostly protons, with a small (~1 percent) admixture of electrons recently discovered. About 15 percent of the primaries are helium, and atomic numbers up to ~90 (and perhaps beyond) have been identified, with an abundance distribution roughly matching that in the universe (see section B). The energy spectra of protons and helium nuclei in the primary cosmic radiation are shown in figure 13-4. The energy spectra per nucleon have the same shape for heavier nuclei also, at least above ~500 MeV per nucleon.

The origin of cosmic rays is still a matter of debate. Most of them probably originate in our galaxy, specifically in supernova explosions, although the highest-energy components may well be of extragalactic origin, since galactic magnetic fields appear to be insufficient to contain particles with energies $\gtrsim 10^{17}$ eV amu^{-1} (M1). The sun contributes significantly to the flux of low-energy ($\lesssim 1$ GeV amu^{-1}) cosmic rays arriving at the earth; the emission of these particles is associated with solar flares. Another important effect of the sun is the decrease in galactic cosmic-ray flux reaching us during intense sun-spot activity, because the low-energy particles in the galactic cosmic rays are deflected by the sun's enhanced magnetic field (N2).

Radionuclides from Cosmic Rays. The impact of the primary and very-high-energy secondary cosmic rays, mostly near the top of the atmosphere, produces violent nuclear reactions in which many neutrons, protons, α particles, and other fragments are emitted. These secondary particles in turn react with nuclei in the lower atmosphere (mostly nitrogen, oxygen, and argon nuclei), and many radioactive products resulting from these interactions have been observed, including ^{3}H, ^{7}Be, ^{26}Al, ^{32}Si, and ^{39}Ar.

Most of the neutrons produced by the cosmic rays are slowed to thermal energies and, by (n, p) reaction with ^{14}N, produce ^{14}C, the 5730-y β^- emitter. From cosmic-ray data the production rate averaged over the whole atmosphere is calculated to be approximately $2\ s^{-1}$ per cm^2 of the earth's surface. The lifetime of ^{14}C is long enough for the radioisotope to become thoroughly mixed with all the carbon in the so-called exchangeable reservoirs (D1): atmospheric CO_2 (1.4 percent of the exchangeable carbon), ocean (95 percent, mostly as dissolved bicarbonate), terrestrial biosphere (1.1 percent), and humus (2.1 percent). The total carbon content of these reservoirs is estimated as 9.66 g per cm^2 of the earth's surface. Therefore the specific activity of ^{14}C in all of this carbon[5] is expected to be $2.4 \times 60/9.66 = 14.9$ dis $min^{-1}\ g^{-1}$. This is a readily measurable activity level.

Radiocarbon Dating. The discovery that all carbon in the world's living cycle is kept uniformly radioactive through the production of ^{14}C by cosmic rays led W. F. Libby to propose and pioneer the ^{14}C dating method that has become such a powerful and widely used technique for determining ages of carbon-containing specimens (L2). The underlying assumption is that cosmic-ray intensity has been constant (apart from short-term fluctuations such as those associated with solar activity) over many thousands of years. Then the specific activity of ^{14}C in the exchangeable reservoir has also been constant, and the time elapsed since a specimen was removed from the exchange reservoir can be determined from its $^{14}C/C$ ratio. The ^{14}C method has been used fairly routinely in many laboratories for dating a wide variety of samples. It has become an essential tool for archaeologists and geologists. The carbon in the samples is usually converted to CO_2, purified, and counted in the gas phase, either as CO_2 or after further conversion to CH_4 or C_2H_2. Typical ^{14}C counters may have volumes of several liters and be operated at pressures of several atmospheres. The range of the method is, of course, limited by the lower limit of detectability of ^{14}C. With radioactivity measurement this is at about 0.1 dis $min^{-1}\ g^{-1}$ of carbon, which corresponds to an age of about 40,000 years. More recently, possibilities for extending the method to perhaps twice that age have been opened up through development of techniques for direct detection of ^{14}C atoms. One approach is selective laser excitation and subsequent ionization

[5] The specific activities in the different reservoirs actually vary somewhat (D1) because the mixing times are not infinitely fast (see below) and there are some isotope effects. In the terrestrial biosphere the value is 13.6 dis $min^{-1}\ g^{-1}$ of carbon.

of ^{14}C-containing molecules; another is acceleration of ^{14}C ions in a cyclotron or tandem Van de Graaff and counting of the accelerated ions (M2, L3).

The basic assumption of the ^{14}C method that the cosmic-ray flux has been constant turns out to be only approximately correct. By dendrochronology (dating through tree-ring counting) and ^{14}C measurements in tree rings it has been possible to calibrate the ^{14}C method back to about 7500 years ago and to establish that there has been a long-term sinusoidal variation in ^{14}C production rate, with an amplitude of about 10 percent and a period of about 10,000 years, as well as many smaller short-term fluctuations. This is shown in figure 13-5. The long-term change is well correlated with changes in the earth's dipole moment as determined from paleomagnetic data. The shorter-term fluctuations are believed to be associated with changes in solar activity (S1). For the last three centuries this correlation is well documented through observations of sun spots: whenever sun spot activity was high (enhanced magnetic activity), ^{14}C production was reduced, presumably because more low-energy cosmic rays were deflected away from the earth. The peak in ^{14}C production rate around the year 1700 coincided with a striking, almost complete absence of sun spots from about 1649 to 1715 (the so-called Maunder minimum).

In addition to its applications for dating archaeological objects and recent geological events, the ^{14}C method has yielded other interesting results. Some years ago a puzzling problem was the apparent absence of isotope effects evidenced by the virtually identical $^{14}C/^{12}C$ ratios found in sea shell carbonates and in wood. The $^{13}C/^{12}C$ ratio in sea shells is about 1.025 times that in wood, presumably as a result of the isotope effect in the exchange equilibrium between CO_2 and HCO_3^-. The $^{14}C/^{12}C$ ratios are therefore expected (chapter 11, section A) to differ by a factor of about 1.05. The experimentally found equality of the $^{14}C/^{12}C$ ratios in shells and wood is interpreted as an apparent age of 400–500 years (^{14}C decay by about 1.05) for sea shells. This, in turn, means that the average residence time of dissolved carbon in ocean surface water is about 400–500 years. The almost exact cancellation between isotope effect and ocean residence time is fortuitous.

The residence time of carbon (as CO_2) in the atmosphere can be estimated from other data. The CO_2 from combustion of fossil fuels (which are old and therefore contain no ^{14}C) has been diluting the ^{14}C concentration in the atmosphere. By 1950 the total amount of "dead" CO_2 added to the atmosphere from this source (mostly since about 1900) amounted to about 12 percent of the total atmospheric CO_2. Yet, as shown in figure 13-5*b*, plants grown in 1950 show a specific ^{14}C activity not 12 percent, but only 2–3 percent lower than wood from the nineteenth century (after correction for decay). Thus it is clear that any given carbon atom remains in the atmosphere for a time short compared with 50 years, and from the data the average residence time of carbon in the atmosphere has been estimated as 5–10 years. Exchange with the oceans is presumably the

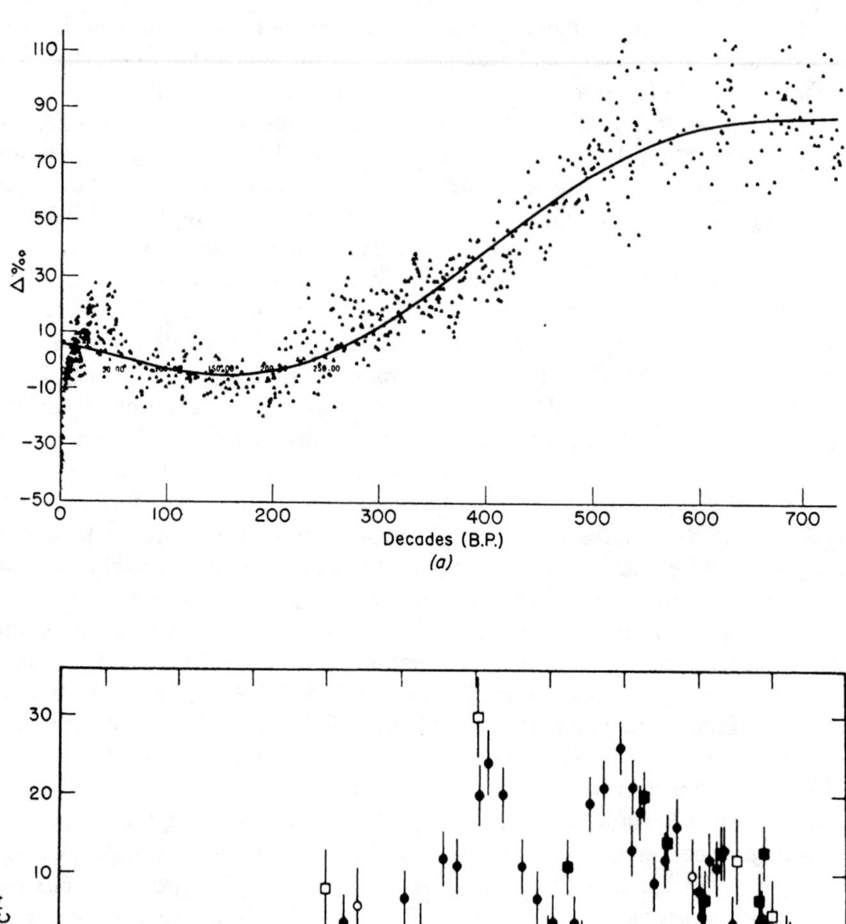

Fig. 13-5 Calibration of ^{14}C dates with tree-ring record. (*a*) Record over the past 7500 years. Note that abscissa is time before the present (B.P.). (*b*) Record since 1000 AD. In both graphs the ordinate is the difference between actual ^{14}C specific activity and ^{14}C specific activity based on a constant production rate (in per mil). [(*a*) From reference D1, reproduced, with permission, from the Annual Review of Earth and Planetary Science, Volume 6. © 1978 by Annual Reviews Inc. (*b*) From H. Suess, *J. Geophys. Res.* **70**, 5937 (1965).]

principal mechanism for the removal of CO_2 from the atmosphere. In the 1950s and 1960s the dilution of atmospheric ^{14}C by "dead" CO_2 was overshadowed by an effect in the opposite direction: the increasing concentration of ^{14}C brought about by the neutrons released in nuclear bomb tests. By 1963 the ^{14}C concentration in the troposphere had approximately doubled, in the stratosphere it had increased by much larger factors, but since the moratorium on atmospheric testing it has been gradually decreasing again as a result of vertical mixing and exchange with ocean bicarbonate (L1). Future generations will certainly be confronted with complex problems in the interpretation of ^{14}C dating of 20th century objects.

Cosmic-Ray Effects in Extraterrestrial Bodies. The cosmic-ray interactions on our planet are largely confined to the atmosphere because (1) the surface of the earth is effectively shielded from the primary cosmic rays and (2) any reaction products that *are* formed on the surface are likely to be removed by weathering in times that are short on a geological scale. On the other hand, much interesting information has been obtained from the study of cosmic-ray-induced nuclear reactions in meteorites and in material on the lunar surface.

Any unshielded body in interplanetary space is bombarded by cosmic rays, and their interaction produces many stable and radioactive spallation products. Since the ranges of typical galactic cosmic-ray protons in solids are only of the order of 1 m, the amounts of spallation products found give information on the length of time the material investigated has been near the surface—the so-called **exposure age**. On the other hand, as discussed below, the spallation products can also give clues to the time dependence of cosmic-ray flux in the past. The interpretation of the data depends strongly on laboratory measurements of cross sections for nuclear reactions by multi-GeV protons (chapter 4, section G) and is greatly aided by the fortunate circumstance that most such cross sections and certainly the *ratios* of cross sections for similar reactions are hardly energy-dependent above about 1 GeV, so that details of the cosmic-ray spectrum are usually not needed for the interpretation of results.

Exposure Ages. Many radioactive products, ranging in half life from a few days to millions of years, as well as some stable products of spallation have been identified in meteorites and in lunar samples. In an object in space, exposed to cosmic rays, any radioactive product should be at saturation (as many atoms decaying per unit time as are being formed), provided the cosmic-ray flux has been constant for a time long compared to the half life of the product. Typical saturation activity levels of radionuclides in meteorites are in the range 10–100 dis min^{-1} kg^{-1}. The saturation activity of a radioactive spallation product depends on the cosmic-ray flux averaged over a time of the order of a half life, whereas the amount of a stable spallation product that has accumulated is a measure of the total integrated cosmic-ray fluence throughout the exposure. If we assume for the moment that the cosmic-ray intensity has been constant in time (see below) and if the relative cross sections for formation of a stable and a radioactive product are known from accelerator data (or from systematics), the ratio of their concentrations in a meteorite at time of fall or in a lunar sample at time of collection immediately gives the exposure age. Isobaric pairs such as ^{36}Cl-^{36}Ar, ^{3}H-^{3}He, and ^{22}Na-^{22}Ne are perhaps most reliable, but others such as ^{26}Al-^{21}Ne, ^{39}Ar-^{38}Ar, and ^{40}K-^{41}K have also been used successfully. The stable nuclide is always determined mass-spectrometrically and must be known to have had a negligible primordial concentration—hence the preference for rare-gas isotopes. Greatest reliance can, of course, be placed on exposure ages determined by more than one method.[6]

The exposure ages of meteorites range from 2×10^4 to 8×10^7 y for stone meteorites and from 4×10^6 to 2.3×10^9 y for iron meteorites. For some classes of meteorites there appears to be a pronounced clustering around certain ages. This is illustrated in figure 13-6 for the large class of stone meteorites called chondrites and for the much rarer iron meteorites. The relatively short and variable exposure ages of meteorites, together with their common formation age of about 4.57×10^9 y, lead to the conclusion that meteorites were formed by breakup of larger bodies, presumably asteroids with diameters $\lesssim 500$ km, through collisional processes. The clustering of exposure ages for certain chemically characterized types of iron meteorites probably dates the breakup of their parent bodies. The much shorter exposure ages of stones could mean that they resulted from different parent bodies or that they were subject to subsequent further breakup by collisions because of their greater fragility.

Exposure ages determined for lunar rocks are harder to interpret. The moon, as we have seen, presumably also formed $(4.5\text{–}4.6) \times 10^9$ y ago and lunar-rock ages range between 3.0×10^9 and 4.5×10^9 y. The much shorter cosmic-ray exposure ages of $1 \times 10^6\text{–}7 \times 10^8$ y measure the time the rocks have spent in the top layer of the lunar surface; there may, however, have been a variety of causes for the exposure of a rock at a certain time, such as meteorite impacts, rock slides, or erosion. Lunar soils (fine particles) show exposure ages of $(1.5\text{–}4.5) \times 10^8$ y, which may be interpreted in terms of a continuous turnover at the rate of a few millimeters per 10^6 y.

Constancy of Cosmic-Ray Flux (S2). The ratios of saturation activities of products of different half lives in iron meteorites should equal the ratios of their production cross sections in iron (and nickel) spallation, if the cosmic-ray flux has been constant for a time of the order of the half-life

[6] In so-called meteorite "finds," that is meteorites whose fall has not been observed, the apparent *disagreement* between different exposure age determinations in the sense of reduced specific activities of shorter-lived radionuclides can be used to deduce the time since the meteorites' fall.

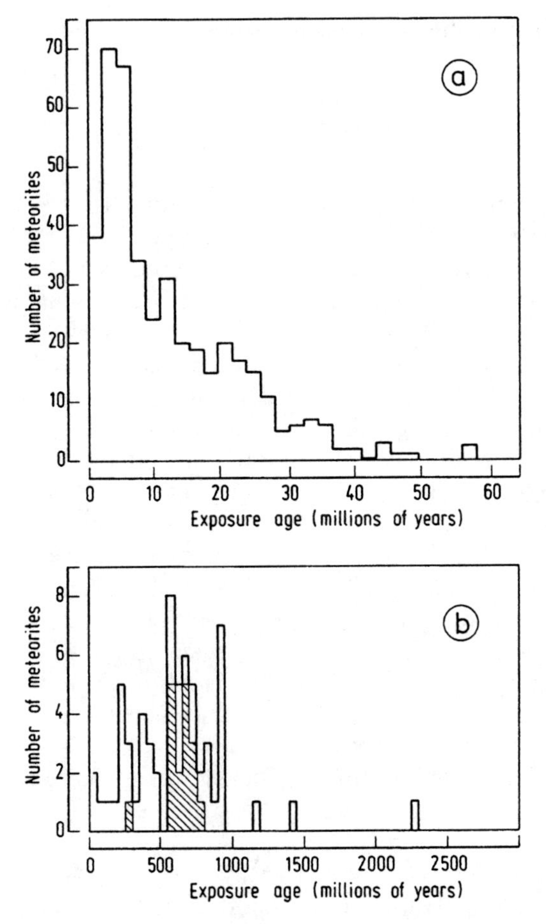

Fig. 13-6 Meteorite exposure ages. (*a*) Histogram for 419 chondrites. (*b*) Histogram for 62 iron meteorites. The shaded area is for a particular class, the medium octaedrites. (Data from reference K1).

range covered. Such comparisons for pairs such as ^{39}Ar(269 y)-^{36}Cl(3×10^5 y), ^{22}Na(2.6 y)-^{26}Al(7.2×10^5 y), and ^{54}Mn(312 d)-^{53}Mn(3.7×10^6 y) have all indicated little or no variation in cosmic-ray intensity over the respective time spans. The level of cosmogenic ^{40}K(1.28×10^9 y) indicates that, even over the last billion years or so, the average cosmic-ray intensity has remained approximately constant. The method is, of course, not sensitive to variations in intensity over times short compared to the shortest of the half lives studied. Deviations of the activity ratio for ^{37}Ar(35 d)-^{39}Ar from that expected from cross-section measurements have been interpreted in terms of a gradient in cosmic-ray flux with distance from the sun due to the sun's magnetic field. This effect presumably comes about because meteorite orbits are quite eccentric and the short-lived ^{37}Ar is formed while the meteorite is near the earth, whereas the ^{39}Ar integrates the flux over the entire orbit.

B. NUCLEAR ASTROPHYSICS

The observed emission rates of radiation from stars require enormous sources of energy. Chemical reactions cannot possibly be of significance as stellar energy sources.[7] Although gravitational energy of contraction plays an important role in certain phases of stellar evolution, it has been recognized since the 1920s that most stars must derive the energy they radiate from exoergic nuclear reactions occurring in their interiors. These reactions not only account for the vast amounts of energy radiated by stars but also constantly change the elementary and isotopic composition of matter in their interiors. We also know that, on the one hand, stars are formed by condensation out of interstellar gas and dust and that, on the other hand, matter is ejected from stars into interstellar space in a variety of processes ranging from the relatively minor emissions of flares as observed on our sun to the violent explosions of supernovae. Therefore the matter from which new stars are formed contains the debris of old stars and thus the products of previous "element-cooking." It is the aim of nuclear astrophysics to account, in terms of nuclear processes, for the observed element and isotope abundances, as well as for energy production in stars. Since nuclear processes appear, in fact, to be involved in most facets of astronomy and cosmology, nuclear astrophysics may be said to be concerned with the entire history of the universe.

The Big Bang. It is now generally believed that our universe is expanding. In 1929 E. Hubble (H2) reported that a number of galaxies are receding from us with velocities proportional to their distances, as evidenced by the amounts of Doppler shifts in their spectra toward the red. This observation, confirmed many times and greatly extended since then, has clearly established the picture of an expanding universe. However, the numerical value of the constant of proportionality between radial distance and velocity, known as the Hubble constant, has undergone repeated revisions because it requires very accurate knowledge of absolute distances. The best value appears to be (S3) in the range of 50–65 km s^{-1} per megaparsec[8] (although recent measurements by a new technique indicate a

[7] The energy release in chemical reactions is only about 10^{13} erg g^{-1}, compared to $\sim 10^{19}$ ergs g^{-1} for nuclear reactions (see p. 111). Thus the sun, which radiates 3.8×10^{33} ergs s^{-1}, would burn up its entire mass of 2.0×10^{33} g in about 10^5 y if chemical reactions were the source of its energy.

[8] One parsec (pc), defined as the distance that gives rise to a parallax of one second of arc when subtended by the radius of the earth's orbit, is the accepted unit of astronomical distances. 1 pc $= 3.0857 \times 10^{16}$ m ≈ 3.26 light years.

value about twice as large—see H3). The time since the galaxies all started out from one "point," which is just the reciprocal of the Hubble constant, is then $(15\text{–}20) \times 10^9$ y.

Beginning in 1946 Gamow and his collaborators championed the theory that not only was the universe born in a gigantic explosion of an extraordinarily hot and dense "singularity," but that in the first stages of this "big bang" the elements were built up to their present abundances (G1, A2). The theory proved untenable in its original form, in part because appreciable element buildup beyond ^{4}He on the time scale involved is made impossible by the nonexistence of particle-stable nuclei at $A = 5$ and $A = 8$. However, the big-bang concept itself is now universally accepted and has found strong support from the observation by A. A. Penzias and R. W. Wilson of an all-pervasive, isotropic microwave radiation corresponding to a 3 K temperature (P3). This blackbody radiation background has been interpreted as the remnant of the big bang (D2, W3).

Leaving aside many interesting problems[9] concerning the big-bang theory (H4, W3), we merely note that element synthesis by nuclear reactions could start only when the temperature had dropped to about 10^9 K, or about 3 minutes after the beginning, and must have practically stopped again after an hour or so, when temperature and pressure had dropped too low to sustain further significant nuclear reaction rates. Furthermore, as already mentioned, no appreciable buildup beyond ^{4}He could have occurred. According to the so-called standard model of the early universe (H3), about 13 percent of the nucleons were neutrons at the time nucleosynthesis started, and the sequence of reactions was:

$$p + n \rightarrow d + \gamma, \tag{13-13}$$

$$d + p \rightarrow {}^3\text{He} + \gamma, \tag{13-14}$$

$${}^3\text{He} + {}^3\text{He} \rightarrow {}^4\text{He} + 2p. \tag{13-15}$$

The net effect is the transformation of two neutrons and two protons into a ^{4}He nucleus. After all the neutrons were used up this would have led to a helium abundance of about 25 percent by weight, in reasonable agreement with observational evidence.

[9] Among these problems are the role of various elementary particles in the earliest stages as well as the mechanisms that account for the great preponderance of matter over antimatter in our universe. Much debated also is the question whether the universe is open or closed, that is, whether the expansion will go on forever or whether it will reach a point at which gravitational contraction will take over, leading eventually to a new singularity or big bang. The answer to this question depends crucially on the average density of matter in the universe. If the density exceeds a critical value ($\sim 5 \times 10^{-30}$ g cm^{-3} provided in the Hubble constant has the value given above), the universe is closed; if the density is less than that value, the universe is open. The actual value of the average density depends very much on the contribution of neutrino masses (if any) to the mass of matter in the universe.

Formation of Galaxies and Stars. For the first 10^6 years or so the radiation density in the universe exceeded the matter density. Only much later, perhaps after about 2×10^8 y, was radiation pressure sufficiently low for permanent large-scale inhomogeneities to form in the expanding gas; these were then subject to gravitational contraction, and thus galaxies and stars within galaxies were presumably born. Further nucleosynthesis processes are intimately connected with stellar evolution.

The current picture of stellar evolution is based on a rich store of observational astronomical material, on laboratory measurements of many nuclear reactions, and on much theoretical work. Without being able to present any of this background material, for which the reader is referred to books (e.g., C2) and review articles (e.g., R1), we outline in the following the major conclusions concerning the connection between stellar evolution and nucleosynthesis.

Hydrogen-Burning Stars—The Main Sequence. As a mass of gas contracts it heats up. Thus a prestellar nebula consisting of hydrogen and helium presumably contracts until the central temperature becomes high enough ($\sim 10^7$ K) to ignite the thermonuclear **pp reaction:**

$$p + p \rightarrow d + e^+ + \nu, \tag{13-16}$$

and its follow-up reactions (13-14) and (13-15).[10] Eventually the energy (heat) production by these nuclear reactions is sufficient to halt the gravitational contraction and the star enters a stable period in which the fusion energy produced in the core just balances the energy radiated from the surface. Energy transport from the interior to the surface is largely by radiative processes. Stars in this stage of their evolution, in which they derive their energy almost exclusively from the transformation of hydrogen into helium, constitute the vast majority of the total star population. On a **Hertzsprung-Russell (or H-R) diagram**, which is a plot of luminosity (or absolute magnitude) versus surface temperature (or some color index related to surface temperature), these hydrogen-burning stars form the so-called main sequence, a diagonal band shown schematically in figure 13-7. The diagram illustrates the fact that the more massive and luminous a main-sequence star is, the higher is its surface temperature (whereas such a simple relationship does not hold for some other types of stars, as we see later). The more massive a star is, the more quickly it will exhaust its hydrogen supply, and the shorter therefore will be its residence time on the main sequence.

The sun is a main-sequence star of modest size, with mass 2.0×10^{33} g,

[10] Note that (13-16) is a β process governed by the weak interaction and therefore much slower than the subsequent steps governed by the strong (nuclear) interaction.

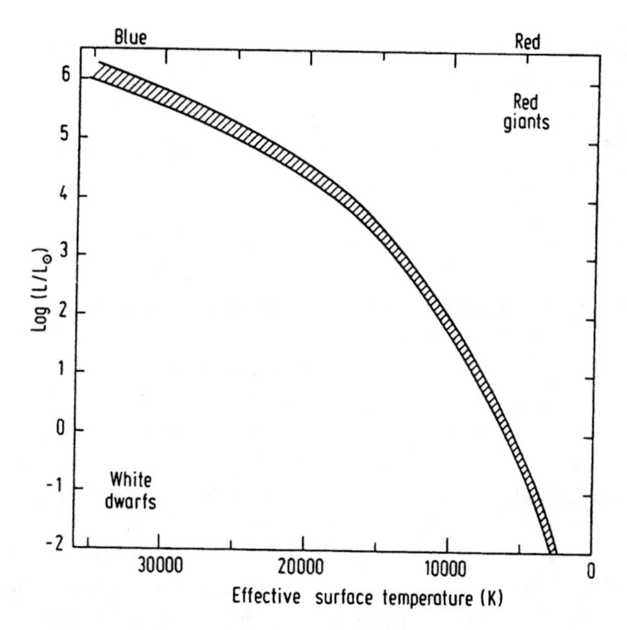

Fig. 13-7 Schematic H-R diagram. The ordinate is luminosity relative to that of the sun, on a logarithmic scale. The diagonal band is the main sequence.

surface temperature 6×10^3 K, estimated central temperature 1.5×10^7 K, and energy production rate 3.8×10^{33} ergs s^{-1}. The main sequence includes stars with luminosities between 10^{-2} and 10^6 times solar luminosity, and surface temperatures ranging from about 2.6×10^3 to 3.5×10^4 K.

Proton-Proton Chains. In a first-generation star on the main sequence, that is, one that evolved out of the original material resulting from the big bang, the energy-producing reactions are principally those already mentioned: (13-16), followed by (13-14) and (13-15). They constitute the main p-p chain and the net result is the transformation of four protons into a ^{4}He nucleus, two positrons, and two neutrinos.

However, some side reactions are also significant. A small fraction (in the present sun estimated at 5 percent) of the ^{3}He reacts not, according to (13-15) with another ^{3}He, but with ^{4}He:

$$^3\text{He} + {}^4\text{He} \to {}^7\text{Be} + \gamma. \qquad (13\text{-}17)$$

The ^{7}Be decays by EC, and the resulting ^{7}Li reacts with another proton to form ^{8}Be, which is unstable:

$$^7\text{Be} + e^- \to {}^7\text{Li} + \nu, \qquad (13\text{-}18)$$

$$^7\text{Li} + p \to {}^8\text{Be}^* \to {}^4\text{He} + {}^4\text{He}. \qquad (13\text{-}19)$$

In an even rarer branch ^7Be can capture a proton instead of an electron, leading to the following set of reactions:

$$^7Be + p \rightarrow {}^8B + \gamma, \tag{13-20}$$

$$^8B \rightarrow {}^8Be + \beta^+ + \nu, \tag{13-21}$$

$$\downarrow$$

$$^4He + {}^4He.$$

The net effect of each of these side branches is again the conversion of four protons to $^4He + 2e^+ + 2\nu$.

The formation of ^8B is of particular interest because, as a result of the relatively high Coulomb barriers for the reactions producing it, its concentration is a very sensitive function of temperature and therefore a determination of its concentration would constitute a good test of the stellar-evolution models used. The only possibility of obtaining direct experimental evidence on the thermonuclear reactions in the sun's interior is via the neutrinos emitted. An extraordinarily difficult and important radiochemical experiment for measuring the flux of the high-energy ($E_{max} \approx 14$ MeV) neutrinos from ^8B decay in the sun has therefore been done by R. Davis and coworkers (B2). It is based on detecting the neutrinos by the inverse β process $^{37}Cl + \nu \rightarrow {}^{37}Ar + e^-$ and measuring the radioactive ^{37}Ar formed. Because of the extremely small cross section for the neutrino capture reaction, a very large detector is required—4×10^5 l of perchlorethylene (C_2Cl_4)—and even so only a few atoms of ^{37}Ar ($t_{1/2} = 35$ d) are formed at saturation. The experiment has aroused particular interest among astrophysicists because it indicates a lower solar-neutrino flux than theoretically predicted. The cause of the discrepancy remains controversial (B2).

We should note that even in the hottest stellar interiors the thermal energies are very low compared to Coulomb barrier heights—10^7 K corresponds to kT of about 1 keV. Thus the thermonuclear reactions in stars typically take place far out in the tails of the Maxwellian energy distributions. The reaction rates therefore involve the product of a rapidly falling exponential—the tail of the Maxwell distribution that goes as $\exp(-E/kT)$—and a rapidly rising exponential—the Coulomb penetration factor that goes as $\exp(-bE^{-1/2})$. This interplay results in a reaction probability sharply peaked at an energy several times kT (known as the Gamow peak). However, even these energies are generally too low to be used in laboratory experiments; to get measurable rates the laboratory experiments have to be done at higher energies and extrapolated.

Carbon-Nitrogen Cycle. Second- and later-generation stars have incorporated, in addition to hydrogen and helium, some heavier elements formed and ejected in later stages of stellar evolution of preceding generations (see below). In such stars another sequence of reactions is possible for the conversion of hydrogen into helium, with ^{12}C acting as a catalyst. This so-called carbon-nitrogen cycle was actually proposed by H. Bethe (B3) before the p-p chains discussed above were worked out. The reaction sequence in the C-N cycle is:

$$
\begin{aligned}
^{12}C + {}^{1}H &\rightarrow {}^{13}N + \gamma, \\
^{13}N &\rightarrow {}^{13}C + \beta^{+} + \nu, \\
^{13}C + {}^{1}H &\rightarrow {}^{14}N + \gamma, \\
^{14}N + {}^{1}H &\rightarrow {}^{15}O + \gamma, \\
^{15}O &\rightarrow {}^{15}N + \beta^{+} + \nu, \\
^{15}N + {}^{1}H &\rightarrow {}^{12}C + {}^{4}He.
\end{aligned}
\tag{13-22}
$$

The net reaction is once again the conversion of four protons into $^{4}He + 2e^{+} + 2\nu$, with an energy release of about 26.7 MeV. Here too side reactions can occur. In a 0.04 percent branch the reaction between ^{15}N and a proton leads to ^{16}O rather than $^{12}C + {}^{4}He$. This is followed by $^{16}O(p, \gamma)^{17}F \rightarrow {}^{17}O + \beta^{+} + \nu$ and $^{17}O(p, \alpha)^{14}N$, thus leading back to the main cycle.

The C-N cycle involves much higher Coulomb barriers than the p-p chains, and its overall reaction rate is therefore an even steeper function of temperature than that of ^{8}B formation. It has a T^{20} dependence and is thus not believed to play a significant role in stars of solar size or smaller, but is presumably the dominant reaction sequence in much larger, hotter main-sequence stars.

Both observational data and theoretical considerations lead to the conclusion that the rate of hydrogen-burning in main-sequence stars depends strongly on their sizes. Luminosity varies roughly as the fourth power of mass. We can thus translate the ordinate of an H-R diagram into an approximate main-sequence lifetime scale. The very large, hot stars near the top of the H-R diagram are extremely young and will use up their hydrogen supply in about 10^{6} y, whereas the smallest stars must be first-generation stars that have survived since the early universe and will not exhaust their hydrogen for 3×10^{10} y or more. The sun is thought to have enough hydrogen supply to remain on the main sequence for at least as long as it has existed already.

Helium-Burning: Red Giants. A star that has exhausted an appreciable fraction of the hydrogen in its core will eventually have a central region that consists largely of helium. Model calculations show that the helium core of such a star will contract, whereas the hydrogen-containing shell may expand considerably. The surface will thus increase substantially and become much cooler, while the luminosity is approximately maintained. On the H-R diagram the star thus moves off the main sequence to the right and becomes what is known as a red giant. The gravitational contraction of the core now leads to further heating until, at temperatures in the neighborhood of 10^8 K, helium-burning reactions become important in the core. In particular, the reaction

$$^4He + ^4He \rightarrow ^8Be \tag{13-23}$$

produces, at 10^8 K, a steady-state concentration of about one ^8Be per 10^9 ^4He nuclei, despite the short (7×10^{-17} s) half life of ^8Be. This concentration is sufficient to support the reaction

$$^8Be + ^4He \rightarrow ^{12}C + \gamma \tag{13-24}$$

at an appreciable rate because this process can take place by resonance capture into a 0^+ state in ^{12}C at 7.65 MeV. Interestingly enough, this state was predicted from the astrophysical arguments for the processes (13-23) and (13-24) before it was found in the laboratory. The combination of reactions (13-23) and (13-24) is usually called the 3α reaction and can be written

$$3\,^4He \rightarrow ^{12}C + \gamma.$$

This reaction is presumably the principal energy source in red giants. In a shell surrounding the helium-burning core hydrogen-burning may continue at the same time. During the helium-burning stage the star is again stable, with the fusion reactions just supplying the energy being radiated from the surface. However, the helium-burning stage is relatively short-lived, typically 10^7–10^8 y.

As ^4He becomes depleted and the ^{12}C concentration builds up, the reaction

$$^{12}C + ^4He \rightarrow ^{16}O + \gamma, \tag{13-25}$$

and subsequently the reaction

$$^{16}O + ^4He \rightarrow ^{20}Ne + \gamma \tag{13-26}$$

become important. Buildup beyond ^{20}Ne by (α, γ) reactions is not thought to be significant because it is strongly inhibited by Coulomb barriers (which get higher with increasing Z) at the temperatures that are reached prior to exhaustion of the ^4He supply.

Element and Isotope Abundances. Before proceeding to trace further transformations of matter in stellar interiors, we pause to consider the observational evidence that any theory of nucleosynthesis must account for, namely the relative abundances of elements and isotopes in the universe.

By far the most complete data on abundances exist, of course, for our own solar system. The information comes from analyses of terrestrial, meteoritic, and lunar material, from spectroscopic observations of the

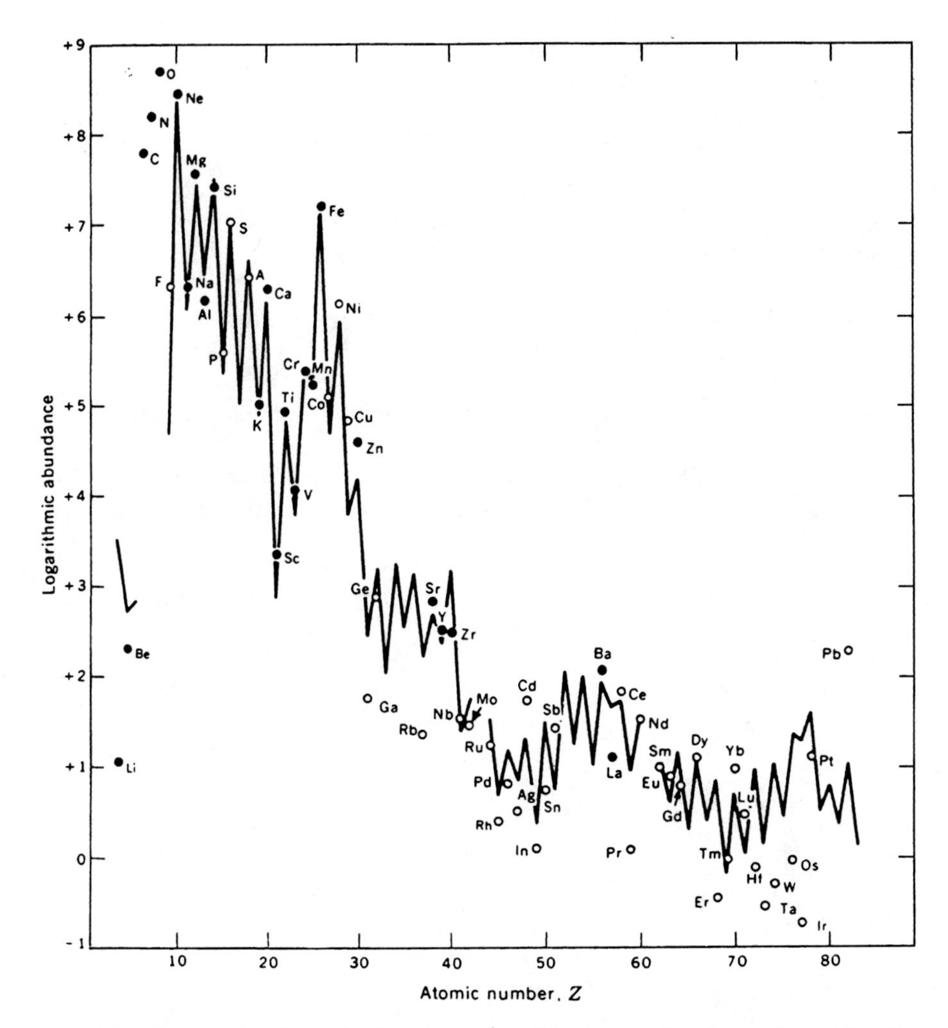

Fig. 13-8 Element abundances in the solar system. The dots are based on solar spectra, the line on analyses of terrestrial and meteorite samples. (From *Principles of Stellar Evolution and Nucleosynthesis* by D. D. Clayton. Copyright © 1968 McGraw-Hill Book Company. Used with permission.)

sun,[11] and from such additional data as the densities of sun and planets. Figure 13-8 shows the solar-system abundances plotted against Z. Some pertinent observations are as follows:

[11] The surface composition of the sun is assumed to represent the original material from which the solar system was formed (except for loss of some highly volatile elements), because transmutations by nuclear reactions have presumably taken place only in the deep interior, and no significant convection is believed to have occurred. The same assumption is made about most other stars.

1. There is a steep, roughly exponential drop in abundances up to $Z \approx 35$, followed by a much more gradual decrease.
2. Superimposed on these general trends is a strong even-odd alternation, even Z being favored.
3. Iron and nickel have "abnormally" high abundances.
4. Lithium, beryllium, and boron abundances are several orders of magnitude lower than one would expect from the general trend.

Additional trends and systematic features are found when *isotopic* abundances are considered. Some of these can be seen in figure 13-9, which is a (somewhat schematic) plot of abundance versus mass number. The iron group peak is even more pronounced in this representation. Double humps appear at $A = 80$ and 90, at $A = 130$ and 138, and at $A = 194$ and 208. Not evident from either graph, but easily verified from any table of abundances (e.g., appendix D) is a striking regularity in the relative isotopic

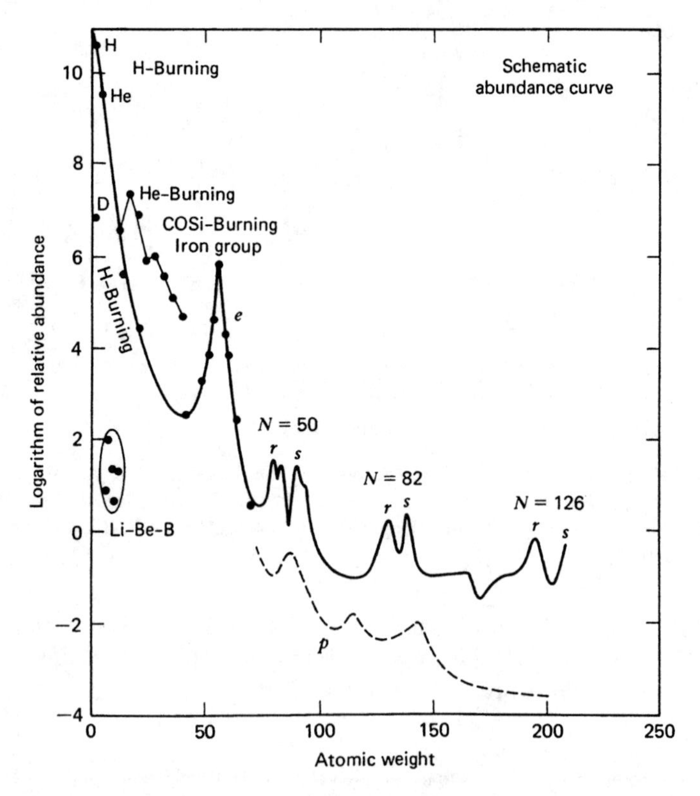

Fig. 13-9 Schematic representation of solar-system abundances as a function of A. Features due to specific processes are indicated. (From reference R1, adapted from reference B4. Reproduced, with permission, from the Annual Review of Nuclear and Particle Science, Volume 28. © 1978 by Annual Reviews Inc.)

abundances in individual even-Z elements. Above $Z = 33$ the lightest isotope is always rare and the heaviest tends to be quite abundant; for $Z < 33$ the lightest isotope is always much more abundant than the heaviest one (except in argon where the radiogenic origin of ^{40}Ar distorts the picture).

The solar-system abundance curves such as those shown in figures 13-8 and 13-9 have often been referred to as cosmic abundance distributions. However this is a gross misnomer. Spectroscopic observations give ample evidence for wide variations in composition among the stars in our galaxy. In particular, the abundances of elements heavier than carbon are two to three orders of magnitude greater, relative to hydrogen, in very young, that is, recently formed stars (so-called population I, typically blue and luminous, often $<10^8$ y old) than in old stars (population II, formed early in the history of the galaxy). This in fact is the most persuasive evidence for the view that the major part of heavy elements ($Z > 6$) has been synthesized in stars during the life of our galaxy. The same conclusion presumably holds for stars in other galaxies.

Many stars with special abundance anomalies are known. For example, He/H ratios as large as several hundred and C/H ratios of about 10 have been reported. Even abnormal *isotopic* abundance ratios, as determined from intensities in band spectra and isotope shifts of spectral lines, have been found. Finally we mention that spectral lines of the unstable element technetium (its longest-lived isotope, ^{97}Tc, has $t_{1/2} = 2.6 \times 10^6$ y) have been discovered in the spectra of certain stars, which most likely means that there *has* been some mixing between the deep interior and the surface in those particular stars.

Further Nucleosynthesis in Stars. The observations on element and isotope abundances, sketched in the preceding paragraphs, together with our earlier discussion of the chronology of the solar system, indicate that the present abundance distribution in the solar system must have been established before the differentiation into sun and planets about 4.6×10^9 y ago and has remained unchanged since then except for the alterations resulting from conversion of hydrogen to helium in the sun and from decay of the radioactive nuclides. On the other hand, it is clear from the widely varying compositions of different types of stars that a common origin for the present matter composition of our entire galaxy cannot be assumed. In fact, the observation of technetium in stars proves rather conclusively that, in addition to the hydrogen- and helium-burning and other light-element reactions already discussed, other nuclear processes that can lead to heavy-element synthesis must be taking place in stellar interiors essentially right up to the present time. To account for the observed abundance distribution in the solar system as well as for the variations in other stars has been a major challenge for nuclear astrophysicists. The treatments have followed the pioneering paper by E. M. Burbidge, G. R. Burbidge, W.

A. Fowler, and F. Hoyle (B4, often referred to as B^2FH), although many modifications and refinements have appeared since. We can only give the briefest sketch of the major points in a very complex story.

We left stellar evolution at the end of the helium-burning stage, when the major nuclides in the core are ^{12}C, ^{16}O, and ^{20}Ne. When helium has been exhausted in the core contraction sets in again and the gravitational energy causes further heating of the core. At this stage the star's path on the H-R diagram is not entirely clear. Luminosity will in general decrease, while surface temperature is expected to go up, so that the path will tend toward the lower left. In small stars (<0.7 solar mass) gravitational contraction will eventually stop as a result of pressure developed because of electron degeneracy, and such stars become **white dwarfs**, located in the lower left corner of the H-R diagram. White dwarfs are very small and dense, have no internal energy sources, and lose their remaining energy by radiation. On the way to the white-dwarf stage much material may be ejected from the outer layers of the star into the interstellar space.

In larger stars the next stage of "element cooking" commences when the central temperature reaches $(6\text{–}7) \times 10^8$ K. The reactions that then become important are the ^{12}C-^{12}C reactions:

$$^{12}C + ^{12}C \rightarrow ^{24}Mg + \gamma \tag{13-27}$$

$$\rightarrow ^{23}Na + p \tag{13-28}$$

$$\rightarrow ^{20}Ne + \alpha \tag{13-29}$$

$$\rightarrow ^{23}Mg + n \tag{13-30}$$

$$\rightarrow ^{16}O + 2\alpha. \tag{13-31}$$

Reactions 13-28 and 13-29 appear to be the most important of these. At still higher temperatures other reactions such as $^{12}C + ^{16}O$ and $^{16}O + ^{16}O$ can lead to a variety of products including isotopes of magnesium, silicon, phosphorus, and sulfur. At the end of the ^{12}C- and ^{16}O-burning stage the most abundant nuclei will be ^{28}Si and ^{32}S.

An entirely new set of reactions now becomes significant at temperatures above $\sim 10^9$ K. Gamma-ray intensities are sufficiently high to cause photonuclear reactions: (γ, n), (γ, p), and (γ, α). The resulting nucleons and α particles (often of much higher energy than "thermal," even at 10^9 K!) are in turn captured, and a complicated network of photodisintegration and particle-capture reactions develops. The effect of all these reactions, which take place on a much shorter time scale than any of the previous processes, is to move nucleons from less tightly bound to more tightly bound nuclei (C2). Whatever the very complicated details, this buildup process must stop when the Fe-Ni region at $A = 56$ is reached, since these nuclei have the maximum binding energy per nucleon. Attempts to account for the observed abundance distribution in the region of $A < 56$ by detailed analyses of the nuclear processes have had considerable but not total success (B4, B5).

The s-Process. The key to the synthesis of nuclides beyond the iron group, which is not possible by exoergic charged-particle reactions, is to be found in neutron-induced reactions. In a second- or later-generation star, that is, one that has condensed out of interstellar matter that already contained debris from previously evolved stars, elements up to iron may be present during the hydrogen- and helium-burning stages. This not only makes the C-N cycle operative in the hydrogen-burning phase, it also makes possible, at about 10^8 K, a number of exoergic neutron-producing reactions, in particular $^{13}C(\alpha, n)^{16}O$, $^{17}O(\alpha, n)^{20}Ne$, $^{21}Ne(\alpha, n)^{24}Mg$, and $^{25}Mg(\alpha, n)^{28}Si$. These reactions will be important principally in red giants.

The neutrons furnished by these reactions can now continue the element-building process beyond iron by successive (n, γ) reactions. This process is slow relative to all but the slowest β-decay processes and is therefore called the *s*-process. It follows a zigzag path up the nuclide chart as illustrated in figure 13-10, with β^- decays (and occasional EC and β^+ branches as in ^{128}I) interspersed between (n, γ) captures. Qualitatively the *s*-process immediately accounts for the prevalence of the heavier isotopes in even-Z elements (the lighter isotopes being depleted by neutron captures), for the relatively flat abundance distribution [(n, γ) cross sections not being a strong function of Z, in contrast to charged-particle cross sections], for the abundance peaks at magic neutron numbers [where (n, γ) cross sections are low], and for the odd-even alternations (because of lower level densities in even-N and even-Z compound nuclei). Quantitative calculations, using all the relevant (n, γ) cross sections at the appropriate "thermal" energies of 10–30 keV (figure 13-11), have been very successful in accounting for the abundance distribution of the majority of nuclides up to bismuth.

Because of its slow time scale—$\sim 10^2$–10^5 y per neutron-capture step—the *s*-process cannot possibly carry the synthesis beyond bismuth to thorium and uranium because of the intervening short-lived species. The *s*-process also bypasses the lightest stable isotopes of some elements, as illustrated in figure 13-10 for ^{124}Xe, ^{126}Xe, ^{130}Ba, and ^{132}Ba. The formation of such (quite rare) nuclides is thought to result from what is called the ***p*-process**. This involves successive (p, γ) reactions and can take place when already synthesized heavy elements are mixed with high concentrations of hydrogen at temperatures of $\sim 2.5 \times 10^9$ K, conditions that presumably prevail in the envelopes of supernovae (see below).

The r-Process. To explain the existence of uranium and thorium as well as the abundance peaks at $A \approx 80$, 130, and 194 and some other abundance features not accounted for by the *s*- and *p*-processes (e.g., the neutron-rich isotopes of even-Z elements not reached by the *s*-process, such as ^{82}Se, ^{96}Zr, ^{110}Pd), a much more rapid neutron-capture chain than the *s*-process has been postulated. It is called the *r*-process. In an enormous neutron flux many successive neutron captures can take place within

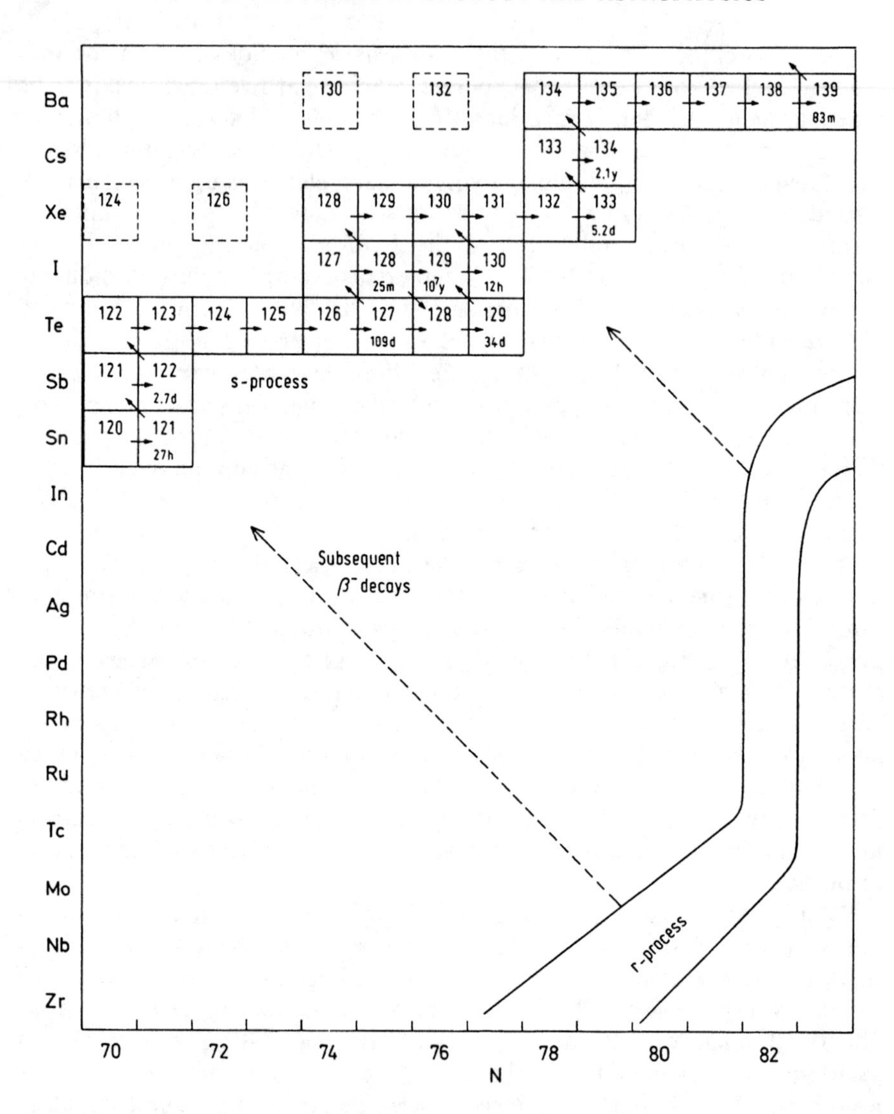

Fig. 13-10 Portion of s- and r-process paths. The s-process path involves (n, γ) reactions, indicated by horizontal arrows, and β decays, shown by diagonal arrows. The squares with only mass numbers are stable nuclei; the β emitters have half lives shown. The squares with dashed borders are nuclides not reached by the s-process, but by the p-process. The r-process path is indicated schematically, with the prominent effect of the $N = 82$ shell shown.

milliseconds to seconds without intermediate α or β decays. Such processes have in fact been observed terrestrially as a result of nuclear explosions and have produced the first man-made samples of the elements einsteinium ($Z = 99$) and fermium ($Z = 100$) through rapid multiple neutron captures in uranium followed by successive β^- decays.

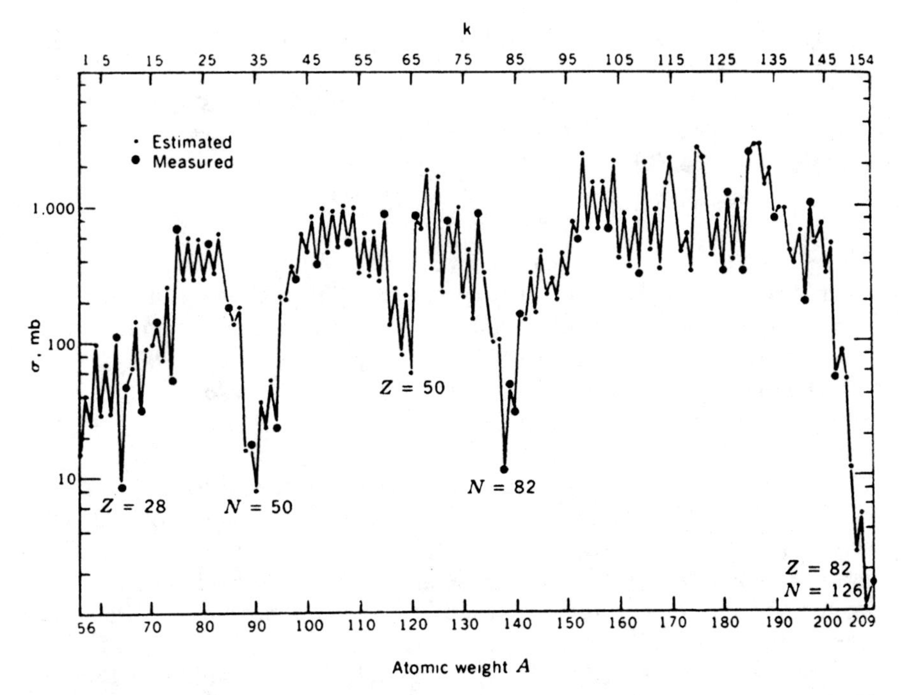

Fig. 13-11 Neutron-capture cross sections at 25 keV for nuclei on the *s*-process path. The odd-even alternation and strong shell effects are evident. (From *Principles of Stellar Evolution and Nucleosynthesis* by D. D. Clayton. Copyright © 1968 McGraw-Hill Book Company. Used with permission.)

The *r*-process path, like the *s*-process, follows a band approximately parallel to the valley of β stability, but far on the neutron-excess side, where β-decay half lives eventually become comparable to the neutron-capture times of milliseconds. A portion of the path is schematically indicated in figure 13-10. Where the *r*-process reaches magic neutron numbers 50, 82, and 126, it moves for a while along these neutron numbers, thus coming closer to stability and leading to a pile-up in these regions when, after a very short interval, the *r*-process is terminated. This magic-number effect far from stability is reflected, after subsequent β^- decays, in the abundance peaks at $A = 80$, 130, 194, which at first sight do not appear to be correlated with closed shells.

Quantitative calculations of the *r*-process depend on knowledge of properties of nuclei far from β stability; this must be largely based on extrapolation from known regions by means of nuclear systematics. Extension of experimental information to nuclides further and further out from the stability valley is therefore of great interest to astrophysics. An interesting question is how far the *r*-process can have carried element synthesis beyond uranium. The evidence cited earlier (p. 491) for the existence of ^{244}Pu in the early solar system clearly shows that the process did not stop at uranium. Whether it could have produced the postulated superheavy elements near $Z = 114$ and $N = 184$ or whether its path would have been halted earlier by nearly instantaneous fission is not clear. It is interesting to note that most of the nuclides used in geo- and cosmochronology—^{244}Pu, ^{238}U, ^{235}U, ^{232}Th, ^{129}I, and ^{87}Rb—are produced entirely or largely by the *r*-process.

Supernovae. The *r*-process is a very plausible mechanism, in fact the only plausible one that has been proposed, for the heaviest terrestrial elements and for a variety of abundance features. But where and how can conditions exist for such a process? There appears to be general agreement that supernovae can provide these conditions. However, the exact mechanisms involved are still matters of discussion. There are, in fact, two types of supernovae, and theories as to which one of these is mostly responsible for *r*-process synthesis have varied (see, e.g., F2 and S4).

Type-I supernovae represent the final, evolutionary stage of old, relatively small stars (1.2–1.5 solar masses), in which the entire star disintegrates in a giant thermonuclear explosion. The time scale for this phase is seconds and the temperatures reached in various regions of the star range from 10^9 to nearly 10^{10} K.

Type-II supernovae, by contrast, occur only in stars with initial masses at least ten times that of the sun. In such stars, as we have seen, element buildup may occur until the core consists almost entirely of iron group elements. At that point gravitational contraction must again set in, and with it further heating. What happens next, however, is quite a new phenomenon: At about 5×10^9 K iron and nickel nuclei are rather suddenly photodisintegrated into α particles and neutrons, the necessary energy being supplied by gravitation. With the accelerating gravitational collapse, ^{4}He is further dissociated into nucleons and the protons almost immediately capture electrons to form neutrons. In other words, the entire core collapses, within a time of the order of a second, into an extraordinarily dense ($\sim 10^{14}$ g cm^{-3}) mass of neutrons—a neutron star is born. This implosion of the core is accompanied by the explosive ejection of the star's outer layers into the interstellar medium, and by massive element synthesis in these various layers that represent different evolutionary stages in the star's history.

Each of the two types of supernova involves the release of enormous amounts of energy (of the order of 10^{51} or 10^{52} ergs), and each type occurs at the rate of one every few hundred years in each galaxy. The most recent supernovae observed in our galaxy occurred in 1572 (observed by Tycho Brahe) and in 1604 (observed by Johannes Kepler).

We have noted (p. 490) the record of extinct radioactivities (^{129}I and ^{244}Pu), which shows that *r*-process nucleosynthesis must have occurred within $(1\text{–}2.5) \times 10^8$ y prior to formation of the solar system 4.6×10^9 y ago. This means that the last supernova explosion that contributed to solar-system abundances must have occurred during that time interval in our region of the galaxy. The average age of the elements in the solar system, however, is likely to be much greater, probably between 6×10^9 and 10×10^9 y (S4), as deduced from the relative abundances of the various long-lived radioactive nuclides produced by the *r*-process—^{235}U, ^{238}U, ^{232}Th, and ^{187}Re. Most likely very many supernova explosions have contributed to the matter from which our solar system formed. What finally caused this cloud to begin to contract and become sun and planets is not certain, but it may well have been the shock wave produced by another supernova approximately 10^8 y after the one that contributed the last nucleosynthesis in that cloud.

REFERENCES

A1 L. T. Aldrich and G. W. Wetherill, "Geochronology by Radioactive Decay," *Ann. Rev. Nucl. Sci.* **8**, 257 (1958).

A2 R. A. Alpher, H. A. Bethe, and G. Gamow, "The Origin of Chemical Elements," *Phys. Rev.* **73**, 803 (1948).

B1 B. B. Boltwood, "On the Ultimate Disintegration Products of the Radioactive Elements," *Am. J. Sci.* **23**, 77 (1907).

B2 J. N. Bahcall and R. Davis Jr., "Solar Neutrinos: A Scientific Puzzle," *Science* **191**, 264 (1976).

B3 H. A. Bethe, "Energy Production in Stars," *Phys. Rev.* **55**, 534 (1939).

*B4 E. M. Burbidge et al., "Synthesis of the Elements in Stars," *Rev. Mod. Phys.* **29**, 547 (1957).

B5 D. Bodansky, D. D. Clayton, and W. A. Fowler, "Nuclear Quasi-Equilibrium during Silicon Burning," *Astrophys. J. Suppl.*, Ser. 16(148), 371 pp. (1968).

C1 R. N. Clayton, "Isotopic Anomalies in the Early Solar System," *Ann. Rev. Nucl. Part. Sci.* **28**, 501 (1978).

*C2 D. D. Clayton, *Principles of Stellar Evolution and Nucleosynthesis*, McGraw-Hill, New York, 1968.

D1 P. E. Damon, J. C. Lerman, and A. Long, "Temporal Fluctuations of ^{14}C: Causal Factors and Implications," *Ann. Rev. Earth Planet. Sci.* **6**, 457 (1978).

D2 R. H. Dicke et al., "Cosmic Black-Body Radiation," *Astrophys. J.* **142**, 414 (1965).

*F1 G. Faure, *Principles of Isotope Geology*, Wiley, New York, 1977.

F2 W. A. Fowler and F. Hoyle, *Nucleosynthesis in Massive Stars and Supernovae*, University of Chicago Press, Chicago, IL, 1965.

G1 G. Gamow, "Expanding Universe and the Origin of the Elements," *Phys. Rev.* **70**, 572 (1946).

*H1 C. T. Harper, *Geochronology*, Dowden, Hutchison, & Ross, Stroudsburg, PA, 1973.

H2 E. Hubble, "A Relation Between Distance and Radial Velocity Among Extragalactic Nebulae," *Proc. Nat. Acad. Sci.* **15**, 168 (1929).

H3 B. K. Hartline, "Double Hubble, Age in Trouble," *Science* **207**, 167 (1980).

H4 E. R. Harrison, "Standard Model of the Early Universe," *Ann. Rev. Astron. Astrophys.* **11**, 155 (1973).

K1 T. Kirsten, "Time and the Solar System," in *Origin of the Solar System* (S. F. Dermott, Ed.), Wiley, London, 1978, pp. 267–346.

L1 D. Lal and H. E. Suess, "The Radioactivity of the Atmosphere and Hydrosphere," *Ann. Rev. Nucl. Sci.* **18**, 407 (1968).

L2 W. F. Libby, *Radiocarbon Dating*, University of Chicago Press, Chicago, IL, 1955.

L3 A. E. Litherland, "Ultrasensitive Mass Spectrometry with Accelerators," *Ann. Rev. Nucl. Part. Sci.* **30**, 437 (1980).

M1 P. Meyer, "Cosmic Rays in the Galaxy," *Ann. Rev. Astron. Astrophys.* **7**, 1 (1969).

M2 R. A. Muller, "Radioisotope Dating with Accelerators," *Phys. Today* **32**(2), 23 (Feb. 1979).

N1 H. V. Neher, "The Primary Cosmic Radiation," *Ann. Rev. Nucl. Sci.* **8**, 217 (1958).

N2 E. P. Ney, "Experiments on Cosmic Rays and Related Subjects During the International Geophysical Year," *Ann. Rev. Nucl. Sci.* **10**, 461 (1960).

P1 C. Patterson, "Age of Meteorites and the Earth," *Geochim. Cosmochim. Acta* **10**, 230 (1956).

P2 F. A. Podosek, "Dating of Meteorites by the High-Temperature Release of Iodine-Correlated ^{129}Xe," *Geochim. Cosmochim. Acta* **34**, 341 (1970).

P3 A. A. Penzias and R. W. Wilson, "A Measurement of Excess Antenna Temperature at 4080 Mc/s," *Astrophys. J.* **142**, 419 (1965).

R1 C. Rolfs and H. P. Trautvetter, "Experimental Nuclear Astrophysics," *Ann. Rev. Nucl. Part. Sci.* **28**, 115 (1978).

S1 M. Stuiver and P. D. Quay, "Changes in Atmospheric Carbon-14 Attributed to a Variable Sun," *Science* **207**, 11 (1980).

S2 O. A. Schaeffer, "Nuclear Chemistry of the Earth and Meteorites," in *Nuclear Chemistry* Vol. II (L. Yaffe, Ed.), Academic, New York, 1968, pp. 371–393.

S3 A. Sandage and G. A. Tammann, "Steps Towards the Hubble Constant VII," *Astrophys. J.* **210**, 7 (1976).

S4 D. Schramm, "Nucleo-Cosmochronology," *Ann. Rev. Astron. Astrophys.* **12**, 383 (1974).

W1 G. W. Wetherill, "Discordant Uranium-Lead Ages I," *Trans. Am. Geophys. Union* **37**, 320 (1956).

W2 G. W. Wetherill, "Radiometric Chronology of the Early Solar System," *Ann. Rev. Nucl. Sci.* **25**, 283 (1975).

W3 S. Weinberg, *The First Three Minutes*, Bantam, New York, 1979.

EXERCISES

1. A 1-kg sample of an ocean sediment is found to contain 1.50 mg of uranium, 4.20 mg of thorium and 6.0×10^{-3} cm^3 of helium (at standard temperature and pressure). Estimate the age of the deposit, assuming complete helium retention.
Answer: 2.0×10^7 y.

2. The lead isolated from a sample of uraninite is mass-spectrometrically analyzed and found to contain ^{204}Pb, ^{206}Pb, and ^{207}Pb in the ratio 1.00:914.7:92.6. Estimate the time since the formation of the uraninite ore.
Answer: 1.33×10^9 y.

3. From data in chapter 1 and appendix D calculate the heat generated per gram of normal uranium per year (a) by ^{238}U, (b) by ^{235}U, each in secular equilibrium with all its decay products. Assume the uranium is embedded in a massive body so that all α, β, and γ rays are absorbed.
Answers: (a) 0.6 cal; (b) 0.03 cal.

4. Assuming the earth's crust contains uranium, thorium, and potassium in the weight ratio 1:4:10^4, estimate the relative contributions of ^{238}U, ^{235}U, ^{232}Th, and ^{40}K to the heat production in the earth's crust (a) now, (b) 2×10^9 y ago, (c) 4.5×10^9 y ago. Use the results of the preceding exercise and the additional information that the corresponding heat production rates (in calories per year per gram of element) are 0.18 for thorium in equilibrium with its decay series and 2.6×10^{-5} for potassium at the present time.
5. In a strontium sample isolated from a rubidium-bearing mineral the $^{87}Sr/^{86}Sr$ ratio is found to be 2.1300. What is the atomic weight of the strontium in this sample? Assume the nonradiogenic strontium isotopes to be present in their normal ratios to each other. *Answer:* 87.531.
6. In the mineral of exercise 5 the *weight* ratio of total rubidium to total strontium is found to be 11.06 ± 0.01. Taking the original isotope ratio $(^{87}Sr/^{86}Sr)_0$ as 0.7020, determine the age of the mineral. *Answer:* 2.65×10^9 y.
7. The following data are obtained on five samples isolated from a suite of granitic rocks:

Sample	Weight Ratio Rb/Sr	Isotope Ratio $^{87}Sr/^{86}Sr$
A	1.029	0.7383
B	6.424	0.8829
C	10.373	0.9913
D	9.066	0.9552
E	3.407	0.8016

Construct an isochron and determine the crystallization age of the rock and the initial $^{87}Sr/^{86}Sr$ ratio at the time of crystallization.

Answers: $t = 0.623 \times 10^9$ y; $(^{87}Sr/^{86}Sr)_0 = 0.7112$.

8. The $^{40}Ar/^{39}Ar$ dating method was applied by L. Husain [*J. Geophys. Res.* **79**, 2588 (1974)] to samples of basalt from Hadley Rille near the Apollo 15 landing site on the moon. For a particular sample the results of stepwise heating following activation in a fast-neutron flux were as follows (after small corrections had been applied for ^{40}Ar from trapped air and for ^{39}Ar from calcium in the rock):

Temperature (°C)	$^{40}Ar/^{39}Ar$	$^{39}Ar \times 10^{-8}$ cm^3 STP g^{-1}
650	14.5	0.94
800	13.3	5.32
950	41.9	4.14
1100	61.7	4.34
1200	58.6	2.49
1300	62.4	1.12
1600	67.5	1.55
1700	56.4	0.09

A standard sample of hornblende of known age $(2.668 \pm 0.018) \times 10^9$ y was irradiated simultaneously with the above samples and gave a plateau value of $^{40}Ar/^{39}Ar = 44.50$. Calculate an "age" for each of the fractions, plot these apparent ages against the cumulative argon release, and estimate the true crystallization age of the rock. *Answer:* $t_{cryst} = 2.9 \times 10^9$ y.

9. A proportional counter of about 6-liter volume, filled to 3 atm with pure CO_2 and surrounded with heavy shielding and anticoincidence counters, is used for ^{14}C-dating measurements. Three samples of CO_2 are measured: sample A is "dead" CO_2 (age greater than 75,000 years) from coal, sample B is from contemporary wood (grown 50 years ago according to a tree-ring count), and sample C is of unknown age. The following counts are accumulated: 11,808 counts in 960 min for sample A, 21,749 counts in 180 min for sample B, and 20,583 counts in 480 min for sample C. Great care is taken to introduce the same weight of CO_2 into the counter at each filling. What is the age (and its standard deviation) of sample C?

10. At sea level the cosmic radiation produces about 2 ion pairs $s^{-1}\,cm^{-3}$ of air. At higher altitudes the intensity depends on the latitude, but for much of the United States it is about 10 ion pairs $s^{-1}\,cm^{-3}$ (of sea level air) at 3,000 m and about 200 ion pairs $s^{-1}\,cm^{-3}$ (of sea level air) at 12,000 m above sea level. Estimate the radiation dosage received per 24 h in rads from this source at (a) sea level, (b) 3,000 m, (c) 12,000 m.

11. A 1-kg sample of a recently fallen iron meteorite is found to contain 16 dis min^{-1} of ^{36}Cl, 14 dis min^{-1} of ^{39}Ar, and $1.88 \times 10^{-4}\,cm^3$ of ^{36}Ar at STP. From bombardments of iron targets with high-energy protons it is known that the cross sections for formation of ^{36}Cl, ^{36}Ar, and ^{39}Ar in such bombardments are in the ratio 1:0.2:0.9, rather independently of proton energy above ~400 MeV. (a) What is the cosmic-ray exposure age of the meteorite? (b) What can you conclude about cosmic-ray intensity as a function of time? (c) How would you interpret the finding that another iron meteorite has nearly the same ^{36}Cl and ^{36}Ar contents as the first but contains only 3.6 dis min^{-1} of ^{39}Ar per kg? *Answer:* (a) 5×10^8 y.

12. The radiation flux from the sun at the top of the earth's atmosphere at normal incidence is $0.139\,J\,cm^{-2}\,s^{-1}$. The earth-sun distance is 1.50×10^8 km. Calculate (a) the energy production in the sun per second, (b) the rate of hydrogen consumption in grams per second. (c) Estimate how long hydrogen-burning can continue in the sun under the assumption that energy production continues at the present rate and that the hydrogen-burning phase will cease when 10 percent of the total hydrogen has been used up. *Answer:* (c) $\sim 8 \times 10^9$ y.

13. Verify the statement on p. 504 that each of the three branches of the *p-p* chain is equivalent to the same net reaction: $4\,^1H \rightarrow {}^4He + 2e^+ + 2\nu$. From the masses of the nuclides involved estimate the fraction of the total energy released that is carried off by neutrinos in each branch.

14. If the sun's energy comes predominantly from the *p-p* chain, and if the neutrinos produced are absorbed only negligibly in the sun, what is the flux at the earth of neutrinos from this source? *Answer:* $7 \times 10^{10}\,cm^{-2}\,s^{-1}$.

15. (a) If the *s*-process (neutron capture on a slow time scale) in a star starts with a mixture of ^{56}Fe and ^{58}Ni, which of the stable nuclides with mass numbers between 56 and 85 do you expect to be formed by this process? (b) Which

stable nuclides in this mass range were presumably formed predominantly by the *r*-process? (c) Which nuclides in the same mass region cannot be accounted for by either of these neutron-capture processes, and what type of reaction might be invoked for their synthesis? *Answer:* (b) ^{70}Zn, ^{76}Ge, ^{82}Se.

16. If the collapsing core of a supernova has a mass equal to that of the sun, what will be the radius of the resulting neutron star? Take the final density as 5×10^{14} g cm^{-3}. *Answer:* 10 km.

Chapter 14

Nuclear Energy

A. BASIC PRINCIPLES OF CHAIN-REACTING SYSTEMS

Speculation about the possible use of nuclear processes for the large-scale production of power dates back to the early years of radioactivity. Only with the discovery of fission did such applications become a real possibility. The unique feature of the fission reaction that makes it suitable as a practical energy source is the emission of several neutrons in each neutron-induced fission; this makes a chain reaction possible.

Chain Reaction. The condition for the maintenance of a chain reaction is that on the average at least one neutron created in a fission process cause another fission. This condition is usually expressed in terms of a **multiplication factor** k, defined as the ratio of the number of fissions produced by a particular generation of neutrons to the number of fissions giving rise to that generation of neutrons. If $k < 1$, no self-sustaining chain reaction is possible; if $k = 1$, a chain reaction is maintained at a steady state; if $k > 1$, the number of neutrons and therefore the number of fissions increases with each generation, and a divergent chain reaction results. An assembly of fissionable material is said to be **critical** if $k = 1$ and **supercritical** if $k > 1$.

Since one neutron per fission is required to propagate the chain reaction, the number of neutrons increases by the fraction $k - 1$ in each generation. Thus the rate of change of the number of neutrons in a chain-reacting system is

$$\frac{dN}{dt} = \frac{N(k-1)}{\tau}$$

where τ is the average time between successive neutron generations. By integration we find that at time t the number of neutrons is

$$N = N_0 e^{(k-1)t/\tau}, \tag{14-1}$$

where N_0 is the number of neutrons at $t = 0$. If τ is very short (as it is when no moderator is used and fission takes place with fast neutrons) and if k is suddenly made to exceed unity by an appreciable amount, the chain reaction can proceed explosively as in a fission bomb.

A nuclear reactor is an assembly of fissile material—^{235}U (either in normal uranium or enriched), ^{239}Pu, or ^{233}U— arranged in such a way that a controlled, self-sustaining chain reaction is maintained. In a nuclear reactor, k is kept equal to unity for steady operation. However, a reactor must be designed in such a way that k can be made slightly larger than one (say 1.01 or 1.02) to allow the neutron flux and therefore the power to be brought up to a desired level. Control of the reactor, for example by the motion of neutron-absorbing control rods, is possible only if τ is not too short. Assume that $\tau = 10^{-3}$ s (approximately the life expectancy of a thermal neutron in graphite or D_2O) and $k = 1.001$. Then, according to (14-1), $N = N_0 e^t$, where t is in seconds, and the neutron level will increase by a factor e every second or by a factor of about 2×10^4 in 10 s. This would be too rapid an increase for safe and convenient control.

Effect of Delayed Neutrons. Fortunately some of the neutrons produced in fission are emitted by highly excited fission products and therefore with time delays controlled by the half lives of their β-decay precursors (see p. 166). These delayed neutrons increase the average time between neutron generations substantially. As long as $k - 1$ is smaller than the fraction of neutrons that are delayed—for thermal-neutron fission of ^{235}U that fraction is 0.0070, for that of ^{239}Pu it is 0.0024—the effective time τ between generations is approximately

$$\tau = \tau_0 + \sum_i \left(\frac{f_i}{\lambda_i}\right),$$

where τ_0 is the generation time without delayed neutrons and the f_i's are the fractions of neutrons delayed with the decay constants λ_i. For the approximately 50 delayed-neutron emitters in ^{235}U fission for which f_i's have been measured[1] $\Sigma(f_i/\lambda_i) = 0.080$ s, which is large compared with τ_0. Hence $\tau \approx 0.08$ s and, with $k = 1.001$, the **period** of the system [the time t required to make $N/N_0 = e$, see (14-1)] is about 80 s, which provides ample time for control. It is worth noting that even in so-called fast reactors, that is, reactors in which the chain reaction is propagated by fast neutrons and in which neutron generation times τ_0 are as short as 10^{-7} s, the reactor period for small values of $k - 1$ is still entirely determined by the delayed neutrons (and these are, in fact, even somewhat more abundant in fast- than in thermal-neutron fission).

Multiplication Factor in Infinite Medium. The multiplication factor in a medium of infinite extent, denoted by k_∞, is given by the product of the number ν of neutrons emitted per fission and the fraction of the neutrons that produce another fission. This fraction is the ratio of the macroscopic

[1] The identification and characterization of practically all the individual nuclides that contribute significantly to delayed-neutron emission is fairly recent. In earlier work the gross decay of neutrons was analyzed into six half-life groups, and reactor engineers still use this type of analysis because it is both simple and adequate for their purposes.

fission cross section ($\sigma_f N_f$, where N_f is the number of fissionable nuclei per cubic centimeter) to the sum of this macroscopic fission cross section and all macroscopic capture cross sections:

$$k_\infty = \nu \frac{\sigma_f N_f}{\sigma_f N_f + \sum_i \sigma_{ci} N_i}, \tag{14-2}$$

where N_i is the number of nuclei of the ith substance per cubic centimeter and σ_{ci} is the (ordinary) capture cross section of that substance.[2] The nonfission capture of the fissile material used must be included in $\Sigma_i \sigma_{ci} N_i$.

Since the competition between radiative capture and fission reactions in the fissile material sets an upper limit to the multiplication factor attainable, regardless of the properties of other materials present, the ratio σ_c/σ_f of capture to fission cross section, generally referred to as α, is an important quantity for fissile nuclides. For a given nuclide α is strongly energy-dependent, going through large fluctuations in the resonance region (~1–10^4 eV) and approaching zero in the MeV range, as might be expected from the energy dependence of (n, γ) cross sections. Because of the finite amount of (n, γ) competition in any practical neutron spectrum, the number of neutrons produced per neutron absorbed in the fissile material, designated as η, is always smaller than the number ν of neutrons per fission; in fact, $\eta = \nu/(1+\alpha)$. In table 14-1 the values of σ_f, σ_c, α, ν, and η are given for the three fissile nuclides ^{235}U, ^{239}Pu, and ^{233}U. The "fast-neutron" quantities listed refer to a representative spectrum as it might

Table 14-1 Some Properties of Fissile Materials at Thermal-Neutron and Fast-Neutron Energies[a]

	^{235}U		^{239}Pu		^{233}U	
Value	Thermal	Fast	Thermal	Fast	Thermal	Fast
σ_f(in barns)	580 ± 2	1.44	742 ± 3	1.78	531 ± 2	2.20
σ_c(in barns)	98 ± 1	0.22	271 ± 3	0.15	47 ± 1	0.15
$\alpha = \sigma_c/\sigma_f$	0.169 ± 0.002	0.15	0.366 ± 0.004	0.084	0.089 ± 0.002	0.068
ν	2.423 ± 0.007	2.52	2.880 ± 0.009	2.98	2.487 ± 0.007	2.59
$\eta = \nu/(1+\alpha)$	2.073 ± 0.006	2.19	2.108 ± 0.007	2.75	2.284 ± 0.006	2.43

[a] The thermal-neutron data are given for $v = 2200\ \mathrm{m\ s^{-1}}$; the fast-neutron data represent weighted averages over a typical reactor neutron spectrum.

[2] Since there is always a spectrum of neutron velocities present and cross sections generally vary with neutron velocity, the expression for k_∞ in (14-2) must in fact be appropriately averaged over the velocity spectrum.

exist in an unmoderated reactor, but it should be clear that the actual quantities in anything but a thoroughly thermalized spectrum will depend rather sensitively on the exact spectrum shape, particularly in the resonance region.

Critical Size. In a reactor of finite extent the multiplication factor k is smaller than k_∞ because of the loss of neutrons by leakage through the surface. The smaller the reactor, the greater its ratio of surface to volume and therefore the greater the loss. Quantitative estimates of neutron losses from a reactor surface are very complicated but are quite essential to any estimate of critical size. As a rough approximation, the fractional loss of neutrons for thermal reactors is proportional to the sum $L_S^2 + L^2$, where L_S and L are the average (crow-flight) distances traveled in the moderating medium (of infinite extent) by a fission neutron before reaching thermal energy (L_S) and after reaching thermal energy (L). For a spherical reactor of radius R the approximate relation is $k_\infty - k \approx \pi^2 R^{-2}(L_S^2 + L^2)$. The critical radius R_c is that radius for which $k = 1$. Thus

$$R_c \approx \pi(L_S^2 + L^2)^{1/2}(k_\infty - 1)^{-1/2}. \quad (14\text{-}3)$$

The **slowing-down length** L_S may be known from measurements on various moderators and generally is not appreciably altered by the addition of fuel to the moderator. The **diffusion length** L in the fuel-moderator mixture will be smaller than that of the pure moderator (L_0) and is given by $L^2 = xL_0^2$, where x is the fraction of neutron absorptions that take place in the moderator. Table 14-2 gives values of L_S, L_0, macroscopic absorption cross section ($\sigma_m N_m$), and density for some popular moderators.

In most reactors x, which is the fraction of neutrons absorbed by the moderator, is kept small for reasons of neutron economy. Therefore a crude approximation to (14-3) that neglects L, namely, $R_c \approx \pi L_S(k_\infty - 1)^{-1/2}$, gives the right order of magnitude of the critical size for practical thermal reactors. To obtain a numerical estimate of R_c we must know k_∞. As a rough approximation we set $k_\infty = \eta$, although it must, in fact, always be

Table 14-2 Properties of Moderators

Moderator	L_S (cm)	L_0 (cm)	$\sigma_m N_m$ (cm^{-1})	Density ($g\ cm^{-3}$)
H_2O	5.6	2.76	0.022	1.00
D_2O	11.0	100	0.000080	1.10
Be	9.2	21	0.0012	1.84
C	18.7	54.2	0.00032	1.70

somewhat less.[3] For a solution of ^{235}U in H_2O we thus estimate the critical radius as

$$R_c \approx \frac{\pi L_S}{(k_\infty - 1)^{1/2}} \approx \frac{\pi \times 5.7}{(2.1-1)^{1/2}} = 17\ \text{cm}.$$

Here we assume that the concentration of the solution is great enough to ensure that neutron reaction with ^{235}U is much more probable than capture by hydrogen. The ratio of the two cross sections is $678/0.332 \approx 2 \times 10^3$, so that this condition is met if the concentration of ^{235}U is a few tenths of a mole per liter. The first homogeneous reactor put into operation, the Los Alamos Water Boiler (1944), consisted of a stainless-steel sphere, 30 cm in diameter, filled with a $\sim 1.2M$ solution of uranyl sulfate in H_2O, the uranium being enriched to 14.6 percent in ^{235}U.

In many reactors the fissile material is not dissolved in the moderator but is separated from it in a heterogeneous arrangement. All the reactors that have been constructed with normal uranium as the fuel have the uranium in lumps or rods arranged in a lattice embedded in graphite or heavy water, which are the only moderators with sufficiently small macroscopic absorption cross sections (see table 14-2). The need for this kind of arrangement arises from the existence of several strong absorption resonances in ^{238}U in the energy range between 6 and 200 eV. In a homogeneous mixture of uranium and moderator the probability that a neutron during the slowing-down process is absorbed by $^{238}U(n, \gamma)$ reaction in the resonance region is quite large and the resonance escape probability p is therefore too small to allow a $k_\infty > 1$ (see footnote 3). If, however, the uranium is arranged in aggregates, a much greater fraction of the neutrons will be slowed down in the moderator to energies below the resonance region before encountering uranium nuclei. The optimum lattice spacing is approximately L_S for the moderator. For a typical lattice of normal uranium embedded in graphite $k_\infty \approx 1.07$ and, according to (14-3), the critical radius of a spherical assembly would be $R_c = \pi \times 18.7 \times (0.07)^{-1/2} \approx$ 220 cm. For a cubic assembly the length of an edge is, in the same approximation, $\sqrt{3}R_c$, or about 3.8 m. The actual critical size for a bare, cubic lattice of normal uranium and graphite with optimal uranium rod size ($\sim$1.4 cm diameter) is about 5.5 m on a side.

[3] Frequently k_∞ is expressed as the product of four factors: $k_\infty = \eta \epsilon p f$, where: ϵ, called the **fast-fission factor**, is the ratio of the total number of fast neutrons slowing down past the ^{238}U fission threshold to the number of fast neutrons produced by thermal-neutron fissions; p, called the **resonance escape probability**, is the fraction of neutrons that escape capture while slowing down; and f, called the **thermal utilization**, is the ratio of thermal neutrons absorbed in fuel to total thermal neutrons absorbed in all materials. Both p and f depend on the nature and arrangement of fuel and moderator. The aim is generally to make them as near to unity as possible. ϵ can actually be slightly greater than 1. For homogeneous mixtures of normal uranium ($\eta = 1.33$) with H_2O, graphite, or beryllium, the product $\epsilon p f$ is always too small for k_∞ to exceed 1.

In all practical reactors the core is surrounded by a neutron reflector that reduces neutron loss. This makes the necessary size of the reactor core slightly smaller, but operating in the other direction are effects of impurities, provisions for cooling and for control, and overdesign. Another important effect of the reflector in power reactors is the increase in neutron flux in the outer parts of the core. Because the power level is likely to be limited by the temperature rise at the center, this makes the outer parts contribute a better share to the overall power output. In addition, the fuel lattice may be altered near the center to flatten the neutron- and power-flux distributions.

Reactivity and Reactor Control. Although steady operation of a reactor implies that $k = 1$, the reactor must always be designed such that k can be made to exceed unity. This is necessary not only to make it possible to bring the reactor up to the desired power level but also to allow for some fuel burnup, for the buildup of neutron-absorbing fission products, and for the deliberate introduction of neutron-absorbing materials, for example, for radionuclide production or radiation effect tests. The quantity $(k-1)/k$ is called the reactivity, and reactivity is zero when $k = 1$.

The usual method of handling the excess reactivity that must be built into a reactor is through the use of control rods made of materials with large neutron-capture cross sections, such as boron, cadmium, or hafnium. These control rods are moved in and out of the reactor to compensate for any changes in reactivity. Other methods of control use motion of fuel elements or of the reflector.

As mentioned above, the possibility of fission product "poisoning" is an important reason for providing reserve reactivity. The most troublesome of the fission products, as regards neutron absorption, is ^{135}Xe, which has a half life of 9.1 h and a cross section for thermal-neutron absorption of 2.6×10^6 b, the largest neutron cross section known. In steady-state operation of a high-flux reactor the presence of this poison can reduce k by about 0.04. Furthermore, the concentration of ^{135}Xe increases after shutdown of the reactor, because it continues to be formed by the decay of its parent, 6.6-h ^{135}I, but is no longer being consumed by (n, γ) reaction. The poisoning effect reaches a maximum about 10 hours after shutdown, and at this time it can cause a reduction of as much as 0.3 in k. Since reactors are not built with that much reserve reactivity, there may be some time period after shutdown during which the reactor cannot be brought back to criticality.

B. REACTORS AND THEIR USES

Early History. Reactors were originally developed during World War II for the production of ^{239}Pu as a nuclear weapons material. The very first man-made chain-reacting system was the famous "pile" constructed by

Fermi and his co-workers under the West Stands of the University of Chicago's Stagg Field and brought to criticality on December 2, 1942. It was literally a pile of graphite blocks, stacked layer by layer to form an approximately spherical assembly, with 40 tons of normal uranium in the form of metal and oxide lumps arranged in a cubic array imbedded in the 385 tons of graphite. No cooling was provided, and the power level was therefore limited to a few kilowatts. Successes in achieving a self-sustaining chain reaction in this experimental device led immediately to the construction (within less than one year!) of an air-cooled 1000-kW graphite-uranium reactor at Oak Ridge, Tennessee—the X-10 reactor, which operated for 20 years as a most successful research tool—and of the large, water-cooled, graphite-moderated plutonium production reactors at Hanford, Washington. In all of these and other early reactors, the heat produced by the fission reaction was entirely wasted. It was only later that production of useful power became one of the major goals of reactor engineers.

In considering power production from nuclear fission, it is useful to remember that the energy released in one fission event is about 200 MeV or 3.2×10^{-4} erg $= 3.2 \times 10^{-11}$ W s. Therefore about 3×10^{10} fissions per second are required to produce one watt of power. In other, more easily remembered terms, this means that 1 megawatt (MW) of reactor power[4] corresponds to the fission of about 1 g of fissile material per day. It also follows that for every megawatt day (MWd) of reactor operation, approximately 1 g of fission products is formed. In a reactor fueled with normal uranium the plutonium production is also of the same order. For example, if the number of neutron captures in ^{238}U is half the number of ^{235}U fissions, which is a typical situation in normal-uranium reactors, 0.5 g of ^{239}Pu is produced per megawatt day.

Reactor Types. A wide variety of reactors are in operation in all parts of the world, and the number is increasing year by year.[5] Existing reactors range from small devices, operating at a few watts and principally used as teaching tools, to power reactors delivering nearly 1200 MW*e*.

Reactors may be classified in a variety of ways. We may distinguish between reactors operating on thermal neutrons, fast neutrons, and intermediate or partially moderated neutrons. The fuel may consist of natural uranium, of uranium enriched to various degrees[6] in ^{235}U, of ^{239}Pu, of ^{233}U,

[4] We are speaking here of *total* or *thermal* power. The *electrical* power output of a reactor is always smaller, typically about one third the thermal power. The two quantities are distinguished by using MW*t* and MW*e* for the power ratings.

[5] A catalog of power reactors is periodically published by the International Atomic Energy Agency (P5). In the 1977 edition 224 operating reactors in 21 countries are listed, 68 of them in the United States, rated at about 49,000 MW*e*. In addition, several hundred research and test reactors are in operation.

[6] Terms frequently used are "slightly enriched" (about 2–5 percent ^{235}U), "highly enriched" (typically 20-30 percent), and "fully enriched" (>90 percent).

or even of some mixture of these. The most widely used moderators are light water, heavy water, and graphite, but other materials such as beryllium, BeO, or organic compounds have also been used. The coolant may be air, helium, CO_2, H_2O, D_2O, or a liquid metal such as sodium. We may distinguish between reactors in which fuel and moderator are homogeneously mixed, and the more prevalent heterogeneous reactors. In terms of purpose, reactors may be designed primarily to produce fissile nuclides (^{239}Pu or ^{233}U), to produce useful power, to serve as test facilities for reactor components, to provide excess neutrons for the production of other nuclides (such as ^3H), to serve as research tools, or for some combination of these purposes.

A more detailed discussion of the various types of reactors is beyond the scope of this book. In chapter 15 we briefly consider some aspects of research reactors as neutron sources. Here we confine ourselves to a few comments on reactors for the generation of electrical energy and for ship propulsion.

Reactors for Electric Power Generation. Since the first small-scale

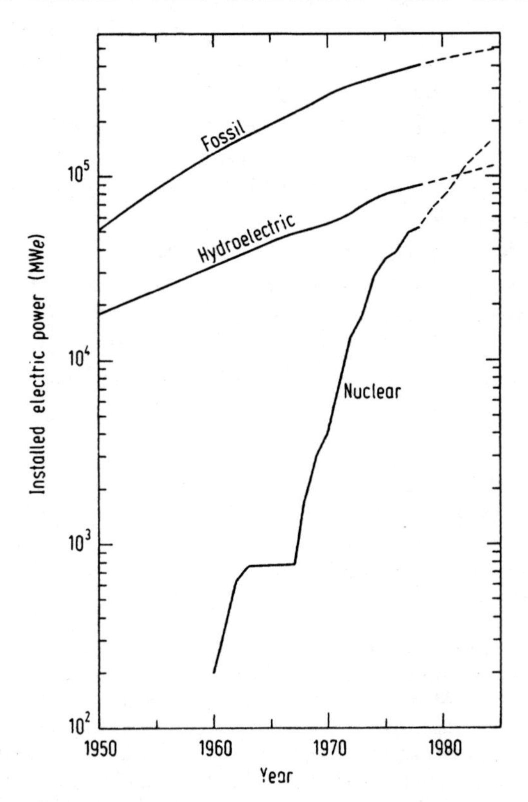

Fig. 14-1 Actual and projected installed electric generating capacity in the United States. (Data from various US government reports.)

application of nuclear reactors to the generation of useful power in the mid-1950s considerable development has taken place in this field. Figure 14-1 shows the growth of nuclear power generation in the United States, together with the corresponding curves for fossil-fueled and hydroelectric plants. In 1978 nuclear-powered plants produced about 14 percent of the total electric power (or about 4 percent of total power) in the United States, and their share is expected to increase.[7] Similar or even more rapid increases in nuclear power generation are occurring and projected in many other countries. Worldwide nuclear generating capacity was 1.1×10^5 MW*e* in 1977 and is expected to increase to 3.5×10^5 MW*e* in 1984 and to $(1\text{–}2.5) \times 10^6$ MW*e* by the year 2000.

In all existing power reactor systems electrical energy is generated by steam-driven turbines, the steam being produced in heat exchangers heated by the primary reactor coolant (except in boiling-light-water reactors, in which steam from the primary coolant is used directly).

Among the desirable characteristics of an efficient power reactor are: high operating temperature to improve the efficiency of converting thermal to electrical energy; large specific power (power per kilogram of fuel) to minimize the inventory of fissile material tied up in the reactor; and large burnup[8] of fissile material to minimize the cost of losses in fuel processing and fuel element refabrication. It also turns out that nuclear power plants in general become more economical with increasing power levels. The design power has therefore tended to increase as the technology has advanced, and at the present time reactors rated at or above 1000 MW*e* are favored.

Most power reactors in operation and under construction as of the late 1970s operate on thermal neutrons and are fueled with uranium, either of natural composition or slightly or highly enriched in ^{235}U. Great Britain, which turned to nuclear power early, based its industry on graphite-moderated, CO_2-cooled reactors fueled with natural uranium, a type refer-

[7] The long time span between initial planning and actual operation of a nuclear power plant should make projections for several years into the future fairly reliable. Most of the plants included in the 1984 forecast were already under construction in 1979. However, increasingly rigorous safety requirements and lengthy licensing procedures have caused many delays in bringing nuclear power plants into operation.

[8] **Burnup** is a measure of the amount of fissile material consumed before fuel must be removed for processing, either because of mechanical deterioration or buildup of neutron-absorbing "poisons". It is usually expressed in megawatt days per ton of initial fuel (MWd ton^{-1}). Recall that 1 MWd corresponds to fission of 1 g of material. Thus in a thermal reactor fueled with pure ^{235}U, burnup of 10^5 MWd ton^{-1} would correspond to the fission of 10^5 g of ^{235}U, and therefore to the disappearance of $10^5(1+\alpha) = 1.17 \times 10^5$ g or 11.7 percent of the ^{235}U. Actually this is not quite correct, because both fission and nonfission reactions in ^{236}U would have to be taken into account. If the fuel contains ^{238}U, the situation becomes even more complicated because of the buildup of ^{239}Pu and its contribution to the fission energy and thus to the burnup. Thus it is possible, in a very efficient reactor, for the burnup in megawatt days per ton to exceed the number of grams of *original* fissile material per ton of fuel.

red to as gas-cooled reactors (GCRs). France initially followed the same path, but later switched to pressurized-water reactors. Most of the plants in the United states, Japan, the Federal Republic of Germany, and several other countries use slightly enriched uranium fuel and are cooled with highly pressurized water (pressurized-water reactors, PWR); the water pressures are typically about 2000 psi ($\sim$140 atm or $\sim 1.4 \times 10^7$ Pa). Less commonly cooling is done with boiling water, which eliminates the need for intermediate heat exchangers (boiling water reactors, BWR). Canada has emphasized heavy-water cooling and natural uranium fuel (pressurized heavy-water reactors, PHWRs also known as CANDUs). The Soviet Union has had a somewhat more diversified program.

In the early versions of most of the reactor types mentioned the fuel elements consisted of uranium metal clad in a corrosion-resistant metal. Later models almost universally use sintered UO_2 pellets clad in stainless steel or zircaloy (zirconium alloyed with small amounts of tin, iron, chromium, and sometimes nickel). The oxide is much less subject to radiation damage than the metal (thus allowing higher burnup), is much less reactive with water (thus minimizing the hazard of cladding failures), and retains gaseous fission products much better, even at high burnup. With PWRs and BWRs efficiencies up to about 33 percent, burnups of about 30,000 MWd ton^{-1}, and specific powers of 20–40 kW per kg uranium have been attained, and modern plants of this type have power ratings of 700–1200 MW*e*. Efficiencies and power levels of GCRs can be comparable, but they have appreciably lower burnups and specific power. However, in a more recently realized gas cooled reactor concept, the high-temperature gas-cooled reactor (HTGR), efficiencies of about 40 percent, burnup of 100,000 MWd ton^{-1}, and specific power of over 900 kW kg^{-1} ^{235}U have been attained. These reactors are fueled with fully enriched ^{235}U in the form of mixed uranium-thorium carbide, moderated by graphite, and cooled with helium, which leaves the reactor core at >700°C.

Converters. The only naturally occurring nuclide that is fissile by thermal neutrons is ^{235}U, which constitutes 0.72 percent of normal uranium. The other two important fissile nuclides, ^{239}Pu and ^{233}U, are produced from ^{238}U and ^{232}Th, respectively, by neutron capture, followed by successive β^- decays:

$$^{238}\mathrm{U}(n, \gamma)^{239}\mathrm{U} \xrightarrow[23.5\ \mathrm{min}]{\beta^-} {}^{239}\mathrm{Np} \xrightarrow[2.35\ \mathrm{d}]{\beta^-} {}^{239}\mathrm{Pu},$$

$$^{232}\mathrm{Th}(n, \gamma)^{233}\mathrm{Th} \xrightarrow[22.2\ \mathrm{min}]{\beta^-} {}^{233}\mathrm{Pa} \xrightarrow[27.0\ \mathrm{d}]{\beta^-} {}^{233}\mathrm{U}.$$

^{238}U and ^{232}Th are called **fertile** nuclides. In any reactor that contains fertile nuclides, either in the core or in the reflecting "blanket," some fraction will be converted to fissile end products. Such reactors are called converters, and the **conversion ratio** is defined as the number of new fissile nuclei (^{239}Pu and ^{233}U) formed divided by the number of ^{235}U nuclei destroyed. Initial conversion ratios in reactors fueled with natural or slightly enriched uranium are typically between 0.4 and 0.8. Conversion ratios of ~0.9 are possible with reactors based on the ^{232}Th-^{233}U system. As a given reactor operates for a while, the conversion ratio tends to decrease to a steady-state value lower than the initial one because some of the newly formed fissile species themselves are consumed by neutron capture or fission. The more highly enriched the fuel, the lower will be the conversion ratio, unless fertile material is used in the blanket or deliberately mixed with the fuel (as is done in the HTGR).

It is interesting to estimate the amount of plutonium production in present-day power reactors. Taking the figure 1.1×10^5 MW*e* for the world wide installed nuclear power and assuming 30 percent efficiency, we get ~3.7×10^5 MW*t*. If the reactors operate on the average 250 days a year, the number of megawatt days per year is about 9×10^7 and, with an average conversion ratio of 0.4, this corresponds to production of ~3.5×10^7 g or 35 tons of ^{239}Pu per year, enough for the fuel requirements of about 30 large power reactors. Whether or not this plutonium is actually recovered and used and, in fact, whether the unburned ^{235}U is recovered and re-enriched is a question not only of economics, but also of safeguards against possible abuses (see section C3 below). The issues involved are vigorously debated in many countries.

Breeders (H1). It has long been recognized that it is in principle possible in a reactor to attain conversion ratios greater than unity, that is, to produce more fissile nuclei than are consumed. Such a reactor is called a breeder, and the necessary condition for breeding is that η, the number of neutrons produced per neutron absorbed in the fissile material, must exceed 2: one neutron to produce another fission so that the chain reaction can proceed, and more than one neutron to be absorbed in fertile material to produce new fissile material. By how much η has to exceed 2 depends on how small other neutron losses (absorption in other materials, resonance absorption in the fissile species, escape from the reactor) can be made. As is readily apparent from the η values in table 14-1, the best candidate for a practical breeder is a ^{239}Pu-fueled reactor operating on fast neutrons, and efforts in a number of countries have indeed been largely concentrated on developing these so-called fast breeders. Attempts also continue towards development of a thermal breeder based on the ^{232}Th-^{233}U system (the η values of ^{235}U and ^{239}Pu at thermal energies are definitely too small for breeding). Such a breeder would make possible the full utilization of the relatively abundant (compared to uranium) resources of thorium. However, appreciable technical problems remain to be solved. The subject of thermal breeders is reviewed in P1.

Fast reactors have advantages as well as disadvantages relative to thermal-neutron reactors. The small fission cross sections for fast neutrons (table 14-1) dictate high fuel concentrations in the core, and the resultant high power densities lead to difficult heat-transfer problems. On the other hand, the choice of structural materials and fuel-alloy components is much wider in fast reactors because most elements have rather small neutron absorption cross sections in the energy range of interest. Similarly, the buildup of fission-product poisons is less serious.

First attempts to design a fast breeder were made as early as 1944, and several prototype fast breeders were built in the United States, the Soviet Union, and Britain prior to 1960. They were fueled with metal—initially uranium, later replaced by plutonium—and cooled with sodium or other liquid metals. The temperatures were only 300–500°C, burnup and specific power were quite low, and most of the breeding was done in the blankets that typically consisted of uranium depleted in ^{235}U (an abundant by-product of the ^{235}U isotope separation plants). These first-generation breeders indeed demonstrated that breeding was possible, but they did not provide an economically viable solution. In particular, it became clear that high burnup had to be achieved in order to minimize the number of times each fissile atom has to go through the chemical-processing and fuel-fabrication cycle with the attendant losses and costs.

Prototypes of second-generation fast breeders, which began to go into operation around 1970 (first in the Soviet Union to be followed by Britain, France, Germany, and the United States) are all quite similar to each other. They use mixed PuO_2-UO_2 (depleted in ^{235}U) as fuel, usually clad in stainless steel, with sodium at 500–600°C as coolant. Burnup is in the range of 50,000–100,000 MWd ton^{-1}, specific power 700–1000 kW*t* per kilogram of fissile material. The efficiencies for conversion from thermal to electrical power are 35–44 percent, and breeding ratios are 1.2–1.5. Most of the prototypes are designed for 250–350 MW*e*, but for economic reasons later models will undoubtedly be designed to deliver at least 1000 MW*e* each.

Since widespread use of breeders would in effect greatly extend the world's uranium resources, it is considered by many a desirable or even an essential option (see, e.g., R1). On the other hand, concerns about the dangers of nuclear proliferation strongly militate against reliance on fast breeders and in the United States have brought about a virtual moratorium on the completion of even a prototype fast breeder. Discussion of the relative merits of nuclear and other (principally coal-fired) electrical generating plants is outside the scope of this book. However, *if* nuclear power is to play a substantial role in the world's energy supply well into the twenty-first century, the use of breeding will almost certainly be dictated simply by the limitations on uranium resources. The world's "reasonably assured reserves" of uranium at a price of $\leq$\$60 per kg of U_3O_8 are about 2×10^6 tons (W1). This is sufficient to fuel perhaps 400 reactors of 1000 MW*e* each for 30 years without fuel reprocessing or breeding. Estimated additional uranium resources would roughly double these figures, which still falls far short of even the most modest forecasts of the world's electrical energy needs in the year 2000. Since breeders can, in principle, convert all ^{238}U and ^{232}Th into the fissile nuclides ^{239}Pu and ^{233}U, the use of breeders would multiply nuclear fuel resources by at least two orders of magnitude.

Reactors for Propulsion. The use of nuclear power for propulsion purposes was suggested almost as soon as chain-reacting systems were developed. The small amount of fuel required per unit energy produced appears immediately as an attractive feature. On the other hand, the necessity of a massive radiation shield around a nuclear reactor makes the weight-to-power ratio for nuclear engines so large that their use in small vehicles such as automobiles is ruled out. Nor has a nuclear-powered aircraft ever been built, although at one time considerable development work was done toward this goal.

For ship propulsion, on the other hand, nuclear reactors have proven very advantageous (P2). For submarines, in particular, nuclear engines are a great boon because they enable subs to remain submerged for almost unlimited periods of time and endow them with greatly extended ranges between refueling stops as well as with higher speeds. Most nuclear-powered ships use PWRs of various designs, with either slightly or highly enriched ^{235}U as fuel. The original development of the PWR was, in fact, for use in submarine engines. Subsequent developments have been aimed at increasing the life of the reactor core in order to extend the cruising range between refueling stops. As in stationary power plants, high burnup is thus desired. Modern submarines are estimated to have ranges of 600,000 km on one reactor core.

Aside from ship propulsion, the most promising area for the application of nuclear engines to transportation is in the space program. For a given weight, nuclear rocket engines can, in principle, produce much larger final velocities than chemical rocket engines. This consideration could be particularly important in space missions undertaken from parking orbits. Intensive research and development on nuclear rockets was underway for a number of years in the United States (G1), but has been discontinued.

Another space application of reactors is not for propulsion, but for power sources aboard satellites and space vehicles. Both in the United States and in the Soviet Union such reactors have been developed and used in various space missions (P3). Those flown up to now have rather low power ratings, but systems producing useful power up to about 1 MW have been designed.[9]

[9] Also important in the space programs have been thermoelectric generators powered by radionuclides, principally the α emitter ^{238}Pu. They have provided power in the 10–500 W range for weather satellites, lunar-surface experiments, and planetary missions (L1). The US program for reactors and radioisotope power sources for space applications is called SNAP (Space Nuclear Auxiliary Power).

Natural Reactors (C1, M1). In 1972, just 30 years after humans first achieved a self-sustaining chain reaction, a group of French scientists made the fascinating discovery that nature had done it long before! The chance discovery that led to this conclusion was the finding that uranium from a rich uraninite ore in the Oklo mine in Gabon, West Africa, was depleted in ^{235}U, with isotopic abundances down to 0.4 percent whereas no naturally occurring uranium had ever before been found to deviate appreciably from the normal 0.720 percent abundance of ^{235}U. Mass-spectrometric analyses of rare earths and other elements in the Oklo ore proved that their isotopic compositions labeled them unmistakably as fission products. Such analyses even made it possible to deduce the total neutron fluence ($\sim 10^{21}\ cm^{-2}$) during the life of the reactor (through the altered isotopic compositions caused by the known high neutron-capture cross sections of some fission products such as ^{143}Nd) and the contribution of ^{239}Pu to the total number of fissions (because of different fission yield distributions for ^{235}U and ^{239}Pu). From other data the duration of the reactor operation—$(6–8) \times 10^5$ y—and the power level—~ 20 kW—could be deduced.

The age of the ore deposit has been determined by various dating methods (see chapter 13, section A) to be $(1.8–2.0) \times 10^9$ y. At that time the ^{235}U abundance was over 3 percent, and all the findings at Oklo are entirely consistent with the picture that, shortly after the formation of that deposit, its uranium-richest parts functioned as PWRs fueled by what was then natural, but would now be called "slightly enriched" uranium. There are, in fact, five such reactor zones at Oklo[10] and the geologic evidence shows that, at the time the reactors were operating, they were buried under several kilometers of rock, which accounts for the high pressures under which they must have been operating. The temperature coefficient of reactivity (due to neutron-capture resonances) provided a natural control mechanism for the reactors (see p. 535).

One of the interesting conclusions from the Oklo phenomenon is that the ore deposit has been so stable as to retain most of the fission products over nearly 2×10^9 y. This fact may have important implications for the problem of radioactive-waste disposal.

C. REACTOR-ASSOCIATED PROBLEMS

1. Reactor Safety

Although, from the points of view of both technology and economics, nuclear reactors are well suited to play a major role in fulfilling the world's energy needs for a long time to come, vigorous debate continues in a

[10] It is more accurate to say that there *were* five such zones. After a moratorium on mining activities to allow thorough exploration of the phenomena, mining was resumed in 1975, and except for one small section preserved as a monument, the ore in the reactor zones has been completely mined out. However, many samples have been retained.

number of countries, including the United States, on whether they can be made sufficiently safe and sufficiently free of environmental effects.

Contrary to some popular misconceptions a reactor cannot, under any circumstances, explode with the force of an atomic bomb. The conditions required for a major explosion of fissile material are very special (S1) and are most unlikely to be achieved accidentally, certainly not in a reactor. For a fission chain reaction to proceed to the intensity of a major explosion, say equivalent to at least 10^3 tons of TNT (which would be a very small fission explosion), about 50 g of fissile material would have to be fissioned before the critical mass has blown itself apart, that is, within the order of a microsecond. With the multiplication factors k achievable in reactors the neutron multiplication is far too slow for such conditions to be reached.

Although a major nuclear explosion cannot happen in a reactor, other kinds of serious reactor accidents are surely conceivable and must be carefully guarded against, which means that they must be made so improbable through multiple and redundant safety devices that the residual risk is considered acceptable in view of the benefits to be gained. Every reactor must be designed, built, and operated in such a way that (1) personnel in the plant are not exposed to any undue hazard during normal operation and maintenance, and (2) neither normal operation nor any credible accident constitutes a danger to the health and safety of the population in the surrounding area.

Radiation- and Thermal-Pollution Hazards. All reactors are shielded sufficiently to keep radiation levels outside the shield well within accepted levels. Since the primary coolant becomes highly radioactive through exposure to the reactor neutrons, the primary cooling circuit and heat exchanger are also shielded, and appropriate precautions are taken against escape of radioactivity through pipe, pump, or valve failures. Planned releases of radioactivity to the atmosphere or in liquid effluents are carefully controlled and monitored (D1, D2).[11] Accidental releases have to be contained.

A much-discussed environmental hazard is thermal pollution. That part of a reactor's thermal power that is not converted to electrical energy is eventually released as heat, often to a body of water. Attention must indeed be given to appropriate dispersal of this heat to prevent ecologically harmful temperature rises. The alternative of using cooling towers to release the heat to the atmosphere is now widely practiced, although somewhat more costly. It should also be pointed out that the thermal pollution problem is not unique to nuclear plants, although it is somewhat less severe for fossil-fueled power plants because of their higher efficiency for conversion from thermal to electrical energy.

[11] In early air- and water-cooled reactors considerable radioactivity was released because the coolant made just one pass through the reactor.

Hazard Control. The best safeguards against reactor accidents are careful design, inherent safety features, highly trained and qualified operators, sound operating procedures, interlocks and controls subject to frequent operational tests, and appropriate caution with regard to untried materials. In the United States each reactor proposal is carefully scrutinized to insure its safety in every respect. The procedure involves Hazards Reports, Environmental Impact Statements, public hearings, and reviews by various governmental bodies before construction of a reactor can commence and again before an operating license is granted.

The geology, hydrology, and meteorology at the reactor site, as well as the population distribution in its vicinity, are factors considered in the hazards evaluation. The structure must be designed to withstand natural disasters such as earthquakes and tornadoes as well as accidents such as airplane crashes. The possibility of sabotage must also be carefully guarded against.

As part of the Hazards Report, the applicant must show convincingly that what is called the **maximum credible accident** will not expose the general public to any significant hazard. Unfortunately there does not appear to be universal agreement on what constitutes convincing evidence in this context. The maximum credible accident typically involves loss of coolant and consequent meltdown of the reactor core, with release of all the volatile and much of the nonvolatile fission products. Consideration of such an eventuality requires that reactors be completely enclosed in containment shells (usually made of steel or prestressed concrete) capable of withstanding the maximum pressure that could develop as a result of such an accident.

In addition to having control rods for normal reactivity control, every reactor must be equipped with several independent safety or "scram" devices designed to shut the reactor down quickly in case of any unforeseen malfunctions, such as a power excursion, loss of coolant, fuel element rupture, or power failure. Such devices, which may, for example, be neutron-absorbing structures that drop into the core by gravity when released by the scram signal, must be so designed as to operate under all circumstances, for example, even in case of an earthquake. The possible consequences of all human and mechanical failures that can be imagined should be carefully analyzed and minimized by incorporation of appropriate measures in the system design.

Safety problems differ markedly in different types of reactors. A negative temperature coefficient of reactivity—decrease in k with increase in temperature—provides some inherent safety since it insures that an increase above normal operating temperature immediately leads to a drop in reactivity and therefore in power. A negative temperature coefficient of reactivity may come about in various ways. For example, in an unpressurized homogeneous aqueous reactor a rise in temperature will lead to expansion of the solution and thus to a lowering of the fuel concentration.

In some reactor types (e.g., the popular TRIGA research reactors) reactivity is reduced as the temperature increases, because additional neutron-capture resonances are brought into play at the higher "thermal" neutron energies.

If coolant and moderator are identical, as is the case with most water-cooled reactors, the production of voids through boiling will tend to reduce reactivity because the moderator becomes less effective. However, high-powered water-cooled reactors (PWRs and BWRs) have one very troublesome safety problem: even though loss of coolant would immediately shut down the reactor, the amount of radioactivity in the fuel elements is so huge that the reactor core would quickly melt unless a backup cooling system were available. So-called emergency core-cooling systems have indeed been incorporated in these reactors, but their adequacy and reliability are matters of considerable controversy.[12]

The actual safety experience with reactor operations has been remarkably good. Various mechanical and human failures have necessitated costly shutdowns and repairs,[13] but have not resulted in significant radiation exposures or other damage to populations. However, we must note that the total number of reactor years of experience is very small compared to what is planned for the future, and much of the controversy centers around the question of "what is safe enough?" Are we willing to tolerate one chance in 10^6 that a given reactor will have a major accident, or does it have to be one in 10^8? One in 10^3 would surely not be acceptable if we expect of the order of 10^3 reactors to be in operation. Furthermore there is lack of agreement on how to evaluate these probabilities.

2. *The Fuel Cycle*

There is much more to the production of nuclear energy than the nuclear reactor itself. In particular, manifold chemical and metallurgical operations are involved in the so-called fuel cycle, and their efficiency and cost profoundly affect the economics of nuclear power generation. By the fuel cycle is meant the path of fuel material from the mine through concentration and purification steps, possibly isotopic enrichment, fuel element fabrication, reactor irradiation, processing of spent fuel to separate fissile and fertile materials from fission products, to the reintroduction of the fissile and fertile species into the cycle, usually again at the fuel element fabrication stage.

[12] In an accident in March 1979 at a PWR at the Three Mile Island plant near Harrisburg, Pennsylvania, the emergency core cooling system came into operation after a loss of coolant, but was subsequently interrupted manually, which led to a very hazardous condition and brought about major damage to the reactor.

[13] In terms of fraction of time available and in terms of ratio of energy produced to maximum capacity, nuclear plants have been quite comparable to coal-fired ones (R2).

Uranium Refining. Uranium occurs in many parts of the world and in a variety of deposits, from relatively high-grade ores in Canada and West Africa containing several percent uranium in the form of pitchblende and uraninite (mixed UO_2-U_3O_8) to low-grade sources such as uranium-bearing phosphates, lignites, and shales that are widespread but contain <0.01 percent uranium.[14] The first steps in the fuel cycle, aimed at concentrating the uranium, are usually carried out near the mine to save on transportation costs. The methods used depend on the nature of the particular ore and may involve mechanical procedures such as crushing, screening, and flotation, followed by acid or alkaline leaching and eventual precipitation, solvent extraction, or ion exchange. The product of these concentration steps, containing perhaps 40–70 percent uranium, is generally shipped to a central processing plant to be further purified either by digestion with nitric acid and extraction of the resulting uranyl nitrate into an organic solvent (such as tributyl phosphate in kerosene), or by conversion to UF_6 and fractional distillation of that volatile compound. The steps leading to UF_6 may be summarized by the following equations:

$$U_3O_8 + 2H_2 \rightarrow 3UO_2 + 2H_2O,$$

$$UO_2 + 4HF \rightarrow UF_4 + 2H_2O,$$

$$UF_4 + F_2 \rightarrow UF_6.$$

Either the distillation of UF_6 or the solvent extraction of $UO_2(NO_3)_2 \cdot 6H_2O$ results in a product that meets the stringent requirements that are set for reactor fuels with respect to neutron-absorbing impurities. The following series of steps indicate the processes used to convert the purified materials to the metal or dioxide as desired.

$$UO_2(NO_3)_2 \cdot 6H_2O \xrightarrow{\text{Heat}} UO_3 \xrightarrow{H_2} UO_2 \xleftarrow[\text{Heat}]{H_2+\text{Steam}} (NH_4)_2(UO_2)_2F_6 \xleftarrow{NH_4OH} UF_6$$

$$UO_2 \xrightarrow[\text{Heat}]{HF} UF_4 \xleftarrow[\text{Heat}]{H_2} UF_6$$

$$UF_4 \xrightarrow[\text{Mg or Ca}]{\text{Heat}} U$$

Isotope Enrichment. If ^{235}U enrichment is required, the purified uranium in appropriate chemical form goes next to an isotope separation plant. The gaseous diffusion process used in the United States and several other countries for uranium isotope enrichment depends on the

[14] Thorium occurs chiefly in monazite, a mixture of rare-earth phosphates with 1–5 percent ThO_2 content. In order not to complicate the discussion unduly, we do not go into the processing of thorium. Many of the steps involved are rather similar to those in the uranium processes.

fact that a gas diffuses through a porous membrane at a rate inversely proportional to the square root of its molecular weight. Thus the optimum separation factor for the separation of $^{235}UF_6$ and $^{238}UF_6$ by gaseous diffusion is $\sqrt{352/349} = 1.0043$. Because this factor is so close to unity, thousands of successive stages of separation are required to obtain highly enriched ^{235}U, and the diffusion plants are huge and very costly. Another method of isotope separation that may be a serious contender for future uranium enrichment plants in the United States and has, in fact, been put into operation in some other countries is centrifugation. The separation factor here depends on the *difference* in the molecular weights of the species to be separated, rather than on the square root of their ratio. Still another approach of great potential that is in the development phase is based on the possibility of selectively exciting, by means of a laser, only one of a pair of isotopic molecules to a vibrationally or electronically excited state, which then either dissociates or is made to undergo a chemical reaction.

Fuel Elements. The great variety of chemical and physical forms that reactor fuels can take makes it impossible to go into any details of fuel element fabrication here. What materials are most suitable for a given reactor type depends on many factors, including neutron spectrum, operating temperature, and coolant. As we already mentioned, uranium dioxide and uranium carbide ceramics are more suitable than the metal for most power reactor applications because they suffer less dimensional change at high burnup, are less subject to radiation damage, are chemically more inert, and retain gaseous fission products better. For fast breeders solid solutions of UO_2-PuO_2 or possibly UC-PuC are among favored fuel materials. In heterogeneous reactors the fuel elements must generally be protected by a cladding or canning material that is inert towards the coolant. Good thermal contact between fuel and cladding is essential, and the cladding as well as the fuel must be as resistant as possible to radiation damage. In high-temperature water-cooled reactors, such as PWRs and BWRs, stainless steel or zirconium alloys are most frequently used, in sodium cooled reactors stainless-steel cladding is most common, and in gas-cooled reactors stainless steel, pyrolytic graphite, or ceramic materials are in use.

Eventually fuel elements must be removed from reactors. Their useful life is limited either by depletion of fissile material, or by the buildup of neutron-absorbing fission product poisons or, most likely, by mechanical deterioration resulting from radiation effects and conversion of fissile material into fission products. Prior to chemical reprocessing the irradiated fuel elements are stored, usually under enough water to absorb all harmful radiation, for periods of at least several months to allow the radioactivity to decrease by a sizable factor. For economic reasons one would try to keep the cooling period reasonably short, since the tying up of storage facilities and valuable fuel adds to the final costs of nuclear power. On the

Table 14-3 Percentage Composition of PWR Fuel Elements[a]

Nuclide	Content (%) Initially	After Burnup of 33,000 MWd ton^{-1}
^{238}U	96.7	94.5
^{235}U	3.3	0.9
^{236}U		0.4
^{239}Pu		0.5
Other plutonium isotopes		0.4
Other actinides		0.1
Fission products		3.2

[a] From Programmstudie *Nukleare Primärenergieträger*, Angewandte Systemanalyse (ASA) in der Arbeitsgemeinschaft der Grossforschungseinrichtungen (AGF), Report ASA-ZE/08/78, Köln, W. Germany, April 1978.

other hand, the concept of **once-through operation** of reactors, that is, of foregoing the reprocessing of fuel completely, has the attraction that plutonium then exists in the form of highly radioactive fuel elements only, and illicit diversion becomes much less likely. For this reason no fuel from commercial nuclear power plants in the United States is at present reprocessed. In the long run this practice is, of course, wasteful of uranium resources since in typical power reactors the fuel elements, at time of discharge from the reactor, still contain 25–30 percent of the initial ^{235}U fuel as well as some ^{239}Pu formed from ^{238}U. The composition of a typical spent fuel is shown in table 14-3.

Fuel Reprocessing. Fuel from plutonium-producing reactors for military purposes has always been reprocessed, and eventually the United States will perhaps also have reprocessing plants for fuel from civilian nuclear power plants, as some other countries already do. The requirements placed on the chemical and metallurgical processing of irradiated fuel elements are quite stringent. Losses of fissile material must be kept very low for economic reasons. Because of the intense radioactivity of the fission products the early process stages must be carried out entirely by remote control in heavily shielded, completely enclosed cells. Discharge of any gaseous or liquid radioactive material must be scrupulously avoided. Finally the processes have to be so designed that it is absolutely impossible for a critical mass of fissile material to be accidentally assembled at any stage.

The separation processes for irradiated fuel can take many forms. Practically all processes have in common as a first step the dissolution of the fuel in hot nitric acid (leaving cladding materials such as zircaloy undissolved.[15] Subsequently the uranium fuel from the wartime plutonium-producing reactors was processed by cycles of precipitation reactions. However, precipitations and filtrations must almost of necessity be carried out as batch processes and are therefore not ideally suited for remote-control operations. Solvent extraction separations in countercurrent columns are much more readily adapted to remote operation and have therefore replaced the precipitation procedures.

Many processes based on extraction of uranium, plutonium, and, if present, thorium into an organic solvent have been investigated. By far the most widely used is the PUREX process (Plutonium Uranium Recovery by EXtraction) in which the organic solvent is a 20–30 percent solution of tri-*n*-butylphosphate (TBP) in kerosene. TBP is particularly suitable because of its high radiation stability.

The efficient extraction (>99.5 percent) of uranium and plutonium in the first step is based on the large partition coefficients for U(VI), Pu(IV), and Pu(VI) in the systems used. At the same time the accompanying total fission product activity (of the order of 10^6 Ci ton^{-1} of fuel in the feed solution) is usually reduced by factors of several hundred, although some specific products, especially ruthenium, are partially extracted into the organic phase. For the subsequent separation of plutonium from uranium advantage is taken of the very low partition coefficient of Pu(III) for extraction into organic solvents. Thus plutonium is reduced to Pu(III) under conditions that leave uranium in the +6 state—such reducing agents as Fe(II), SO_2, or hydrazine can be used—and another countercurrent solvent extraction then separates plutonium and uranium. Further remote-control purification of the fractions by additional solvent extraction steps or by other techniques such as ion exchange is normally required before the material can finally be handled without γ-ray shielding. Even then, some protective measures are always required when the highly α-active and toxic nuclide ^{239}Pu is processed. A flowsheet of the PUREX process is shown in figure 14-2.

To prevent any possibility of plutonium solutions reaching criticality, geometric configurations are carefully designed and controlled and, in addition, neutron-absorbing materials may be used in the walls of the vessels (e.g., hafnium) or as additives in the solutions [e.g., $Gd(NO_3)_3$].

Some significant advantages in efficiency and cost may accrue from the use of fuel-processing methods that do not involve aqueous solution chemistry. The initial dissolving step, for example, could then be avoided and the final reconversion to metal or oxides could be greatly simplified. Intensive development work has therefore been carried out on a number of nonaqueous processes. Among them are volatility methods that accomplish

[15] The radioactive gases set free during dissolution (Kr, Xe, ^3H, I) must be contained in absorbers or cold traps.

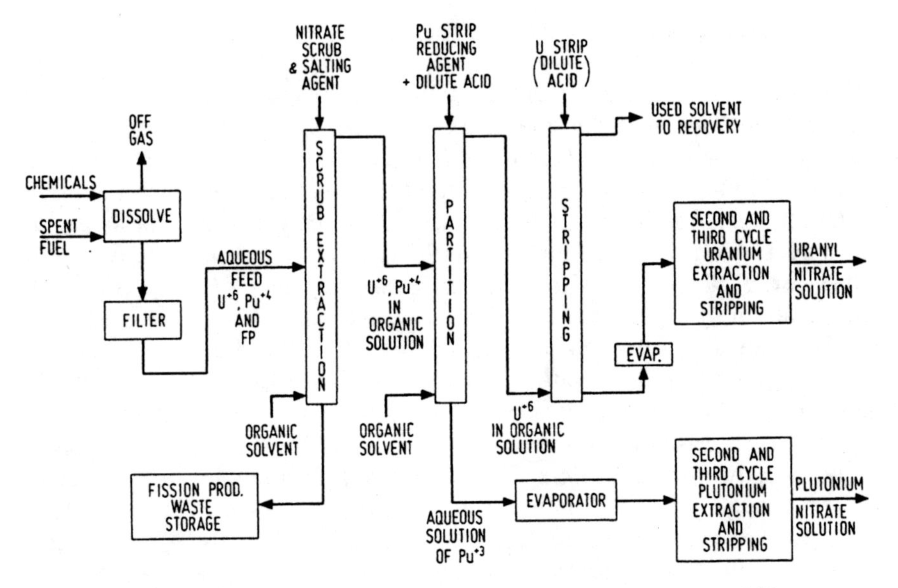

Fig. 14-2 Flow sheet for the PUREX process. (From reference G2.)

separation of uranium from fission products by volatilization of UF_6.[16] Pyrometallurgical processes are of considerable interest also. Among them is a melt-refining method that has actually been in use for reprocessing the fuel of one of the US experimental breeder reactors (EBR-II). In this process uranium fuel elements are melted down in zirconium oxide crucibles at 1300°C in an inert atmosphere. Many fission products, such as the rare gases, alkalis, alkaline earths, and cadmium, distil out; others form oxides (by reduction of the zirconium oxide to a suboxide) and separate in a layer of slag. Still other fission products, such as the noble metals and molybdenum, remain alloyed with the uranium.[17] This alloy, with the addition of some fresh fuel to make up for burnup, is recast into new fuel elements (by remote control) and returned to the reactor. The relative simplicity of this type of process is clearly an asset.

[16] In principle PuF_6, which sublimes at 62°C, can also be separated by volatilization, but it is extremely reactive and difficult to handle. UF_6 volatilization is therefore particularly interesting for fully enriched ^{235}U fuels, where little ^{239}Pu is involved. Otherwise, the plutonium may have to be separated from the fission products by aqueous methods after the UF_6 volatilization step.

[17] The equilibrium mixture of fission products, which remains alloyed with uranium (or plutonium), has been termed "fissium." Uranium-fissium alloys have been found to possess desirable mechanical properties for fuel-element use.

3. *Nuclear Materials Safeguards*

One of the serious concerns about the rapidly expanding nuclear industry is the possibility of theft or diversion of what are called special nuclear materials (SNM)—^{239}Pu, fully enriched ^{235}U, and ^{233}U— for purposes of nuclear blackmail, illicit bomb manufacture, or other forms of terrorism. Much attention has been devoted to providing adequate safeguards against this eventuality, but continued and increased vigilance is undoubtedly called for. Stringent physical security measures (including locks, guards, fences, alarm systems, etc.) are mandatory wherever SNMs are present. In addition, however, it is most important to have reliable accountability and control procedures for keeping track of the flow of SNM throughout the fuel cycle, with provisions for establishing material balance at each step.[18]

It is thus imperative that methods be available for the assay of the fissile nuclides in the various chemical and physical forms in which they occur in the fuel cycle (S2). Sampling, followed by chemical and mass-spectrometric analyses, is widely used, but such analyses are not only time-consuming, they also have definite limitations, for example when it comes to assaying solid scrap or intact fuel elements. Much effort has been and is being devoted to developing techniques and instrumentation for nondestructive assays in all phases of the fuel cycle. Depending on the particular application, one may strive for methods with a combination of some of the following attributes: speed, accuracy, sensitivity, low cost, suitability for remote control, easy use by inexperienced personnel. The assay methods make use of various characteristic "signatures" for the fissile nuclides—specific radiations emitted either spontaneously (passive methods) or under irradiation from an external source (active methods).

Among the radiations useful for passive detection are the 186-keV γ rays of ^{235}U, the 129-keV and 375-keV γ rays in ^{239}Pu decay, and the neutrons from the spontaneous fission of ^{240}Pu (which always accompanies ^{239}Pu and whose assay can serve as a measure of ^{239}Pu content provided the isotopic composition is known). Highly enriched ^{235}U in the form of UF_6 can be accurately determined via the neutrons emitted as a result of the (α, n) reaction on fluorine, the α particles originating almost entirely from the ^{234}U that always accompanies ^{235}U in the isotope enrichment process.

Active interrogation methods use neutron or photon irradiation to induce fission and are based on quantitative measurements of some resulting radiations, again neutrons or γ rays. For example, slow neutrons may be used for irradiation, and prompt fast neutrons measured; ^{252}Cf neutron sources are useful for this purpose. Alternatively, delayed neutrons resulting from fission induced by a modulated neutron source, for example from a 14-MeV neutron generator or Van de Graaff, may be measured. This

[18] Siting reactors, reprocessing plant, and fuel fabrication in one location somewhat alleviates the safeguard difficulties by avoiding repeated transportation.

latter method is, for example, applicable to the assay of fissile nuclides in irradiated fuel elements even though they may contain thousands of curies of fission product activity and be encased in heavy lead shields.

A large variety of techniques has already been developed for many different aspects of the safeguards problem, and active research in this field will undoubtedly continue apace with the increasing variety of physical and chemical forms in which special nuclear materials will occur in the future.

4. *Management of Radioactive Wastes*

Wherever radioactive nuclides are produced or used, safe disposal of radioactive wastes is a problem to be reckoned with. Radioactive materials should not be released into the environment unless dilution to harmless levels can be guaranteed, and in establishing such levels one must take into account the possibility of reconcentration of specific elements in biological systems. Appropriate measures for safeguarding the environment must thus be taken in research laboratories, industrial establishments, and medical institutions using radionuclides. However, by far the largest problem of radioactive-waste management is connected with the fuel cycle of nuclear reactors. Even here the mining, milling, purification, and fuel fabrication steps produce wastes that are not too difficult to manage because they contain only naturally occurring radioactivities at rather low levels.[19] The major concern is with the highly radioactive wastes from the processing of irradiated fuels (M2).

To put the problem in perspective we note that the power reactors operating in the world in 1978 produced of the order of 10^9 Ci of long-lived ($t_{1/2} > 1$ y) fission products per year. Although, as mentioned before, no fuel elements from commercial power reactors have been reprocessed in the United States since 1972, there already exist about 3×10^8 l of high-level liquid wastes from reprocessed fuel of military-purpose and early power reactors. Much of this waste is in temporary storage in large underground tanks. This is certainly not a satisfactory long-range solution, since the useful life of a tank is probably measured in decades, whereas the required storage times are orders of magnitude longer. Leaks into the ground have already been reported from some storage tanks.

Storage of wastes in solid form has long been considered the method of choice. Processes have been developed for drying the liquids, calcining the residues, and incorporating them in glasses, clays, or ceramics. The blocks of glass or ceramic are then sealed in metal canisters to serve as additional barriers against moisture possibly reaching and leaching the radionuclides.

[19] Long-term disposition of the so-called "tailings," solid wastes from the ore-concentrating mills, presents something of a problem since they contain all the radium that has been in equilibrium with ^{238}U. So far the tailings have merely been stored in controlled areas.

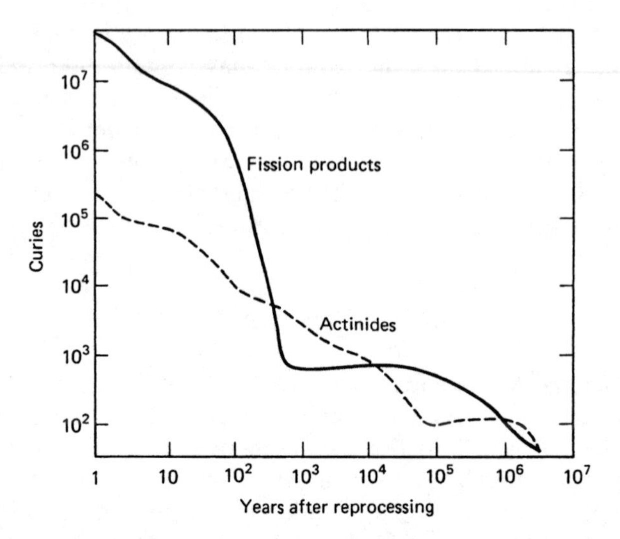

Fig. 14-3 Decay of total fission product and actinide activities in waste originating from 30 metric tons of original uranium subjected to burnup of 33,000 MWd ton^{-1} in a PWR and reprocessed 6 months after discharge from the reactor. 99.5 percent of uranium and plutonium and 99.9 percent of halogens are assumed to have been removed in reprocessing. (Data were kindly supplied by B. L. Cohen; see also reference C2.)

Salt mines are favored as permanent storage sites because they are known to have long been quite free of water; however, doubts about their suitability have not been completely dispelled. The same is true of bedrock storage, which has also been explored.

To give a more quantitative idea of the waste-storage problem, we show in figure 14-3 the decay of total fission product and actinide[20] activities resulting from reprocessing of 30 tons of light-water reactor fuel, which is approximately the amount of fuel that would be processed annually from a 1000 MW*e* PWR. The curves are drawn on the assumption that reprocessing occurs six months after discharge from the reactor and leaves 0.5 percent of the uranium and plutonium and 0.1 percent of iodine and bromine, but no krypton or xenon in the wastes. As the figure shows, the level of fission product activity drops by five orders of magnitude from 1 to 800 y after reprocessing; during all but the first decade of that time span the fission product decay curve represents almost exclusively the decay of ^{90}Sr ($t_{1/2} = 29$ y) and ^{137}Cs ($t_{1/2} = 30$ y). Fortuitously there are no significant fission products with half lives between 30 and 10^5 y.[21] Furthermore, although after a few hundred years fission products and actinides have

[20] The actinide activities include, besides unrecovered ^{239}Pu, principally ^{237}Np, ^{238}Pu, ^{240}Pu, ^{241}Am, ^{243}Am, ^{244}Cm, and some of their descendants. All these nuclides are formed by combination of neutron-induced reactions and radioactive decay processes.

[21] The fission products with $t_{1/2} > 100$ y and fission yields greater than 10^{-4} are ^{126}Sn (1×10^5 y), ^{99}Tc (2.1×10^5 y), ^{93}Zr (1.5×10^6 y), ^{135}Cs (3×10^6 y), ^{107}Pd (6.5×10^6 y), and ^{129}I (1.6×10^7 y).

comparable activity levels, the α-emitting actinides pose much greater health hazards if released into the environment. Some benefit would therefore be derived from much more complete chemical separation of actinides and fission products from each other, particularly if followed by reintroduction of the waste actinide fraction into reactors where they would then largely be "burned" to fission products.

Although burial of solidified wastes in geologic formations seems to be the favored method of ultimate waste disposal, more exotic schemes have been suggested, including rocketing into space and injection under the earth's crust, but at present such methods appear to be neither technically nor economically feasible.

A detailed discussion of the high-level waste problem may be found in C2.

D. CONTROLLED THERMONUCLEAR REACTIONS

So far we have dealt in this chapter with energy derived from the fission of heavy elements. As is evident from a consideration of nuclear binding energies as a function of mass number (see figure 2-1), another potential source of nuclear energy exists in fusion reactions of the lightest nuclei. In terms of energy release per nucleon, or therefore per gram of material, some of these light-element reactions are even more powerful energy sources than the fission reactions.

Fusion reactions are the source from which the energy of the sun and of other stars is derived (see chapter 13, section B). The first man-made application of the energy release accompanying light-element fusion reactions came with the development of the thermonuclear or hydrogen bomb. However, in parallel with that development went the beginnings of intensive efforts towards achieving *controlled* thermonuclear reactions (CTR) in a number of countries. In contrast to the situation with fission, the attainment of useful power from controlled fusion is very much more difficult than the explosive release of fusion energy and, more than 25 years after the first successful hydrogen bomb test (1952), even the scientific feasibility of controlled fusion power remains to be experimentally established, although that goal now seems likely to be achieved within a few years (P4).

Fusion Reaction. The reactions of prime interest for controlled fusion are the following:

$$^{2}\text{H} + {}^{2}\text{H} \rightarrow {}^{3}\text{He} + {}^{1}n + 3.3\ \text{MeV}, \tag{14-4}$$

$$^{2}\text{H} + {}^{2}\text{H} \rightarrow {}^{3}\text{H} + {}^{1}\text{H} + 4.0\ \text{MeV}, \tag{14-5}$$

$$^{2}\text{H} + {}^{3}\text{H} \rightarrow {}^{4}\text{He} + {}^{1}n + 17.6\ \text{MeV}, \tag{14-6}$$

$$^{2}\text{H} + {}^{3}\text{He} \rightarrow {}^{4}\text{He} + {}^{1}\text{H} + 18.3\ \text{MeV}, \tag{14-7}$$

$$^{1}H + {}^{6}Li \rightarrow {}^{3}He + {}^{4}He + 4.0\ \text{MeV}, \qquad (14\text{-}8)$$

$$^{1}H + {}^{7}Li \rightarrow {}^{4}He + {}^{4}He + 17.3\ \text{MeV}. \qquad (14\text{-}9)$$

Of these reactions, (14-6), the deuteron-triton or d-t reaction, has received the greatest amount of attention and is almost certainly the reaction of choice for early fusion reactor designs, since it has by far the highest cross section at low energies (see figure 14-4) and one of the highest Q values. In this reaction, approximately 80 percent of the energy (14.1 MeV) is carried off by the neutron, the remainder by the ^{4}He. Since tritium, which does not exist in nature in appreciable abundance, is one of the reactants, provisions must be made for its regeneration in a breeding cycle. The scheme therefore includes a breeding blanket of lithium that regenerates tritium by the reactions:

$$n + {}^{6}Li \rightarrow {}^{3}H + {}^{4}He, \qquad \text{and} \qquad n + {}^{7}Li \rightarrow {}^{3}H + {}^{4}He + n.$$

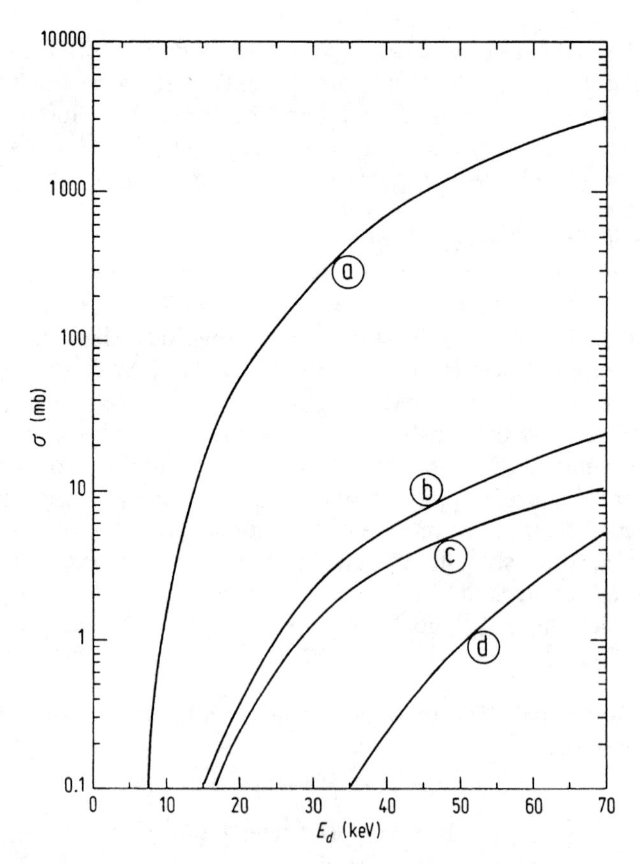

Fig. 14-4 Excitation functions of some light-element fusion reactions. Curve (a) is for the d-t reaction (14-6), curves (b) and (c) are for the d-d reactions (14-5) and (14-4), respectively, and curve (d) is for the $d + {}^{3}He$ reaction (14-7). (Data mostly from reference P4.)

A certain amount of neutron multiplication will be achieved through $(n, 2n)$ reactions in the blanket walls, which might be made of niobium or possibly molybdenum. Tritium breeding ratios up to about 1.3 appear quite feasible. Most of the fusion energy would appear in the form of heat in the lithium blanket, and useful power would presumably be extracted by circulating the lithium through a heat exchanger, thus producing steam for driving a turbine.

At a later stage fusion reactors based on (14-7)–(14-9) may become practical. These reactions are of particular interest because the products are all charged, which makes it possible to devise schemes for direct conversion of the reaction energy to electricity. Since (14-7) uses ^3He as a fuel, it would presumably have to be combined with (14-4), which produces ^3He. Reaction 14-5, the other d-d reaction, would of course also take place, but a fuel cycle involving the two d-d reactions and the d-^3He reaction would result in over 90 percent of the fusion energy going into charged particles. One disadvantage of both schemes mentioned so far is that they involve production of massive amounts of neutrons and of tritium, with attendant potential environmental hazards. From this point of view, (14-8) and (14-9) are attractive, since they produce no radioactive products and no neutrons. However, the Coulomb barriers for these reactions are much higher than for the others, and they are therefore only remote possibilities.

Basic Requirements for CTR. The major problems of achieving controlled fusion arise from the fact that the relevant reactions, in contrast to fission, involve charged particles only. They can therefore proceed at appreciable rates only when the relative velocities of the reaction partners are high enough to overcome, or effectively tunnel through, the Coulomb barriers. This condition is easy to achieve with modest accelerators, but to obtain *useful* fusion power, the reactions must be made self-sustaining, and this requires sufficiently high temperatures for the *thermal* velocities to bring about appreciable reaction rates. The temperatures of interest are in the neighborhood of 10^8 K. At these temperatures gases are completely ionized, and fusion research therefore deals with the behavior of high-density gases of charged particles, known as **plasmas.** The goals of fusion research thus are (1) to achieve the requisite plasma temperatures and (2) to keep the hot plasma together long enough for useful amounts of energy to be produced by the nuclear reactions. A minimum condition for the production of practical fusion power is that the energy extracted from the thermonuclear reactions must exceed the energy required to heat and confine the plasma.

The two requirements of high temperature and adequate confinement time can be put into somewhat more quantitative terms. For each of the reactions there is a so-called ignition temperature, defined as that temperature at which the rate of energy production by fusion overtakes the rate of energy loss by bremsstrahlung, that is, by radiation in close collisions between electrons and nuclei. This crossover between energy release and radiation loss comes about because the bremsstrahlung losses vary as the square root of the electron temperature, whereas the reaction cross sections have much steeper temperature dependence (see figure 14-4).[22] As already indicated, the *d-t* reaction has the lowest ignition temperature, about 4×10^7 K.[23]

The second condition, confinement of the plasma until the **break-even** point (energy output equal to energy input) has been reached, can be expressed, according to an analysis by J. D. Lawson, as a minimum value for the product of plasma density and confinement time. This product, which is an approximate constant for a given thermonuclear reaction, is called the **Lawson criterion**, and an enormous amount of plasma research has been devoted to reaching this condition. For the *d-t* reaction the Lawson criterion is about 10^{14} s cm^{-3}, which could be achieved in various ways, for example, by keeping a plasma of 10^{14} particles per cubic centimeter together for at least 1 s, or by attaining a density of 10^{20} particles per cubic centimeter for 10^{-6} s, and so on. In each case the temperature would also have to exceed the ignition temperature.

Magnetic Confinement (P4). It is evident that a plasma at $\sim 10^8$ K cannot be confined in any known material. The major thrust over the years has been toward achieving confinement by magnetic fields, and a considerable variety of magnetic confinement schemes, referred to collectively as "magnetic bottles" has been developed. Some configurations are linear, with magnetic "mirrors" to reflect escaping particles back into the plasma region, others are toroidal. In some devices, such as the Princeton Stellarator, stabilizing magnetic fields are produced by external current-carrying coils; in others, such as the Tokamak (whose concept originated in the Soviet Union), such fields are generated by currents flowing in the plasma itself. So far none of these devices has yet achieved the Lawson criterion, but several have come within one to two orders of magnitude of it, and there appears to be good reason to believe that this goal will be reached before long. Limitations on attainable magnetic fields, as well as on the rates at which one can hope to remove power from the plasma, limit plasma densities in magnetic-confinement devices to relatively low values—in the range of 10^{12}–10^{17} cm^{-3}. Although collisional relaxation can be sufficiently slow relative to the requisite confinement times, other loss and leakage mechanisms, resulting from various plasma instabilities, have plagued plasma researchers. However, much progress has been made toward overcoming these problems (F1).

[22] Electron and ion temperatures may not be identical in a plasma, which makes the concept of the ignition temperature a little less precise, but still qualitatively correct.

[23] Note that kT of 1 keV corresponds to a temperature of 1.16×10^7 K. Note also that the mean kinetic energy in a Maxwellian distribution at temperature T is $\frac{3}{2}kT$. Furthermore, because of the steeply rising excitation functions, most of the reactions take place at several times kT.

Magnetic confinement, although long the most vexing problem in fusion research, is by no means the only one. Heating the plasma to the requisite temperature is also not a simple matter. However, this has been successfully accomplished in a number of devices, either by magnetic compression or by injection of energetic neutral beams, although it is fair to say that the machines in which the ignition temperature (for d-t) has been reached or exceeded are not the ones that have come closest to reaching the Lawson criterion. Among other major technological problems that have had to be solved we mention the extreme vacuum and plasma purity requirements, as well as the need for very high magnetic fields. All of these problems appear to have practical solutions; fnr example, very high magnetic fields can be obtained by the introduction of superconducting magnet coils.

Inertial Confinement Fusion (M3). A concept very different from magnetic confinement is based on intertial confinement of an exceedingly dense plasma for a very short period of time. The idea here is to use high-powered, pulsed beams of lasers, electrons, or heavy ions to heat and compress small pellets of deuterium tritide (DT). Laser fusion (S3) has been under development longer and more intensively than the other schemes; the general principle is similar for all of them.

The aim is to direct simultaneous, pulsed beams at the fuel pellet, causing it to implode and reach densities several orders of magnitude greater than normal, say 10^{25}–10^{26} atoms per cubic centimeter. To achieve the Lawson criterion such densities would then have to be maintained for about 10^{-11} s, just about the length of time the pellet would stay together through inertial forces.

Lasers of various types (neodymium, CO_2, iodine), each capable of delivering about 10^{12} W, have been developed, and elaborate systems for directing a number of such laser beams at a reaction chamber containing the fuel pellets have been built. For example, in an installation called Shiva at the Lawrence Livermore Laboratory, 20 neodymium laser trains, each delivering 1.2×10^{12} W for 0.1 ns, are focused on a fuel pellet. The total energy is thus 2.4×10^3 J, but this is still nearly two orders of magnitude below what is estimated to be necessary for an operating fusion power reactor.

In an eventual power plant of this type the power would be generated in microexplosions, perhaps 100 of them every second, each producing of the order of 10^7 J in a time of about 10^{-11} s. The neutrons generated would, as in the magnetic confinement schemes, be absorbed in a blanket of lithium. Needless to say, there are still many unsolved problems, some of them relating to the mechanical and structural requirements on materials.

One of the disadvantages of lasers for the achievement of inertial confinement is their low efficiency (0.2 percent for neodymium, a few percent for CO_2). On the other hand, advances in the technology of pulsed particle accelerators have made them quite efficient (~50 percent) and relatively inexpensive. Work on the possible use of electron and heavy-ion beams for inertial confinement fusion has therefore been intensified since the early 1970s (Y1).

REFERENCES

C1 G. A. Cowan, "A Natural Fission Reactor," *Sci. Am.* **235** (1), 36 (July 1976).

C2 B. L. Cohen, "High-Level Radioactive Waste from Light-Water Reactors," *Rev. Mod. Phys.* **49**, 1 (1977).

D1 T. R. Decker, "Radioactive Materials Released from Nuclear Power Plants in 1977," *Nucl. Saf.* **20**, 476 (1979).

D2 W. Davis, Jr., "Radioactive Effluents from Nuclear Power Stations and Fuel Reprocessing Plants in Europe, 1972–1976," *Nucl. Saf.* **20**, 468 (1979).

F1 H. P. Furth, "Progress toward a Tokamak Fusion Reactor," *Sci. Am.* **241**, (2), 51 (Aug. 1979).

G1 D. S. Gabriel and I. Helms, "The New Status of Space Nuclear Propulsion in the United States of America," *At. En. Rev.* **12**, 801 (1974).

*G2 S. Glasstone, *Source Book on Atomic Energy*, 3rd ed., Van Nostrand, Princeton, NJ, 1967.

*G3 S. Glasstone and A. Sesonske, *Nuclear Reactor Engineering*, Van Nostrand, Princeton, NJ, 1963.

H1 W. Haefele et al., "Fast Breeder Reactors," *Ann. Rev. Nucl. Sci.* **20**, 393 (1970).

L1 B. Lubarsky, "Nuclear Power Systems for Space Applications," *Adv. Nucl. Sci. Tech.* **5**, 223 (1969).

*L2 S. E. Liverhant, *Elementary Introduction to Reactor Physics*, Wiley, New York, 1960.

M1 M. Maurette, "Fossil Nuclear Reactors," *Ann. Rev. Nucl. Sci.* **26**, 319 (1976).

M2 "*Management of Wastes from the LWR Fuel Cycle*," in *Proc. Int. Symp.* (Denver, July 1976), Report Conf-76-0701. Available from National Technical Information Service, US Dept. of Commerce, Washington, DC.

M3 J. A. Maniscalco, "Inertial Confinement Fusion," *Ann. Rev. En.* **5**, 33 (1980).

P1 A. M. Perry and A. M. Weinberg, "Thermal Breeder Reactors," *Ann. Rev. Nucl. Sci.* **22**, 317 (1972).

P2 R. F. Pocock, *Nuclear Ship Propulsion*, Ian Allan, London, 1970.

P3 A. S. Pushkarsky and A. S. Okhotin, "Methods of Thermal-to-Electric Energy Conversion in On-Board Nuclear Power Plants for Space Applications," *At. En. Rev.* **13**, 479 (1975).

P4 R. F. Post, "Controlled-Fusion Research and High-Temperature Plasmas," *Ann. Rev. Nucl. Sci.* **20**, 509 (1970).

*P5 *Power Reactors in Member States—1980 Edition*, Report ISP 423-80, International Atomic Energy Agency, Vienna, 1980.

R1 C. L. Rickard and R. C. Dahlberg, "Nuclear Power: A Balanced Approach," *Science* **202**, 581 (1978).

R2 A. D. Rossin and T. A. Rieck, "Economics of Nuclear Power," *Science* **201**, 582 (1978).

*S1 H. D. Smyth, *Atomic Energy for Military Purposes*, Princeton University Press, Princeton, NJ, 1945.

S2 "Safeguards Techniques," in *Proc. Symp. Progress in Safeguards Techniques* (Karlsruhe, July 1970), International Atomic Energy Agency, Vienna, 1970.

S3 C. M. Stickley, "Laser Fusion," *Phys. Today* **31** (5), 50 (May 1978).

W1 C. L. Wilson, *Energy: Global Prospects 1985–2000*, Workshop on Alternative Energy Strategies, Nimrod, Boston, 1977.

Y1 G. Yonas, "Fusion Power with Particle Beams,' *Sci. Am.* **239**, (5), 40 (Nov. 1978).

EXERCISES

1. Assume that natural uranium metal is dispersed in heavy water at a concentration of 0.36 g per gram of D_2O. (a) Estimate the value of k_∞ for this mixture. (b) Estimate the radius of the sphere that is just critical.
Answers: (a) 1.3; (b) ~64 cm.
2. The mixture in exercise 1 corresponds approximately to that used in an early research reactor at Kjeller, Norway, which had 2.5 tons of uranium metal in the form of 2.5-cm diameter slugs dispersed in 7 tons of D_2O in a cylindrical tank. If that reactor is operated at 100 kW, estimate the average thermal-neutron flux in the core. *Answer:* 1.4×10^{11} cm^{-2} s^{-1}.
3. Suppose a submarine reactor operating on highly enriched ^{235}U contains 75 kg of ^{235}U in its fuel elements. If the reactor operates at 300 MW*t*, how long can the submarine run before 10 percent of the fuel is used up?
4. Estimate the flux of antineutrinos 20 m from the reactor of exercise 3.
Answer: 3.6×10^{11} cm^{-2} s^{-1}.
5. A water-cooled uranium-graphite reactor operates at a power level of 200 MW*t*. The reactor core is a cube 5 m on the side. The cooling water enters at 20°C, flows through the reactor at the rate of 80,000 liters min^{-1}, and resides in the reactor core for an average of 2 s. Estimate (a) the exit temperature of the water; (b) its radioactivity (in curies per liter) as it leaves the reactor, assuming the water to be pure; (c) the radioactivity of the water 1 h after leaving the reactor if it contains 1.2 ppm phosphorus, 1.8 ppm sodium, and 0.9 ppm chlorine.
6. Verify the statement of p. 525 that ^{135}Xe poisoning in a high-flux reactor ($\geq 10^{14}$ cm^{-2} s^{-1}) reaches a maximum about 10 hours after shutdown.
7. After a reactor has operated at 2000 MW*t* for 1 y its fuel elements are discharged and stored for 6 months. They are then processed and the fission product wastes are stored. Estimate (a) the total weight of fission products, (b) the activity of the fission product wastes (in curies) 10 y after processing, (c) the activity 100 y after processing. (d) What will be the major contributors to the activity after 100 y?
8. Estimate (a) the equilibrium quantity of ^{140}Ba present in a reactor operating at 50 MW*t*, (b) the total amount of stable ^{140}Ce accumulated in the same reactor after 1-y operation followed by 2 months of shutdown. *Answer:* (a) ~40 g.
9. From (14-2) estimate the minimum ratio of ^{235}U atoms to moderator molecules that is required to make a thermal-neutron chain reaction possible in an infinitely large homogeneous mixture of (a) ^{235}U and H_2O, (b) ^{235}U and D_2O.
Answer: (a) 1 : 1100.
10. Show the sequence of processes by which each of the actinide nuclides listed in footnote 20 on p. 544 may be produced in a reactor.

Chapter 15

Sources of Nuclear Bombarding Particles

A. CHARGED-PARTICLE ACCELERATORS

From the discovery of nuclear transmutations in 1919 until 1932 the only known sources of particles which could induce nuclear reactions were the natural α emitters. In fact, the only type of nuclear reaction known during that period of 13 years was the (α, p) reaction. Today the use of natural α emitters to induce nuclear reactions is largely of historical interest because of the much higher intensities as well as higher energies available from man-made accelerators for charged particles. Common to all accelerators is the use of electric fields for the acceleration of the charged particles; however, the manner in which the fields are applied varies widely.

1. *Direct-Voltage Accelerators (A1, M1)*

Cascade Rectifiers and Transformers. The most straightforward type of accelerator results from the direct application of a voltage between two terminals. To obtain more than about 200 kV of accelerating voltage it is necessary to use one or more stages of voltage-doubling circuits. The first such voltage-multiplying rectifier device for nuclear research was built by J. D. Cockcroft and E. T. S. Walton in 1932 and was used for the first transmutation experiments with artificially accelerated particles (protons). **Cockcroft-Walton accelerators** are still widely used, especially as injectors for higher-energy accelerators and as neutron generators. Voltages up to about 4 MV and dc currents up to ~10 mA of protons are obtainable. Cascade rectifiers and cascade transformers of various types are produced commercially by several firms.

Electrostatic (Van de Graaff) Generator. The adaptation of the electrostatic machine to the production of high potentials for the acceleration of positive ions was pioneered by R. J. Van de Graaff, beginning in 1929. In the Van de Graaff machine a high potential is built up and maintained on a smooth conducting surface by the continuous transfer of static charges from a moving belt to the surface. This is illustrated in figure 15-1 where

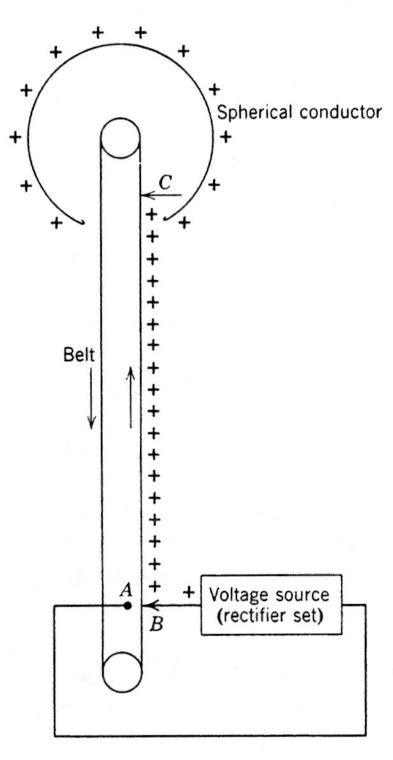

Fig. 15-1 Schematic representation of the charging mechanism of a Van de Graaff generator.

the surface is a sphere.[1] The belt, made of a suitable insulator, is driven by a motor and pulley system. It passes through the gap *AB*, which is connected to a high-voltage source (10–30 kV dc) and adjusted so that a continuous discharge is maintained from the sharp point *B*. Thus positive (or negative) charges are sprayed from *B* onto the belt, which carries them to the interior of the insulated metal sphere. There another sharp point or sharp-toothed comb *C*, connected to the sphere, takes off the charges and distributes them to the surface of the sphere. The sphere will continue to charge up until the loss of charge from the surface by corona discharge and by leakage along its insulating support balances the rate of charge transfer from the belt. The continuous current that can be maintained with an electrostatic generator depends on the rate at which charge can be supplied to the sphere. An improved charging system in which the belt is replaced by a chain of tubular steel segments separated by nylon spacers was developed in the early 1970s by R. G. Herb. These so-called **pelletrons** (and the similar "laddertrons" used in England) have much longer lives than belts and avoid

[1] A sphere was the usual shape of high-voltage electrodes in early machines; in modern Van de Graaffs nearly cylindrical shapes are used.

the dust problem inherent in belt drives, which often leads to electrical breakdown.

An ion source (or electron gun) is located inside the high-voltage terminal, and the ions (or electrons) originating from it are guided through focusing electrodes into an evacuated accelerating tube (see below) along which the electric field is applied.

Since the voltage of an electrostatic generator is limited by the breakdown of the gas surrounding the charged electrode, it is desirable to use conditions under which the breakdown potential is as high as possible. Most electrostatic generators are therefore enclosed in steel tanks pressurized to ten or more atmospheres with an insulating gas such as N_2 or SF_6. These gases have the additional virtue that they will not support combustion following a spark to some combustible material. Terminal voltages up to ~15 MV have been attained and still higher ones are in prospect.

A variety of models for positive-ion and electron acceleration up to about 6 MeV are commercially available. Proton currents of about 100 μA and even larger electron currents are common. The chief application of electrostatic generators is in nuclear physics work requiring high precision because, unlike other machines such as cyclotrons, they supply ions of precisely controllable energies (constant to about 0.1 percent and with an energy spread of about the same order of magnitude).

Accelerating Tubes. Any machine for the acceleration of ions by the application of a high potential requires an accelerating tube across which the potential is applied. A source of ions near the high-voltage end, a system of accelerating electrodes, and a target at the low-voltage end must be provided and enclosed in a vacuum tube connected to the necessary pumping system. The ion source is essentially an arrangement for ionizing the proper gas (hydrogen, deuterium, helium) in an arc or electron beam; the ions are drawn through an opening into the accelerating system. In electron accelerators an electron gun is used as the source.

A typical accelerating tube (figure 15-2) is built of glass or porcelain sections *S*. Inside this tube, sections of metal tube *T* define the path of the ion beam. Each metal section is supported on a disk that passes between two sections of insulator out into the gas-filled space to a corona ring *R* equipped with corona points *P*. The purpose of the corona rings and points is to carry the corona discharge from the high- to the low-voltage end of the tube and to distribute the voltage drop uniformly along the tube. Depending on the number of sections used, a potential difference somewhere between 10 and several hundred kilovolts exists between successive sections. Each gap between successive sections has both a focusing and a defocusing action on the ions traveling down the tube. The ions tend to travel along the electric lines of force (see figure 15-2 for the pattern of these lines between a pair of sections). In entering the gap the ions are

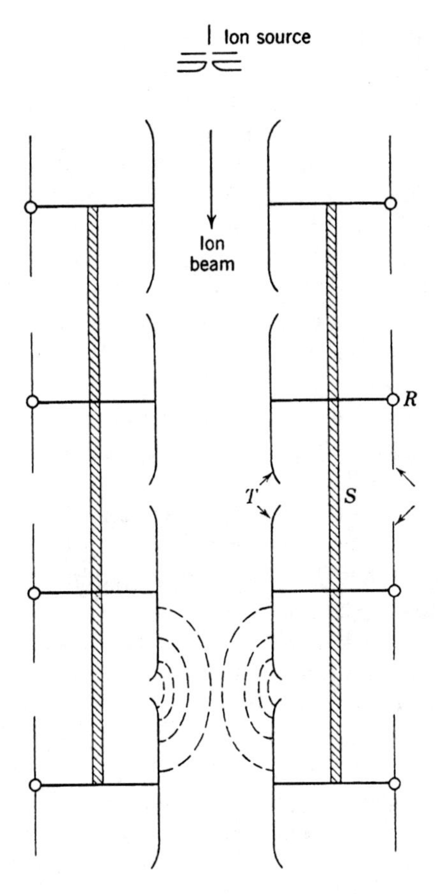

Fig. 15-2 Schematic cross-sectional diagram of a portion of an accelerating tube.

focused and in leaving it they are defocused, but because the ions move more slowly on entering the gap than on leaving it, the focusing effect is stronger than the subsequent defocusing. Well-focused beams (cross-sectional area less than 0.1 cm^2) can be obtained. It should be mentioned that from hydrogen gas in an ion source not only protons but also hydrogen molecule ions (H_2^+) and H_3^+ ions are obtained. These are also accelerated in the tube but can be separated from the protons before striking the target by means of a magnetic analyzer. One of the magnetically analyzed beams is usually used to obtain precise automatic energy control. The position of this deflected beam along a slit system depends on its energy, and a signal from this slit system can be fed back to devices at the high-voltage terminal, which will adjust the accelerating potential in the desired direction.

Tandem Van de Graaff. The energies attainable with electrostatic generators have been greatly increased by the application of the "tandem" principle, an ingenious idea first suggested in 1936 but not put into practice until more than 20 years later. In the two-stage tandem Van de Graaff negative ions (such as H^-) are produced by electron bombardment and are accelerated *toward* the positive high-voltage terminal, which is located in the center of the pressure tank. Inside the terminal the negative ions, which now have an energy in MeV equal in magnitude to the terminal voltage in MV, pass through either a foil or a gas-filled canal and are thus stripped of electrons. The positive-ion beam so produced is further accelerated toward ground potential in the usual way. Two-stage tandems producing tens of microamperes of protons at energies up to about 25 MeV are commercially available. Tandems with still higher terminal voltages are under construction, principally for heavy-ion research (see below).

A further increase in energy can be achieved in the three-stage version of tandem Van de Graaff machine. This requires two separate tanks, one with a negative, the other with a positive high-voltage terminal. Figure 15-3 is a schematic drawing. Ions are produced in an ordinary positive-ion source, magnetically analyzed, and then neutralized by electron bombardment. The neutral beam so produced is allowed to drift to the negative high-voltage electrode in the first tank, where further electron addition produces negative ions, which are then accelerated to ground potential. At this point the negative-ion beam passes from the first to the second tank and receives two additional stages of acceleration analogous to the two-stage tandem operation. A number of three-stage tandem Van de Graaff machines are in operation, with maximum proton energies up to about 45 MeV.

Tandem Van de Graaff machines have become the principal tools for precise nuclear physics research. Great stability, excellent energy resolution, continuous energy variability, and wide choice of ion beams are the major assets.

Tandems for Heavy-Ion Acceleration. Any element for which a negative-ion source can be devised is suitable for acceleration in a tandem Van de Graaff. In recent years much emphasis has been on heavy-ion research

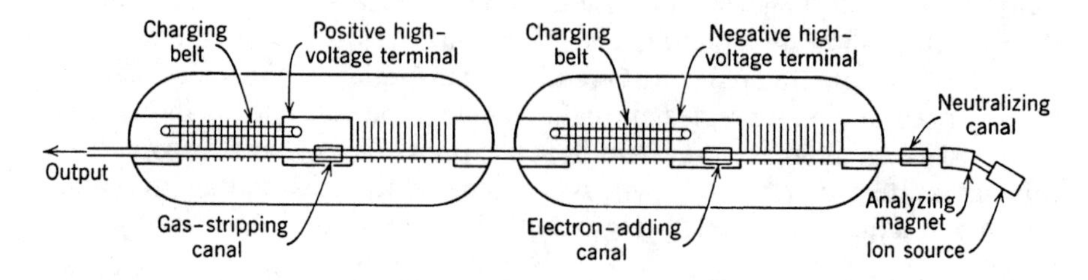

Fig. 15-3 Schematic sketch of a three-stage tandem Van de Graaff generator. [From *Methods of Experimental Physics*, Vol. 5B, *Nuclear Physics* (L. C. L. Yuan and C. S. Wu, Eds.), Academic, New York, 1963.]

with tandems. Development of suitable, efficient ion sources is crucial to obtaining good beam intensities.

The final ion energy attainable depends sensitively on the charge state to which the ion can be stripped in the high-voltage terminal, and this in turn is governed by the terminal voltage. Actually the ions emerge from the stripper with a distribution of charge states (see chapter 6, section A), and the average charge state as well as the width of the distribution can be estimated by (6-19)–(6-21). In the second stage of the tandem the different charge states are then accelerated to different energies, and the final energy can be selected by magnetic analysis. Usually a compromise between energy and intensity has to be made.

The foregoing may be illustrated by reference to figure 15-4, which shows the distribution of charge states for ^{79}Br ions of 15 and 20 MeV, stripped in argon gas. Suppose a $^{79}Br^-$ beam is accelerated to 15 MeV and then stripped in the high-voltage terminal. The final energy attained by the Br^{+9} charge state will be $15 + (9 \times 15) = 150$ MeV, whereas that of the Br^{+7} charge will be $15 + (7 \times 15) = 120$ MeV, but according to figure 15-4 the intensity of the +7 beam will be nearly 5 times that of the +9 beam. If the terminal voltage could be raised to 20 MV, the intensity of the Br^{+9} beam would be approximately tripled and its energy would reach 200 MeV.

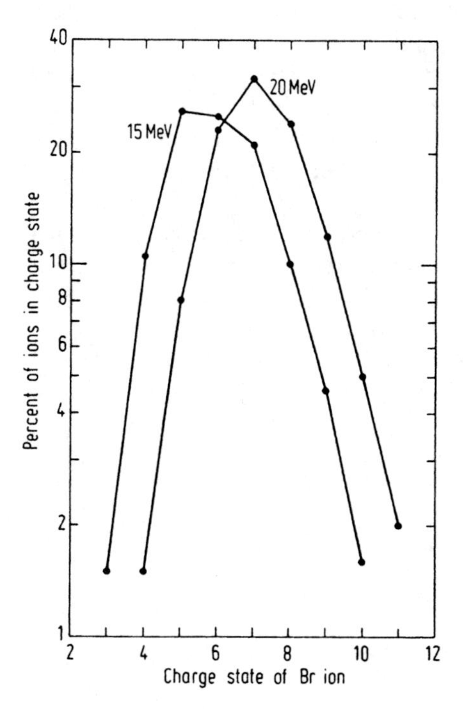

Fig. 15-4 Charge state distribution of ^{79}Br ions after stripping by argon gas. [From data in C. D. Moak et al., *Phys. Rev.* **176**, 427 (1968).]

A 25-MV tandem facility (Holifield Heavy Ion Research Facility at Oak Ridge, Tennessee) and a 30-MV facility (Daresbury, England) will be in operation in the early 1980s.

2. *Linear Accelerators*

Principle. In the types of accelerators mentioned so far the full high potential corresponding to the final energy of the ions must be provided, and the limitations of this type of device are introduced by the insulation problems. These problems are very much reduced in machines that employ repeated acceleration of ions through relatively small potential differences. The linear accelerator was the first device developed in which advantage was taken of this possibility. In the early versions of this machine a beam of ions from an ion source was injected into an accelerating tube containing a number of coaxial cylindrical sections. (See figure 15-5 for a schematic diagram.) Alternate sections were connected, and a high-frequency alternating voltage from an oscillator was applied between the two groups of electrodes. An ion traveling down the tube will be accelerated at a gap between electrodes if the voltage is in the proper phase. By choosing the frequency and the lengths of successive sections correctly one can arrange the system so that the ions arrive at each gap at the proper phase for acceleration across the gap. The successive electrode lengths have to be such that the ions spend just one half cycle in each electrode. Acceleration takes place at each gap, but the focusing action described for the accelerating tubes of direct high-voltage machines is for most types of linear accelerators replaced by a net defocusing effect because the radio frequency field is rising while the particles cross the gap. Special focusing devices such as grids or magnetic lenses must then be provided.

History. The first linear accelerator on record was a two-stage device built in 1928 by R. Wideröe (W1); it accelerated positive ions to about 50 keV. By 1931 E. O. Lawrence and D. H. Sloan had succeeded in accelerating mercury ions to 1.26 MeV in an accelerating system having 30 gaps. Intensive work on linear accelerators was carried out in many laboratories in the early 1930s. However, because the cyclotron was

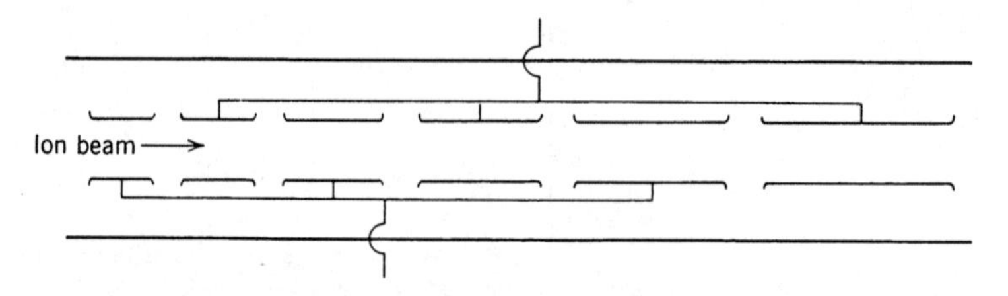

Fig. 15-5 Schematic diagram of the accelerating tube of a linear accelerator.

developed almost simultaneously and had obvious advantages, the linear accelerator did not receive much further attention from about 1934 until after World War II, when the availability of high-power microwave oscillators made possible acceleration to high energies in relatively small linear accelerators. Since then a sizable number of linear accelerators, or linacs, have come into operation, both for electron and proton acceleration, as well as several heavy-ion linacs or "hilacs."

Electron Linacs (S1, B1). Since electrons, even at relatively low energies, travel with essentially the velocity of light, electron linacs use traveling rf waves for acceleration, and the electrons are kept in phase with the traveling wave all the way down the wave guide. The dimensional tolerances for the wave guides are extraordinarily exacting because of the necessity of precisely maintaining the phase velocity of the traveling wave. The largest electron linac in operation is the 2-mile-long machine of the Stanford Linear Accelerator Center (SLAC) at Stanford, California that accelerates electrons up to 20×10^9 eV (20 GeV). Like all linacs, this is a pulsed machine, with up to 360 pulses per second and peak currents of about 50 mA. In a number of electron linacs, including the one at SLAC, positron as well as electron acceleration has been achieved by placing a converter foil in the beam at a point part way down the accelerator; the positrons travel 180° out of phase with the electrons. At SLAC positrons have been accelerated to about 15 GeV.

Electron linacs at intermediate energies (100–1000 MeV) are of increasing importance for electron-scattering research, and the principal design efforts have been in the direction of increased intensity, higher duty cycle, and improved energy definition.

In parallel with the development of high-energy electron linacs for particle and nuclear physics, high-intensity machines of lower energies have been developed for a variety of purposes. Particularly important for nuclear science and technology is the use of electron linacs as pulsed-neutron sources (see section C). Some machines such as the Oak Ridge Electron Linear Accelerator, ORELA (energy up to 120 MeV, peak current 20 A, pulse rate 10^3 s^{-1}, pulse widths 20 ns) are dedicated to this use. Other applications for low-energy linacs are in photonuclear-reaction studies, radiation therapy, and industrial radiation processing.

Proton Linacs. Since protons and other positive ions have much smaller velocities than electrons of comparable energies and speed up markedly as they gain energy, they cannot be accelerated with traveling waves. Standing-wave acceleration is used instead, and the accelerator structure contains a series of drift tubes of increasing length. Their purpose is to shield the particles during the "wrong" phase of the rf cycle, and acceleration takes place by the electric field at each gap between drift tubes. The principle is schematically illustrated in figure 15-5. The necessary focusing inside the drift tubes is provided by quadrupole lenses. Proton linacs with energies up to 200 MeV have been built on the basis of this so-called Alvarez design (A2), and linacs of this type serve as injectors of 50–200-MeV protons into various proton synchrotrons (see below).

The largest and most powerful proton linac in operation is the 800-MeV Los Alamos Meson Physics Facility (LAMPF) at Los Alamos, New Mexico, which, as its name implies, is designed primarily as a "meson factory." In operation since 1972, it has a design intensity of 1 mA, delivered in 120 pulses per second, each of 500-μs duration, that is with a 6 percent duty factor. The 0.25-ns micropulses are separated by 5 ns. Only the first 100 MeV of energy are achieved in an Alvarez linac. The remaining acceleration takes place in so-called side-coupled cavities (C1); here the cylindrical wave guide is loaded with disks and the successive resonant cavities between these disks are coupled by additional small cavities alongside the main accelerating structure. In contrast to the Alvarez mode, the standing waves in successive cavities are not in phase, but out of phase by $\pi/2$.

Although for electron acceleration to multi-GeV energies linacs have clear advantages over circular machines because they avoid the huge energy losses by radiation inherent in the latter (so called synchrotron radiation, see p. 580), this is not true for proton machines. Economic considerations greatly favor circular accelerators for achieving multi-GeV proton energies.

Hilacs (G1). Linear accelerators for heavy ions are similar to those for protons. However, to obtain high energies it is necessary to accelerate high charge states as already discussed under tandem Van de Graaffs, and hilacs are therefore always built in at least two sections, with a foil or gas stripper between them. In the first stage, or prestripper, ions in low charge states as obtained in an ion source[2] are accelerated to an energy of 1 or 1.5 MeV amu^{-1}. After stripping at this energy the ions have charge states q several times the prestripper value and are then accelerated to their final energy.

Among the existing hilacs the most important for nuclear research are the SuperHILAC at the Lawrence Berkeley Laboratory and the UNILAC of the Gesellschaft für Schwerionenforschung (GSI) near Darmstadt, Germany. The SuperHILAC produces microamperes of light heavy ions such as carbon, nitrogen, oxygen, and neon and somewhat smaller currents of ions up to xenon,[3] with energies of 8.5 MeV amu^{-1}. The UNILAC, the most advanced system in operation, can accelerate any ion up to uranium

[2] The linac is, in fact, always preceded by a dc accelerator, usually of the Cockcroft-Walton type, that boosts the ions emerging from the ion source to an energy of 0.3–3 MeV for injection into the linac.

[3] An improvement program underway will permit acceleration of ions up to uranium.

Fig. 15-6 Two views of the UNILAC. (*a*) The Wideröe section as seen from the injection region. (*b*) Experimental area as seen from the end of the accelerator. (Courtesy Gesellschaft für Schwerionenforschung and G. Herrmann.)

to energies up to 10 MeV amu^{-1} (lighter ions to higher energies)—hence the name Universal Linear Accelerator. In this machine injection from one of two 320-kV dc generators is followed by acceleration to 1.4 MeV amu^{-1} in a Wideröe section (rf power delivered to the drift tubes by conductors rather than by means of resonant cavities as in the Alvarez design). Then comes stripping, followed by magnetic analysis to select a single charge state that is further accelerated to 5.9 MeV amu^{-1} in Alvarez tanks. Subsequently 20 single-gap cavities driven independently bring the beam to final energy. These last cavities make it possible to achieve almost continuously variable energy; they can even by used to decelerate the beam. Views of the injection region and Wideröe section and of the experimental area of the UNILAC are shown in figure 15-6.

Development work is underway in a number of laboratories on improved linac structures for heavy ions. In particular, the use of superconducting cavities is being pursued intensively.

An important advance in linac technology in the 1970s was the development of rf quadrupoles for simultaneous focusing and acceleration of particles. These devices promise to make possible the design of very compact, high-current linacs for a variety of applications, including high-intensity fast-neutron sources and possible future fusion reactors (see chapter 14, section D).

3. *Cyclotrons*

The best-known, and one of the most successful of all the devices for the acceleration of positive ions to millions of electron volts is the cyclotron proposed by Lawrence in 1929. A remarkable development has taken place (H1) from the first working model, which produced 80-keV protons in 1930, to the giant synchrocyclotrons now in operation, which accelerate protons to energies as high as 700 MeV.

Principle of Operation. In the cyclotron, as in the linac, multiple acceleration by an rf potential is used. But the ions, instead of traveling along a straight tube, are constrained by a magnetic field to move in a spiral path consisting of a series of semicircles with increasing radii. The principle of operation is illustrated in figure 15-7. Ions are produced in an arc

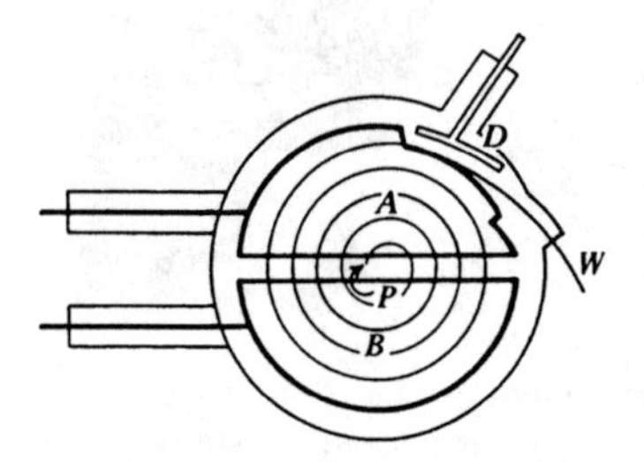

Fig. 15-7 Schematic sketch of cyclotron operation. The ions originate at the ion source *P* and follow a spiral path. The "dees" *A* and *B*, the deflector *D*, and the exit window *W* are indicated. (Reproduced from E. Pollard and W. L. Davidson, *Applied Nuclear Physics*, 2nd ed., Wiley, New York, 1951.)

ion source P near the center of the gap between two hollow semicircular electrode boxes A and B, called "dees." The dees are enclosed in a vacuum tank, which is located between the circular pole faces of an electromagnet and is connected to the necessary vacuum pumping system. An rf potential supplied by an oscillator is applied between the dees. A positive ion starting from the ion source is accelerated toward the dee that is at negative potential at the time. As soon as it reaches the field-free interior of the dee, the ion is no longer acted on by electric forces, but the magnetic field perpendicular to the plane of the dees constrains the ion to a semicircular path. If the frequency of the alternating potential is such that the field has reversed its direction just at the time the ion again reaches the gap between dees, the ion again is accelerated, this time toward the other dee. Now its velocity is greater than before, and it therefore describes a semicircle of larger radius; however, as we shall see from the equations of motion, the time of transit for each semicircle is independent of radius. Therefore, although the ion describes larger and larger semicircles, it continues to arrive at the gap when the oscillating voltage is at the right phase for acceleration. At each crossing of the gap the ion acquires an amount of kinetic energy equal to the product of the ion charge and the voltage difference between the dees. Finally, as the ion reaches the periphery of the dee system, it is removed from its circular path by a negatively charged deflector plate D and is allowed to emerge through a window W and to strike a target.

The equation of motion of an ion of mass M, charge e, and velocity v in a magnetic field H is given by the necessary equality of the centripetal magnetic force Hev and the centrifugal force Mv^2/r, where r is the radius of the ion's orbit:

$$Hev = \frac{Mv^2}{r} \quad \text{and} \quad r = \frac{Mv}{He}. \tag{15-1}$$

Remembering that the angular velocity $\omega = v/r$, we see that

$$\omega = \frac{He}{M}. \tag{15-2}$$

Standard Fixed-Frequency Cyclotron. From (15-2) it is evident that the angular velocity is independent of radius and ion velocity and that the time required for half a revolution is constant for ions of the same e/M, provided that the magnetic-field strength is constant. In practice the magnetic field is kept constant, e/M is a characteristic of the type of ion used, and therefore ω is constant. The radio frequency has to be chosen so that its period equals the time it takes for the ions to make one revolution. For $H = 15{,}000$ G and e/M for a proton the revolution frequency $\omega/2\pi$, and therefore the necessary oscillator frequency, turns out (from 15-2) to be about 23×10^6 Hz. For deuterons or helium ions (He^{2+}) at the same H the frequency is half that value.

It is clear from (15-2) that in a given cyclotron both the magnetic field and the oscillator frequency can be left unchanged when different ions of the same e/M, such as deuterons and α particles, are accelerated. Equation 15-1 shows that the velocity reached at a given radius is the same for ions of the same e/M. Therefore α particles are accelerated to the same velocity, hence twice the energy, as deuterons. To accelerate protons in a cyclotron designed for deuterons either the frequency must be approximately doubled (which is usually impractical) or H must be about halved. Although the latter method makes inefficient use of the magnet, it is often used, and the final velocity is again the same as for deuterons (15-1); therefore protons are accelerated to half the energy available for deuterons.

By squaring and rearranging (15-1) we see that the final energy attainable for a given ion varies with the square of the radius of the cyclotron. With $H = 15{,}000$ G, the deuteron energy $E = 5.4 \times 10^{-3} r^2$ MeV if r is in centimeters. The size of a cyclotron is usually given in terms of its pole-face diameter.

From the equations of motion it is clear that an ion can reach the dee gap at any phase of the dee potential and still be in resonance with the radio frequency. As we have just derived, the final energy acquired by an ion is entirely independent of the energy increment the ion receives at each crossing of the dee gap. However, in practice only ions that enter the first gap in a favorable phase of the radio frequency (perhaps during about one third of the cycle) contribute to the beam current. To avoid difficulties due to excessive phase differences between beam and radio frequency as well as to excessively long paths for the ions, rather high dee voltages (50–500 kV) are generally used.

A very important feature of the cyclotron is the focusing action it provides for the ion beam. The electrostatic focusing at the dee gap is entirely analogous to that in high-voltage accelerating tubes. However, as the energy of the ions increases, this effect becomes almost negligible. Fortunately, a magnetic focusing effect becomes more and more pronounced as the ions travel toward the periphery. This can be seen from the shape of the magnetic field as shown in figure 15-8. Near the edge of the pole faces the magnetic lines of force are curved, and therefore the field has a horizontal component that provides a restoring force toward the

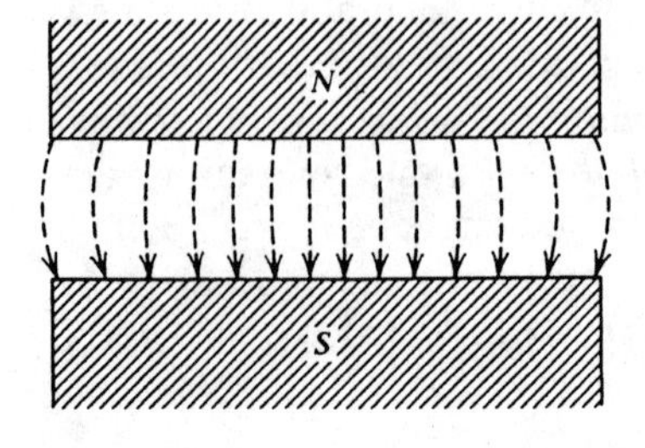

Fig. 15-8 Shape of magnetic field in the gap of a cyclotron magnet. The curvature of the lines of force gives rise to the focusing action.

median plane to an ion either below or above that plane. The focusing is so good that a cyclotron beam is generally less than 1 cm high at the target.

The maximum energy to which ions can be accelerated in standard cyclotrons is limited by their relativistic mass increase. It is clear from (15-2) that if the revolution frequency is to be kept constant the increase in mass must be compensated by a proportional increase in field strength. When the relativity effects are small, this increase of the magnetic field toward the periphery can be readily achieved by slight radial shaping or shimming of the pole faces.[4] Notice, however, that this shaping of the field creates regions of magnetic defocusing. For moderate relativistic mass increases this difficulty has been overcome, mainly by the use of higher dee voltages and correspondingly shorter ion paths. The practical limits for acceleration in standard fixed-frequency cyclotrons are about 25 MeV for protons and deuterons and 50 MeV for ^{4}He.

Cyclotrons have rather high beam intensities but relatively large energy spreads. Typically we may find circulating beams of hundreds of microamperes; the deflected external beams are somewhat smaller. The large beam currents available have made target cooling a rather severe problem. The power dissipation in a target receiving 100 μA of 20-MeV particles is 2 kW, and even iron targets are melted unless water cooling is provided.

Sector-Focused or Azimuthally-Varying-Field (AVF) Cyclotron. A method for overcoming the relativistic-energy limitation of cyclotron acceleration was suggested as early as 1938 by L. H. Thomas (T1) but was not put to use until almost two decades later (see H1 and L1). Thomas showed that azimuthal variations of the magnetic field can result in axial focusing (i.e., focusing in the direction perpendicular to the pole faces). It is thus possible to let the *average* field increase with radius (as required to compensate for the relativistic mass increase), yet to achieve focusing by means of azimuthal field variations. It has been shown that, with this type of design, particles can be accelerated to a kinetic energy approximately equal to their rest energy. The periodic azimuthal field variations are obtained by the use of pole faces that have alternate "hill" and "valley" sectors (figure 15-9). In most, but not all, designs the sectors have spiral rather than radial contours. Therefore the name spiral-ridge cyclotron is frequently used, although the terms sector-focused, isochronous, or azimuthally-varying-field (AVF) cyclotron are more generally applicable. The designation isochronous is meant to convey that in contrast to the synchrocyclotron (see later) the time per revolution (or the revolution

[4] In a given cyclotron the field should be shaped slightly differently for protons and for deuterons because of the different relativity effects. For this reason deuteron cyclotrons do not give very good proton beams without major readjustments. Better proton beams can be obtained by acceleration of H_2^+ ions at full magnetic field; the required field shapes for H_2^+ and D^+ acceleration are almost identical. The final proton energy is the same whether H_2^+ ions are accelerated at full field or H^+ ions at half field.

Fig. 15-9 The TRIUMF variable-energy isochronous cyclotron while under construction (in 1972). This view clearly shows the six spiral magnet sectors supported on the center post of the machine. The many holes through the sectors accomodate tie rods, which restrain the vacuum tank from collapse under atmospheric load. This cyclotron accelerates H^- ions to 500 MeV, producing beam currents of over 100 μA. (Photo courtesy of TRIUMF, University of British Columbia, Vancouver, B.C. and B.D. Pate.)

frequency) stays constant just as in the ordinary cyclotron [cf. (15-2)], although both magnetic field H and ion mass M vary over the ion path. Note that as in all cyclotrons the magnetic field stays constant with time.

In existing machines three, four, or six pairs of hill and valley sectors are used. The ratio of hill gap to valley gap varies widely in different designs but typically may be between 0.25 and 0.7. The pole faces must be radially shaped also (usually in both hill and valley regions) to provide the radial rise of magnetic field required by the relativistic mass increase. Resulting pole-face contours and field shapes can become quite intricate, as shown in figures 15-9 and 15-10. A combination of model measurements and computer calculations is usually required to arrive at the final pole shapes. The dee structures often take on odd shapes also, because they have to fit the pole designs.

Several dozen AVF cyclotrons are in operation in many parts of the world. In contrast to standard cyclotrons, they have the great virtue of allowing easy and continuous energy variation over wide ranges as well as

Fig. 15-10 Contour map of the magnetic field at the median plane of the 50-MeV sector-focused cyclotron at UCLA. The numbers on the contour lines are field strengths in kilogauss. (From D. J. Clark, J. R. Richardson, and B. T. Wright, *Nucl. Instr. Methods* **18–19**, 1 (1962).

great flexibility in the ions that can be accelerated (^{2}H, ^{3}He, ^{4}He, and heavier ions, as well as protons). Appropriate correcting coils have to be provided to achieve this flexibility, and the currents in these coils as well as in the main windings are programmed to give the field shape required for the acceleration of a particular ion to a particular energy. The advantages of AVF operation have prompted the conversion of a number of standard cyclotrons to sector-focused design. Existing AVF cyclotrons span a wide range of energies, the largest one, at the Swiss Institute for Nuclear Research (SIN), near Zurich, accelerating protons to about 580 MeV and serving as a "meson factory." Large circulating beam currents (up to about 1 mA of protons) can be achieved in AVF machines, and efficient beam extraction, although more difficult than in standard cyclotrons, has been accomplished.

Synchrocyclotron. Another way of overcoming the relativity limitation in cyclotrons is by modulation of the oscillator frequency. Although this was, in a sense, an obvious solution that followed from the basic cyclotron equations, it was not seriously considered until about 1945 because the difficulty of maintaining synchronism between oscillator frequency and revolution frequency seemed formidable. The discovery of the principle of phase stability in 1944–1945 (independently found by V. Veksler in the Soviet Union and E. M. McMillan in the United States led to the realization that, as McMillan put it (M1), "Nature had already provided the means" for overcoming this difficulty.

Without going into the quantitative theory we can easily see how phase stability in any circular resonance accelerator comes about qualitatively.[5] Suppose a particle at a particular crossing of the accelerating gap is given more than its proper amount of energy. Its next orbit will then have a larger radius than if the particle had just the "right" amount of energy, and the transit time for that revolution will be too long (longer than one rf wavelength). If the accelerator is designed so that the particles cross the gap when the sinusoidal rf voltage is in the 90–180° phase—decreasing with time—the particle that has previously received too much energy and therefore arrives late at the gap now receives less energy. Conversely, a particle with too little energy follows an orbit of smaller than equilibrium radius, arrives at the gap early, and is given more energy than before. Thus the particles will perform phase oscillations (called **synchrotron oscillations**) around a stable phase. The frequency of these phase oscillations is usually much (perhaps several hundred times) lower than the revolution frequency.

As soon as phase stability was recognized, frequency modulation was applied to cyclotrons, and a number of frequency-modulated (FM) cyclotrons, or synchrocyclotrons as they are usually called, were built in several countries in the 1940s and 1950s. Many of them have since been shut down; others have been upgraded to deliver both higher energies and larger beam currents than was possible with the original designs. The AVF experience has led to the incorporation of a certain amount of sector focusing in recent FM cyclotron redesigns. The highest proton energies achieved in synchrocyclotrons are around 700 MeV.

In most synchrocyclotrons the frequency modulation is brought about by means of a rotating condenser in the oscillator circuit. Obviously, for successful acceleration ions have to start their spiral path at or near the time of maximum frequency. Because ions are accepted into stable orbits only during about 1 percent of the FM cycle, the beam consists of successive pulses. Beam currents are therefore lower than in standard and sector-focused cyclotrons. Average beam currents of one to several microamperes at pulse rates between 50 and 500 s^{-1} are typical. Focusing presents less of a problem than in standard cyclotrons because the relativity effects need not be compensated for by the shape of the magnetic

[5] In modern linacs for protons and other heavy particles phase stability is also important but the details are different; in linacs stability is obtained on the rising (0–90°) rather than the falling (90–180°) part of the rf sine wave.

field. The magnetic field can actually be decreased near the edges of the pole faces to increase the magnetic focusing. Most FM cyclotrons have but one working dee, and the dee voltage is relatively low, typically 10–30 kV. Acceleration to hundreds of MeV thus requires of the order of 10^4 revolutions, or a total path length of the order of 10^7 cm; an acceleration cycle then takes about 10^{-3} s. The low dee voltages have as a consequence small differences in radius between successive turns in the spiral orbits. Beam extraction is therefore more difficult than in standard cyclotrons, requiring the application of perturbing magnetic fields near the periphery in order to induce appropriate oscillations that eventually bring the beam into a magnetically shielded channel or deflector. Extraction of more than about 10 percent of the circulating beam is difficult, and many synchrocyclotron experiments are done with target probes inserted into the circulating beam. The bombarding energy can then be varied by choice of the radius of interception.

Cyclotrons in Accelerator Combinations (G1, B2). We already mentioned that acceleration of heavy ions to high energies always involves at least two stages of acceleration, with a stripper between them. In addition to tandem Van de Graaffs and linacs, cyclotrons have found important application as poststripper accelerators for heavy ions. Injection may be from a tandem (as in several proposed facilities), a linac (as, for example, in the accelerator ALICE in Orsay, France), or a cyclotron (as in the machines at Indiana University, at Dubna, USSR, and in the large GANIL[6] facility under construction at Caen, France).

An essential difference between linacs and cyclotrons as poststrippers lies in the very different dependence of ion energy on mass in the two types of machine. In linacs the energy per amu (E/A) is approximately independent of mass number A and charge state q. For a cyclotron

$$\frac{E}{A}=\frac{Kq^2}{A^2}. \tag{15-3}$$

Since, for $A > 40$, q/A decreases with increasing A (even for highly stripped ions), E/A is a strongly decreasing function of A. Heavy-ion cyclotrons are characterized by the constant K. For example, the ALICE cyclotron, injected from a linac with ions of $E/A = 1.1$ MeV amu^{-1}, has $K = 100$, whereas the GANIL facility uses $K = 400$ cyclotrons for both the prestripper and the poststripper.

New variants of cyclotron design have been developed to fit the particular requirements of poststrippers for heavy ions. The separated-sector cyclotron, with open spaces between the four or six magnet sections, makes injection and extraction easier and gives improved ion optics. The Indiana University and GANIL machines use this design (see figure 15-11).

[6] Grand Accélérateur National d'Ions Lourds.

Fig. 15-11 Top view of separated-sector isochronous cyclotron at Indiana University. The four magnet sectors are clearly visible. On the left-hand side is shown the extraction port, on the right-hand side one of the two 250-kV rf accelerating cavities. (Photo by Berglund, courtesy Indiana University Cyclotron Facility and T. Ward.)

The technology of superconductivity has advanced to the point where superconducting magnets offer important economic advantages in construction and operation, and a pair of superconducting cyclotrons with $K = 500$ and $K = 800$ will be used in the major heavy-ion facility under construction at Michigan State University.

Microtron. This is a device for *electron* acceleration somewhat akin to the cyclotron in that it uses a constant, uniform magnetic field and a constant rf accelerating voltage. However, the particle orbits are circles of increasing radius with a common tangent point at which the accelerating voltage is applied. The time per revolution must be an integral number of rf periods, with this number increasing (usually by one) in each successive turn. The microtron was proposed in Veksler's original paper on phase stability. A few microtrons have been built, most of them in the Soviet Union, with maximum electron energies of about 30 MeV. The requirements on magnet homogeneity are very severe. An asset of microtrons that makes them quite suitable as injectors into higher-energy machines (such as high-current synchrotrons and linacs) is the ease with which the beam can be extracted because successive turns are well separated at a point 180° from the accelerating cavity.

4. *Betatrons*

The betatron was the first device developed for accelerating electrons to energies above a few million electron volts. The basic principle had been suggested by several investigators, but the first practical betatron was developed by D. W. Kerst in 1940. A betatron may be thought of as a transformer in which the secondary winding is replaced by a stream of electrons in a vacuum "doughnut." The acceleration is supplied by the electromotive force induced at the position of the doughnut by a steadily increasing magnetic flux perpendicular to and inside the electron orbit. In order for the electrons to move in a fixed orbit, it is necessary that the field at the orbit change proportionally with the momentum of the electrons. This condition is fulfilled if the field at the orbit increases at just half the rate at which the average magnetic flux inside the orbit increases; this may be achieved by the proper tapering of the pole faces, as indicated schematically in figure 15-12. The radial variation of the field in the region of the electron orbit is important in the focusing problem. It turns out that, if the field falls off as the inverse nth power of the radius in the region of the orbit and if $0 < n < 1$, the electrons will describe damped oscillations about the equilibrium orbit. If the revolution frequency is f, the frequency of the vertical **betatron oscillations** about the equilibrium orbit is $f_v = f\sqrt{n}$, that of the radial oscillations $f_r = f\sqrt{1-n}$. Betatrons are generally designed with $n = 0.75$, and the good focusing permits the electrons to make hundreds of thousands of revolutions.

Electrons are injected into the doughnut from an electron gun when the field at the orbit is very small. The ac magnets (60–1800 Hz) of betatrons are made of laminated iron. The energy obtainable with a given betatron is limited by the saturation of the central flux bars. Auxiliary coils can be used to steer the beam at any chosen electron energy onto a target placed in the doughnut.

Betatrons with energies up to 320 MeV have been built. Few of them remain in operation because synchrotrons and linacs have proved to be superior machines for electron acceleration. The ultimate limitation on the energy attainable in a betatron or in any other circular electron accelerator is presumably set by the radiation of energy by electrons under centripetal

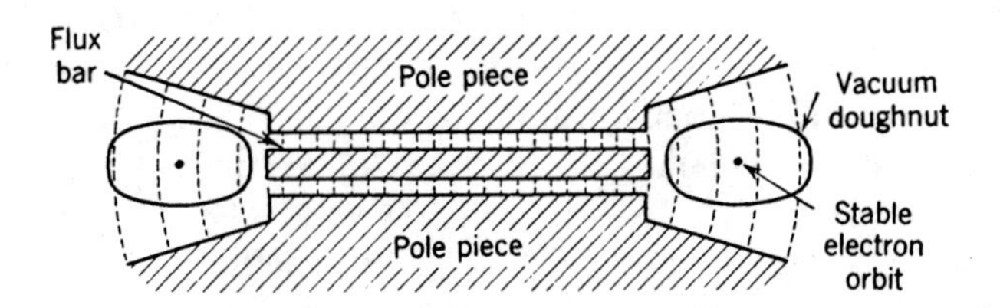

Fig. 15-12 Cross section through central region of a betatron (schematic), with the magnetic lines of force indicated.

acceleration; the radiative energy loss at a given radius increases with the fourth power of the electron energy.

5. Synchrotrons

Both Veksler and McMillan, along with the discovery of the principle of phase stability (see the discussion of the synchrocyclotron above), proposed a new type of accelerator, called synchrotron by McMillan and synchrophasotron by Veksler. The two names have survived, one in Western countries, the other in the Soviet Union, and the device has been enormously successful for the acceleration of both electrons and protons.

Principle of Constant-Gradient Synchrotron[7]. In the synchrotron, as in the betatron, the radius of the orbit is kept approximately constant by a magnetic field that increases proportionally with the momentum of the particles. However, the acceleration (or rather the increase in energy, since the velocity must remain essentially constant at $v \approx c$) is provided not by a changing central flux but (more nearly as in the cyclotron) by a rf oscillator that supplies an energy increment every time the particles cross a gap in a resonator that forms part of the vacuum doughnut.

As in the synchrocyclotron (see p. 568), phase stability results when the acceleration takes place during the decreasing part of the rf cycle (phase between 90 and 180°). The phase stability considerations are qualitatively the same as for the synchrocyclotron, although the variation of the magnetic field with time makes the detailed phase relations somewhat more complex. As the magnetic field is steadily increased, a slight phase difference is maintained between the particle orbits and the resonator voltage, and on the average the particles gain some energy at each passage of the gap. As in the betatron, the particles perform radial and vertical oscillations around their equilibrium orbit. The radial field index

$$n = -\frac{dB/B}{dr/r} = -\frac{d \ln B}{d \ln r} \tag{15-4}$$

has to be carefully chosen to avoid resonances between radial and vertical betatron oscillations or between either of these and the revolution frequency. At such resonances the beam would be lost. From the relations between the oscillation and revolution frequencies (see section A, 4 above) it is clear, for example, that n values of exactly 0.2, 0.5, and 0.8 must be avoided because they would lead to ratios of vertical to radial oscillation frequencies of exactly 0.5, 1, and 2, respectively. The actual choice of n is dictated by economic factors. For a given angular deviation from the equilibrium orbit the amplitude of an oscillation is inversely proportional to

[7] The term "constant gradient" means that the field gradient n does not vary azimuthally.

its frequency, and therefore small n corresponds to large vertical, large n to large radial amplitudes. A value of n between the 0.5 and 0.8 resonances results in the most economical shape of magnetic gap.

Proton Synchrotrons. For proton (or other positive-ion) acceleration to energies in the GeV range the synchrotron has proved to be the first practical device. Because it requires a ring-shaped magnet only, a proton synchrotron is much cheaper to construct for these energies than a synchrocyclotron with its solid magnet structure. Since protons do not approach the speed of light until they have energies of billions of electron volts ($v = 0.98c$ at 3.8 GeV), the revolution frequency of the protons changes by a large factor during the acceleration (a factor of 12 for acceleration from 4 MeV to the limiting velocity). The frequency of the rf accelerating voltage must be modulated over this wide range, and in most of the present machines this is accomplished electronically rather than by rotating condensers, as in the FM cyclotrons.

Proton synchrotrons are generally built with field-free straight sections between magnet sectors to facilitate injection, rf acceleration, and targeting. The stored energy in the magnets of these accelerators is enormous—peak power inputs range from a few to more than 100 MW—and for this reason most of them use rather low pulse rates (5 to 30 pulses per minute) and have provisions for storage of most of the energy in flywheels between magnet pulses.[8] Injection is usually from linacs or Van de Graaffs.

With maximum magnetic fields of 12–20 kG, proton synchrotrons have orbit radii of many meters (e.g., 18 m for the 6-GeV Bevatron at Berkeley). In order to accommodate oscillations around the equilibrium particle orbits, the vacuum chambers and therefore the magnet gaps have to be fairly large (30×170 cm in the Bevatron). The magnets are thus rather massive, although still much lighter than they would be for synchrocyclotrons of comparable energy rating.

Targeting. Typical beam intensities in proton synchrotrons are 10^{11}–10^{12} per pulse. For most radiochemical studies of nuclear reactions such intensities are quite ample, especially since the fractional energy loss of the protons in going through a thin target is so small that a given proton can make many target traversals in successive revolutions. The product of actual target thickness and number of traversals is approximately constant for all target thicknesses up to some maximum, of the order of 1–5 g cm^{-2}, the exact value differing from machine to machine and also depending on the atomic number of the target because protons are lost from the beam by multiple Coulomb scattering. Targeting in proton synchrotrons can be accomplished in a variety of ways: the target may be rammed into or

[8] The decommissioned 3-GeV Princeton-Pennsylvania Accelerator (PPA) was unique in that it used a choke-capacitor system for energy storage and operated at 20 pulses per second.

flipped through the beam at the end of each acceleration cycle, or the beam orbit may be allowed to collapse into a target, which is achieved by switching off the rf accelerating field while the magnetic field is rising. The energy of the protons striking the target can be varied at will up to the maximum energy of the machine.

External Beams. For many types of experimentation it is advantageous to have external proton beams as well as the beams of secondary particles, such as neutrons, π mesons, K-mesons, and antiprotons, which originate in targets struck by the primary proton beam. Several proton synchrotron installations have external-beam systems based on a scheme for beam extraction, developed by O. Piccioni et al. (P1), that is capable of yielding external-beam intensities up to 50 percent of circulating beams. The beam is first made to go through an energy-loss target, which causes the beam orbit to shrink so that on the next revolution the protons pass through a deflecting magnet placed at a smaller radius in a straight section. This magnet bends the beam into a trajectory that, at some distance downstream from the magnet, emerges from the periphery of the machine. Several such external proton beams may be available at a machine, as well as a number of beams of secondary particles. It is characteristic for these large accelerators to have a multiplicity of complex configurations of experimental equipment (deflecting, analyzing, and focusing magnets, bubble chambers, spark chambers, counter telescopes) simultaneously set up in the experimental areas outside the main accelerator shielding. The proton synchrotrons themselves and the external-beam facilities require very bulky shielding (L2) because the primary proton beams as well as some of the secondary particles are so penetrating.

Bevalac. The majority of the constant-gradient proton synchrotrons, mostly built in the 1950s, have been shut down, and the more recent, higher-energy machines use the alternating-gradient principle (see below). One of the remaining older synchrotrons, the Bevatron, is now largely used for heavy-ion acceleration. A 250-m transfer line was built for injection of heavy ions (at about 8.5 MeV amu^{-1}) from the SuperHILAC into the Bevatron. Prior to further acceleration in the Bevatron the ions are fully stripped and these fully stripped ions are then brought to relativistic energies (2.6 GeV amu^{-1} if $q/A = \frac{1}{2}$, somewhat less for lower q/A values). Ions up to $^{56}Fe^{+26}$ have been accelerated in this accelerator combination known as the Bevalac. An improvement program underway, which includes both a new preaccelerator for the SuperHILAC and a greatly improved vacuum system for the Bevatron, is expected to make possible the acceleration of ions up to $A \approx 238$ to relativistic energies.

Alternating-Gradient Synchrotron. A novel design principle that makes great reductions in the cost of high-energy synchrotrons possible was suggested by H. Christophilos in 1950 and independently rediscovered by E. D. Courant, M. S. Livingston, and H. S. Snyder in 1952 (C2). To understand qualitatively how this so-called alternating-gradient focusing or strong focusing works we recall that, for a given angular deviation from the orbit, the vertical oscillation amplitude is proportional to $n^{-1/2}$, the radial amplitude proportional to $(1-n)^{1/2}$ (see section A, 4 above). Thus in an ordinary synchrotron the vertical dimension of the aperture can be reduced only at the expense of the radial dimension and vice versa. The important new discovery was that a ring magnet design with alternate sections of large positive and large negative n values leads to strong focusing in both dimensions (and thus to small aperture requirements) while preserving phase stability for synchrotron acceleration. One undesirable consequence of the large n values is the existence of many resonances that can lead to beam loss. The range of operating conditions under which stable operation is possible is therefore rather small, and very careful control of a number of parameters is necessary (L1, L3).

Proton synchrotrons with alternating-gradient focusing have become the major tools of particle physics since about 1960 when the 28-GeV machine at CERN (European Center for Nuclear Research, Geneva, Switzerland) and the 33-GeV AGS (alternating-gradient synchrotron) at Brookhaven National Laboratory (Upton, New York) came into operation. They were followed by the 70-GeV accelerator at Serpukhov, USSR (initial operation 1967) and the 500-GeV machine at the Fermi National Accelerator Laboratory (FNAL), Batavia, Illinois (1972). These are enormous machines, the FNAL accelerator (S2) having an orbit radius of 1 km and external beams extending over 2 km in length. Yet, thanks to the strong-focusing principle, the magnet gaps can be kept to very modest sizes (e.g., 7×15 cm in the Brookhaven AGS and even smaller at FNAL) and hence the magnets are relatively small—the total weight of iron in the 240 magnets of the AGS is 4000 tons, equal to the weight of iron in the Berkeley 184-in. synchrocyclotron, which accelerates protons to only about $\frac{1}{50}$ the energy. Typical magnet cross sections are shown in figure 15-13. The field index (n) values are of the order of several hundred. Acceleration to full energy takes typically 0.5–1 s, and repetition rates are 10–60 min^{-1}. Injection into most of the machines is from linacs (e.g., at 200 MeV in the AGS). At FNAL a fast-cycling 8-GeV synchrotron "booster" is used between the linac and the main ring, injecting pulses into the main accelerator until its circumference is filled with protons. Through increase in the injection energies (which reduces space charge limitations) and other improvements, the circulating-beam intensities of alternating-gradient synchrotrons have steadily increased and are now of the order of $10^{13}\ \text{s}^{-1}$.

Targeting problems are similar to those in constant-gradient synchrotrons. But the strong focusing causes what is known as momentum compaction: a great reduction in the radius change caused by a given momentum change. This leads to large numbers of multiple traversals in targets. In

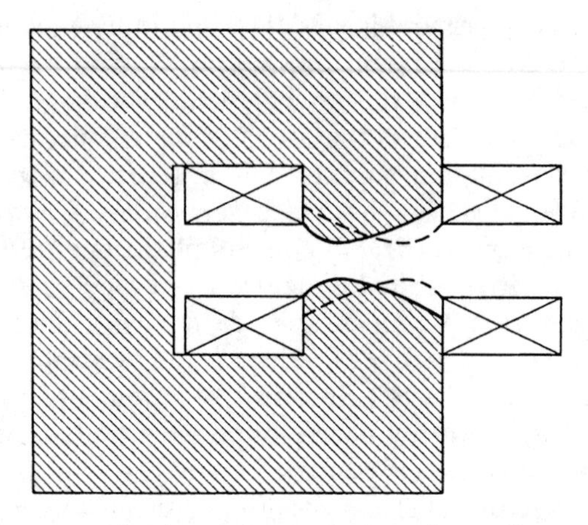

Fig. 15-13 Typical cross section for a magnet of alternating-gradient synchrotron. Magnets with poles shown by the solid lines are alternated with magnets whose poles are shaped as shown by the dashed lines. The boxes with crosses indicate the positions of the magnet coils. In the Brookhaven AGS the external dimensions of the steel laminations are about 84 cm wide by 99 cm high; the pole width is 32 cm and the gap height at the central orbit position is 8.9 cm. [From *Methods of Experimental Physics*, Vol. 5B, *Nuclear Physics*, (L. C. L. Yuan and C. S. Wu, Eds.), Academic, New York, 1963.]

fact, a light-element target tends to be traversed by the beam until a large fraction ($\geq\frac{1}{2}$) of the protons have made nuclear interactions. Beam extraction is usually achieved either by kicking the beam out of its normal orbit by means of pulsed magnets (for fast extraction, as is needed, e.g., for spark and bubble chambers) or by gradually increasing the betatron oscillations around the equilibrium orbit (for slow extraction, up to 1 s in duration, as required for counter experiments). Nearly 100 percent extraction has been achieved and is, in fact, essential to avoid radiation damage and severe activation of accelerator components by the intense beams.

Electron Synchrotrons. For acceleration of electrons to high energies synchrotrons are much more economical than betatrons because no large central magnetic flux is required in synchrotrons, and magnet costs are therefore much lower. If electrons are injected at moderately high energies, no frequency modulation is required ($v/c = 0.99$ at 3 MeV and 0.999 at 11 MeV).

As with proton machines, the constant-gradient electron synchrotrons have been largely superseded by alternating-gradient designs. The 6.5-GeV machine at DESY (Deutsches Elektronen Synchrotron), Hamburg, Germany and a 12-GeV accelerator at Cornell University are among the highest-energy alternating-gradient electron synchrotrons built. As pointed out in connection with betatrons in section A, 4 above, the ultimate limitation for circular electron accelerators is set by the radiation losses; to offset these, very large energy increments per turn are required in these synchrotrons (e.g., about 10 MeV per turn at 10 GeV). Nevertheless, even larger machines are under serious consideration to answer some outstanding questions in particle physics. The energy losses can also be turned into a virtue: the emitted radiation, known as synchrotron radiation, has become a very important research tool, and synchrotron radiation facilities are now being built to serve as dedicated photon sources (see section B).

Storage Rings and Colliding Beams. Although there is no indication of technical barriers to further increases in accelerator energies, there are clearly economic limitations. No new principle comparable in potential savings to the discoveries of phase stability and alternating-gradient focusing have appeared on the horizon, except to some degree the advent of superconducting magnets, which can lead to smaller machine dimensions for a given energy. Even if costs did not become prohibitive, it is not clear that the best way to gain new insights in particle physics is to accelerate particles to even higher energies and let them collide with stationary targets. The problem is that, in such collisions, conservation of momentum requires that a certain fraction of the incident momentum go to momentum of the products, and this fraction increases rapidly with increasing incident energy, more so as the velocities approach the velocity of light. From relativistic mechanics (see appendix B) and conservation of momentum we can derive the following expression for the total center-of-mass energy W_{cm} corresponding to the collision of a particle of rest mass m_1 and total (rest + kinetic) energy W_1 with a stationary particle of rest mass m_2:

$$W_{cm}^2 = m_1^2c^4 + m_2^2c^4 + 2W_1m_2c^2. \tag{15-5}$$

For the case of $m_1 = m_2$ (15-5) reduces to the simple form

$$W_{cm} = mc^2\left[2\left(1 + \frac{W_1}{mc^2}\right)\right]^{1/2}. \tag{15-6}$$

Thus the total center-of-mass energy in the collision of a 30-GeV proton with a stationary proton is only 7.6 GeV, of which only 5.9 GeV is available for particle production, the remainder being accounted for by the rest energy of the two protons. Equation 15-6 shows that, for $W_1 \gg mc^2$, the center-of-mass energy increases approximately as the square root of the laboratory energy.

Because of these considerations, various colliding-beam accelerators have been proposed since the mid-1950s, and a number of such machines (mostly for electrons) have been put into operation (P2). Clearly when two proton beams of equal kinetic energy collide head-on no energy is lost to center-of-mass motion. The drawback of such schemes is the relatively small number of interactions that can be achieved. This problem has been greatly relieved by the construction of storage rings in which many pulses from the primary accelerator are stored so that currents of the order of amperes can be achieved and stored for hours. This development was made possible through great improvements in high-vacuum technique; storage rings typically have pressures of the order of 10^{-10} torr.

The parameter used to characterize the effectiveness of a colliding-beam system is the **luminosity** L (in units of $cm^{-2}\,s^{-1}$), defined as the ratio of the number of events of a given kind per second to the cross section of that process in square centimeters. Colliding-beam devices for p-p as well as for e^+-e^- collisions[9] are in operation, with luminosities up to $2\times 10^{31}\,cm^{-2}\,s^{-1}$. With $L = 10^{31}\,cm^{-2}\,s^{-1}$ a process with a cross section of 1 nb ($10^{-33}\,cm^2$) would be observed at the rate of about one event every two minutes. Colliding-beam machines are clearly not of great interest for radiochemical experiments.

Among the largest colliding-beam facilities in operation are the ISR (intersecting storage rings) at CERN, Geneva, Switzerland, and the PETRA facility at Hamburg, Germany. The two concentric ISR rings are injected with 28-GeV protons from the CERN proton synchrotron; p-p collisions at a center-of-mass energy of 56 GeV can be achieved at 8 crossing points. PETRA is injected from DESY in a somewhat complicated scheme that involves intermediate storage in a smaller ring and additional acceleration; the ultimate electron energy is 19 GeV and the design luminosity is $10^{32}\,cm^{-2}\,s^{-1}$. Another e^+-e^- facility of comparable specifications is the PEP facility at Palo Alto, California, injected from the SLAC electron linac. A p-p colliding-beam facility for 800-GeV center-of-mass energy and $10^{33}\,cm^{-2}\,s^{-1}$ luminosity is under construction at Brookhaven National Laboratory; it will be injected by the AGS at 30 GeV and is called ISABELLE.

As already mentioned, electron synchrotrons are important as radiation sources and, for obvious reasons, the high currents achievable in storage rings make these particularly attractive for this application (see section B).

B. PHOTON SOURCES

Photons for nuclear research are produced usually as secondary beams in electron accelerators. Only a few radioactive decay processes result in γ rays of sufficiently high energy to induce nuclear reactions of low threshold (such as the photodisintegration of ^2H and ^9Be). Of somewhat broader interest are energetic γ rays emitted in some light-element reactions, particularly the reaction $^3H(p, \gamma)^4He$, which has a Q value of 19.8 MeV. This large energy release is emitted in the form of a single γ ray so that monoenergetic γ rays over a limited energy range can be obtained by variation of the proton energy up to a few million electron volts.

The principal mechanisms by which photon beams are produced in

[9] Even the storage of appreciable intensities of antiprotons has been accomplished, thus making possible the study of $p\bar{p}$ collisions at high center-of-mass energies.

electron accelerators are bremsstrahlung and emission of synchrotron radiation.

Bremsstrahlung. The continuous X rays produced when electrons are decelerated in the Coulomb fields of atomic nuclei are called bremsstrahlung (German for "slowing-down radiation"). This type of radiation is produced whenever fast electrons pass through matter and, as discussed on p. 222, the efficiency of conversion of kinetic energy into bremsstrahlung goes up with increasing electron energy and with increasing atomic number of the material in which the conversion takes place. In tungsten, for example, the fraction of energy lost by radiation increases from 0.5 for 10-MeV electrons to >0.9 for 100-MeV electrons.

The spectrum of bremsstrahlung from a monoenergetic electron source extends from the electron energy down to zero, with approximately equal amounts of energy in equal energy intervals. In other words, the number of quanta in a narrow energy interval is about inversely proportional to the mean energy of the interval.

The stopping of fast electrons in matter thus produces a continuous spectrum of X rays, and any electron accelerator can also serve as an X-ray source. The higher the energy of the electrons, the more the X-ray emission is concentrated in the forward direction; about half the intensity of the X-ray beam from a 100-MeV electron accelerator is contained in a 2° cone.

A serious disadvantage of bremsstrahlung sources for nuclear work is their continuous energy spectrum. Measurements have to be made at a series of closely spaced electron energies, and the resulting yield curve for the nuclear reaction studied has to be differentiated to obtain cross sections. This procedure requires accurate knowledge of bremsstrahlung intensity and spectrum shape.

Monochromatic Photon Beams (B3). To circumvent the difficulties just mentioned two schemes have been developed for producing monochromatic beams from bremsstrahlung. In the first of these the bremsstrahlung photons are "tagged" as follows. Bremsstrahlung is produced in a thin target of high Z and the degraded electrons are analyzed in an electron spectrometer. Those bremsstrahlung photons in coincidence with electrons in a particular energy range can then be selected. This type of monochromator is useful only for experiments that can be gated by a coincidence pulse, not for activation experiments.

In the second technique monoenergetic photons are produced by in-flight annihilation of positrons. Electrons from an accelerator interact in a thick, high-Z target, producing there not only bremsstrahlung, but also an appreciable intensity of electron-positron pairs. The positrons may be further accelerated (e.g., in a linac) or simply momentum-analyzed in an electron spectrograph of high transmission and are then allowed to strike a thin, low-Z target. Positron annihilation in flight produces a beam of monoenergetic photons in the forward direction, contaminated by a small amount of bremsstrahlung.[10] Photon intensities of about $10^7\ \mathrm{s}^{-1}$ with 1 percent energy resolution have been obtained by this scheme, at energies up to about 30 MeV.

Synchrotron Radiation. As mentioned on p. 560, a continuous spectrum of electromagnetic radiation is emitted whenever relativistic electrons are bent in a magnetic field. Large circular electron accelerators can thus be used as sources of this synchrotron radiation, which is emitted tangentially in the plane of the electron orbit. The spectrum is usually described in terms of a characteristic wavelength given by

$$\lambda_c = \frac{5.59R}{E^3} = \frac{18.64}{BE^2}, \tag{15-7}$$

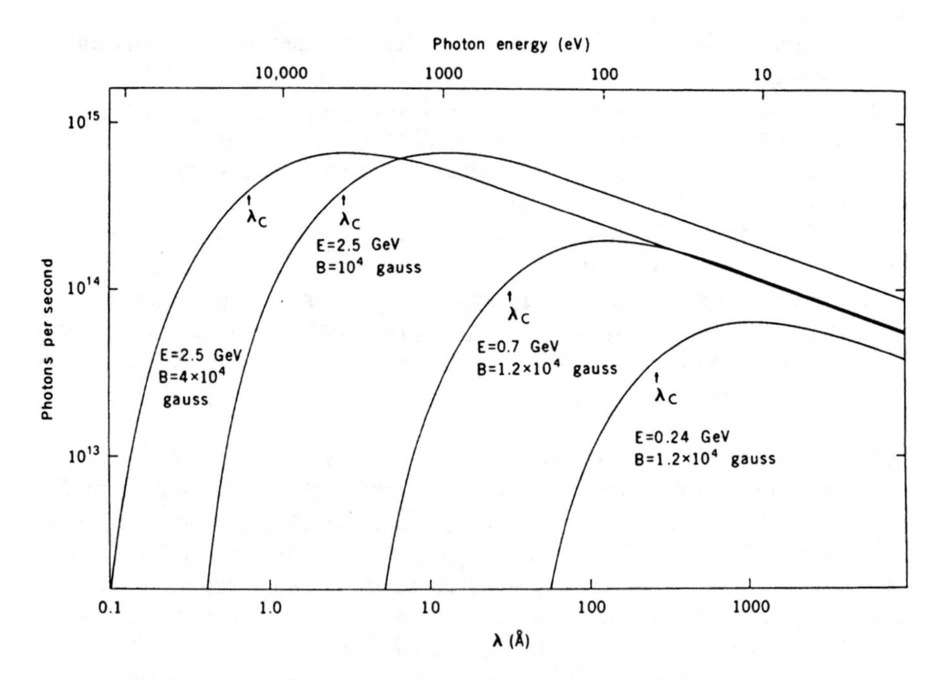

Fig. 15-14 Synchrotron radiation spectra for different electron energies and bending magnetic fields. The ordinate gives intensity within a 1 percent wavelength band and a 10 mrad horizontal acceptance angle, and for a 100-mA electron current. (From reference W2, copyright 1978 by the American Association for the Advancement of Science.)

[10] The bremsstrahlung component is minimized by the low Z of the target. Its contribution to the process under investigation can be determined by a separate experiment with electrons in place of positrons hitting the low-Z target.

where λ_c is in angstroms, R is the orbit radius in meters, E the electron energy in GeV, and B the magnetic field in tesla. Toward the short-wavelength side of λ_c the intensity drops rapidly, whereas it rises slightly to a peak at about $4\lambda_c$ and then decreases slowly toward still longer wavelengths (figure 15-14). In present-day synchrotrons (energies of a few billion electron volts, orbit radii of the order of 10 m) λ_c values are in the range of 1–10 Å (corresponding to photon energies of between 12 and 1 keV); synchrotron radiation is thus of no particular interest for nuclear research. However, it has become a very important tool in other fields including solid-state physics, radiation chemistry, photoelectron spectroscopy, and X-ray crystallography (W2). Electron storage rings for energies up to several billion electron volts and designed to serve as dedicated sources of synchrotron radiation are being built in several laboratories. Intensities of the order of 10^{13} photons s^{-1} Å^{-1} mrad^{-1} per milliampere of circulating current can be achieved, and some of the designs call for circulating beams up to 1 A. Photons in otherwise unattainable energy and intensity ranges thus become available.[11] Furthermore, the fact that photons are emitted tangentially all around the azimuth of a machine makes it possible to perform many experiments simultaneously.

C. NEUTRON SOURCES

Radioactive Sources. Neutrons are produced in nuclear reactions and decay. Several naturally occurring and several artificially produced α and γ emitters can be combined with a suitable light element to make useful neutron sources (O1). Because of the short ranges of the α particles, α emitters must be intimately mixed with the light element (usually beryllium because it gives the highest yield). Such sources necessarily give neutrons with energies spread over a wide range. A γ emitter may be enclosed in a capsule surrounded by a beryllium or deuterium oxide target. (Only beryllium and deuterium have (γ, n) thresholds below 5 MeV.) Some of these sources can, in principle, give monoenergetic neutrons. However, because of neutron and γ-ray scattering in targets of practical thickness, the actual spectrum usually has an energy spread of about 30 percent, and the average energy is roughly 20 percent below the expected maximum value. Some useful sources are listed in table 15-1.

Since spontaneous fission, like any fission process, is always accompanied by neutron emission, a sample of a nuclide that undergoes spontaneous fission can serve as a neutron source. At present by far the most

[11] Further enhancement of intensities in particular wavelength regions is possible by means of so-called wigglers, which produce local regions of smaller radius of curvature. Particularly interesting are helical wigglers (K1); they produce photon spectra that are much more sharply peaked than those shown in figure 15-14.

Table 15-1 Alpha- and Gamma-Ray Neutron Sources

Source	Main Reaction	Q (MeV)	Neutron Energy (MeV)	Neutron Yield (per 10^6 dis)
Ra + Be(mixed)	$^9Be(\alpha, n)^{12}C$	5.65	up to 13	460
Po + Be (mixed	$^9Be(\alpha, n)^{12}C$	5.65	up to 11, av. 4	80
Ra + B (mixed)	$^{11}B(\alpha, n)^{14}N$	0.28	up to 6	180
^{239}Pu + Be (mixed)	$^9Be(\alpha, n)^{12}C$	5.65	up to 11	60
Ra + Be (separated)	$^9Be(\gamma, n)^8Be^a$	−1.67	<0.6	0.9^b
Ra + D_2O (separated)	$^2H(\gamma, n)^1H$	−2.23	0.1	0.03^b
^{24}Na + Be	$^9Be(\gamma, n)^8Be^a$	−1.67	0.8	3.8^b
^{24}Na + D_2O	$^2H(\gamma, n)^1H$	−2.23	0.2	7.8^b
^{88}Y + Be	$^9Be(\gamma, n)^8Be^a$	−1.67	0.16	2.7^b
^{88}Y + D_2O	$^2H(\gamma, n)^1H$	−2.23	0.3	0.08^b
^{124}Sb + Be	$^9Be(\gamma, n)^8Be^a$	−1.67	0.02	5.1^b
^{140}La + Be	$^9Be(\gamma, n)^8Be^a$	−1.67	0.6	0.06^b
^{140}La + D_2O	$^2H(\gamma, n)^1H$	−2.23	0.15	0.2^b

[a] The product ^8Be is unstable and decomposes in less than 10^{-14} s into two ^4He nuclei.
[b] The photoneutron yields are given for 1 g of target (D_2O or Be) at 1 cm from the γ-ray source.

practical nuclide for this purpose is ^{252}Cf, which has a half life of 2.64 y, decays 96.9 percent by α emission, 3.1 percent by spontaneous fission, and emits on the average 3.76 neutrons per fission. Thus ^{252}Cf sources emit about 2.3×10^9 neutrons mg^{-1} s^{-1}, along with approximately ten times that many α particles.

Neutron-Producing Reactions with Accelerators. Much more copious sources of neutrons than can be obtained with radioactive α and γ emitters are available with ion accelerators. The reaction $^2H(d, n)^3He$ (often called a *d-d* reaction) is exoergic ($Q = +3.27$ MeV), and, because the potential barrier is low, good neutron yields can be obtained with deuteron energies as low as 100–200 keV. With thick targets of solid D_2O, the yields are about 0.7, 3, and 80 neutrons per 10^7 deuterons at 100 keV, 200 keV, and 1 MeV deuteron energy, respectively. Direct-voltage accelerators are often used to produce the *d-d* reaction. The neutrons are monoenergetic if monoenergetic deuterons of moderate energies (up to a few million electron volts) fall on a sufficiently thin target.

Even more widely used is the *d-t* reaction: $^3H(d, n)^4He$. Tritium ($t_{1/2}$ = 12.33 y) has become available in large quantities. For use as a target it is usually adsorbed on zirconium or titanium. Neutron yields of about 150 per 10^7 deuterons are obtained at 200 keV. The reaction has a strong resonance at 100 keV deuteron energy and can be a remarkable source of neutrons from very-low-energy deuterons. The reaction is exoergic with $Q = 17.6$ MeV, and monoenergetic neutrons of about 14 MeV are produced from a thin target.

For a controlled source of monoenergetic neutrons of very low energy (down to about 30 keV) the ^{7}Li(p, n)^{7}Be reaction is suitable, especially when produced with the protons of well-defined energy available from electrostatic generators. The reaction is endoergic ($Q = -1.644$ MeV) and has a threshold of 1.88 MeV. Advantage may be taken of the differences in neutron energy in the forward and backward (and intermediate) directions.

With X rays from electron accelerators neutrons can be produced by means of the ^{9}Be (γ, n) or ^{2}H(γ, n) reactions. The yields of these reactions go up quite sharply with energy. With an electrostatic generator operating at 2.5 MeV with 100 μA electron current the neutron yield per gram of beryllium is 7×10^7 s^{-1}; at 3.2 MeV the corresponding figure is 4×10^8 s^{-1}.

When deuteron and proton beams above a few million electron volts are available a number of reactions can be used to produce copious quantities of neutrons. The ^{2}H(d, n)^{3}He reaction discussed previously and the ^{9}Be(d, n)^{10}B reaction are especially favorable for neutron production. The latter reaction has a positive Q value of 3.80 MeV, but the neutrons are far from monoenergetic; deuterons of E MeV produce a distribution of neutron energies up to about $E + 3.5$ MeV. The neutron yield goes up rapidly with deuteron energy, from 10^8 s^{-1}(μA)$^{-1}$ at 1 MeV to 10^{10} s^{-1} (μA)$^{-1}$ at 8 MeV and 3×10^{10} s^{-1}(μA)$^{-1}$ at 14 MeV. With both reactions mentioned the neutrons are emitted largely in the forward direction and both reactions have been used extensively for neutron therapy and neutron radiobiology.

Considerably higher neutron energies than with beryllium targets are obtained by deuteron bombardment of lithium targets, since the reaction ^{7}Li(d, n)^{8}Be is exoergic by 15 MeV.[12] The neutron yield is only about one third that of the ^{9}Be(d, n)^{10}B reaction. Neutrons are also obtained in the bombardment of almost any element with fast protons, deuterons, or α particles. The yields and energies vary from reaction to reaction, but if a neutron bombardment is needed for the activation of some substance it is often sufficient to place the sample near a cyclotron target that is being bombarded by deuterons, even if the target is not beryllium or lithium.

In the bombardment of targets with deuterons of much higher energy (>100 MeV), high-energy neutrons are emitted in a rather narrow cone in the forward direction as a result of deuteron stripping (see p. 147). The energy distribution of these neutrons is approximately Gaussian, with the maximum at half the deuteron energy.

When high-energy protons strike target nuclei they produce neutrons in

[12] This reaction is to be used for the production of very intense fast-neutron beams for testing fusion reactor materials; a high-current deuteron linac for this purpose will be operating at Hanford, Washington, in the 1980s.

the forward direction by elastic or nearly elastic collisions. Useful neutron beams of modest energy spread are thus made at a number of high-energy proton accelerators. They lend themselves particularly to the production and study of neutron-rich nuclides by such reactions as $(n, 2p)$ and $(n, 2pn)$.

Neutrons from all nuclear reactions initially are fast neutrons. Their slowing down and some properties of thermal neutrons are discussed in chapter 6, section D.

Nuclear Chain Reactors. By far the most prolific sources of neutrons known are the nuclear chain reactors. The general characteristics of reactors are discussed in chapter 14; there, however, the principal emphasis is on energy production whereas, in the present context, we are concerned only with reactors as neutron sources. Several hundred research and test reactors of a variety of designs are in operation, spanning a range of power levels from 0.1 W to about 100 MW. Apart from a handful of experimental plutonium-fueled reactors and a small number using natural uranium, these reactors are fueled with uranium enriched to various degrees in ^{235}U. The fuel may be: in the form of plates, rods, bars, or some other shape, made of the metal or of some alloy; in the form of uranium oxide pellets; or homogeneously dispersed in a moderator (either an aqueous solution or a solid medium such as zirconium hydride or polyethylene).

The vast majority of research reactors use thermal neutrons for the propagation of the chain reaction and therefore have moderators to slow the fast neutrons emitted in fission (average energy $\approx$2.5 MeV, most probable energy $\approx$0.6 MeV) to thermal energies. Ordinary water, D_2O, and graphite (in that order) are the favored moderator materials. All but the lowest-power reactors (power $\lesssim$1 kW, neutron flux $\lesssim 10^9\ cm^{-2}\ s^{-1}$) require cooling and, since high fluxes imply high specific power, that is, high power per unit volume, the limiting problem in achieving maximum neutron flux is the design of the requisite cooling system for carrying away the enormous amounts of heat developed. Water, heavy water, CO_2, helium, air, and liquid metals appear as coolants in research reactors, the most economical designs being those in which the same substance, usually H_2O or D_2O, serves as both moderator and coolant. Most widespread are the so-called **pool-type** or **swimming pool** reactors, which have the entire reactor core suspended in the bottom of an open pool, with some 5–7 m of water above the core for shielding, and with water also serving as moderator, coolant, and reflector. Pool-type reactors give neutron fluxes up to several times 10^{13} neutrons $cm^{-2}\ s^{-1}\ MW^{-1}$ and have been built with power levels up to about 50 MW. Higher fluxes can be achieved if the reactor core, instead of being suspended in the bottom of an open pool, is enclosed in a sealed tank, with H_2O or D_2O under pressure serving as moderator and coolant.

Maximum thermal-neutron fluxes available in research reactors range up to several times $10^{15}\ cm^{-2}\ s^{-1}$. Some reactors such as the rather popular

Table 15-2 Representative Thermal Neutron Fluxes

A. Reactors

Reactor Type[a]	Fuel, Moderator, Coolant	Power	Maximum Neutron Flux ($cm^{-2}\,s^{-1}$)
Aerojet General Corporation's Teaching Reactor AGN-201	20% ^{235}U as U_3O_8 homogeneously dispersed in polyethylene; uncooled	0.1 W	5×10^6
Argonaut Reactor (Argonne Nuclear Assembly for University Teaching)	80–93% ^{235}U as U_3O_8 dispersed in aluminum; graphite moderator; H_2O cooling	10 kW	1.7×10^{11}
General Atomic's TRIGA Reactor Mark II	20% ^{235}U as uranium zirconium hydride (solid homogeneous fuel-moderator system); H_2O cooled	250–2000 kW steady; up to 6.4×10^3 MW pulsed	$(1\text{–}8) \times 10^{13}$ steady; $\sim 2 \times 10^{17}$ pulsed
Pool-Type Reactors	10–93% ^{235}U as U-Al alloy or UO_2 pellets; H_2O moderated and cooled	10 kW–40 MW	$(0.5\text{–}2) \times 10^{13}$ per MW
Brookhaven HFBR	93% ^{235}U as U-Al alloy; D_2O moderated and cooled	60 MW	1×10^{15}

B. Other Sources

Source	Conditions	Thermal Flux ($cm^{-2}\,s^{-1}$)
1 g Ra mixed with Be	Immersed in a large volume of water or paraffin; flux measured 4 cm from source	1×10^5
Po + Be, 3.7×10^{10} α particles s^{-1}	Immersed in a large volume of water or paraffin; flux estimated 4 cm from source	2×10^4
Van de Graaff, 10 μA of 1-MeV deuterons on Be	Target backed up with large paraffin block; flux estimated in paraffin near target	1×10^7
Cyclotron, 100 μA or 8-MeV deutrons on Be	Target backed up with large paraffin block; flux estimated in paraffin near target	1×10^9
Cyclotron, 100 μA of 14-MeV deuterons on Be	Target backed up with large paraffin block; flux estimated in paraffin near target	3×10^9

[a]The first four reactors listed are general types, each represented by many individual units that may differ somewhat among themselves; typical data are given.

TRIGA reactors built by the General Atomic Company are designed such that they can be pulsed to give brief (~10 ms) bursts with fluxes up to more than 10^{17} neutrons $cm^{-2}\,s^{-1}$, although at steady operation they produce only up to 8×10^{13} neutrons $cm^{-2}\,s^{-1}$. In table 15-2 we compare some typical reactor fluxes with those available from other neutron sources.

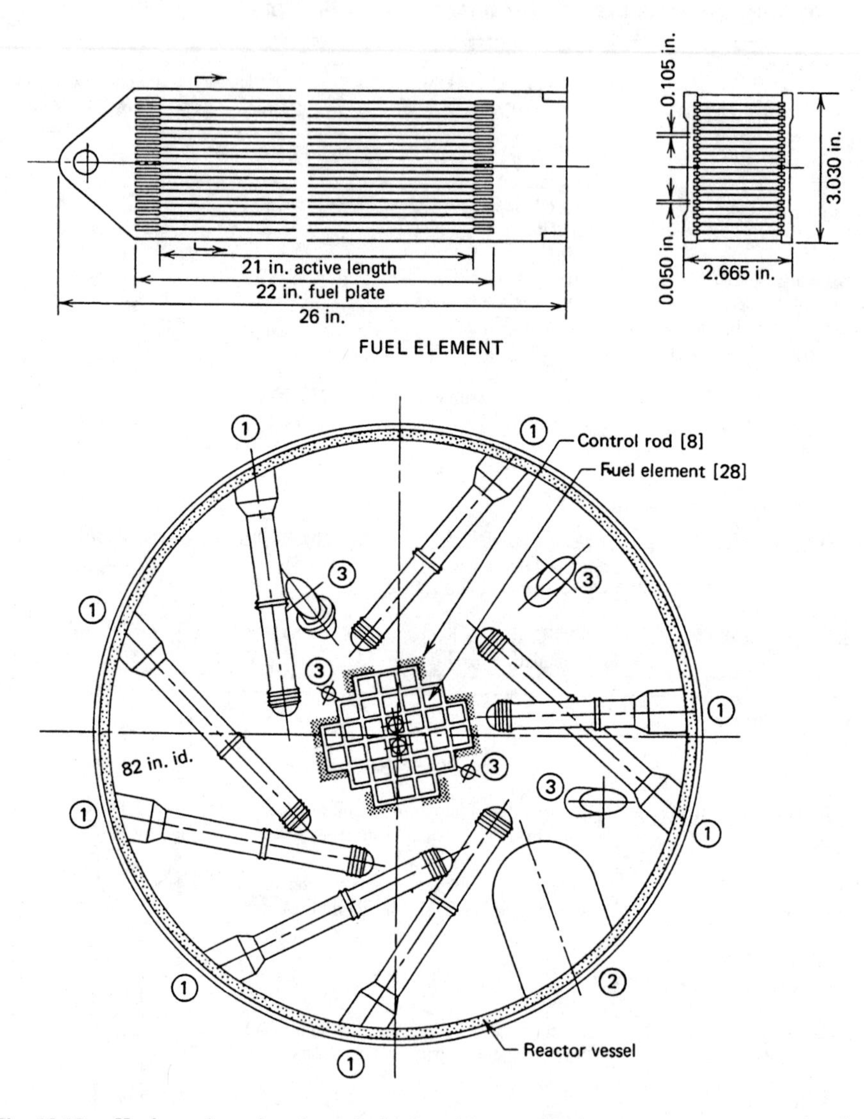

Fig. 15-15a Horizontal section through the Brookhaven HFBR. Beam tubes are labeled ①, a special large-diameter tube for a cold neutron moderator is designated ②, and irradiation facilities are labeled ③. Details of the fuel elements are shown above. (Courtesy Brookhaven National Laboratory.)

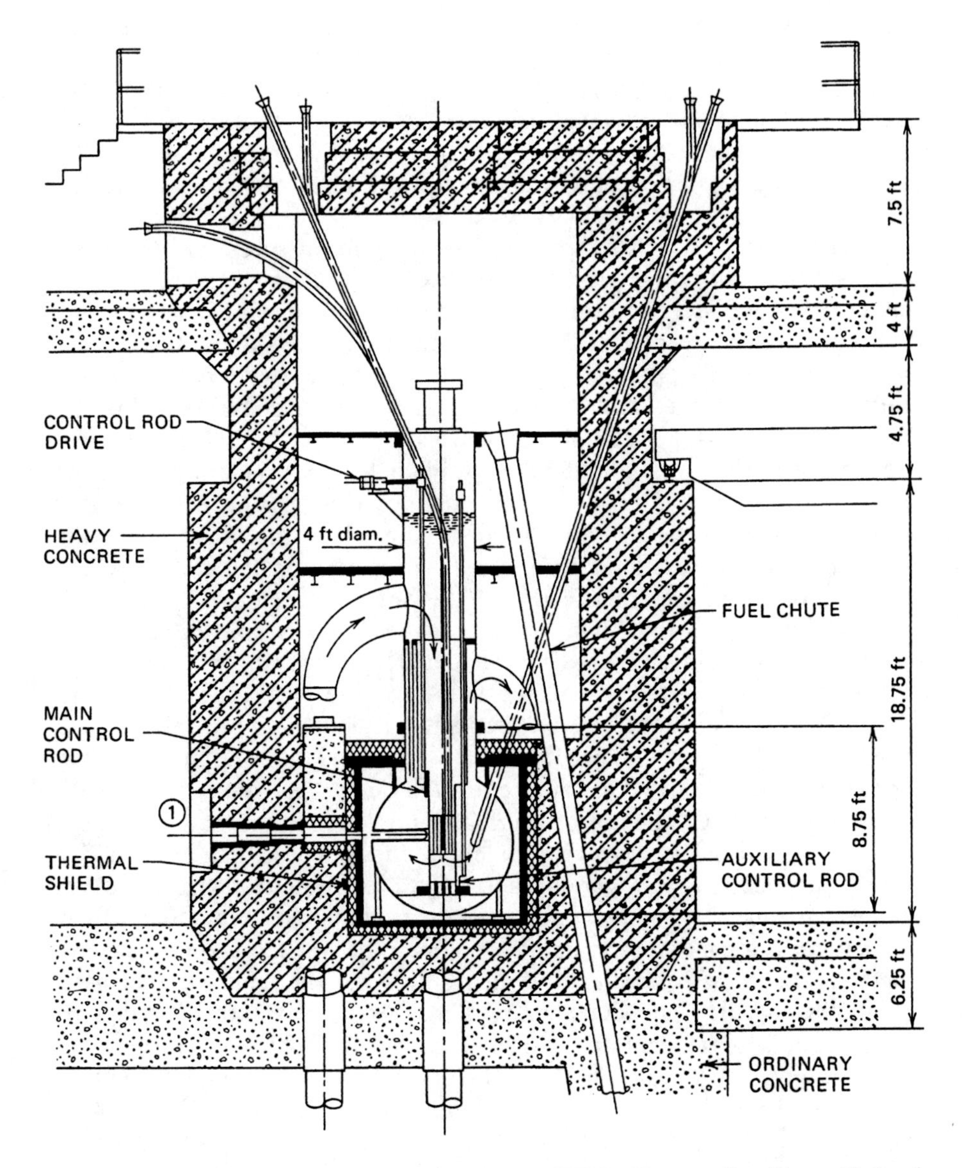

Fig. 15-15b Vertical section through the Brookhaven HFBR. (Courtesy Brookhaven National Laboratory.)

Two general types of facilities for neutron use are desirable: (1) means (including pneumatic tubes) for the irradiation of samples in high-flux regions, (2) tubes and channels for bringing neutron beams outside the reactor for such purposes as structure studies by neutron diffraction, capture-γ-ray spectroscopy, or neutron cross-section measurements. Not all types of reactors are equally well suited for both classes of applications. Pool-type reactors have flexibility for all sorts of irradiation studies but do not lend themselves well to the installation of beam tubes. On the other hand, a few tank-type reactors have been especially designed to give multiple, high-intensity neutron beams. The most advanced of these are the High-Flux Beam Reactor (HFBR) at Brookhaven National Laboratory (fueled with 93 percent ^{235}U, cooled and moderated with D_2O) and the similar reactor at the Laue-Langevin Laboratory at Grenoble, France. Figure 15-15 shows horizontal and vertical sections through the HFBR. The reactor core is only about 50 cm in diameter and 53 cm high, and the design is such that the maximum thermal-neutron fluxes are in the D_2O reflector, where the beam tubes originate and where the facilities for the highest-intensity irradiations are located. What may be thought of as the opposite concept is used in the High-Flux Isotope Reactor (HFIR) at Oak Ridge, Tennessee. Here one of the highest fluxes anywhere ($\sim 5 \times 10^{15}$ cm^{-2} s^{-1}) is produced by having the fuel elements arranged in an annulus surrounding a central region called a flux trap. The HFIR uses 93 percent ^{235}U for fuel, H_2O as moderator and coolant, beryllium as a reflector. One of its main functions is the production of ^{252}Cf and other transplutonium nuclides by multiple neutron capture in such targets as ^{242}Pu.

Although we have spoken of thermal-neutron reactors, the actual neutron energy distributions can vary widely in different reactor types and also in different locations in a given reactor. To provide pure thermal-neutron sources, so-called **thermal columns** are often attached to reactors. A thermal column is a column of graphite (or some other moderator) of sufficient length to ensure a thermal-energy distribution for the neutrons that have passed through it. The neutron flux at the end of a thermal column is several orders of magnitude smaller than that available inside the associated reactor. Especially *large* ratios of fast-neutron to slow-neutron fluxes can be obtained inside uranium-walled containers placed in a reactor.

Neutron Monochromators. A number of means have been devised for the conduct of experiments with neutrons selected to have a particular energy. One of these, the **crystal spectrometer**, is analogous to an optical-grating monochromator. A thermal neutron with a velocity of 2.2×10^5 cm s^{-1} (the most probable velocity at 20°C) has a wavelength $\lambda = h/mv = 1.8 \times 10^{-8}$ cm. This length is in the range of common X-ray wavelengths (1.54×10^{-8} cm for copper K_α radiation), and the spacing between crystal planes is about the proper "grating" spacing for slow-neutron diffraction as it is for X-ray diffraction. Neutrons of considerably higher energy (i.e., shorter wavelength) may be diffracted sucessfully by the crystal at grazing incidence angles. With an intense source of slow neutrons available, such as a nuclear reactor, the crystal and slit system may be arranged to select neutrons from the spectrum with good resolution from about 0.02 to about 10 eV.

The other common means for selecting monoenergetic neutrons from a spectrum depend on control of the **time of flight** (TOF) of the neutrons over a measured course. A burst of neutrons, containing all energies in the spectrum of the source, may be selected mechanically by interposing in the beam a rotating **chopper**, that is, a disk made of neutron-absorbing material[13] with one or more radial slits to let neutrons through. Timing devices actuate the detector circuits at a chosen time in each chopping cycle, and different neutron energies may be selected by varying that time. Only processes that produce instantaneous response in a detector can be studied by the chopper technique—it is not applicable, for example, to the study of activation cross sections.

A neutron of energy $E = 0.025$ eV has a velocity $v = 1.38 \times 10^6\ E^{1/2} = 2.2 \times 10^5\ \text{cm s}^{-1}$ and so traverses a distance of 10 m in 4.55×10^{-3} s. Monochromators with burst times of a few microseconds give very good energy resolution for neutrons in that energy range. Choppers are useful from about 10^{-3} to about 10^4 eV. At the higher energies long flight paths are needed to obtain good resolution.

If the source of neutrons is an accelerator, the bursts of neutrons for TOF studies can be generated directly by suitable modulation of the accelerator ion beam. The pulsed beams from linear accelerators, including electron linacs, are particularly well suited for this application. Even with an electron energy of only 15 MeV the bremsstrahlung gives, by (γ, n) reaction, more than one neutron per 10^4 electrons and, with such sources, monochromators with wide useful ranges have been built (e.g., one at Harwell, England, that is used from 10^{-3} to $>10^3$ eV).

D. MEASUREMENT OF BEAM ENERGIES AND INTENSITIES

In almost any investigation of a nuclear reaction it is necessary to know either the energy or the intensity of the bombarding particles or, more frequently, both. The accuracy with which these quantities are required can vary widely, depending on the problem at hand. The techniques available for energy and intensity determinations differ for different energy ranges. In the following paragraphs a brief account is given of the principal

[13] The chopper material must be chosen according to the energy range to be studied; for slow neutrons it is usually cadmium, for fast neutrons often iron or nickel.

methods used by nuclear chemists for the measurement of these important parameters, and attention is called to some of the problems encountered.

Determination of Beam Energy. In general, the most accurate methods for the determination of the energy of charged-particle beams use deflection in magnetic or electric fields. Beam-deflection equipment is commonly employed in the external beams of low-energy accelerators (such as Van de Graaff machines, cyclotrons, and linear accelerators) not only for energy determinations but also to achieve energy analysis of initially inhomogeneous beams. Such analyzed, highly monoenergetic particle beams (energy spread typically of the order of 0.1 percent) are often essential for scattering experiments, nuclear spectroscopy, and so on, but they are not always practical for certain types of experiments of interest to nuclear chemists (such as excitation function determinations) because the magnetic or electrostatic analysis generally results in greatly lowered beam intensities.

In any circular accelerator the magnetic field of the accelerator itself accomplishes some energy analysis, and a knowledge of field strength and orbit radius in principle gives the beam energy. For many purposes this type of information on beam energy is sufficient. In both standard cyclotrons and synchrocyclotrons the maximum beam energies are usually known from the machine characteristics within perhaps 2 or 3 percent. The actual inhomogeneity in beam energy is usually somewhat less (typically ~1 percent) for the full-energy beam. But, since much of this inhomogeneity results from nonconcentric orbits (which, in turn, are brought about by ion source optics), the percentage energy spread increases with decreasing radius. Therefore excitation functions determined by beam interception at different radii in cyclotrons or synchrocyclotrons can be subject to rather serious distortion because of this energy spread. Yet at synchrocyclotron energies, radius variation is often the only practical method for selecting different energies.

In a synchrotron the particle energy at any particular time in the acceleration cycle is uniquely determined if the radio frequency at that time and the radius of the equilibrium orbit are known. Accurate frequency measurements can be made more easily than accurate magnetic-field measurements; therefore particle energies in synchrotrons are readily known to about 1 percent. Furthermore, energy variation is easily achieved by variation of the time within the acceleration cycle at which the rf field is turned off. A calibration of energy versus rf-turnoff time is usually available for a synchrotron.

At sufficiently low beam energies the beam particles can be stopped in a semiconductor detector to achieve accurate energy measurements. To use this technique it is usually necessary to reduce beam intensity significantly and this is best accomplished by Rutherford scattering from a thin, high-Z film through a large angle. A less accurate, but still often useful method (at energies below 50 or 100 MeV) is the measurement of range, in conjunction with a range-energy relation. The range measurement can be done in a variety of ways but basically always involves the use of absorbers and of some detector. The detector may be a Faraday cup (particularly if a beam-intensity measurement is to be coupled with the energy determination) or almost any other radiation-sensitive device; the bleaching of blue cellophane gives a very convenient beam indication. It is extremely useful to be able to do the range measurements remotely. A convenient device for this purpose is a wheel that, by means of a servomechanism, can be rotated to interpose, in the beam, absorbers of various thicknesses mounted on its periphery.

As discussed on p. 124, absorption in foils is used not only to *measure*, but also to *degrade* beam energies, especially in the stacked-foil technique for excitation function measurement. The secondary particles produced in the absorbers and the energy spread caused by straggling (see p. 220) can cause trouble, increasingly so with increasing beam energy. If reasonably monoenergetic degraded beams are wanted and a considerable loss in intensity (factor 10–100) can be tolerated, magnetic analysis after degradation is recommended.

Neutron Flux Measurement. The thermal-neutron flux, for example in a reactor, is usually determined by the activation of a substance of known activation cross section under the exact conditions for which the flux is desired. The most frequently used flux monitor is gold (^{197}Au), from which, by the capture of thermal neutrons, 2.696-d ^{198}Au is formed with a cross section $\sigma_a = 98.8$ b. The number of ^{198}Au atoms formed from W mg of ^{197}Au at the end of an irradiation of t s in a flux of thermal neutrons is

$$N_{198} = nv \times \sigma_a \times \frac{W}{197} \times 6.02 \times 10^{20} \frac{1 - e^{-\lambda t}}{\lambda},$$

where λ is the decay constant of ^{198}Au in s^{-1}, nv is the neutron flux in cm^{-2} s^{-1} (n is the neutron density per cubic centimeter and $v = 2.2 \times 10^5$ cm s^{-1}). Since W and t are readily measured, the measurement of the thermal flux nv reduces to the problem of determining N_{198}, the number of ^{198}Au atoms, that is, to an absolute disintegration rate measurement. Fortunately the decay scheme of ^{198}Au is simple enough in its main features (cf. figure 8-4) to lend itself to the use of the coincidence method described in chapter 8, section G. Other capture reactions, for example ^{59}Co(n, γ)^{60}Co, can also be used for thermal-neutron flux measurements. A general requirement for a convenient thermal-neutron flux monitor is that the cross section follow a $1/v$ law, so that the Maxwellian velocity distribution can be replaced by the single velocity $v = 2200$ m s^{-1}.

Primary Monitoring of Charged-Particle Fluxes (C3). The most widely used instrument for absolute determination of charged-particle fluxes is the **Faraday cup.** This is essentially an insulated electrode designed to stop all the beam particles striking it as well as any charged secondaries produced in it by the beam. The total charge built up on the Faraday cup divided by the charge per particle (e for protons, $2e$ for α particles, etc.) thus gives the total number of particles that have fallen on the cup. Commercially obtainable electronic current integrators may be used to measure the current flowing to the cup, with the cup kept near ground potential at all times.

The design of Faraday cups presents a number of problems. Good insulation is essential, and the cup should be operated in a high vacuum, since gas ionization in the vicinity of the cup can lead to erroneous results. The principal concern in Faraday-cup design is usually the retention of charged secondaries, chiefly secondary electrons. Proper design of the cup-shaped electrode, with an entrance aperture small compared to the depth of the cup, is used to minimize the solid angle for escape of secondary electrons. Magnetic fields of a few hundred gauss are also useful for preventing electron escape. On the other hand, secondary electrons coming from any other objects in the beam path (windows, collimators, etc.) must be prevented from reaching the cup. With increasing beam energy the difficulty of retaining the charged secondaries increases, and Faraday cups for beams of several hundred MeV become quite unwieldy.

Another method that for practical reasons is limited to relatively low energies is calorimetry. If a beam and all its secondaries are completely stopped in a calorimeter, the product of beam energy and beam current is measured. One advantage of calorimetry over the Faraday-cup method is that it can be used for the circulating beams inside accelerators, where strong magnetic and electric fields would interfere with the operation of a Faraday cup. In external beams Faraday-cup measurements are to be preferred.

At higher energies ($\gg$100 MeV) absolute measurements of beam intensities are usually based on counting individual beam particles by means of counter telescopes or nuclear emulsions (C4). Both methods are limited to use at fairly low beam intensities, and secondary devices suitable for higher intensities are usually calibrated in terms of these absolute primary monitors. Emulsion monitoring of near-minimum ionizing particles can be done with total intensities up to $\sim 5 \times 10^6$ particles cm^{-2} (independent of time distribution). Measurements in a counter telescope are limited by the requirement that the dead-time losses should be small; therefore the maximum average intensity that can be monitored with a given telescope and associated circuitry depends on the time distribution of the beam, which in synchrotrons tends to be strongly bunched.

Secondary Beam Monitors. Once a primary beam monitor is available, any other device can, in principle, be calibrated in terms of it. Caution is required in the use of secondary monitors in intensity ranges outside the range of primary calibration; linearity of response must be checked. This is particularly true for any instruments based on ionization measurements or on light collection (scintillation and Čerenkov counters). A large variety of secondary monitoring devices, each with its own virtues and shortcomings, has been described (C3).

Secondary beam monitors particularly useful for nuclear chemists and essentially free of nonlinearity problems are nuclear reactions of known cross section. The absolute cross section does not, in fact, have to be known, so long as the activity of the reaction product, preferably without chemical separation from the target foil, is measured in the same arrangement in which it was determined when its production was calibrated against an absolute beam monitor. Radioactivity induced by nuclear reactions can be useful as a beam monitor in almost any energy range but becomes of paramount importance at high energies (>100 MeV) for several reasons:

1. Scattering and absorption in thin foils become relatively unimportant, and monitor foil and target foil can thus be made to intercept virtually the same number of beam particles of the same energy (which is not the case at lower energies).
2. In the circulating beams of synchrocyclotrons and proton synchrotrons the effective particle fluxes through targets may greatly exceed the circulating beam intensities (multiple traversals, cf. p. 573); since the number of traversals depends on particle energy, accelerator characteristics, and thickness and composition of target, activation of a monitor foil incorporated in the target stack is the only reliable method for the measurement of the effective particle flux through the target. In spite of this difficulty, circulating-beam irradiations are often attractive just because the multiple traversals raise the effective beam intensities by large factors over those that might be available in the more readily monitored external beams.

The nuclear reactions that have been found most useful for monitoring high-energy proton beams are shown in table 15-3. The reaction $^{12}C(p, pn)^{11}C$ is the one that has been most thoroughly calibrated against absolute monitors over a wide energy range (50 MeV–300 GeV). However, because of the 20-min half life of ^{11}C, this reaction is useful for relatively short irradiations only. For this reason the production of ^{24}Na from aluminum is the most widely used monitor reaction. The cross section of this reaction is almost independent of proton energy from 100 MeV to 30 GeV (cf. table 15-3), and the ^{24}Na activity is readily measured in aluminum foils without chemical separation. About 24 hours after irradiation, when the shorter-lived activities have died out, ^{24}Na can be measured by β counting without interference from other products. A disadvantage is the fact that ^{24}Na can also be made by low-energy secondary neutrons

Table 15-3 Monitor Reactions for High-Energy Proton Beams

Reaction[a]	Product Half-Life	Principal Radiation Detected	Cross Section (mb)			Remarks
			300 MeV	3 GeV	30 GeV	
$^{12}C(p, pn)^{11}C$	20.4 min	β^+	35.8	27.1	26.8	Best monitor for short irradiations
$^{27}Al(p, 3pn)^{24}Na$	15.0 h	β^-, γ	10.1	9.1	8.6	Sensitive to low-energy secondaries
$^{27}Al(p, spall)^{18}F$	110 min	β^+	6.6	6.8	6.2	
$^{12}C(p, spall)^{7}Be$	53.3 d	γ	10.0	10.3	9.2	Useful for long irradiations
$^{197}Au(p, spall)^{149}Tb$	4.15 h	α	0.0	8.1	5.8	Threshold at ~600 MeV

[a] The notation (p, spall) indicates a spallation reaction.
[b] The cross section values are mostly taken from reference C4. The cross section values for ^{149}Tb production are based on 17 percent of its decays proceeding by α emission.

produced in the target, according to the reaction $^{27}Al(n, \alpha)^{24}Na$. Production of ^{18}F in aluminum is less sensitive to secondaries and is therefore a preferable monitor under some circumstances. The last reaction listed in table 15-3, the production of the α emitter ^{149}Tb in gold, has the advantage of a very high effective threshold (~600 MeV), so that it is very insensitive to secondaries. Also, the α activity of ^{149}Tb ($t_{1/2} = 4.15$ h) is readily measured in irradiated gold foils after some short-lived α emitters have decayed out. All these monitor reactions are discussed in detail in C4. Analogous reactions can be used to monitor beams of particles other than protons, such as ^4He, deuterons, pions, and so on, once proper calibration measurements against primary monitors have been made.

REFERENCES

A1 K. W. Allen, "Electrostatic Accelerators," in *Nuclear Spectroscopy and Reactions*, Part A (J. Cerny, Ed.), Academic, New York, 1974, pp. 3–34.

A2 L. W. Alvarez et al., "Berkeley Proton Linear Accelerator," *Rev. Sci. Inst.* **26**, 111 (1955).

B1 M. H. Blewett, "Characteristics of Typical Accelerators," *Ann. Rev. Nucl. Sci.* **17**, 427 (1967).

B2 R. Bock, "Heavy Ion Accelerators," in *Nuclear Spectroscopy and Reactions*, Part A (J. Cerny, Ed.), Academic, New York, 1974, pp. 79–111.

B3 B. L. Berman, "Photonuclear Reactions," in *Nuclear Spectroscopy and Reactions,* Part C (J. Cerny, Ed.), Academic, New York, 1974, pp. 377–416.

*B4 J. P. Blewett, "Recent Advances in Particle Accelerators," *Adv. Electron. Electron Phys.* **29,** 233 (1970).

C1 E. D. Courant, "Accelerators for High Intensities and High Energies," *Ann. Rev. Nucl. Sci.* **18,** 435 (1968).

C2 E. D. Courant, M. S. Livingston, and H. S. Snyder, "The Strong-Focusing Synchrotron—A New High-Energy Accelerator," *Phys. Rev.* **88,** 1190 (1952).

C3 O. Chamberlain, "Determination of Flux of Charged Particles," in *Methods of Experimental Physics,* Vol. 5B, *Nuclear Physics* (L. C. L. Yuan and C. S. Wu, Eds.), Academic, New York, 1963, pp. 485–507.

C4 J. B. Cumming, "Monitor Reactions for High-Energy Proton Beams," *Ann. Rev. Nucl. Sci.* **13,** 261 (1963).

G1 H. A. Grunder and F. B. Selph, "Heavy Ion Accelerators," *Ann. Rev. Nucl. Sci.* **27,** 353 (1977).

H1 B. G. Harvey, "The Cyclotron," in *Nuclear Spectroscopy and Reactions,* Part A (J. Cerny, Ed.), Academic, New York, 1974, pp. 35–77.

K1 B. M. Kincaid, "A Short-Period Helical Wiggler as an Improved Source of Synchrotron Radiation," *J. Appl. Phys.* **48,** 2684 (1977).

*L1 M. S. Livingston, *The Development of High-Energy Accelerators,* Dover, New York, 1966.

L2 S. J. Lindenbaum, "Shielding of High-Energy Accelerators," *Ann. Rev. Nucl. Sci.* **11,** 213 (1961).

*L3 M. S. Livingston and J. P. Blewett, *Particle Accelerators,* McGraw-Hill, New York, 1962.

*M1 E. M. McMillan, "Particle Accelerators," in *Experimental Nuclear Physics,* Vol. III (E. Segre, Ed.), Wiley, New York, 1959, pp. 639–785.

O1 G. D. O'Kelley, "Radioactive Sources," in *Methods of Experimental Physics,* Vol. 5B, *Nuclear Physics* (L. C. L. Yuan and C. S. Wu, Eds.), Academic, New York, 1963, pp. 555–580.

P1 O. Piccioni et al., "External Proton Beam of the Cosmotron," *Rev. Sci. Inst.* **26,** 232 (1955).

P2 C. Pellegrini, "Colliding-Beam Accelerators," *Ann. Rev. Nucl. Sci.* **22,** 1 (1972).

S1 H. A. Schwettman, "Electron Linear Accelerators," in *Nuclear Spectroscopy and Reactions,* Part A (J. Cerny, Ed.), Academic, New York, 1974, pp. 129–147.

S2 J. R. Sanford, "The Fermi National Accelerator Laboratory," *Ann. Rev. Nucl. Sci.* **26,** 151 (1976).

T1 L. H. Thomas, "The Paths of Ions in the Cyclotron," *Phys. Rev.* **54,** 580, 588 (1938).

W1 R. Wideröe, "Über ein Neues Prinzip zur Herstellung Hoher Spannungen," *Arch. Elektrotech.* **21,** 387 (1928).

W2 R. E. Watson and M. L. Perlman, "Seeing with a New Light: Synchrotron Radiation," *Science,* **199,** 1295 (1978).

EXERCISES

1. A standard cyclotron of 120-cm pole diameter is operated with a 10-MHz oscillator. (a) What magnetic field is required for the acceleration of deuterons? (b) What will be the final deuteron energy? (c) With the same rf frequency, what is the maximum kinetic energy to which ^{3}He ions could be accelerated in this cyclotron? (d) Under the assumption that the magnetic field calculated in (a) is the maximum available, what oscillator frequency would be required to obtain the highest possible ^3H energy and what is that energy?

Answers: (b) 14.8 MeV; (d) 6.67 MHz, 9.9 MeV.

2. A linac for the acceleration of protons to 45.3 MeV is designed so that, between any pair of accelerating gaps, the protons spend one complete rf cycle inside a drift tube (field-free region). The rf frequency used is 200 MHz. (a) What is the length of the final drift tube? (b) If the first drift tube is 5.35 cm long, at what kinetic energy are the protons injected into the linac?

   Answer: (b) 0.60 MeV.

3. Estimate (a) the percentage frequency modulation and (b) the pole diameter required for an FM cyclotron designed to accelerate protons to 350 MeV. Assume $H = 16{,}000$ G. *Answer:* (b) 3.7 m.

4. The H^+ and H_2^+ ions accelerated in a Van de Graaff generator to 5 MeV are to be magnetically separated from one another. Approximately over what distance must a 10,000-G field be applied if the two beams are to diverge by 20°?

   Answer: ~34 cm.

5. A synchrotron is to be designed to accelerate protons to 12-GeV kinetic energy. (a) Assuming a maximum field strength of 14,300 G, estimate the radius of curvature of the proton orbit. (b) If about 25 percent of the protons' path is spent in field-free straight sections, what is the final revolution frequency? (c) Assuming one revolution per rf cycle, at what kinetic energy must the protons be injected if the frequency over the entire acceleration cycle is to vary by a factor of 5? What type of device would you suggest for the injector?

   Answers: (a) 30 m; (b) 1.2 MHz.

6. In the synchrotron of the preceding problem the protons receive an energy increment of 7.5 keV per revolution. Approximately how long does it take to accelerate them from injection to full energy? What is the total path length?

   Answer: ~1.4 s.

7. A Cockcroft-Walton accelerator can be used to accelerate either deuterons or tritons to 500 keV. What is the maximum neutron energy attainable with (a) deuterons incident on a tritium target, (b) tritons incident on a deuterium target?

8. From the data given on p. 569 and from information about charge states of ions in chapter 6, estimate the final kinetic energy to which (a) ^{40}Ar, (b) ^{84}Kr can be accelerated in the cyclotron ALICE. Assume that stripping at injection from the linac into the cyclotron occurs in a solid stripper foil and that the most probable charge state resulting from this stripping is accelerated.

9. What is the energy available in the center-of-mass system when a 500-GeV proton from the FNAL synchrotron hits a stationary proton?

10. What would be the minimum (n, γ) cross section detectable by means of the product activity in a sample of 10-cm^2 area containing 1 mg-equivalent of target isotope, with a mixed Ra-Be source containing 1 g radium? Assume that the bombardment is continued to saturation and that 1 percent of the neutrons emitted by the source strike each square centimeter of the target sample as slow neutrons. Consider 30 dis min^{-1} as the minimum detectable activity.

11. If the fuel loading of a research reactor consists of 2 kg of uranium enriched to

90 percent ^{235}U, how long can the reactor be operated without reloading at a power level of 1 MW before burnup reaches 20 percent?

12. In a neutron chopper, a neutron-absorbing rotor with a narrow slit near its periphery rotates in front of a stationary collimator with a similar slit to produce short bursts of neutrons. The rotor is 20 cm in diameter and rotates at 15,000 rpm; the slits are 0.03 cm wide and 1 cm long. (a) What is the burst length? (b) With the neutron detector actuated for a time equal to the burst length, what flight path is required to achieve ± 10 percent energy resolution for 5-keV neutrons? *Answers:* (a) 4 μs; (b) $\sim$80 m.

13. A 0.70-mg sample of cobalt in the form of a fine wire is used to monitor the thermal-neutron flux in a reactor. It is exposed to the flux for exactly 5 min and a few hours later it is placed inside a well-type scintillation counter for a determination of the amount of ^{60}Co formed. The counter has 93% efficiency for ^{60}Co. The total γ-counting rate in the well counter is 180,800 cpm. What was the neutron flux to which the cobalt was exposed? Ignore any variation of flux within the small sample and consider the counting rates given as net counting rates after background subtraction and correction for coincidence losses. *Answer:* 1×10^{13} cm^{-2} s^{-1}.

14. A 3-GeV proton bombardment is monitored by means of the ^{24}Na activity induced in a 25 μm (6.85 mg cm^{-2}) aluminum foil (surrounded by two other aluminum foils in order to compensate any recoil losses of ^{24}Na from the monitor by equal recoil gains). Exactly 20 h after the end of the 15-min irradiation the ^{24}Na activity in the aluminum monitor is measured with a calibrated end-window counter (efficiency 0.037 in the geometrical arrangement used). The net counting rate is 27,430 min^{-1}. What was the average proton flux through the sample during the irradiation?
Answer: 1.17×10^{14} min^{-1}.

Appendix A

Constants and Conversion Factors

The values in table A-1 are taken from the consistent set of least-squares-adjusted physical constants published by E. R. Cohen and B. N. Taylor in *J. Phys. Chem. Ref. Data* **2**, 663 (1973). Values are given both in SI and cgs units; the SI (Système International d'Unités or International System of Units) is described in the National Bureau of Standards Publication 330 (1977) and the most important units are listed in footnote *b*.

Table A-1 Fundamental Constants

Quantity	Symbol	Value[a]	SI[b]	cgs
Speed of light in vacuum	c	2.997924580(12)	10^{8} m s^{-1}	10^{10} cm s^{-1}
Planck's constant	h	6.626176(36)	10^{-34} J s	10^{-27} erg s
	$\hbar = h/2\pi$	1.0545887(57)	10^{-34} J s	10^{-27} erg s
Boltzmann constant	k	1.380662(44)	10^{-23} J K^{-1}	10^{-16} erg K^{-1}
Electronic charge	e	4.803242(14)		10^{-10} esu
	e/c	1.6021892(46)	10^{-19} C	10^{-20} emu
Avogadro's number	N	6.022045(31)	10^{23} mol^{-1}	10^{23} mol^{-1}
Faraday constant	$F = Ne/c$	9.648456(27)	10^{4} C mol^{-1}	10^{3} emu mol^{-1}
Fine structure constant	$\alpha = e^2/\hbar c$	1/137.03604(11)		
Atomic mass unit (amu)	u	1.6605655(86)	10^{-27} kg	10^{-24} g
Electron rest mass	m_e	5.4858026(21)	10^{-4} u	10^{-4} u
Hydrogen atom rest mass	m_{H}	1.007825037(10)	u	u
Neutron rest mass	m_{N}	1.008665012(37)	u	u
Bohr magneton	$\mu_B = \hbar e/2m_e c$	9.274078(36)	10^{-24} J T^{-1}	10^{-21} erg G^{-1}
Nuclear magneton	$\mu_N = \hbar e/2m_{\mathrm{H}} c$	5.050824(20)	10^{-27} J T^{-1}	10^{-24} erg G^{-1}
Compton wavelength of electron	$h/m_e c$	2.4263089(40)	10^{-12} m	10^{-10} cm

[a] The number in parentheses gives the uncertainty (standard deviation) in the last digit.

[b] The basic SI units are as follows:

Length—meter (m)	Temperature—kelvin (K)
Mass—kilogram (kg)	Amount of substance—mole (mol)
Time—second (s)	Luminous intensity—candela (cd)
Current—ampere (A)	

(*Table* A-1 *footnotes continued on p.* 600)

Table A-2 Some Useful Conversion Factors

One electron volt (eV)	= $1.6021892(46) \cdot 10^{-19}$ J
	= $1.6021892(46) \cdot 10^{-12}$ erg
	= $3.829324(28) \cdot 10^{-20}$ cal
Energy equivalent of:	
Atomic mass unit (u)	931.5016(26) MeV
Electron rest mass	0.5110034(14) MeV
Hydrogen atom rest mass	938.7906(27) MeV
Neutron rest mass	939.5731(27) MeV
Temperature corresponding to 1 eV	$1.160450(36) \cdot 10^4$ K
Photon wavelength associated with 1 eV	$1.239852(3) \cdot 10^{-6}$ m
Number of seconds per day	$8.6400 \cdot 10^4$
Number of seconds per sidereal year	$3.1558150 \cdot 10^7$
One standard atmosphere (atm)	= $1.01325 \cdot 10^5$ Pa
	= 1.01325 bar
	= 760 torr

(*Footnote b continued from p. 599*)

Among derived SI units are the following, some used in the table:

Frequency—hertz (Hz = s^{-1})
Force—newton (N = m kg s^{-2})
Pressure—pascal (Pa = m^{-1} kg s^{-2})
Energy—joule (J = m^2 kg s^{-2}) [1 J = 10^7 erg]
Power—watt (W = J s^{-1})
Electric charge—coulomb (C = A s) [1 C = 10^{-1} emu]
Electric potential—volt (V = W A^{-1})
Capacitance—farad (F = C V^{-1})
Resistance—ohm (Ω = V A^{-1})
Magnetic flux—weber (Wb = V s)
Magnetic flux density—tesla (T = Wb m^{-2}) [1 T = 10^4 G]
Radioactive disintegrations—becquerel (Bq = s^{-1})
Absorbed dose—gray (Gy = J kg^{-1})
Atomic mass—unified atomic mass unit ($u = \frac{1}{12}$ of atomic mass of ^{12}C)

Appendix B

Relativistic Relations

Consider a particle of rest mass m_0 moving at a velocity v. If $\beta = v/c$, where c is the velocity of light, then the particle has

$$\text{mass} \equiv m = \frac{m_0}{\sqrt{1-\beta^2}}, \tag{B-1}$$

$$\text{momentum} \equiv p = mv = \frac{m_0 v}{\sqrt{1-\beta^2}} = \frac{m_0 \beta c}{\sqrt{1-\beta^2}}, \tag{B-2}$$

$$\text{kinetic energy} \equiv T = m_0 c^2\left(\frac{1}{\sqrt{1-\beta^2}} - 1\right) = mc^2 - m_0 c^2, \tag{B-3}$$

$$\text{total energy} \equiv E = mc^2 = \frac{m_0 c^2}{\sqrt{1-\beta^2}}. \tag{B-4}$$

A useful relation between momentum and total energy may be obtained by squaring and rearranging (B-1):

$$m^2c^2 - m_0^2c^2 = m^2v^2 = p^2.$$

Dividing by $m_0^2c^2$, we have

$$\left(\frac{p}{m_0 c}\right)^2 = \left(\frac{mc}{m_0 c}\right)^2 - 1 = \left(\frac{mc^2}{m_0 c^2}\right)^2 - 1 = \left(\frac{E}{m_0 c^2}\right)^2 - 1. \tag{B-5}$$

Thus if momentum is expressed in units of $m_0 c$ ($\eta = p/m_0 c$) and total energy in units of $m_0 c^2$ ($W = E/m_0 c^2$), the following relation holds:

$$\eta^2 = W^2 - 1. \tag{B-6}$$

Equation B-5 may also be rearranged to yield the relation

$$E^2 = E_0^2 + p^2c^2, \tag{B-7}$$

where $E_0 = m_0 c^2$.

If a system A moves with a velocity v ($= \beta c$) relative to another system B, a time interval Δt_A measured in system A will in system B appear as

$$\Delta t_B = \frac{\Delta t_A}{\sqrt{1-\beta^2}}. \tag{B-8}$$

For a particle of zero rest mass (such as a photon or a neutrino) the following relations hold:

$$p = \frac{E}{c}, \tag{B-9}$$

$$E = h\nu, \tag{B-10}$$

where ν is the frequency and h is Planck's constant.

Table B-1 lists, for a number of values of β, the corresponding kinetic energies of electrons, π mesons, and protons. For each value of β the ratio of moving mass to rest mass is also given.

Table B-1 Relativistic Mass and Energy Relationships

		Kinetic Energy (MeV)		
β	m/m_0	Electron	π Meson	Proton
0.10	1.005	0.00257	0.710	4.72
0.20	1.021	0.0107	2.96	19.7
0.30	1.048	0.0247	6.81	45.3
0.40	1.091	0.0465	12.8	85.5
0.50	1.155	0.0791	21.8	145
0.60	1.250	0.128	35.2	235
0.70	1.400	0.205	56.4	375
0.75	1.512	0.262	72.2	480
0.80	1.667	0.341	94.0	625
0.85	1.898	0.459	127	843
0.90	2.294	0.661	182	1.21×10^3
0.95	3.203	1.13	311	2.07×10^3
0.96	3.571	1.31	363	2.41×10^3
0.97	4.114	1.59	439	2.92×10^3
0.98	5.025	2.06	568	3.78×10^3
0.99	7.089	3.11	859	5.71×10^3
0.995	10.01	4.61	1.27×10^3	8.45×10^3
0.999	22.37	10.9	3.01×10^3	2.00×10^4
0.9999	70.71	35.6	9.83×10^3	6.54×10^4
0.99999	223.6	114	3.14×10^4	2.09×10^5

Appendix C

Center-of-Mass System

In discussing nuclear reactions it is often convenient to use the center-of-mass rather than the laboratory system.

The velocity that characterizes a reaction is the *relative* velocity of the two reacting particles (masses m_a and m_b), and the relevant mass is the **reduced mass** of the system:

$$\mu = \frac{m_a m_b}{m_a + m_b}. \tag{C-1}$$

The kinetic energy that characterizes the reaction is then given by

$$T = \tfrac{1}{2}\mu v^2. \tag{C-2}$$

This is another way of expressing what was stated in chapter 4, p. 113: if a is the bombarding particle and b the stationary target nucleus, then the fraction of the incident kinetic energy available to make the reaction go is $m_b/(m_a + m_b)$; conversely, if a is stationary and b is the bombarding particle, the fraction of the kinetic energy of b available for the reaction is $m_a/(m_a + m_b)$.

The kinetic energy given by (C-2) is the kinetic energy in a coordinate system whose origin O' is at the center of mass of the two particles a and b and moves with a velocity v_{cm} with respect to the fixed laboratory origin O. The relationship between the velocities of a particle in the two coordinate systems is illustrated in figure C-1 for the general situation in which both particles, a and b, are in motion in the laboratory.

The velocities of the two particles in the center-of-mass system, v'_a and v'_b, must necessarily be in exactly opposite directions, with a ratio of magnitudes given by the inverse of the ratio of their masses:

$$\frac{v'_a}{v'_b} = \frac{m_b}{m_a}.$$

The momenta of the two particles, then, are exactly equal in magnitude but opposite in direction in the new coordinate system and their vectorial sum vanishes. It is this property of the center-of-mass system that makes the energy given by (C-2) the energy *available* for the nuclear reaction since in this system the products of the reaction may have *zero kinetic energy*.

The over-all conservation of momentum is assured by the motion of the

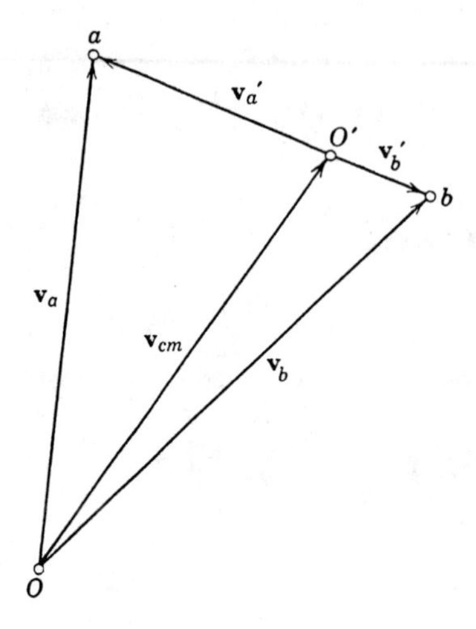

Fig. C-1 Relationship between velocities v_a and v_b of particles a and b in the laboratory system and velocities v_a' and v_b' of the same particles in the center-of-mass system. The velocity of the center of mass with respect to the laboratory frame is v_{cm}.

center of mass with respect to the fixed origin:

$$(m_a + m_b)\mathbf{v}_{\text{cm}} = m_a\mathbf{v}_a + m_b\mathbf{v}_b. \tag{C-3}$$

What (C-3) expresses is the replacement of the reacting system of two particles by a fictitious complex particle of mass $m_a + m_b$ moving with a velocity $\mathbf{v}_{\text{cm}}$; any interaction between a and b is considered as occurring in the internal coordinates of the fictitious complex particle. A velocity diagram for the nuclear reaction

$$a + b \rightarrow c + d,$$

as seen in the laboratory system and in the center-of-mass system, is shown in figure C-2.

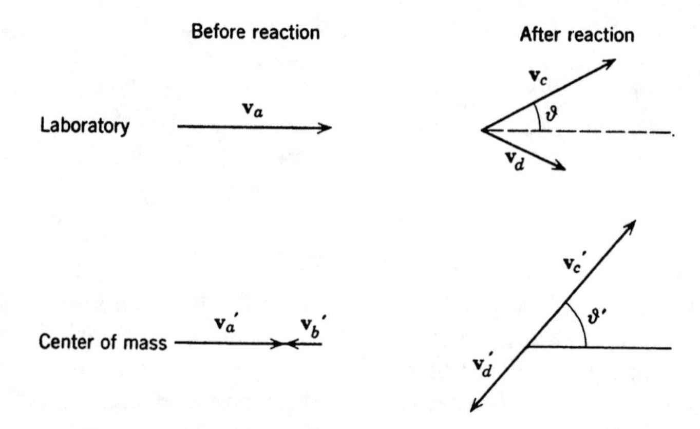

Fig. C-2 Velocity diagram of nuclear reaction in laboratory system and in center-of-mass system.

Corresponding to the kinetic energy of relative motion given by (C-2), the relative momentum[1] of the two particles is:

$$\mathbf{p} = \mu \mathbf{v}, \tag{C-4}$$

$$\mathbf{v} = \mathbf{v}_a - \mathbf{v}_b = \mathbf{v}'_a - \mathbf{v}'_b. \tag{C-5}$$

By combining (C-1), (C-4), and (C-5) we obtain the relation between the *magnitude* of the relative momentum and the momenta in the laboratory system:

$$p = \left| \frac{m_b}{m_a + m_b} \mathbf{p}_a - \frac{m_a}{m_a + m_b} \mathbf{p}_b \right|. \tag{C-6}$$

The magnitude of the relative momentum in the laboratory system is the same as the magnitude of the momentum of *either* particle in the center-of-mass system.

[1] The relative momentum **p** of the two particles is the proper quantity to use in evaluating the density of translational quantum states for the reacting system discussed in footnote 17 of chapter 3.

Appendix D

Table of Nuclides

V. S. Shirley and C. M. Lederer
Isotopes Project
Lawrence Berkeley Laboratory

This table presents properties of nuclides, both stable and radioactive, adopted from the seventh edition of the *Table of Isotopes* (L1). The data are based on experimental results reported in the literature, with the cutoff date varying from January to December, 1977. (The earliest date refers to the lightest nuclides, and vice versa.) Most mass excesses are from the 1977 Atomic Mass Evaluation (W1), with some recent experimental values added. For a few of the very unstable nuclides for which no values were reported in the 1977 Atomic Mass Evaluation, estimates are taken from the tables of W. D. Myers (M1). Natural isotopic abundances (H1) and neutron cross sections (H2) are taken from compilations by N. E. Holden. For other references, original data, and information on the data measurements the reader is referred to L1.

Column 1, Nuclide

Nuclides are listed in order of increasing atomic number Z, and are subordered by increasing mass number A. All isotopic species with half lives longer than about 1 s are included, as are the few shorter-lived ground states, fission isomers, and "historic" isomers (e.g., $^{24}Na^{m}$). Isotopes in L1 with ambiguous or very uncertain assignments, or whose assignments are probably in error (class "G"), have been omitted. Also not included are those nuclides identified in nuclear reactions, but for which radioactive decay has not been observed (class "R" in L1). Isomeric states are denoted by the conventional symbols m, m_1, m_2, and so on. Identical mass assignments (with no m) for several species indicate that the relative positions of the isomers are unknown.

Column 2, Abundance and Half Life

Half lives are given in plain type, natural isotopic abundances in italics. Half lives are rounded so that the uncertainty is ≤5 units in the last place. A question mark following the half life indicates that the assignment of the half life (and other measured decay properties) to the listed values of Z and A is rather uncertain (nuclides with class "F" in L1).

Abundances (in atom percent) are also rounded to an uncertainty of ≤ 5 units in the last place, although the uncertainties are not well known. (Note that, because of the rounding, the abundances for an element do not always add to exactly 100 percent.) For additional information on abundances observed in specific sources and for variations in abundances the reader is referred to L1 and H1.

Column 3, Decay Mode

β^-	negative beta decay.
β^+, EC or EC, β^+	both positive beta decay and electron capture have been experimentally shown to occur, with the first-named mode dominant from theoretical considerations; percentage branchings are given when known, e.g., EC 90%, β^+ 10%.
β^+ or EC	β^+ (or EC) has been observed or inferred from genetic relationships, with the other decay mode probably $\leqslant 1$ percent from theoretical considerations.
$\beta^+ +$ EC or EC $+ \beta^+$	the first-named mode has been observed or inferred from genetic relationships; the second mode is probably $\geqslant 1$ percent from theoretical considerations.
IT	isomeric transition (γ-ray and conversion-electron decay).
α	alpha decay.
SF	spontaneous fission (listed only if branching by this mode is $\gtrsim 1$ percent).
p	direct proton decay ($^{53}Co^m$).
$\beta^-\beta^-$	double negatron emission.
$\beta^- n$	"delayed" neutron emission following β^- decay to unbound states. Other delayed particle-emission modes include $\beta^-\alpha$, $\beta^+ p$, $\beta^+\alpha$, β^+SF, and so on.

Decay modes inferred from the means of production are enclosed in square brackets. For nuclides that decay by more than one mode, branching ratios are given if known; they are rounded so that the uncertainty is ≤ 5 units in the last place.

Column 4, Mass Excess Δ

Mass excesses are given in MeV, with $\Delta(^{12}C)$ defined as zero. Values are quoted to the number of significant figures implied in W1, except that very precise values have been rounded to the nearest keV. An appended s denotes a mass excess estimated from systematic considerations.

Column 5, Spin J and Parity π

Spin and parity assignments without parentheses are definite; assignments in parentheses are probable. Values enclosed in square brackets are inferred from systematics.

Column 6, Neutron Cross Section σ_n

Neutron cross sections are given in barns ($b \equiv 10^{-24}\ cm^2$), and, in the absence of additional notation, refer to thermal-neutron capture cross sections [$\sigma_c = \sigma(n, \gamma)$] at a neutron velocity of $2200\ m\ s^{-1}$ ($E = 0.0253$ eV, or $T =$ 293 K). A superscript sc following the value indicates a cross-section measurement with "subcadmium" neutrons, those with energy $\lesssim 0.5$ eV to which a cadmium absorber is "opaque"; the superscript rs refers to "reactor spectrum" neutrons, with an energy spectrum that is not well defined but that is approximately characteristic of a "thermal" irradiation position in a reactor. The subscripts f (fission), a (total absorption, $n\alpha$, and np identify cross sections other than the capture cross section. The symbols m and g as subscripts stand for "metastable" and "ground," and are used wherever separate cross sections are reported for capture to ground and isomeric states. For those cases for which a single total cross section includes both direct capture and indirect capture via the isomeric states, the subscript g + m is used. For additional details the reader is referred to L1.

To use the table for the calculation of the rate of a nuclear reaction in a sample placed in a nuclear reactor, it must be realized that the magnitude of the flux of neutrons at any point in a reactor is given by the expression

$$n \int_0^\infty vP(v)\, dv = n\bar{v}, \qquad \text{(D-1)}$$

where n is the density of neutrons at that point, $P(v)\, dv$ is the probability that a neutron will have a velocity between v and $v + dv$, and $\bar{v}$ is the average velocity. If the neutron-reaction cross section is in the $1/v$ region (see chapter 4, section D), then the rate of the reaction is

$$\mathbf{R} = Nnv_0\sigma_0, \qquad \text{(D-2)}$$

where N is the number of target nuclei in the sample and σ_0 is the cross section at v_0.

REFERENCES

H1 N. E. Holden, "Isotopic Composition of the Elements and Their Variation in Nature: A Preliminary Report," Brookhaven National Laboratory Report No. BNL-NCS-50605, 1977 (unpublished, available from Nat. Tech. Info. Service, Springfield, VA).

H2 N. E. Holden, private communication to C. M. Lederer and V. S. Shirley, 1977.

L1 C. M. Lederer and V. S. Shirley (Eds.); E. Browne, J. M. Dairiki, and R. E. Doebler (principal authors); A. A. Shihab-Eldin, L. J. Jardine, J. K. Tuli, and A. B. Buyrn (authors), *Table of Isotopes*, 7th ed., Wiley, New York, 1978.

M1 W. D. Myers, *Droplet Model of Atomic Nuclei*, IFI/Plenum, New York, 1977; see also *At. Data Nucl. Data Tables* **17**, 474 (1976).

W1 A. H. Wapstra and K. Bos, "The 1977 Atomic Mass Evaluation," *At. Data Nucl. Data Tables* **19**, 175 (1977); *At. Data Nucl. Data Tables* **20**, 1 (1977); Errata: *At. Data Nucl. Data Tables* **20**, 126 (1977).

TABLE OF NUCLIDES

Z	El	A	Abundance or $t_{1/2}$	Decay Mode	Δ(MeV)	Jπ	σ_n(b)
0	n	1	10.6 m	β^-,no γ	8.071	1/2+	
1	H	1	*99.985%*		7.289	1/2+	0.332
		2	*0.0148%*		13.136	1+	5.2×10^{-4}
		3	12.33 y	β^-,no γ	14.950	1/2+	$<6\times10^{-6\,sc}$
2	He	3	*1.38×10^{-4}%*		14.931	1/2+	$5.33\times10^{3}{}_{np}$
		4	*99.99986%*		2.425	0+	
		6	0.808 s	β^-,no γ	17.597	0+	
		8	0.122 s	β^-,β^-n 12%	31.609	0+	
3	Li	6	*7.5%*		14.087	1+	$942_{n\alpha}$
		7	*92.5%*		14.908	3/2−	0.045^{rs}
		8	0.84 s	$\beta^-2\alpha$	20.947	2+	
		9	0.178 s	β^-,β^-n2α 35%	24.955	(3/2)−	
		11	8.5 ms	β^-,β^-n 61%	40.94		
4	Be	7	53.3 d	EC	15.770	3/2−	$5\times10^{4\,rs}_{np}$
		9	*100%*		11.348	3/2−	0.008
		10	1.6×10^{6} y	β^-,no γ	12.608	0+	$<0.001^{rs}$
		11	13.8 s	β^-,$\beta^-\alpha$ 3%	20.176	1/2+	
		12	11.4 ms	β^-,β^-n	25.03	0+	
5	B	8	0.769 s	$\beta^+2\alpha$	22.922	2+	
		10	*19.8%*		12.052	3+	$3838_{n\alpha}$
		11	*80.2%*		8.668	3/2−	0.005^{rs}
		12	20.4 ms	β^-,$\beta^-3\alpha$ 1.6%	13.370	1+	
		13	17.4 ms	β^-,β^-n 0.28%	16.562	3/2−	
		14	16 ms	β^-	23.657	2−	
6	C	9	0.1265 s	β^+p2α	28.912	(3/2−)	
		10	19.2 s	β^+	15.703	0+	
		11	20.38 m	β^+ 99.76%, EC 0.24%,no γ	10.650	3/2−	
		12	*98.89%*		0	0+	0.0034
		13	*1.11%*		3.125	1/2−	9×10^{-4}
		14	5730 y	β^-,no γ	3.020	0+	$<1\times10^{-6\,rs}$
		15	2.449 s	β^-	9.873	1/2+	
		16	0.75 s	β^-n >98.8%	13.693	0+	
7	N	12	11.0 ms	β^+,$\beta^+3\alpha$ 3.5%	17.338	1+	
		13	9.96 m	β^+,no γ	5.346	1/2−	
		14	*99.63%*		2.863	1+	1.82^{sc}_{np}
		15	*0.366%*		0.102	1/2−	$4\times10^{-5\,rs}$
		16	7.13 s	β^-,$\beta^-\alpha$ 0.0012%	5.682	2−	
		17	4.17 s	β^-,β^-n 95%	7.870	1/2−	
		18	0.63 s	β^-	13.274	0,1,2−	
8	O	13	8.9 ms	β^+p	23.105	3/2−	
		14	70.60 s	β^+	8.008	0+	
		15	122 s	β^+ 99.89%, EC 0.11%,no γ	2.855	1/2−	
		16	*99.76%*		−4.737	0+	$1.8\times10^{-4\,rs}$
		17	*0.038%*		−0.810	5/2+	$0.235^{sc}_{n\alpha}$
		18	*0.204%*		−0.783	0+	1.6×10^{-4}
		19	26.9 s	β^-	3.331	5/2+	
		20	13.5 s	β^-	3.799	0+	
9	F	17	64.5 s	β^+,no γ	1.952	5/2+	
		18	109.8 m	β^+ 96.9%, EC 3.1%,no γ	0.872	1+	
		19	*100%*		−1.487	1/2+	0.010^{rs}
		20	11.0 s	β^-	−0.017	2+	
		21	4.32 s	β^-	−0.047	5/2+	
		22	4.23 s	β^-	2.826	4+	
		23	2.2 s	β^-	3.35	(5/2)+	
10	Ne	17	0.109 s	β^+p	16.478	1/2−	
		18	1.67 s	β^+	5.319	0+	
		19	17.3 s	β^+ 99+%,EC 0.102%	1.751	1/2+	
		20	*90.51%*		−7.043	0+	0.038^{rs}
		21	*0.27%*		−5.733	3/2+	0.7^{rs}

TABLE OF NUCLIDES

Z	El	A	Abundance or $t_{1/2}$	Decay Mode	Δ(MeV)	Jπ	σ_n(b)
10	Ne	22	*9.22%*		−8.026	0+	0.05^{rs}
		23	37.6 s	β^-	−5.155	5/2+	
		24	3.38 m	β^-	−5.949	0+	
		25	0.60 s	β^-	−2.15	(1/2)+	
11	Na	20	0.446 s	β^+,$\beta^+\alpha$ 21%	6.844	2+	
		21	22.47 s	β^+	−2.186	3/2+	
		22	2.602 y	β^+ 90.5%,EC 9.5%	−5.184	3+	3.2×10^4
		23	*100%*		−9.530	3/2+	0.43_m 0.10_g
		24	15.02 h	β^-	−8.418	4+	
		24m	20.2 ms	IT,β^-(weak)	−7.945	1+	
		25	60 s	β^-	−9.357	5/2+	
		26	1.07 s	β^-	−6.888	3+	
		27	0.30 s	β^-,β^-n 0.08%	−5.63	3/2,5/2+	
		28	31 ms	β^-,β^-n 0.6%	−1.13	1+	
		29	43 ms	β^-,β^-n 15%	2.66		
		30	54 ms	β^-,β^-n 33%	8.38		
		31	17 ms	β^-,β^-n 30%	10.61		
		32	14.5 ms	β^-	16.41		
		33	0.02 s	β^-			
12	Mg	21	123 ms	β^+p	10.912	5/2+	
		22	3.86 s	β^+	−0.394	0+	
		23	11.3 s	β^+	−5.471	3/2+	
		24	*78.99%*		−13.931	0+	0.053^{rs}
		25	*10.00%*		−13.191	5/2+	0.18^{rs}
		26	*11.01%*		−16.212	0+	0.038
		27	9.46 m	β^-	−14.585	1/2+	0.15^{rs}
		28	21.0 h	β^-	−15.016	0+	
		29	1.4 s	β^-	−10.75	(3/2+)	
		30	1.2 s	β^-	−9.79 s	0+	
13	Al	23	0.47 s	β^+,β^+p	6.768		
		24	2.07 s	β^+,$\beta^+\alpha$ 0.0077%	−0.052	4+	
		24m	0.13 s	IT 93%,β^+ 7%,$\beta^+\alpha$	0.387	1+	
		25	7.18 s	β^+	−8.913	5/2+	
		26	7.2×10^5 y	β^+ 82%,EC 18%	−12.208	5+	
		26m	6.36 s	β^+,no γ	−11.979	0+	
		27	*100%*		−17.194	5/2+	0.231
		28	2.24 m	β^-	−16.848	3+	
		29	6.6 m	β^-	−18.212	5/2+	
		30	3.69 s	β^-	−15.89	(2,3)+	
		31	0.64 s	β^-	−15.10	5/2,3/2+	
14	Si	25	0.22 s	β^+,β^+p	3.824	3/2,5/2+	
		26	2.21 s	β^+	−7.143	0+	
		27	4.13 s	β^+	−12.385	5/2+	
		28	*92.23%*		−21.491	0+	0.17^{rs}
		29	*4.67%*		−21.894	1/2+	0.10^{rs}
		30	*3.10%*		−24.432	0+	0.108
		31	2.62 h	β^-	−22.949	3/2+	0.5^{rs}
		32	≈650 y	β^-,no γ	−24.092	0+	
		33	6.2 s	β^-	−20.57		
		34	2.8 s	β^-	−19.85	0+	
15	P	28	270 ms	β^+	−7.160	3+	
		29	4.1 s	β^+	−16.949	1/2+	
		30	2.50 m	β^+,EC	−20.204	1+	
		31	*100%*		−24.440	1/2+	0.18^{rs}
		32	14.28 d	β^-,no γ	−24.305	1+	
		33	25.3 d	β^-,no γ	−26.337	1/2+	
		34	12.4 s	β^-	−24.55	1+	
		35	47 s	β^-	−24.94	(1/2,3/2)+	
16	S	29	0.19 s	β^+,β^+p	−3.16	5/2+	
		30	1.2 s	β^+	−14.062	0+	
		31	2.6 s	β^+	−19.044	1/2+	
		32	*95.02%*		−26.015	0+	0.53^{rs}
		33	*0.75%*		−26.586	3/2+	$0.09^{rs}_{n\alpha}$
		34	*4.21%*		−29.931	0+	0.24^{rs}
		35	87.4 d	β^-,no γ	−28.846	3/2+	

TABLE OF NUCLIDES

Z	El	A	Abundance or $t_{1/2}$	Decay Mode	Δ(MeV)	Jπ	σ_n(b)
16	S	36	*0.017%*		−30.666	0+	0.15 ^{rs}
		37	5.0 m	β^-	−26.908	5/2,7/2−	
		38	170 m	β^-	−26.862	0+	
17	Cl	32	298 ms	β^+,β^+p ≈0.007%, $\beta^+\alpha$ ≈0.01%	−13.329	1+	
		33	2.51 s	β^+	−21.003	3/2+	
		34	1.526 s	β^+,no γ	−24.438	0+	
		34m	32.0 m	β^+ 53%,IT 47%	−24.292	3+	
		35	*75.77%*		−29.014	3/2+	43
		36	3.00×10^5 y	β^- 98.1%,EC 1.9%, β^+ 0.0017%,no γ	−29.522	2+	<1 ^{rs}
		37	*24.23%*		−31.762	3/2+	0.428 $_g$ 0.005 $_m$
		38	37.3 m	β^-	−29.798	2−	
		38m	0.715 s	IT	−29.127	5−	
		39	56 m	β^-	−29.803	3/2+	
		40	1.35 m	β^-	−27.54	2−	
		40	0.10 s?	β^-?			
		41	34 s	β^-	−27.40	(1/2,3/2)+	
18	Ar	33	0.18 s	β^+,β^+p 34%	−9.385	1/2+	
		34	0.844 s	β^+	−18.379	0+	
		35	1.78 s	β^+	−23.049	3/2+	
		36	*0.337%*		−30.231	0+	5 ^{rs}
		37	35.0 d	EC,no γ	−30.948	3/2+	
		38	*0.063%*		−34.715	0+	0.8 ^{rs}
		39	269 y	β^-,no γ	−33.241	7/2−	600 ^{rs}
		40	*99.60%*		−35.040	0+	0.64
		41	1.83 h	β^-	−33.068	7/2−	0.5 ^{rs}
		42	33 y	β^-,no γ	−34.42	0+	
		43	5.4 m	β^-	−31.98		
		44	11.9 m	β^-	−32.271	0+	
19	K	36	0.34 s	β^+	−17.426	2+	
		37	1.23 s	β^+	−24.799	3/2+	
		38	7.61 m	β^+	−28.802	3+	
		38m	0.93 s	β^+,no γ	−28.671	0+	
		39	*93.26%*		−33.806	3/2+	2.1 ^{rs}
		40	*0.0117%* 1.28×10^9 y	β^- 89.3%,EC 10.7%, β^+ 0.0010%	−33.535	4−	70 ^{rs}
		41	*6.73%*		−35.560	3/2+	1.46
		42	12.36 h	β^-	−35.023	2−	
		43	22.3 h	β^-	−36.588	3/2+	
		44	22.1 m	β^-	−35.807	2−	
		45	20 m	β^-	−36.611	3/2+	
		46	115 s	β^-	−35.420	(2−)	
		47	17.5 s	β^-	−35.698	1/2+	
		48	6.8 s	β^-	−32.22	(2−)	
		49	≧2 s?	β^-			
		50	≈0.3 s?	β^-	−23.57 s		
20	Ca	37	0.173 s	β^+,β^+p	−13.164	3/2+	
		38	0.44 s	β^+	−22.060	0+	
		39	0.86 s	β^+,no γ	−27.282	3/2+	
		40	*96.94%*		−34.847	0+	0.4 ^{rs}
		41	1.0×10^5 y	EC,no γ	−35.138	7/2−	
		42	*0.647%*		−38.544	0+	0.7 ^{rs}
		43	*0.135%*		−38.405	7/2−	6 ^{rs}
		44	*2.09%*		−41.466	0+	0.88
		45	165 d	β^-	−40.810	7/2−	
		46	*0.0035%*		−43.138	0+	0.7 ^{rs}
		47	4.536 d	β^-	−42.343	7/2−	
		48	*0.187%*		−44.216	0+	1.1 ^{rs}
		49	8.72 m	β^-	−41.286	(3/2)−	
		50	14 s	β^-	−39.572	0+	
21	Sc	40	182 ms	β^+,β^+p	−20.527	4−	
		41	0.596 s	β^+,no γ	−28.644	7/2−	
		42	682 ms	β^+,no γ	−32.121	0+	
		42m	62.0 s	β^+	−31.503	7+	

TABLE OF NUCLIDES

Z	El	A	Abundance or $t_{1/2}$	Decay Mode	Δ(MeV)	Jπ	σ_n(b)
21	Sc	43	3.89 h	β^++EC	−36.185	7/2−	
		44	3.93 h	β^+ 95%,EC 5%	−37.811	2+	
		44m	2.44 d	IT 98.61%,EC 1.39%	−37.540	6+	
		45	*100%*		−41.066	7/2−	17^{rs}_{g} 9^{rs}_{m}
		45m	0.31 s	IT	−41.054	3/2+	
		46	83.80 d	β^-	−41.756	4+	8^{sc}
		46m	18.7 s	IT	−41.613	1−	
		47	3.42 d	β^-	−44.330	7/2−	
		48	43.7 h	β^-	−44.498	6+	
		48	3 h?	?			
		49	57.0 m	β^-	−46.555	(7/2)−	
		50	1.71 m	β^-	−44.539	(5)+	
		50m	0.35 s	IT	−44.282	(2)+	
		51	12.4 s	β^-	−43.220	(7/2)−	
22	Ti	41	80 ms	β^+p	−15.78	3/2+	
		42	0.20 s	β^+	−25.122	0+	
		43	0.49 s	β^+,no γ	−29.324		
		44	47 y	EC	−37.546	0+	
		45	3.09 h	β^+,EC	−39.004	7/2−	
		46	*8.2%*		−44.123	0+	0.6^{rs}
		47	*7.4%*		−44.931	5/2−	1.7^{rs}
		48	*73.7%*		−48.488	0+	7.9^{rs}
		49	*5.4%*		−48.559	7/2−	2.1^{rs}
		50	*5.2%*		−51.432	0+	0.179
		51	5.80 m	β^-	−49.733	3/2−	
		52	1.7 m	β^-	−49.469	0+	
		53	33 s	β^-	−46.84	(3/2)−	
23	V	44	0.09 s	[β^+],$\beta^+\alpha$	−23.85 s		
		46	0.423 s	β^+,no γ	−37.071	0+	
		47	32.6 m	β^++EC	−42.001	3/2−	
		48	15.976 d	EC 50.4%,β^+ 49.6%	−44.473	4+	
		49	330 d	EC,no γ	−47.957	7/2−	
		50	*0.250%*		−49.219	6+	50
		51	*99.750%*		−52.199	7/2−	4.88
		52	3.76 m	β^-	−51.439	3+	
		53	1.6 m	β^-	−51.863	7/2−	
		54	43 s	β^-	−49.93		
24	Cr	45	0.05 s	β^+p	−19.46	[7/2−]	
		46	0.26 s	β^+,no γ	−29.461	0+	
		48	21.56 h	EC	−42.818	0+	
		49	41.9 m	β^+,EC	−45.329	5/2−	
		50	*4.35%*		−50.258	0+	15.9
		51	27.70 d	EC	−51.448	7/2−	
		52	*83.79%*		−55.415	0+	0.8^{rs}
		53	*9.50%*		−55.284	3/2−	18^{rs}
		54	*2.36%*		−56.931	0+	0.38^{rs}
		55	3.55 m	β^-	−55.106	3/2−	
		56	5.9 m	β^-	−55.265	0+	
25	Mn	50	0.283 s	β^+,no γ	−42.626	0+	
		50m	1.74 m	β^+	−42.40	5+	
		51	46.2 m	β^++EC	−48.240	5/2−	
		52	5.59 d	EC 72%,β^+ 28%	−50.704	6+	
		52m	21.1 m	β^++EC 98.25%, IT 1.75%	−50.326	2+	
		53	3.7×10^6 y	EC,no γ	−54.687	7/2−	70^{rs}
		54	312 d	EC	−55.554	3+	$<10^{rs}$
		55	*100%*		−57.710	5/2−	13.3
		56	2.579 h	β^-	−56.909	3+	
		57	1.6 m	β^-	−57.487	5/2−	
		58	65 s	β^-	−55.802	3+	
		58	3.0 s	β^-,no γ	−55.832	(0+)	
26	Fe	49	0.07 s	[β^+],β^+p	−24.47		
		52	8.27 h	β^+ 57%,EC 43%	−48.332	0+	
		53	8.51 m	β^+,EC	−50.944	7/2−	
		53m	2.53 m	IT	−47.904	19/2−	

Nuclide Z El	A	Abundance or $t_{1/2}$	Decay Mode	Δ(MeV)	Jπ	σ_n(b)
26 Fe	54	*5.8%*		−56.251	0+	2.2 rs
	55	2.7 y	EC,no γ	−57.479	3/2−	
	56	*91.8%*		−60.604	0+	2.6 rs
	57	*2.15%*		−60.179	1/2−	2.4 rs
	58	*0.29%*		−62.152	0+	1.14
	59	44.6 d	β^-	−60.661	3/2−	
	60	3×10^5 y	β^-	−61.437	0+	
	61	6.0 m	β^-	−59.01	(3/2)−	
	62	68 s	β^-	−58.86	0+	
27 Co	53	0.26 s	β^+,no γ	−42.640	[7/2−]	
	53m	0.25 s	β^+ ≈98.5%,p ≈1.5%	−39.453	[19/2−]	
	54	193.2 ms	β^+,no γ	−48.010	0+	
	54m	1.46 m	β^+	−47.811	(7+)	
	55	17.5 h	β^+ 77%,EC 23%	−54.024	7/2−	
	56	78.8 d	EC 81%,β^+ 19%	−56.037	4+	
	57	271 d	EC	−59.342	7/2−	
	58	70.8 d	EC 85.00%,β^+ 15.00%	−59.844	2+	1.9×10^3
	58m	9.2 h	IT	−59.819	5+	1.4×10^5
	59	*100%*		−62.226	7/2−	19_m 18_g
	60	5.271 y	β^-	−61.647	5+	2.0 sc
	60m	10.5 m	IT 99.75%,β^- 0.25%	−61.588	2+	58 sc
	61	1.65 h	β^-	−62.897	7/2−	
	62(g)	1.50 m	β^-	−61.430	(2)+	
	62(m)	13.9 m	β^-	−61.408	(5)+	
	63	27.5 s	β^-	−61.850	7/2,5/2−	
	64	0.3 s	β^-	−59.791	(1+)	
28 Ni	53	0.05 s	[β^+],β^+p	−29.41	[7/2−]	
	56	6.10 d	EC	−53.902	0+	
	57	36.0 h	EC 60%,β^+ 40%	−56.077	3/2−	
	58	*68.3%*		−60.224	0+	4.6 rs
	59	7.5×10^4 y	EC 99+%, β^+ 1.5×10^{-5}%,no γ	−61.153	3/2−	92_a
	60	*26.1%*		−64.470	0+	2.8 rs
	61	*1.13%*		−64.219	3/2−	2 rs
	62	*3.59%*		−66.745	0+	14.2
	63	100 y	β^-,no γ	−65.513	1/2−	23 rs
	64	*0.91%*		−67.098	0+	1.49
	65	2.520 h	β^-	−65.124	5/2−	24 sc
	66	54.8 h	β^-,no γ	−66.021	0+	
	67	18 s	β^-	−63.47		
29 Cu	58	3.20 s	β^+	−51.662	1+	
	59	82 s	β^+	−56.352	3/2−	
	60	23.4 m	β^+ 93%,EC 7%	−58.343	2+	
	61	3.41 h	β^+ 62%,EC 38%	−61.981	3/2−	
	62	9.73 m	β^+ 97.8%,EC 2.2%	−62.796	1+	
	63	*69.2%*		−65.578	3/2−	4.4
	64	12.70 h	EC 41%,β^+ 19%, β^- 40%	−65.423	1+	
	65	*30.8%*		−67.262	3/2−	2.17
	66	5.10 m	β^-	−66.257	1+	140 sc
	67	61.9 h	β^-	−67.305	3/2−	
	68	31 s	β^-	−65.39	1+	
	68m	3.8 m	IT 86%,β^- 14%	−64.66	(6−)	
	69	3.0 m	β^-	−65.94	(3/2)−	
	70(g)	5 s	β^-	−63.39	1+	
	70(m)	47 s	β^-	−63.25	(5−)	
30 Zn	57	0.04 s	[β^+],β^+p	−32.61	[7/2−]	
	60	2.4 m	β^+ ≈97%,EC ≈3%	−54.184	0+	
	61	89.1 s	β^+ ≈99%,EC ≈1%	−56.58	3/2−	
	62	9.2 h	EC 93%,β^+ 7%	−61.169	0+	
	63	38.1 m	β^+ 93%,EC 7%	−62.211	3/2−	
	64	*48.6%*		−66.001	0+	0.78
	65	244.1 d	EC 98.54%,β^+ 1.46%	−65.910	5/2−	
	66	*27.9%*		−68.898	0+	1 rs

TABLE OF NUCLIDES

Z	El	A	Abundance or $t_{1/2}$	Decay Mode	Δ(MeV)	Jπ	σ_n(b)
30	Zn	67	*4.10%*		−67.880	5/2−	7^{rs}
		68	*18.8%*		−70.006	0+	0.81_g 0.072_m
		69	56 m	β^-	−68.417	1/2−	
		69m	14.0 h	IT 99+%, β^- 0.033%	−67.978	9/2+	
		70	*0.62%*		−69.560	0+	0.09_g 0.0082_m
		71	2.4 m	β^-	−67.324	1/2−	
		71m	3.9 h	β^-	−67.167	(9/2)+	
		72	46.5 h	β^-	−68.134	0+	
		73	24 s	β^-	−65.03		
		74	95 s	β^-	−65.67	0+	
		75	10.2 s	β^-	−62.46 s		
		76	5.7 s	β^-	−62.55	0+	
		77	1.4 s	β^-	−58.91 s		
		79	2.6 s?	β^-n			
31	Ga	62	118 ms	β^+	−51.77 s	(0+)	
		63	32 s	β^+	−56.69	3/2,5/2−	
		64	2.62 m	β^++EC	−58.836	0+	
		65	15.2 m	β^+ 86%,EC 14%	−62.654	3/2−	
		66	9.4 h	β^+ 56.5%,EC 43.5%	−63.723	0+	
		67	78.3 h	EC	−66.878	3/2−	
		68	68.1 m	β^+ 90%,EC 10%	−67.085	1+	
		69	*60.1%*		−69.322	3/2−	1.7
		70	21.1 m	β^- 99.8%,EC 0.2%	−68.905	1+	
		71	*39.9%*		−70.142	3/2−	4.6
		72	14.10 h	β^-	−68.591	3−	
		73	4.87 h	β^-	−69.73	(3/2)−	
		74	8.1 m	β^-	−68.02	(4)−	
		74m	10 s	IT	−67.96	1+	
		75	2.10 m	β^-	−68.56	(3/2−)	
		76	27.1 s	β^-	−66.44	(3−)	
		77	13 s	β^-	−66.41 s		
		78	5.1 s	β^-	−63.68		
		79	3.0 s	β^-	−62.80		
		80	1.66 s	β^-,β^-n	−59.53 s		
		81	1.2 s	β^-,β^-n			
		82	0.60 s?	[β^-],β^-n			
		83	0.31 s	[β^-],β^-n			
32	Ge	64	64 s	β^++EC	−54.43	0+	
		65	31 s	β^++EC, (β^++EC)p 0.013%	−56.41	3/2,5/2−	
		66	2.3 h	EC 73%,β^+ 27%	−61.621	0+	
		67	19.0 m	β^+ 96%,EC 4%	−62.45	(1/2)−	
		68	288 d	EC,no γ	−66.972	0+	
		69	39.0 h	EC 64%,β^+ 36%	−67.096	5/2−	
		70	*20.5%*		−70.561	0+	3.2^{rs}
		71	11.2 d	EC,no γ	−69.906	1/2−	
		72	*27.4%*		−72.583	0+	1.0^{rs}_{g+m}
		73	*7.8%*		−71.294	9/2+	15^{rs}
		73m	0.50 s	IT	−71.227	1/2−	
		74	*36.5%*		−73.422	0+	0.4^{rs}_g 0.16^{rs}_m
		75	82.8 m	β^-	−71.856	1/2−	
		75m	48 s	IT 99.97%,β^- 0.03%	−71.716	7/2+	
		76	*7.8%*		−73.214	0+	0.10^{rs}_m 0.06^{rs}_g
		77	11.30 h	β^-	−71.214	7/2(+)	
		77m	53 s	β^- 80%,IT 20%	−71.055	1/2−	
		78	1.45 h	β^-	−71.76	0+	
		79	19 s?	β^-			
		79	42 s	β^-	−69.56	(1/2)−	
		80	29 s	β^-	−69.43	0+	
		81	10 s	β^-	−66.34 s		
		82	4.6 s	β^-	−65.99 s	0+	
		83	1.9 s	β^-	−≈62.5 s		

TABLE OF NUCLIDES

Nuclide Z	El	A	Abundance or $t_{1/2}$	Decay Mode	Δ(MeV)	Jπ	σ_n(b)
32	Ge	84	1.2 s	β^-		0+	
33	As	68	2.6 m	β^+	−58.77 s		
		69	15 m	β^+ 98%,EC 2%	−63.12	(5/2)−	
		70	53 m	β^+ 84%,EC 16%	−64.339	4(+)	
		71	61 h	EC 68%,β^+ 32%	−67.893	5/2−	
		72	26.0 h	β^+ 77%,EC 23%	−68.232	2−	
		73	80.3 d	EC	−70.949	3/2−	
		74	17.78 d	EC 37%,β^+ 31%, β^- 32%	−70.860	2−	
		75	*100%*		−73.034	3/2−	4.4
		76	26.3 h	β^-	−72.291	2−	
		77	38.8 h	β^-	−73.916	3/2−	
		78	91 m	β^-	−72.74	(2−)	
		79	9.0 m	β^-	−73.71	3/2−	
		80	16 s	β^-	−72.06	1(+)	
		81	33 s	β^-	−72.64	(3/2)−	
		82	14 s	β^-		(5−)	
		82	19 s	β^-	−70.39	(1+)	
		83	13 s	β^-	−69.87		
		84	0.6 s	β^-	−66.16 s		
		84	5.3 s	β^-,β^-n 0.1%	−66.16 s	(1−)	
		85	2.03 s	β^-,β^-n 23%	−63.52 s		
		86	0.9 s	β^-,β^-n ≈4%	−59.7 s		
		87	0.6 s	β^-	−≈56.2 s		
34	Se	68	1.6 m	β^+	−54.17 s	0+	
		69	27.4 s	β^+,β^+p 0.07%	−56.30		
		70	41.1 m	β^++EC	−61.74 s	0+	
		70m	4 m?	β^++EC			
		71	4.9 m	β^++EC	−63.46	(5/2)−	
		72	8.4 d	EC	−67.894	0+	
		73	7.1 h	β^+ ≈65%,EC ≈35%	−68.209	7/2+	
		73m	41 m	IT 73%,(β^+,EC) 27%	−68.183	1/2−	
		74	*0.87%*		−72.213	0+	52
		75	118.5 d	EC	−72.169	5/2+	
		76	*9.0%*		−75.259	0+	64_g 21_m^{rs}
		77	*7.6%*		−74.606	1/2−	42^{rs}
		77m	17.4 s	IT	−74.444	7/2+	
		78	*23.5%*		−77.032	0+	0.4_{g+m} 0.3_m^{rs}
		79	≦6.5×10^4 y	β^-,no γ	−75.911	7/2+	
		79m	3.90 m	IT	−75.815	1/2−	
		80	*49.8%*		−77.761	0+	0.6_g 0.07_m
		81	18.5 m	β^-	−76.391	(1/2)−	
		81m	57.3 m	IT 99+%,β^- 0.058%	−76.288	(7/2)+	
		82	*9.2%* 1.4×10^{20} y	$\beta^-\beta^-$	−77.586	0+	0.04_m 0.006_g
		83	22.5 m	β^-	−75.333	(9/2)+	
		83m	70 s	β^-	−75.105	(1/2)−	
		84	3.3 m	β^-	−75.942	0+	
		85	31 s	β^-	−72.57 s		
		85	19 s?	β^-			
		86	16 s	β^-	−70.86 s	0+	
		87	5.8 s	β^-,β^-n 0.16%	−≈66.2 s		
		88	1.5 s	β^-,β^-n 0.8%	−64.09 s	0+	
		89	0.41 s	[β^-],β^-n 5%	−59.89 s		
		91	0.27 s	β^-,β^-n ≈21%			
35	Br	70	23 s?	β^+p	−51.29 s		
		71	<1 m?	β^++EC	−56.86 s		
		72	1.31 m	β^+	−58.93 s	(3)	
		73	3.4 m	β^++EC	−63.67	(3/2−)	
		74	25.3 m	β^++EC	−65.295	(0,1−)	
		74	4 m?	β^+			
		74m	41 m	β^++EC	−≈65.1	(4−)	
		75	98 m	β^+ 76%,EC 24%	−69.159	(3/2−)	
		76	16.1 h	β^+ 57%,EC 43%	−70.303	1−	

TABLE OF NUCLIDES

Nuclide Z El	A	Abundance or $t_{1/2}$	Decay Mode	Δ(MeV)	Jπ	σ_n(b)
35 Br	77	57.0 h	EC 99.26%,β^+ 0.74%	−73.242	3/2−	
	77m	4.3 m	IT	−73.136	9/2+	
	78	6.46 m	β^+ 92%,EC 8%, β^- ≤0.01%	−73.458	1+	
	79	*50.69%*		−76.070	3/2−	10.8_g 2.4_m
	79m	4.9 s	IT	−75.863	9/2+	
	80	17.6 m	β^- 91.7%, EC 5.7%,β^+ 2.6%	−75.891	1+	
	80m	4.42 h	IT	−75.805	5−	
	81	*49.31%*		−77.976	3/2−	2.7_{g+m}
	82	35.34 h	β^-	−77.498	5−	
	82m	6.1 m	IT 97.6%,β^- 2.4%	−77.452	2−	
	83	2.39 h	β^-	−79.025	(3/2)−	
	84	31.8 m	β^-	−77.759	2−	
	84m	6.0 m	β^-	−77.46	(6−)	
	85	2.9 m	β^-	−78.67	3/2−	
	86	56 s	β^-	−75.96	(2−)	
	86	4.5 s?	[IT]			
	87	55.6 s	β^-,β^-n 2.3%	−74.21 s	(3/2−)	
	88	16.6 s	β^-,β^-n 6%	−71.09 s	(1−)	
	89	4.4 s	β^-,β^-n 13%	−69.09 s		
	90	1.9 s	β^-,β^-n 23%	−≈65.2 s		
	91	0.54 s	β^-,β^-n 9%			
	92	0.36 s	β^-,β^-n 16%	−≈57.8 s		
36 Kr	72	17 s	β^++EC	−53.87 s	0+	
	73	27 s	β^+,β^+p 0.7%	−56.98		
	74	11.5 m	β^+,EC	−62.02	0+	
	75	4.3 m	β^++EC	−64.16 s		
	76	14.8 h	EC	−69.10	0+	
	77	75 m	β^+ ≈80%,EC ≈20%	−70.236	(5/2+)	
	78	*0.356%*		−74.150	0+	5_g 0.21_m
	79	35.0 h	EC 93%,β^+ 7%	−74.439	1/2−	
	79m	50 s	IT	−74.309	7/2+	
	80	*2.27%*		−77.897	0+	12_g 5_m
	81	2.1×10^5 y	EC	−77.654	7/2+	
	81m	13 s	IT	−77.464	1/2−	
	82	*11.6%*		−80.591	0+	23_g 20_m
	83	*11.5%*		−79.985	9/2+	200
	83m	1.83 h	IT	−79.943	1/2−	
	84	*57.0%*		−82.432	0+	0.09_m 0.042_g
	85	10.7 y	β^-	−81.472	9/2+	1.7
	85m	4.48 h	β^- 79%,IT 21%	−81.167	1/2−	
	86	*17.3%*		−83.263	0+	0.06^{rs}
	87	76 m	β^-	−80.707	(5/2)+	
	88	2.84 h	β^-	−79.689	0+	
	89	3.18 m	β^-	−76.79		
	90	32.3 s	β^-	−75.18	0+	
	91	8.6 s	β^-	−71.77		
	92	1.84 s	β^-,β^-n 0.032%	−69.15	0+	
	93	1.29 s	β^-,β^-n 1.9%	−65.6		
	94	0.20 s	β^-,β^-n 6%	−≈61.32 s	0+	
	95	0.78 s	[β^-]			
37 Rb	74	65 ms	β^+,no γ	−51.43 s	(0+)	
	75	18 s	β^+	−57.51		
	76	39 s	β^+	−60.61		
	77	3.9 m	β^++EC	−65.11	(5/2−)	
	78	18 m	β^++EC	−68.8		
	78m	6 m	β^++EC,IT	−68.7		
	79	23.0 m	β^+ 84%,EC 16%	−70.86	(3/2,5/2−)	
	80	34 s	β^+	−72.190	1+	
	81	4.58 h	EC 73%,β^+ 27%	−75.392	3/2−	
	81m	32 m	β^++EC,IT	−75.307	9/2+	

Nuclide Z	El	A	Abundance or $t_{1/2}$	Decay Mode	Δ(MeV)	Jπ	σ_n(b)
37	Rb	82	1.25 m	β^+ 96%,EC 4%	−76.213	1+	
		82m	6.2 h	EC 74%,β^+ 26%	−≈76.1	5−	
		83	86.2 d	EC	−78.914	5/2−	
		84	32.9 d	EC 75%,β^+ 22%, β^- 3.0%	−79.752	2−	12^{rs}_{np}
		84m	20.5 m	IT	−79.288	(6+)	
		85	*72.17%*		−82.159	5/2−	0.40_g 0.047_m
		86	18.8 d	β^- 99+%,EC 0.005%	−82.738	2−	
		86m	1.02 m	IT	−82.182	6−	
		87	*27.83%* 4.8×10^{10} y	β^-,no γ	−84.596	3/2−	0.12^{rs}
		88	17.8 m	β^-	−82.602	2−	1.0^{rs}
		89	15.2 m	β^-	−81.717	(3/2−)	
		90	153 s	β^-	−79.57	(1−)	
		90m	258 s	β^-,IT	−79.46	(4−)	
		91	58 s	β^-	−77.97		
		92	4.52 s	β^-,β^-n 0.012%	−75.12	(1−)	
		93	5.85 s	β^-,β^-n 1.3%	−72.92		
		94	2.72 s	β^-,β^-n 10%	−68.82		
		95	0.38 s	β^-,β^-n 8.4%	−66.55		
		96	0.201 s	β^-,β^-n 13%	−62.77 s		
		97	0.170 s	β^-,β^-n 27%			
		98	0.13 s	β^-,β^-n 13%			
		99	76 ms	[β^-]			
38	Sr	77	9 s	β^+,β^+p ≤0.25%	−57.96		
		78	31 m	β^++EC	−65.5 s	0+	
		79	8.1 m	β^++EC	−65.46 s		
		79	4 m?	β^++EC			
		80	106 m	EC+β^+	−70.39 s	0+	
		81	26 m	β^+ ≈87%,EC ≈13%	−71.40	(1/2−)	
		82	25.0 d	EC,no γ	−75.999	0+	
		83	32.4 h	EC 76%,β^+ 24%	−76.664	7/2+	
		83m	5.0 s	IT	−76.405	1/2−	
		84	*0.56%*		−80.641	0+	0.6^{sc}_m 0.3_g
		85	64.8 d	EC	−81.095	9/2+	
		85m	68 m	IT 87%,EC 13%	−80.856	1/2−	
		86	*9.8%*		−84.512	0+	0.84^{sc}_m
		87	*7.0%*		−84.869	9/2+	
		87m	2.80 h	IT 99.7%,EC 0.3%	−84.480	1/2−	
		88	*82.6%*		−87.911	0+	0.0057^{rs}
		89	50.5 d	β^-	−86.203	5/2+	0.42^{rs}
		90	28.8 y	β^-,no γ	−85.935	0+	0.8^{rs}
		91	9.5 h	β^-	−83.666	5/2+	
		92	2.71 h	β^-	−82.892	0+	
		93	7.4 m	β^-	−80.28		
		94	75 s	β^-	−78.96	0+	
		95	24.4 s	β^-	−75.14		
		96	1.1 s	β^-	−73.07	0+	
		97	0.40 s	β^-	−69.08 s		
		98	0.7 s	β^-	−67.38 s	0+	
		99	0.6 s	β^-,β^-n 3%			
39	Y	81	5 m	β^++EC			
		82	12 m?	[β^+]	−67.91 s		
		83	7.1 m	β^+ ≈95%,EC ≈5%	−72.36 s	(9/2+)	
		83	2.85 m	β^++EC	−72.36 s	(1/2)−	
		84	39 m	β^++EC	−73.692		
		84	4.6 s	β^++EC		(1+)	
		85(g)	2.7 h	β^+ 55%,EC 45%	−77.855	(1/2)−	
		85(m)	4.9 h	β^+ 70%,EC 30%	−77.835	(9/2)+	
		86	14.74 h	EC 66%,β^+ 34%	−79.239	4−	
		86m	48 m	IT 99.31%, β^++EC 0.69%	−79.021	8+	
		87	80.3 h	EC 99.8%,β^+ 0.2%	−83.007	1/2−	
		87m	13 h	IT 98%, EC ≈2%,β^+ 0.75%	−82.626	9/2+	

Z	El	A	Abundance or $t_{1/2}$	Decay Mode	Δ(MeV)	Jπ	σ_n(b)
39	Y	88	106.6 d	EC 99+%,β^+ 0.210%	−84.298	4−	
		89	*100%*		−87.695	1/2−	1.2_g 0.0010_m
		89m	16.1 s	IT	−86.786	9/2+	
		90	64.1 h	β^-	−86.481	2−	$<6.5^{rs}_g$
		90m	3.19 h	IT 99+%,β^- 0.0021%	−85.799	7+	
		91	58.5 d	β^-	−86.350	1/2−	1.4^{rs}
		91m	49.7 m	IT	−85.794	9/2+	
		92	3.54 h	β^-	−84.822	2−	
		93	10.2 h	β^-	−84.227	1/2−	
		93m	0.82 s	IT	−83.468	9/2+	
		94	18.7 m	β^-	−82.382	2−	
		95	10.3 m	β^-	−81.233	(1/2)−	
		96	9.8 s	β^-			
		96	6.0 s	β^-	−78.43	(0−)	
		97(g)	3.7 s	β^-	−76.28	(1/2−)	
		97(m)	1.21 s	β^- ≧99.3%, IT(?) ≤0.7%	−75.61	(9/2)+	
		98	0.6 s	β^-	−73.19 s	(1+)	
		98	2.0 s	β^-	−73.19 s		
		99	1.4 s	β^-,β^-n 1%	−71.50		
		100	0.8 s	β^-	−67.96 s		
		102	0.9 s?	[β^-]	−63.36 s		
40	Zr	81	≈11 m	[β^+]			
		82	10 m	[β^++EC]		0+	
		83	≈8 m	[β^+]	−≈65.4 s		
		83	0.7 m	β^+	−≈65.4 s		
		84	5 m	EC+β^+	−71.44 s	0+	
		85	7.9 m	β^++EC	−73.16 s		
		85	1.4 h?	[β^++EC]			
		85m	10.9 s	IT,β^++EC	−72.87 s		
		86	16.5 h	EC	−77.94 s	0+	
		87	1.6 h	β^+,EC	−79.43	(9/2+)	
		87m	14 s	IT	−79.09	(1/2−)	
		88	83.4 d	EC	−83.621	0+	
		89	78.4 h	EC 77.7%,β^+ 22.3%	−84.860	9/2+	
		89m	4.18 m	IT 93.8%, EC 4.7%,β^+ 1.5%	−84.272	1/2−	
		90	*51.5%*		−88.765	0+	0.03^{rs}
		90m	809 ms	IT	−86.446	5−	
		91	*11.2%*		−87.892	5/2+	1.1^{rs}
		92	*17.1%*		−88.456	0+	0.2^{rs}
		93	1.5×10^6 y	β^-	−87.117	5/2+	1^{rs}
		94	*17.4%*		−87.264	0+	0.06
		95	64.0 d	β^-	−85.663	5/2+	
		96	*2.80%*		−85.445	0+	0.020
		97	16.9 h	β^-	−82.954	1/2+	
		98	31 s	β^-,no γ	−81.292	0+	
		99	2.1 s	β^-	−77.89	(1/2+)	
		100	7.1 s	β^-	−76.60	0+	
		101	2.0 s	β^-	−73.05 s		
		102	2.9 s	β^-	−72.36 s	0+	
41	Nb	86	1.4 m	β^+	−69.34 s		
		87	2.6 m	β^++EC	−74.43 s	(9/2+)	
		87	3.9 m	β^++EC	−74.43 s	(1/2−)	
		88	7.8 m	β^++EC	−76.42 s	(4−)	
		88	14.3 m	β^++EC	−76.42 s	(8+)	
		89	2.0 h	β^++EC	−80.621	(9/2+)	
		89	66 m	EC 74%,β^+ 26%	−80.621	(1/2)−	
		90	14.6 h	β^+ 53%,EC 47%	−82.654	8+	
		90m	18.8 s	IT	−82.529	4−	
		91	long	[EC]	−86.637	9/2+	
		91m	62 d	IT 96.6%,EC 3.4%	−86.532	1/2−	
		92	*≈2×10⁻¹¹%* 3.2×10^7 y	EC	−86.448	7+	
		92m	10.15 d	EC 99.94%,β^+ 0.06%	−86.313	2+	
		93	*100%*		−87.209	9/2+	1.1_{g+m}
		93m	13.6 y	IT	−87.179	1/2−	

TABLE OF NUCLIDES

Z	El	A	Abundance or $t_{1/2}$	Decay Mode	Δ(MeV)	Jπ	σ_n(b)
41	Nb	94	2.0×10^4 y	β^-	−86.367	6+	15^{sc}_{g} 0.59^{rs}_{m}
		94m	6.26 m	IT 99.5%,β^- 0.5%	−86.326	3+	
		95	35.0 d	β^-	−86.786	9/2+	$<7^{rs}$
		95m	87 h	IT 97.5%,β^- 2.5%	−86.552	1/2−	
		96	23.4 h	β^-	−85.608	(6)+	
		97	72 m	β^-	−85.612	9/2+	
		97m	1.0 m	IT	−84.868	1/2−	
		98	2.9 s	β^-	−83.530	1+	
		98m	51 m	β^-	−83.446	(5+)	
		99	15.0 s	β^-	−82.346	(9/2)+	
		99m	2.6 m	β^-,IT?(weak)	−81.981	(1/2)−	
		100	1.5 s	β^-	−79.96		
		100	3.1 s	β^-			
		101	7.0 s	β^-	−78.95		
		101	1.0 m?	β^-			
		102	4.3 s	β^-	−76.36 s		
		102	1.3 s	β^-	−76.36 s		
		103	1.5 s	β^-	−75.41 s		
		104	0.8 s	β^-	−72.65 s		
		104	4.8 s	β^-			
		105	2 s	[β^-]	−≈70.14 s		
		106	≈1 s	β^-			
42	Mo	88	27 m?	β^++EC	−72.92 s		
		88	8 m	β^++EC	−72.92 s		
		90	5.67 h	EC 75%,β^+ 25%	−80.167	0+	
		91	15.49 m	β^+ 94.1%,EC 5.9%	−82.199	9/2+	
		91m	65 s	(β^+,EC) 50%,IT 50%	−81.546	1/2−	
		92	*14.8%*		−86.807	0+	0.3^{rs}_{a}
		93	3×10^3 y	EC	−86.803	5/2+	
		93m	6.9 h	IT 99.88%,EC 0.12%	−84.378	21/2+	
		94	*9.3%*		−88.412	0+	
		95	*15.9%*		−87.712	5/2+	14^{rs}
		96	*16.7%*		−88.795	0+	1^{rs}
		97	*9.6%*		−87.544	5/2+	2^{rs}
		98	*24.1%*		−88.115	0+	0.13
		99	66.02 h	β^-	−85.970	1/2+	
		100	*9.6%*		−86.189	0+	0.20
		101	14.6 m	β^-	−83.516	1/2+	
		102	11.0 m	β^-	−83.562	0+	
		103	60 s	β^-	−80.61 s		
		104	1.0 m	β^-	−81.65 s	0+	
		105	36 s	β^-	−77.14 s		
		106	9.5 s	β^-	−≈76.1 s	0+	
		107	≈5 s	β^-			
		108	1.1 s	β^-	−≈70.9 s	0+	
43	Tc	90	50 s	β^+			
		90	7.9 s	β^+	−71.3	(1+)	
		91	3.14 m	β^++EC	−75.98	(9/2+)	
		91	3.3 m	β^++EC		(1/2)−	
		92	4.4 m	β^++EC	−78.936	(8)+	
		93	2.7 h	EC 87%,β^+ 13%	−83.610	9/2+	
		93m	43 m	IT 80%,EC 20%	−83.217	1/2−	
		94	293 m	EC 89%,β^+ 11%	−84.156	7+	
		94m	52 m	β^+ 72%,EC 28%	−84.081	(2)+	
		95	20.0 h	EC	−86.013	9/2+	
		95m	61 d	EC 95.8%,β^+ 0.31%, IT 3.9%	−85.974	1/2−	
		96	4.3 d	EC	−85.821	7+	
		96m	52 m	IT 98%,EC 2%, β^+ ≈0.01%	−85.787	4+	
		97	2.6×10^6 y	EC,no γ	−87.224	9/2+	
		97m	90 d	IT	−87.128	1/2−	
		98	4.2×10^6 y	β^-	−86.434	(6)+	3^{rs}_{m}
		99	2.14×10^5 y	β^-	−87.326	9/2+	19
		99m	6.02 h	IT 99+%,β^- ≥9×10^{-5}%	−87.184	1/2−	
		100	15.8 s	β^-	−86.019	1+	

TABLE OF NUCLIDES

Z	El	A	Abundance or $t_{1/2}$	Decay Mode	Δ(MeV)	Jπ	σ_n(b)
43	Tc	101	14.3 m	β^-	−86.327	9/2+	
		102	5.3 s	β^-	−84.60 s	1+	
		102m	4.4 m	β^- ≈98%,IT ≈2%	−≈84.3 s	(5)	
		103	50 s	β^-	−84.91		
		104	18.1 m	β^-	−83.85	(3)	
		105	7.6 m	β^-	−82.54		
		106	36 s	β^-	−80.03 s		
		107	21.2 s	β^-	−79.51 s		
		108	5.1 s	β^-	−≈75.8 s		
		109	1.4 s	β^-			
		110	0.82 s	β^-	−≈71 s		
44	Ru	92	3.7 m	β^++EC	−≈74.7 s	0+	
		93	60 s	β^++EC	−77.31 s	(9/2)+	
		93m	10.8 s	β^++EC 79%,IT 21%	−76.58 s	(1/2)−	
		94	52 m	EC	−82.571	0+	
		95	1.65 h	EC 85%,β^+ 15%	−83.452	5/2+	
		96	*5.5%*		−86.075	0+	0.25
		97	2.88 d	EC	−86.07	5/2+	
		98	*1.86%*		−88.226	0+	<8 rs
		99	*12.7%*		−87.620	5/2+	4
		100	*12.6%*		−89.222	0+	6
		101	*17.0%*		−87.952	5/2+	5
		102	*31.6%*		−89.100	0+	1.3
		103	39.4 d	β^-	−87.261	5/2+	
		104	*18.7%*		−88.099	0+	0.47
		105	4.44 h	β^-	−85.938	(3/2+)	0.30
		106	367 d	β^-,no γ	−86.333	0+	0.12
		107	4.2 m	β^-	−83.71		
		108	4.5 m	β^-	−83.82	0+	
		109	34 s	β^-	−80.81 s		
		109	13 s	β^-	−80.81 s		
		110	16 s	β^-	−≈80.3 s	0+	
		111	1.5 s	β^-			
		111	≈1 m?	[β^-]			
		112	0.7 s?	[β^-]		0+	
45	Rh	94	25 s	β^+			
		94	80 s	β^+			
		95	5.0 m	β^++EC	−78.34	(9/2)+	
		95m	1.96 m	IT 88%,β^++EC 12%	−77.80	(1/2)−	
		96	9.9 m	β^++EC	−79.633	(5+)	
		96m	1.51 m	IT 60%,β^++EC 40%	−79.581	(2+)	
		97	31 m	β^+,EC	−82.56	(9/2)+	
		97m	44 m	(β^+,EC) 95%,IT 5%	−82.30	(1/2)−	
		97	1 m?	?			
		98(g)	8.7 m	β^+,EC	−83.168	(2+)	
		98(m)	3.5 m	β^++EC	−83.162	(5+)	
		99	15.0 d	EC 97.4%,β^+ 2.6%	−85.517	(1/2−)	
		99m	4.7 h	EC ≈90%,β^+ ≈10%	−85.452	9/2+	
		100	20.8 h	EC 95%,β^+ 5%	−85.592	1−	
		100m	4.7 m	IT 93%,EC+β^+ 7%	−85.252	(5+)	
		101	3.3 y	EC	−87.410	1/2−	
		101m	4.34 d	EC 92.8%,IT 7.2%	−87.253	9/2+	
		102	2.9 y	EC		(6+)	
		102m	206 d	EC 62%,β^+ 14%, β^- 19%,IT 5%	−86.777	(2−)	
		103	*100%*		−88.024	1/2−	134_g 11_m
		103m	56.1 m	IT	−87.984	7/2+	
		104	42.3 s	β^- 99.6%,EC 0.4%	−86.952	1+	40^{rs}_{g+m}
		104m	4.34 m	IT 99.87%,β^- 0.13%	−86.823	5+	800^{rs}_{g+m}
		105	35.4 h	β^-	−87.855	(7/2)+	$1.1\times10^4{}_g$ $5\times10^3{}_m$
		105m	45 s	IT	−87.725	1/2−	
		106	29.8 s	β^-	−86.372	1+	
		106m	130 m	β^-	−86.235	4,5,6+	
		107	21.7 m	β^-	−86.86	(5/2)+	
		108	16.8 s	β^-	−85.02	1+	

TABLE OF NUCLIDES

Z	El	A	Abundance or $t_{1/2}$	Decay Mode	Δ(MeV)	Jπ	σ_n(b)
45	Rh	108	6.0 m	β^-	−85.09		
		109	80 s	β^-	−85.11 s	(5/2,3/2)+	
		110	3 s	β^-	−82.8		
		110	28 s	β^-	−82.93		
		111	11 s	[β^-]	−82.53 s		
		112	4.6 s	β^-	−80.3 s		
		113	0.9 s	[β^-]			
		114	1.7 s?	β^-			
46	Pd	97	3.3 m	β^++EC	−77.76 s		
		98	18 m	EC+β^+	−81.27 s	0+	
		99	21.4 m	β^+,EC	−86.112	(5/2+)	
		100	3.6 d	EC	−85.230	0+	
		101	8.5 h	EC 93.6%,β^+ 6.4%	−85.428	(5/2)+	
		102	*1.0%*		−87.925	0+	5^{rs}
		103	17.0 d	EC	−87.478	5/2+	
		104	*11.0%*		−89.400	0+	
		105	*22.2%*		−88.422	5/2+	
		106	*27.3%*		−89.913	0+	0.28_g 0.013_m
		107	6.5×10^6 y	β^-,no γ	−88.371	5/2+	
		107m	21.3 s	IT	−88.156	11/2−	
		108	*26.7%*		−89.523	0+	11_g^{rs} 0.19_m^{rs}
		109	13.43 h	β^-	−87.606	5/2+	
		109m	4.69 m	IT	−87.417	11/2−	
		110	*11.8%*		−88.335	0+	0.36_g^{rs} 0.02_m^{rs}
		111	22 m	β^-	−86.03	(5/2+)	
		111m	5.5 h	IT 71%,β^- 29%	−85.86	(11/2−)	
		112	21.1 h	β^-	−86.326	0+	
		113	1.5 m	β^-,no γ	−83.64 s		
		114	2.4 m	β^-,no γ	−83.76 s	0+	
		115	37 s	β^-			
		116	14 s	β^-	−≈80.12 s	0+	
		117	5 s	[β^-]			
		118	3.1 s	β^-	−76.21 s	0+	
47	Ag	99	1.8 m?	β^++EC	−76.51 s		
		100	2.3 m	β^++EC	−77.93		
		100	8 m?	β^++EC			
		101	10.8 m	β^++EC	−81.33 s	(9/2+)	
		102	13.0 m	β^+ ≈68%,EC ≈32%	−82.33	5+	
		102m	7.7 m	(β^+,EC) 51%,IT 49%	−82.32	2+	
		103	1.10 h	EC ≈58%,β^+ ≈42%	−84.80	7/2+	
		103m	5.7 s	IT	−84.67	(1/2)−	
		104	69 m	β^+,EC	−85.150	5+	
		104m	33 m	(β^+,EC) 67%,IT 33%		.2+	
		105	41.3 d	EC 99+%,β^+ 9×10^{-4}%	−87.075	1/2−	
		105m	7.2 m	IT 99.7%,EC 0.3%	−87.049	(7/2)+	
		106	24.0 m	(EC,β^+) ≧99%, β^-? ≦1%	−86.929	1+	
		106m	8.5 d	EC	−86.841	6+	
		107	*51.83%*		−88.404	1/2−	37_g 0.3_m^{sc}
		107m	44.3 s	IT	−88.311	7/2+	
		108	2.4 m	β^- 97.7%,EC 2.1%, β^+ 0.24%	−87.602	1+	
		108m	127 y	EC+β^+ 91%,IT 9%	−87.492	6+	
		109	*48.17%*		−88.722	1/2−	88_g 4_m
		109m	39.8 s	IT	−88.634	7/2+	
		110	24.4 s	β^- 99.7%,EC 0.3%	−87.456	1+	
		110m	252 d	β^- 98.5%,IT 1.5%	−87.338	6+	80_{g+m}
		111	7.45 d	β^-	−88.226	1/2−	3^{rs}
		111m	65 s	IT 99.7%,β^- 0.3%	−88.166	(7/2+)	
		112	3.14 h	β^-	−86.620	2(−)	

TABLE OF NUCLIDES

Z	El	A	Abundance or $t_{1/2}$	Decay Mode	Δ(MeV)	Jπ	σ_n(b)
47	Ag	113	1.15 m	β^-	−86.82		
		113	5.37 h	β^-	−87.040	1/2(−)	
		114	4.5 s	β^-	−85.16	1+	
		115	18 s	β^-			
		115	20 m	β^-	−84.91	(1/2−)	
		116	2.68 m	β^-	−82.62 s		
		116m	10.5 s	β^- ≈98%,IT ≈2%	−82.54 s		
		117	1.21 m	β^-	−82.24		
		117	5.3 s	β^-	−82.24		
		118	3.7 s	β^-	−80.21 s		
		118m	2.8 s	β^- 59%,IT 41%	−80.08 s		
		119	2.1 s	β^-	−79.31 s	(7/2+)	
		120	1.2 s	β^-	−≈78.0 s	{3+}	
		120m	0.32 s	β^- ≈63%,IT ≈37%	−≈77.8 s	{6−}	
		121	≲3 s	[β^-]			
		122	1.5 s	[β^-]	−≈70.5 s		
		123	0.39 s	[β^-],β^-n			
48	Cd	100	1.1 m	β^++EC	−73.43 s	0+	
		101	1.2 m	β^++EC	−75.53 s		
		102	5.5 m	EC,β^+	−79.43 s	0+	
		103	7.3 m	β^+,EC	−80.60		
		104	58 m	EC 99.2%,β^+ 0.8%	−83.57	0+	
		105	56.0 m	EC,β^+	−84.336	5/2+	
		106	*1.25%*		−87.131	0+	1^{rs}
		107	6.50 h	EC 99.77%,β^+ 0.23%	−86.987	5/2+	
		108	*0.89%*		−89.251	0+	1.2^{rs}
		109	453 d	EC	−88.540	5/2+	700^{rs}
		110	*12.5%*		−90.349	0+	11^{rs}_{g} 0.10^{rs}_{m}
		111	*12.8%*		−89.254	1/2+	24^{rs}
		111m	48.6 m	IT	−88.858	11/2−	
		112	*24.1%*		−90.578	0+	2^{rs}_{g}
		113	*12.2%* 9×10^{15} y	β^-,no γ	−89.050	1/2+	1.98×10^{4}
		113m	14 y	β^- 99.9%,IT 0.1%	−88.787	11/2−	
		114	*28.7%*		−90.020	0+	0.30^{sc}_{g} 0.04^{sc}_{m}
		115	53.4 h	β^-	−88.093	1/2+	
		115m	44.8 d	β^-	−87.920	11/2−	
		116	*7.5%*		−88.718	0+	0.05^{sc}_{g} 0.025^{sc}_{m}
		117	2.4 h	β^-	−86.416	1/2+	
		117m	3.4 h	β^-	−86.29	11/2−	
		118	50.3 m	β^-,no γ	−86.707	0+	
		119	2.7 m	β^-	−84.23	1/2+	
		119m	1.9 m	β^-	−84.08	11/2−	
		120	50.8 s	β^-	−83.981	0+	
		121	12.8 s	β^-	−≈81.3 s		
		121	4.8 s	β^-	−≈81.3 s		
		122	5.8 s	β^-	−≈80.0 s	0+	
		124	0.9 s	β^-	−≈76.4 s	0+	
49	In	104	1.5 m	β^++EC	−75.57 s		
		105	5.1 m	β^++EC	−79.34 s		
		105m	55 s?	IT?			
		106	5.3 m	β^++EC	−80.586	(3)	
		106	6.3 m	β^++EC			
		107	32.4 m	EC 65%,β^+ 35%	−83.50	9/2+	
		107m	50 s	IT	−82.82	1/2−	
		108	40 m	EC,β^+	−84.10	3+	
		108	58 m	EC,β^+	−84.13	(5,6+)	
		109	4.2 h	EC 94%,β^+ 6%	−86.524	9/2+	
		$109m_1$	1.3 m	IT	−85.874	1/2−	
		$109m_2$	0.21 s	IT	−84.41	(19/2+)	
		110	4.9 h	EC		7+	
		110	69 m	β^+,EC	−86.409	2+	
		111	2.83 d	EC	−88.405	9/2+	

Nuclide Z	El	A	Abundance or $t_{1/2}$	Decay Mode	Δ(MeV)	Jπ	σ_n(b)
49	In	111m	7.6 m	IT	−87.869	1/2−	
		112	14.4 m	β^- 44%,EC 34%, β^+ 22%	−88.000	1+	
		112m	20.9 m	IT	−87.845	4+	
		113	*4.3%*		−89.372	9/2+	5_m 3_g
		113m	99.5 m	IT	−88.980	1/2−	
		114	71.9 s	β^- 98.1%,EC 1.9%, β^+ 0.004%	−88.576	1+	
		114m	49.51 d	IT 96.7%,EC 3.3%	−88.386	5+	
		115	*95.7%* 5.1×10^{14} y	β^-,no γ	−89.541	9/2+	91_{m2} 70_{m1} 41_g
		115m	4.49 h	IT 95%,β^- 5%	−89.205	1/2−	
		116	14.10 s	β^-	−88.253	1+	
		$116m_1$	54.1 m	β^-	−88.126	5+	
		$116m_2$	2.16 s	IT	−87.963	8−	
		117	42 m	β^-	−88.944	9/2+	
		117m	1.93 h	β^- 53%,IT 47%	−88.629	1/2−	
		118	5.0 s	β^-	−87.45	1+	
		118	4.4 m	β^-	−87.37	(5)+	
		118	8.5 s	IT 98.5%,β^- 1.5%	−87.23	(8)−	
		119	2.1 m	β^-	−87.730	9/2+	
		119m	18.0 m	β^- 95%,IT 5%	−87.419	1/2−	
		120	44 s	β^-	−85.8	(5)+	
		120	3.0 s	β^-	−85.5	1+	
		121	30.0 s	β^-	−85.842	9/2+	
		121m	3.8 m	β^- 98.8%,IT 1.2%	−85.528	1/2−	
		122	9.2 s	β^-	−83.4		
		122	1.5 s	β^-	−83.5	(1+)	
		123(g)	6.0 s	β^-	−83.44	(9/2)+	
		123(m)	48 s	β^-	−83.12	(1/2)−	
		124	2.4 s	β^-			
		124	3.2 s	β^-	−81.10	(2+)	
		125	2.32 s	β^-	−80.50	(9/2)+	
		125	12.2 s	β^-			
		126	1.53 s	β^-	−77.90		
		127	1.3 s	β^-	−77.36		
		127	3.7 s	β^-,β^-n	−77.36		
		128	12 s?	[β^-],β^-n			
		129	2.5 s	β^-,β^-n			
		129	0.99 s	β^-,β^-n	−73.12		
		130	0.58 s	β^-,β^-n	−70.08 s		
		131	0.29 s	β^-,β^-n	−69.8 s	(9/2+)	
		132	0.12 s	β^-,β^-n	−≈65 s		
50	Sn	106	1.9 m	EC+β^+	−76.99 s	0+	
		107	2.90 m	β^++EC	−78.40 s		
		108	10.5 m	EC	−81.90 s	0+	
		109	18.0 m	β^+,EC	−82.62 s	7/2+	
		109	1.5 m?	?			
		110	4.1 h	EC	−85.834	0+	
		111	35 m	EC 71%,β^+ 29%	−85.941	7/2+	
		112	*1.01%*		−88.658	0+	0.4_g^{rs} 0.3_m^{rs}
		113	115.1 d	EC	−88.332	1/2+	
		113m	21 m	IT 91%,EC 9%	−88.253	7/2+	
		114	*0.67%*		−90.560	0+	
		115	*0.38%*		−90.035	1/2+	50^{rs}
		116	*14.8%*		−91.526	0+	0.006_m^{rs}
		117	*7.75%*		−90.399	1/2+	3^{rs}
		117m	14.0 d	IT	−90.084	11/2−	
		118	*24.3%*		−91.654	0+	0.08_m^{rs}
		119	*8.6%*		−90.067	1/2+	2
		119m	≈250 d	IT	−89.977	11/2−	
		120	*32.4%*		−91.102	0+	0.16_g 0.001_m

Z	El	A	Abundance or $t_{1/2}$	Decay Mode	Δ(MeV)	Jπ	σ_n(b)
50	Sn	121	27.1 h	β^-, no γ	−89.202	3/2+	
		121m	55 y	β^-	−89.196	(11/2)−	
		122	*4.56%*		−89.946	0+	0.15_m 0.001_g
		123	129 d	β^-	−87.821	11/2−	
		123m	40.1 m	β^-	−87.796	(3/2)+	
		124	*5.64%*		−88.240	0+	0.13_m 0.005_g
		125	9.62 d	β^-	−85.903	11/2−	
		125m	9.5 m	β^-	−85.876	3/2+	
		126	≈1×10^5 y	β^-	−86.024	0+	
		127	2.1 h	β^-	−83.79	(11/2−)	
		127m	4.1 m	β^-	−83.78	(3/2)+	
		128	59.3 m	β^-	−83.44	0+	
		129	2.2 m	β^-	−80.64	(3/2+)	
		129m	7.5 m	β^-	−80.60	(11/2−)	
		130	3.7 m	β^-	−80.38	0+	
		130m	1.7 m	β^-		(7−)	
		131	63 s	β^-	−77.48 s	(3/2+)	
		132	40 s	β^-	−76.60	0+	
		133	1.47 s	β^-, β^-n	−71.5		
		134	1.04 s	β^-, β^-n ≈17%		0+	
51	Sb	108	7.0 s	β^+	−72.40 s	(3+)	
		109	18.3 s	β^++EC	−76.12 s		
		110	23 s	β^+ ≈92%, EC ≈8%	−76.75	(3)+	
		111	75 s	β^+, EC	−81.47	(5/2)+	
		112	54 s	β^+, EC	−81.63	(3+)	
		113	6.7 m	EC, β^+	−84.443	(5/2)+	
		114	3.5 m	β^+, EC	−84.14	(3)+	
		114	8 m	?			
		115	31.8 m	EC 67%, β^+ 33%	−87.005	5/2+	
		116	16 m	EC 72%, β^+ 28%	−86.93	3+	
		116m	60.4 m	EC 81%, β^+ 19%	−86.32	8−	
		117	2.80 h	EC 97.5%, β^+ 2.5%	−88.654	5/2+	
		118	3.5 m	EC, β^+	−87.967	1+	
		118	0.87 s	?			
		118m	5.00 h	EC 99.84%, β^+ 0.16%	−87.747	8−	
		119	38.0 h	EC	−89.483	5/2+	
		120	15.8 m	EC 56%, β^+ 44%	−88.421	.1+	
		120	5.76 d	EC		8−	
		121	*57.3%*		−89.588	5/2+	6.1_g 0.06_m
		122	2.68 d	β^- 97.0%, EC 3.0%, β^+ 0.0063%	−88.323	2−	
		122m	4.2 m	IT	−88.160	(8−)	
		123	*42.7%*		−89.218	7/2+	4.0_g 0.04_{m1}
		124	60.20 d	β^-	−87.613	3−	7^{rs}
		124m_1	93 s	IT 80%, β^- 20%	−87.603	(5)+	
		124m_2	20.2 m	IT	−87.578		
		125	2.7 y	β^-	−88.252	7/2+	
		126	12.4 d	β^-	−86.402	(8−)	
		126m	19.0 m	β^- 86%, IT 14%	−86.384	(5)+	
		127	3.9 d	β^-	−86.704	7/2+	
		128(g)	9.1 h	β^-	−≈84.75	8−	
		128(m)	10.0 m	β^- 96.4%, IT 3.6%	−84.73	5+	
		129	4.4 h	β^-	−84.630	7/2+	
		130	40 m	β^-	−82.38	(8−)	
		130	6.5 m	β^-		(4,5)+	
		131	23.03 m	β^-	−82.10 s	(7/2+)	
		132	2.8 m	β^-	−79.68	(4+)	
		132	4.2 m	β^-		(8−)	
		133	2.7 m	β^-	−78.98		
		134	10.4 s	β^-, β^-n 0.09%	−73.87 s		
		134	0.8 s	β^-, no γ	−73.87 s		
		135	1.70 s	β^-, β^-n 20%	−70.44 s	(7/2+)	
		136	0.82 s	β^-, β^-n 32%			
52	Te	107	2.1 s	α			

TABLE OF NUCLIDES

Z	El	A	Abundance or $t_{1/2}$	Decay Mode	Δ(MeV)	Jπ	σ_n(b)
52	Te	108	5.3 s	α,[β^++EC],(β^++EC)p	−65.32 s	0+	
		109	4.2 s	[β^++EC],(β^++EC)p,α	−67.47 s		
		111	19 s	β^++EC,(β^++EC)p	−74.10 s		
		113	2.0 m	β^++EC	−78.96		
		114	17 m	EC+β^+	−81.46 s	0+	
		115	6.0 m	β^+ ≈75%,EC ≈25%	−82.58	(7/2+)	
		115	7.5 m	β^++EC		(1/2+)	
		116	2.50 h	EC,β^+	−85.37	0+	
		117	62 m	EC 70%,β^+ 30%	−85.164	1/2+	
		118	6.00 d	EC,no γ	−87.671	0+	
		119	16.05 h	EC 97.2%,β^+ 2.8%	−87.189	1/2+	
		119m	4.68 d	EC	−≈86.89	11/2−	
		120	*0.091%*		−89.404	0+	2.0^{rs}_{g} 0.3^{rs}_{m}
		121	16.8 d	EC	−88.486	1/2+	
		121m	154 d	IT 90%,EC 10%, β^+ 0.002%	−88.192	11/2−	
		122	*2.5%*		−90.304	0+	3^{rs}_{g+m}
		123	*0.89%*		−89.166	1/2+	400^{rs}
		123m	119.7 d	IT	−88.918	11/2−	
		124	*4.6%*		−90.518	0+	7^{rs}_{g} 0.05^{rs}_{m}
		125	*7.0%*		−89.019	1/2+	1.6^{rs}
		125m	58 d	IT	−88.874	11/2−	
		126	*18.7%*		−90.066	0+	0.9^{rs}_{g} 0.13^{rs}_{m}
		127	9.4 h	β^-	−88.285	3/2+	
		127m	109 d	IT 97.6%,β^- 2.4%	−88.197	11/2−	
		128	*31.7%* 1.5×10^{24} y	$\beta^-\beta^-$	−88.992	0+	0.20_{g} 0.016_{m}
		129	69 m	β^-	−87.007	3/2+	
		129m	33.5 d	IT 63%,β^- 37%	−86.901	11/2−	
		130	*34.5%* 2×10^{21} y	$\beta^-\beta^-$	−87.348	0+	0.2^{rs}_{g} 0.03^{rs}_{m}
		131	25.0 m	β^-	−85.201	3/2+	
		131m	30 h	β^- 78%,IT 22%	−85.019	11/2−	
		132	78 h	β^-	−85.213	0+	
		133	12.4 m	β^-	−82.93	(3/2+)	
		133m	55.4 m	β^- 83%,IT 17%	−82.60	(11/2−)	
		134	42 m	β^-	−82.67 s	0+	
		135	19.2 s	β^-	−77.60		
		136	17.5 s	β^-,β^-n 0.7%	−74.83 s	0+	
		137	4 s	β^-,β^-n 2.5%			
		138	1.4 s	β^-,β^-n 6%		0+	
53	I	115	1.3 m	β^++EC	−76.78 s		
		116	2.9 s	β^++EC	−77.61	1+	
		117	2.2 m	EC 54%,β^+ 46%	−80.85		
		118	14.3 m	β^+ 54%,EC 46%	−80.60	(2−)	
		118m	8.5 m	β^+,EC,IT	−80.50		
		119	19.3 m	β^+ 51%,EC 49%	−83.82		
		120	1.35 h	EC 54%,β^+ 46%	−83.789	2−	
		120m	53 m	β^+,EC	−82.86		
		121	2.12 h	EC 94%,β^+ 6%	−86.12	5/2+	
		122	3.6 m	β^+ 77%,EC 23%	−86.16	1+	
		123	13.0 h	EC	−87.97	5/2+	
		124	4.2 d	EC 75%,β^+ 25%	−87.361	2−	
		125	60.2 d	EC	−88.841	5/2+	900^{sc}
		126	13.0 d	EC 53%,β^+ 1.0%, β^- 46%	−87.911	2−	$6\times10^{3\,rs}$
		127	*100%*		−88.980	5/2+	6.1
		128	24.99 m	β^- 94%,EC 6%, β^+ 0.003%	−87.734	1+	
		129	1.6×10^{7} y	β^-	−88.505	7/2+	18_{m} 9_{g}

TABLE OF NUCLIDES

Z	El	A	Abundance or $t_{1/2}$	Decay Mode	Δ(MeV)	Jπ	σ_n(b)
53	I	130	12.36 h	β^-	−86.897	5+	18^{rs}
		130m	9.2 m	IT 83%,β^- 17%	−86.849	2+	
		131	8.040 d	β^-	−87.451	7/2+	0.7_{g+m}
		132	2.28 h	β^-	−85.706	4+	
		132m	83 m	IT 86%,β^- 14%	−≈85.59	(8−)	
		133	20.9 h	β^-	−85.902	7/2+	
		133m	9 s	IT	−84.268	(19/2−)	
		134	52.6 m	β^-	−83.97	(4)+	
		134m	3.5 m	IT 98%,β^- 2%	−83.65	(8−)	
		135	6.61 h	β^-	−83.796	7/2+	
		136	46 s	β^-		(5,6−)	
		136	83 s	β^-	−79.43	(2−)	
		137	24.5 s	β^-,β^-n 6%	−76.72		
		138	6.5 s	β^-,β^-n 5%	−71.85 s	(2,3−)	
		139	2.3 s	β^-,β^-n 10%	−≈68.8 s		
		140	0.8 s	β^-,β^-n 14%			
		141	0.5 s	[β^-],β^-n ≈60%			
54	Xe	113	2.8 s	[β^++EC],(β^++EC)p	−71.86 s		
		115	18 s	β^+,EC,(β^++EC)p 0.3%	−68.87 s		
		116	57 s	β^++EC	−73.27	0+	
		117	61 s	EC 65%,β^+ 35%, (EC+β^+)p 0.003%	−74.48 s		
		118	6 m	EC 86%,β^+ 14%	−77.30 s	0+	
		119	6 m	EC 82%,β^+ 18%	−78.83		
		120	40 m	EC 97%,β^+ 3%	−81.84	0+	
		121	39 m	EC 92%,β^+ 8%	−82.33		
		122	20.1 h	EC	−85.16 s	0+	
		123	2.08 h	EC 87%,β^+ 13%	−85.29	(1/2+)	
		124	*0.096%*		−87.45	0+	100_g 20_m
		125	17 h	EC 99.7%,β^+ 0.3%	−87.11	(1/2)+	
		125m	57 s	IT	−86.86	(9/2)−	
		126	*0.090%*		−89.162	0+	3_g 0.4_m
		127	36.41 d	EC	−88.316	(1/2+)	
		127m	69 s	IT	−88.019	(9/2−)	
		128	*1.92%*		−89.861	0+	0.4_m $<8_a^{rs}$
		129	*26.4%*		−88.698	1/2+	20^{rs}
		129m	8.89 d	IT	−88.461	11/2−	
		130	*4.1%*		−89.881	0+	0.4_m $<26_a$
		131	*21.2%*		−88.421	3/2+	90^{rs}
		131m	11.77 d	IT	−88.257	11/2−	
		132	*26.9%*		−89.286	0+	0.4_g 0.03_m
		133	5.25 d	β^-	−87.662	3/2+	190_g^{rs}
		133m	2.19 d	IT	−87.429	11/2−	
		134	*10.4%*		−88.125	0+	0.25_g 0.003_m
		134m	0.29 s	IT	−86.160	(7−)	
		135	9.10 h	β^-	−86.506	3/2+	2.6×10^6
		135m	15.6 m	IT 99+%,β^- 0.004%	−85.979	11/2−	
		136	*8.9%*		−86.425	0+	0.16
		137	3.82 m	β^-	−82.215	(7/2)−	
		138	14.1 m	β^-	−80.15	0+	
		139	39.7 s	β^-	−75.75	(7/2−)	
		140	14 s	β^-	−73.18	0+	
		141	1.73 s	β^-,β^-n 0.05%	−69.00		
		142	1.2 s	β^-,β^-n 0.41%	−66.05	0+	
		143	0.30 s	β^-			
		143	0.96 s	β^-			
		144	1.2 s	β^-		0+	
		145	0.9 s	β^-			
55	Cs	116	3.9 s	β^++EC,(β^++EC)p 0.3%	−62.63 s		

TABLE OF NUCLIDES

Nuclide Z	El	A	Abundance or $t_{1/2}$	Decay Mode	Δ(MeV)	Jπ	σ_n(b)
55	Cs	117	8 s	β^++EC	−66.85 s		
		118	16 s	β^++EC, (β^++EC)p 0.04%, (β^++EC)α 0.0024%	−67.89 s		
		119	38 s	β^++EC	−72.53 s		
		120	60 s	β^++EC, (β^++EC)α 2.0×10^{-5}%, (β^++EC)p 7×10^{-6}%	−73.4		
		121	126 s	β^+,EC	−77.13 s		
		122	4.5 m	β^++EC			
		122	21 s	β^++EC	−78.01 s	(2,3+)	
		122	0.4 s	?			
		123	5.9 m	β^+,EC	−81.19	(1/2+)	
		123m	1.6 s	IT			
		124	31 s	β^+ ≈92%,EC ≈8%	−81.53	(1+)	
		125	45 m	EC 61%,β^+ 39%	−84.04	1/2+	
		126	1.64 m	β^+ 82%,EC 18%	−84.33	1+	
		127	6.2 h	EC 96.5%,β^+ 3.5%	−86.226	1/2+	
		128	3.6 m	β^+ 61%,EC 39%	−85.935	1+	
		129	32.3 h	EC 99+%,β^+ 0.0030%	−87.493	1/2+	
		130	29.9 m	(EC,β^+) 98.4%, β^- 1.6%	−86.863	1+	
		131	9.688 d	EC,no γ	−88.066	5/2+	
		132	6.47 d	EC 96.5%,β^+ 1.5%, β^- 2.0%	−87.175	2(−)	
		133	*100%*		−88.089	7/2+	27_g 2.5_m
		134	2.062 y	β^- 99+%,EC 3×10^{-4}%	−86.909	4+	140^{rs}_g
		134m	2.90 h	IT	−86.770	8−	
		135	3×10^6 y	β^-,no γ	−87.665	7/2+	9^{sc}
		135m	53 m	IT	−86.038	(19/2−)	
		136	13.1 d	β^-	−86.358	5+	
		136m	19 s	IT			
		137	30.17 y	β^-	−86.560	7/2+	0.11^{rs}_{g+m}
		138	32.2 m	β^-	−82.98	3−	
		138m	2.9 m	IT 75%,β^- 25%	−82.90	(6−)	
		139	9.5 m	β^-	−80.63	(7/2+)	
		140	65 s	β^-	−77.24	1,2−	
		141	24.9 s	β^-,β^-n 0.05%	−75.00		
		142	1.69 s	β^-,β^-n 0.28%	−70.95		
		143	1.78 s	β^-,β^-n 1.7%	−68.36 s		
		144	1.00 s	β^-,β^-n 3.0%	−63.93 s		
		145	0.58 s	β^-,β^-n 12%	−61.72 s		
		146	0.34 s	β^-,β^-n 14%			
56	Ba	117	1.9 s	[β^++EC],(β^++EC)p			
		119	5.3 s	(β^++EC)p	−64.53 s		
		120	32 s	β^++EC	−68.8 s	0+	
		121	30 s	β^++EC, (β^++EC)p 0.02%	−70.55 s		
		122	2.0 m	β^++EC	−74.26 s	0+	
		122	≈4 s	[β^++EC]			
		123	2.7 m	β^++EC	−75.69 s		
		124	11 m	EC+β^+	−78.75 s	0+	
		125	3.5 m	β^+,EC	−79.53		
		125	8 m	β^++EC			
		126	100 m	EC+β^+	−82.56 s	0+	
		127	13 m	β^+ ≈51%,EC ≈49%	−82.78	(1/2+)	
		127	18 m	β^++EC			
		128	2.43 d	EC	−85.482	0+	
		129	2.2 h	EC,β^+	−85.046	1/2+	
		129m	2.1 h	EC+β^+	−84.769	(11/2)−	
		130	*0.106%*		−87.303	0+	8^{rs}_g 2.5^{rs}_m
		131	12.0 d	EC	−86.726	1/2+	
		131m	14.6 m	IT	−86.538	9/2−	

Z	El	A	Abundance or $t_{1/2}$	Decay Mode	Δ(MeV)	Jπ	σ_n(b)
56	Ba	132	*0.101%*		−88.453	0+	7^{rs}_{g} 0.6^{rs}_{m}
		133	10.7 y	EC	−87.569	1/2+	
		133m	38.9 h	IT 99+%,EC 0.011%	−87.281	11/2−	
		134	*2.42%*		−88.968	0+	0.16^{rs}_{m} $\approx 2_{a}$
		135	*6.59%*		−87.870	3/2+	6^{rs}_{g} 0.014^{rs}_{m}
		135m	28.7 h	IT	−87.602	11/2−	
		136	*7.85%*		−88.906	0+	0.011^{rs}_{m} 0.4_{a}
		136m	0.31 s	IT	−86.876	7−	
		137	*11.2%*		−87.733	3/2+	5.1^{rs}
		137m	2.551 m	IT	−87.071	11/2−	
		138	*71.7%*		−88.273	0+	0.4^{rs}
		139	82.9 m	β^-	−84.925	(7/2)−	6^{rs}
		140	12.79 d	β^-	−83.285	0+	1.6
		141	18.2 m	β^-	−79.98		
		142	10.6 m	β^-	−77.82	0+	
		143	13.5 s	β^-	−74.01 s		
		144	11.9 s	β^-	−72.03 s	0+	
		145	5 s	β^-	−67.82 s		
		146	1.7 s	β^-	−65.56 s	0+	
		148	0.5 s	β^-		0+	
57	La	125	<1 m?	?			
		126	1.0 m	β^++EC			
		127	3.8 m	β^++EC	−77.78 s		
		128	4.6 m	β^++EC	−78.68 s		
		129	10 m	β^++EC	−81.05 s	(3/2+)	
		129m	0.56 s	IT	−80.88 s	(11/2−)	
		130	8.7 m	β^+,EC	−81.60 s	(3+)	
		131	61 m	EC 76%,β^+ 24%	−83.77	3/2+	
		132	4.8 h	β^+,EC	−83.74	2−	
		132m	24.3 m	IT 76%,EC+β^+ 24%	−83.55	6−	
		133	3.91 h	EC,β^+	−85.57 s	5/2+	
		134	6.67 m	β^+ 62%,EC 38%	−85.268	1+	
		135	19.4 h	EC 99+%,β^+ 0.009%	−86.670	5/2+	
		136	9.87 m	EC 64%,β^+ 36%	−86.04	1+	
		137	6×10^4 y	EC,no γ	−87.13 s	7/2+	
		138	*0.089%* 1.1×10^{11} y	EC 68%,β^- 32%	−86.524	5+	57
		139	*99.911%*		−87.231	7/2+	9.2
		140	40.3 h	β^-	−84.320	3−	2.7^{sc}
		141	3.90 h	β^-	−83.008		
		142	93 m	β^-	−80.018	2−	
		143	14.0 m	β^-	−78.31		
		144	40 s	β^-	−74.93 s		
		145	30 s	β^-	−72.92 s		
		146	11 s	β^-	−69.46 s		
		148	1.3 s	β^-	−63.99 s		
58	Ce	128	≈6 m	[EC+β^+]		0+	
		129	3.5 m	β^++EC			
		130	25 m	EC+β^+		0+	
		131	5 m	EC+β^+	−79.47 s		
		131	10 m	EC 89%,β^+ 11%	−79.47 s		
		132	3.5 h	EC	−82.34 s	0+	
		133	97 m	EC+β^+	−82.17 s	1/2(+)	
		133	5.4 h	EC,β^+	−82.17 s	9/2−	
		134	76 h	EC	−84.77 s	0+	
		135	17.8 h	EC 99%,β^+ 1%	−84.55	1/2(+)	
		135m	20 s	IT	−84.10	(11/2−)	
		136	*0.190%*		−86.50	0+	6^{rs}_{g} 1.0^{rs}_{m}
		137	9.0 h	EC 99+%,β^+ 0.014%	−85.91 s	3/2+	

Nuclide Z	El	A	Abundance or $t_{1/2}$	Decay Mode	Δ(MeV)	Jπ	σ_n(b)
58	Ce	137m	34.4 h	IT 99.2%,EC 0.8%	−85.66 s	11/2−	
		138	*0.254%*		−87.565	0+	1.0_g^{rs} 0.015_m^{rs}
		139	137.2 d	EC	−86.966	3/2+	
		139m	56 s	IT	−86.212	11/2−	
		140	*88.5%*		−88.081	0+	0.56^{rs}
		141	32.5 d	β^-	−85.438	7/2−	29^{rs}
		142	*11.1%*		−84.535	0+	0.95
		143	33.0 h	β^-	−81.610	3/2−	6^{rs}
		144	284 d	β^-	−80.431	0+	1.0
		145	3.0 m	β^-	−77.12		
		146	14 m	β^-	−75.76	0+	
		147	56 s	β^-	−72.24 s		
		148	48 s	β^-	−70.81 s	0+	
		149	5.0 s	β^-	−67.47 s		
		150	4 s	β^-	−65.3 s	0+	
		151	1.0 s	β^-	−62.68 s		
59	Pr	121	1 s	[β^++EC],(β^++EC)p			
		129	24 s	β^++EC			
		130	28 s	β^+,EC			
		132	1.6 m	β^++EC	−75.34 s		
		133	6.5 m	β^+,EC	−77.97 s	5/2(+)	
		134	17 m	β^++EC	−78.47 s	2+	
		134	≈11 m	β^++EC	−78.47 s		
		135	25 m	EC ≈75%,β^+ ≈25%	−80.99	3/2(+)	
		136	13.1 m	β^+,EC	−81.40	2+	
		137	1.28 h	EC 75%,β^+ 25%	−83.21 s	5/2+	
		138	1.4 m	β^+,EC	−83.128	1+	
		138m	2.1 h	EC 77%,β^+ 23%	−82.765	7−	
		139	4.4 h	EC 92%,β^+ 8%	−84.854	5/2+	
		139	≈6 m?	?			
		140	3.39 m	EC 51%,β^+ 49%	−84.693	1+	
		141	*100%*		−86.018	5/2+	8_g 3.9_m
		142	19.2 h	β^- 99+%,EC 0.016%	−83.790	2−	20^{rs}
		142m	14.6 m	IT	−83.786	5−	
		142	1.6 m?	?			
		143	13.58 d	β^-	−83.065	7/2+	90_{g+m}
		144	17.3 m	β^-	−80.750	0−	
		144m	7.2 m	IT 99.96%,β^- 0.04%	−80.691	3−	
		145	5.98 h	β^-	−79.625	(7/2+)	
		146	24.0 m	β^-	−76.84	(1,2−)	
		147	13 m	β^-	−75.44		
		148	2.30 m	β^-	−72.61	(3)	
		149	2.3 m	β^-	−71.37	(5/2+)	
		150	6.2 s	β^-	−68.0		
		150	30 s?	β^-			
		151	4 s	β^-	−67.44 s		
60	Nd	129	6 s	[β^++EC],(β^++EC)p			
		130	28 s	β^+,EC		0+	
		132	1.8 m	β^++EC		0+	
		133	1.2 m	β^++EC			
		134	8 m	EC+β^+		0+	
		135	12 m	β^++EC	−76.29 s	9/2(−)	
		135	5.5 m	[β^++EC]	−76.29 s		
		136	50.6 m	EC 94%,β^+ 6%	−79.19	0+	
		137	38 m	β^+,EC	−79.41 s	1/2+	
		137m	1.6 s	IT	−78.89 s	11/2−	
		137	≈22 m	?			
		138	5.1 h	EC	−82.03 s	0+	
		139	30 m	EC 74.4%,β^+ 25.6%	−82.05	3/2+	
		139m	5.5 h	EC 87%,β^+ ≈1%, IT 12%	−81.82	11/2−	
		140	3.37 d	EC,no γ	−84.22	0+	
		141	2.5 h	EC 97.3%,β^+ 2.7%	−84.203	3/2+	
		141m	61 s	IT 99.97%, EC+β^+ 0.03%	−83.446	11/2−	

TABLE OF NUCLIDES

Z	El	A	Abundance or $t_{1/2}$	Decay Mode	Δ(MeV)	Jπ	σ_n(b)
60	Nd	142	*27.2%*		−85.949	0+	19^{rs}
		143	*12.2%*		−84.000	7/2−	320
		144	*23.8%* 2.1×10^{15} y	α	−83.746	0+	4^{rs}
		145	*8.3%*		−81.430	7/2−	41
		146	*17.2%*		−80.923	0+	1.3
		147	11.0 d	β^-	−78.144	5/2−	440^{rs}
		148	*5.7%*		−77.407	0+	2.5
		149	1.73 h	β^-	−74.374	5/2−	
		150	*5.6%*		−73.682	0+	1.2
		151	12.4 m	β^-	−70.945	(3/2+)	
		152	11.4 m	β^-	−70.146	0+	
		154	40 s	β^-		0+	
61	Pm	132	4 s	β^++EC			
		133	12 s	β^++EC			
		134	24 s	β^++EC			
		135	0.9 m	β^++EC		(11/2−)	
		136	107 s	β^++EC	−71.36 s	(5+)	
		137	2.4 m	β^++EC	−74.21 s	(11/2−)	
		138	3.5 m	β^++EC	−75.03 s	(3+)	
		139	4.15 m	β^+,EC	−77.60	(5/2)+	
		140	9.2 s	β^+,EC	−78.18	1+	
		140m	5.9 m	EC 58%,β^+ 42%	−77.78	(7−)	
		141	20.9 m	β^+ 57%,EC 43%	−80.47	5/2+	
		142	40.5 s	β^+ 69%,EC 31%	−81.06	1+	
		143	265 d	EC	−82.959	5/2+	
		144	349 d	EC	−81.416	5−	
		145	17.7 y	EC 99+%,α 2.8×10^{-7}%	−81.270	5/2+	
		146	5.5 y	EC 63%,β^- 37%	−79.442	3−	$8\times10^{3\,rs}$
		147	2.6234 y	β^-	−79.040	7/2+	97^{g} 85^{m}
		148	5.37 d	β^-	−76.870	1−	$<3\times10^{3\,rs}$
		148m	41.3 d	β^- 95%,IT 5%	−76.733	6−	1.06×10^{4}
		149	53.1 h	β^-	−76.063	7/2+	$1.4\times10^{3\,rs}$
		150	2.68 h	β^-	−73.55	(1−)	
		151	28.4 h	β^-	−73.386	5/2+	$<700^{rs}$
		152	4.1 m	β^-	−71.29	(1+)	
		152	7.5 m	β^-		(4)	
		152	15 m	β^-		(≥6)	
		153	5.4 m	β^-	−70.76	(5/2−)	
		154	1.7 m	β^-	−68.45	(0,1)	
		154	2.7 m	β^-		(3,4)	
62	Sm	133	32.0 s	β^++EC,(β^++EC)p			
		134	12 s	β^++EC		0+	
		135	10 s	β^++EC,(β^++EC)p			
		137	44 s	β^++EC			
		138	3.0 m	β^++EC		0+	
		139	2.5 m	β^+,EC	−72.40		
		139m	10 s	IT 93.7%,β^++EC 6.3%	−71.94	(11/2)−	
		140	14.8 m	EC,β^+	−75.48 s	0+	
		141	10.2 m	EC 53%,β^+ 47%	−75.91	1/2+	
		141m	22.5 m	(β^+,EC) 99.69%, IT 0.31%	−75.73	11/2−	
		142	72.49 m	EC 90%,β^+ 10%	−78.978	0+	
		143	8.83 m	EC 54%,β^+ 46%	−79.511	3/2+	
		143m	66 s	IT 99.80%, β^++EC 0.20%	−78.757	11/2−	
		144	*3.1%*		−81.964	0+	0.7^{rs}
		145	340 d	EC	−80.656	7/2−	110^{rs}
		146	*<2×10⁻⁷%* 1.03×10^{8} y	α	−80.984	0+	
		147	*15.1%* 1.06×10^{11} y	α	−79.265	7/2−	60
		148	*11.3%* 8×10^{15} y	α	−79.335	0+	4.7
		149	*13.9%*		−77.135	7/2−	4.2×10^{4}
		150	*7.4%*		−77.049	0+	104

Z	El	A	Abundance or $t_{1/2}$	Decay Mode	Δ(MeV)	Jπ	σ_n(b)
62	Sm	151	90 y	β^-	−74.574	5/2−	1.5×10^4 rs
		152	*26.6%*		−74.761	0+	204
		153	46.8 h	β^-	−72.557	3/2+	
		154	*22.6%*		−72.454	0+	5
		155	22.4 m	β^-	−70.196	3/2−	
		156	9.4 h	β^-	−69.368	0+	
		157	8.0 m	β^-	−66.86		
63	Eu	138	1.5 s	β^+			
		138	35 s	β^+			
		139	22 s	β^++EC			
		140	1.3 s	β^+,EC			
		140	≈20 s	β^++EC			
		141	40 s	β^+,EC	−69.88	(5/2+)	
		141m	3.3 s	β^++EC 67%,IT 33%	−69.78	(11/2−)	
		142	2.4 s	β^+,EC	−71.48 s	1+	
		142	1.22 m	β^+,EC	−71.48 s	(7−)	
		143	2.61 m	β^+ ≈72%,EC ≈28%	−74.41	(5/2)+	
		144	10.2 s	β^+ ≈80%,EC ≈20%	−75.636	1+	
		145	5.93 d	EC 98%,β^+ 2%	−77.936	5/2+	
		146	4.62 d	EC 96.1%,β^+ 3.9%	−77.111	4−	
		146	38 h?	?			
		147	22 d	EC 99.5%,β^+ 0.5%, α 0.002%	−77.535	5/2+	
		148	54 d	EC 99.8%,β^+ 0.2%, α 9×10^{-7}%	−76.235	5−	
		149	93.1 d	EC	−76.439	5/2+	
		150	36 y	EC		(4,5−)	
		150	12.6 h	β^- 89%,EC 10.6%, β^+ ≈0.6%	−74.756	0(−)	
		151	*47.9%*		−74.650	5/2+	5.8×10^3 g 3.2×10^3 m1 4 m2
		152	13 y	EC 73.0%, β^+ 0.019%,β^- 27.0%	−72.884	3−	
		$152m_1$	9.3 h	β^- 76%,EC 24%, β^+ 0.011%	−72.836	0−	<3 rs
		$152m_2$	96 m	IT	−72.736	8−	
		153	*52.1%*		−73.363	5/2+	380 g
		154	8.5 y	β^- 99.98%,EC 0.02%	−71.726	3−	
		154m	46 m	IT	−≈71.57	(8−)	
		155	4.9 y	β^-	−71.825	5/2+	4.0×10^3
		156	15 d	β^-	−70.083	0+	
		157	15.13 h	β^-	−69.465	(5/2+)	
		158	45.9 m	β^-	−67.24	(1−)	
		159	18.1 m	β^-	−65.93	(5/2+)	
		160	0.8 m	β^-	−63.54 s	(0−)	
		160	≈2.5 m?	β^-			
64	Gd	143	1.83 m	β^++EC	−68.51 s	(11/2,13/2−)	
		143	39 s?	?			
		144	4.5 m	β^++EC	−71.94 s	0+	
		145	22 m	β^+,EC	−72.94 s	1/2+	
		145m	85 s	IT 95.3%,β^++EC 4.7%	−72.19 s	11/2−	
		146	48.3 d	EC 99.93%,β^+ 0.07%	−75.361	0+	
		146	7 h?	α,EC			
		147	38.1 h	EC 99.74%,β^+ 0.26%	−75.207	7/2−	
		148	98 y	α	−76.268	0+	
		149	9.3 d	EC 99+%,α 5×10^{-4}%	−75.131	7/2−	
		150	1.8×10^6 y	α	−75.765	0+	
		151	120 d	EC 99+%,α ≈8×10^{-7}%	−74.168	7/2−	
		152	*0.20%* 1.1×10^{14} y	α	−74.703	0+	1.1×10^3
		153	241.6 d	EC	−73.119	3/2−	
		154	*2.1%*		−73.704	0+	90
		155	*14.8%*		−72.071	3/2−	6.1×10^4
		156	*20.6%*		−72.536	0+	2
		157	*15.7%*		−70.825	3/2−	2.55×10^5
		158	*24.8%*		−70.691	0+	2.4
		159	18.6 h	β^-	−68.562	3/2−	

TABLE OF NUCLIDES

Nuclide Z	El	A	Abundance or $t_{1/2}$	Decay Mode	Δ(MeV)	Jπ	σ_n(b)
64	Gd	160	*21.8%*		−67.943	0+	0.77
		161	3.7 m	β^-	−65.507	5/2−	4×10^4 rs
		162	9 m	β^-	−64.36	0+	
65	Tb	146	23 s	β^++EC	−67.26 s	(4−)	
		147	1.6 h	EC 95%,β^+ 5%	−70.51 s	5/2+	
		147	1.9 m	EC,β^+	−70.51 s	11/2−	
		148	2.2 m	EC+β^+		(9+)	
		148	60 m	EC 80%,β^+ 20%	−70.64	2−	
		149	4.15 h	EC 79%,β^+ 4.0%, α 17%	−71.434	(3/2,5/2+)	
		149m	4.2 m	(EC,β^+) 99+%, α 0.020%	−71.394	(11/2−)	
		150	3.3 h	EC 90%,β^+ 10%, α ≤0.05%	−71.098	(2)−	
		150	6.0 m	EC+β^+		(8,9+)	
		151	17.6 h	EC 99%,β^+ ≈1%, α 0.009%	−71.608	1/2(+)	
		152	17.5 h	EC 87%,β^+ 13%	−70.853	2−	
		152m	4.2 m	IT 78%,EC 22%	−70.351	(8+)	
		153	2.30 d	EC 99+%,β^+ 0.04%	−71.329	5/2+	
		154	21 h	EC 98%,β^+ 2%	−70.24	0(+)	
		$154m_1$	9 h	EC+β^+ 78%,IT 22%		3(−)	
		$154m_2$	23 h	EC 98%,IT 2%		(7,8−)	
		155	5.3 d	EC	−71.256	3/2+	
		156	5.3 d	EC	−70.098	3−	
		156m	5.0 h	IT,EC,β^+ 0.02%, β^-(weak)	−70.010	(0)+	
		156	24 h	IT			
		157	150 y	EC	−70.767	3/2+	
		158	150 y	EC 82%,β^- 18%	−69.475	3−	
		158m	10.5 s	IT	−69.365	0−	
		159	*100%*		−69.536	3/2+	23
		160	72.1 d	β^-	−67.840	3−	500 rs
		161	6.90 d	β^-	−67.466	3/2+	
		162	7.7 m	β^-	−65.76	(1)−	
		163	19.5 m	β^-	−64.68	3/2+	
		164	3.0 m	β^-	−62.11	(5+)	
66	Dy	147m	59 s	IT,EC?	−63.46 s		
		148	3.1 m	EC+β^+	−67.77 s	0+	
		149	4.1 m	β^++EC	−67.53 s	(7/2−)	
		150	7.17 m	(EC,β^+) 69%,α 31%	−69.14 s	0+	
		151	17 m	EC+β^+ 94%,α 6%	−68.601	7/2−	
		152	2.37 h	EC 99.91%,α 0.09%	−70.116	0+	
		153	6.3 h	(EC,β^+) 99+%, α 0.010%	−69.155	7/2(−)	
		154	≈1×10^7 y	α	−70.392	0+	
		155	10.0 h	EC 97%,β^+ 3%	−69.157	3/2−	
		156	*0.057%*		−70.527	0+	33 sc
		157	8.1 h	EC	−69.425	3/2−	
		158	*0.100%*		−70.410	0+	70 rs
		159	144.4 d	EC	−69.171	3/2−	
		160	*2.3%*		−69.674	0+	60
		161	*19.0%*		−68.056	5/2+	570
		162	*25.5%*		−68.181	0+	160
		163	*24.9%*		−66.382	5/2−	130
		164	*28.1%*		−65.967	0+	1.8×10^3 m 900 g
		165	2.33 h	β^-	−63.611	7/2+	4.0×10^3 rs
		165m	1.26 m	IT 97.8%,β^- 2.2%	−63.503	1/2−	2.1×10^3 rs
		166	81.5 h	β^-	−62.583	0+	
		167	6.2 m	β^-	−59.97	(1/2−)	
67	Ho	150	40 s	β^++EC	−62.04 s	(8,9+)	
		151	47 s	β^++EC 90%,α 10%	−63.44 s		
		151	35.6 s	β^++EC 80%,α 20%	−63.44 s		
		152	52 s	β^++EC 94%,α 6%		(9+)	
		152	2.4 m	β^++EC 98.3%,α 1.7%	−63.71	(3+)	

TABLE OF NUCLIDES

Z	El	A	Abundance or $t_{1/2}$	Decay Mode	Δ(MeV)	Jπ	σ_n(b)
67	Ho	153	2.0 m	EC+β^+ 99.96%, α 0.04%			
		153	9.3 m	EC+β^+ 99.9%,α 0.1%	−64.954		
		153	27 m?	α			
		154	12 m	β^++EC 99+%,α 0.017%	−64.635	1	
		154	3.2 m	EC+β^+ 99+%, α <0.002%			
		155	49 m	EC,β^+,α	−66.055	5/2	
		156	56 m	β^++EC,IT		1	
		156	2 m	β^++EC	−65.43 s	(5+)	
		156	7.4 m?	β^++EC			
		157	12.6 m	β^+,EC	−66.89	7/2−	
		158	11.5 m	EC,β^+	−66.433	5+	
		$158m_1$	27 m	IT 65%,EC+β^+ 35%	−66.366	2−	
		158(m_2)	21 m	EC+β^+		(9+)	
		159	33 m	EC	−67.318	7/2−	
		159m	8.3 s	IT	−67.112	1/2+	
		160	25.6 m	EC 99+%,β^+ ≈0.4%	−66.388	5+	
		160m	5.02 h	IT 65%,(EC,β^+) 35%	−66.328	2−	
		160	3 s	?			
		160	7 m	?		(1+)	
		160	≈1 h	?		(9+)	
		161	2.48 h	EC	−67.203	7/2−	
		161m	6.7 s	IT	−66.992	1/2+	
		162	15 m	EC 95%,β^+ 5%	−66.047	1+	
		162m	68 m	IT 61%,EC+β^+ 39%	−≈65.94	6−	
		163	≈33 y	EC,no γ	−66.379	7/2−	
		163m	1.09 s	IT	−66.081	1/2+	
		164	29.0 m	EC 58%,β^- 42%	−64.937	1+	
		164m	37 m	IT	−64.797	6(−)	
		165	*100%*		−64.896	7/2−	62_g 3_m
		166	26.80 h	β^-	−63.067	0−	
		166m	1.2×10^3 y	β^-	−63.062	(7−)	
		167	3.1 h	β^-	−62.316	(7/2−)	
		168	3.0 m	β^-	−60.27	3+	
		169	4.6 m	β^-	−58.793	(7/2−)	
		170	43 s	β^-	−56.10		
		170	2.8 m	β^-	−56.09		
68	Er	151	23 s	β^++EC	−58.20 s		
		152	10 s	α ≈90%,EC+β^+ ≈10%	−60.41 s	0+	
		153	36 s	EC+β^+ ≈62%,α ≈38%	−60.31 s		
		154	3.8 m	EC+β^+ 99.5%,α 0.5%	−62.44 s	0+	
		155	5 m	EC+β^+ 99+%,α ≥0.02%	−62.057		
		156	20 m	EC+β^+	−63.93 s	0+	
		157	24 m	β^+,EC	−63.09 s	3/2−	
		158	2.4 h	EC,β^+	−65.03 s	0+	
		159	36 m	EC,β^+	−64.39	3/2−	
		160	28.6 h	EC,no γ	−66.052	0+	
		161	3.24 h	EC 99.96%,β^+ 0.04%	−65.197	3/2−	
		162	*0.14%*		−66.335	0+	19
		163	75.1 m	EC 99+%,β^+ 0.004%	−65.168	5/2−	
		164	*1.56%*		−65.940	0+	13
		165	10.4 h	EC,no γ	−64.518	5/2−	
		166	*33.4%*		−64.921	0+	15_m^{rs} 5_g^{rs}
		167	*22.9%*		−63.286	7/2+	650^{rs}
		167m	2.28 s	IT	−63.078	1/2−	
		168	*27.1%*		−62.985	0+	2.0
		169	9.40 d	β^-	−60.917	1/2−	
		170	*14.9%*		−60.104	0+	5.7
		171	7.52 h	β^-	−57.714	5/2−	300^{rs}
		172	49.5 h	β^-	−56.491	0+	
		173	1.4 m	β^-	−53.73	(7/2−)	
		173	12 m	β^-			
69	Tm	153	1.6 s	α	−53.87 s		
		154	5 s	α	−54.53 s		
		154	3.0 s	α	−54.53 s		

TABLE OF NUCLIDES

Z	El	A	Abundance or $t_{1/2}$	Decay Mode	Δ(MeV)	Jπ	σ_n(b)
69	Tm	155	39 s	α	−56.45 s		
		156	80 s	α,β^++EC	−56.94 s		
		156	19 s	α	−56.94 s		
		157	3.6 m	EC+β^+	−58.49 s		
		158	4.0 m	β^++EC	−58.43 s		
		159	9.0 m	EC,β^+	−60.19 s	5/2(+)	
		160	9.2 m	EC 85%,β^+ 15%	−60.13	1−	
		161	30 m	EC,β^+	−61.68 s	7/2(+)	
		161	7 m?	?			
		162	22 m	EC 93%,β^+ 7%	−61.54	1−	
		162m	24 s	IT 90%,EC+β^+ 10%		(5+)	
		163	1.8 h	EC 99.8%,β^+ 0.2%	−62.99	1/2+	
		163m	11 m	IT?,EC?			
		164	2.0 m	EC 61%,β^+ 39%	−61.978	1+	
		164m	5.1 m	IT ≈80%,EC+β^+ ≈20%		6(−)	
		165	30.06 h	EC 99+%,β^+ 0.007%	−62.924	1/2+	
		166	7.7 h	EC 98%,β^+ 2%	−61.874	2+	
		167	9.25 d	EC	−62.537	1/2+	
		168	93.1 d	EC ≈98%,β^-? ≈2%	−61.306	3(+)	
		169	*100%*		−61.269	1/2+	98_{g}
		170	128.6 d	β^- 99+%,EC 0.144%	−59.791	1−	92^{sc}
		171	1.92 y	β^-	−59.205	1/2+	4.5^{sc}
		172	63.6 h	β^-	−57.380	2−	
		173	8.2 h	β^-	−56.226	(1/2+)	
		174	5.4 m	β^-	−53.85	(4−)	
		175	15 m	β^-	−52.29	(1/2+)	
		176	1.9 m	β^-	−49.59 s	(4+)	
		176	1.5 m?	β^-			
70	Yb	154	0.39 s	α	−50.05 s	0+	
		155	1.6 s	α	−50.45 s		
		156	24 s	α	−53.06 s	0+	
		157	34 s	α	−53.27 s		
		158	1.1 m	EC+β^+	−55.53 s	0+	
		160	4.8 m	EC+β^+	−57.55 s	0+	
		161	4.2 m	β^++EC	−57.40 s		
		162	18.9 m	EC ≥98%,β^+ ≤2%	−59.34 s	0+	
		163	11.0 m	EC+β^+	−59.62	(3/2−)	
		164	76 m	EC	−60.88 s	0+	
		165	10 m	EC,β^+	−60.161	(5/2)−	
		166	56.7 h	EC	−61.582	0+	
		167	17.5 m	EC 99.6%,β^+ 0.4%	−60.583	5/2−	
		168	*0.135%*		−61.565	0+	$3.5\times10^{3}_{g+m}$
		169	32.0 d	EC	−60.361	7/2+	
		169m	46 s	IT	−60.337	1/2−	
		170	*3.1%*		−60.759	0+	10
		171	*14.4%*		−59.302	1/2−	53
		172	*21.9%*		−59.250	0+	1
		173	*16.2%*		−57.546	5/2−	17
		174	*31.6%*		−56.940	0+	19_{g}
		175	4.19 d	β^-	−54.691	7/2−	
		176	*12.6%*		−53.490	0+	2.4_{g+m}
		176m	11.7 s	IT	−52.439	(8)−	
		177	1.9 h	β^-	−50.986	9/2+	
		177m	6.4 s	IT	−50.655	1/2−	
		178	74 m	β^-	−49.66	0+	
71	Lu	155	0.07 s	α	−42.60 s		
		156	0.23 s	α	−43.81 s		
		156	≈0.5 s	α	−43.81 s		
		162	1.4 m?	β^++EC	−52.34 s		
		164	3.17 m	β^++EC	−54.58 s		
		165	11.8 m	EC+β^+	−56.16 s	1/2	
		166	2.6 m	EC,β^+	−56.10	(6−)	
		$166m_1$	1.4 m	EC+β^+ 58%,IT 42%	−56.07	(3−)	
		$166m_2$	2.1 m	EC+β^+ >80%	−56.06	(0−)	
		167	52 m	EC 98.2%,β^+ 1.8%	−57.45	7/2+	
		168	5.5 m	EC,β^+	−57.10	(6)−	
		168m	6.7 m	EC ≈88%,β^+ ≈12%	−56.88	3+	
		169	34.1 h	EC 99.3%,β^+ 0.7%	−57.881	7/2+	
		169m	2.7 m	IT	−57.852	1/2−	

TABLE OF NUCLIDES

Z	El	A	Abundance or $t_{1/2}$	Decay Mode	Δ(MeV)	Jπ	σ_n(b)
71	Lu	170	2.02 d	EC,β^+	−57.319	0+	
		170m	0.7 s	IT	−57.226	4−	
		171	8.25 d	EC 99+%,β^+ ≈0.005%	−57.821	7/2+	
		171m	79 s	IT	−57.750	1/2−	
		172	6.70 d	EC	−56.726	4−	
		172m	3.7 m	IT	−56.684	1−	
		173	1.37 y	EC	−56.871	7/2+	
		174	3.3 y	EC 99+%,β^+ 0.025%	−55.562	(1−)	
		174m	142 d	IT 99.3%,EC 0.7%	−55.391	(6−)	
		175	*97.39%*		−55.159	7/2+	16_m 10_g
		176	*2.61%* 3.6×10^{10} y	β^-	−53.381	7−	$2.0\times10^3{}_g$ 7_m
		176m	3.68 h	β^-	−53.254	1−	
		177	6.71 d	β^-	−52.382	7/2+	
		177m	160.5 d	β^- 78%,IT 22%	−51.412	23/2−	
		178	28.4 m	β^-	−50.30	1+	
		178m	23 m	β^-	−≈50.00	(9)−	
		178	5 m?	β^-			
		179	4.6 h	β^-	−49.11	(7/2+)	
		180	5.7 m	β^-	−46.68		
72	Hf	157	0.12 s	α	−38.96 s		
		158	3.0 s	α	−42.22 s	0+	
		159	5.6 s	α	−42.80 s		
		160	≈12 s	α	−45.75 s	0+	
		161	17 s	α	−46.13 s		
		166	6.8 m	EC+β^+	−53.48 s	0+	
		167	2.05 m	EC,β^+	−53.15 s	(5/2−)	
		168	25.9 m	EC ≈98%,β^+? ≈2%	−55.10 s	0+	
		169	3.3 m	EC 86%,β^+ 14%	−54.53	(5/2)−	
		170	16.0 h	EC	−56.12 s	0+	
		171	12.1 h	EC+β^+	−55.30 s	7/2+	
		172	1.87 y	EC	−56.33 s	0+	
		173	24.0 h	EC	−55.27 s	1/2−	
		174	*0.16%* 2.0×10^{15} y	α	−55.830	0+	400
		175	70 d	EC	−54.548	5/2−	
		176	*5.2%*		−54.567	0+	30_g 390_g
		177	*18.6%*		−52.879	7/2−	1.0_{m1} $2\times10^{-7}{}^{rs}_{m2}$
		177m_1	1.1 s	IT	−51.564	23/2+	
		177m_2	51 m	IT	−50.139	37/2−	
		178	*27.1%*		−52.434	0+	50_{m1} 40_g
		178m_1	4.0 s	IT	−51.287	8−	
		178m_2	31 y	IT	−49.987	16+	
		179	*13.7%*		−50.462	9/2+	50_g 0.4^{sc}_m
		179m_1	18.7 s	IT	−50.087	1/2−	
		179m_2	25.1 d	IT	−49.356	25/2−	
		180	*35.2%*		−49.779	0+	14
		180m	5.5 h	IT	−48.637	8−	
		181	42.4 d	β^-	−47.403	1/2−	30^{rs}_g
		182	9×10^6 y	β^-	−45.99	0+	
		182m	62 m	β^- 54%,IT 46%	−44.82	(8−)	
		183	64 m	β^-	−43.269	(3/2−)	
		184	4.12 h	β^-	−41.48	0+	
73	Ta	166	32 s	β^++EC	−46.10 s		
		167	3 m	β^++EC	−47.95 s		
		168	2.4 m	β^++EC	−48.40 s	(2−,3+)	
		169	5 m	EC+β^+	−50.03 s		
		170	6.8 m	β^+,EC	−50.12 s	(3+)	
		171	24 m	EC+β^+	−51.60 s		
		171	2.0 m?	?			
		171	6.3 m?	?			

TABLE OF NUCLIDES

Z	El	A	Abundance or $t_{1/2}$	Decay Mode	Δ(MeV)	Jπ	σ_n(b)
73	Ta	172	37 m	EC ≈85%,β^+ ≈15%	−51.41 s	(3−)	
		173	3.6 h	EC,β^+	−52.37 s	(5/2−)	
		174	1.1 h	EC,β^+	−51.98	3(+)	
		175	10.5 h	EC,β^+	−52.35 s	7/2+	
		176	8.1 h	EC 99.3%,β^+ 0.7%	−51.47	(1−)	
		177	56.6 h	EC 99+%, β^+ 2.9×10^{-4}%	−51.721	7/2+	
		178	9.3 m	EC 98.9%,β^+ 1.1%	−50.52	1+	
		178	2.4 h	EC		(7)−	
		179	1.7 y	EC,no γ	−50.347	(7/2+)	
		180(g)	*0.0123%* $\gtrsim1\times10^{13}$ y			(8+)	700
		180(m)	8.1 h	EC 87%,β^- 13%	−48.914	1	
		181	*99.9877%*		−48.425	7/2+	21_g 0.010_{m2}
		182	115 d	β^-	−46.417	3−	8.2×10^3
		$182m_1$	0.28 s	IT	−46.400	5+	
		$182m_2$	15.8 m	IT	−45.897	10−	
		183	5.1 d	β^-	−45.279	7/2+	
		184	8.7 h	β^-	−42.821	(5−)	
		185	49 m	β^-	−41.360	(7/2+)	
		186	10.5 m	β^-	−38.60	(3−)	
74	W	160	≤0.2 s?	α		0+	
		162	<0.25 s	α	−34.13 s	0+	
		163	2.5 s	α,(β^++EC)?	−35.31 s		
		164	6 s	α	−38.04 s	0+	
		165	5.1 s	α	−38.67 s		
		166	16 s	α	−41.48 s	0+	
		170	4 m	EC+β^+	−46.92 s	0+	
		171	9 m	EC+β^+	−46.90 s		
		172	6.7 m	EC,β^+	−48.81 s	0+	
		173	16 m	EC+β^+	−48.47 s		
		174	29 m	EC	−50.08 s	0+	
		175	34 m	EC+β^+	−49.45 s	(1/2−)	
		176	2.3 h	EC	−50.57 s	0+	
		177	135 m	EC+β^+	−49.72 s	(1/2−)	
		178	21.5 d	EC,no γ	−50.43	0+	
		179	38 m	EC	−49.283	(7/2−)	
		179m	6.4 m	IT 99.69%,EC 0.31%	−49.061	(1/2−)	
		180	*0.13%*		−49.624	0+	$\approx10^{rs}$
		181	121 d	EC	−48.237	9/2+	
		182	*26.3%*		−48.228	0+	21_{g+m}
		183	*14.3%*		−46.347	1/2−	10.1
		183m	5.3 s	IT	−46.038	(11/2)+	
		184	*30.7%*		−45.687	0+	1.8_g 0.002_m
		185	75.1 d	β^-	−43.370	3/2−	
		185m	1.66 m	IT	−43.173	11/2+	
		186	*28.6%*		−42.498	0+	38
		187	23.9 h	β^-	−39.893	3/2−	70
		188	69.4 d	β^-	−38.657	0+	
		189	11.5 m	β^-	−35.47		
		190	30 m	β^-	−34.22	0+	
75	Re	170	8.0 s	β^++EC	−38.92 s		
		172	30 s	β^++EC	−41.51 s		
		174	2.3 m	β^++EC	−43.58 s		
		175	4.6 m	EC+β^+	−45.15 s		
		176	5.2 m	EC+β^+	−44.97 s		
		177	14 m	EC+β^+	−46.12 s	(5/2−)	
		178	13.2 m	EC 89%,β^+ 11%	−45.77	(3)	
		179	19.7 m	EC 99.1%,β^+ 0.9%	−46.59	(5/2+)	
		180	2.4 m	EC 92%,β^+ 8%	−45.829	(1)−	
		181	20 h	EC	−46.44 s	5/2+	
		182	64 h	EC	−45.43 s	7+,6±	
		182	12.7 h	EC 99.8%,β^+ 0.2%	−45.43 s	2+	
		183	71 d	EC	−45.791	(5/2)+	
		184	38 d	EC	−44.191	3−	
		184m	169 d	IT 75%,EC 25%	−44.003	8+	
		184	2.2 d?	?			

Z	El	A	Abundance or $t_{1/2}$	Decay Mode	Δ(MeV)	Jπ	σ_n(b)
75	Re	185	*37.40%*		−43.802	5/2+	110_g 0.3^{sc}_m
		186	90.6 h	β^- 92.2%,EC 7.8%	−41.910	1−	
		186m	2×10^5 y	IT	−≈41.76	[8+]	
		187	*62.60%* 4×10^{10} y	β^-,no γ	−41.205	5/2+	74_g 1.0_m
		188	16.9 h	β^-	−39.006	1−	
		188m	18.7 m	IT	−38.834	(6)−	
		189	24.3 h	β^-	−37.970	(5/2+)	
		189	4.3 d?	β^-			
		190	3.1 m	β^-	−35.52	(2−)	
		190m	3.2 h	β^- ≈51%,IT ≈49%	−≈35.30	(6−)	
		191	9.8 m	β^-	−34.343		
		192	16 s	β^-	−≈31.9 s		
76	Os	169	3.0 s	α	−30.55 s		
		170	7.1 s	α	−33.53 s	0+	
		171	8 s	α	−34.16 s		
		172	19 s	EC+β^+ 99+%,α ≤0.3%	−36.84 s	0+	
		173	16 s	β^++EC 99.98%, α 0.02%	−37.41 s		
		174	45 s	EC+β^+ 99.98%, α 0.02%	−39.62 s	0+	
		175	1.4 m	[EC+β^+]	−39.71 s		
		176	3.6 m	EC+β^+	−41.81 s	0+	
		177	4 m	EC+β^+	−41.62 s		
		178	5.0 m	EC+β^+	−43.35 s	0+	
		179	7 m	EC+β^+	−42.89 s		
		180	22 m	EC	−44.22 s	0+	
		181	105 m	EC+β^+		(1/2−)	
		181	2.7 m	EC,β^+	−43.41 s	(7/2−)	
		182	22.0 h	EC	−44.58 s	0+	
		183	13 h	EC 99.91%,β^+ 0.09%	−43.49 s	9/2+	
		183m	9.9 h	EC 89%,IT 11%	−43.32 s	1/2−	
		184	*0.018%*		−44.233	0+	3×10^3
		185	93.6 d	EC	−42.787	1/2−	
		186	*1.6%* 2×10^{15} y	α	−42.987	0+	80
		187	*1.6%*		−41.208	1/2−	330
		188	*13.3%*		−41.125	0+	$\leq5_g$
		189	*16.1%*		−38.978	3/2−	20_g $2.6\times10^{-4}{}_m$
		189m	5.7 h	IT	−38.947	9/2−	
		190	*26.4%*		−38.699	0+	9_m 4_g
		190m	9.9 m	IT	−36.994	10−	
		191	15.4 d	β^-	−36.388	9/2−	
		191m	13.1 h	IT	−36.314	3/2−	
		192	*41.0%*		−35.875	0+	2.0
		192m	6.1 s	IT	−33.860	(10−)	
		193	30.6 h	β^-	−33.387	(3/2−)	$1.5\times10^{3\,rs}$
		194	6.0 y	β^-	−32.417	0+	
		195	6.5 m	β^-	−29.69		
		196	35.0 m	β^-		0+	
77	Ir	171	1.0 s	α	−26.18 s		
		172	1.7 s	α	−27.32 s		
		173	3.0 s	α	−29.91 s		
		174	4 s	α	−30.89 s		
		175	4 s	α	−33.16 s		
		176	8 s	α	−33.84 s		
		177	21 s	α	−35.82 s		
		178	12 s	β^++EC	−36.27 s		
		179	4 m	EC+β^+	−37.89 s		
		180	1.5 m	EC+β^+	−37.93 s		
		181	5 m	EC+β^+	−39.34 s		
		182	15 m	EC+β^+	−38.98 s		
		183	0.9 h	EC+β^+	−40.09 s	(9/2−)	
		184	3.0 h	EC,β^+	−39.51	5	

TABLE OF NUCLIDES

Nuclide Z	El	A	Abundance or $t_{1/2}$	Decay Mode	Δ(MeV)	Jπ	σ_n(b)
77	Ir	185	14 h	EC+β^+	−40.29 s	5/2(−)	
		186	16 h	EC 98%,β^+ 2%	−39.156	5(+)	
		186	1.7 h	EC,β^+		(2−)	
		187	10.5 h	EC	−39.71 s	3/2+	
		188	41.5 h	EC 99.6%,β^+ 0.4%	−38.323	2−	
		189	13.1 d	EC	−38.48 s	3/2+	
		190	11.8 d	EC	−36.70	(4+)	
		$190m_1$	1.2 h	IT	−36.67	(7+)	
		$190m_2$	3.2 h	EC 95%,IT 5%	−36.52	(11−)	
		191	*37.3%*		−36.698	3/2+	540_g 400_{m1} 0.10_{m2}
		191m	4.9 s	IT	−36.527	11/2−	
		192	74.2 d	β^- 95.4%,EC 4.6%	−34.826	4(−)	$1.0\times10^{3\,rs}_{a}$
		$192m_1$	1.45 m	IT 99+%,β^- 0.017%	−34.768	1(+)	
		$192m_2$	241 y	IT	−34.665	9(+)	
		193	*62.7%*		−34.519	3/2+	110_g
		193m	10.6 d	IT	−34.439	11/2−	0.05_m
		194	19.2 h	β^-	−32.514	1−	
		194m	0.47 y	β^-		(11)	
		195	2.5 h	β^-	−31.692	(3/2+)	
		195m	3.8 h	β^-	−31.57	(11/2−)	
		196	52 s	β^-	−29.44	(0,1−)	
		196m	1.40 h	β^-	−29.01	(10,11)	
		197	9.8 m	β^-	−28.43		
		198	8 s	β^-	−25.52		
78	Pt	173	≤1 s	α	−21.79 s		
		174	0.7 s	α ≈80%	−24.93 s	0+	
		175	2.4 s	α ≈75%	−25.64 s		
		176	6.3 s	α 42%	−28.54 s	0+	
		177	7 s	α 9%	−29.35 s		
		178	21 s	α 7%	−31.63 s	0+	
		179	33 s	α 0.27%	−32.01 s	1/2	
		180	52 s	α ≈0.3%	−34.12 s	0+	
		181	51 s	α ≈0.06%	−34.06 s		
		182	2.6 m	EC+β^+ 99+%,α ≈0.02%	−35.98 s	0+	
		183	7 m	EC+β^+ 99+%, α ≈0.0013%	−35.63 s		
		184	17.3 m	EC+β^+ 99+%, α ≈0.001%	−37.21 s	0+	
		184	42 m?	EC			
		185	71 m	EC+β^+	−36.49 s		
		185	33 m	EC+β^+	−36.49 s		
		186	2.0 h	EC 99+%, α ≈1.4×10^{-4}%	−37.83 s	0+	
		187	2.35 h	EC+β^+	−36.81 s	3/2	
		188	10.2 d	EC 99+%,α 3×10^{-5}%	−37.788	0+	
		189	10.9 h	EC,β^+	−36.57 s	3/2−	
		190	*0.013%* 6×10^{11} y	α	−37.318	0+	800
		191	2.9 d	EC	−35.698	3/2−	
		192	*0.78%*		−36.283	0+	10_{g+m} 2^{rs}_{m}
		193	50 y	EC,no γ	−34.458	(1/2)−	
		193m	4.3 d	IT	−34.308	(13/2)+	
		194	*32.9%*		−34.765	0+	$\approx1^{rs}_{g}$ 0.09^{rs}_{m}
		195	*33.8%*		−32.802	1/2−	27^{rs}
		195m	4.02 d	IT	−32.543	13/2+	
		196	*25.3%*		−32.652	0+	0.7^{rs}_{g} 0.05^{rs}_{m}
		197	18.3 h	β^-	−30.431	1/2−	
		197m	94 m	IT 97%,β^- 3%	−30.032	13/2+	
		198	*7.2%*		−29.921	0+	3.7_g 0.027_m

TABLE OF NUCLIDES

Z	El	A	Abundance or $t_{1/2}$	Decay Mode	Δ(MeV)	Jπ	σ_n(b)
78	Pt	199	30.8 m	β^-	−27.420	(5/2−)	≈15 [rs]
		199m	14 s	IT	−26.996	(13/2+)	
		200	12.6 h	β^-	−26.60 s	0+	
		201	2.5 m	β^-	−23.74		
79	Au	175	≈0.14 s	α	−17.16 s		
		176	1.2 s	α	−18.40 s		
		177	1.3 s	α	−21.19 s		
		178	2.6 s	α	−22.41 s		
		179	7.5 s	α	−24.75 s		
		181	11 s	α 1.1%	−27.64 s		
		182	21 s	β^++EC 99+%, α? ≈0.04%	−28.18 s		
		183	42 s	α 0.30%	−30.01 s		
		184	53 s	β^++EC 99+%,α 0.022%	−30.22 s		
		185	4.3 m	EC+β^+ 99.91%, α 0.09%	−31.73 s		
		185	6.8 m	EC+β^+	−31.73 s		
		186	11 m	EC+β^+	−31.69 s	3	
		186	≤2 m	EC+β^+	−31.69 s		
		187	8 m	EC+β^+,α?	−32.87 s	1/2	
		188	8.8 m	EC+β^+	−32.49 s	1	
		189	28.7 m	EC+β^+	−33.41 s	1/2+	
		189m	4.6 m	EC+β^+,IT?	−33.16 s	11/2−	
		190	43 m	EC 98%,β^+ 2%	−32.876	1−	
		191	3.2 h	EC	−33.87	3/2+	
		191m	0.9 s	IT	−33.60	(11/2−)	
		192	5.0 h	EC ≈99%,β^+ ≈1%	−32.768	1−	
		193	17.5 h	EC,β^+?	−33.36 s	3/2+	
		193m	3.9 s	IT 99.97%,EC 0.03%	−33.07 s	11/2−	
		194	39.5 h	EC ≈97%,β^+ ≈3%	−32.256	1−	
		195	183 d	EC	−32.572	3/2+	
		195m	30.6 s	IT	−32.253	11/2−	
		196	6.18 d	EC 93.0%,β^+ 5×10^{-5}%, β^- 7.0%	−31.162	2−	
		196m_1	8.2 s	IT	−31.077	5+	
		196m_2	9.7 h	IT	−30.567	12−	
		197	100%		−31.150	3/2+	98.8
		197m	7.7 s	IT	−30.741	11/2−	
		198	2.696 d	β^-	−29.591	2−	2.5×10^4
		198m	2.30 d	IT	−28.779	(12−)	
		199	3.14 d	β^-	−29.104	3/2+	≈30 [rs] [g]
		200	48.4 m	β^-	−27.30	1(−)	
		200m	18.7 h	β^- ≈84%,IT ≈16%	−≈26.3	12(−)	
		201	26 m	β^-	−26.40	(3/2+)	
		202	29 s	β^-	−23.86	(1−)	
		203	53 s	β^-	−22.98 s		
		204	4 s?	β^-			
		204	40 s	β^-			
80	Hg	177	≈0.2 s	α	−12.65 s		
		178	0.5 s	α ≈84%,[EC+β^+] ≈16%	−15.93 s	0+	
		179	1.09 s	α ≈53%,EC+β^+ ≈47%, (EC+β^+)p	−16.80 s		
		179	3.5 s?	α	−16.80 s		
		180	2.9 s	α	−19.86 s	0+	
		180	5.9 s?	α			
		181	3.6 s	β^++EC 74%,α 26%, (β^++EC)p 0.014%, (β^++EC)α 9×10^{-6}%	−20.79 s	1/2(−)	
		182	11 s	EC+β^+ 91%,α 9%	−23.21 s	0+	
		183	8.8 s	EC 61%,β^+ 27%, α 12%, (EC+β^+)p 3×10^{-4}%	−23.69 s	1/2	
		184	30.6 s	EC+β^+ 98.7%,α 1.3%	−26.04 s	0+	
		185(g)	48 s	EC+β^+ ≲95%,α ≳5%	−26.14 s	1/2−	
		185(m)	17 s	α,IT?			
		185	155 s	?			
		186	1.4 m	EC 96%,β^+ 4%, α 0.016%	−28.35 s	0+	

Nuclide Z	El	A	Abundance or $t_{1/2}$	Decay Mode	Δ(MeV)	Jπ	σ_n(b)
80	Hg	187	1.6 m	EC+β^+,α >2.5×10^{-4}%	−28.06 s		
		187	2.4 m	EC+β^+,α >1.2×10^{-4}%	−28.06 s	3/2	
		187	3.0 m	EC+β^+,α			
		188	3.3 m	EC+β^+	−29.88 s	0+	
		189	8.7 m	EC+β^+	−29.21 s	(13/2+)	
		189	7.5 m	EC+β^+	−29.21 s	3/2−	
		190	20 m	EC	−30.96	0+	
		191	≈49 m	EC+β^+	−30.48	(3/2−)	
		191m	51 m	EC+β^+	−≈30.34	(13/2+)	
		192	4.9 h	EC	−31.97 s	0+	
		193	4 h	EC,β^+?	−31.02 s	3/2−	
		193m	11 h	EC 92%,β^+ 0.34%, IT 8%	−30.88 s	13/2+	
		194	260 y	EC,no γ	−32.206	0+	
		194	0.40 s	?			
		195	10 h	EC	−31.05	1/2−	
		195m	41 h	EC 50%,IT 50%	−30.87	13/2+	
		196	*0.15%*		−31.846	0+	$3.0\times10^{3}\,{}^{sc}_{g}$ 120^{sc}_{m}
		197	64.1 h	EC	−30.735	1/2−	
		197m	23.8 h	IT 93%,EC 7%	−30.436	13/2+	
		198	*10.0%*		−30.964	0+	0.018_{m}
		199	*16.8%*		−29.557	1/2−	2×10^{3}
		199m	42.6 m	IT	−29.025	13/2+	
		200	*23.1%*		−29.514	0+	<60
		201	*13.2%*		−27.672	3/2−	<60
		202	*29.8%*		−27.356	0+	5.0
		203	46.8 d	β^-	−25.277	5/2−	
		204	*6.9%*		−24.703	0+	0.4^{rs}
		205	5.2 m	β^-	−22.299	1/2−	
		206	8.1 m	β^-	−20.955	0+	
81	Tl	184	11 s	β^++EC 98%,α 2%	−16.90 s		
		185m	1.7 s	α,IT	−18.66 s	(9/2−)	
		186	28 s	(β^+,EC) 99+%, α? ≈0.006%	−19.86 s		
		186m	3 s	IT	−19.49 s		
		187m	16 s	α,IT	−≈21.60 s	(9/2−)	
		188	71 s	β^+,EC	−22.29 s	(7)	
		189	1.4 m	β^++EC	−24.02 s		
		189	2.3 m	β^++EC	−24.02 s		
		190	2.6 m	β^+,EC	−24.16	(2−)	
		190	3.7 m	β^+,EC		(7+)	
		191	5.2 m	EC 98%,β^+ 2%	−25.67		
		192	10.8 m	EC+β^+	−25.59 s	(7+)	
		192	10.6 m	EC+β^+	−25.59 s	(2−)	
		193	22 m	EC ≧96%,β^+ ≦4%	−27.02 s	1/2+	
		193m	2.1 m	IT		(9/2−)	
		194	33.0 m	EC+β^+	−26.81 s	2−	
		194m	32.8 m	EC+β^+	−≈26.51 s	(7+)	
		195	1.16 h	EC 99.3%,β^+ 0.7%	−27.85	1/2+	
		195m	3.6 s	IT	−27.37	9/2−	
		196	1.84 h	EC+β^+	−27.35 s	2(−)	
		196m	1.41 h	EC+β^+ 96.2%,IT 3.8%	−26.95 s	(7+)	
		197	2.84 h	EC 99.5%,β^+ 0.5%	−28.33 s	1/2+	
		197m	0.54 s	IT	−27.72 s	9/2−	
		198	5.3 h	EC ≈99.3%,β^+ ≈0.7%	−27.50	2−	
		198m	1.87 h	EC+β^+ 56%,IT 44%	−26.96	7+	
		199	7.4 h	EC	−28.08	1/2+	
		200	26.1 h	EC 99.65%,β^+ 0.35%	−27.060	2−	
		201	73 h	EC	−27.185	1/2+	
		202	12.2 d	EC	−25.988	2−	
		203	*29.5%*		−25.769	1/2+	10
		204	3.77 y	β^- 97.4%,EC 2.6%	−24.353	2−	22^{rs}
		205	*70.5%*		−23.837	1/2+	0.10^{rs}_{g}
		206	4.20 m	β^-	−22.269	0−	
		206m	3.6 m	IT	−19.626	(12−)	
		207	4.77 m	β^-	−21.041	1/2+	
		207m	1.3 s	IT	−19.700	11/2−	

TABLE OF NUCLIDES

Z	El	A	Abundance or $t_{1/2}$	Decay Mode	Δ(MeV)	Jπ	σ_n(b)
81	Tl	208	3.053 m	β^-	−16.768	(5+)	
		209	2.2 m	β^-	−13.650	(1/2+)	
		210	1.30 m	β^-,β^-n ≈0.007%	−9.251		
82	Pb	185	≈2 s	α	−11.74 s		
		186	8 s	α ≈2.4%	−14.33 s	0+	
		187	17 s	α ≈2.0%	−14.94 s		
		188	25 s	EC+β^+ 97%,α 3%	−17.50 s	0+	
		189	51 s	EC+β^+ 99+%,α ≈0.4%	−17.86 s		
		190	1.2 m	EC+β^+ 99.8%,α 0.2%	−20.22 s	0+	
		191	1.3 m	EC+β^+ 99+%,α 0.013%	−20.23 s		
		192	2.3 m	EC+β^+ 99+%,α 0.007%	−22.29 s	0+	
		193	5.8 m	EC+β^+	−22.07 s	(13/2+)	
		194	11 m	EC+β^+	−23.81 s	0+	
		195	16.4 m	EC+β^+	−23.55 s	(13/2+)	
		196	37 m	EC	−25.15 s	0+	
		197	?	EC+β^+	−24.63 s	(3/2−)	
		197m	42 m	EC+β^+ 81%,IT 19%	−24.31 s	(13/2+)	
		198	2.4 h	EC	−25.90 s	0+	
		199	1.5 h	EC ≈98.6%,β^+ ≈1.4%	−25.28	5/2−	
		199m	12.2 m	IT 93%,EC+β^+ 7%	−24.86	13/2+	
		200	21.5 h	EC	−26.16 s	0+	
		201	9.4 h	EC 99+%,β^+ ≲0.034%	−25.327	5/2−	
		201m	61 s	IT	−24.699	13/2+	
		202	≈3×10^5 y	EC,no γ	−25.942	0+	
		202m	3.62 h	IT 90.5%,EC 9.5%	−23.772	9−	
		203	52.0 h	EC	−24.794	5/2−	
		203m$_1$	6.1 s	IT	−23.969	13/2+	
		203m$_2$	0.48 s	IT	−21.844	29/2−	
		204	*1.42%*		−25.117	0+	0.7
		204m	66.9 m	IT	−22.932	9−	
		205	1.4×10^7 y	EC,no γ	−23.777	5/2−	3.8^{rs}
		206	*24.1%*		−23.795	0+	0.03
		207	*22.1%*		−22.463	1/2−	0.71^{g}
		207m	0.81 s	IT	−20.830	13/2+	
		208	*52.3%*		−21.759	0+	5.0×10^{-4}
		209	3.25 h	β^-,no γ	−17.624	9/2+	
		210	22.3 y	β^- 99+%,α 1.7×10^{-6}%	−14.738	0+	0.5
		211	36.1 m	β^-	−10.492	(9/2)+	
		212	10.64 h	β^-	−7.562	0+	
		213	10.2 m	β^-	−3.14 s		
		214	26.8 m	β^-	−0.185	0+	
83	Bi	189	<1.5 s	α	−9.87 s		
		190	5.4 s	α ≈90%	−10.85 s		
		191	13 s	α ≈40%	−13.05 s		
		191m	≈20 s	α			
		192	42 s	α ≈20%	−13.67 s		
		193	64 s	α ≈60%	−15.56 s		
		193m	3.5 s	α ≈25%			
		194	1.7 m	β^++EC 99+%,α <0.2%	−15.98	(10−)	
		195	2.8 m	α <0.2%	−17.68		
		195m	90 s	α 4%			
		196	4.5 m	β^++EC	−17.76 s		
		197(m)	8 m	β^++EC 99.89%, α 0.11%			
		198	11.8 m	EC+β^+	−19.30 s	(7+)	
		198m	7.7 s	IT	−19.05 s	(10−)	
		199(g)	27 m	EC	−20.61 s	9/2−	
		199(m)	24.7 m	α	−≈20.00 s		
		200	36 m	EC,β^+(weak)	−20.46 s	7(+)	
		200m	0.40 s	IT	−20.03 s	10(−)	
		201	1.8 h	EC+β^+	−21.41 s	9/2−	
		201m	59 m	EC+β^+,IT,α ≳0.02%	−20.56 s	(1/2+)	
		202	1.7 h	EC 99.5%,β^+ 0.5%	−21.04 s	5(+)	
		203	11.8 h	EC ≈99.7%,β^+ ≈0.3%	−21.60	9/2−	
		204	11.2 h	EC	−20.82 s	6+	
		205	15.3 d	EC 99.90%,β^+ 0.10%	−21.070	9/2−	
		206	6.243 d	EC,β^+? 8×10^{-4}%	−20.033	6+	
		207	38 y	EC 99+%,β^+ 0.012%	−20.058	9/2−	
		208	3.68×10^5 y	EC	−18.879	(5)+	

TABLE OF NUCLIDES

Z	El	A	Abundance or $t_{1/2}$	Decay Mode	Δ(MeV)	Jπ	σ_n(b)
83	Bi	209	*100%*		−18.268	9/2−	0.019^{rs}_{g} 0.014^{rs}_{m}
		210	5.01 d	β^- 99+%,α 1.3×10^{-4}%	−14.801	1−	0.050^{sc}
		210m	3.0×10^{6} y	α	−14.530	9−	
		211	2.15 m	α 99.72%,β^- 0.28%	−11.865	(9/2)−	
		212	60.60 m	β^- 64.0%, $\beta^-\alpha$ 0.014%,α 36.0%	−8.135	1(−)	
		$212m_1$	25 m	α ≦93%,β^- ≧7%	−7.88	[9−]	
		$212m_2$	9 m	β^- ≦100%		[15−]	
		213	45.6 m	β^- 97.8%,α 2.2%	−5.243	(9/2−)	
		214	19.7 m	β^- 99+%, $\beta^-\alpha$ 0.0031%, α 0.021%	−1.209	(1−)	
		215	7 m	β^-	1.71		
84	Po	193	≲1 s	α	−8.31 s		
		194	0.6 s	α	−10.81 s	0+	
		195(g)	4.5 s	α	−11.06 s		
		195(m)	2.0 s	α			
		196	5 s	α	−13.21 s	0+	
		197	58 s	α 90%	−13.23 s		
		197m	26 s	α			
		198	1.78 m	α 70%,EC+β^+ 30%	−15.07 s	0+	
		199	5.2 m	EC+β^+ 88%,α 12%	−15.05 s	(3/2−)	
		199m	4.2 m	EC+β^+ 61%,α 39%		(13/2+)	
		200	11.4 m	EC+β^+ 86%,α 14%	−16.74 s	0+	
		201	15.2 m	EC+β^+ 98.4%,α 1.6%	−16.41 s	3/2(−)	
		201m	8.9 m	IT 53%,EC+β^+ 44%, α 2.9%	−15.98 s	(13/2+)	
		202	44 m	EC+β^+ 98.0%,α 2.0%	−17.78 s	0+	
		203	33 m	EC+β^+ 99.89%, α 0.11%	−17.36	5/2−	
		203m	1.2 m	IT 96%,EC+β^+ 4%	−16.72	(13/2+)	
		204	3.57 h	EC 99.4%,α 0.6%	−18.25 s	0+	
		205	1.80 h	EC+β^+ 99.5%,α 0.5%	−17.576	5/2−	
		206	8.8 d	EC 94.5%,α 5.5%	−18.190	0+	
		207	5.7 h	EC 99.5%,β^+ 0.5%, α 0.008%	−17.150	5/2−	
		207m	2.8 s	IT	−15.766	19/2−	
		208	2.90 y	α 99+%,EC 0.0018%	−17.475	0+	
		209	102 y	α 99.74%,EC 0.26%	−16.373	1/2−	
		210	138.38 d	α	−15.963	0+	$<0.03^{rs}_{g}$ $<5\times10^{-4}{}^{rs}_{m}$ $<0.002_{o}$
		211	0.516 s	α	−12.444	(9/2+)	
		211m	25 s	α	−10.982	(25/2+)	
		212	0.30 μs	α	−10.381	0+	
		212m	45 s	α	−7.476	[16+]	
		213	4 μs	α	−6.663	9/2+	
		214	164 μs	α	−4.479	0+	
		215	1.78 ms	α 99+%,β^- 2.3×10^{-4}%	−0.540	(9/2)+	
		216	0.15 s	α	1.769	0+	
		217	<10 s	α	5.96 s		
		218	3.05 m	α 99+%,β^- 0.018%	8.355	0+	
85	At	196	0.3 s	α	−4.05 s		
		197	0.4 s	α	−6.03 s		
		198	4.9 s	α	−6.67		
		198m	1.5 s	α			
		199	7.2 s	α	−8.47		
		200(g)	42 s	α 53%,EC+β^+ 47%	−8.67 s		
		200(m)	4.3 s	α			
		201	1.5 m	α 71%,EC+β^+ 29%	−10.52 s		
		202	3.0 m	EC+β^+ 85%,α 15%	−10.52 s		
		203	7.3 m	EC+β^+ 69%,α 31%	−11.97 s		
		204	9.1 m	EC+β^+ 95.6%,α 4.4%	−11.97 s	(5+)	
		205	26 m	EC 87%,β^+ 3%,α 10%	−12.96 s	9/2−	
		206	31 m	EC 82%,β^+ 17%, α 1.0%	−12.73 s	(5+)	

Nuclide Z El	A	Abundance or $t_{1/2}$	Decay Mode	Δ(MeV)	Jπ	σ_n(b)
85 At	207	1.8 h	EC+β^+ ≈90%,α ≈10%	−13.31	9/2−	
	208	1.63 h	EC+β^+ 99.4%,α 0.6%	−12.64 s	(6+)	
	209	5.4 h	EC 95.9%,α 4.1%	−12.888	9/2−	
	210	8.3 h	EC+β^+ 99.82%, α 0.18%	−11.976	5+	
	211	7.21 h	EC 58.1%,α 41.9%	−11.653	9/2−	
	212	0.315 s	α	−8.625		
	212m	0.12 s	α	−8.403		
	213	0.11 μs	α,no γ	−6.589	9/2−	
	214	≈2 μs	α	−3.389		
	215	0.10 ms	α	−1.262	(9/2)−	
	216	0.30 ms	α	2.237	1(−)	
	217	32.3 ms	α 99+%,β^- 0.012%	4.382	(9/2−)	
	218	≈2 s	α 99.9%,β^- 0.1%	8.099		
	219	0.9 m	α ≈97%,β^- ≈3%	10.53		
86 Rn	200	1 s	α	−3.74 s	0+	
	201(g)	7.0 s	α	−3.95 s		
	201(m)	3.8 s	α			
	<202	<1 s?	α			
	202	9.9 s	α >70%	−5.88 s	0+	
	203	45 s	α 65%,EC+β^+ 35%	−6.00 s		
	203m	28 s	α	−≈5.95 s		
	204	75 s	α ≈72%,EC+β^+ ≈28%	−7.77 s	0+	
	205	170 s	EC+β^+ 77%,α 23%	−7.60 s		
	206	5.7 m	α 64%,EC+β^+ 36%	−8.97 s	0+	
	207	9.3 m	EC+β^+ 77%,α 23%	−8.69	5/2−	
	208	24 m	α 52%,EC+β^+ 48%	−9.56 s	0+	
	209	29 m	EC 80%,β^+ 3%,α 17%	−8.994	5/2−	
	210	2.4 h	α 96%,EC 4%	−9.608	0+	
	211	14.6 h	EC+β^+ 74%,α 26%	−8.761	1/2−	
	212	23 m	α	−8.666	0+	
	213	25.0 ms	α	−5.706	(9/2+)	
	214	0.27 μs	α	−4.328	0+	
	215	2.3 μs	α,no γ	−1.179	(9/2+)	
	216	45 μs	α	0.245	0+	
	217	0.54 ms	α	3.649	9/2+	
	218	35 ms	α	5.212	0+	
	219	3.96 s	α	8.831	(5/2)+	
	220	55.6 s	α	10.599	0+	$<0.2^{rs}$
	221	25 m	β^- ≈80%,α ≈20%	14.38 s		
	222	3.8235 d	α	16.370	0+	0.73^{rs}
	223	43 m	β^-			
	224	1.8 h	β^-	22.26 s	0+	
	225	4.5 m	β^-	27.59 s		
	226	6.0 m	β^-		0+	
87 Fr	203	0.7 s	α,β^++EC	1.23		
	204	2.1 s	α	0.92 s		
	205	3.7 s	α	−1.04 s		
	206	16.0 s	α 85%,EC+β^+ 15%	−1.18 s		
	207	14.8 s	α 93%,EC+β^+ 7%	−2.65 s	(9/2−)	
	208	58.0 s	α 74%,EC+β^+ 26%	−2.77 s		
	209	50.0 s	α 89%,EC+β^+ 11%	−3.76 s	9/2−	
	210	3.2 m	α,EC+β^+	−3.64 s		
	211	3.1 m	α,EC+β^+	−4.22		
	212	19.3 m	EC+β^+ 56%,α 44%	−3.69 s		
	213	34.7 s	α 99.45%,EC 0.55%	−3.556	(9/2−)	
	214	5.0 ms	α	−0.965	(1−)	
	214m	3.4 ms	α	−0.843	(9−)	
	215	0.12 μs	α,no γ	0.309	9/2−	
	216	0.70 μs	α,no γ	2.975		
	217	22 μs	α,no γ	4.307	9/2−	
	218	≈0.7 ms	α	7.050		
	219	0.020 s	α	8.617	(9/2)−	
	220	27.4 s	α 99.65%,β^- 0.35%	11.470		
	221	4.8 m	α	13.265	(5/2−)	
	222	14.4 m	β^- 99+%,α 0.01−0.1%	16.338		
	223	21.8 m	β^- 99+%,α ≈0.005%	18.382	(3/2)	
	224	2.7 m	β^-	21.71 s		
	225	3.9 m	β^-	23.79 s		
	226	48 s	β^-	27.46		

TABLE OF NUCLIDES

Z	El	A	Abundance or $t_{1/2}$	Decay Mode	Δ(MeV)	Jπ	σ_n(b)
87	Fr	227	2.4 m	β^-	29.58		
		228	39 s	β^-			
		229	0.8 m	β^-			
88	Ra	206	0.4 s	α,(EC+β^+)?	3.96 s	0+	
		207	1.3 s	α,(EC+β^+)?	3.70 s		
		208	1.5 s	α	1.93 s	0+	
		209	4.7 s	α	1.97 s		
		210	3.7 s	α	0.61 s	0+	
		211	14 s	α,(EC+β^+)?	0.78	(5/2−)	
		212	13.0 s	α,(EC+β^+)?	−0.11 s	0+	
		213	2.7 m	α 80%,EC+β^+ 20%	0.290	(1/2−)	
		213m	2.1 ms	IT ≈99%,α ≈1%	2.060	(17/2−,13/2+)	
		214	2.46 s	α 99+%,EC 0.059%	0.090	0+	
		215	1.6 ms	α	2.531		
		216	0.18 μs	α	3.285	0+	
		217	1.6 μs	α	5.881		
		218	14 μs	α	6.644	0+	
		219	10 ms	α	9.377		
		220	23 ms	α	10.263	0+	
		221	30 s	α	12.957		
		222	38 s	α	14.312	0+	
		223	11.435 d	α	17.235	1/2+	134 [rs]
		224	3.66 d	α	18.813	0+	12 [rs]
		225	14.8 d	β^-	21.987	(3/2)+	
		226	1.60×10^3 y	α	23.666	0+	8
		227	42.2 m	β^-	27.185	(3/2+)	
		228	5.76 y	β^-	28.941	0+	36 [sc]
		229	4.0 m	β^-	32.72 s		
		230	93 m	β^-	34.56 s	0+	
89	Ac	209	0.10 s	α	9.12 s		
		210	0.35 s	α	8.86 s		
		211	0.25 s	α,EC+β^+	7.40 s		
		212	0.93 s	α	7.18 s		
		213	0.80 s	α,no γ	6.17 s	(9/2−)	
		214	8.2 s	α ≥86%,EC ≤14%	6.14 s		
		215	0.17 s	α 99.91%, EC+β^+ 0.09%	5.95		
		216	≈0.33 ms	α	7.98 s		
		216m	0.33 ms	α	8.02 s		
		217	0.11 μs	α,no γ	8.701	(9/2−)	
		218	0.27 μs	α,no γ	10.837		
		219	7 μs	α,no γ	11.560	(9/2−)	
		220	26 ms	α	13.747		
		221	52 ms	α	14.518		
		222	5 s	α	16.617		
		222m	66 s	α ≳90%, IT <10%,EC ≈1%			
		223	2.2 m	α 99%,EC 1%	17.825	(5/2−)	
		224	2.9 h	EC ≈90%,α ≈10%	20.219		
		225	10.0 d	α	21.626	(3/2−)	
		226	29 h	β^- 83%,EC 17%, α 0.006%	24.301	(1−)	
		227	21.773 y	β^- 98.62%,α 1.38%	25.850	3/2−	900 [sc]
		228	6.13 h	β^-	28.895	(3+)	
		229	62.7 m	β^-	30.72	(3/2+)	
		230	122 s	β^-	33.76 s		
		231	7.5 m	β^-	35.91	(1/2+)	
		232	35 s	[β^-]	39.15 s		
90	Th	215	1.2 s	α	10.87	(1/2−)	
		216	0.028 s	α	10.39 s	0+	
		217	0.25 ms	α	12.141		
		218	0.10 μs	α	12.362	0+	
		219	1.05 μs	α	14.470		
		220	10 μs	α	14.663	0+	
		221	1.7 ms	α	16.934		
		222	2.8 ms	α	17.197	0+	
		223	0.66 s	α	19.256		
		224	1.04 s	α	19.993	0+	

TABLE OF NUCLIDES

Nuclide Z	El	A	Abundance or $t_{1/2}$	Decay Mode	Δ(MeV)	Jπ	σ_n(b)
90	Th	225	8.0 m	α ≈90%,EC ≈10%	22.303	(3/2+)	
		226	30.9 m	α	23.189	0+	
		227	18.718 d	α	25.806	3/2+	200_f^{rs}
		228	1.9131 y	α	26.758	0+	100^{rs}
		229	7.3×10^3 y	α	29.581	5/2+	30_f
		230	8.0×10^4 y	α	30.861	0+	40
		231	25.52 h	β^-	33.812	5/2+	
		232	*100%* 1.41×10^{10} y	α	35.447	0+	7.4
		233	22.3 m	β^-	38.732	1/2+	$1.4\times10^{3\ sc}$
		234	24.10 d	β^-	40.612	0+	2.0
		235	6.9 m	β^-	44.15 s		
		236	37 m	β^-	46.64 s	0+	
91	Pa	216	0.20 s	α			
		217	≈10 ms	α			
		222	5.7 ms	α	21.959		
		223	6 ms	α	22.330		
		224	0.9 s	α	23.798		
		225	1.8 s	α	24.320		
		226	1.8 m	α 74%,EC 26%	26.029		
		227	38.3 m	α ≈85%,EC ≈15%	26.832	(5/2−)	
		228	22 h	EC ≈98%,α ≈2%	28.870	3+	
		229	1.4 d	EC 99.75%,α 0.25%	29.887	(5/2+)	
		230	17.7 d	EC 90%,β^- 10%, α 0.0032%	32.166	(2−)	$1.5\times10^3{}_f^{rs}$
		231	3.28×10^4 y	α	33.423	3/2−	200
		232	1.31 d	β^-	35.934	(2−)	800^{rs} 700_f^{rs}
		233	27.0 d	β^-	37.487	3/2−	20_m 19_g
		234	6.75 h	β^-	40.349	4(+)	$<5\times10^3{}_f^{rs}$
		234m	1.175 m	β^- 99.87%,IT 0.13%	≈40.43	(0−)	$<500_f^{rs}$
		235	24.2 m	β^-	42.32	(3/2−)	
		236	9.1 m	β^-	45.54	(1−)	
		237	8.7 m	β^-	47.64	(1/2+)	
		238	2.3 m	β^-	51.27	(3−)	
92	U	226	0.5 s	α	27.186	0+	
		227	1.1 m	α	28.88 s		
		228	9.1 m	α ≥95%,EC ≤5%	29.221	0+	
		229	58 m	EC ≈80%,α ≈20%	31.201	(3/2+)	
		230	20.8 d	α	31.607	0+	20_f^{rs}
		231	4.2 d	EC 99+%,α 0.0055%	33.78	(5/2)	$\approx300_f^{rs}$
		232	72 y	α	34.597	0+	74 76_f
		233	1.592×10^5 y	α	36.915	5/2+	530_f 46
		234	*0.0054%* 2.45×10^5 y	α	38.143	0+	100_{g+m}
		235	*0.720%* 7.038×10^8 y	α	40.916	7/2−	580_f 98
		235m	26 m	IT	40.916	1/2+	
		235f	20 ns?	SF			
		236	2.342×10^7 y	α	42.442	0+	5.1
		236f	0.12 μs	SF	44.79		
		237	6.75 d	β^-	45.389	1/2+	400
		238	*99.275%* 4.468×10^9 y	α	47.307	0+	2.7
		238f	0.19 μs	IT ≈96%,SF ≈4%	49.866	(0+)	
		239	23.5 m	β^-	50.572	5/2+	22^{rs} 15_f^{rs}
		240	14.1 h	β^-	52.712	0+	
93	Np	229	4.0 m	α ≥50%,EC ≤50%	33.758		
		230	4.6 m	α 99+%,EC+β^+ ≤0.97%	35.232		
		231	48.8 m	EC <99%,α >1%	35.626	(5/2)	

TABLE OF NUCLIDES

Z	El	A	Abundance or $t_{1/2}$	Decay Mode	Δ(MeV)	Jπ	σ_n(b)
93	Np	232	14.7 m	EC	37.29 s		
		233	36.2 m	EC 99+%,α ≈0.001%	38.01 s	(5/2+)	
		234	4.4 d	EC 99.95%,β^+ 0.05%	39.951	(0+)	$1.0\times10^{3}{}_{f}^{rs}$
		235	396 d	EC 99+%,α 0.0016%	41.040	5/2+	$160^{rs}_{22.5\ h}$
		236	1.1×10^5 y	EC 91%,β^- 9%		(6−)	$3\times10^{3}{}_{f}$
		236	22.5 h	EC 50%,β^- 50%	43.426	1(−)	
		237	2.14×10^6 y	α	44.869	5/2+	180
		237f	45 ns	SF	47.57		
		238	2.117 d	β^-	47.453	2+	$2.1\times10^{3}{}_{f}^{sc}$
		239	2.35 d	β^-	49.306	5/2+	32^{rs}_{m} $\approx20^{rs}_{g}$
		240	67 m	β^-	52.21	(5+)	
		240m	7.5 m	β^- 99.89%,IT 0.11%		1(−)	
		241	16.0 m	β^-	54.31		
94	Pu	232	34 m	EC ≥80%,α ≤20%	38.362	0+	
		233	20.9 m	EC 99.88%,α 0.12%	40.042		
		234	8.8 h	EC 94%,α 6%	40.342	0+	
		235	25.6 m	EC 99+%,α 0.003%	42.16	(5/2)+	
		235f	30 ns	SF	43.86		
		236	2.85 y	α	42.889	0+	150^{rs}_{f}
		$236f_1$	0.03 ns	SF			
		$236f_2$	0.03 μs	SF	46.39		
		237	45.4 d	EC 99+%,α 0.0033%	45.087	7/2−	$2.1\times10^{3}{}_{f}^{rs}$
		237m	0.18 s	IT	45.233	1/2+	
		$237f_1$	0.11 μs	SF	47.39		
		$237f_2$	1.1 μs	SF	47.69		
		238	87.74 y	α	46.161	0+	500^{rs} 17^{rs}_{f}
		$238f_1$	0.6 ns	SF	48.56		
		$238f_2$	6 ns	SF	49.86		
		239	2.41×10^4 y	α	48.585	1/2+	742_{f} 271
		$239f_1$	8 μs	SF	50.79		
		$239f_2$	0.01 μs?	SF			
		240	6.57×10^3 y	α	50.123	0+	290
		240f	3.8 ns	SF	52.52	(0+)	
		241	14.4 y	β^- 99+%,α 0.0024%	52.953	5/2+	$1.01\times10^{3}{}_{f}$ 370
		$241f_1$	24 μs	SF	54.95		
		$241f_2$	30 ns?	SF			
		242	3.76×10^5 y	α	54.715	0+	19 $<0.2_{f}$
		$242f_1$	4 ns	SF			
		$242f_2$	28 ns	SF			
		243	4.956 h	β^-	57.752	7/2+	200^{rs}_{f} 100^{rs}
		243f	0.05 μs	SF	59.55		
		244	8.1×10^7 y	α	59.803	0+	1.7
		244f	0.4 ns	SF			
		245	10.5 h	β^-	63.157	(9/2−)	150
		246	10.85 d	β^-	65.29	0+	
95	Am	232	1.4 m?	[EC+β^+],(EC+β^+)SF			
		234	2.6 m	[EC+β^+],(EC+β^+)SF	44.46 s		
		235f	?	SF			
		236f	?	SF			
		237	1.22 h	EC 99+%,α 0.025%	46.64 s	5/2(−)	
		237f	5 ns	SF	48.74 s		
		238	1.63 h	EC 99+%,α 1.0×10^{-4}%	48.417	1+	
		238f	35 μs	SF	50.72		
		239	11.9 h	EC 99+%,α 0.010%	49.389	5/2−	
		239f	0.16 μs	SF	51.89		
		240	50.8 h	EC 99+%,α 1.9×10^{-4}%	51.443	(3−)	
		240f	0.9 ms	SF	54.04		

TABLE OF NUCLIDES

Nuclide Z	El	A	Abundance or $t_{1/2}$	Decay Mode	Δ(MeV)	Jπ	σ_n(b)
95	Am	241	433 y	α	52.932	5/2−	562_g 62_m 3.2_f
		241f	1.5 μs	SF	55.13		
		242	16.01 h	β^- 82.7%,EC 17.3%	55.463	1−	$2.1\times10^3{}_f^{rs}$
		242m	152 y	IT 99.52%,α 0.48%	55.511	5−	$7.4\times10^3{}_f^{rs}$ $1.6\times10^3{}^{rs}$
		242f	14.0 ms	SF	57.76		
		243	7.37×10^3 y	α	57.170	5/2−	80_m^{rs} 6_g^{rs}
		243f	5 μs	SF	59.17		
		244	10.1 h	β^-	59.877	(6−)	$2.2\times10^3{}_f^{rs}$
		244m	26 m	β^- 99+%,EC 0.036%	59.948	(1−)	$1.6\times10^3{}_f^{rs}$
		244f	1.1 ms	SF	61.48		
		245	2.05 h	β^-	61.897	(5/2)+	
		245f	0.6 μs	SF			
		246	39 m	β^-		(7−)	
		246	25.0 m	β^-	64.92	(2−)	
		246f	0.07 ms	SF			
		247	24 m	β^-	67.13 s	(5/2)	
96	Cm	238	2.3 h	EC <90%,α >10%	49.398	0+	
		239	2.9 h	EC	51.09 s		
		240	27 d	α	51.712	0+	
		240f	10 ps	SF			
		241	32.8 d	EC 99.0%,α 1.0%	53.696	1/2+	
		241f	15 ns	SF	55.70		
		242	162.8 d	α	54.802	0+	20 $<5_f^{rs}$
		242f_1	0.04 ns	SF			
		242f_2	0.2 μs	SF	57.60		
		243	28.5 y	α 99.74%,EC 0.26%	57.177	5/2+	$1.0\times10^3{}_a^{rs}$ 610_f 130
		243f	0.04 μs	SF	58.68		
		244	18.11 y	α	58.450	0+	14 1.0_f
		244(f_1)	<5 ps?	[SF]			
		244(f_2)	>100 ns	SF	61.45		
		245	8.5×10^3 y	α	61.001	7/2+	$2.0\times10^3{}_f$ 350
		245f	13 ns	SF	62.70		
		246	4.7×10^3 y	α	62.616	0+	1.3 0.2_f
		247	1.6×10^7 y	α	65.530	9/2−	100_f^{sc} 60
		248	3.5×10^5 y	α 91.74%,SF 8.26%	67.389	0+	4^{sc} 0.3_f
		249	65 m	β^-	70.748	1/2+	2^{rs}
		250	$\leq1.1\times10^4$ y	SF	72.986	0+	$\approx80^{rs}$
		251	16.8 m	β^-	76.67 s	(1/2+)	
97	Bk	240	5 m	EC+β^+, (EC+β^+)SF 0.001%	55.71 s		
		242	7 m	EC	57.80 s		
		242f_1	0.6 μs	SF			
		242f_2	10 ns	SF			
		243	4.5 h	EC 99.85%,α 0.15%	58.685	(3/2−)	
		243f	5 ns	SF	60.88		
		244	4.4 h	EC 99+%,α 0.006%	60.646		
		244f	0.8 μs	SF			
		245	4.90 d	EC 99.88%,α 0.12%	61.811	3/2−	
		245f	2 ns	SF			
		246	1.80 d	EC	64.02 s	(2−)	
		247	1.4×10^3 y	α	65.484	(3/2−)	

TABLE OF NUCLIDES

Nuclide Z	El	A	Abundance or $t_{1/2}$	Decay Mode	Δ(MeV)	Jπ	σ_n(b)
97	Bk	248	23.5 h	β^- 70%,EC 30%	67.99 s	(1−)	
		248	>9 y	?	67.99 s	(6+)	
		249	0.88 y	β^- 99+%,α 0.0015%	69.848	7/2+	$1.0\times10^{3}{}^{rs}_{\alpha}$
		250	3.22 h	β^-	72.950	2−	$1.0\times10^{3}{}^{rs}_{f}$
		251	56 m	β^-	75.25 s	(3/2−)	
98	Cf	240	1.1 m	α	58.03 s	0+	
		241	4 m	α	59.19 s		
		242	3.5 m	α	59.332	0+	
		243	11 m	[EC] ≈86%,α ≈14%	60.91 s		
		244	19 m	α	61.465	0+	
		245	44 m	EC ≈70%,α ≈30%	63.377		
		246	35.7 h	α	64.096	0+	
		246f	0.05 μs	SF			
		247	3.15 h	EC 99.96%,α 0.04%	66.15 s	(7/2+)	
		248	333 d	α	67.243	0+	
		249	351 y	α	69.722	9/2−	$1.63\times10^{3}{}_{f}$ 480^{sc}
		250	13.1 y	α	71.170	0+	$2.0\times10^{3}{}^{sc}$ $<350{}^{rs}_{f}$
		251	9.0×10^{2} y	α	74.127	1/2+	$4\times10^{3}{}^{sc}_{f}$ $2.9\times10^{3}{}^{sc}$
		252	2.64 y	α 96.91%,SF 3.09%	76.031	0+	$32{}^{sc}_{f}$ 20
		253	17.8 d	β^- 99.69%,α 0.31%	79.299	(7/2+)	$1.3\times10^{3}{}^{rs}_{f}$
		254	60.5 d	SF 99.69%,α 0.31%	81.342	0+	$100{}^{rs}_{\alpha}$
		255	≈2 h?	[β^-]			
		256	12 m	SF		0+	
99	Es	243	21 s	α	64.80 s		
		244	37 s	EC+β^+ 96%,α 4%	65.97 s		
		245	1.3 m	EC 60%,α 40%	66.38 s		
		246	7.7 m	EC+β^+ 90%,α 10%	67.93 s		
		247	4.7 m	EC ≈93%,α ≈7%	68.550		
		248	28 m	EC ≈99.7%,α ≈0.3%	70.22 s		
		249	1.70 h	EC 99.4%,α 0.6%	71.116	(7/2+)	
		250	8.6 h	EC	73.17 s	(6+)	
		250	2.1 h	EC	73.17 s	(1−)	
		251	33 h	EC 99.5%,α 0.5%	74.507	(3/2−)	
		252	472 d	α 78%,EC 22%	77.15 s	(5−)	
		253	20.47 d	α	79.012	7/2+	$160{}_{m}$ $<3{}_{g}$ $<60{}^{rs}_{f}$
		254	276 d	α	81.992	(7+)	$2.8\times10^{3}{}_{f}$
		254m	39.3 h	β^- 99.59%, α 0.33%,EC 0.08%	82.070	2+	$1.8\times10^{3}{}^{rs}_{f}$
		255	38.3 d	β^- 92.0%,α 8.0%, SF 0.004%	84.12 s	(7/2+)	65^{rs}
		256	7.6 h	β^-	87.26 s	(7,8)	
		256	22 m	β^-	87.26 s		
100	Fm	242	0.8 ms?	SF		0+	
		244	3.3 ms	SF	68.77 s	0+	
		245	4 s	α	70.02 s		
		246	1.2 s	α 92%,SF 8%	70.131	0+	
		247	9 s	α	71.54 s		
		247	35 s	α ≳50%,EC ≲50%	71.54 s		
		248	36 s	α 99.9%,SF 0.1%	71.891	0+	
		249	3 m	α	73.50 s		
		250	30 m	α,EC?	74.069	0+	
		250m	1.8 s	IT			
		251	5.3 h	EC 98.2%,α 1.8%	76.00 s	(9/2−)	
		252	25.4 h	α	76.822	0+	
		253	3.0 d	EC 88%,α 12%	79.346	1/2+	
		254	3.240 h	α 99+%,SF 0.0590%	80.899	0+	
		255	20.1 h	α	83.793	7/2+	$3.3\times10^{3}{}^{sc}_{f}$

TABLE OF NUCLIDES

Nuclide Z	El	A	Abundance or $t_{1/2}$	Decay Mode	Δ(MeV)	Jπ	σ_n(b)
100	Fm	256	2.63 h	SF 91.9%,α 8.1%	85.481	0+	
		257	100.5 d	α 99.79%,SF 0.21%	88.588	(9/2+)	$6\times10^{3}\,{}^{sc}_{\alpha}$ $3.0\times10^{3}\,{}^{rs}_{f}$
		258	0.4 ms	SF		0+	
		259	1.5 s	SF			
101	Md	248	7 s	EC+β^+ 80%,α 20%	77.00 s		
		249	24 s	EC+β^+ ≲80%,α ≳20%	77.26 s		
		250	0.9 m	EC+β^+ 94%,α 6%	78.60 s		
		251	4.0 m	EC ≳90%,α ≲10%	79.03 s		
		252	2 m	EC+β^+	80.50 s		
		254	10 m	EC	83.39 s		
		254	28 m	EC	83.39 s		
		255	27 m	EC 92%,α 8%	84.843	(7/2−)	
		256	75 m	EC 90.1%,α 9.9%	87.42 s		
		257	5.0 h	EC 90%,α 10%	89.04 s	(7/2−)	
		258	56 d	α	91.82 s		
		258	43 m	EC(?)	91.82 s		
		259	1.6 h	SF			
102	No	250	0.25 ms?	SF		0+	
		251	0.8 s	α			
		252	2.3 s	α 73%,SF 27%	82.867	0+	
		253	1.7 m	α	84.33 s		
		254	55 s	α	84.729	0+	
		254m	0.28 s	IT			
		255	3.1 m	α 62%,EC 38%	86.87 s	(1/2+)	
		256	3.2 s	α ≈99.7%,SF ≈0.3%	87.801	0+	
		257	26 s	α	90.223		
		258	1.2 ms	SF	91.52 s	0+	
		259	58 m	α ≈78%,EC ≈22%	94.012		
103	Lr	255	22 s	α	90.25 s		
		256	27 s	α	91.82 s		
		257	0.65 s	α	92.97 s		
		258	4.3 s	α	94.82 s		
		259	5 s	α	95.97 s		
		260	3.0 m	α	98.14 s		
104		253	1.8 s?	SF ≈50%(?)			
		254	0.5 ms?	SF		0+	
		255	2 s?	SF ≈50%			
		256	≈5 ms?	SF		0+	
		257	5 s	α	95.95 s		
		258	11 ms?	SF	96.55 s	0+	
		259	3 s	α	98.50 s		
		260	0.08 s?	SF	99.23 s	0+	
		261	1.1 m	α	101.25 s		
105		255	≈1.2 s?	SF ≈20%			
		257	5 s?	SF ≈20%			
		260	1.5 s	α 90%,SF 10%	103.65 s		
		261	2 s	α ≈75%,SF ≈25%	104.46 s		
		262	0.7 m	α ≈40%, SF or EC(?) ≈60%	106.04 s		
106		259	4 to 10 ms?	SF ≈70%(?)			
		263	0.9 s	α			
107		261	1 to 2 ms?	SF ≈20%			

Appendix E

Gamma-Ray Sources

In this table are listed the energies and intensities of γ rays emitted in the decay of nuclides with half lives greater than 12 h and having intensities of 10 percent or more. Gamma rays emitted by daughter nuclides are included only if the daughter half life is ≤ 10 min; they are designated by *D* preceding the energy.

The intensity I_γ of each γ ray is given in percentage of disintegrations. Note that I_γ refers to γ-*ray* intensity, that is, it does not include the intensity of conversion electrons. Where only relative intensities of several γ rays are known, the I_γ entries are preceded by *R*.

Half lives are included for orientation and correspond to those in Appendix D. However, they are given here to no more than three significant figures and no more than one decimal place. The notation 3.2E7 y means 3.2×10^7 years.

The energy and intensity data for this table are taken with kind permission of the authors from the *Gamma-Ray Catalog* compiled by U. Reus, W. Westmeier, and I. Warnecke and issued as GSI Report 79-2 (February 1979) by the Gesellschaft für Schwerionenforschung, Darmstadt, West Germany.

Table E-1 Gamma-Ray Sources

Nuclide	$t_{1/2}$	E_γ (keV)	I_γ (%)
^{7}Be	53.3 d	477.6	10.4
^{22}Na	2.6 y	1274.6	99.9
^{24}Na	15.0 h	1368.5	100
		2753.9	99.9
^{28}Mg	21.0 h	30.6	95.0
		400.6	35.9
		941.7	35.9
		1342.2	54.0
		D 1778.9	100
^{26}Al	7.2E5 y	1808.7	99.7
^{40}K	1.3E9 y	1460.8	10.7
^{47}Ca	4.5 d	1297.1	74.9
$^{44}Sc^{m}$	2.4 d	271.2	77.8
^{46}Sc	83.8 d	889.2	100
		1120.5	100
^{47}Sc	3.4 d	159.4	68
^{48}Sc	43.7 h	983.5	100
		1037.5	97.5
		1312.1	100
^{44}Ti	47 y	67.8	91
		78.4	96
^{48}V	16.0 d	983.5	100
		1312.1	97.5
^{48}Cr	21.6 h	112.5	95
		308.3	99
^{52}Mn	5.6 d	744.2	90.0
		935.5	94.5
		1434.1	100
^{54}Mn	312 d	834.8	100
^{59}Fe	44.6 d	1099.3	56.5
		1291.6	43.2
^{55}Co	17.5 h	477.2	20.3
		931.5	75
		1408.7	16.5
^{56}Co	78.8 d	846.8	99.9
		1037.8	14.1
		1238.3	67.0
		1771.4	15.5
		2598.6	16.8
^{57}Co	271 d	122.1	85.6
		136.5	10.6
^{58}Co	70.8 d	810.8	99.4
^{60}Co	5.3 y	1173.2	100
		1332.5	100
^{56}Ni	6.1 d	158.4	98.8
		269.5	36.5
		480.4	36.5
		750.0	49.5
		811.9	86.0
		1561.8	14.0
^{57}Ni	36.0 h	127.2	12.9
		1377.6	77.9
		1919.4	14.7
^{67}Cu	61.9 h	93.3	16.1
		184.6	48.7
^{65}Zn	244 d	1115.7	50.7
$^{69}Zn^{m}$	14.0 h	438.6	94.8
^{72}Zn	46.5 h	144.7	83.0
^{67}Ga	78.3 h	93.3	37.0
		184.6	20.4
		300.2	16.6
^{72}Ga	14.1 h	629.9	25.2
		834.0	95.6
		2201.7	25.6
		2507.8	12.7
^{69}Ge	39.0 h	573.9	11.0
		1106.4	25.7
^{71}As	61 h	174.9	83.6
^{72}As	26.0 h	834.0	80.1
^{73}As	80.3 d	53.4	10.5
^{74}As	17.8 d	595.8	60.3
		634.8	15.1
^{76}As	26.3 h	559.1	44.7
^{72}Se	8.4 d	46.0	58.8
^{75}Se	118 d	121.1	16.3
		136.0	55.6
		264.6	58.2
		279.5	24.6
		400.6	11.1
^{76}Br	16.1 h	559.1	72.3
		657.0	15.5
		1853.7	14.0

Table E-1 Gamma-Ray Sources

Nuclide	$t_{1/2}$	E_γ (keV)	I_γ (%)
^{77}Br	57.0 h	239.0	23.1
		520.7	22.4
^{82}Br	35.3 h	554.3	70.6
		619.1	43.1
		698.3	27.9
		776.5	83.4
		827.8	24.2
		1044.0	27.4
		1317.5	26.9
		1474.8	16.6
^{76}Kr	14.8 h	44.5	18
		270.2	21
		315.7	40
		406.5	12
^{79}Kr	35.0 h	261.3	12.7
^{83}Rb	86.2 d	520.4	46.1
		529.5	30.0
		552.5	16.3
^{84}Rb	32.9 d	881.6	75.3
^{83}Sr	32.4 h	381.6	19.6
		762.7	29.7
^{85}Sr	64.8 d	514.0	100
^{86}Y	14.7 h	443.1	16.9
		627.7	32.6
		703.3	15.4
		777.4	22.4
		1076.6	82.5
		1153.0	30.5
		1854.4	17.2
		1920.7	20.8
^{87}Y	80.3 h	484.8	90.7
^{87}Y^m	13 h	381.1	78.5
^{88}Y	107 d	898.0	94.0
		1836.0	99.4
^{86}Zr	16.5 h	28.0	21
		243	96
^{88}Zr	83.4 d	392.9	97.3
^{89}Zr	78.4 h	*D* 909.2	99.9
^{95}Zr	64.0 d	724.2	43.7
		756.7	55.4
^{97}Zr	16.9 h	*D* 743.4	92.6
^{90}Nb	14.6 h	141.2	69.0
		1129.1	92.0
		2186.4	17.5
		2319.1	82.8
^{92}Nb	3.2E7 y	561.1	100
		934.5	100
^{92}Nb^m	10.2 d	934.5	99.2
^{94}Nb	2.0E4 y	702.6	100
		871.1	100
^{95}Nb	35.0 d	765.8	99.9
^{95}Nb^m	87 h	235.7	25.1
^{96}Nb	23.4 h	460.0	28.2
		568.9	55.7
		778.2	96.9
		849.9	20.7
		1091.3	49.5
		1200.2	20.1
^{99}Mo	66.0 h	739.4	14.0
^{95}Tc	20.0 h	765.8	93.9
^{95}Tc^m	61 d	204.1	66.5
		582.1	31.5
		835.1	28.0
^{96}Tc	4.3 d	778.2	10.0
		812.5	82.2
		849.9	97.8
		1126.8	15.2
^{98}Tc	4.2E6 y	652.4	100
		745.3	100
^{97}Ru	2.9 d	215.7	85.8
		324.5	10.2
^{103}Ru	39.4 d	497.1	86.4
^{106}Ru	367 d	*D* 511.8	20.6
^{99}Rh	15.0 d	89.4	30.9
		353.0	31.9
		527.7	40.7
^{100}Rh	20.8 h	446.2	11
		539.6	78.4
		822.5	20.1
		1107.1	13.2
		1362.1	15.0
		1553.4	20.5

Table E-1 Gamma-Ray Sources

Nuclide	$t_{1/2}$	E_γ (keV)	I_γ (%)
		1929.7	12.2
		2376.1	35.0
^{101}Rh	3.3 y	127.2	65.6
		198.0	63.6
		325.2	12.0
^{101}Rh^m	4.3 d	306.8	86.8
^{102}Rh	2.9 y	475.1	94.0
		631.3	55.5
		697.5	43.2
		766.8	33.8
		1046.6	33.8
		1112.8	18.8
^{102}Rh^m	206 d	475.1	44.0
^{105}Rh	35.4 h	318.9	19.2
^{100}Pd	3.6 d	74.8	R 98
		84.0	R100
		126.1	R 11
^{105}Ag	41.3 d	64.0	10.5
		280.4	29.5
		344.5	40.9
		443.4	11.4
		644.5	11.4
^{106}Ag^m	8.5 d	406.2	13.5
		429.6	13.2
		451.0	28.4
		511.8	88.2
		616.2	21.7
		717.3	29.1
		748.4	20.7
		804.3	12.4
		824.7	15.4
		1045.8	29.7
		1128.0	11.8
		1199.4	11.3
		1527.7	16.4
^{108}Ag^m	127 y	433.9	90.7
		614.4	90.7
		722.9	91.5
^{110}Ag^m	252 d	657.7	94.7
		677.6	10.7
		706.7	16.7
		763.9	22.4
		884.7	72.9
		937.5	34.3
		1384.3	24.3
		1505.0	13.1
^{115}Cd	53.4 h	527.9	27.5
^{111}In	2.8 d	171.3	90.3
		245.4	94.0
^{114}In^m	49.5 d	191.6	18.2
^{117}Sn^m	14.0 d	158.6	86.3
^{119}Sn^m	250 d	23.9	16.4
^{126}Sn	1E5 y	87.6	37.0
^{119}Sb	38.0 h	23.9	16.4
^{120}Sb	5.8 d	89.8	80.0
		197.3	88.0
		1023.3	99.0
		1171.7	100
^{122}Sb	2.7 d	564.0	70.8
^{124}Sb	60.2 d	602.7	98.4
		722.8	11.3
		1691.0	49.0
^{125}Sb	2.7 y	427.9	29.4
		463.4	10.5
		600.6	17.8
		635.9	11.3
^{126}Sb	12.4 d	414.8	83.6
		666.3	100
		695.0	100
		697.0	30
		720.5	54.0
		856.7	17.7
^{127}Sb	3.9 d	473.0	25.0
		685.7	35.7
		783.7	14.7
^{119}Te	16.0 h	644.0	84.4
		699.8	10.1
^{119}Te^m	4.7 d	153.6	67.1
		270.5	28.3
		1212.7	67.0
^{121}Te	16.8 d	507.6	17.5
		573.1	79.7
^{121}Te^m	154 d	212.2	83.9

Table E-1 Gamma-Ray Sources

Nuclide	$t_{1/2}$	E_γ (keV)	I_γ (%)	Nuclide	$t_{1/2}$	E_γ (keV)	I_γ (%)
$^{123}Te^m$	120 d	159.0	83.9	^{128}Ba	2.4 d	273.4	14.5
$^{131}Te^m$	30 h	773.7	38.1	^{131}Ba	12.0 d	123.8	29.2
		793.8	13.8			216.1	19.9
		852.2	21.0			373.2	14.1
		1125.5	11.4			496.3	47.1
^{132}Te	78 h	49.7	14.4	^{133}Ba	10.7 y	81.0	32.8
		228.2	88.1			302.9	18.6
^{123}I	13.0 h	159.0	83.2			356.0	62.3
^{124}I	4.2 d	602.7	61.0	$^{133}Ba^m$	38.9 h	276.1	17.5
		722.8	10.1	$^{135}Ba^m$	28.7 h	268.2	15.6
		1691.0	10.5	^{140}Ba	12.8 d	30.0	14
^{126}I	13.0 d	388.6	32.2			537.3	19.9
		666.3	31.3	^{140}La	40.3 h	328.8	18.5
^{130}I	12.4 h	418.0	34.2			487.0	43.0
		536.1	99.0			815.8	22.4
		668.5	96.1			1596.5	95.5
		739.5	82.3	^{135}Ce	17.8 h	265.6	42.4
		1157.5	11.3			300.1	22.9
^{131}I	8.0 d	364.5	81.2			518.1	13.6
^{133}I	20.9 h	529.9	87.0			572.3	10.6
^{122}Xe	20.1 h	*D* 564.0	17.7			606.8	19.5
						783.6	10.6
^{125}Xe	17 h	188.4	55.1	$^{137}Ce^m$	34.4 h	254.3	10.9
		243.4	28.9	^{139}Ce	137 d	165.8	78.9
^{127}Xe	36.4 d	172.1	24.7	^{141}Ce	32.5 d	145.4	48.4
		202.8	68.1	^{143}Ce	33.0 h	57.4	~12
		375.0	17.4			293.3	~51
^{133}Xe	5.2 d	81.0	35.9	^{144}Ce	284 d	133.5	10.8
^{129}Cs	32.3 h	371.9	31.1	^{147}Nd	11.0 d	91.1	27.9
		411.5	22.7			531.0	13.3
^{132}Cs	6.5 d	667.5	97.4	^{143}Pm	265 d	742.0	38.3
^{134}Cs	2.1 y	569.3	15.4	^{144}Pm	349 d	476.8	42.0
		604.7	97.6			618.0	98.6
		795.8	85.4			696.5	99.5
^{136}Cs	13.1 d	66.9	12.5	^{146}Pm	5.5 y	453.8	62.3
		176.6	13.6			736.2	22.4
		273.6	12.7			747.4	35.9
		340.6	46.9	^{148}Pm	5.4 d	550.3	22.0
		818.5	99.8			914.9	11.5
		1048.1	79.8			1465.1	22.2
		1235.3	19.8				
^{137}Cs	30.2 y	*D* 661.6	85.1	$^{148}Pm^m$	41.3 d	288.1	12.6

Table E-1 Gamma-Ray Sources

Nuclide	$t_{1/2}$	E_γ (keV)	I_γ (%)
		414.1	18.7
		550.3	95.6
		599.7	12.6
		630.0	89.1
		725.7	32.9
		915.3	17.2
		1013.8	20.3
^{151}Pm	28.4 h	340.1	22.4
^{145}Sm	340 d	61.3	12.7
^{153}Sm	46.8 h	103.2	28.3
^{145}Eu	5.9 d	653.6	16.0
		893.8	63.9
		1658.7	15.4
^{146}Eu	4.6 d	633.2	43
		634.1	37
		747.2	98.0
^{147}Eu	22 d	121.3	19.7
		197.3	22.8
^{148}Eu	54 d	413.9	18.6
		550.3	99.0
		553.2	17.1
		611.3	19.3
		629.9	70.9
		725.7	13.0
^{150}Eu	36 y	334.0	94.0
		439.4	78.7
		584.3	51.5
^{152}Eu	13 y	121.8	28.4
		344.3	26.6
		778.9	13.0
		964.1	14.6
		1085.9	10.2
		1112.1	13.6
		1408.0	20.8
^{154}Eu	8.5 y	123.1	40.5
		723.3	19.7
		873.2	11.5
		996.3	10.3
		1004.8	17.4
		1274.5	35.5
^{155}Eu	4.9 y	86.5	32.7
		105.3	21.8
^{156}Eu	15 d	811.8	10.3
^{157}Eu	15.1 h	64	22
		373	10
		413	17
^{146}Gd	48.3 d	114.7	43.5
		115.5	44.1
		154.6	43.0
^{147}Gd	38.1 h	229.2	57.3
		370.0	13.2
		396.5	26.2
		765.6	10.0
		928.3	18
^{149}Gd	9.3 d	149.6	41.7
		298.5	22.6
		346.5	17.9
^{153}Gd	242 d	97.4	30.1
		103.2	21.8
^{159}Gd	18.6 h	363.6	10.3
^{151}Tb	17.6 h	108.3	25
		180.4	11
		251.7	26
		287.0	25
		443.7	10
		479.0	16
		587.3	17
		616.6	10
^{152}Tb	17.5 h	344.3	67.2
^{153}Tb	2.3 d	211.9	32.5
^{154}Tb	21 h	123.1	28
		1274.4	11.4
		2187.2	10.8
$^{154}Tb^{m2}$	23 h	123.1	43
		226.1	27.0
		248.0	79.3
		346.7	69.6
		426.8	17.4
		993.0	16.4
		1419.9	46.5
^{155}Tb	5.3 d	86.5	29.3
		105.3	23.0
^{156}Tb	5.3 d	89.0	18.0

Table E-1 Gamma-Ray Sources

Nuclide	$t_{1/2}$	E_γ (keV)	I_γ (%)
		199.2	37.4
		356.5	12.4
		534.3	60.3
		1065.1	10.0
		1222.4	29.0
		1421.6	11.6
$^{156}Tb^{m1}$	24 h	49.6	72.8
^{158}Tb	150 y	79.5	11.3
		944.2	43.0
		962.1	19.9
^{160}Tb	72.1 d	86.8	13.4
		298.6	27.4
		879.4	30.0
		962.3	10.0
		966.2	25.5
		1177.9	15.5
^{161}Tb	6.9 d	25.6	21.0
		48.9	14.8
^{166}Dy	81.5 h	82.5	13.1
$^{166}Ho^{m}$	1.2E3 y	80.6	12.6
		184.4	73.9
		280.5	30.1
		410.9	11.7
		529.8	10.3
		711.7	59.3
		752.3	13.2
		810.3	63.3
		830.6	10.7
^{172}Er	49.5 h	407.3	44.8
		610.1	47.0
^{165}Tm	30.1 h	242.9	35.1
		297.4	17.5
^{167}Tm	9.2 d	207.8	41.0
^{168}Tm	93.1 d	79.8	11.0
		184.3	16.4
		198.2	50.0
		447.5	21.9
		720.3	10.9
		741.3	11.3
		815.9	46.3
		821.1	11.1
^{166}Yb	56.7 h	82.3	15.2
^{169}Yb	32.0 d	63.1	45.0
		109.8	18.0
		130.5	11.5
		177.2	22.0
		198.0	36.0
		307.7	11.1
^{169}Lu	34.1 h	191.3	22.4
		960.3	23.7
^{171}Lu	8.2 d	19.4	~13
		739.8	36.7
^{172}Lu	6.7 d	78.7	10.9
		181.5	19.9
		810.0	15.8
		900.7	28.7
		912.0	14.7
		1093.6	63.6
^{173}Lu	1.4 y	272.0	13.0
$^{174}Lu^{m}$	142 d	44.7	12.4
^{176}Lu	3.6E10 y	88.3	13.1
		201.8	84.7
		306.9	93.3
^{177}Lu	6.7 d	208.4	11.0
$^{177}Lu^{m}$	160 d	*D* 105.4	12.2
		D 113.0	21.8
		D 128.5	15.5
		D 153.2	18.3
		D 174.4	12.8
		D 204.1	14.5
		D 208.4	62.2
		D 228.4	37.8
		D 281.8	14.3
		319.0	10.5
		D 327.7	17.8
		D 378.5	28.3
		413.7	16.7
		D 418.5	20.4
^{170}Hf	16.0 h	120.2	19.2
		164.7	33.5
		572.9	18.5
		620.7	22.9
^{171}Hf	12.1 h	122.1	12.6
		662.0	14.8
		1071.4	11.9

Table E-1 Gamma-Ray Sources

Nuclide	$t_{1/2}$	E_γ (keV)	I_γ (%)
^{172}Hf	1.9 y	24.0	20.4
^{173}Hf	24.0 h	123.7	82.7
		139.7	12.8
		297.0	33.8
		311.3	10.7
^{175}Hf	70 d	343.4	86.6
^{178}Hf^{m2}	31 y	*D* 88.9	62.0
		D 93.2	17.3
		D 213.4	80.9
		216.7	63.7
		257.6	16.6
		D 325.6	93.9
		D 426.4	96.9
		454.0	16.3
		495.0	68.7
		574.2	83.6
^{179}Hf^{m2}	25.1 d	122.7	26.9
		146.1	26.3
		169.8	18.9
		192.8	20.9
		236.6	18.3
		268.9	11.0
		316.0	19.7
		362.6	38.5
		409.8	20.9
		453.7	66.0
^{181}Hf	42.4 d	133.0	43.0
		345.9	14.0
		482.0	85.5
^{182}Hf	9E6 y	270.4	80.0
^{182}Ta	115 d	67.7	41.3
		100.1	14.1
		1121.3	35.0
		1189.0	16.4
		1221.4	27.4
		1231.0	11.6
^{183}Ta	5.1 d	107.9	10.8
		246.1	26.7
		354.0	11.4
^{187}W	23.9 h	72.5	12.9
		134.2	10.3
		479.5	25.3
		865.8	31.6
^{181}Re	20 h	360.7	12.0
		365.5	56.4
^{182}Re	64 h	67.8	24.6
		100.1	16.2
		169.2	12.2
		229.3	27.9
		256.4	10.3
		351.1	11.2
		1076.2	11.4
		1121.3	23.9
		1221.4	18.9
		1231.0	16.2
		1427.3	10.6
^{182}Re	12.7 h	67.7	38.0
		100.1	14.4
		1121.4	31.9
		1189.2	15.2
		1221.5	25.1
^{183}Re	71 d	162.3	23.5
^{184}Re	38 d	111.2	17.0
		792.1	37.1
		894.8	15.4
		903.3	37.5
^{184}Re^{m}	169 d	104.7	13.5
^{186}Re^{m}	2E5 y	59.0	18.6
^{188}Re	16.9 h	155.0	16.0
^{182}Os	22.0 h	180.2	36.8
		510.0	55.0
^{183}Os	13 h	114.4	20.8
		381.8	77.0
^{185}Os	93.6 d	646.1	79.5
^{191}Os	15.4 d	*D* 129.4	25.9
^{185}Ir	14 h	254.3	13.0
^{186}Ir	16 h	137.2	41.3
		296.8	62.2
		434.8	33.8
^{188}Ir	41.5 h	155.0	29.8
		478.0	14.8
		633.0	18.0
		2214.6	18.8
^{190}Ir	11.8 d	186.7	48.2
		361.1	12.6

Table E-1 Gamma-Ray Sources

Nuclide	$t_{1/2}$	E_γ (keV)	I_γ (%)
		371.2	22.0
		407.2	27.5
		518.6	32.8
		558.0	29.0
		569.3	27.5
		605.1	38.5
^{192}Ir	74.2 d	295.9	28.7
		308.4	29.7
		316.5	82.9
		468.1	48.1
^{194}Ir	19.2 h	328.4	13.0
^{194}Ir^m	170 d	328.4	92.8
		338.8	55.1
		390.8	35.1
		482.9	96.9
		562.4	69.9
		600.5	62.3
		687.8	59.1
^{188}Pt	10.2 d	187.6	19
		195.1	18
^{191}Pt	2.9 d	538.1	13.7
^{195}Pt^m	4.0 d	98.9	11.1
^{197}Pt	18.3 h	77.3	17.0
^{200}Pt	12.6 h	76.2	14.4
^{193}Au	17.5 h	186.2	10.1
^{194}Au	39.5 h	293.5	11.0
		328.4	63.0
^{195}Au	183 d	98.9	11.0
^{196}Au	6.2 d	333.0	23.1
		355.7	87.7
^{198}Au	2.7 d	411.8	95.5
^{198}Au^m	2.3 d	97.2	69.0
		180.3	50.8
		204.1	41.5
		214.9	76.9
^{199}Au	3.1 d	158.4	36.9
^{200}Au^m	18.7 h	181.2	56
		255.9	71
		332.8	12
		367.9	74.5
		497.8	73.5
		579.3	72
		759.5	67
^{197}Hg	64.1 h	77.3	18.1
^{197}Hg^m	23.8 h	133.9	34.1
^{203}Hg	46.8 d	279.2	81.5
^{200}Tl	26.1 h	367.9	87.5
		579.3	13.8
		828.3	10.9
		1205.7	30.1
^{202}Tl	12.2 d	439.6	90.0
^{200}Pb	21.5 h	147.6	37.9
^{203}Pb	52.0 h	279.2	80.1
^{205}Bi	15.3 d	703.5	31.1
		987.7	16.1
		1764.3	32.5
^{206}Bi	6.2 d	184.0	15.8
		343.5	23.4
		398.0	10.7
		497.1	15.3
		516.2	40.7
		537.5	30.4
		803.1	98.9
		881.0	66.2
		895.1	15.7
		1098.3	13.5
		1718.7	31.8
^{207}Bi	38 y	569.7	97.8
		1063.6	74.9
^{208}Bi	3.7E5 y	2614.6	100
^{210}Bi^m	3.0E6 y	265.7	51.0
		304.8	28.0
^{206}Po	8.8 d	286.4	22.8
		338.4	18.4
		511.4	23.0
		522.5	15.2
		807.4	21.9
		1032.3	32.2
^{211}Rn	14.6 h	442.2	22.5
		674.1	44.2
		678.4	28.3
		947.4	18

Table E-1 Gamma-Ray Sources

Nuclide	$t_{1/2}$	E_γ (keV)	I_γ (%)
		1126.7	21.6
		1362.9	31.8
^{223}Ra	11.4 d	269.4	13.6
^{225}Ra	14.8 d	40	29
^{225}Ac	10.0 d	*D* 217.6	12.5
^{226}Ac	29 h	158.0	17.3
		230.0	26.6
^{227}Th	18.7 d	236.0	11.2
^{231}Th	25.5 h	25.6	14.8
^{228}Pa	22 h	463.0	13.2
		911.2	16.0
		964.6	10.0
		969.1	13.0
^{230}Pa	17.7 d	952.0	28.1
^{232}Pa	1.3 d	150.1	10.8
		894.3	19.8
		969.3	41.6
^{233}Pa	27.0 d	311.9	36.0
^{231}U	4.2 d	25.6	12
^{235}U	7.0E8 y	143.8	10.5
		185.7	54.0
^{237}U	6.8 d	59.5	33
		208.0	21.7
^{240}U	14.1 h	*D* 554.6	22.6
		D 597.4	12.6
^{234}Np	4.4 d	1527.5	12
		1558.7	19
^{236}Np	1.1E5 y	160.3	27.5
^{237}Np	2.1E6 y	29.4	14
		86.5	12.6
^{238}Np	2.1 d	984.5	23.8
		1028.5	17.3
^{239}Np	2.4 d	106.1	22.7
		228.2	10.7
		277.6	14.1
^{246}Pu	10.8 d	43.8	30
		179.9	12
		223.7	28
^{240}Am	50.8 h	888.8	25.1
		987.8	73.2
^{241}Am	433 y	59.5	35.9
^{243}Am	7.4E3 y	74.7	66.0
^{241}Cm	32.8 d	471.8	71.3
^{243}Cm	28.5 y	228.2	10.6
		277.6	14.0
^{247}Cm	1.6E7 y	402.4	76
^{245}Bk	4.9 d	252.8	29.1
^{246}Bk	1.8 d	798.7	61.4
^{247}Bk	1.4E3 y	84.0	~40
		267	~30
^{249}Cf	351 y	333.4	15.5
		387.9	66.0
^{251}Cf	9.0E2 y	176.6	17.7
^{252}Es	472 d	139.0	12
		785.1	16
^{254}Es^{m}	39.3 h	648.8	28.4
		688.7	12.2
		693.8	24.3

Appendix F

Selected References to Nuclear Data Compilations

In the following we give references to a small number of selected publications that tabulate or compile data of interest to nuclear scientists. No attempt has been made to be complete. Much more extensive listings may be found in the following two publications:

T. W. Burrows and N. E. Holden, "A Source List of Nuclear Data Bibliographies, Compilations, and Evaluations," Report BNL-NCS-50702, 2nd ed. (Oct. 1978). Available from National Technical Information Service, 5285 Pt. Royal Road, Springfield, VA 22161.

F. Ajzenberg-Selove, "A Guide to Nuclear Compilations," in *Nuclear Spectroscopy and Reactions*, Vol. C (J. Cerny, Ed.), Academic, New York, 1974, pp. 551–559.

A large variety of tabulations appear in the journal *Atomic Data and Nuclear Data Tables* and its predecessor, *Nuclear Data Tables.*

General Nuclear Properties

Nuclear Data Sheets (Academic Press, New York).
This periodic publication provides the most extensive compilation of evaluated data on the properties of ground and excited states of nuclei, including masses, half lives, abundances, decay modes, radiations, spins, parities, and many other data. Revisions are periodically published for each mass chain.

C. M. Lederer and V. S. Shirley, Eds., *Table of Isotopes*, 7th ed., Wiley, New York, 1978.
This is a much more compact, but still quite extensive compilation of level and decay properties. An abbreviated version very similar to Appendix D is available as wallet cards.

F. W. Walker et al., *General Electric Chart of the Nuclides*, 12th ed., General Electric Company, Schenectady, NY, 1977.
A very handy reference, available as a wall chart and in booklet form.

Masses, Spins, and Moments

A. H. Wapstra and K. Bos, "The 1977 Atomic Mass Evaluation," *At. Data Nucl. Data Tables* **19**, 177 (1977); **20**, 1 (1977).

G. H. Fuller, "Nuclear Spins and Moments," *J. Phys. Chem. Ref. Data* **5**, 835 (1976).

Half Lives

The Lederer and Shirley *Table of Isotopes* and the *Nuclear Data Sheets* listed above are major references for half lives. The following listing is often useful.

N. R. Large and R. J. Bullock, "Table of Radioactive Nuclides Arranged in Ascending Order of Half Life," *Nucl. Data Tables* **A7**, 477 (1970).

Alpha Decay

A. Rytz, "Catalogue of Recommended Alpha Energy and Intensity Values," *At. Data Nucl. Data Tables* **12**, 479 (1973).

Beta Decay

H. Behrens and J. Jaenecke, "Numerical Tables for Beta Decay and Electron Capture," *Landolt-Börnstein I*, Vol. 4, Springer, Berlin, 1969.

N. B. Gove and M. J. Martin, "Log f Tables for β Decay," *Nucl. Data Tables* **A10**, 205 (1971).

Gamma Rays

R. L. Heath, *Gamma-Ray Spectrum Catalogue*, Vol. 2, Report ANCR-1000-2, Ed. 3 (1975). Available from National Technical Information Service, 5285 Pt. Royal Road, Springfield, VA 22161.

J. B. Marion, "Gamma-Ray Calibration Energies," *Nucl. Data Tables* **A4**, 301 (1968).

U. Reus, W. Westmeier, and I. Warnecke, *Gamma-Ray Catalog*, GSI Report 79-2, Gesellschaft für Schwerionenforschung, Darmstadt, West Germany 1979.

S. A. Lis et al., "Gamma-Ray Tables for Neutron, Fast-Neutron, and Photon Activation Analysis," *J. Radioanal. Chem.* **24**, 125 (1975); **25**, 303 (1975).

D. Duffey et al., "Thermal Neutron Capture Gamma Rays," *Nucl. Instr. Methods* **80**, 149 (1970); **93**, 425 (1971).

X Rays

W. Bambynek et al., "X-Ray Fluorescence Yields, Auger, and Coster-Kronig Transition Probabilities", *Rev. Mod. Phys.* **44**, 716 (1972).

Internal Conversion Coefficients

F. Rösel et al., "Internal Conversion Coefficients for all Atomic Shells," *At. Data Nucl. Data Tables* **21**, 91 (1978).

Cross Sections and Excitation Functions

S. F. Mughabghab and D. I. Garber, *Neutron Cross Sections*, Report BNL-325, Vol. 1, 3rd ed. (1973). Available from National Technical Information Service, 5285 Pt. Royal Road, Springfield, VA 22161.

D. I. Garber and R. R. Kinsey, *Neutron Cross Sections*, Report BNL-325, Vol. 2, 3rd ed. (1976). Available as above.

W. E. Alley and R. M. Lessler, "Neutron Activation Cross Sections," *Nucl. Data Tables* **A11**, 621 (1973).

K. A. Keller, "Excitation Functions for Charged-Particle-Induced Reactions," *Landolt-Börnstein I*, Vol. 5, Pt. B, Springer, Berlin, 1973.

Fission

E. A. C. Crouch, "Fission Product Yields from Neutron-Induced Fission," *At. Data Nucl. Data Tables* **19**, 419 (1977).

J. Blachot and C. Fiche, "Gamma Ray and Half Life Data for the Fission Products," *At. Data Nucl. Data Tables* **20**, 241 (1977).

Ranges and Stopping Powers

L. C. Northcliffe and R. F. Schilling, "Range and Stopping-Power Tables for Heavy Ions," *Nucl. Data Tables* **A7**, 233 (1970).

L. Pages et al., "Energy Loss, Range, and Bremsstrahlung Yield for 10-keV to 100-MeV Electrons in Various Elements and Chemical Compounds," *At. Data* **4**, 1 (1972).

Shielding

P. F. Sauermann, *Tables for the Calculation of Gamma Radiation Shielding*, Thiemig, Munich, West Germany, 1976.

J. C. Courtney, Ed., *A Handbook of Radiation Shielding Data*, Am. Nucl. Soc. rep. ANS-SD-14.

Health Physics

Radiological Health Handbook, United States Public Health Service, Rockville, MD 20852, 1970.

Standards for Protection against Radiation, Title 10, *Code of Federal Regulations*, Part 20 (published annually).

Name Index

Subject Index